Handbook of Research Methods in Health Social Sciences

Pranee Liamputtong
Editor

Handbook of Research Methods in Health Social Sciences

Volume 2

With 192 Figures and 81 Tables

Springer

Editor
Pranee Liamputtong
School of Science and Health
Western Sydney University
Penrith, NSW, Australia

ISBN 978-981-10-5250-7 ISBN 978-981-10-5251-4 (eBook)
ISBN 978-981-10-5252-1 (print and electronic bundle)
https://doi.org/10.1007/978-981-10-5251-4

Library of Congress Control Number: 2018960888

© Springer Nature Singapore Pte Ltd. 2019
This work is subject to copyright. All rights are reserved by the Publisher, whether the whole or part of the material is concerned, specifically the rights of translation, reprinting, reuse of illustrations, recitation, broadcasting, reproduction on microfilms or in any other physical way, and transmission or information storage and retrieval, electronic adaptation, computer software, or by similar or dissimilar methodology now known or hereafter developed.
The use of general descriptive names, registered names, trademarks, service marks, etc. in this publication does not imply, even in the absence of a specific statement, that such names are exempt from the relevant protective laws and regulations and therefore free for general use.
The publisher, the authors, and the editors are safe to assume that the advice and information in this book are believed to be true and accurate at the date of publication. Neither the publisher nor the authors or the editors give a warranty, express or implied, with respect to the material contained herein or for any errors or omissions that may have been made. The publisher remains neutral with regard to jurisdictional claims in published maps and institutional affiliations.

This Springer imprint is published by the registered company Springer Nature Singapore Pte Ltd.
The registered company address is: 152 Beach Road, #21-01/04 Gateway East, Singapore 189721, Singapore

To my mother:
Yindee Liamputtong
and
To my children:
Zoe Sanipreeya Rice and Emma Inturatana Rice

Preface

Research is defined by the Australian Research Council as "the creation of new knowledge and/or the use of existing knowledge in a new and creative way so as to generate new concepts, methodologies, inventions and understandings." Research is thus the foundation for knowledge. It produces evidence and informs actions that can provide wider benefit to a society. The knowledge that researchers cultivate from a piece of research can be adopted for social and health programs that can improve the health and well-being of the individuals, their communities, and the societies in which they live. As we have witnessed, in all corners of the globe, research has become an endeavor that most of us in the health and social sciences cannot avoid. This Handbook is conceived to provide the foundation to readers who wish to embark on a research project in order to form knowledge that they need. The Handbook comprises four main parts: Traditional Research Methods in Health and Social Sciences, Innovative Research Methods in Health Social Sciences, Doing Cross-Cultural Research in Health Social Sciences, and Sensitive Research Methodology and Approach. This Handbook attests to the diversity and richness of research methods in the health and social sciences. It will benefit many readers, particularly students and researchers who undertake research in health and social science areas. It is also valuable for the training needs of postgraduate students who wish to undertake research in cross-cultural settings, with special groups of people, as it provides essential knowledge not only on the methods of data collection but also salient issues that they need to know if they wish to succeed in their research endeavors.

Traditionally, there are several research approaches and practices that researchers in the health social sciences have adopted. These include qualitative, quantitative, and mixed-methods approaches. Each approach has its own philosophical foundations, and the ways researchers go about to form knowledge can be different. But all approaches do share the same goal: to acquire knowledge that can benefit the world. This Handbook includes many chapters that dedicate to the traditional ways of conducting research. These chapters provide the "traditional ways of knowing" that many readers will need.

As health and social science researchers, we are now living in a moment that needs our imagination and creativity when we carry out our research. Indeed, we are now living "in the new age" where we will see more and more "new experimental works" being invented by researchers. And in this new age, we have witnessed many innovative and creative forms of research in the health and social sciences. In this Handbook, I also bring together a unique group of health and social science researchers to present their innovative and creative research methods that readers can adopt in their own research. The Handbook introduces many new ways of doing research. It embraces "methodological diversity," and this methodological diversity will bring "new ways of knowing" in the health and social sciences. Chapters in this Handbook will help to open up our ideas about doing research differently from the orthodox research methods that we have been using or have been taught to do.

Despite the increased demands on cross-cultural research, discussions on "culturally sensitive methodologies" are still largely neglected in the literature on research methods. As a result, researchers who are working in cross-cultural settings often confront many challenges with very little information on how to deal with these difficulties. Performing cross-cultural research is exciting, but it is also full of ethical and methodological challenges. This Handbook includes a number of chapters written by researchers who have undertaken their research in cross-cultural settings. They are valuable to many readers who wish to embark on doing a cross-cultural research in the future.

Globally too, we have witnessed many people become vulnerable to health and social issues. It will be difficult, or even impossible, for health and social science researchers to avoid carrying out research regarding vulnerable and marginalized populations within the "moral discourse" of the postmodern world, as it is likely that these population groups will be confronted with more and more problems in their private and public lives as well as in their health and well-being. Similar to undertaking cross-cultural research, the task of conducting research with the "vulnerable" and/or the "marginalized" presents researchers with unique opportunities and yet dilemmas. The Handbook also includes chapters that discuss research that involves sensitive and vulnerable/marginalized people.

This Handbook cannot be born without the help of others. I would like to express my gratitude to many people who helped to make this book possible. I am grateful to all the contributors who worked very hard to get their chapter done timely and comprehensively. I hope that the process of writing your chapter has been a rewarding endeavor to you as well. My sincere appreciation is given to Mokshika Gaur who believes in the value of the volume on research methods in health social sciences and has given me a contract to edit the Handbook. I also thank Tina Shelton, Vasowati Shome, and Ilaria Walker of Springer who helped to bring this book to life.

I dedicate this book to my mother Yindee Liamputtong, who has been a key person in my life. It was my mother who made it possible for me to continue my education amidst poverty. Without her, I would not have been where I am now. I also dedicate the book to my two daughters, Zoe Sanipreeya Rice and Emma Inturatana Rice, who have formed an important part of my personal and professional lives in Australia.

Sydney, Australia   Pranee Liamputtong

Contents

Volume 1

Section I Traditional Research Methods in Health and Social Sciences ... 1

1. Traditional Research Methods in Health and Social Sciences: An Introduction ... 3
   Pranee Liamputtong

2. Qualitative Inquiry ... 9
   Pranee Liamputtong

3. Quantitative Research ... 27
   Leigh A. Wilson

4. The Nature of Mixed Methods Research ... 51
   Cara Meixner and John D. Hathcoat

5. Recruitment of Research Participants ... 71
   Narendar Manohar, Freya MacMillan, Genevieve Z. Steiner, and Amit Arora

6. Ontology and Epistemology ... 99
   John D. Hathcoat, Cara Meixner, and Mark C. Nicholas

7. Social Constructionism ... 117
   Viv Burr

8. Critical Theory: Epistemological Content and Method ... 133
   Anastasia Marinopoulou

9. Positivism and Realism ... 151
   Priya Khanna

10. Symbolic Interactionism as a Methodological Framework ... 169
    Michael J. Carter and Andrea Montes Alvarado

11	**Hermeneutics: A Boon for Cross-Disciplinary Research** .... Suzanne D'Souza	189
12	**Feminism and Healthcare: Toward a Feminist Pragmatist Model of Healthcare Provision** .... Claudia Gillberg and Geoffrey Jones	205
13	**Critical Ethnography in Public Health: Politicizing Culture and Politicizing Methodology** .... Patti Shih	223
14	**Empathy as Research Methodology** .... Eric Leake	237
15	**Indigenist and Decolonizing Research Methodology** .... Elizabeth F. Rix, Shawn Wilson, Norm Sheehan, and Nicole Tujague	253
16	**Ethnomethodology** .... Rona Pillay	269
17	**Community-Based Participatory Action Research** .... Elena Wilson	285
18	**Grounded Theory Methodology: Principles and Practices** ...... Linda Liska Belgrave and Kapriskie Seide	299
19	**Case Study Research** .... Pota Forrest-Lawrence	317
20	**Evaluation Research in Public Health** .... Angela J. Dawson	333
21	**Methods for Evaluating Online Health Information Systems** ... Gary L. Kreps and Jordan Alpert	355
22	**Translational Research: Bridging the Chasm Between New Knowledge and Useful Knowledge** .... Lynn Kemp	367
23	**Qualitative Interviewing** .... Sally Nathan, Christy Newman, and Kari Lancaster	391
24	**Narrative Research** .... Kayi Ntinda	411
25	**The Life History Interview** .... Erin Jessee	425
26	**Ethnographic Method** .... Bonnie Pang	443
27	**Institutional Ethnography** .... Michelle LaFrance	457

28	**Conversation Analysis: An Introduction to Methodology, Data Collection, and Analysis** .... Sarah J. White	471
29	**Unobtrusive Methods** .... Raymond M. Lee	491
30	**Autoethnography** .... Anne Bunde-Birouste, Fiona Byrne, and Lynn Kemp	509
31	**Memory Work** .... Lia Bryant and Katerina Bryant	527
32	**Traditional Survey and Questionnaire Platforms** .... Magen Mhaka Mutepfa and Roy Tapera	541
33	**Epidemiology** .... Kate A. McBride, Felix Ogbo, and Andrew Page	559
34	**Single-Case Designs** .... Breanne Byiers	581
35	**Longitudinal Study Designs** .... Stewart J. Anderson	603
36	**Eliciting Preferences from Choices: Discrete Choice Experiments** .... Martin Howell and Kirsten Howard	623
37	**Randomized Controlled Trials** .... Mike Armour, Carolyn Ee, and Genevieve Z. Steiner	645
38	**Measurement Issues in Quantitative Research** .... Dafna Merom and James Rufus John	663
39	**Integrated Methods in Research** .... Graciela Tonon	681
40	**The Use of Mixed Methods in Research** .... Kate A. McBride, Freya MacMillan, Emma S. George, and Genevieve Z. Steiner	695

Volume 2

41	**The Delphi Technique** .... Jane Chalmers and Mike Armour	715
42	**Consensus Methods: Nominal Group Technique** .... Karine Manera, Camilla S. Hanson, Talia Gutman, and Allison Tong	737

43	Jumping the Methodological Fence: Q Methodology .... 751 Tinashe Dune, Zelalem Mengesha, Valentina Buscemi, and Janette Perz
44	Social Network Research .... 769 Janet C. Long and Simon Bishop
45	Meta-synthesis of Qualitative Research .... 785 Angela J. Dawson
46	Conducting a Systematic Review: A Practical Guide .... 805 Freya MacMillan, Kate A. McBride, Emma S. George, and Genevieve Z. Steiner
47	Content Analysis: Using Critical Realism to Extend Its Utility .. 827 Doris Y. Leung and Betty P. M. Chung
48	Thematic Analysis .... 843 Virginia Braun, Victoria Clarke, Nikki Hayfield, and Gareth Terry
49	Narrative Analysis .... 861 Nicole L. Sharp, Rosalind A. Bye, and Anne Cusick
50	Critical Discourse/Discourse Analysis .... 881 Jane M. Ussher and Janette Perz
51	Schema Analysis of Qualitative Data: A Team-Based Approach .... 897 Frances Rapport, Patti Shih, Mia Bierbaum, and Anne Hogden
52	Using Qualitative Data Analysis Software (QDAS) to Assist Data Analyses .... 917 Pat Bazeley
53	Sequence Analysis of Life History Data .... 935 Bram Vanhoutte, Morten Wahrendorf, and Jennifer Prattley
54	Data Analysis in Quantitative Research .... 955 Yong Moon Jung
55	Reporting of Qualitative Health Research .... 971 Allison Tong and Jonathan C. Craig
56	Writing Quantitative Research Studies .... 985 Ankur Singh, Adyya Gupta, and Karen G. Peres
57	Traditional Academic Presentation of Research Findings and Public Policies .... 999 Graciela Tonon
58	Appraisal of Qualitative Studies .... 1013 Camilla S. Hanson, Angela Ju, and Allison Tong

59	**Critical Appraisal of Quantitative Research**	1027
	Rocco Cavaleri, Sameer Bhole, and Amit Arora	
60	**Appraising Mixed Methods Research**	1051
	Elizabeth J. Halcomb	

Section II   Innovative Research Methods in Health Social Sciences **1069**

61	**Innovative Research Methods in Health Social Sciences: An Introduction**	1071
	Pranee Liamputtong	
62	**Personal Construct Qualitative Methods**	1095
	Viv Burr, Angela McGrane, and Nigel King	
63	**Mind Maps in Qualitative Research**	1113
	Johannes Wheeldon and Mauri Ahlberg	
64	**Creative Insight Method Through Arts-Based Research**	1131
	Jane Marie Edwards	
65	**Understanding Health Through a Different Lens: Photovoice Method**	1147
	Michelle Teti, Wilson Majee, Nancy Cheak-Zamora, and Anna Maurer-Batjer	
66	**IMAGINE: A Card-Based Discussion Method**	1167
	Ulrike Felt, Simone Schumann, and Claudia G. Schwarz-Plaschg	
67	**Timeline Drawing Methods**	1183
	E. Anne Marshall	
68	**Semistructured Life History Calendar Method**	1201
	Ingrid A. Nelson	
69	**Calendar and Time Diary Methods**	1219
	Ana Lucía Córdova-Cazar and Robert F. Belli	
70	**Body Mapping in Research**	1237
	Bronwyne Coetzee, Rizwana Roomaney, Nicola Willis, and Ashraf Kagee	
71	**Self-portraits and Maps as a Window on Participants' Worlds** ..	1255
	Anna Bagnoli	
72	**Walking Interviews**	1269
	Alexandra C. King and Jessica Woodroffe	
73	**Participant-Guided Mobile Methods**	1291
	Karen Block, Lisa Gibbs, and Colin MacDougall	

74	**Digital Storytelling Method** .... 1303 Brenda M. Gladstone and Elaine Stasiulis	
75	**Netnography: Researching Online Populations** .... 1321 Stephanie T. Jong	
76	**Web-Based Survey Methodology** .... 1339 Kevin B. Wright	
77	**Blogs in Social Research** .... 1353 Nicholas Hookway and Helene Snee	
78	**Synchronous Text-Based Instant Messaging: Online Interviewing Tool** .... 1369 Gemma Pearce, Cecilie Thøgersen-Ntoumani, and Joan L. Duda	
79	**Asynchronous Email Interviewing Method** .... 1385 Mario Brondani and Rodrigo Mariño	
80	**Cell Phone Survey** .... 1403 Lilian A. Ghandour, Ghinwa Y. El Hayek, and Abla Mehio Sibai	
81	**Phone Surveys: Introductions and Response Rates** .... 1417 Jessica Broome	
82	**The Freelisting Method** .... 1431 Marsha B. Quinlan	
83	**Solicited Diary Methods** .... 1447 Christine Milligan and Ruth Bartlett	
84	**Teddy Diaries: Exploring Social Topics Through Socially Saturated Data** .... 1465 Marit Haldar and Randi Wærdahl	
85	**Qualitative Story Completion: A Method with Exciting Promise** .... 1479 Virginia Braun, Victoria Clarke, Nikki Hayfield, Naomi Moller, and Irmgard Tischner	

Volume 3

Section III Doing Cross-Cultural Research in Health Social Sciences **1497**

86	**Doing Cross-Cultural Research: An Introduction** .... 1499 Pranee Liamputtong	
87	**Kaupapa Māori Health Research** .... 1507 Fiona Cram	

| 88 | Culturally Safe Research with Vulnerable Populations (Māori) | 1525 |

Denise Wilson

| 89 | Using an Indigenist Framework for Decolonizing Health Promotion Research | 1543 |

Karen McPhail-Bell, Alison Nelson, Ian Lacey, Bronwyn Fredericks, Chelsea Bond, and Mark Brough

| 90 | Engaging Aboriginal People in Research: Taking a Decolonizing Gaze | 1563 |

Emma Webster, Craig Johnson, Monica Johnson, Bernie Kemp, Valerie Smith, and Billie Townsend

| 91 | Space, Place, Common Wounds and Boundaries: Insider/Outsider Debates in Research with Black Women and Deaf Women | 1579 |

Chijioke Obasi

| 92 | Researcher Positionality in Cross-Cultural and Sensitive Research | 1601 |

Narendar Manohar, Pranee Liamputtong, Sameer Bhole, and Amit Arora

| 93 | Considerations About Translation: Strategies About Frontiers | 1617 |

Lía Rodriguez de la Vega

| 94 | Finding Meaning: A Cross-Language Mixed-Methods Research Strategy | 1639 |

Catrina A. MacKenzie

| 95 | An Approach to Conducting Cross-Language Qualitative Research with People from Multiple Language Groups | 1653 |

Caroline Elizabeth Fryer

| 96 | The Role of Research Assistants in Qualitative and Cross-Cultural Social Science Research | 1675 |

Sara Stevano and Kevin Deane

| 97 | Indigenous Statistics | 1691 |

Tahu Kukutai and Maggie Walter

| 98 | A Culturally Competent Approach to Suicide Research with Aboriginal and Torres Strait Islander Peoples | 1707 |

Monika Ferguson, Amy Baker, and Nicholas Procter

| 99 | Visual Methods in Research with Migrant and Refugee Children and Young People | 1723 |

Marta Moskal

| 100 | Participatory and Visual Research with Roma Youth | 1739 |

Oana Marcu

101 Drawing Method and Infant Feeding Practices Among Refugee Women 1757
June Joseph, Pranee Liamputtong, and Wendy Brodribb

102 Understanding Refugee Children's Perceptions of Their Well-Being in Australia Using Computer-Assisted Interviews 1777
Jeanette A. Lawrence, Ida Kaplan, and Agnes E. Dodds

103 Conducting Focus Groups in Terms of an Appreciation of Indigenous Ways of Knowing 1795
Norma R. A. Romm

104 Visual Depictions of Refugee Camps: (De)constructing Notions of Refugee-ness? 1811
Caroline Lenette

105 Autoethnography as a Phenomenological Tool: Connecting the Personal to the Cultural 1829
Jayne Pitard

106 Ethics and Research with Indigenous Peoples 1847
Noreen D. Willows

107 Conducting Ethical Research with People from Asylum Seeker and Refugee Backgrounds 1871
Anna Ziersch, Clemence Due, Kathy Arthurson, and Nicole Loehr

108 Ethical Issues in Cultural Research on Human Development 1891
Namrata Goyal, Matthew Wice, and Joan G. Miller

Section IV Sensitive Research Methodology and Approach: Researching with Particular Groups in Health Social Sciences 1905

109 Sensitive Research Methodology and Approach: An Introduction 1907
Pranee Liamputtong

110 "With Us and About Us": Participatory Methods in Research with "Vulnerable" or Marginalized Groups 1919
Jo Aldridge

111 Inclusive Disability Research 1935
Jennifer Smith-Merry

112 Understanding Sexuality and Disability: Using Interpretive Hermeneutic Phenomenological Approaches 1953
Tinashe Dune and Elias Mpofu

113	Ethics and Practice of Research with People Who Use Drugs ... Julaine Allan	1973
114	Researching with People with Dementia .... Jane McKeown	1991
115	Researching with Children .... Graciela Tonon, Lia Rodriguez de la Vega, and Denise Benatuil	2007
116	Optimizing Interviews with Children and Youth with Disability .... Gail Teachman	2023
117	Participant-Generated Visual Timelines and Street-Involved Youth Who Have Experienced Violent Victimization .... Kat Kolar and Farah Ahmad	2041
118	Capturing the Research Journey: A Feminist Application of Bakhtin to Examine Eating Disorders and Child Sexual Abuse .... Lisa Hodge	2061
119	Feminist Dilemmas in Researching Women's Violence: Issues of Allegiance, Representation, Ambivalence, and Compromise .... Lizzie Seal	2079
120	Animating Like Crazy: Researching in the Animated Visual Arts and Mental Welfare Fields .... Andi Spark	2093
121	Researching Underage Sex Work: Dynamic Risk, Responding Sensitively, and Protecting Participants and Researchers ...... Natalie Thorburn	2111
122	The Internet and Research Methods in the Study of Sex Research: Investigating the Good, the Bad, and the (Un)ethical .... Lauren Rosewarne	2127
123	Emotion and Sensitive Research .... Virginia Dickson-Swift	2145
124	Doing Reflectively Engaged, Face-to-Face Research in Prisons: Contexts and Sensitivities .... James E. Sutton	2163
125	Police Research and Public Health .... Jyoti Belur	2179

126	**Researching Among Elites** 2197
	Neil Stephens and Rebecca Dimond
127	**Eliciting Expert Practitioner Knowledge Through Pedagogy and Infographics** 2213
	Robert H. Campbell

Index 2225

About the Editor

Pranee Liamputtong is a medical anthropologist and Professor of Public Health at the School of Science and Health, Western Sydney University, Australia. Previously, Pranee held a Personal Chair in Public Health at the School of Psychology and Public Health, College of Science, Health and Engineering, La Trobe University, Melbourne, Australia, until January 2016. She has also previously taught in the School of Sociology and Anthropology and worked as a Public Health Research Fellow at the Centre for the Study of Mothers' and Children's Health, La Trobe University. Pranee has a particular interest in issues related to cultural and social influences on childbearing, childrearing, and women's reproductive and sexual health. She works mainly with refugee and migrant women in Melbourne and with women in Asia (mainly in Thailand, Malaysia, and Vietnam). She has published several books and a large number of papers in these areas.

Some of her books in the health and social sciences include *The Journey of Becoming a Mother Among Women in Northern Thailand* (Lexington Books, 2007); *Population Health, Communities and Health Promotion* (with Sansnee Jirojwong, Oxford University Press, 2009); *Infant Feeding Practices: A Cross-Cultural Perspective* (Springer, 2011); *Motherhood and Postnatal Depression: Narratives of Women and Their Partners* (with Carolyn Westall, Springer, 2011); *Health, Illness and Wellbeing: Perspectives and Social Determinants* (with Rebecca Fanany and Glenda Verrinder, Oxford University Press, 2012); *Women, Motherhood and Living with HIV/AIDS: A Cross-Cultural Perspective* (Springer, 2013); *Stigma, Discrimination and Living with HIV/AIDS: A Cross-Cultural*

Perspective (Springer, 2013); *Contemporary Socio-Cultural and Political Perspectives in Thailand* (Springer, 2014; *Children and Young People Living with HIV/AIDS: A Cross-Cultural Perspective* (Springer, 2016); *Public Health: Local and Global Perspectives* (Cambridge University Press, 2016, 2019); and *Social Determinants of Health: Individuals, Communities and Healthcare* (Oxford University Press, 2019).

Pranee is a Qualitative Researcher and has also published several method books. Her most recent method books include *Researching the Vulnerable: A Guide to Sensitive Research Methods* (Sage, 2007); *Performing Qualitative Cross-Cultural Research* (Cambridge University Press, 2010); *Focus Group Methodology: Principle and Practice* (Sage, 2011); *Qualitative Research Methods, 4th Edition* (Oxford University Press, 2013); *Participatory Qualitative Research Methodologies in Health* (with Gina Higginbottom, Sage, 2015); and *Research Methods in Health: Foundations for Evidence-Based Practice*, which is now in its third edition (Oxford University Press, 2017).

About the Contributors

Mauri K. Ahlberg was Professor of Biology and Sustainability Education at the University of Helsinki (2004–2013). On his 69th birthday, January 01, 2014, he had to retire. He has since the 1980s studied research methodology. He is interested in theory building and its continual testing as a core of scientific research. He has developed an integrating approach to research. He is an expert in improved concept mapping.

Farah Ahmad is an Associate Professor at the School of Health Policy and Management, Faculty of Health, York University. Applying equity perspective, she conducts mixed-method research to examine and improve the healthcare system for psychosocial health, vulnerable communities, access to primary care, and integration of eHealth innovations. The key foci in her research are mental health, partner violence, and cancer screening.

Jo Aldridge is a Professor of Social Policy and Criminology at Loughborough University, UK. She specializes in developing and using participatory research methods with "vulnerable" or marginalized people, including children (young carers), people with mental health problems and learning difficulties and women victims/survivors of domestic violence.

Julaine Allan is a Substance Use Researcher and Practitioner with over 30 years' experience in social work. Julaine's research is grounded in a human rights approach to working with stigmatized and marginalized groups. Julaine is also Senior Research Fellow and Deputy CEO at Lyndon, a substance treatment, research, and training organization. She holds conjoint positions at the National Drug and Alcohol Research Centre at UNSW and Charles Sturt University.

Jordan M. Alpert received his Ph.D. in Communication from George Mason University and is currently an Assistant Professor in the Department of Advertising at the University of Florida.

Andrea Montes Alvarado is an M.A. candidate in Sociology at California State University, Northridge. Her main research interests are in education, immigration, demography, and social inequality. Her current research explores the interpersonal relations and lived experiences of drag queens, specifically how drag queens construct and maintain families of choice within the drag community.

Stewart Anderson is a Professor of Biostatistics and Clinical and Translational Medicine at the University of Pittsburgh, Graduate School of Public Health.

His areas of current methodological research interests include (1) general methodology in longitudinal and survival data analysis, (2) modern regression techniques, and (3) methods in the design and analysis of clinical trials. He has over 25 years of experience in cancer clinical trial research in the treatment and prevention of breast cancer and in mental health in mid- to late-life adults.

Mike Armour is a Postdoctoral Research Fellow at NICM, Western Sydney University. Mike's research focus is on implementing experimental designs that can replicate complex clinical interventions that are often seen in the community. Mike has extensive experience in the design and conduct of clinical trials using a mixed-methods approach to help shape trial design, often including clinicians and community groups.

Amit Arora is a Senior Lecturer and National Health and Medical Research Council (NHMRC) Research Fellow in Public Health at Western Sydney University, Australia, where he teaches public health to undergraduate and postgraduate students. His research focuses on developing interventions to improve maternal and child health and oral health. Amit's research expertise includes mixed-methods research, health promotion, and life course approach in health research.

Kathy Arthurson is the Director of Neighborhoods, Housing and Health at Flinders Research Unit, Flinders University of South Australia. Her past experiences as a Senior Policy Analyst in a range of positions including public health, housing, and urban policy are reflected in the nature of her research, which is applied research grounded in broader concepts concerning social inclusion, inequality, and social justice.

Anna Bagnoli is an Associate Researcher in the Sociology Department of the University of Cambridge. Anna has a distinctive interest in methodological innovation and in visual, arts-based, and other creative and participatory approaches. She teaches postgraduates on qualitative analysis and CAQDAS and supervises postgraduates engaged with qualitative research projects. Her current work looks at the identity processes of migrants, with a focus on the internal migrations of Europeans, particularly young Italians.

Amy Baker is an academic in the Occupational Therapy Program at the University of South Australia. Her teaching and research focuses on mental health and suicide prevention, particularly working with people of culturally and linguistically diverse backgrounds and research approaches that are qualitative and participatory in nature.

Ruth Bartlett is an Associate Professor based in the Faculty of Health Sciences, University of Southampton, with a special interest in people with dementia and participatory research methods. Ruth has designed and conducted several funded research projects using innovative qualitative techniques, including diary and visual methods. Ruth has published widely in the health and social sciences and teaches and supervises postgraduate students.

Pat Bazeley has 25 years' experience in exploring, teaching, and writing about the use of software for qualitative and mixed-methods analysis for social and health research. Having previously provided training, consulting, and retreat facilities to researchers through Research Support P/L, she is now focusing on writing and researching in association with an Adjunct Professorial appointment at Western Sydney University.

Linda Liska Belgrave is an Associate Professor of Sociology in the Department of Sociology at the University of Miami. Her scholarly interests include grounded theory, medical sociology, social psychology, and social justice. She has pursued research on the daily lives of African-American caregivers of family with Alzheimer's disease, the conceptualization of successful aging, and academic freedom. She is currently working on research into the social meanings of infectious disease.

Robert F. Belli is a Professor of Psychology at the University of Nebraska–Lincoln. He received his Ph.D. in Experimental Psychology from the University of New Hampshire. Robert's research interests focus on the role of memory in applied settings, and his published work includes research on autobiographical memory, eyewitness memory, and the role of memory processes in survey response. The content of his work in surveys focuses on methods that can improve retrospective reporting accuracy.

Jyoti Belur is a Lecturer in the UCL Department of Security and Crime Science. She worked as a Lecturer before joining the Indian Police Service as a Senior Police Officer. She has undertaken research for the UK Home Office, College of Policing, ESRC, and the Met. Police. Her research interests include countering terrorism, violence against women and children, crime prevention, and police-related topics such as ethics and misconduct, police deviance, use of force, and investigations.

Denise Benatuil obtained her Ph.D. in Psychology from Universidad de Palermo, Argentina. She has a Licenciatura in Psychology from UBA, Argentina. She is the Director of the Psychology Program and Professor of Professional Practices in Psychology at Universidad de Palermo, Argentina. She is member of the Research Center in Social Sciences (CICS) at Universidad de Palermo, Argentina.

Sameer Bhole is the Clinical Director of Sydney Dental Hospital and Sydney Local Health District Oral Health Services and is also attached to the Faculty of Dentistry, the University of Sydney, as the Clinical Associate Professor. He has dedicated his career to address improvement of oral health for the disadvantaged populations with specific focus on health inequities, access barriers, and social determinants of health.

Mia Bierbaum is a Research Assistant at Macquarie University's Centre for Healthcare Resilience and Implementation Science, within the Australian Institute of Health Innovation. Mia had worked on the Healthy Living after Cancer project, providing a coaching service for cancer survivors, and conducted population monitoring research to examine the perceptions of cancer risk factors and behavioral change for patients. Her interests include health systems enhancement, behavioral research, and the prevention of chronic disease.

Simon Bishop is an Associate Professor of Organizational Behavior at Nottingham University Business School and a founding member of the Centre for Health Innovation and Learning. His research focuses around the relationship between public policy, organizational arrangements, and frontline practice in healthcare. His work has been published in a number of leading organization and policy and health sociology journals including *Human Relations*, *Journal of Public Administration Research and Theory*, and *Social Science & Medicine*.

Karen Block is a Research Fellow in the Jack Brockhoff Child Health and Wellbeing Program in the Centre for Health Equity at the University of Melbourne. Karen's research interests are social inclusion and health inequalities with a focus on children, young people, and families and working in collaborative partnerships with the community.

Chelsea Bond is a Munanjali and South Sea Islander Australian woman and Health Researcher with extensive experience in Indigenous primary healthcare, health promotion, and community development. Chelsea is interested in the emancipatory possibilities of health research for Indigenous peoples by examining the capabilities rather than deficiencies of Indigenous peoples, cultures, and communities.

Virginia Braun is a Professor in the School of Psychology at the University of Auckland, Aotearoa/New Zealand. She is a feminist and Critical Psychologist whose research has explored the intersecting areas of gender, bodies, sex/sexuality, and health/well-being across multiple topics and the possibilities, politics, and ethics of lives lived in neoliberal times.

Wendy Brodribb is an Honorary Associate Professor in the Primary Care Clinical Unit at the University of Queensland in Brisbane, Australia. She has a background as a medical practitioner focusing on women's health. Her Ph.D. investigated the breastfeeding knowledge of health professionals. Wendy's research interests are focused on infant feeding, especially breastfeeding, and postpartum care of mother and infant in the community following hospital discharge.

Mario Brondani is an Associate Professor and Director of Dental Public Health, University of British Columbia. Mario has developed a graduated program in dental public health (as a combined M.P.H. with a Diploma) in 2014 in which he currently directs and teaches. His areas of research include dental public health, access to care, dental geriatrics, psychometric measures, policy development, and dental education.

Jessica Broome has designed and applied qualitative and quantitative research programs for clients including top academic institutions, Fortune 500 companies, and grassroots community groups, since 2000. Jessica holds a Ph.D. in Survey Methodology from the University of Michigan.

Mark Brough is a Social Anthropologist with extensive experience in social research and teaching. Mark specializes in the application of qualitative methodologies to a wide range of health issues in diverse social contexts. He has a particular focus on strength-based approaches to community health.

Katerina Bryant completed a Bachelor of Laws and Bachelor of Arts degree in 2016. She is currently completing an honors degree in the Department of English and Creative Writing at Flinders University. Her nonfiction has been published widely, including in the *Griffith Review*, *Overland*, *Southerly*, and *The Lifted Brow*.

Lia Bryant is an Associate Professor in the School of Psychology, Social Work and Social Policy, University of South Australia, and is Director of the Centre for Social Change. She has vast experiences in working with qualitative methods and has published several books and refereed journal articles. She has authored two books and edited books. One of these relates specifically to research methodologies and methods. Lia also teaches innovation in research to social science students.

Anne Bunde-Birouste is Director of Yunus Social Business for Health Hub and Convener of the Health Promotion Program at the Graduate School of Public Health and Community Medicine, University of New South Wales in Sydney Australia. She specializes in innovative health promotion approaches for working with disadvantaged groups, particularly community-based research in sport for development and social change, working with vulnerable populations particularly youth.

Vivien Burr is a Professor of Critical Psychology at the University of Huddersfield. Her publications include *Invitation to Personal Construct Psychology* (2nd edition 2004, with Trevor Butt) and *Social Constructionism* (3rd edition 2015). Her research is predominantly qualitative, and she has a particular interest in innovative qualitative research methods arising from Personal Construct Psychology.

Valentina Buscemi is an Italian Physiotherapist with a strong interest in understanding and managing chronic pain conditions. She completed her M.Sc. in Neurorehabilitation at Brunel University London, UK, and is now a Ph.D. student at the Brain Rehabilitation and Neuroplasticity Unit, Western Sydney University. Her research aims to investigate the role of psychological stress in the development of chronic low back pain.

Rosalind Bye is the Director of Academic Program for Occupational Therapy in the School of Science and Health at Western Sydney University. Rosalind's research interests include adaptation to acquired disability, family caregiving experiences, and palliative care. Rosalind employs qualitative research methods that allow people to tell their story in an in-depth and meaningful way.

Breanne Byiers is a Researcher in the Department of Educational Psychology at the University of Minnesota. Her research interests include the development of measurement strategies to document changes in cognitive, health, and behavioral function among individuals with severe disabilities, assessment and treatment of challenging behavior, and single-subject experimental design methodology.

Fiona Byrne is a Research Officer in the Translational Research and Social Innovation (TReSI) Group, part of the School of Nursing and Midwifery at Western Sydney University. She works in the international support team for the Maternal Early Childhood Sustained Home-visiting (MECSH) program.

Robert Campbell is a Reader in Information Systems and Research Coordinator for the School of Creative Technologies at the University of Bolton. He holds a Ph.D. in this area and is on the board of the UK Academy for Information Systems. He has been professionally recognized as a Fellow of the British Computer Society and as a Senior Fellow of the Higher Education Academy. For many years, he was an Information Technology Specialist in the UK banking sector.

Michael J. Carter is an Associate Professor of Sociology at California State University, Northridge. His main research interests are in social psychology and microsociological theory, specifically the areas of self and identity. His current work examines how identities motivate behavior and emotions in face-to-face versus virtual environments. His research has appeared in a variety of academic journals, including *Social Psychology Quarterly* and *American Sociological Review*.

Rocco Cavaleri is a Ph.D. candidate and Associate Lecturer at Western Sydney University. He has been the recipient of the Australian Physiotherapy Association (APA) Board of Director's Student Prize and the New South Wales APA Award for academic excellence. Rocco has been conducting his research with the Brain Rehabilitation and Neuroplasticity Unit at Western Sydney since 2014. His research interests include chronic illnesses, continuing education, and exploring the nervous system using noninvasive technologies.

Jane Chalmers is a Lecturer in Physiotherapy at Western Sydney University and is also undertaking her Ph.D. through the University of South Australia. Jane's research focus has been on the assessment, management, and pathology of pelvic pain and in particular vulvodynia. She has extensive research experience using a range of methodologies and has recently used the Delphi technique in a novel way to create a new tool for assessing pelvic pain in women by using a patient-as-expert approach.

Nancy Cheak-Zamora is an Assistant Professor in the Department of Health Sciences at the University of Missouri. She conducts innovative research to inform policy-making, advocacy, service delivery, and research for youth with special healthcare needs and adults with chronic medical conditions.

Betty P.M. Chung, a graduate from the University of Sydney, Australia, has worked in the field of palliative and end-of-life care as practicing nurse and researcher. Her focus is particularly in psychosocial care for the dying and their families, family involvement to care delivery, good "quality of death," and well-being at old age. She has methodological expertise in qualitative interpretation and close-to-practice approaches in health sciences research.

Victoria Clarke is an Associate Professor in Sexuality Studies at the University of the West of England, Bristol, UK. She has published three prize-winning books, including most recently *Successful Qualitative Research: A Practical Guide for Beginners* with Virginia Braun (Sage), and over 70 papers in the areas of LGBTQ and feminist psychology and relationships, appearance psychology, human sexuality, and qualitative methods.

Bronwyne Coetzee is a Lecturer in the Department of Psychology at Stellenbosch University. She teaches introductory psychology and statistics to undergraduate students and neuropsychology to postgraduate students. Bronwyne's main interests are in health psychology, specifically in behavioral aspects of pediatric HIV and strengthening caregiver and child relationships to improve adherence to chronic medications.

Ana Lucía Córdova-Cazar holds a Ph.D. in Survey Research and Methodology from the University of Nebraska–Lincoln and is a Professor of Quantitative Methods and Measurement at the School of Social and Human Sciences of Universidad San Francisco de Quito, Ecuador. She has particular interests in the use of calendar and time diary data collection methods and the statistical analysis of complex survey data through multi-level modeling and structural equation modeling.

Jonathan C. Craig is a Professor of Clinical Epidemiology at the Sydney School of Public Health, the University of Sydney. His research aims to improve healthcare and clinical outcomes particularly in the areas of chronic kidney disease (CKD) and more broadly in child health through rigorous analysis of the evidence for commonly used and novel interventions in CKD, identifying gaps/inconsistency in the evidence, conducting methodologically sound clinical trials, and application of the research findings to clinical practice and policy.

Fiona Cram has tribal affiliations with Ngāti Pahauwera (Indigenous Māori, Aotearoa New Zealand) and is the mother of one son. She has over 20 years of Kaupapa Māori (by Māori, for Māori) research and evaluation experience with Māori (Indigenous peoples of Aotearoa New Zealand) tribes, organizations, and communities, as well as with government agencies, district health boards, and philanthropic organizations.

Anne Cusick is Professor Emeritus at Western Sydney University, Australia, Professor and Chair of Occupational Therapy at the University of Sydney, Editor in Chief of the *Australian Occupational Therapy Journal*, and inaugural Fellow of the Occupational Therapy Australia Research Academy. She has widely published in social, medical, health, and rehabilitation sciences.

Angela Dawson is a Public Health Social Scientist with expertise in maternal and reproductive health service delivery to priority populations in Australian and international settings. Angela has also undertaken research into innovative approaches to deliver drug and alcohol services to Aboriginal people. Angela is nationally recognized by the association for women in science, technology, engineering, mathematics, and medicine and is a recipient of the 2016 Sax Institute's Research Action Awards.

Kevin Deane is a Senior Lecturer in International Development at the University of Northampton, UK. His educational background is in development economics, but his research draws on a range of disciplines including political economy, development studies, economics, public health, and epidemiology. His research interests continue to focus on mobility and HIV risk, local value chains, transactional sex, and female economic empowerment in relation to HIV prevention.

Virginia Dickson-Swift was a Senior Lecturer in the La Trobe Rural Health School, La Trobe University, Bendigo, Australia. She has a wealth of experience in teaching research methods to undergraduate and postgraduate students throughout Australia. Her research interests lie in the practical, ethical, and methodological challenges of undertaking qualitative research on sensitive topics, and she has published widely in this area.

Rebecca Dimond is a Medical Sociologist at Cardiff University. Her research interests are patient experiences, clinical work, the classification of genetic syndromes and their consequences, and reproductive technologies. Her work is currently funded by an ESRC Future Research Leaders Award.

Agnes Dodds is an Educator and Evaluator and Associate Professor in the Department of Medical Education, the University of Melbourne. Agnes' current research interests are the developmental experiences of young people. Her current projects include studies of the development of medical students, selection into medical school, and school-related experiences of refugee children.

Suzanne D'Souza is a literacy tutor with the School of Nursing and Midwifery at Western Sydney University. Suzanne is keenly interested in the academic writing process and is actively involved with designing literacy strategies and resources to strengthen students' writing. Suzanne is also a doctoral candidate researching the hybrid writing practices of nursing students and their implications for written communicative competence.

Joan L. Duda is a Professor of Sport and Exercise Psychology in the School of Sport, Exercise and Rehabilitation Sciences at the University of Birmingham, UK. Joan is one of the most cited researchers in her discipline and is internationally known for her expertise on motivational processes and determinants of adherence and optimal functioning within physical and performance-related activities such as sport, exercise, and dance.

Clemence Due is a Postdoctoral Research Fellow in the Southgate Institute for Health, Society and Equity at Flinders University. She is the author of more than 50 peer-reviewed academic articles or book chapters, with a primary focus on research with adults and children with refugee backgrounds. Her research has focused on trauma, child well-being, housing and health, access to primary healthcare, and oral health.

Tinashe Dune is a Clinical Psychologist and Health Sociologist with significant expertise in sexualities and sexual and reproductive health. Her work focuses on the experiences of the marginalized and hidden populations using mixed methods and participatory action research frameworks. She is a recipient of a Freilich Foundation Award (2015) and one of Western's *Women Who Inspire* (2016) for her work on improving diversity and inclusivity in health and education.

Jane Edwards is a Creative Arts Therapy Researcher and Practitioner and is currently Associate Professor for Mental Health in the Faculty of Health at Deakin University. She has conducted research into the uses of the arts in healthcare in a range of areas. She is the Editor in Chief for *The Arts in Psychotherapy* and edited *The Oxford Handbook of Music Therapy* (2016). She is the inaugural President of the International Association for Music & Medicine.

Carolyn Ee is a GP and Research Fellow at NICM, Western Sydney University. Carolyn has significant experience in randomized controlled trials having conducted a large NHMRC-funded trial on acupuncture for menopausal hot flushes for her Ph. D. Carolyn combines clinical practice as a GP and acupuncturist with a broad range of research skills including health services and translational research as well as mixed-methods approaches to evaluating the effectiveness of interventions in the field of women's health.

Ulrike Felt is a Professor in the Department of Science and Technology Studies and Dean of the Faculty of Social Sciences at the University of Vienna. Her research focuses on public engagement and science communication; science, democracy, and governance; changing knowledge cultures and their institutional dimensions; as well as development in qualitative methods. Her work is often comparative between national context and technoscientific fields (especially life sciences, biomedicine, sustainability research, and nanotechnologies).

Monika Ferguson is a Research Associate in the Mental Health and Substance Use Research Group at the University of South Australia. Her current program of research focuses on educational interventions to reduce stigma and improve support for people at risk of suicide, both in the health sector and in the community.

Pota Forrest-Lawrence is a University Lecturer in the School of Social Science and Psychology, Western Sydney University. Pota holds a Ph.D. in Criminology from the University of Sydney Law School. Her research interests include drug law and policy, criminological and social theory, criminal law, and people and risk. She has published in the areas of cyber security, illicit drugs, and media.

Bronwyn Fredericks is an Indigenous Australian woman from Southeast Queensland (Ipswich/Brisbane) and is Professor and Pro Vice-Chancellor (Indigenous Engagement) and BHP Billiton Mitsubishi Alliance Chair in Indigenous Engagement at Central Queensland University, Australia. Bronwyn is a member of the National Indigenous Research and Knowledges Network and the Australian Institute of Aboriginal and Torres Strait Islander Studies.

Caroline Fryer is a Physiotherapist and a Lecturer in the School of Health Sciences at the University of South Australia. Her interest in equity in healthcare for people from culturally diverse backgrounds began as a clinician and translated to a doctoral study investigating the experience of healthcare after stroke for older people with limited English proficiency.

Emma George is a Lecturer in Health and Physical Education at Western Sydney University. Emma teaches across a range of health science subjects with a focus on physical activity, nutrition, health promotion, and evidence-based research methodology. Her research aims to promote lifelong physical activity and improve health outcomes. Emma's research expertise includes men's health, intervention design, implementation and evaluation, mixed-methods research, and community engagement.

Lilian A. Ghandour is an Associate Professor at the Faculty of Health Sciences, American University of Beirut (AUB). She holds a Ph.D. from the Johns Hopkins Bloomberg School of Public Health and a Master of Public Health (M.P.H.) from the American University of Beirut. Her research focuses on youth mental health and substance use, and Dr. Ghandour has been involved in the implementation and analyses of several local and international surveys. She has over 40 publications in high-tier peer-reviewed international journals.

Lisa Gibbs is an Associate Professor and Director of the Jack Brockhoff Child Health and Wellbeing Program at the University of Melbourne. She leads a range of complex community-based public health studies exploring sociocultural and environmental influences on health and well-being.

Claudia Gillberg is a Research Associate with the Swedish National Centre for Lifelong Learning (ENCELL) at Jonkoping University, Sweden. Her research interests include feminist philosophy, social citizenship, lifelong learning, participation, methodology, and ethics. Claudia divides her time between the UK and Sweden, closely following developments in health politics in both countries and worldwide. In her work, she has expressed concerns about the rise in human rights violations of the sick and disabled.

Brenda Gladstone is Associate Director of the Centre for Critical Qualitative Health Research and Assistant Professor at Dalla Lana School of Public Health,

University of Toronto. Brenda's research uniquely examines intergenerational experiences and effects of mental health and illness, focusing on parental mental illnesses. She teaches graduate-level courses on qualitative methodologies and uses innovative visual and participatory research methods to bring young people's voices into debates about their mental/health and social care needs.

Namrata Goyal is a Postdoctoral Fellow at the New School for Social Research. She received her Ph.D. in Psychology from the New School for Social Research and completed her B.A. at York University. Her research interests include cultural influences on reciprocity, gratitude, and social support norms, as well as developmental perspectives on social expectations and motivation.

Adyya Gupta is a Research Scholar in the School of Public Health at the University of Adelaide, Australia. She has an expertise in applying mixed methods. Her interests are in social determinants of health, oral health, and behavioral epidemiology.

Talia Gutman is a Research Officer at the Sydney School of Public Health, the University of Sydney. She has an interest in patient and caregiver involvement in research in chronic kidney disease and predominantly uses qualitative methods to elicit stakeholder perspectives with the goal of informing patient-centered programs and interventions. She has conducted focus groups with nominal group technique both around Australia and overseas as part of global studies.

Elizabeth Halcomb is a Professor of Primary Healthcare Nursing at the University of Wollongong. Elizabeth has taught research methods at an undergraduate and postgraduate level and has been an active supervisor of doctoral candidates. Her research interests include nursing workforce in primary care, chronic disease management, and lifestyle risk factor reduction, as well as mixed-methods research.

Marit Haldar is a Professor of Sociology. Her studies are predominantly on childhood, gender, and families. Her most recent focus is on the vulnerable subjects of the welfare state and inequalities in healthcare. Her perspectives are to understand the culturally and socially acceptable in order to understand what is unique, different, or deviant. She has a special interest in the development of new methodologies and analytical strategies.

Camilla S. Hanson is a Research Officer at the Sydney School of Public Health, University of Sydney. Her research interest is in psychosocial outcomes for patients with chronic disease. She has extensive experience in conducting systematic reviews and qualitative and mixed-methods studies involving pediatric and adult patients, caregivers, and health professionals worldwide.

John D. Hathcoat is an Assistant Professor of Graduate Psychology and Associate Director of University Learning Outcomes Assessment in the Center for Assessment and Research Studies at James Madison University. John has taught graduate-level courses in educational statistics, research methods, measurement theory, and performance assessment. His research focuses on instrument development, validity theory, and measurement issues related to "authentic" assessment practices in higher education.

Nikki Hayfield is a Senior Lecturer in Social Psychology in the Department of Health and Social Sciences at the University of the West of England, Bristol,

UK. Her Ph.D. was an exploration of bisexual women's visual identities and bisexual marginalization. In her research she uses qualitative methodologies to explore LGB and heterosexual sexualities, relationships, and (alternative) families.

Lisa Hodge is a Lecturer in Social Work in the College of Health and Biomedicine, Victoria University, and an Adjunct Research Fellow in the School of Nursing and Midwifery, University of South Australia. Her primary research interests include eating disorders and sexual trauma in particular and mental health more broadly, as well as self-harm, the sociology of emotions, and the use of creative expression in research methodologies.

Anne Hogden is a Research Fellow at Macquarie University's Centre for Healthcare Resilience and Implementation Science, within the Australian Institute of Health Innovation, and a Visiting Fellow at the Centre for Health Stewardship, Australian National University (ANU). Her expertise is in healthcare practice and research. Her research uses qualitative methodology, and she is currently focusing on patient-centered care, patient decision-making, and healthcare communication in Motor Neurone Disease.

Nicholas Hookway is a Lecturer in Sociology in the School of Social Sciences at the University of Tasmania, Australia. Nick's principle research interests are morality and social change, social theory, and online research methods. He has published recently in *Sociology* and *The British Journal of Sociology*, and his book *Everyday Moralities: Doing It Ourselves in an Age of Uncertainty* (Routledge) is forthcoming in 2017. Nick is current Co-convener of the Australian Sociological Association Cultural Sociology group.

Kirsten Howard is a Professor of Health Economics in the Sydney School of Public Health at the University of Sydney. Her research focuses on methodological and applied health economics research predominantly in the areas of the assessment of patient and consumer preferences using discrete choice experiment (DCE) methods as well as in economic evaluation and modeling. She has conducted many discrete choice experiments of patient and consumer preferences in diverse areas.

Martin Howell is a Research Fellow in Health Economics in the Sydney School of Public Health at the University of Sydney. His research focuses on applied health economics predominantly in the areas of assessment of preferences using discrete choice experiment (DCE) methods applied to complex health questions including kidney transplant and the trade-offs to avoid adverse outcomes of immunosuppression, preferences, and priorities for the allocation of deceased donor organs and relative importance of outcomes in nephrology.

Erin Jessee is a Lord Kelvin Adam Smith Research Fellow on Armed Conflict and Trauma in the Department of Modern History at the University of Glasgow. She works primarily in the fields of oral history and genocide studies and has extensive experience conducting fieldwork in conflict-affected settings, including Rwanda, Bosnia-Hercegovina, and Uganda. Her research interests also include the ethical and methodological challenges that surround conducting qualitative fieldwork in highly politicized research settings.

James Rufus John is a Research Assistant and Tutor in Epidemiology at Western Sydney University. He has published quantitative research articles on community perception and attitude toward oral health, utilization of dental health services, and public health perspectives on epidemic diseases and occupational health.

Craig Johnson is from the Ngiyampaa tribe in western NSW. Craig did over 30 years of outdoor and trade-based work; he moved to Dubbo in 1996 and became a trainee health worker in 2010. Craig is now a registered Aboriginal Health Practitioner and qualified diabetes educator and valued member of the Dubbo Diabetes Unit at Dubbo Base Hospital. Craig is dedicated to lifelong learning and improving healthcare for Aboriginal people.

Monica Johnson is from the Ngiyampaa tribe in western NSW. Monica has always been interested in health and caring for people and has worked in everything from childhood immunization to dementia. Monica works for Marathon Health (previously Western NSW Medicare Local) in an audiology screening program in western NSW. Monica has qualifications in nursing and has been an Aboriginal Health Worker for the past 8 years.

Stephanie Jong is a Ph.D. candidate and a sessional academic staff member in the School of Education at Flinders University, South Australia. Her primary research interests are in social media, online culture, and health. Stephanie's current work adopts a sociocultural perspective using netnography to understand how online interactions influence health beliefs and practices.

June Mabel Joseph is a Ph.D. candidate in the University of Queensland. She has a keen interest in researching vulnerable groups of populations – with a keen interest on mothers from refugee background in relation to their experiences of displacement and identity. She is also passionate in the field of breastfeeding, maternal and infant health, sociology, and Christian theology.

Angela Ju is a Research Officer at the Sydney School of Public Health, the University of Sydney. Her research interest is in the development and validation of patient-reported outcome measures. She has experience in conducting primary qualitative studies with patients and health professionals and in systematic review of qualitative health research in the areas of chronic disease including chronic kidney disease and cardiovascular disease.

Yong Moon Jung, with a long engagement in quantitative research, has strong expertise and extensive experience in statistical analysis and modeling. He has employed quantitative research skills for the development and evaluation purposes at either a policy or program level. He has also been teaching quantitative methods and data analysis for the undergraduate and postgraduate courses. Currently, he is working as part of the Quality and Analytics Group of the University of Sydney.

Ashraf Kagee is a Distinguished Professor of Psychology at Stellenbosch University. His main interests are in health psychology, especially HIV and mental health, health behavior theory, and stress and trauma. He has an interest in global mental health and lectures on research methods, cognitive psychotherapy, and psychopathology.

Ida Kaplan is a Clinical Psychologist and Director of Direct Services at the Victorian Foundation for Survivors of Torture. Ida is a specialist on research and practice in the area of trauma, especially for refugees and asylum seekers.

Bernie Kemp was born in Wilcannia and is a member of the Barkindji people. Bernie has spent most of the past 25 years working in Aboriginal health in far western NSW providing health checks and preventing and treating chronic disease. Bernie has qualifications in nursing and diabetes education and is a registered Aboriginal Health Practitioner. Bernie recently relocated to Dubbo to be closer to family and works at the Dubbo Regional Aboriginal Health Service.

Lynn Kemp is a Professor and Director of the Translational Research and Social Innovation (TReSI), School of Nursing and Midwifery, Western Sydney University. Originally trained as a Registered Nurse, Lynn has developed a significant program of community-based children and young people's research that includes world and Australian-first intervention trials. She is now leading an international program of translational research into the implementation of effective interventions at population scale.

Priya Khanna holds a Ph.D. in Science Education and Master's degrees in Education and Zoology. She has been working as a Researcher in medical education for more than 10 years. Presently, she is working as an Associate Lecturer, Assessment at the Sydney Medical School, University of Sydney, New South Wales.

Alexandra King graduated from the University of Tasmania in December 2014 with a Ph.D. in Rural Health. Her doctoral thesis is entitled "Food Security and Insecurity in Older Adults: A Phenomenological Ethnographic Study." Alexandra's research interests include social gerontology, social determinants of health, ethnographic phenomenology, and qualitative research methods.

Kat Kolar is a Ph.D. candidate, Teaching Assistant, and Research Analyst in the Department of Sociology, University of Toronto. She is also completing the Collaborative Program in Addiction Studies at the University of Toronto, a program held in collaboration with the Centre for Addiction and Mental Health, the Canadian Centre on Substance Abuse, and the Ontario Tobacco Research Unit.

Gary L. Kreps received his Ph.D. in Communication from the University of Southern California. He currently serves as a University Distinguished Professor and Director of the Center for Health and Risk Communication at George Mason University.

Tahu Kukutai belongs to the Waikato, Ngāti Maniapoto, and Te Aupouri tribes and is an Associate Professor at the National Institute of Demographic and Economic Analysis, University of Waikato. Tahu specializes in Māori and Indigenous demographic research and has written extensively on issues of Māori and tribal population change, identity, and inequality. She also has an ongoing interest in how governments around the world count and classify populations by ethnic-racial and citizenship criteria.

Ian Lacey comes from a professional sporting background and was contracted to the Brisbane Broncos from 2002 to 2007. Ian has expertise in smoking cessation, completing the University of Sydney Nicotine Addiction and Smoking Cessation

Training Course and the Smoke Check Brief Intervention Training. He has recently completed a Certificate IV in Frontline Management and Diploma in Management.

Michelle LaFrance teaches graduate and undergraduate courses in writing course pedagogy, ethnography, cultural materialist, and qualitative research methodologies. She has published on peer review, preparing students to write across the curriculum, e-portfolios, e-research, writing center and WAC pedagogy, and institutional ethnography. She has written several texts and her upcoming book, *Institutional Ethnography: A Theory of Practice for Writing Studies Researchers*, will be published by Utah State Press in 2018.

Kari Lancaster is a Scientia Fellow at the Centre for Social Research in Health, UNSW, Sydney. Kari is a Qualitative Researcher who uses critical policy study approaches to contribute to contemporary discussions about issues of political and policy significance in the fields of drugs and viral hepatitis. She has examined how policy problems and policy knowledges are constituted and the dynamics of "evidence-based" policy.

Jeanette Lawrence is a Developmental Psychologist and Honorary Associate Professor of the Melbourne School of Psychological Sciences. Jeanette's current research specializes in developmental applications to cultural and refugee studies. With Ida Kaplan, she is developing age- and culture-appropriate computer-assisted interviews to assist refugee and disadvantaged children to express their thoughts, feelings, and activities.

Eric Leake is an Assistant Professor of English at Texas State University. His areas of research include the intersection of rhetoric and psychology as well as civic literacies and writing pedagogy.

Raymond M. Lee is Emeritus Professor of Social Research Methods at Royal Holloway University of London. He has written extensively about a range of methodological topics including the problems and issues involved in research on "sensitive" topics, research in physically dangerous environments, and the use of unobtrusive measures. He also provides support and advice to the in-house researchers at Missing People, a UK charity that works with young runaways, missing and unidentified people, and their families.

Caroline Lenette is a Lecturer of Social Research and Policy at the University of New South Wales, Australia. Caroline's research focuses on refugee and asylum seeker mental health and well-being, forced migration and resettlement, and arts-based research in health, particularly visual ethnography and community music.

Doris Leung graduated from the University of Toronto and worked there as a Mental Health Nurse and Researcher focused on palliative end-of-life care. Since moving to Hong Kong, she works at the School of Nursing, the Hong Kong Polytechnic University. Her expertise is in interpretive and post-positivistic qualitative research approaches.

Nicole Loehr is a Ph.D. candidate in the School of Social and Policy Studies at Flinders University. Her doctoral work examines the integration of the not-for-profit, governmental, and commercial sectors in meeting the long-term housing needs of resettled refugees and asylum seekers in the private rental market. She has worked as a Child and Family Therapist and in community development and youth work roles.

Janet Long is a Health Services Researcher at the Australian Institute of Health Innovation, Macquarie University, Australia. She has a background in nursing and biological science. She has published a number of studies using social network analysis of various health and medical research settings to demonstrate silos, key players, brokerage and leadership, and strategic network building. Her other research interests are in behavior change, complexity, and implementation science.

Colin MacDougall is concerned with equity, ecology, and healthy public policy. He is currently Professor of Public Health at Flinders University and Executive member of the Southgate Institute of Health, Society and Equity. He is a Principal Fellow (Honorary) at the Jack Brockhoff Child Health and Wellbeing Program at the University of Melbourne. He studies how children experience/act on their worlds.

Catrina A. MacKenzie is a Mixed-Methods Researcher investigating how incentives, socioeconomic conditions, and household well-being influence conservation attitudes and behaviors, with a particular focus on tourism revenue sharing, loss compensation, and resource extraction. For the last 8 years, she has worked in Uganda studying the spatial distribution of perceived and realized benefits and losses accrued by communities as a result of the existence of protected areas.

Freya MacMillan is a Lecturer in Health Science at Western Sydney University, where she teaches in health promotion and interprofessional health science. Her research focuses on the development and evaluation of lifestyle interventions for the prevention and management of diabetes in those most at risk. Particularly relevant to this chapter, she has expertise in working with community to develop appropriate and appealing community-based interventions.

Wilson Majee is an Assistant Professor in the Master of Public Health program and Health Sciences Department at the University of Missouri. He is a Community Development Practitioner and Researcher whose primary research focuses on community leadership development in the context of engaging and empowering local residents in resource-limited communities in improving their health and well-being.

Karine E. Manera is a Research Officer at the Sydney School of Public Health, the University of Sydney. She uses qualitative and quantitative research methods to generate evidence for improving shared decision-making in the area of chronic kidney disease. She has experience in conducting focus groups with nominal group technique and has applied this approach in global and multi-language studies.

Narendar Manohar is a National Health and Medical Research Council Postgraduate Research candidate and a Research Assistant at Western Sydney University, Australia. He teaches public health and evidence-based practice to health science students and has published several quantitative research studies on several areas of public health. His interests include evidence-based practice, quantitative research, and health promotion.

Oana Marcu is a Researcher at the Faculty of Political and Social Sciences of the Catholic University in Milan. Her interests include qualitative methods (the ethnographic method, visual and participatory action research). She has extensively worked with Roma in Europe on migration, transnationalism, intersectionality, and health. She teaches social research methodology and sociology of migration.

Anastasia Marinopoulou is teaching political philosophy, political theory, and epistemology at the Hellenic Open University and is the coeditor of the international edition *Philosophical Inquiry*. Her publications include the recent monograph *Critical Theory and Epistemology: The Politics of Modern Thought and Science* (Manchester University Press, 2017). Her latest research award was the Research Fellowship at the University of Texas at Austin (2017).

Kate McBride is a Lecturer in Population Health in the School of Medicine, Western Sydney University. Kate teaches population health, basic and intermediate epidemiology, and evidence-based medicine to undergraduate and postgraduate students and has also taught at Sydney University. Kate's research expertise is in epidemiology, public health, and the use of mixed methods to improve health at a population level through the prevention of and reduction of chronic and non-communicable disease prevalence.

Angela McGrane is Director of Review and Approvals at Newcastle Business School, Northumbria University. She is currently investigating the effect of placement experience on student perceptions of themselves in a work role through a longitudinal study following participants from first year to graduation.

Jane McKeown is a Mental Health Nurse specializing in dementia research, practice, and education, working for the University of Sheffield and Sheffield Health and Social Care NHS Foundation Trust. Jane has a special interest in developing and implementing approaches that enable people with dementia to have their views and experiences heard. She also has an interest in the ethical aspects of research with people with dementia.

Karen McPhail-Bell is a Qualitative Researcher whose interest lies in the operation of power in relation to people's health. Karen facilitates strength-based and reciprocal processes in support of community-controlled and Aboriginal and Torres Strait Islander-led agendas. She has significant policy and program experience across academic, government and non-government roles in health equity, health promotion, and international development.

Joan Miller is Professor of Psychology at the New School for Social Research and Director of Undergraduate Studies in Psychology at Lang College. Her research interests center on culture and basic psychological theory, with a focus on interpersonal motivation, theory of mind, social support, moral development, as well as family and friend relationships.

Rodrigo Mariño is a Public Health Dentist and a Principal Research Fellow at the Oral Health Cooperative Research Centre (OH-CRC), the University of Melbourne. Rodrigo has an excellent publication profile and research expertise in social epidemiology, dental workforce issues, public health, migrant health, information and communication technology, gerontology, and population oral health. Rodrigo has also been a consultant to the Pan American Health Organization/World Health Organization in Washington, DC.

Anna Maurer-Batjer is a Master of Social Work and Master of Public Health student at the University of Missouri. She has a special interest in maternal and child health, particularly in terms of neurocognitive development. Through her graduate research assistantship, she has explored the use of photovoice and photo-stories among youth with Autism Spectrum Disorder.

Cara Meixner is an Associate Professor of Psychology at James Madison University, where she also directs the Center for Faculty Innovation. A scholar in brain injury advocacy, Cara has published mixed-methods studies and contributed to methodological research on this genre of inquiry. Also, Cara teaches a doctoral-level course on mixed-methods research at JMU.

Zelalem Mengesha has Bachelor's and Master's qualifications in Public Health with a special interest in sexual and reproductive health. He is passionate about researching the sexual and reproductive health of culturally and linguistically diverse communities in Australia using qualitative and mixed-methods approaches.

Dafna Merom is a Professor in the School of Science and Health, Western Sydney University, Australia. She is an expert in the area of physical activity measurement, epidemiology, and promotion. She has recently led several clinical trials comparing the incidence of falls, heart health, and executive functions of older adults participating in complex motor skills such as dancing and swimming versus simple functional physical activity such as walking.

Magen Mhaka-Mutepfa is a Senior Lecturer in the Department of Psychology, Faculty of Social Sciences, the University of Botswana. Previously, she was a Teaching Assistant in the School of Public Health at the University of Sydney (NSW) and was a recipient of Australia International Postgraduate Research Scholarship. Her main research interests are on health and well-being, HIV and AIDS, and ageing.

Christine Milligan is a Professor of Health and Social Geography and Director of the Centre for Ageing Research at Lancaster University, UK. With a keen interest in innovative qualitative techniques, Christine is an active researcher who has led on both national and international research projects. She has published over a hundred books, journal articles, and book chapters – including a recently published book (2015) with Ruth Bartlett on *What Is Diary Method?*

Naomi Moller is a Chartered Psychologist and Lecturer in Psychology at the Open University. Trained as a Counseling Psychologist, she recently coedited with Andreas Vossler *The Counselling and Psychotherapy Research Handbook* (Sage). Her research interests include perceptions and understandings of counselors and counseling, relationship, and family research including infidelity.

Marta Moskal is a Sociologist and Human Geographer based at the University of Glasgow, UK. Her research lies within the interdisciplinary area of migration and mobility studies, with a particular interest in family, children, and young people's experiences and in using visual research methods.

Elias Mpofu is a Rehabilitation Counseling Professional with a primary research focus in community health services intervention design, implementation, and evaluation applying mixed-methods approaches. His specific qualitative inquiry orientation is interpretive phenomenological analysis to understand meanings around and actions toward health-related quality of community living with chronic illness or disability.

Sally Nathan is a Senior Lecturer at the School of Public Health and Community Medicine, UNSW, Sydney. She has undertaken research into consumer and community participation in health as well as research approaches which engage and partner directly with vulnerable and marginalized communities and the organizations that represent and advocate for them, including a focus on adolescent drug and alcohol treatment and Aboriginal health and well-being.

Alison Nelson is an occupational therapist with extensive research, teaching, and practice experience working alongside urban Aboriginal and Torres Strait Islander people. Alison has a particular interest in developing practical strategies which enable non-Indigenous students, researchers, and practitioners to understand effective ways of working alongside Aboriginal and Torres Strait Islander Australians.

Ingrid A. Nelson is an Assistant Professor of Sociology at Bowdoin College, in Brunswick, Maine, USA. Her research examines the ways that families, schools, communities, and community-based organizations support educational attainment among marginalized youth. Her current work examines the educational experiences of rural college graduates, through the lens of social capital theory.

Christy E. Newman is an Associate Professor and Social Researcher of health, sexuality, and relationships at the Centre for Social Research in Health, UNSW, Sydney. Her research investigates social aspects of sexual health, infectious disease, and chronic illness across diverse contexts and communities. She has a particular passion for promoting sociological and qualitative approaches to understanding these often culturally and politically sensitive areas of health and social policy.

Mark C. Nicholas is Executive Director of Institutional Assessment at Framingham State University. He has published qualitative research on program review, classroom applications of critical thinking, and its assessment.

Kayi Ntinda is a Lecturer of Educational Counseling and Mixed-Methods Inquiry Approaches at the University of Swaziland. Her research interests are in assessment, learner support services service learning, and social justice research among vulnerable populations. She has authored and coauthored several research articles and book chapters that address assessment, inclusive education, and counseling practices in the diverse contexts.

Chijioke Obasi is a Principal Lecturer in Equality and Diversity at the University of Wolverhampton, England. She is a qualified British Sign Language/English Interpreter of Nigerian Origin and was brought up in a bilingual and bicultural household.

Felix Ogbo is a Lecturer in the School of Medicine, Western Sydney University. Felix studied at the University of Benin (M.B.B.S.), Benue State University (M.H.M.), and Western Sydney University – M.P.H. (Hons.) and Ph.D. He is an active collaborator in the landmark Global Burden of Disease Study – the world's largest scientific effort to quantify the magnitude of health loss from all major diseases, injuries, and risk factors by year-age-sex-geography and cause, which are essential to health surveillance and policy decision-making.

Andrew Page is a Professor of Epidemiology in the School of Medicine, Western Sydney University. He has been teaching basic and intermediate epidemiology and population health courses to health sciences students for 10 years and has published over 150 research articles and reports across a diverse range of population health topics. He has been a Research Associate at the University of Bristol and has also worked at the University of Queensland and University of Sydney in Australia.

Bonnie Pang is a Lecturer at the Western Sydney University and a school-based member of the Institute for Culture and Society. As a Sociocultural Researcher, she specializes in ethnographic methods, social theories, and youth health and physical activity. She has over 10 years of research experiences in exploring Chinese youth communities in school, familial, and neighborhood environments in Australia and Hong Kong.

Gemma Pearce is a Chartered Psychologist currently working in the Health Behaviour and Interventions Research (BIR) group at Coventry University, UK. Her research focuses on women's health, public health, and self-management support of people with long-term conditions. She specializes in systematic reviews and pragmatic research methods (including qualitative, quantitative, and mixed-methods research). She developed a synchronous method of online interviewing when conducting her Ph.D.

Karen Peres is an Associate Professor of Population Oral Health and Director of the Dental Practice Education Research Unit (DPERU) at the University of Adelaide. Her main research interests are in the epidemiology of oral diseases, particularly in social inequality in oral health, mother and child oral health-related issues, and life course and with over 100 publications in peer-reviewed journals.

Janette Perz is a Professor of Health Psychology and Director of the Translational Health Research Institute, Western Sydney University. She researches in the field of reproductive and sexual health with a particular focus on gendered experiences, subjectivity, and identity. She has demonstrated expertise in research design and analysis and mixed-methods research.

Rona Pillay (nee' Tranberg) is a Lecturer and Academic Course Advisor BN Undergraduate. She has unit coordinated research units and teaches primarily in research. Her research interests are in health communication, teams and teamwork, decision-making, and patient safety based on ethnomethodology and conversational analysis. Her additional interests lie in communication of cancer diagnosis in Aboriginal people and treatment decision-making choices.

Jayne Pitard is a Researcher in education at Victoria University, Melbourne, Australia. She has completed a thesis focused on her teaching of a group of students from Timor-Leste. She has held various teaching and research positions within Victoria University for the last 27 years.

Jennifer Prattley is a Research Associate at the Cathie Marsh Institute for Social Research, University of Manchester, UK. She works across a variety of ageing and life course studies, with research questions relating to women's retirement, frailty, and social exclusion. Jennifer has expertise in the analysis of longitudinal data and an interest in applying and appraising new and innovative quantitative methods. She also maintains an interest in statistics and mathematics education and the teaching of advanced methods to nonspecialists.

Nicholas Procter is a Professor and Chair of Mental Health Nursing and Convener of the Mental Health and Substance Use Research Group at the University of

South Australia. Nicholas has over 30 years' experience as a mental health clinician and academic. He works with various organizations within the mental health sector providing education, consultation, and research services in the suicide prevention and trauma-informed practice areas.

Marsha B. Quinlan is a Medical Anthropologist focusing on the intersection of health with ethnobiology and an Associate Professor in the Department of Anthropology and affiliate of the School for Global Animal Health at Washington State University, Pullman, WA, USA. Her research concentrates on ethnomedical concepts including ethnophysiology, ethnobotany, and ethnozoology, along with health behavior in families, and psychological anthropology.

Frances Rapport is a Professor of Health Implementation Science at Macquarie University's Centre for Healthcare Resilience and Implementation Science within the Australian Institute of Health Innovation. She is a social scientist with a background in the arts. Frances has won grants within Australia and the United Kingdom, including the Welsh Assembly Government, the National Institute for Health Research in England, and Cochlear Ltd. to examine the role of qualitative and mixed methods in medical and health services research.

Liz Rix is a Researcher and Academic working with Gnibi College of Indigenous Australian Peoples at Southern Cross University, Lismore. She is also a practicing Registered Nurse, with clinical expertise in renal and aged care nursing. Liz currently teaches Indigenous health to undergraduate and postgraduate health and social work professionals. Her research interests include Indigenous health and chronic disease, reflexive practice, and improving the cultural competence of non-Indigenous clinicians.

Norma R. A. Romm is a Research Professor in the Department of Adult Education and Youth Development at the University of South Africa. She is author of *The Methodologies of Positivism and Marxism: A Sociological Debate* (1991); *Accountability in Social Research: Issues and Debates* (2001); *New Racism* (2010); *People's Education in Theoretical Perspective: Towards the Development* (with V. McKay 1992); *Diversity Management: Triple Loop Learning* (with R. Flood 1996); and *Assessment of the Impact of HIV and AIDS in the Informal Economy of Zambia* (with V. McKay 2006).

Rizwana Roomaney is a Research Psychologist, Registered Counselor, and Lecturer in the Psychology Department at Stellenbosch University. She teaches research methods and quantitative data analysis to undergraduate and postgraduate students. Rizwana's main interests are health psychology and research methodology. She is particularly interested in women's health and reproductive health.

Lauren Rosewarne is a Senior Lecturer at the University of Melbourne, Australia, specializing in gender, sexuality, media, and popular culture. Her most recent books include *American Taboo: The Forbidden Words, Unspoken Rules, and Secret Morality of Popular Culture* (2013); *Masturbation in Pop Culture: Screen, Society, Self* (2014); *Cyberbullies, Cyberactivists, Cyberpredators: Film, TV, and Internet Stereotypes* (2016); and *Intimacy on the Internet: Media Representations of Online Connections* (2016).

Simone Schumann is a doctoral candidate and former Lecturer and Researcher in the Department of Science and Technology Studies, University of Vienna. Her main research interest is in public engagement with emerging technologies, especially on questions of collective sense-making and the construction of expertise/power relations within group settings. In her dissertation project, she focuses on the case of nano-food to understand how citizens encounter and negotiate an emerging food technology in Austria.

Claudia G. Schwarz is a Postdoc Researcher at the University of Vienna and Lecturer and former Researcher in the Department of Science and Technology Studies. Her current research interests lie mainly in the public understanding of and engagement with science and technology. In her dissertation project, she examined how laypeople use analogies as discursive devices when talking about nanotechnology.

Lizzie Seal is a Senior Lecturer in Criminology at the University of Sussex. Her research interests are in the areas of feminist, historical, and cultural criminology. She is the author of *Women, Murder and Femininity: Gender Representations of Women Who Kill* (Palgrave, 2010); *Transgressive Imaginations: Crime, Deviance and Culture* (with Maggie O'Neill, Palgrave, 2012); and *Capital Punishment in the Twentieth-Century Britain: Audience, Justice, Memory* (Routledge, 2014).

Kapriskie Seide is a doctoral student in the Department of Sociology at the University of Miami with concentrations in Medical Sociology and Race, Ethnicity, and Immigration. Her current research explores the impacts of infectious disease epidemics on everyday life and human health and their overlap with lay epidemiology, social justice, and grounded theory.

Nicole Sharp is a Lecturer in the School of Science and Health at Western Sydney University. Nicole's research interests include the in-depth experiences of people with disability, particularly at times of significant life transition. Nicole is focused on using inclusive research methods that give voice to people who may otherwise be excluded from participation in research.

Norm Sheehan is a Wiradjuri man and a Professor. He is currently Director of Gnibi College SCU. Basing on his expertise in Indigenous Knowledge and Education that employs visual and narrative principles to activate existing strengths within Indigenous education contexts for all individuals and learning communities, he has led the development of two new degrees: the Bachelor of Indigenous Knowledge and the Bachelor of Aboriginal Health and Wellbeing.

Patti Shih is a Postdoctoral Research Fellow at Macquarie University's Centre for Healthcare Resilience and Implementation Science, within the Australian Institute of Health Innovation. She teaches qualitative research methods in public health and has conducted qualitative and mixed-methods projects across a number of healthcare settings including HIV prevention in Papua New Guinea, e-learning among medical students, and professional training in the aged care sector.

Abla Mehio Sibai is a Professor of Epidemiology at the Faculty of Health Sciences, American University of Beirut (AUB). She has led a number of population-based national surveys, including the WHO National Burden of Disease

Study and the Nutrition and Non-communicable Disease Risk Factor (NNCD-BRF) STEPWise study. She is the author of over 150 scholarly articles and book chapters. Abla holds a degree in Pharmacy from AUB and a Ph.D. in Epidemiology from the London School of Hygiene and Tropical Medicine.

Ankur Singh is a Research Fellow in Social Epidemiology at the University of Melbourne, Australia. Ankur has published in different areas of public health including oral epidemiology, tobacco control, and health promotion. He is an invited reviewer for multiple peer-reviewed journals.

Valerie Smith was born in Dubbo and is a Wiradjuri woman. Val started in the healthcare industry in 2012, becoming an Aboriginal Health Worker with the Dubbo Regional Aboriginal Health Service in 2014. Val has recently moved to Port Macquarie and is currently taking a career break to spend time with her family.

Jennifer Smith-Merry is an Associate Professor of Qualitative Health Research in the Faculty of Health Sciences at the University of Sydney. She has methodological expertise in a range of qualitative methods including inclusive research, process evaluation of policy and services, narrative analysis, and critical discourse analysis. Her research focuses on mental health service and policy and consumer experiences of health and healthcare.

Helene Snee is a Senior Lecturer in Sociology at Manchester Metropolitan University, UK. Her research explores stratification with a particular focus on youth and class. Helene is the author of *A Cosmopolitan Journey? Difference, Distinction and Identity Work in Gap Year Travel* (Ashgate, 2014), which was short-listed for the BSA's Philip Abrams Memorial Prize for the best first and sole-authored book within the discipline of Sociology.

Andi Spark led the animation program at the Griffith Film School for 10 years following two decades in various animation industry roles. Working as an animation director and producer, she has been involved with projects ranging from large-scale outdoor projection installations to micro-looping animations for gallery exhibitions and small-screen devices and to children's television series and feature-length films and supervised the production of more than 200 short animated projects.

Elaine Stasiulis is a community-based Research Project Manager and Research Fellow at the Hospital for Sick Children and a doctoral candidate at the University of Toronto. Her work has involved an extensive range of qualitative and participatory arts-based health research projects with children and young people experiencing mental health difficulties and other health challenges.

Genevieve Steiner is a multi-award-winning NHMRC-ARC Dementia Research Development Fellow at NICM, Western Sydney University. Gen uses functional neuroimaging and physiological research methods to investigate the biological bases of learning and memory processes that will inform the prevention, diagnosis, and treatment of cognitive decline in older age. She also conducts many rigorous clinical trials including herbal and lifestyle medicines that may reduce the risk of dementia.

Neil Stephens is a sociologist and Science and Technology Studies (STS) scholar based at Brunel University London. He uses qualitative methods, including ethnography, to research innovation in biomedical contexts and explore the cultural and political aspects of the setting. He has researched topics including human embryonic stem cell research, biobanking, robotic surgery, and cultured meat.

Sara Stevano is a Postdoctoral Fellow in Economics at SOAS. She has a background in development economics, with interdisciplinary skills in the field of political economy and anthropology. Her research interests include the political economy of food and nutrition, agrarian change, and labor markets, with a focus on sub-Saharan Africa (Mozambique and Ghana in particular).

James Sutton is an Associate Professor of Anthropology and Sociology at Hobart and William Smith Colleges in Geneva, New York, USA, where he serves as a member of the Institutional Review Board. His research is focused in the areas of criminology and criminal justice, and his substantive research emphases include prisons, gangs, sexual assault, white-collar crime, and criminological research methods.

Roy Tapera is an Epidemiologist and Medical Informatician with more than 10 years of experience in the field of public health. He is currently a Lecturer of Epidemiology and Health Informatics at the University of Botswana, School of Public Health.

Gail Teachman is a Researcher and an Occupational Therapist. She is currently completing postdoctoral studies at McGill University, Montreal, Canada, and was a trainee with the Critical Disability and Rehabilitation Studies Lab at Bloorview Research Institute in Toronto, Canada. Gail's interdisciplinary research draws on critical qualitative inquiry, rehabilitation science, social theory, and social studies of childhood.

Gareth Terry is a Senior Research Fellow at the Centre for Person Centred Research at the Auckland University of Technology. He comes from a background in critical health and critical social psychologies and currently works in critical rehabilitation studies. His research interests are in men's health, gendered bodies, chronic health conditions, disability and accessibility, and reproductive decision-making.

Cecilie Thøgersen-Ntoumani is an Associate Professor in the School of Psychology and Speech Pathology at Curtin University, Western Australia. She conducts mixed-methods research with a particular emphasis on the development, implementation, and evaluation of theory-based health behavior interventions.

Natalie Thorburn is a Ph.D. candidate at the University of Auckland, New Zealand. Natalie works as Policy Advisor in the community sector, and her background is in sexual violence and intimate partner violence. Her research interests focus on sexual violence, sexual exploitation, and trafficking, and she sits on the board of Child Alert, an organization committed to ending sexual exploitation of children through child trafficking, child prostitution, and child pornography.

Irmgard Tischner is a Senior Lecturer in Social Psychology at the University of the West of England, Bristol, UK, and member of the Centre for Appearance Research. Focusing on poststructuralist, feminist, and critical psychological approaches, her research interests include issues around embodiment and subjectivity, particularly in relation to (gendered) discourses of body size, health, and physical activity in contemporary western industrialized societies.

Allison Tong is an Associate Professor at the Sydney School of Public Health, the University of Sydney. She developed the consolidated criteria for reporting qualitative health research (COREQ) and the enhancing transparency in reporting the synthesis of qualitative health research [ENTREQ], which are both endorsed as key reporting guidelines by leading journals and by the EQUATOR Network for promoting the transparency of health research.

Graciela Tonon is the Director of the Master Program in Social Sciences and the Research Center in Social Sciences (CICS) at Universidad de Palermo, as well as the Director of UNICOM, Faculty of Social Sciences, Universidad Nacional de Lomas de Zamora, Argentina. She is Vice-President of External Affairs of the International Society for Quality of Life Studies (ISQOLS) and Secretary of the Human Development and Capability Association (HDCA).

Billie Townsend is a non-Aboriginal woman born in Dubbo with family connections to the local area. Billie graduated with a double degree in International Studies and History in 2015 from the University of Wollongong. Billie worked on the Aboriginal people's stories of diabetes care study as a Volunteer Research Assistant for the University of Sydney, School of Rural Health, prior to returning to Wollongong to complete her honors year.

Nicole Tujague is a descendant of the Gubbi Gubbi nation from Mt Bauple, Queensland, and the South Sea Islander people from Vanuatu and the Loyalty Islands. Nicole has extensive experience interpreting intercultural issues for both Indigenous and non-Indigenous stakeholders. She has the ability to engage with Aboriginal and Torres Strait Islanders effectively to enable and support changes on an individual and family level.

Jane M. Ussher is Professor of Women's Health Psychology, in the Centre for Health Research at the Western Sydney University. Her research focuses on examining subjectivity in relation to the reproductive body and sexuality and the gendered experience of cancer and cancer care. Her current research include sexual health in CALD refugee and migrant women, sexuality and fertility in the context of cancer, young women's experiences of smoking, and LGBTI experiences of cancer.

Bram Vanhoutte is Simon Research Fellow in the Department of Sociology of the University of Manchester, UK. His research covers a wide field both in terms of topics and methods used. He has investigated aspects of the political socialization of youth and network influences on community-level social cohesion to more recently life course perspectives on ageing. Methodologically, he is interested in how to improve measurement and modeling using survey and administrative data and has recently developed an interest in life history data.

Lía Rodriguez de la Vega is a Professor and Researcher of the University of Palermo and Lomas de Zamora National University (Argentina). She obtained her doctoral degree in International Relations from El Salvador University, Argentina, in 2006. She is also a postdoctoral stay at the Psychology Faculty, Universidade Federal de Rio Grande do Sul (UFRGS, Porto Alegre, Brazil, in the context of the Bilateral Program Scholarship MINCYT-CAPES, 2009).

Randi Wærdahl is an Associate Professor of Sociology. Her work has an everyday perspective on transitions and trajectories in childhood and for families, in consumer societies, in times of social and economic change, and in changing contexts due to migration. She combines traditional methods of interviews and surveys with participatory methods such as photography or diary writing.

Morten Wahrendorf is a Senior Researcher at the Centre for Health and Society, University of Duesseldorf, Germany, with substantial expertise in sociology and research methodology. His main research interest are health inequalities in ageing populations and underlying pathways, with a particular focus on psychosocial working conditions, patterns of participation in paid employment and social activities in later life, and life course influences.

Maggie Walter is a member of the Palawa Briggs/Johnson Tasmanian Aboriginal family. She is Professor of Sociology and Pro Vice-Chancellor of Aboriginal Research and Leadership at the University of Tasmania. She has published extensively in the field of race relations and inequality and is passionate about Indigenous statistical engagement.

Emma Webster is a non-Aboriginal woman who has lived in Dubbo over 20 years and has family connections to the area. Emma joined the University of Sydney, School of Rural Health, in 2015 after working in the NSW public health system for 21 years. She has extensive experience in guiding novice researchers through their first research study having worked in the NSW Health Education and Training Institute's Rural Research Capacity Building Program.

Johannes Wheeldon has more than 15 years' experience managing evaluation and juvenile justice projects. He has worked with the American Bar Association, the Open Society Foundations, and the World Bank. He has published 4 books and more than 25 peer-reviewed papers on aspects of criminal justice, restorative justice, organizational change, and evaluation. He is an Adjunct Professor at Norwich University.

Sarah J. White is a Qualitative Health Researcher and Linguist with a particular interest in using conversation analysis to understand communication in surgical practice. She is a Senior Lecturer at the Faculty of Medicine and Health Sciences at Macquarie University, Sydney.

Matthew Wice is a doctoral student at the New School for Social Research where he studies Developmental and Social Psychology. His research interests include the influence of culture on moral reasoning, motivation, and the development of social cognition.

Nicola Willis is a Pediatric HIV Nurse Specialist. She is the Founder and Director of AfricAid, Zvandiri, and has spent 14 years developing and scaling up differentiated HIV service delivery models for children and adolescents in Zimbabwe. Nicola's work has been extensively influenced by the engagement of HIV-positive young people in creative, participatory approaches as a means of identifying and informing their policy and service delivery needs.

Noreen Willows uses anthropological, qualitative, and quantitative methodologies for exploring the relationships between food and health, cultural meanings of food and health, how food beliefs and dietary practices affect the well-being of communities, and how sociocultural factors influence food intake and food selection.

Noreen aims to foster an understanding among academics and nonacademics of the value of community-based participatory research.

Denise Wilson is a Professor in Māori Health and the Director of Taupua Waiora Centre for Māori Health Research at Auckland University of Technology. Her research and publications focus on Māori health, health services access and use, family violence, cultural responsiveness, and workforce development.

Elena Wilson is a Ph.D. candidate in the rural health research program, "Improving the Health of Communities through Participation," in the Rural Health School at La Trobe University, Bendigo. Elena's research interests are research methods, research ethics, and community participation with a focus on rural health and well-being.

Leigh Wilson is a Senior Lecturer in the Ageing Work and Health Research Group at the University of Sydney. Leigh comes from a public health and behavioral science background and has worked as an Epidemiologist and Health Service Manager in the NSW Health system. Her key research interests are the epidemiology of ageing, the impact of climate change (particularly heatwaves) on the aged, and the impact of behavior and perceptions of age on health.

Shawn Wilson has worked with Indigenous people internationally to apply Indigenist philosophy within the contexts of Indigenous education, health, and counselor education. His research focuses on the interrelated concepts of identity, health and healing, culture, and well-being. His book, *Research is Ceremony: Indigenous Research Methods*, has been cited as bridging understanding between mainstream and Indigenist research and is used as a text in many universities.

Jess Woodroffe works in research, academic supervision, teaching, and community engagement. Her research interests include community partnerships and engagement, social inclusion, health promotion, inter-professional learning and education, health sociology, qualitative research, and evaluation.

Kevin B. Wright is a Professor of Health Communication in the Department of Communication at George Mason University. His research focuses on social support processes and health outcomes in both face-to-face and online support groups/communities, online health information seeking, and the use of technology in provider-patient relationships.

Anna Ziersch is an Australian Research Council Future Fellow based at the Southgate Institute for Health, Society and Equity at Flinders University. She has an overarching interest in health inequities, in particular multidisciplinary and multi-method approaches to understanding the social determinants of health, and in particular for refugees and asylum seekers.

Contributors

Mauri Ahlberg Department of Teacher Education, University of Helsinki, Helsinki, Finland

Farah Ahmad School of Health Policy and Management, York University, Toronto, ON, Canada

Jo Aldridge Department of Social Sciences, Loughborough University, Leicestershire, UK

Julaine Allan Lyndon, Orange, NSW, Australia

Jordan Alpert Department of Advertising, University of Florida, Gainesville, FL, USA

Stewart J. Anderson Department of Biostatistics, University of Pittsburgh Graduate School of Public Health, Pittsburgh, PA, USA

E. Anne Marshall Educational Psychology and Leadership Studies, Centre for Youth and Society, University of Victoria, Victoria, BC, Canada

Mike Armour NICM, Western Sydney University (Campbelltown Campus), Penrith, NSW, Australia

Amit Arora School of Science and Health, Western Sydney University, Sydney, NSW, Australia

Discipline of Paediatrics and Child Health, Sydney Medical School, Sydney, NSW, Australia

Oral Health Services, Sydney Local Health District and Sydney Dental Hospital, NSW Health, Sydney, NSW, Australia

COHORTE Research Group, Ingham Institute of Applied Medical Research, Liverpool, NSW, Australia

Kathy Arthurson Southgate Institute for Health, Society and Equity, Flinders University, Adelaide, SA, Australia

Anna Bagnoli Department of Sociology, Wolfson College, University of Cambridge, Cambridge, UK

Amy Baker School of Health Sciences, University of South Australia, Adelaide, SA, Australia

Ruth Bartlett Centre for Innovation and Leadership in Health Sciences, Faculty of Health Sciences, University of Southampton, Southampton, UK

Pat Bazeley Translational Research and Social Innovation Group, Western Sydney University, Liverpool, NSW, Australia

Linda Liska Belgrave Department of Sociology, University of Miami, Coral Gables, FL, USA

Robert F. Belli Department of Psychology, University of Nebraska-Lincoln, Lincoln, NE, USA

Jyoti Belur Department of Security and Crime Science, University College London, London, UK

Denise Benatuil Master Program in Social Sciences and CICS, Universidad de Palermo, Buenos Aires, Argentina

Sameer Bhole Sydney Dental School, Faculty of Medicine and Health, The University of Sydney, Surry Hills, NSW, Australia

Oral Health Services, Sydney Local Health District and Sydney Dental Hospital, NSW Health, Surry Hills, NSW, Australia

Mia Bierbaum Centre for Healthcare Resilience and Implementation Science, Australian Institute of Health Innovation (AIHI), Macquarie University, Sydney, NSW, Australia

Simon Bishop Centre for Health Innovation, Leadership and Learning, Nottingham University Business School, Nottingham, UK

Karen Block Melbourne School of Population and Global Health, The University of Melbourne, Melbourne, VIC, Australia

Chelsea Bond Aboriginal and Torres Strait Islander Studies Unit, The University of Queensland, St Lucia, QLD, Australia

Virginia Braun School of Psychology, The University of Auckland, Auckland, New Zealand

Wendy Brodribb Primary Care Clinical Unit, Faculty of Medicine, The University of Queensland, Herston, QLD, Australia

Mario Brondani University of British Columbia, Vancouver, BC, Canada

Jessica Broome University of Michigan, Ann Arbor, MI, USA
Sanford, NC, USA

Contributors

Mark Brough School of Public Health and Social Work, Queensland University of Technology, Kelvin Grove, Australia

Katerina Bryant Department of English and Creative Writing, Flinders University, Bedford Park, SA, Australia

Lia Bryant School of Psychology, Social Work and Social Policy, Centre for Social Change, University of South Australia, Magill, Australia

Anne Bunde-Birouste School of Public Health and Community Medicine, UNSW, Sydney, NSW, Australia

Viv Burr Department of Psychology, School of Human and Health Sciences, University of Huddersfield, Huddersfield, UK

Valentina Buscemi Western Sydney University, Sydney, NSW, Australia

Rosalind A. Bye School of Science and Health, Western Sydney University, Campbelltown, NSW, Australia

Breanne Byiers Department of Educational Psychology, University of Minnesota, Minneapolis, MN, USA

Fiona Byrne Translational Research and Social Innovation (TReSI) Group, School of Nursing and Midwifery, Ingham Institute for Applied Medical Research, Western Sydney University, Liverpool, NSW, Australia

Robert H. Campbell The University of Bolton, Bolton, UK

Michael J. Carter Sociology Department, California State University, Northridge, Northridge, CA, USA

Rocco Cavaleri School of Science and Health, Western Sydney University, Campbelltown, NSW, Australia

Jane Chalmers Western Sydney University (Campbelltown Campus), Penrith, NSW, Australia

Nancy Cheak-Zamora Department of Health Sciences, University of Missouri, Columbia, MO, USA

Betty P. M. Chung School of Nursing, The Hong Kong Polytechnic University, Hong Kong, SAR, China

Victoria Clarke Department of Health and Social Sciences, Faculty of Health and Applied Sciences, University of the West of England (UWE), Bristol, UK

Bronwyne Coetzee Department of Psychology, Stellenbosch University, Matieland, South Africa

Ana Lucía Córdova-Cazar Colegio de Ciencias Sociales y Humanidades, Universidad San Francisco de Quito, Diego de Robles y Vía Interoceánica, Quito, Cumbayá, Ecuador

Jonathan C. Craig Sydney School of Public Health, The University of Sydney, Sydney, NSW, Australia

Centre for Kidney Research, The Children's Hospital at Westmead, Sydney, NSW, Australia

Fiona Cram Katoa Ltd, Auckland, Aotearoa, New Zealand

Anne Cusick Faculty of Health Sciences, Sydney University, Sydney, Australia

Angela J. Dawson Australian Centre for Public and Population Health Research, University of Technology Sydney, Sydney, NSW, Australia

Kevin Deane Department of Economics, International Development and International Relations, The University of Northampton, Northampton, UK

Lía Rodriguez de la Vega Ciudad Autónoma de Buenos Aires, University of Palermo, Buenos Aires, Argentina

Virginia Dickson-Swift LaTrobe Rural Health School, College of Science, Health and Engineering, LaTrobe University, Bendigo, VIC, Australia

Rebecca Dimond School of Social Sciences, Cardiff University, Cardiff, UK

Agnes E. Dodds Melbourne Medical School, The University of Melbourne, Melbourne, VIC, Australia

Suzanne D'Souza School of Nursing and Midwifery, Western Sydney University, Sydney, NSW, Australia

Joan L. Duda School of Sport and Exercise Sciences, University of Birmingham, Birmingham, West Midlands, UK

Clemence Due Southgate Institute for Health, Society and Equity, Flinders University, Adelaide, SA, Australia

Tinashe Dune Western Sydney University, Sydney, NSW, Australia

Jane Marie Edwards Deakin University, School of Health and Social Development, Geelong, VIC, Australia

Carolyn Ee NICM, Western Sydney University (Campbelltown Campus), Penrith, NSW, Australia

Ghinwa Y. El Hayek Department of Epidemiology and Population Health, Faculty of Health Sciences, American University of Beirut, Beirut, Lebanon

Ulrike Felt Department of Science and Technology Studies, Research Platform Responsible Research and Innovation in Academic Practice, University of Vienna, Vienna, Austria

Monika Ferguson School of Nursing and Midwifery, University of South Australia, Adelaide, SA, Australia

Pota Forrest-Lawrence School of Social Sciences and Psychology, Western Sydney University, Milperra, NSW, Australia

Bronwyn Fredericks Office of Indigenous Engagement, Central Queensland University, Rockhampton, Australia

Caroline Elizabeth Fryer Sansom Institute for Health Research, University of South Australia, Adelaide, SA, Australia

Emma S. George School of Science and Health, Western Sydney University, Sydney, NSW, Australia

Lilian A. Ghandour Department of Epidemiology and Population Health, Faculty of Health Sciences, American University of Beirut, Beirut, Lebanon

Lisa Gibbs Melbourne School of Population and Global Health, The University of Melbourne, Melbourne, VIC, Australia

Claudia Gillberg Swedish National Centre for Lifelong Learning (ENCELL), Jonkoping University, Jönköping, Sweden

Geoffrey Jones Centre for Welfare Reform, Sheffield, UK

Brenda M. Gladstone Dalla Lana School of Public Health, Centre for Critical Qualitative Health Research, University of Toronto, Toronto, ON, Canada

Namrata Goyal Department of Psychology, New School for Social Research, New York, NY, USA

Adyya Gupta School of Public Health, The University of Adelaide, Adelaide, SA, Australia

Talia Gutman Sydney School of Public Health, The University of Sydney, Sydney, NSW, Australia
Centre for Kidney Research, The Children's Hospital at Westmead, Westmead, NSW, Australia

Elizabeth J. Halcomb School of Nursing, University of Wollongong, Wollongong, NSW, Australia

Marit Haldar Department of Social Work, Child Welfare and Social Policy, Oslo and Akershus University College of Applied Sciences, Oslo, Norway

Camilla S. Hanson Sydney School of Public Health, The University of Sydney, Sydney, NSW, Australia
Centre for Kidney Research, The Children's Hospital at Westmead, Westmead, NSW, Australia

John D. Hathcoat Department of Graduate Psychology, Center for Assessment and Research Studies, James Madison University, Harrisonburg, VA, USA

Nikki Hayfield Department of Health and Social Sciences, Faculty of Health and Applied Sciences, University of the West of England (UWE), Bristol, UK

Lisa Hodge College of Health and Biomedicine, Victoria University, Melbourne, VIC, Australia

Anne Hogden Centre for Healthcare Resilience and Implementation Science, Australian Institute of Health Innovation (AIHI), Macquarie University, Sydney, NSW, Australia

Nicholas Hookway University of Tasmania, Launceston, Tasmania, Australia

Kirsten Howard School of Public Health, University of Sydney, Sydney, NSW, Australia

Martin Howell School of Public Health, University of Sydney, Sydney, NSW, Australia

Erin Jessee Modern History, University of Glasgow, Glasgow, UK

James Rufus John Translational Health Research Institute, School of Medicine, Western Sydney University, Penrith, NSW, Australia

Capital Markets Cooperative Research Centre, Sydney, NSW, Australia

Craig Johnson Dubbo Diabetes Unit, Dubbo, NSW, Australia

Monica Johnson Marathon Health, Dubbo, NSW, Australia

Stephanie T. Jong School of Education, Flinders University, Adelaide, SA, Australia

June Joseph Primary Care Clinical Unit, Faculty of Medicine, The University of Queensland, Herston, QLD, Australia

Angela Ju Sydney School of Public Health, The University of Sydney, Sydney, NSW, Australia

Centre for Kidney Research, The Children's Hospital at Westmead, Westmead, NSW, Australia

Yong Moon Jung Centre for Business and Social Innovation, University of Technology Sydney, Ultimo, NSW, Australia

Ashraf Kagee Department of Psychology, Stellenbosch University, Matieland, South Africa

Ida Kaplan The Victorian Foundation for Survivors of Torture Inc, Brunswick, VIC, Australia

Bernie Kemp Dubbo Regional Aboriginal Health Service, Dubbo, NSW, Australia

Lynn Kemp Translational Research and Social Innovation (TReSI) Group, School of Nursing and Midwifery, Ingham Institute for Applied Medical Research, Western Sydney University, Liverpool, NSW, Australia

Priya Khanna Sydney Medical Program, University of Sydney, Camperdown, NSW, Australia

Alexandra C. King Rural Clinical School, Faculty of Health, University of Tasmania, Burnie, Tasmania, Australia

School of Pharmacy, Faculty of Health, University of Tasmania, Hobart, Tasmania, Australia

Nigel King Department of Psychology, University of Huddersfield, Huddersfield, UK

Kat Kolar Department of Sociology, University of Toronto, Toronto, ON, Canada

Gary L. Kreps Department of Communication, George Mason University, Fairfax, VA, USA

Tahu Kukutai University of Waikato, Hamilton, New Zealand

Ian Lacey Deadly Choices, Institute for Urban Indigenous Health, Brisbane, Australia

Michelle LaFrance George Mason University, Fairfax, VA, USA

Kari Lancaster Centre for Social Research in Health, Faculty of Arts and Social Sciences, UNSW, Sydney, NSW, Australia

Jeanette A. Lawrence Melbourne School of Psychological Science, The University of Melbourne, Melbourne, VIC, Australia

Eric Leake Department of English, Texas State University, San Marcos, TX, USA

Raymond M. Lee Royal Holloway University of London, Egham, UK

Caroline Lenette Forced Migration Research Network, School of Social Sciences, University of New South Wales, Kensington, NSW, Australia

Doris Y. Leung School of Nursing, The Hong Kong Polytechnic University, Hong Kong, SAR, China

The Lawrence S. Bloomberg Faculty of Nursing, University of Toronto, Toronto, ON, Canada

Pranee Liamputtong School of Science and Health, Western Sydney University, Penrith, NSW, Australia

Nicole Loehr School of Social and Policy Studies, Flinders University, Adelaide, SA, Australia

Janet C. Long Australian Institute of Health Innovation, Macquarie University, Sydney, NSW, Australia

Colin MacDougall Health Sciences Building, Flinders University, Bedford Park, SA, Australia

Catrina A. MacKenzie Department of Geography, McGill University, Montreal, QC, Canada

Department of Geography, University of Vermont, Burlington, VT, USA

Freya MacMillan School of Science and Health and Translational Health Research Institute (THRI), Western Sydney University, Penrith, NSW, Australia

Wilson Majee Department of Health Sciences, University of Missouri, Columbia, MO, USA

Karine Manera Sydney School of Public Health, The University of Sydney, Sydney, NSW, Australia

Centre for Kidney Research, The Children's Hospital at Westmead, Westmead, NSW, Australia

Narendar Manohar School of Science and Health, Western Sydney University, Sydney, NSW, Australia

Oana Marcu Università Cattolica del Sacro Cuore, Milan, Italy

Rodrigo Mariño University of Melbourne, Melbourne, VIC, Australia

Anastasia Marinopoulou Department of European Studies, Hellenic Open University, Patra, Greece

Anna Maurer-Batjer Department of Health Sciences, University of Missouri, Columbia, MO, USA

Kate A. McBride School of Medicine and Translational Health Research Institute, Western Sydney University, Sydney, NSW, Australia

Angela McGrane Newcastle Business School, Northumbria University, Newcastle-upon-Tyne, UK

Jane McKeown School of Nursing and Midwifery, The University of Sheffield, Sheffield, UK

Karen McPhail-Bell University Centre for Rural Health, University of Sydney, Camperdown, NSW, Australia

Poche Centre for Indigenous Health, Sydney Medical School, The University of Sydney, Camperdown, NSW, Australia

Abla Mehio Sibai Department of Epidemiology and Population Health, Faculty of Health Sciences, American University of Beirut, Beirut, Lebanon

Cara Meixner Department of Graduate Psychology, Center for Faculty Innovation, James Madison University, Harrisonburg, VA, USA

Zelalem Mengesha Western Sydney University, Sydney, NSW, Australia

Dafna Merom School of Science and Health, Western Sydney University, Penrith, Sydney, NSW, Australia

Translational Health Research Institute, School of Medicine, Western Sydney University, Penrith, NSW, Australia

Joan G. Miller Department of Psychology, New School for Social Research, New York, NY, USA

Christine Milligan Division of Health Research, Lancaster University, Lancaster, UK

Naomi Moller School of Psychology, Faculty of Social Sciences, The Open University, Milton Keynes, UK

Andrea Montes Alvarado Sociology Department, California State University, Northridge, Northridge, CA, USA

Marta Moskal Durham Univeristy, Durham, UK

Elias Mpofu University of Sydney, Lidcombe, NSW, Australia

Educational Psychology and Inclusive Education, University of Johannesburg, Johannesburg, South Africa

Magen Mhaka Mutepfa Department of Psychology, University of Botswana, Gaborone, Botswana

Sally Nathan School of Public Health and Community Medicine, Faculty of Medicine, UNSW, Sydney, NSW, Australia

Alison Nelson Allied Health and Workforce Development, Institute for Urban Indigenous Health, Brisbane, Australia

Ingrid A. Nelson Sociology and Anthropology Department, Bowdoin College, Brunswick, ME, USA

Christy Newman Centre for Social Research in Health, Faculty of Arts and Social Sciences, UNSW, Sydney, NSW, Australia

Mark C. Nicholas Framingham State University, Framingham, MA, USA

Kayi Ntinda Discipline of Educational Counselling and Mixed-Methods Inquiry Approaches, Faculty of Education, Office C.3.5, University of Swaziland, Kwaluseni Campus, Manzini, Swaziland

Chijioke Obasi University of Wolverhampton, Wolverhampton, UK

Felix Ogbo School of Medicine and Translational Health Research Institute, Western Sydney University, Campbelltown, NSW, Australia

Andrew Page School of Medicine and Translational Health Research Institute, Western Sydney University, Campbelltown, NSW, Australia

Bonnie Pang School of Science and Health and Institute for Culture and Society, University of Western Sydney, Penrith, NSW, Australia

Gemma Pearce Centre for Advances in Behavioural Science, Coventry University, Coventry, West Midlands, UK

Karen G. Peres Australian Research Centre for Population Oral Health (ARCPOH), Adelaide Dental School, The University of Adelaide, Adelaide, SA, Australia

Janette Perz Translational Health Research Institute, School of Medicine, Western Sydney University, Sydney, NSW, Australia

Rona Pillay School of Nursing and Midwifery, Western Sydney University, Sydney, NSW, Australia

Jayne Pitard College of Arts and Education, Victoria University, Melbourne, VIC, Australia

Jennifer Prattley Department of Social Statistics, University of Manchester, Manchester, UK

Nicholas Procter School of Nursing and Midwifery, University of South Australia, Adelaide, SA, Australia

Marsha B. Quinlan Department of Anthropology, Washington State University, Pullman, WA, USA

Frances Rapport Centre for Healthcare Resilience and Implementation Science, Australian Institute of Health Innovation (AIHI), Macquarie University, Sydney, NSW, Australia

Elizabeth F. Rix Gnibi Wandarahn School of Indigenous Knowledge, Southern Cross University, Lismore, NSW, Australia

Lia Rodriguez de la Vega Ciudad Autónoma de Buenos Aires, Buenos Aires, Argentina

Norma R. A. Romm Department of Adult Education and Youth Development, University of South Africa, Pretoria, South Africa

Rizwana Roomaney Department of Psychology, Stellenbosch University, Matieland, South Africa

Lauren Rosewarne School of Social and Political Sciences, University of Melbourne, Melbourne, VIC, Australia

Simone Schumann University of Vienna, Vienna, Austria

Claudia G. Schwarz-Plaschg Research Platform Nano-Norms-Nature, University of Vienna, Vienna, Austria

Lizzie Seal University of Sussex, Brighton, UK

Kapriskie Seide Department of Sociology, University of Miami, Coral Gables, FL, USA

Nicole L. Sharp School of Science and Health, Western Sydney University, Campbelltown, NSW, Australia

Norm Sheehan Gnibi Wandarahn School of Indigenous Knowledge, Southern Cross University, Lismore, NSW, Australia

Patti Shih Centre for Healthcare Resilience and Implementation Science, Australian Institute of Health and Innovation (AIHI), Macquarie University, Sydney, NSW, Australia

Ankur Singh Centre for Health Equity, Melbourne School of Population and Global Health, The University of Melbourne, Melbourne, VIC, Australia

Valerie Smith Formerly with Dubbo Regional Aboriginal Health Service, Dubbo, NSW, Australia

Jennifer Smith-Merry Faculty of Health Sciences, The University of Sydney, Sydney, Australia

Helene Snee Manchester Metropolitan University, Manchester, UK

Andi Spark Griffith Film School, Queensland College of Art, Griffith University, Brisbane, QLD, Australia

Elaine Stasiulis Child and Youth Mental Health Research Unit, SickKids, Toronto, ON, Canada

Institute of Medical Science, University of Toronto, Toronto, ON, Canada

Genevieve Z. Steiner NICM and Translational Health Research Institute (THRI), Western Sydney University, Penrith, NSW, Australia

Neil Stephens Social Science, Media and Communication, Brunel University London, Uxbridge, UK

Sara Stevano Department of Economics, University of the West of England (UWE) Bristol, Bristol, UK

James E. Sutton Department of Anthropology and Sociology, Hobart and William Smith Colleges, Geneva, NY, USA

Roy Tapera Department of Environmental Health, University of Botswana, Gaborone, Botswana

Gail Teachman McGill University, Montreal, QC, Canada

Gareth Terry Centre for Person Centred Research, School of Clinical Sciences, Auckland University of Technology, Auckland, New Zealand

Michelle Teti Department of Health Sciences, University of Missouri, Columbia, MO, USA

Cecilie Thøgersen-Ntoumani Health Psychology and Behavioural Medicine Research Group, School of Psychology and Speech Pathology, Curtin University, Perth, WA, Australia

Natalie Thorburn The University of Auckland, Auckland, New Zealand

Irmgard Tischner Faculty of Sport and Health Sciences, Technische Universität München, Lehrstuhl Diversitätssoziologie, Munich, Germany

Allison Tong Sydney School of Public Health, The University of Sydney, Sydney, NSW, Australia

Centre for Kidney Research, The Children's Hospital at Westmead, Westmead, Australia

Graciela Tonon Master Program in Social Sciences and CICS-UP, Universidad de Palermo, Buenos Aires, Argentina

UNICOM- Universidad Nacional de Lomas de Zamora, Buenos Aires, Argentina

Billie Townsend School of Rural Health, Sydney Medical School, University of Sydney, Dubbo, NSW, Australia

Nicole Tujague Gnibi Wandarahn School of Indigenous Knowledge, Southern Cross University, Lismore, NSW, Australia

Jane M. Ussher Translational Health Research Institute, School of Medicine, Western Sydney University, Sydney, NSW, Australia

Bram Vanhoutte Department of Sociology, University of Manchester, Manchester, UK

Randi Wærdahl Department of Social Work, Child Welfare and Social Policy, Oslo and Akershus University College of Applied Sciences, Oslo, Norway

Morten Wahrendorf Institute of Medical Sociology, Centre of Health and Society (CHS), Heinrich-Heine-University Düsseldorf, Medical Faculty, Düsseldorf, Germany

Maggie Walter University of Tasmania, Hobart, Tasmania, Australia

Emma Webster School of Rural Health, Sydney Medical School, University of Sydney, Dubbo, NSW, Australia

Johannes Wheeldon School of Sociology and Justice Studies, Norwich University, Northfield, VT, USA

Sarah J. White Macquarie University, Sydney, NSW, Australia

Matthew Wice Department of Psychology, New School for Social Research, New York, NY, USA

Nicola Willis Zvandiri House, Harare, Zimbabwe

Noreen D. Willows Faculty of Agricultural, Life and Environmental Sciences, University of Alberta, Edmonton, AB, Canada

Denise Wilson Auckland University of Technology, Auckland, New Zealand

Elena Wilson Rural Health School, College of Science, Health and Engineering, La Trobe University, Melbourne, VIC, Australia

Leigh A. Wilson School of Science and Health, Western Sydney University, Penrith, NSW, Australia

Faculty of Health Science, Discipline of Behavioural and Social Sciences in Health, University of Sydney, Lidcombe, NSW, Australia

Shawn Wilson Gnibi Wandarahn School of Indigenous Knowledge, Southern Cross University, Lismore, NSW, Australia

Jessica Woodroffe Access, Participation, and Partnerships, Academic Division, University of Tasmania, Launceston, Australia

Kevin B. Wright Department of Communication, George Mason University, Fairfax, VA, USA

Anna Ziersch Southgate Institute for Health, Society and Equity, Flinders University, Adelaide, SA, Australia

The Delphi Technique

Jane Chalmers and Mike Armour

Contents

1	Introduction	716
2	What Questions Can Be Answered Using a Delphi Technique?	717
	2.1 Forecasting	718
	2.2 Reaching Consensus	718
3	Delivery of the Delphi	719
	3.1 Anonymity	719
	3.2 Postal Delphi	719
	3.3 Online Delphi	720
	3.4 Other and Combination Approaches	720
4	Using Experts	720
	4.1 Defining "Expert"	720
	4.2 Size of Panel	721
5	Round One	722
	5.1 Provision of Pre-existing Information	723
	5.2 Modified Delphi	724
6	Subsequent Rounds	724
	6.1 Purpose of Round Two	724
	6.2 Developing a Structured Questionnaire for Round Two	725
	6.3 Purpose of Rounds Three and Beyond	725
	6.4 Providing Feedback	725
7	Delphi Rounds and Reaching Consensus	727
	7.1 Determining Number of Rounds	727
	7.2 How Much Agreement Is Needed for Consensus?	728
	7.3 Stability Indicating Termination	729
8	Reporting Results	729

J. Chalmers (✉)
Western Sydney University (Campbelltown Campus), Penrith, NSW, Australia
e-mail: j.chalmers@westernsydney.edu.au

M. Armour
NICM, Western Sydney University (Campbelltown Campus), Penrith, NSW, Australia
e-mail: m.armour@westernsydney.edu.au

© Springer Nature Singapore Pte Ltd. 2019
P. Liamputtong (ed.), *Handbook of Research Methods in Health Social Sciences*,
https://doi.org/10.1007/978-981-10-5251-4_99

9	Limitations of the Method	730
	9.1 Lack of Standard Guidelines for the Delphi Technique	730
	9.2 Sample Size	730
	9.3 The Use of "Experts"	731
	9.4 Non-responders	731
	9.5 Determining When Questioning Should End	732
10	Conclusion and Future Directions	732
References		732

Abstract

This chapter introduces the Delphi technique and explores its applications relevant to the health field. The Delphi technique is a method of gaining consensus on a particular topic through the use of rounds of questioning of experts in the field. It has three characteristics that make it distinct from other group interaction methods: (1) anonymous group interactions and responses, (2) multiple rounds of questioning, and (3) the provision of feedback to the group between each round. Each characteristic is designed to reduce bias in gaining consensus, such as removing the influence of societal and peer pressures and encouraging the convergence of ideas while still using an anonymous group setting. This chapter will facilitate decision-making in setting up and executing a Delphi study and covers question design, delivery method, employing experts, determining the point of study termination, and reporting of results. Finally, the limitations of the method, particularly the lack of guidelines for researchers, are highlighted to encourage researchers to make a priori decisions which will assist in reducing bias and improving the validity of the Delphi technique.

Keywords

Delphi technique · Consensus · Forecasting · Experts · Stability · Modified Delphi · Panel

1 Introduction

The Delphi technique was originally named after the famous Greek oracle, but despite its name, the technique has no oracular or prophetic powers. Rather, it uses current knowledge and opinions to make predictions about the future or to reach decisions on present questions. Developed by Dalkey and Helmer (1963) for the RAND Corporation, the Delphi technique was originally used to forecast military priorities, particularly with the evolution of technologies. The technique is now much more widely used to make decisions, particularly where hard data to support decision-making cannot be obtained (Linstone and Turoff (1975). Instead, experts in the relevant field are consulted for their knowledge and experience to make group decisions. It is now commonly used across many areas including computer science, education, psychology, retail, and healthcare, for the purpose of forecasting future events, goal setting, problem-solving, and developing policies.

The Delphi technique generally has three characteristics that make it distinct from other group interaction methods: (1) anonymous group interactions and responses, (2) multiple rounds of questioning, and (3) the provision of feedback to the group between each round (Murry Jr and Hammons 1995). Each characteristic provides the Delphi with a unique approach to gaining consensus on a topic. Anonymous group interactions and responses are thought to better represent the group's true position, as it removes the effect of powerful personalities, individual's status, social pressures of conformity, and peer pressure from group members (Dalkey and Helmer 1963). Multiple rounds of questioning are used on the rationale that "two heads are better than one, or...n heads are better than one" (Dalkey 1972, p. 15). That is, input from multiple individuals working as a group is more likely to result in solving a complex issue than from one individual alone. The final characteristic of providing feedback is thought to increase the convergence of ideas within the group to aid in problem-solving. That feedback is given after each round means group members are able to gradually mold their opinion and are more likely to be "swayed by persuasively stated opinions of others" (Dalkey and Helmer 1963, p. 459).

Since the development of the technique, the Delphi has undergone several modifications to allow a wider application. In particular, advancements in technology have enabled the undertaking of online Delphi studies, which greatly improve response times and response rates. The accessibility of literature on the internet has also modified the Delphi. Often, experts are now consulted to provide their input and opinion on a predefined set of variables extracted from the literature, rather than being asked traditional open-ended questions. In healthcare, the Delphi technique is commonly used to determine, predict, and explore group attitudes, needs, and priorities. Many studies consult health professionals as experts in healthcare, but there is a growing body of literature using Delphi studies to problem solve from the health-user perspective, employing patients as experts.

This chapter outlines the key characteristics of the Delphi technique and provides a simple guide on how to employ the technique in a health context. The chapter also provides ongoing examples of how the technique can be used and how decisions about the technique can be made.

2 What Questions Can Be Answered Using a Delphi Technique?

The Delphi technique and its modifications are often used to answer two main types of questions: forecasting and generating consensus. It is important at the outset to understand which category your own research question falls in to. Forecasting is around determining the direction of future trends based on past and present data and soliciting opinions of the expert panel on where a particular area or topic of interest is moving, usually with the aim of directing policy decisions. Generating consensus, most commonly in an area with conflicting evidence or a small evidence base, is used in healthcare to determine practice guidelines, assessment tools, and treatment strategies (Junger et al. 2017), while in research it is often used to design treatment protocols.

2.1 Forecasting

One of the first applications for the Delphi technique since its design at RAND has been forecasting, with initial studies being in the area of defense (Dalkey and Helmer 1963) using experts to determine "from the point of view of a Soviet strategic planner – an optimal U.S. industrial target system, with a corresponding estimation of the number of atomic bombs required to reduce munitions output by a prescribed amount." When using Delphi to perform forecasting, it is not meant to replace other forms of forecasting, such as statistical models, but rather provide the human input required when current or past data is not comprehensive enough to allow for accurate model forecasting (Rowe and Wright 1999). In a healthcare context, forecasting using the Delphi technique can be used in a wide range of healthcare settings, and forecasting in this context is usually found in the area of policy development, where having an outlook on future directions, trends, and probabilities is crucial to allocating resources to areas where they may be, for example, areas of future unmet need.

Examples of using forecasting to understand future trends in healthcare are diverse and include determining likely rates of illicit drug use and misuse of medical drugs (Lintonen et al. 2014), examining the likely effect of different alcohol control policies (Nelson et al. 2013), and forecasting trends in dentistry techniques and materials needed for tooth restoration (Seemann et al. 2014). The importance of using Delphi to undertake forecasting is that it allows desirable areas or projects (such as successful alcohol reduction initiatives) to be developed; likely but undesirable trends (such as an increase in the misuse of prescription painkillers) can be counteracted, and training for future likely treatments planned.

2.2 Reaching Consensus

Consensus building is one of the primary roles of the Delphi technique in modern-day healthcare research. Consensus guidelines can be very valuable in providing care that is founded on evidence-based "best practice" for a specific clinical problem. By combining the latest healthcare evidence from the scientific literature with expert opinion, this kind of systematic consensus can help form guidelines for areas where the evidence is insufficient or controversial (Boulkedid et al. 2011). For interventions where the complexity of practice can make evidence-based guidelines difficult, such as herbal medicine or psychotherapy, Delphi can provide expert guidance in the creation of practice guidelines and is appropriate where a high-quality evidence base may be lacking (Flower et al. 2007). When using modified Delphi techniques (such as face-to-face discussions at the conclusion of the process) that allow more clarification and discussion, clinically useful tools such as clinical pathway algorithms can be developed more easily (Eubank et al. 2016). In addition to clinical pathways, consensus building via Delphi can provide expert input into the design of clinical trial protocols (Cochrane et al. 2011; Cotchett et al. 2011; Smith et al. 2012) and evaluation of the quality of evidence in clinical trials (Smith et al. 2011). This is especially useful in the design of trials into complex interventions, interventions where there are multiple likely synergistic components that affect the clinical outcome, due to the unclear nature of which of these may be responsible. However, when developing guidelines or trial protocols, in the absence of strong evidence, caution must be taken as expert opinion on treatment choice may be radically different from how clinicians treat in daily practice (Alraek et al. 2011). It is important to clarify at the outset that one very common misconception with Delphi is that its sole goal is to reach a single consensus, where all members of the panel agree on an issue. This is incorrect; a single consensus should not be the primary goal, and a bimodal distribution of consensus (e.g., where many panelists how two diverging viewpoints) can be an important finding, especially in areas where there is significant controversy in the literature or evidence base (Linstone and Turoff 2011). Discussion around consensus, and what defines consensus, occurs in more depth in Sect. 7.2.

3 Delivery of the Delphi

3.1 Anonymity

One of the key features of a traditional Delphi is the anonymity of individual responses within the panel (Linstone and Turoff 1975). Anonymity stimulates focus, reflection, and imagination of the individuals (Linstone and Turoff 1975) and removes biases introduced by the effects of status, personalities, and group pressures (Thangaratinam and Redman 2005). The identity of participants should never be revealed, even after the completion of the study; therefore, the method of delivery should be carefully chosen to ensure anonymity is upheld.

3.2 Postal Delphi

Prior to the introduction and wide availability of the internet, Delphi questionnaires were delivered in paper format through post. A postal Delphi is easily administered by the research team and easy to complete by respondents, as well as allowing respondents to fill in the questionnaire at their own pace. However, the administration time of a postal Delphi is much longer than other methods and often incurs low response rates. Salant et al. (1994) recommend a four-step method to improve response rates of a postal Delphi: (1) an advance notice letter; (2) approximately 1 week later, a packet containing a cover letter, the round one questionnaire, and a stamped return envelope; (3) approximately 1 week later, a follow-up reminder notice; and (4) approximately 2 weeks later, a packet containing a new cover letter, a replacement questionnaire, and another stamped envelope sent to those who have not yet responded. With this method, the overall time of administering the postal Delphi will take around 16–20 weeks for four rounds of questioning.

3.3 Online Delphi

The accessibility of the internet has made online Delphi studies much more appealing. An online Delphi allows for faster collection of data, with some studies more than halving the time using an online Delphi versus a postal Delphi (Whyte 1992; Young and Ross 2000). It is also much easier to include a worldwide panel, with turnaround times between subsequent rounds being reduced online and through the post. An online approach also allows for a broader range of data collection methods – online survey, chat room or forum questioning, or even real-time conferencing (Hasson and Keeney 2011; see ► Chaps. 76, "Web-Based Survey Methodology," ► 78, "Synchronous Text-Based Instant Messaging: Online Interviewing Tool," and ► 79, "Asynchronous Email Interviewing Method").

The availability of free online software has made the online Delphi process much simpler for both researchers and respondents alike. However, there are some factors to take into consideration when using an online process. For example, the security of these freely available software should be taken into consideration. As respondents in a Delphi should remain anonymous, the security of the software is paramount. The availability of a Delphi survey online leaves open the possibility that nonparticipants may access the survey through the URL. One way to avoid this scenario is to provide respondents with a password to be used to access the survey (Young and Jamieson 2001). An online Delphi also limits the inclusion of certain respondents, for example, those people who do not have access to a computer or the internet or those people who are not proficient at using a computer.

3.4 Other and Combination Approaches

In order to improve response rates, many Delphi studies are delivered using multiple platforms of communication. For example, researchers may contact potential panel members via telephone asking for their involvement in the project. The project, however, may be run using a different platform, such as online.

Many Delphi studies are also now delivered using a combined approach; that is, respondents are contacted using various modes and can choose to respond in various modes. This process is thought to improve response times as respondents can choose a method most suitable to them (Okoli and Pawlowski 2004). The most common combined approach is for respondents to be given the option of email, fax, or online submission of their responses (Okoli and Pawlowski 2004). In the interest of time, postal and online methods are rarely combined when delivering a Delphi study.

4 Using Experts

4.1 Defining "Expert"

The Oxford Dictionary defines an expert as "a person who is very knowledgeable about or skilful in a particular area" (https://en.oxforddictionaries.com/definition/expert). Unfortunately, broad definitions of experts provide little guidance for selection of a panel to use within a Delphi study. As such, there is limited consensus as to what standards an expert should meet for inclusion in a Delphi study; however, Mead and Moseley (2001) suggest experts may be defined by their position in a hierarchy, such as those seen within a university structure; by public acknowledgement, such as those who have published widely on a topic; or by their experiences, such as patients who have undergone a certain type of treatment.

Undoubtedly, the core characteristic sought after in an expert is a certain level of knowledge. The difficulty, however, comes with ruling the amount of knowledge necessary for inclusion. When consulting health professionals, knowledge is often benchmarked by level of experience. This is often in the form of a certain qualification or a length time spent working within an area. The danger of these benchmarks is that experience may not always translate to knowledge in the area. That is, to become knowledgeable in a particular area, an individual must go above and beyond simply achieving a certain qualification or undergoing experience: they must have a drive to continue learning within the field. This is of particular relevance in the health sector where advancements in the assessment and treatment of patients are continually evolving, and to maintain knowledge in that area, an expert must actively seek to update their understanding throughout this evolvement. When consulting health consumers, knowledge is often benchmarked by the length of time an individual has spent using a service or through their exposure to a certain event, such as surgery. The definition of an expert may vary according to the objectives of a Delphi study. Therefore, it is imperative that researchers develop a set of strictly defined inclusion criteria for potential experts prior to undertaking the study (Mead and Moseley 2001).

When selecting a panel of experts, the homogeneity or heterogeneity of the panel should be considered (Baker et al. 2006). There is merit to both types of panels. A homogenous group of experts may share the same attributes required to reach a consensus, while a heterogeneous group may find it more difficult to reach consensus. The inclusion of a heterogeneous group is thought to increase the validity of findings, because if a varied panel reaches consensus then the findings must be worthwhile (Baker et al. 2006). Of course, obtaining a heterogeneous group may also require a larger panel size which may increase the difficulty of reaching consensus, so is not always advantageous in a Delphi method (Table 1).

4.2 Size of Panel

The number of experts included in a Delphi panel can vary greatly. For example, studies may include as few as 4 participants to as many as 3000 (Thangaratinam and Redman 2005). Linstone (1985) suggests the minimum suitable panel size is seven, which is a generally accepted guide when planning and conducting Delphi studies. However, the final determinant of the size of a Delphi panel is often determined by several pragmatic factors, including, but not limited to, the question to be answered, delivery method, access to experts and resources, timeframe of study, and expenses. A further consideration should be the expectation of a dropout rate between 20% and 30% between subsequent rounds of questioning (Bardecki 1984). Motivating panel

Table 1 Two examples of predefined criteria for determining expert panel member inclusion

Health-user inclusion criteria (Chalmers et al. 2017)	Health professional inclusion criteria (Whitehead 2008)
Aim of study: to determine the areas of life most greatly impacted in women with pelvic pain	*Aim of study:* to arrive at an expert consensus in relation to health promotion and health education constructs as they apply to nursing practice, education, and policy

Inclusion
  Female
  Aged over 18 years
  Self-reports pelvic pain, defined as pain in the pelvic region and associated structures
	Inclusion
	  Senior clinician or academic serving in a nursing area
	  Involved with health promotion policy formation
	  Established public domain research publication record
	  Possess a health promotion-related higher degree qualification

Comment: The health-user inclusion criteria were broad and hence attracted a heterogeneous group for the study. This was in line with the aims of the study. The health professional inclusion criteria were specific and hence attracted a homogeneous group. Again, this was in line with the aims of the study. Notably, the health-user study required three rounds of questioning to reach consensus, while the health professional study with a homogeneous group achieved consensus with only two rounds of questioning

members and ensuring a quick turnaround between rounds of questioning may help to reduce dropout rates (Hsu and Sandford 2007a).

Consensus is likely to be reached in a faster timeframe with smaller panel sizes. However, a small panel size is more likely to represent a homogenous group, limiting the validity of expanding the results beyond that group. A larger panel size will allow for greater heterogeneity among experts, making the results more meaningful to a more varied population. As panel size increases, reliability of findings also improves, and error is reduced (Cochran 1983). However, there are also suggestions that no new ideas are generated nor improvement in results achieved in Delphi studies with panel sizes greater than 25–30 (Delbecq et al. 1975; Brooks 1979). When choosing a panel, the number of experts as well as their quality should be considered.

5 Round One

The initial round of Delphi aims to identify broad statements around each of various areas where consensus is sought. This can be achieved in various ways depending on the specific question(s) that researchers desire to answer. One common technique that is used when the expert opinion (rather than say, previous literature) is paramount is using an open-ended questionnaire. The open-ended questionnaire allows the researchers to solicit information from the panel of experts by providing some open-ended questions. An example of this may be a question such as "List all the components that influence how your patients transition from acute to chronic pain." In this instance, the researchers are not limiting the information that the experts can provide (Thangaratinam and Redman 2005). This may be more common in areas where evidence-based practice is limited or where individual opinion guides clinical practice strongly. However, the shortcoming with this open-ended approach is that there may be a wide variety of answers, which may need to be combined or collapsed to allow for the second round to occur, or the researchers risk "survey fatigue" from the large number of items, causing panel attrition and slow response (Custer et al. 1999).

5.1 Provision of Pre-existing Information

In healthcare, especially when designing consensus statements around clinical problems, often researchers wish to use pre-existing research as the basis for round one, then soliciting the panel opinion on the validity or importance of these statements. Figure 1 outlines one way in which this can be done, where a tick-box approach allows a "Yes/No" answer for each of these symptoms. Another similar option would be to have an "Agree/Disagree" option for various statements, as used in this study for developing treatment guidelines for rotator cuff pathology based on a systematic review of the literature (Eubank et al. 2016). In both styles, there is an "other" option, or similar, that allows addition of items that the researchers may not have missed, allowing the panel to feedback their own items. These may then be incorporated into round two, if there is a sufficient common theme that occurs. A final, and very common, way to structure pre-existing information, either from previous literature or the researchers own interest, is to use a Likert, or similar, scale where a numeric rating, usually from 1 to 5, allows the panel member to numerically rate how strongly they agree, or disagree, on a particular statement.

Please select which aspects of your life have been impacted by your pelvic pain:

☐ Jogging or running
☐ Getting to sleep at night
☐ Household activities such as cooking or cleaning
☐ Sitting for longer than 20 minutes
☐ Riding bicycle
☐ Staying asleep during the night
☐ Wearing tight fitting clothes

☐ Other (please elaborate):

Fig. 1 An example of round one questioning in a Delphi study identifying significant quality of life impact areas in women with pelvic pain. Here, eight aspects are shown. In the actual study, 52 aspects were identified through the literature and were included in round one. *Comment*: The availability of the "other" section allows for a qualitative addition by respondents if their experience was not reflected in the impact areas provided. This is essential when round one questionnaires have been developed in consultation with the literature to ensure that real-life experiences of the experts can be added to the information published within the literature

An example of this may be around how important it is to explain side effects of a medication to patients, with responders rating from 1, not at all important, to 5, vitally important. There are, therefore, a number of question styles that can be used, depending on the quantity and importance of any pre-existing information.

5.2 Modified Delphi

There are numerous modifications to the Delphi technique, although all of these maintain the same overall intent as the original technique (to predict future events or arrive at consensus) and the same overall procedure (multiple rounds using a panel of selected experts). The original Delphi technique is designed to be used with an open-ended questionnaire, as outlined above. The most common modification is that instead of an open-ended questionnaire, the researchers use a carefully selected set of preselected items (such as in Sect. 5.1 above). Two significant advantages of this modification are an increase in the response rate and a solid foundation in pre-existing knowledge (Custer et al. 1999). Other modifications can include the inclusion of a "face-to-face" discussion at the end of the initial anonymous rounds (Eubank et al. 2016; Schneider et al. 2016). This final "face-to-face" round can serve a number of purposes including allowing easier discussion and presentations of clarification and justification for certain points and allowing easier clarification of disagreements. However, this also means that one of the key principles of Delphi is broken, allowing one or more dominant personalities to overpower others (Boulkedid et al. 2011). Another variant can be that these initial recommendations (developed during the anonymous rounds) can be given to a separate panel of experts, for example, physicians, to comment on, and then their recommendations can be returned to the anonymous panel for further round(s) of consensus. This ensures that the expert recommendations are feasible for clinical practice (van Vliet et al. 2016) and avoid the issue of discrepancy between experts and clinicians (Alraek et al. 2011).

6 Subsequent Rounds

6.1 Purpose of Round Two

The aims of round two are to prioritize or rank items discussed in round one and provide justification for their ranking (Hsu and Sandford 2007a). In this process, a more structured questionnaire is created based on the responses received in the first round, and respondents are asked to quantitatively rate or rank order their responses in order to prioritize them (Ludwig 1994). Respondents can also then use qualitative responses to supplement their ratings, providing justification for their chosen ratings. As a result of round two, areas of agreement and disagreement between respondents begin to emerge, and a level of consensus can be identified (Hsu and Sandford 2007a).

6.2 Developing a Structured Questionnaire for Round Two

On the completion of round one, responses are then used to develop a more structured questionnaire for further rounds of questioning and clarification. As most round one surveys are conducted in a qualitative manner, there are few guidelines on the exact procedure to then develop a structured questionnaire. Research teams should aim to identify themes from the round one qualitative responses, remove duplicates of any responses, consolidate themes and responses, and unify terminology used (Okoli and Pawlowski 2004). Often, there is a need to collapse responses in order to shorten the subsequent round questionnaires (Keeney et al. 2001). An inclusive approach where responses have not been collapsed can result in large questionnaires for subsequent rounds, which may put off panel members participating.

Once key themes have been identified and the overall content of the questionnaire has been finalized, the method of response from the expert panel should be considered. Whether respondents are asked to rank or rate items will influence the type of data gathered and the subsequent data analysis. Rating of items indicates how respondents feel about individual items within the questionnaire, whereas ranking of items asks respondents to compare all items to one another and order according to their preference. Linear numerical scales, such as those described by Likert (1932), are commonly used as a means of collecting rated responses. Care should be taken to select a validated scale to collect quantitative responses (Hasson and Keeney 2011). While closely related, rating and ranking are separate entities, and the choice to use one or the other or both should be considered in conjunction with the aims of the Delphi study (Figs. 2 and 3).

6.3 Purpose of Rounds Three and Beyond

Subsequent rounds of questioning serve to allow respondents to revise their judgements or to provide justification for remaining outside of the consensus (Pfeiffer 1968). In rounds three and beyond, feedback on the group and the individual respondents' position is provided from the previous round structured questionnaire, and respondents can clarify their rationale for differing from the group. Often, there is only a slight increase in the degree of consensus in subsequent rounds of questioning (Dalkey and Rourke 1971; Weaver 1971), and as such, Delphi studies often include only three rounds of questioning. Theoretically, Delphi studies could continue for many number of rounds before reaching the required consensus, although three is often enough to obtain sufficient consensus (Hsu and Sandford 2007a).

6.4 Providing Feedback

The provision of feedback between each round will be dependent on the aims of the study and the format of questions included in the Delphi. Qualitative content

In the past month, how much has your pelvic pain affected your:	Not at all (0)	A little bit (1)	Somewhat (2)	Quite a bit (3)	A great deal (4)
energy levels?	☐	☐	☐	☐	☐
mood?	☐	☐	☐	☐	☐
sleep?	☐	☐	☐	☐	☐
stomach and intestinal function?	☐	☐	☐	☐	☐
ability to sit for longer than 20 minutes?	☐	☐	☐	☐	☐
ability to perform and function normally in your everyday role?	☐	☐	☐	☐	☐
ability to take part in physical activity?	☐	☐	☐	☐	☐
ability to wear certain clothes?	☐	☐	☐	☐	☐

Fig. 2 An example of structured questioning in round two in a Delphi study identifying significant quality of life impact areas in women with pelvic pain. *Comment*: In this example, respondents use a qualitative scale to provide answers. The qualitative aspect of the scale was also piloted on the expert panel members to ensure its ease of use. The scale has also been validated as a numerical Likert scale (see Chalmers et al. 2017), with numbers 0–4 corresponding with the qualitative responses

In the past month, how much has your pelvic pain affected your:	Your last score	Group's last score	Your new score					Comment
			Not at all (0)	A little bit (1)	Somewhat (2)	Quite a bit (3)	A great deal (4)	
energy levels?	4	3	☐	☐	☐	☐	☐	
mood?	3	3	☐	☐	☐	☐	☐	
sleep?	3	2	☐	☐	☐	☐	☐	
stomach and intestinal function?	4	4	☐	☐	☐	☐	☐	
ability to sit for longer than 20 minutes?	4	2	☐	☐	☐	☐	☐	
ability to perform and function normally in your everyday role?	3	3	☐	☐	☐	☐	☐	
ability to take part in physical activity?	2	2	☐	☐	☐	☐	☐	
ability to wear certain clothes?	3	2	☐	☐	☐	☐	☐	

Fig. 3 An example of feedback after round two in a Delphi study identifying significant quality of life impact areas in women with pelvic pain. Feedback delivered to respondent one (Tables 2 and 3). *Comment:* Feedback received should include the respondent's previous rating along with the group's previous rating, for the respondent to consider their position in relation to the group's. Respondents also have the opportunity to provide further comments and are instructed to provide comments on times where ratings have changed or where they do not wish to change their rating but remain an outlier in comparison to the group's median rating

analysis techniques are usually used after the first round to identify key themes from the unstructured questionnaire (see also ▶ Chap. 47, "Content Analysis: Using Critical Realism to Extend Its Utility"). These are then used to create the structured questionnaire in subsequent rounds. In rounds three and beyond, feedback can be numerical or statistical, providing an overview of the group position, with supplemental qualitative data such as rationale responses or other comments. Quantitative feedback generally consists of central tendencies (mean, median, or mode) and level of dispersion (standard deviation or interquartile ranges) (Thangaratinam and Redman 2005; Hsu and Sandford 2007a). Respondents are also given specific feedback on their individual responses from the previous round in relation to the group. This gives respondents the opportunity to revise their responses in light of the group responses (Powell 2003).

7 Delphi Rounds and Reaching Consensus

7.1 Determining Number of Rounds

Delphi studies can continue for any number of rounds, and there have been up to 25 rounds of questioning reported (Couper 1984). However, Delphi studies generally use a maximum of three or four rounds of questioning before termination. The termination of rounds should occur when either consensus is reached, when "there is enough convergence to justify using the results without complete consensus" (Whitman 1990, p. 378), or when stability of responses is achieved (Dajani et al. 1979). With successive rounds of questioning, the variability in results will generally stabilize. This can either indicate a convergence of responses (Murry Jr and Hammons 1995) toward consensus or that the panel members are unmoving in their responses between rounds (Dajani et al. 1979). In these instances, consensus on the topic may never be reached, but the stability of responses should indicate that terminating the survey is appropriate. See Tables 2, 3, and 4 for examples of consensus and stability.

Another consideration for determining the number of rounds in a Delphi study is the dropout rate of expert panel members. Some dropout between rounds of

Table 2 Results of respondents in round two of a Delphi study identifying significant quality of life impact areas in women with pelvic pain

		\multicolumn{8}{c}{Impact area}							
		Energy	Mood	Sleep	Stomach/ intestinal function	Sitting	Daily functioning	Physical activity	Wearing certain clothes
Respondent	1	4	3	3	4	4	3	2	3
	2	4	4	2	3	2	3	3	0
	3	3	3	2	4	2	4	4	3
	4	1	2	2	2	0	1	1	0
	5	4	3	4	4	4	3	2	3
	6	3	3	3	4	3	3	2	3
	7	1	1	0	2	0	1	0	0
Median		3	3	2	4	2	3	2	2
IQR		2	0.5	1	1.5	2.5	1	1	3

IQR: interquartile range. Colour code: IQR indicates consensus not reached   IQR indicates consensus has been reached

Comment: Overall agreement between respondents in round two was 23.2%. However, inspection of the IQR for each impact area shows that consensus appears to be reached on four out of eight areas. To obtain a higher level of consensus, a third round of questioning was used (Table 3)

Table 3 Results of respondents in round three of a Delphi study identifying significant quality of life impact areas in women with pelvic pain

		Impact area						
		Energy	Mood	Sleep	Stomach/intestinal function	Sitting	Daily functioning	Physical activity	Wearing certain clothes
Respondent	1	4	3	3	3	4	3	2	1
	2	4	4	2	3	2	3	3	0
	3	3	3	2	4	2	4	4	2
	4	4	3	2	2	2	1	1	0
	5	4	3	4	3	4	3	2	3
	6	3	3	3	4	3	3	2	2
	7	4	3	2	2	2	2	0	0
Median		4	3	2	3	2	3	2	1
IQR		0.5	0	1	1	1.5	0.5	1	2

IQR: interquartile range. Colour code: IQR indicates consensus not reached   IQR indicates consensus has been reached

Comment: Overall agreement between respondents in round three was greater than round two, at 34.5%. Although this is seemingly low, inspection of the IQR for each impact area shows that consensus appears to be reached on six out of eight areas

Table 4 Wilcoxon signed-rank test results comparing responses from round two and three of a Delphi study identifying significant quality of life impact areas in women with pelvic pain

	Impact area							
	Energy	Mood	Sleep	Stomach/intestinal function	Sitting	Daily functioning	Physical activity	Wearing certain clothes
Z-score	−1.414	−1.342	−1.00	−1.414	−1.414	−1.00	0.000	−1.633
Significance	0.157	0.180	0.317	0.157	0.157	0.317	1.000	0.102

questioning is to be expected (Bardecki 1984), and the research team should anticipate this in their consideration of the size of the expert panel. The makeup of the expert panel should be assessed after each round to ensure the uniformity of the panel remains as was intended. Dropouts have the tendency to modify a once heterogeneous group into a more homogenous group, which can influence findings (Wheeller et al. 1990). Where the panel size significantly diminishes or the uniformity of the panel changes, the survey should be terminated.

7.2 How Much Agreement Is Needed for Consensus?

Consensus is essentially the amount of responses that fall within a prescribed range (Miller 2006), and it is imperative that consensus is determined prior to the beginning of the Delphi study. However, there is little guidance from the literature that will help researchers to determine a level of consensus, and different levels may be set depending on the aspirations of the research team, the question being asked, the type of responses given, and the number of expert panel members consulted.

Many Delphi studies define consensus as a certain percentage of respondents being in agreement. The particular level of agreement is often arbitrarily chosen and may range from 100% down to as little as 51% (Williams and Webb 1994;

Heiko 2012). Other studies are even less specific with the set level of agreement, using unclear values (Beech 1997) or terms such as "good enough" (Butterworth and Bishop 1995, p. 30) agreement. A more objective and rigorous method of determining consensus is through the use of interquartile range (IQR). The IQR is the measure of dispersion around the median, consisting of the middle 50% of responses. The range of the IQR is usually dependent on the number of response choices available, with more response choices usually resulting in larger IQRs. There are some basic guidelines that may assist with interpreting consensus using IQRs: where the number of response choices is ten or more, an IQR of two or less can be considered consensus (Sheibe et al. 1975); where the number of response choices is four or five, an IQR of one or less can be considered consensus (Raskin 1994; Rayens and Hahn 2000). Tables 2 and 3 outline an example of determining consensus using level of agreement and IQRs.

7.3 Stability Indicating Termination

Stability of a Delphi study occurs when there is little shift in the distribution of responses from one round to another. Whether responses are stable or instable is determined through statistical means. Where the statistical analysis indicates that responses have not differed between two subsequent rounds, the Delphi survey should be terminated. Similar to level of agreement, there are no set guidelines for how this statistical analysis should be conducted. The choice of statistical method to determine stability will depend on the questions asked in the survey and the type of responses provided. More information on different techniques of determining stability can be found in reviews by von der Gracht (2008, 2012). Some common examples include the Wilcoxon sign-ranked test, Chi-squared test, intra-class correlation coefficient, and Kendall's W coefficient of concordance. Table 4 provides an example of determining stability levels using a Wilcoxon sign-ranked test.

8 Reporting Results

A common issue with Delphi is that due to the wide variety of modifications, there is often a lack of clear reporting around both the technique itself, including vital information such as the definition of consensus, how the experts were selected, and which, if any, modifications to the original Delphi technique were used (Boulkedid et al. 2011). This lack of clarity in reporting the methods often undermines the conclusions from then Delphi as it is unclear if the process was followed correctly. To assist with standardize reporting of Delphi, there have been attempts to generate guidelines for reporting of Delphi in the past (Hasson et al. 2000), but these have not focused on development a checklist approach similar to other reporting guidelines such as SPIRIT (Chan et al. 2013) which has limited their usefulness. The use of a checklist can improve not only the reporting but, if used at an early stage, the design of the study as it helps identify decisions that need to be made with design choices early on. The Guidance on Conducting and REporting DElphi Studies (CREDES) guideline provides the first checklist approach for evaluating study quality (Junger et al. 2017). This checklist covers the rationale for using Delphi. Some key areas include the design process including how consensus will be designed and what to do if consensus is not reached, how the initial information is generated (sources and how it was synthesized), how the experts are selected, and a flow chart outlining the number of rounds including the preparation phases. The use of a checklist such as this, or other checklists that are currently in development such as the "Reporting E-Delphi Studies (REDS) in health research checklist," is strongly encouraged to improve design.

9 Limitations of the Method

This chapter has outlined some of the uses and benefits of employing the Delphi technique; however, there are a number of limitations of this technique that should be highlighted.

9.1 Lack of Standard Guidelines for the Delphi Technique

The mixed qualitative-quantitative design of the Delphi technique does not allow for studies to follow a standardized scientific approach as is possible with other study methodologies. As such, there is wide variation in the ways the Delphi technique has been implemented in the literature. The only general guidelines that exist for the technique are the requirement to provide feedback to respondents between rounds and to include at least two rounds of questioning to reach consensus (Keeney et al. 2010). However, even the method of feedback provided can range from a single number to group distributions or qualitative responses of the respondents. The lack of guidelines for the Delphi technique has resulted in some researchers questioning the validity of the methodology (Sackman 1975), although others view it as beneficial, with the flexibility of the technique allowing researchers to answer a broad range of questions (Keeney et al. 2011). While efforts have been made to create guidelines and recommendations for the Delphi technique, there is still no standardized method for undertaking a Delphi study (Keeney et al. 2001; Hasson and Keeney 2011).

9.2 Sample Size

As discussed in Sect. 4.2, the number of expert panel members in a Delphi study can vary greatly. There are no guidelines for selecting the number of expert panel members with the Delphi technique, and so the decision is often made empirically and pragmatically based on several factors (Thangaratinam and Redman 2005).

The size and type of the expert panel may limit the findings of a Delphi study: a smaller, homogeneous panel may reach consensus quickly but has limitations on the expansion of results to a heterogeneous population; a heterogeneous panel may take longer or indeed fail to reach consensus but allows for results to be validly extrapolated to a broader population. The size and makeup of the panel should be determined with the aims of the study in mind.

9.3 The Use of "Experts"

The definition of an expert suitable for inclusion in a Delphi study is not well defined. Often, the definition of an expert being included is based on the objectives of the study being run. With healthcare professionals, experts may be classified according to their level of qualification, although this has limitations in assuming that a qualification equals experience and expertise. With healthcare users, experts may be classified as patients or groups who have a certain condition or have undergone the same procedure or treatment. That the makeup of a panel can vary greatly highlights the need for a clear set of inclusion criteria prior to the beginning of a Delphi study.

While the Delphi technique is used to reach consensus, there is also an argument that consensus in a Delphi study does not necessarily mean a correct answer has been achieved (Keeney et al. 2011). Unfortunately, the use of statistical analysis methods often means that extreme opinions can be masked (Rudy 1996), and respondents who hold opinions that stray away from the mean are most likely to drop out of Delphi studies (Bardecki 1984). The anonymity of the Delphi technique theoretically allows respondents with minority views to retain their opinion, but some researchers also question the issue of social pressure and conformity despite this anonymity (Keeney et al. 2011).

9.4 Non-responders

As with all study designs with repeated measures, there is a risk of dropouts or non-responders in the Delphi technique. There is no set percentage of attrition that is acceptable in Delphi studies, but a dropout rate of 20–30% can be expected between rounds (Bardecki 1984), and this should be taken into consideration when determining the panel sample size.

Hsu and Sandford (2007b) recommend several strategies for dealing with non-respondents in a Delphi study. First, the use of follow-up reminder strategies should be implemented to nudge nonrespondents into filling in their questionnaires. Second, setting deadlines can help researchers avoid nonrespondents slowing down the Delphi process. If several follow-up reminders have been sent in the days or weeks following a round of questioning, it is reasonable to proceed with the study and classify any nonrespondents at that point dropouts. Finally, the provision of incentives for respondents can help to reduce dropouts. Incentives may be provided at the submission of each round to encourage participation or given at the conclusion of the final round of questioning as a gesture of thanks.

9.5 Determining When Questioning Should End

Both stability and consensus provide some general outlines for when questioning should end. However, there are no set guidelines on the cutoff level that should be used to determine the end of a Delphi study. Often, this figure is set arbitrarily and decided upon post hoc or omitted entirely from studies (Keeney et al. 2011). The particular level of consensus should be determined prior to a study beginning. Setting the consensus level may depend on several factors, including the questions posed, the expected outcomes, the homogeneity or heterogeneity of the panel, and the size of the panel. In studies where consensus is not reached should not necessarily be viewed negatively. Results not reaching consensus indicate that expert opinion differs among panel members, which can be interesting and clinically meaningful results in themselves.

10 Conclusion and Future Directions

The Delphi technique, when properly designed and reported, can be a powerful tool to answer questions for healthcare researchers around future trends and in decision-making in areas where there is a lack of, or conflicting, evidence. Due to the lack of standard guidelines around the Delphi technique, careful consideration by the researchers must be given to ensure that the criteria for "experts," the presynthesis of any information, and definition of consensus (if required) are clearly thought out a priori and clearly reported in any publications, preferably using one of the emerging guidelines for Delphi reporting such as CREDES. Given the new focus on ensuring healthcare outcomes from clinical trials are able to be translated from the more "ideal" clinical trial context into practice, the use of Delphi, including its modifications of face-to-face meetings, to ensure that trial design both reflects and is able to influence clinical practice, through the use of key stakeholders, is an important consideration moving forward. This has the potential to change the landscape of clinical trial design, especially when considering complex interventions or multidisciplinary teams working on complex, and often chronic, healthcare issues.

References

Alraek T, Borud E, White A. Selecting acupuncture treatment for hot flashes: a delphi consensus compared with a clinical trial. J Altern Complement Med. 2011;17(1):33–8. https://doi.org/10.1089/acm.2010.0070.

Baker J, Lovell K, Harris N. How expert are the experts? An exploration of the concept of 'expert' within Delphi panel techniques. Nurs Res. 2006;14(1):59–70.

Bardecki MJ. Participants' response to the Delphi method: an attitudinal perspective. Technol Forecast Soc Chang. 1984;25(3):281–92.

Beech B. Studying the future: a Delphi survey of how multi-disciplinary clinical staff view the likely development of two community mental health centres over the course of the next two years. J Adv Nurs. 1997;25(2):331–8.

Boulkedid R, Abdoul H, Loustau M, Sibony O, Alberti C. Using and reporting the Delphi method for selecting healthcare quality indicators: a systematic review. PLoS One. 2011;6(6):e20476. https://doi.org/10.1371/journal.pone.0020476.

Brooks KW. Delphi technique: expanding applications. N Cent Assoc Q. 1979;53(3):377–85.

Butterworth T, Bishop V. Identifying the characteristics of optimum practice: findings from a survey of practice experts in nursing, midwifery and health visiting. J Adv Nurs. 1995;22(1):24–32.

Chalmers KJ, Catley MJ, Evans SF, Moseley GL. Clinical assessment of the impact of pelvic pain on women. Pain. 2017;158(3):498–504.

Chan AW, Tetzlaff JM, Altman DG, Laupacis A, Gotzsche PC, Krleza-Jeric K, Moher D. SPIRIT 2013 statement: defining standard protocol items for clinical trials. Ann Intern Med. 2013;158(3):200–7. https://doi.org/10.7326/0003-4819-158-3-201302050-00583.

Cochran SW. The Delphi method: formulating and refining group judgements. J Hum Sci. 1983;2(2):111–7.

Cochrane S, Smith CA, Possamai-Inesedy A. Development of a fertility acupuncture protocol: defining an acupuncture treatment protocol to support and treat women experiencing conception delays. J Altern Complement Med. 2011;17(4):329–37. https://doi.org/10.1089/acm.2010.0190.

Cotchett MP, Landorf KB, Munteanu SE, Raspovic AM. Consensus for dry needling for plantar heel pain (plantar fasciitis): a modified delphi study. Acupunct Med. 2011;29(3):193–202. https://doi.org/10.1136/aim.2010.003145.

Couper MR. The Delphi technique: characteristics and sequence model. Adv Nurs Sci. 1984;7(1):72–7.

Custer RL, Scarcella JA, Stewart BR. The modified Delphi technique-A rotational modification. J Career Tech Educ. 1999;15(2).

Dajani JS, Sincoff MZ, Talley WK. Stability and agreement criteria for the termination of Delphi studies. Technol Forecast Soc Chang. 1979;13(1):83–90.

Dalkey NC. The Delphi method: an experimental study of group opinion. In: Dalkey NC, Rourke DL, Lewis R, Snyder D, editors. Studies in the quality of life: Delphi and decision-making. Lexington: Lexington Books; 1972. p. 13–54.

Dalkey N, Helmer O. An experimental application of the Delphi method to the use of experts. Manag Sci. 1963;9(3):458–67.

Dalkey NC, Rourke DL. Experimental assessment of Delphi procedures with group value judgments. In: Dalkey NC, Rourke DL, Lewis R, Snyder D, editors. Studies in the quality of life: Delphi and decision-making. Lexington: Lexington Books; 1971.

Delbecq AL, Van de Ven AH, Gustafson DH. Group techniques for program planning: a guide to nominal group and Delphi processes. Illinois: Scott Foresman; 1975.

Eubank BH, Mohtadi NG, Lafave MR, Wiley JP, Bois AJ, Boorman RS, Sheps DM. Using the modified Delphi method to establish clinical consensus for the diagnosis and treatment of patients with rotator cuff pathology. BMC Med Res Methodol. 2016;16:56. https://doi.org/10.1186/s12874-016-0165-8.

Flower A, Lewith GT, Little P. Seeking an oracle: using the Delphi process to develop practice guidelines for the treatment of endometriosis with Chinese herbal medicine. J Altern Complement Med. 2007;13(9):969–76.

Hasson F, Keeney S. Enhancing rigour in the Delphi technique research. Technol Forecast Soc Chang. 2011;78(9):1695–704.

Hasson F, Keeney S, McKenna H. Research guidelines for the Delphi survey technique. J Adv Nurs. 2000;32(4):1008–15.

Heiko A. von der Gracht. Consensus measurement in Delphi studies. Technological Forecasting and Social Change. 2012; 79(8):1525–1536.

Hsu C-C, Sandford BA. The Delphi technique: making sense of consensus. Pract Assess Res Eval. 2007a;12(10):1–8.

Hsu C-C, Sandford BA. Minimizing non-response in the Delphi process: how to respond to non-response. Pract Assess Res Eval. 2007b;12(17):62–78.

Junger S, Payne SA, Brine J, Radbruch L, Brearley SG. Guidance on Conducting and REporting DElphi Studies (CREDES) in palliative care: recommendations based on a methodological systematic review. Palliat Med. 2017;31(8):684–706. https://doi.org/10.1177/0269216317690685.

Keeney S, Hasson F, McKenna HP. A critical review of the Delphi technique as a research methodology for nursing. Int J Nurs Stud. 2001;38(2):195–200.

Keeney S, McKenna H, Hasson F. The Delphi technique in nursing and health research. Oxford: Wiley; 2010.

Keeney S, Hasson F, McKenna H. Debates, criticisms and limitations of the Delphi. In: Keeney Ss, Hasson F, McKenna H, editors. Delphi Technique in Nursing and Health Research. Sussex: John Wiley & Sons; 2011.

Likert R. A technique for the measurement of attitudes. Arch Psychol. 1932;22(140):55.

Linstone HA. The Delphi technique environmental impact assessment, technology assessment, and risk analysis. Berlin: Springer; 1985. p. 621–49.

Linstone HA, Turoff M. The Delphi method: techniques and applications, vol. 29. Reading: Addison-Wesley; 1975.

Linstone HA, Turoff M. Delphi: a brief look backward and forward. Technol Forecast Soc Chang. 2011;78(9):1712–9.

Lintonen T, Konu A, Rönkä S, Kotovirta E. Drugs foresight 2020: a Delphi expert panel study. Subst Abus Treat Prev Policy. 2014;9(1):18.

Ludwig BG. Internationalizing extension: an exploration of the characteristics evident in a state university extension system that achieves internationalization. PhD Dissertation, The Ohio State University; 1994.

Mead D, Moseley L. The use of the Delphi as a research approach. Nurs Res. 2001;8(4):4–23.

Miller L. Determining what could/should be: the Delphi technique and its application. Paper presented at the meeting of the 2006 annual meeting of the Mid-Western Educational Research Association, Columbus; 2006.

Murry JW Jr, Hammons JO. Delphi: a versatile methodology for conducting qualitative research. Rev High Educ. 1995;18(4):423–36.

Nelson TF, Xuan Z, Babor TF, Brewer RD, Chaloupka FJ, Gruenewald PJ, et al. Efficacy and the strength of evidence of US alcohol control policies. Am J Prev Med. 2013;45(1):19–28.

Okoli C, Pawlowski SD. The Delphi method as a research tool: an example, design considerations and applications. Inf Manag. 2004;42(1):15–29.

Pfeiffer J. New look at education. Poughkeepsie: Odyssey Press; 1968.

Powell C. The Delphi technique: myths and realities. J Adv Nurs. 2003;41(4):376–82.

Raskin MS. The Delphi study in field instruction revisited: expert consensus on issues and research priorities. J Soc Work Educ. 1994;30(1):75–89.

Rayens MK, Hahn EJ. Building consensus using the policy Delphi method. Policy Polit Nurs Pract. 2000;1(4):308–15.

Rowe G, Wright G. The Delphi technique as a forecasting tool: issues and analysis. Int J Forecast. 1999;15(4):353–75.

Sackman H. Delphi critique. Lexington: Lexington Books; 1975.

Salant P, Dillman I, Don A. How to conduct your own survey. New York: Wiley; 1994.

Schneider P, Evaniew N, Rendon JS, McKay P, Randall RL, Turcotte R, Investigators P. Moving forward through consensus: protocol for a modified Delphi approach to determine the top research priorities in the field of orthopaedic oncology. BMJ Open. 2016;6(5):e011780. https://doi.org/10.1136/bmjopen-2016-011780.

Seemann R, Flury S, Pfefferkorn F, Lussi A, Noack MJ. Restorative dentistry and restorative materials over the next 20 years: a Delphi survey. Dent Mater. 2014;30(4):442–8.

Sheibe M, Skutsch M, Schofer J. Experiments in Delphi methodology. In: Linstone HA, Turoff M, editors. The Delphi method: techniques and applications, vol. 29. Reading: Addison-Wesley; 1975. p. 262–87.

Smith CA, Zaslawski CJ, Zheng Z, Cobbin D, Cochrane S, Lenon GB, Bensoussan A. Development of an instrument to assess the quality of acupuncture: results from a delphi process. J Altern Complement Med. 2011;17(5):441–52. https://doi.org/10.1089/acm.2010.0457.

Smith CA, Grant S, Lyttleton J, Cochrane S. Using a Delphi consensus process to develop an acupuncture treatment protocol by consensus for women undergoing Assisted Reproductive Technology (ART) treatment. BMC Complement Altern Med. 2012;12:88. https://doi.org/10.1186/1472-6882-12-88.

Thangaratinam S, Redman CW. The Delphi technique. Obstet Gynaecol. 2005;7(2):120–5.

van Vliet DCR, van der Meij E, Bouwsma EVA, Vonk Noordegraaf A, van den Heuvel B, Meijerink WJHJ, Anema JR. A modified Delphi method toward multidisciplinary consensus on functional convalescence recommendations after abdominal surgery. Surg Endosc. 2016;30(12):5583–95. https://doi.org/10.1007/s00464-016-4931-9.

von der Gracht. The future of logistics: scenarios for 2025. Springer Science & Business Media; 2008.

von der Gracht. Consensus measurement in Delphi studies: review and implications for future quality assurance. Technol Forecast Soc Chang. 2012;79(8):1525–36.

Weaver WT. The Delphi forecasting method. Phi Delta Kappan. 1971;52(5):267–71.

Wheeller B, Hart T, Whysall P. Application of the Delphi technique: a reply to Green, Hunter and Moore. Tour Manag. 1990;11(2):121–2.

Whitehead D. An international Delphi study examining health promotion and health education in nursing practice, education and policy. J Clin Nurs. 2008;17(7):891–900.

Whitman NI. The Delphi technique as an alternative for committee meetings. J Nurs Educ. 1990;29(8):377–9.

Whyte DN. Key trends and issues impacting local government recreation and park administration in the 1990s: a focus for strategic management and research. J Park Recreat Adm. 1992;10(3):89–106.

Williams PL, Webb C. The Delphi technique: a methodological discussion. J Adv Nurs. 1994;19(1):180–6.

Young SJ, Jamieson LM. Delivery methodology of the Delphi: a comparison of two approaches. J Park Recreat Adm. 2001;19(1).

Young SJ, Ross CM. Recreational sports trends for the 21st century: results of a Delphi study. NIRSA J. 2000;24(2):24–37.

Consensus Methods: Nominal Group Technique

42

Karine Manera, Camilla S. Hanson, Talia Gutman, and Allison Tong

Contents

1	Introduction to Nominal Group Technique (NGT)	738
2	When and Why to Use NGT	738
	2.1 Guideline Development	739
	2.2 Identifying Research Priorities	739
	2.3 Establishing Core Outcomes	740
3	Methods	741
	3.1 Participant Selection and Recruitment	741
	3.2 The Facilitator	741
	3.3 Conducting the Nominal Group Technique	742
	3.4 The Modified NGT	743
	3.5 Data Analysis	744
4	Advantages and Disadvantages of the NGT	745
	4.1 Considerations for Involving Patients in NGT	746
5	Conclusion and Future Directions	747
References		748

Abstract

Nominal group technique uses structured small group discussion to achieve consensus among participants and has been used for priority setting in healthcare and research. A facilitator asks participants to individually identify and contribute

K. Manera (✉) · C. S. Hanson · T. Gutman
Sydney School of Public Health, The University of Sydney, Sydney, NSW, Australia

Centre for Kidney Research, The Children's Hospital at Westmead, Westmead, NSW, Australia
e-mail: karine.manera@sydney.edu.au; camilla.hanson@sydney.edu.au; talia.gutman@sydney.edu.au

A. Tong
Sydney School of Public Health, The University of Sydney, Sydney, NSW, Australia

Centre for Kidney Research, The Children's Hospital at Westmead, Westmead, Australia
e-mail: allison.tong@sydney.edu.au

© Springer Nature Singapore Pte Ltd. 2019
P. Liamputtong (ed.), *Handbook of Research Methods in Health Social Sciences*,
https://doi.org/10.1007/978-981-10-5251-4_100

ideas to generate a list. The group discusses, elaborates, clarifies, and adds new ideas as appropriate. Each participant independently prioritizes the ideas, for example, by voting, rating, or ranking. The facilitator may summarize the scores to ascertain the overall group priorities. This method is useful for generating a diverse range of views and ideas in a structured manner, prevents participants from dominating the discussion, and promotes input from all members.

Keywords
Nominal group technique · Consensus methods · Focus group · Ranking

1 Introduction to Nominal Group Technique (NGT)

The nominal group technique (NGT) is a structured group process used to achieve consensus among participants. The process involves participants identifying and contributing ideas toward a topic or question specified by the facilitator. Participants then discuss and individually prioritize the ideas. There are variations on how the prioritization can be done, for example, by voting, rating, ranking, or a combination of methods. The facilitator may summarize the individual scores to produce a set of prioritized ideas that represent the group's preferences. Using the NGT as a method of consensus generates a diverse range of views and ideas in a structured manner, ensures that the opinions of all group members are taken into account, and prevents the discussion and process being dominated by an individual participant.

The NGT was developed by Delbecq and Van de Ven in 1968 and has since been applied in a variety of fields, particularly in health and social research (Delbecq et al. 1975). This chapter will discuss when and why to use NGT, present some examples of NGT studies in health, provide an overview of the methods involved in conducting a NGT, outline the advantages and disadvantages of using NGT, and discuss future directions in terms of potential applications and developments in NGT.

2 When and Why to Use NGT

The NGT is used to generate and prioritize ideas and to achieve group consensus on a particular topic. The James Lind Alliance (Cowan and Oliver 2013), an organization that facilitates research priority setting partnerships, suggests that NGT may be useful when:

- There is concern about participants not contributing.
- Some participants are more outspoken than others.
- The group does not easily generate many ideas.
- Some participants think better in silence.
- The topic is contentious or there is serious disagreement.

In health and social research, the NGT has been applied in numerous studies across a broad range of topics and with diverse groups of participants, including patients, caregivers, and health professionals. Some applications of the NGT include research priority setting (Knight et al. 2016; Ghisoni et al. 2017; Lavigne et al. 2017; Williams et al. 2017), eliciting patient priorities and treatment outcomes (Sanderson et al. 2010; Howell et al. 2012; Urquhart-Secord et al. 2016), identifying patient and caregiver challenges and needs related to chronic conditions (Miller et al. 2000; Dewar et al. 2003; Sav et al. 2015), identifying preferences for end-of-life care (Aspinal et al. 2006; Dening et al. 2013), and identifying exemplars of patient-centered professionalism in nursing and pharmacy (Hutchings et al. 2010, 2012). Some detailed examples of NGT studies used in guideline development, research priority setting, and establishing core outcomes will be provided in the following section.

2.1 Guideline Development

The World Health Organization (2014) has recognized the need to use formal consensus methods in the development of clinical practice guidelines and often use the NGT to identify priorities of stakeholders for integration into healthcare recommendations. For example, the NGT may be used to inform guideline development through the generation and prioritization of guideline topics, with participants being asked questions such as, "What are the main issues that need to be addressed by guidelines for this disease/condition/treatment?" The NGT has been successfully used to inform the development of guidelines for dementia (Trickey et al. 1998), intensive care (Rolls and Elliott 2008), and treatment of vulvovaginal candidiasis (Bond and Watson 2003), among others.

Rolls and Elliott (2008) used the NGT in combination with other group processes to develop state-based clinical guidelines for six common intensive care practices – eye care, oral care, endotracheal tube management, suctioning, arterial line management, and central venous catheter management. Clinicians, academics, and staff from the Intensive Care Coordination and Monitoring Unit participated in the guideline development process which included a preparatory educational seminar, formation of working groups, and identification of scope of practice for each guideline. A consensus development conference was conducted, and the NGT was used to develop recommendations for practice.

2.2 Identifying Research Priorities

Research priority setting partnerships are increasingly being conducted to involve stakeholders in a transparent process to prioritize health research, to help ensure that resources are directed toward areas of high priority and address important needs. The James Lind Alliance (JLA) was established in 2004 to facilitate a priority setting partnership among patients, family members, and health professionals to identify and prioritize unanswered questions about treatment that they agree are important (Cowan and Oliver 2013). The process involves consultation, and the collation and prioritization of research questions (treatment uncertainties). The JLA recommends the NGT to prioritize research topics, and it has been used in a many of their partnership initiatives including in asthma, prostate cancer, urinary incontinence, schizophrenia, and kidney disease.

Rees et al. (2017) conducted a research priority setting exercise in gestational diabetes mellitus using the JLA approach. The steering committee short-listed research priorities obtained through a survey and review of clinical practice guidelines. They used the NGT in the final workshop to identify the top 10 research priorities.

2.3 Establishing Core Outcomes

In clinical trials, treatments are developed and tested by researchers to make sure they are effective and safe. Researchers look at the effects those treatments have on patients by measuring outcomes, which are things that can arise or change because of a health condition or treatment (see ▶ Chap. 37, "Randomized Controlled Trials"). However, outcomes that are measured in research are often not appropriate or meaningful to end users (i.e., patients) (Yudkin et al. 2011). In addition, there is large heterogeneity in the reporting of outcomes, resulting in the inability to compare the effectiveness across trials, ultimately contributing to research waste (Manera et al. 2017). In view of this, there is growing recognition of the need for core outcomes informed by all stakeholders. Core outcomes are defined as outcomes that should be measured and reported, as a minimum, in all clinical trials in specific areas of health or healthcare. Developing a standardized set of core outcomes will improve the consistency in reporting of outcomes that are critically important to patients, caregivers, and their clinicians. A number of global and discipline-specific initiatives develop core outcome sets, including the Outcome Measures in Rheumatology (OMERACT) initiative and the Standardized Outcomes in Nephrology (SONG) initiative. Both initiatives recommend and implement the NGT as a method to reach consensus on core outcomes (Boers et al. 2014; SONG Initiative 2017).

As part of the SONG initiative, an international nominal group technique study was conducted to identify patient and caregiver priorities for outcomes in hemodialysis (Urquhart-Secord et al. 2016). In total, 82 patients on hemodialysis and their families from Australia and Canada were recruited to participate in 12 nominal groups. The study used a combined focus group/nominal group technique to identify and rank outcomes, and to discuss the reasons behind the ranking decisions. Qualitative data analysis of the group discussions was done using thematic analysis to generate themes and subthemes. Quantitative data analysis used the individual rank scores of each participant to determine a mean rank score for the top 10 most important outcomes, based on the number of people who ranked each outcome. Across all groups they identified 68 different outcomes. The top ranked outcomes were fatigue/energy, survival, ability to travel, dialysis-free time, impact on family, ability to work, sleep, anxiety/stress, decrease in blood pressure, and lack of appetite/taste. Four themes were identified which explained their choice and prioritization of outcomes: living well, ability to control outcomes, tangible and experiential relevance, and severity and intrusiveness.

3 Methods

3.1 Participant Selection and Recruitment

The NGT seeks to generate a range of ideas or solutions to a specified topic; thus key informants (i.e., people with relevant expertise or experience) should be selected to participate. For example, a NGT study aimed at identifying important issues for palliative care patients and their families conducted groups with patients, bereaved relatives, and healthcare professionals with direct experience and interest in palliative care (Aspinal et al. 2006). As it may not be feasible to recruit all eligible participants, a purposive sampling strategy can be applied, whereby participants with a diversity of demographic characteristics are selected to obtain a range of perspectives.

The number of participants per group usually ranges from six to eight, although groups have been run with as few as two and as many as 14 participants (McMillan et al. 2015). However, too few participants may reduce the potential for idea generation and discussion if the group dynamic is lacking, while very large groups may be difficult to manage, and it may not be possible to elicit detailed input from all participants.

3.2 The Facilitator

The facilitator (also known as a moderator) plays a key role in the NGT as their task is to encourage and help participants contribute their ideas while guiding the group through the NGT process. The facilitator therefore acts as the group leader, and it is important for them to be trained in conducting NGT, as an untrained facilitator may influence the quality of the data collected. The facilitator must also have sufficient knowledge of the topic under investigation and must understand the informational goals which are trying to be achieved. These factors are important because the way in which the questions are framed by the facilitator will impact the quality, breadth, and depth of the responses (Elliott and Shewchuk 2002).

In addition to having adequate content knowledge, the facilitator must be able to build rapport and provide a comfortable setting for participants to support and stimulate thoughtful and relevant discussions. This includes being respectful and nonjudgmental of participants' views, being patient and sensitive to their needs, and ensuring the confidentiality of the discussion. If participants are unknown to each other, the facilitator may also prepare an icebreaker exercise to build rapport so that participants can have open, free-flowing discussions.

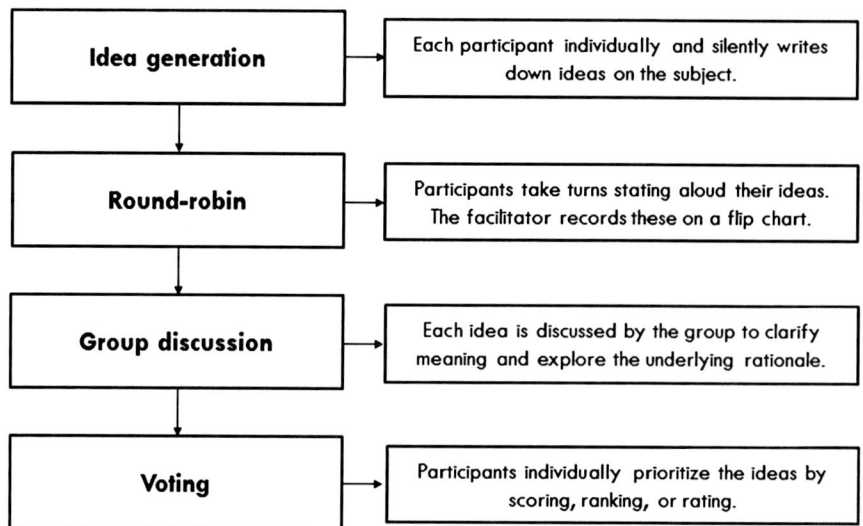

Fig. 1 The four phases of the nominal group technique

3.3 Conducting the Nominal Group Technique

The NGT typically lasts between 60 and 90 minutes and involves four phases detailed in the following section and summarized in Fig. 1:

1. *Silent generation of ideas:* The facilitator presents the question to the group and directs the participants to individually and silently write down their responses and ideas. Participants are given anywhere from 5 (Aspinal et al. 2006) to 20 minutes (Claxton et al. 1980) to generate ideas, and during this time the participants and facilitator remain silent.

2. *Round-robin recording of ideas:* Participants are asked one at a time to contribute a single idea to the group which the facilitator then records on a flip chart (or similar), which is visible to the whole group. The round-robin proceeds until all participants have stated their ideas and there are no new ideas being generated, or when the group determines that a sufficient number of ideas have been produced. Comments and discussion from participants at this stage are not recommended but may be appropriate or necessary depending on the group (McMillan et al. 2014, 2016).

3. *Discussion of ideas:* This phase of the NGT involves a group discussion on each of the ideas listed on the flip chart. It provides an opportunity for participants to clarify and express their understanding and opinions of the ideas, as well as to explore the rationale underlying their perspectives. Throughout the discussion, the group may decide to add, alter, or group similar ideas together. The facilitator

must ensure that all ideas are discussed and that participants have a full understanding of each of the ideas, to enable them to make informed decisions in the voting process. Moreover, the facilitator should emphasize that participants do not have to agree with all the ideas and that their personal preferences will be recorded in the next phase of the NGT. Researchers should note that the use of audio recordings during this phase is usually required for qualitative data analysis but may be refused by some participants.

4. *Voting:* The voting phase consists of participants individually prioritizing the ideas based on what they believe to be most important. There are a number of ways to conduct this process, which include scoring, ranking, or rating the ideas. For example, participants may be asked to select between five and ten of their most important ideas from the list and then to rank them in order of importance. The number of items to be ranked depends on a range of factors, including the study topic and aims, total number of ideas generated, and participant ability and willingness; however the literature commonly reports the use of five items (Delbecq et al. 1975; Dening et al. 2013; McMillan et al. 2014). Variations of prioritizing may include each participant ranking the entire list of ideas, with the most important idea ranked one, the next ranked two, and so on. Other variations may involve scoring, such as each idea being scored a number from one to five (Vander Laenen 2015). Depending on the desired anonymity of the voting process, participants may share their votes with the group and the facilitator may tabulate these on the flip chart to be discussed by the group.

3.4 The Modified NGT

The NGT is a highly adaptable method and is commonly used by researchers to inform, or in combination with, other qualitative (e.g., interviews or focus groups) and quantitative (e.g., discrete choice experiment) methods (see also ▶ Chaps. 23, "Qualitative Interviewing," and ▶ 36, "Eliciting Preferences from Choices: Discrete Choice Experiments"). A combined nominal group/focus group technique has frequently been used (Howell et al. 2012; Urquhart-Secord et al. 2016) and combines the prioritization process of a standard NGT with the in-depth discussion of a focus group. This approach is considered beneficial as it allows for further exploration and discussion of ideas (Varga-Atkins et al. 2017).

Hiligsmann et al. (2013) conducted a NGT to identify and select attributes for inclusion in a discrete choice experiment. Participants were asked to rank the list of attributes in order of importance from 1 (most important) to 12 (least important). New or missing attributes could also be included by participants. In this study, the traditional first phase of silent generation of ideas by participants was not included because the researchers had already identified ideas from the literature.

Typically, the phases most commonly modified in the NGT are phase one (silent generation of ideas) and phase four (voting). As demonstrated in the study by

Hiligsmann et al. (2013) described above, phase one may not involve the generation of ideas by participants. Instead, ideas may be obtained by the researchers prior to the NGT from a literature review or survey (Vella et al. 2000). Examples of variations in phase four have been highlighted in the section above.

Some studies have also conducted phases two (round-robin) and three (discussion of ideas) simultaneously. For example, McMillan et al. (2014) conducted a NGT with Aboriginal and Torres Strait Islander people and found it was culturally appropriate for discussion among participants to occur during the round-robin, as it helped to create a safe and supportive environment for participants. Ultimately, the NGT may need to be modified to suit the participant needs, logistics, and research aims.

3.5 Data Analysis

The NGT elicits both qualitative and quantitative data through the group discussions and voting process, respectively. Qualitative data analysis for NGT typically aims to provide context and rationale for the generated and prioritized ideas. This involves an interpretive and iterative process of data reduction in which meaningful sections of data (i.e., the transcribed group discussions) are coded into concepts. Codes serve as a way to summarize and organize the data and may be grouped together to develop themes (see ▶ Chaps. 47, "Content Analysis: Using Critical Realism to Extend Its Utility," and ▶ 48, "Thematic Analysis"). Researchers may use qualitative data management software (e.g., HyperRESEARCH, NVivo, ATLAS.ti) to code, organize, and retrieve data (see ▶ Chap. 52, "Using Qualitative Data Analysis Software (QDAS) to Assist Data Analyses"). The qualitative data can be compared to the priority scores (quantitative data) and used to contextualize and justify group priorities (McMillan et al. 2014). Quotations from individual participants can be presented to help provide explanations for individual and group priorities (Potter et al. 2004).

Quantitative data analysis depends on the way the voting process was conducted (i.e., ranking, rating, or scoring). In general, the data processing and analysis of NGT quantitative data is relatively simple, as a summary of the group priorities may be generated by the facilitator during the session (Aspinal et al. 2006; Vander Laenen 2015). Following the NGT, researchers will usually input the raw data (i.e., individual participant votes) into an Excel spreadsheet to calculate the priority of each item. For example, this may be done by summing the scores for each item (Aspinal et al. 2006), or calculating a mean rank score based on the number of participants who ranked the item (Urquhart-Secord et al. 2016). It is suggested that items should be assessed in terms of the score as well as the frequency of votes, as this may be more representative of priorities (McMillan et al. 2014). If multiple NGT sessions are conducted for the same research question, group scores can be collated to give an overall priority of items (Dening et al. 2013). If participant details are collected and can be linked to the individuals' votes, analysis can be performed to compare across participant characteristics (e.g., gender, age, ethnicity) (Vander Laenen 2015).

In a proposal to use NGT to develop core outcomes in polycystic kidney disease, Cho et al. (2017) report methods of quantitative data analysis to generate a measure of importance (i.e., importance score). The importance score is a summary measure of the importance of each outcome that incorporates the consistency of being nominated and the rankings given by participants. To calculate this measure, the distribution of the ranking for each outcome was obtained by calculating the probability of each rank for each outcome: $P(O_j$ in rank $i)$, i.e., the probability of the outcome O_j being assigned the rank i. Hence for each outcome, they obtained the probability of being ranked in first place, second place, and so on. By the total law of probabilities, the probabilities are decomposed as:

$$P(O_j \text{ in rank } i) = = P(O_j \text{ in rank } i \mid O_j \text{ is nominated}) \times P(O_j \text{ is nominated}) + P(O_j \text{ in rank } i \mid O_j \text{ not nominated}) \times P(O_j \text{ not nominated})$$

where "nominated" means that the outcome was ranked by the participant. This assumes that the $P(O_j$ in rank $i \mid O_j$ not nominated) is 0. The reasoning behind this is if the participant did not mention the outcome O_j, then the probability of any rank is 0. Therefore, the expression above simplifies to

$$P(O_j \text{ in rank } i) = P(O_j \text{ in rank } i \mid O_j \text{ is nominated}) \times P(O_j \text{ is nominated})$$

From this expression, the probability has two components: (1) the importance given to the outcome by the ranking and (2) the consistency of being nominated by the participants. These probabilities and the computed weighted sum of the inverted ranking $\left(\frac{1}{i}\right)$ are used to obtain the importance score (IS).

$$IS = \sum_{i=1}^{\text{nr of outcomes}} P(O_j \text{ in rank } i) \times \frac{1}{i}$$

The reason for inverting the ranks is to give more weight to top ranks and less to lower ranks. Higher values of the score identify outcomes that are more valued by the participants. The standard errors for the importance score could be obtained through bootstrapping. This measure proposed by Cho et al. (2017) has a similar motivation to the Expected Reciprocal Rank Evaluation Metric proposed in a different context (Chapelle et al. 2009).

4 Advantages and Disadvantages of the NGT

Using the NGT allows researchers to explore a range of ideas generated by participants with expertise or experience in the proposed topic or question in a relatively short timeframe with little or no preparation required by participants. The combination of silent idea generation followed by discussion enables broad generation of ideas while reducing direct criticism and simultaneously allowing for clarification and definition of ideas. The primary strength of the NGT is the structure of the group which ensures equal participation by each member and prevents dominant or outspoken participants from controlling the discussion, resulting in a balance of influence (Carney et al. 1996). This strength, however, relies on the skill of the facilitator to effectively moderate the group discussion as explained above. The NGT may also be more suitable compared to other consensus methods, such as the Delphi technique, for people who may feel more comfortable participating in an in-person meeting than in a relatively complex multi-round survey (McMillan et al. 2016). Unlike the Delphi technique, the NGT enables the opportunity to explore reasons for disagreement in opinions through group discussion, and participants can revise their views by exploring and considering the opinions of others (Allen et al. 2004; see also ▶ Chap. 41, "The Delphi Technique"). Finally, the NGT process can be adapted to accommodate participant needs such as low health literacy, and group moderation by facilitators can ensure that all voices are heard (McMillan et al. 2014).

One disadvantage of NGT is that this method is suited to addressing only one idea or question at a time, thereby limiting the ability to foster new ideas which participants may consider to be important. Additionally, the structured format may be less stimulating than other group techniques and may minimize discussion and generation of ideas. Moreover, while this technique is effective in generating consensus among a group, this does not always equate with the "correct" answer to the research question. This technique, therefore, may be best suited to eliciting expert perspectives on their own experiences and priorities. Finally, this technique requires extensive planning and preparation. Participants must be recruited and attend in-person, which may be burdensome and ultimately lead to nonattendance. For example, Rupert et al. (2017) conducted a study comparing differences in recruitment and logistics between in-person and online focus groups. The authors found that in-person groups were less geographically diverse, and more likely to be white, more educated, and healthier. Despite the fact that sampling methods – specifically purposive sampling – can be applied to capture a broad range of participants, the study by Rupert et al. (2017) reveals that some limitations remain which may have implications for the group dynamics as well as the findings. In view of this, some authors propose that the findings from consensus methods are not a definitive end point but rather an exploratory initial step requiring further investigation (Jones and Hunter 1995). The advantages and disadvantages of the NGT are summarized in Table 1.

4.1 Considerations for Involving Patients in NGT

While it is widely acknowledged that it is important to involve patients and caregivers in research, the impact of this involvement on the participant is sometimes overlooked. Some participants may value the opportunity to share their personal experiences and to meet other people with similar conditions. However, other participants may feel confronted by topics or outcomes relating to their condition

Table 1 Advantages and disadvantages of the nominal group technique

Advantages	Disadvantages
Utilizes informed expert panel	Addresses a single idea/question
Neutralizes dominant participants	Potential for selection bias
Encourages equal representation	Consensus does not equal "correct" answer
Reduces direct criticism and rejection	Benefits rely on facilitator skills
Allows time to generate ideas individually in silence	Resource and time intensive
Allows for clarification and definition of ideas	
Time efficient	

that they may not have been aware of previously. Due to the heterogeneity of symptom presentation in some health conditions, it may be helpful to select participants who are at similar stages of the disease to ensure that they feel the process is relevant to them. As participants may discuss issues relating to medical treatments, it is important to clarify that the NGT is not a forum for medical advice or recommendations and that participants should clarify any concerns that may have been raised in the discussion with their treating physicians. Researchers should consider how the research question and participant selection may impact participants emotionally and psychologically and ensure that participants have access to support following the NGT to address any issues that may have emerged in the discussions.

5 Conclusion and Future Directions

The purpose of this chapter was to outline the utility of the NGT, to review the advantages and disadvantages of using this technique, and to provide an overview of methods for conducting NGT, including participant recruitment and sampling, data collection, and analysis. This section briefly discusses potential future applications of this research method.

Online group techniques are commonly used throughout health and social research (see, e.g., Tates et al. 2009; Thomas et al. 2013; Howells et al. 2017), though very few have involved NGT. In view of its adaptability, the NGT could be applied online to address the challenges inherent to in-person groups. Benefits include the ability to access hard-to-reach populations, such as those who live outside of metropolitan areas, or those with severe disabilities who are unable to travel. Furthermore, online and anonymous NGT may be more appropriate and elicit greater participation than in-person groups for sensitive topics such as sexual health and behavior, illicit drug use, and violence. For example, Woodyatt et al. (2016) conducted a study comparing the quality and depth of data generated from in-person versus online focus groups. The study recruited gay and bisexual men and compared the data from two online and two in-person focus group discussions on the topic of intimate partner violence. The findings revealed that the online groups provided responses which were more succinct and relevant to the research question and were without the interruption experienced during in-person groups. However, the frequency of intragroup conflicts (including disagreement and insults) was greater in the online groups. The facilitator, therefore, has an important role to ensure a safe and supportive environment for participants in this setting. Overall, the content and quality of the data from the two settings was found to be equivalent (see also Kramish Campbell et al. 2001; Synnot et al. 2014).

The NGT is a useful method to generate a large amount of qualitative and quantitative data regarding the preferences and priorities of its participants. The flexibility of the technique enables it to be modified to suit the needs of participants and researchers, and it functions both as a stand-alone technique and within a mixed methods study design.

References

Allen J, Dyas J, Jones M. Building consensus in health care: a guide to using the nominal group technique. Br J Community Nurs. 2004;9(3):110–4.

Aspinal F, Hughes R, Dunckley M, Addington-Hall J. What is important to measure in the last months and weeks of life?: a modified nominal group study. Int J Nurs Stud. 2006;43(4):393–403.

Boers M, Kirwan J, Tugwell P, Beaton D, Bingham III C, Conaghan P, et al. The OMERACT handbook [online]. OMERACT. 2014.

Bond CM, Watson MC. The development of evidence-based guidelines for over-the-counter treatment of vulvovaginal candidiasis. Pharm World Sci. 2003;25(4):177–81.

Carney O, McIntosh J, Worth A. The use of the nominal group technique in research with community nurses. J Adv Nurs. 1996;23(5):1024–9.

Chapelle O, Metlzer D, Zhang Y, Grinspan P. Expected reciprocal rank for graded relevance. Paper presented at the Proceedings of the 18th ACM conference on Information and knowledge management. Hong Kong. 2009.

Cho Y, Sautenet B, Rangan G, Craig JC, Ong ACM, Chapman A, et al. Standardised outcomes in nephrology—polycystic kidney disease (SONG-PKD): study protocol for establishing a core outcome set in polycystic kidney disease. Trials. 2017;18(1):560.

Claxton JD, Ritchie JRB, Zaichkowsky J. The nominal group technique: its potential for consumer research. J Consum Res. 1980;7(3):308–13.

Cowan K, Oliver S. James Lind alliance guidebook (version 5). Southampton: James Lind Alliance; 2013.

Delbecq AL, Van de Ven AH, Gustafson DH. Group techniques for program planning: a guide to nominal group and Delphi processes. Scott Foresman Company. Glenview, Illinois; 1975.

Dening KH, Jones L, Sampson EL. Preferences for end-of-life care: a nominal group study of people with dementia and their family carers. Palliat Med. 2013;27(5):409–17.

Dewar A, White M, Posade ST, Dillon W. Using nominal group technique to assess chronic pain, patients' perceived challenges and needs in a community health region. Health Expect. 2003;6(1):44–52.

Elliott TR, Shewchuk RM. Using the nominal group technique to identify the problems experienced by persons living with severe physical disabilities. J Clin Psychol Med Settings. 2002;9(2):65–76.

Ghisoni M, Wilson CA, Morgan K, Edwards B, Simon N, Langley E, et al. Priority setting in research: user led mental health research. Res Involv Engagem. 2017;3:4.

Hiligsmann M, van Durme C, Geusens P, Dellaert BG, Dirksen CD, van der Weijden T, et al. Nominal group technique to select attributes for discrete choice experiments: an example for drug treatment choice in osteoporosis. Patient Prefer Adherence. 2013;7:133–9.

Howell M, Tong A, Wong G, Craig JC, Howard K. Important outcomes for kidney transplant recipients: a nominal group and qualitative study. Am J Kidney Dis. 2012;60(2):186–96.

Howells LM, Chalmers JR, Cowdell F, Ratib S, Santer M, Thomas KS. 'When it goes back to my normal I suppose': a qualitative study using online focus groups to explore perceptions of 'control' among people with eczema and parents of children with eczema in the UK. BMJ Open. 2017;7(11).

Hutchings H, Rapport FL, Wright S, Doel MA, Wainwright P. Obtaining consensus regarding patient-centred professionalism in community pharmacy: nominal group work activity with professionals, stakeholders and members of the public. Int J Pharm Pract. 2010;18(3):149–58.

Hutchings H, Rapport F, Wright S, Doel M, Jones A. Obtaining consensus about patient-centred professionalism in community nursing: nominal group work activity with professionals and the public. J Adv Nurs. 2012;68(11):2429–42.

Jones J, Hunter D. Consensus methods for medical and health services research. BMJ. 1995;311 (7001):376–80.

Knight SR, Metcalfe L, O'Donoghue K, Ball ST, Beale A, Beale W, et al. Defining priorities for future research: results of the UK kidney transplant priority setting partnership. PLoS One. 2016;11(10):e0162136.

Kramish Campbell M, Meier A, Carr C, Enga Z, James AS, Reedy J, et al. Health behavior changes after colon cancer: a comparison of findings from face-to-face and on-line focus groups. Fam Community Health. 2001;24(3):88–103.

Lavigne M, Birken CS, Maguire JL, Straus S, Laupacis A. Priority setting in paediatric preventive care research. Arch Dis Child. 2017;102(8):748–53.

Manera KE, Tong A, Craig JC, Brown EA, Brunier G, Dong J, et al. Standardized outcomes in nephrology-peritoneal dialysis (SONG-PD): study protocol for establishing a core outcome set in PD. Perit Dial Int. 2017;37(6):639–47.

McMillan SS, Kelly F, Sav A, Kendall E, King MA, Whitty JA, et al. Using the nominal group technique: how to analyse across multiple groups. Health Serv Outcome Res Methodol. 2014;14(3):92–108.

McMillan SS, Kelly F, Sav A, Kendall E, King MA, Whitty JA, et al. Consumers and carers versus pharmacy staff: do their priorities for Australian pharmacy services align? Patient Patient Centered Outcomes Res. 2015;8(5):411–22.

McMillan SS, King M, Tully MP. How to use the nominal group and Delphi techniques. Int J Clin Pharm. 2016;38(3):655–62.

Miller D, Shewchuk R, Elliot TR, Richards S. Nominal group technique: a process for identifying diabetes self-care issues among patients and caregivers. Diabetes Educ. 2000;26(2):305–10. 12, 14.

Potter M, Gordon S, Hamer P. The nominal group technique: a useful consensus methodology in physiotherapy research. N Z J Physiother. 2004;32(2):70–5.

Rees SE, Chadha R, Donovan LE, Guitard AL, Koppula S, Laupacis A, et al. Engaging patients and clinicians in establishing research priorities for gestational diabetes mellitus. Can J Diabetes. 2017;41(2):156–63.

Rolls KD, Elliott D. Using consensus methods to develop clinical practice guidelines for intensive care: the intensive care collaborative project. Aust Crit Care. 2008;21(4):200–15.

Rupert DJ, Poehlman JA, Hayes JJ, Ray SE, Moultrie RR. Virtual versus in-person focus groups: comparison of costs, recruitment, and participant logistics. J Med Internet Res. 2017;19(3):e80.

Sanderson T, Morris M, Calnan M, Richards P, Hewlett S. Patient perspective of measuring treatment efficacy: the rheumatoid arthritis patient priorities for pharmacologic interventions outcomes. Arthritis Care Res. 2010;62(5):647–56.

Sav A, McMillan SS, Kelly F, King MA, Whitty JA, Kendall E, et al. The ideal healthcare: priorities of people with chronic conditions and their carers. BMC Health Serv Res. 2015;15:551.

SONG Initiative. The SONG handbook version 1.0. Sydney; 2017.

Synnot A, Hill S, Summers M, Taylor M. Comparing face-to-face and online qualitative research with people with multiple sclerosis. Qual Health Res. 2014;24(3):431–8.

Tates K, Zwaanswijk M, Otten R, van Dulmen S, Hoogerbrugge PM, Kamps WA, et al. Online focus groups as a tool to collect data in hard-to-include populations: examples from paediatric oncology. BMC Med Res Methodol. 2009;9:15.

Thomas C, Wootten A, Robinson P. The experiences of gay and bisexual men diagnosed with prostate cancer: results from an online focus group. Eur J Cancer Care (Engl). 2013;22(4):522–9.

Trickey H, Harvey I, Wilcock G, Sharp D. Formal consensus and consultation: a qualitative method for development of a guideline for dementia. Quality in Health Care : QHC. 1998;7(4):192–9.

Urquhart-Secord R, Craig JC, Hemmelgarn B, Tam-Tham H, Manns B, Howell M, et al. Patient and caregiver priorities for outcomes in hemodialysis: an international nominal group technique study. Am J Kidney Dis. 2016;68(3):444–54.

Vander Laenen F. Not just another focus group: making the case for the nominal group technique in criminology. Crime Science. 2015;4(1):5.

Varga-Atkins T, McIsaac J, Willis I. Focus group meets nominal group technique: an effective combination for student evaluation? Innov Educ Teach Int. 2017;54(4):289–300.

Vella K, Goldfrad C, Rowan K, Bion J, Black N. Use of consensus development to establish national research priorities in critical care. BMJ. 2000;320(7240):976–80.

Williams A, Sell D, Oulton K, Wilson N, Wray J, Gibson F. Identifying research priorities with nurses at a tertiary children's hospital in the United Kingdom. Child Care Health Dev. 2017;43(2):211–21.

Woodyatt CR, Finneran CA, Stephenson R. In-person versus online focus group discussions: a comparative analysis of data quality. Qual Health Res. 2016;26(6):741–9.

World Health Organization. WHO handbook for guideline development. 2nd ed. Geneva: World Health Organization; 2014.

Yudkin JS, Lipska KJ, Montori VM. The idolatry of the surrogate. BMJ. 2011;343.

Jumping the Methodological Fence: Q Methodology

43

Tinashe Dune, Zelalem Mengesha, Valentina Buscemi, and Janette Perz

Contents

1	Introduction	752
2	What Is Q Methodology?	753
3	Developing a Study Using Q Methodology	754
	3.1 Concourse Development	754
	3.2 Item Sampling: Q Set	755
	3.3 Selection of Participants: The P Set	755
	3.4 Q Sorting: Techniques for Collecting Q Data	756
	3.5 Process of Analysis and Factor Interpretation	757
4	Strengths and Limitations of Q Methodology	762
5	Q Methodology in Action	763
6	Conclusion and Future Directions	764
References		766

Abstract

Mixed methods research is consistently used quantitatively and qualitatively to understand and explore the many facets of a range of phenomena. Generally, mixed methods research involves the use of qualitative and quantitative methods simultaneously or concurrently, yet for the most part independently. What if these methods could be truly mixed? This chapter introduces readers to a methodology that aims to address this question – Q methodology. Q methodology allows for the sampling of subjective viewpoints and can assist in identifying patterns,

T. Dune (✉) · Z. Mengesha · V. Buscemi
Western Sydney University, Sydney, NSW, Australia
e-mail: t.dune@westernsydney.edu.au; z.mengesha@westernsydney.edu.au; v.buscemi@westernsydney.edu.au

J. Perz
Translational Health Research Institute, School of Medicine, Western Sydney University, Sydney, NSW, Australia
e-mail: j.perz@westernsydney.edu.au

© Springer Nature Singapore Pte Ltd. 2019
P. Liamputtong (ed.), *Handbook of Research Methods in Health Social Sciences*,
https://doi.org/10.1007/978-981-10-5251-4_101

including areas of difference or overlap, across various perspectives on a given phenomenon. Q methodology can be described as "'qualiquantilogical' combining elements from qualitative and quantitative research traditions" (Perz et al. BMC Cancer 13: 270, 2013, p. 13). This chapter will outline the five steps involved in conducting a Q methodology study: (1) developing the concourse, (2) developing the Q set, (3) selection of the P set, (4) Q sorting, and (5) Q analysis and interpretation. In order to contextualize and demonstrate how Q methodology can be used, we will present reflections on the use of this methodology with respect to constructions of sexual and reproductive health, chronic low back pain, and culturally and linguistically diverse people. These examples demonstrate how Q methodology can provide a unique and truly mixed way of studying human subjectivity.

Keywords
Q methodology · Social research · Health research · Mixed methods · Subjectivity

1   Introduction

Mixed methods research involves the collecting, analysis, and interpretation of both quantitative and qualitative data in a singular study or series of studies (Creswell and Plano Clark 2018). The use of both qualitative and quantitative methods to answer a research question is central to the concept of mixed methods research which maximizes the strengths and minimizes the weaknesses of employing a single approach (2011). In order for a study to be considered mixed methods, there should be a clear integration of approaches at the design, analysis, or interpretation stages of the project (Fetters et al. 2013).

A fundamental assumption behind mixed methods is that it can address research questions more comprehensively than employing qualitative or quantitative approaches alone (Creswell and Plano Clark 2018). Broad and complex research questions with multiple aspects that are difficult to address using either a qualitative or quantitative approach are appropriate for a mixed methods investigation (Tashakkori and Teddlie 2010). Usually, quantitative research is used to assess the magnitude of concepts or constructs or test a hypothesis, while qualitative studies are associated with exploring the meanings of complex social or cultural concepts including how and why people think in a certain way (see ▶ Chaps. 2, "Qualitative Inquiry," and ▶ 3, "Quantitative Research"). Combining the two approaches in a study has the potential to allow for a broader examination of the study phenomenon (see ▶ Chap. 4, "The Nature of Mixed Methods Research").

Mixed methods researchers suggest several reasons for combining qualitative and quantitative data which include:

- Complementary: Findings from one method can be used to explain results from another. For example, Tariq et al. (2012), in their mixed methods study of the

impact of African ethnicity and migration on pregnancy, identified potential disparities in health-care access and outcomes by analyzing a national surveillance data. Then, semi-structured interviews with pregnant African women allowed them to understand factors driving the disparity.
- Development: Results from one study can be used to inform the development of the other. For instance, Stoller et al. (2009) conducted a qualitative study first to identify factors contributing to the reduction of alcohol use in hepatitis C-positive patients. A quantitative survey was then conducted to quantify and test the identified factors.
- Triangulation: Data obtained from both methods can be used to corroborate findings. Mengesha et al. (2017) used a survey to measure health-care professionals' knowledge, skills, and attitude to provide sexual health care to refugee and migrant women. Semi-structured interviews with health-care providers were conducted to confirm survey findings.
- Expansion: Examine several research questions in a study using different methodologies. For instance, Montgomery et al. (2010) conducted a randomized control trial to assess the efficacy of a vaginal microbicidal gel on vaginal HIV transmission and in-depth interviews with trial participants to assess the acceptability of the gel.

Although mixed methods research has several benefits, there are also some challenges that should be noted (Tariq and Woodman 2013). For instance, the collection, analysis, and interpretation of multiple forms of data require extensive resources and time and involve skills and experiences in both qualitative and quantitative approaches (Tariq and Woodman 2013; Creswell and Plano Clark 2018). This suggests that a well-working team from different disciplines is required for a rigorous conduct of mixed methods research. Interpreting the findings of a mixed methods study can also be challenging as investigators may place unequal weight in relation to the depth of the data and accuracy, validity, and theoretical emphasis of each data set (Creswell et al. 2011; Tariq and Woodman 2013). (See also ▶ Chaps. 4, "The Nature of Mixed Methods Research," ▶ 39, "Integrated Methods in Research," and ▶ 40, "The Use of Mixed Methods in Research.")

2 What Is Q Methodology?

Q methodology is derived from a range of earlier and similar ranking techniques including Kelly's repertory grid methods (Robbins and Krueger 2000) to aid researchers in "revealing the subjectivity involved in any situation" (Brown 1993, p. 561). The term Q methodology was coined by Stephenson in the 1950s. Q methodology allows for the sampling of subjective viewpoints and can assist in identifying patterns, including areas of difference or overlap, across various perspectives on a given phenomenon (Watts and Stenner 2012). Q methodology can be described as "'qualiquantilogical' combining elements from qualitative and quantitative research traditions" (Perz et al. 2013, p. 13). The data is composed of participant's constructions of a specific topic, obtained by ranking a set of predefined items according to their experiences, understandings, and perspectives. The factors resulting from factor analysis represent constructions of subjectivity that are present within the rankings (Brown 1993). In Q methodology, a factor analysis is performed to identify associations between patterns expressed by participants, a procedural inversion to conventional factor analysis that is used to identify associations between variables (Lazard et al. 2011). As such, the focus in Q methodology is not the "constructors" (the participants) but the "'constructions" themselves (Rogers 1995).

Q methodology has been described as a more robust technique than alternative methods for the measurement of subjective opinions and has been recommended in the study of attitudes within the health field (Cross 2005). In the area of health and well-being, previous authors have also used this approach to examine constructions of sex and intimacy after cancer (Perz et al. 2013), explore constructions of bulimia (Churruca et al. 2014), investigate understandings of irritable bowel syndrome (Stenner et al. 2000), examine beliefs about the sexuality of the intellectually disabled (Brown and Pirtle 2008), and explore experiences of postnatal perineal morbidity (Herron-Marx et al. 2007).

3 Developing a Study Using Q Methodology

Watts and Stenner (2012) have captured five primary steps in conducting a Q methodology study. These include (1) developing the concourse, (2) developing the Q set, (3) selection of the P set, (4) Q sorting, and (5) Q analysis and interpretation.

3.1 Concourse Development

The first step in conducting a Q methodological study is generating a "collection of all the possible items the respondents can make about the subject at hand" called a concourse (Van Exel and de Graaf 2005). A concourse represents all the things people think or say about the subject being studied. Researchers commonly use statements of opinion (verbal concourse), although other items such as paintings, photographs, musical selections, and pieces of art can be used in Q methodology (Brown 1993; Paige 2014). Developing an effective concourse can begin with a comprehensive review of literature or media from different sources. A literature review can be conducted to explore and analyze existing evidence concerning the phenomena in question. Online and other media can also be explored to identify relevant constructions of phenomena to help draw out statements to add to the initial concourse (Brown and Pirtle 2008).

Concourse can also be developed by eliciting statements from key informants who can provide their perspectives on the issue in question to make sure that the statements within the concourse represent the views of people who may be part of the sample. This approach also allows the researcher to develop a concourse that covers perspectives from a range of stakeholders (Perz et al. 2013). Such stakeholder can include experts in academic or practical areas surrounding the phenomena in question. Qualitative methods are helpful in this regard as they assist in drawing out the why, how, and what of a particular topic of interest (Liamputtong 2013; see also ▶ Chap. 2, "Qualitative Inquiry"). In the case of interviews, for example, these can be audio-recorded, transcribed verbatim, and then the transcripts can be examined for statements to be included in the concourse (see ▶ Chap. 23, "Qualitative Interviewing"). Key informants can also assist in reviewing the statements collated in the concourse and give feedback on how the statements could be amended in order to better reflect a representative sample of perspectives.

3.2 Item Sampling: Q Set

The Q set is a representative sample of the concourse provided to participants for rank ordering (Brown 1993). The Q set results from deconstructing relevant topics into a series of sub-themes and in line with the research question (Watts and Stenner 2012). This helps to ensure that all aspects of the subject of interest are covered and items are not inclined toward a particular viewpoint (Coogan and Herrington 2011). The refinement process of statements within each of the themes continues until a final set of items is ready for administration. Through the process, statements can be revised, duplicate items discarded, and additional items added. The resulting set can then be piloted with key and expert informants. Based on their feedback, the pool is then reduced to a final set of statements that broadly represents a range of viewpoints. Although the number of statements can vary depending on the study question, sample, and other logistical variables, Rogers (1995) has suggested that a Q set with more than 60 statements can result in participant fatigue and therefore disengagement with statements after this point.

3.3 Selection of Participants: The P Set

Unlike survey studies (see ▶ Chaps. 32, "Traditional Survey and Questionnaire Platforms," and ▶ 76, "Web-Based Survey Methodology"), the emphasis of a Q methodological study is not on the number of people who think in certain ways; rather it is focused on why and how people may present a particular way of thinking about the study topic. Therefore, studies involving Q methodology use relatively small sample sizes (Brown 1993; McKeown and Thomas 1988). According to Rogers (1995), a group of 40–60 participants is enough to establish the existence of particular viewpoints. Watts and Stenner (2012) suggest strategic selection of the participants in order to achieve a broad sample that can give the best insights about the topic.

3.4 Q Sorting: Techniques for Collecting Q Data

Once the Q set is developed and the response format selected (e.g., type of grid distribution – see Fig. 1), participants can start rating and ranking the statements – a process called Q sorting. This can be done either manually or electronically.

3.4.1 Manual Techniques

To manually collect Q sorts, the researcher can meet with the participant or mail the materials to participants with detailed instructions. Participants receiving the materials by mail can send their sorts back to the researchers or take a digital picture of their sort and email it to the researcher. A study comparing these two manual data collection tools have demonstrated no difference in the results and interpretations (Tubergen and Olins 1978). The Q grid and statement can also be sent through email to participants with detailed instructions, and participants can complete the sort and email their results to the researcher (Perz et al. 2013). The following section explains how physical cards can be used to complete a Q sort (Brown 1993).

Sorting begins by organizing the statements into three clusters: agree, disagree, and neutral. Participants then rank-order the statements by beginning with those they most agree with (+4) to those they most disagree with (−4) based on their subjective viewpoints. They then review each cluster of statements and sort them into a quasi-normal distribution sorting grid (see Fig. 1). This stage is more challenging as participants have to compare statements to each other and make a "forced" choice based on what they felt most strongly about (Paige 2015). The sorting activity is complete once the participant feels that the statements positioned into the grid accurately represent their view on the topic of interest. Once Q sort data is collected via manual methods, they need to be manually entered and then analyzed.

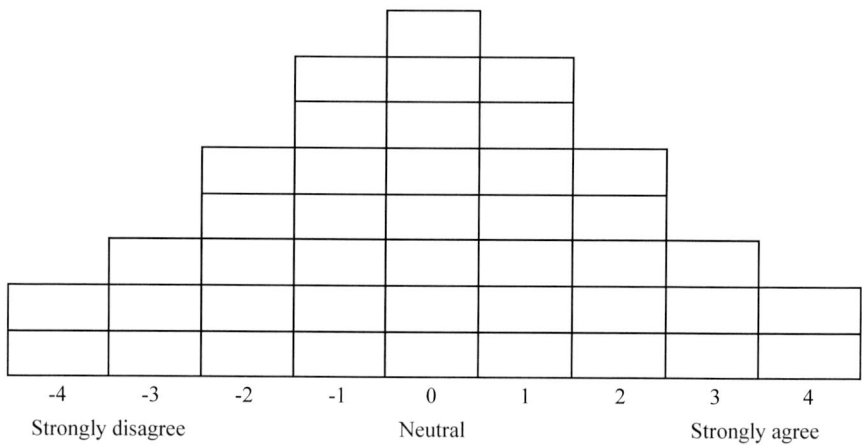

Fig. 1 Q sorting grid

3.4.2 Electronic Techniques

Online Q sorting generally requires participants to complete the sorting in three phases – similar to that of the manual process described above. Online or software-based Q data collection methods can reduce the time and expense sometimes required for manual data collection. Further, online programs like Q-Assessor reduced the difficulty and complexity of manual analysis for researchers. From the mid-1990s, some primary data collection solutions have been developed; in particular WebQ helps to collect the Q sorts online, and Q-Assessor allows for the collection and analysis of Q sort data (Reber and Kaufman 2011).

The Example of Q-Assessor

Q-Assessor is a paid service for a simple web-based sorting technique, suitable for users who can access the Internet (Omeri et al. 2006). It helps the investigator to configure the different parts of the Q study, collect and analyze data, and engage with participants (Reber and Kaufman 2011). It has been validated against the more traditional manual Q sort collections (Reber et al. 2000), and its use is increasingly growing for the following reasons: (i) it allows easily to first cluster the statements into the "agree," "neutral," and "disagree" bins and then to proceed with the final ranking on the grid; (ii) allows participants to change their mind at any time; (iii) helps to sort the statements starting from the extremes, firstly ranking the "mostly agree," then "mostly disagree," and then "neutral" statements; (iv) collects other information, such as post-sort comments; (v) automatically sends results to the investigator who can check in real time the ongoing development of the study (i.e., if the subject has started and/or completed the Q sort); (vi) manages participant recruitment, including automatic reminders and post-completion communication; and importantly (vii) performs data analysis.

3.5 Process of Analysis and Factor Interpretation

The main objective of Q analysis is to identify viewpoints or perspectives shared by participants where these shared views are represented by factors. This is achieved by performing a by-person factor analysis – a commonly applied statistical method in Q methodology. Factor analysis begins with the calculation of correlation matrix that reveals the degree of agreement or disagreement between the sorts or the similarity or dissimilarity in the views of the Q sorters. From this matrix, the centroid method can be used to extract initial sets of factors. This method is preferred by most Q methodologists (Brown 1993; Watts and Stenner 2012) as it allows the researcher to explore the data through rotation until the best factor solution is achieved (Watts and Stenner 2012). The extracted factors should then be subjected to varimax rotation: a statistically and theoretically sound method for factor rotation. In Q analysis, varimax positions factors so that the overall solution maximizes the amount of study variance explained (Watts and Stenner 2012). It also gives a very workable factor solution considering the majority viewpoints of the study participants (Watts and Stenner 2012). Following rotation, factors for interpretation will be selected if they fulfilled the stringent criteria of at least two sorts significantly loading upon a factor and eigenvalues greater than one (Watts and Stenner 2005). At this stage, factors should also be examined if they reveal a distinct and meaningful viewpoint. The nature of the research topic may also influence our decision regarding the number of factors that should be retained for interpretation. For instance, Mengesha et al. (2017) chose a broad factor solution that gives a larger number of factors to capture the complexity of engaging refugee and migrant women in sexual and reproductive health in salient and clear ways.

Before starting factor interpretation, a factor array (see Table 1 Q-set statements and factor array) will be created for the factors to facilitate factor interpretation. A factor array is where each of the factors is represented by its own "best-estimate Q sort or 'factor array'...These factor arrays or best-estimate Q sorts can then be subjected to interpretation" (Watts and Stenner 2005, p. 82). The interpretation can be done by reading the statements thematically and analyzing their position relative to all other statements within individual factors. In the process of interpretation, attention is given to the whole configuration of items in a factor with the aim of achieving holistic factor interpretations. Statements that have statistically different factor scores across factor arrays (distinguishing) are then examined to clarify the difference between factors. For those participants whose responses load significantly onto a factor, responses to auxiliary questions and demographic data can be examined to assist in identifying the meaning of each factor.

3.5.1 Tools to Assist with Analysis

Traditional statistical software (i.e., SPSS) can be used for analysis but are not preferred due to syntax adjustments required during factor analysis. There are other Q methodological packages available that can be used for data analysis and generate the by-person correlation matrix and factor extraction (Watts and Stenner 2005). These also offer the centroid method as default choice that provides an unlimited number of rotated solutions (Watts and Stenner 2005). PCQ for Windows and PQMethod are the most recommended and dedicated packages for Q analysis, and the latest versions can be downloaded from (http://schmolck.userweb.mwn.de/qmethod/).

3.5.2 Data Analysis with Q-Assessor

Once all data has been collected, the Q-Assessor performs the analysis based on data extraction techniques described by Brown (1980) but also allows investigators to manipulate their data on their own computers. Q-Assessor analysis conducts correlation matrix, provides unrotated and rotated factors flagging the significant responses, and provides the possibility to download the reports: (i) rank statement totals for each factor, (ii) normalized factor scores, (iii) comparisons between factor scores, (iv) Q sort values for each statement, (v) distinguishing and consensus statements for factors, and (vi) factor characteristics. Notably, Q-Assessor allows also to detect the participant linked to his/her Q sort and to the open-ended comments. Information regarding how to use Q-Assessor is available at q-assessor.com (http://q-assessor.com).

Table 1 Q-set statements and factor array for a study on the sexual and reproductive health and help-seeking among 1.5 generation migrants

	Q-set statements							
	Factor							
		A	B	C	D	E	F	G
1	If my family or community found out that I had a sexually transmitted infection, they would not be very supportive	2	0	−1	0	1	1	−2
2	I would never let my family or community know that I had sex outside of marriage	1	3	−2	−2	1	4	1
3	If I had a sexual and/or reproductive health issue, I would have to find a way to go to a clinic without my family or community knowing	−3	2	0	−1	0	2	−2
4	If my family or community found out that I was involved in an unplanned pregnancy, they would not be very supportive	−1	−2	−2	−1	2	−1	2
5	If I had a sexual and/or reproductive health issue, other people's perceptions about it would impact how and when I got professional help	2	1	2	−3	1	0	2
6	Going through adolescence and puberty was sometimes difficult because the things I was taught at school or in the media about sexual and reproductive health were different to what my family or community believe	1	0	−2	0	−3	3	0
7	People who move from one country as children and grow up in Australia are often confused about sexual and reproductive health	0	0	4	0	−1	0	−1
8	I want to pass on to future generations the values about sexual and reproductive health held by my culture of origin	−3	−3	3	0	2	−1	2
9	The way that sexual and reproductive health is dealt with in Australia is very different than the way it is understood in the country where I was born	1	−3	1	0	1	3	1
10	In my origin culture, openly discussing sexual and reproductive health is encouraged	−2	−4	2	2	−3	−3	−1
11	In my origin culture, women have control over their sexual and reproductive health	−4	0	1	1	1	−3	3
12	In Australia people are encouraged to discuss sexual and reproductive health	−2	−1	3	3	4	3	4
13	People who were born in Australia have an easier time with sexual and reproductive health than migrants like me	1	−2	1	2	4	−1	−1

(*continued*)

Table 1 (continued)

	Q-set statements							
	Factor							
		A	B	C	D	E	F	G
14	Australians can have intimate relationships with whomever they like, and no one would mind	0	2	2	3	−1	0	3
15	Culture plays a large role in how people experience sexual and reproductive health	1	3	−1	4	2	4	0
16	Sexual and reproductive health in my culture of origin is a taboo subject	4	1	1	−3	0	1	−4
17	Australia is very conservative about sexual and reproductive health	−1	1	−3	−2	−2	−1	3
18	In my origin culture, sexual and reproductive health is perceived of more in terms of women's or men's health	0	1	−1	0	1	1	−1
19	I avoid casual sexual encounters because my family or community would think I was disrespecting my origin culture	0	2	−2	−4	3	2	0
20	Australian men and women think of sexual and reproductive health in the same ways	3	−2	−1	1	−1	−1	−3
21	Australian values lead my understanding of sexual and reproductive health	−3	−1	−4	2	−1	2	−2
22	Sexual and reproductive health refers mostly to the prevention of and protection from disease	0	4	−3	1	0	−2	−2
23	Sexual and reproductive health refers mostly to means prevention of unplanned pregnancy	−2	−4	0	−1	−2	−3	0
24	Sexual and reproductive health refers mostly to contraception	2	0	−4	−1	−2	−4	−3
25	Sexual and reproductive health is usually something only promiscuous people have to deal with	−4	1	0	−3	−4	−4	1
26	The way that sexual and reproductive health is understood in Australia is very different than the way it is understood in the country where I was born	1	−1	3	−1	3	2	0
27	There are no words in my culture of origin for sexual and reproductive health	−1	−1	2	−1	−1	0	−4
28	Culture plays a large role in how people understand sexual and reproductive health	3	−1	4	4	2	2	2
29	Health-care workers are well equipped to deal with the sexual and reproductive health needs of people from my background	1	0	1	1	0	−2	−1
30	Australians can more easily get help for sexual and reproductive health issues than people from my culture of origin	2	−1	−2	1	3	0	1

(*continued*)

Table 1 (continued)

	Q-set statements							
	Factor							
		A	B	C	D	E	F	G
31	Migrants need more assistance from health services with sexual and reproductive health than people born in Australia	−2	0	1	3	−1	1	0
32	Health-care workers have very little knowledge of the beliefs related to sexual and reproductive health within my culture	0	−3	−3	1	0	0	−2
33	Health services provide the anonymity needed to cater to migrants sexual and reproductive health needs	−1	2	0	1	2	1	0
34	Health services provide clients with a choice between a male or female health-care provider	−1	2	0	0	0	0	−3
35	Health services cannot do much else to better cater to the sexual and reproductive health needs of migrants	−2	1	0	0	−1	−2	−1
36	Migrants who identify most as being Australian have more sexual and reproductive health issues than other Australians (excluding Aboriginal and Torres Strait Islander peoples)	−1	3	−1	−2	0	−1	0
37	Migrants who identify most with their culture of origin have more sexual and reproductive health issues than other Australians (excluding Aboriginal and Torres Strait Islander peoples)	0	0	1	−2	−4	−1	1
38	Migrants from certain cultures are carriers of sexually transmitted infections	2	−2	2	−4	−2	−2	−1
39	Migrants who do not take on Australian ways of understanding sexual and reproductive health have failed to assimilate	−1	−1	0	−2	−2	−2	1
40	Australians should take on more values about sexual and reproductive health from migrant cultures	0	4	0	2	−3	1	1
41	Migrant sexual and reproductive health needs are quite different from those of nonmigrants	3	−2	−1	−1	0	0	2
42	Australians may think that some migrant groups have outdated ideas about sexual and reproductive health	4	1	−1	2	1	1	4

4   Strengths and Limitations of Q Methodology

Q methodology is a unique research method that identifies commonalities and divergences within and between groups. It combines strengths from quantitative and qualitative research methods (Ellingsen et al. 2010), using a systematic approach to analyze the Q sorts but providing an interpretation that is typical of qualitative research methods. Q methodology reveals subjectivity that is particularly important in complex contexts, such as social and health fields.

In Q methodology, user participation is the main focus and strength for different reasons. Firstly, the concourse is developed using information deriving from interviews and group discussions provided by participants on the topic of interest. As a consequence, researcher's preconception is minimized (Corr 2006). Secondly, shared viewpoints are gained through factor analysis of the Q sorts based on participants' perspectives, and the forced distribution, used to collect Q sorts, helps to reduce the number of uncertain statements and decide the most important ones (Cross 2005).

Q methodology is an easy-to-use method where participants do not necessarily need to verbalize about sensitive topics. Therefore, it can be considered as a non-threatening means to gather personal viewpoints. Q methodology has also the advantage to enhance the level of participants' inclusion. In fact, as statements can be pictures, images, or even single words, individuals with a wide range of ages and social and physical conditions (including children or people with learning disabilities) can actively take part and provide their stories (Ellingsen et al. 2010).

Q methodology also has some limitations. One of the main ones is that developing the concourse can be time-consuming, for example, when interviews and group discussions needed to be conducted to gather more information on the topic of interest. Secondly, the sorting activity and the extensive explanations provided to instruct participants on how to use the method are also time-consuming, and the lack of comprehension on how to participate can affect the validity of results.

Further, Q methodology has received some critics regarding how results are obtained: reliability has been questioned as repeating the method on the same participant does not necessarily yield the same viewpoints (Cross 2005). However, few studies have demonstrated that administering the same Q sample to the same participants at two different time points resulted in correlation coefficients between 0.80 and 0.95 (Fairweather 1981).

Another limitation is that the expression of participants' subjectivity is limited by sorting predetermined statements, chosen by the researcher. However, using interviews or group discussion to gather information for the statements' development helps to alleviate this limitation. Participants' free expression is also limited by the "forced choice" of the sorting activity. On the other hand, some authors argue that the unforced choice risks to produce less discrimination than forced methods, which would make assessment and interpretation more difficult (Dziopa and Ahern 2011).

Finally, regardless the sample size, in Q methodology, results cannot be generalized in that they are limited to the participants' viewpoints of that particular study, nor researchers can claim that all potential viewpoints on the topic of interest are elicited (Dziopa and Ahern 2011).

5   Q Methodology in Action

We have conducted studies using Q methodology as a primary investigator, and these example studies are used to illustrate how Q methodological study can be done including the pros and cons for each project. Dune et al. (2017) investigated the role of culture in constructions of SRH from the perspective of 1.5 generation migrants (those who migrate as children or adolescents). Forty-two adults from various ethno-cultural backgrounds rank-ordered 42 statements about constructions of SRH and SRH help-seeking behavior. A by-person factor analysis was then conducted, with factors extracted using the centroid technique and a varimax rotation. A seven-factor solution provided the best conceptual fit for constructions of SRH. Factor A compared SRH values within Australia and migrants' culture of origin. Factor B highlighted the influence of culture on SRH values. Factor C explored migrant understandings of SRH in the context of culture. Factor D explained the role of culture in migrants' intimate relationships, beliefs about migrant SRH, and engagement of health-care services. Factor E described the impact of culture on SRH-related behavior. Factor F presented the messages migrant youth are given about SRH. Lastly, Factor G compared constructions of SRH across cultures.

Mengesha et al. (2017) also conducted a Q methodological study to identify the challenges of providing sexual and reproductive health care to refugee and migrant women. Forty-seven health professionals rank-ordered 42 statements and commented on their rankings in subsequent open-ended questions. Seven factors each with a distinct and meaningful viewpoint were identified. These factors are "communication difficulties – hurdles to counseling," "the lack of access to culturally appropriate care," "navigating SRH care," "cultural constraints on effective communication," "effects of the lack of cultural competency," "impacts of low income and language barrier," and "SRH services are accessible, but not culturally relevant."

Buscemi et al. (2018) used Q methodology to identify shared perspectives on day-to-day stressors in people experiencing chronic low back pain. Fifty statements representing the broad experience of psychosocial stress were ranked and sorted using a predefined quasi-normally distributed grid by 61 participants. A by-person correlation and factor analysis revealed seven factors corresponding to seven different group viewpoints (Groups 1–7). Viewpoints were enriched using standardized questionnaires to identify the presence of distress (DASS-21), stress (PSS), pain self-efficacy (PSEQ), and pain catastrophizing (PCS). Only in one group (Group 1) did participants demonstrate a sense of mastery over their life and chronic pain condition ("I've learned to live with the pain, so it doesn't stress me out"). In Groups 2 and 3, the main focus of stress was the perceived limitation and life disruption caused by living with CLBP ("I feel unwell all the time"). In Group 4, stress manifested as an overwhelming force, associated with a feeling of being trapped by having too many responsibilities in life ("The more stress, the more often I have back pain"). In Groups 5 and 6, participants felt stressed by worries and demands from the outside world including family, work, and recent changes in living conditions ("Children, health problems, full time work"). In Group 7, stress resulted from personal fears and worries coupled with low self-esteem that impacted on participants' perceptions of themselves and the world around them ("Not feeling appreciated or valued").

Finally, Perz et al. (2013) explored the complex perspectives that people with personal and professional experience with cancer hold about sexuality in the context of cancer. Participants were asked to rank-order 56 statements about sexuality and intimacy after cancer and asked to comment on their rankings in a subsequent semi-structured interview. A by-person factor analysis was performed with factors extracted according to the centroid method with a varimax rotation. A three-factor solution provided the best conceptual fit for the perspectives regarding intimacy and sexuality post-cancer: Factor 1, entitled "communication – dispelling myths about sex and intimacy;" Factor 2, "valuing sexuality across the cancer journey;" and Factor 3, "intimacy beyond sex."

Table 2 presents the pros and cons of using Q methodology for the two example studies (Mengesha et al. 2017; Dune et al. 2017).

6   Conclusion and Future Directions

This chapter provides readers with an introduction to Q methodology, a mixed methods approach to exploring phenomena using quantitative and qualitative principles within an interdependent data collection, analysis, and interpretive strategy. This chapter outlined the five steps involved in conducting a Q methodology study (i.e., (1) developing the concourse, (2) developing the Q set, (3) selection of the P set, (4) Q sorting, and (5) Q analysis and interpretation). Examples from our own work with Q methodology across a range of fields, including constructions of sexual and reproductive health, chronic low back pain, and culturally and linguistically diverse people, demonstrate the breadth and potential applicability of Q methodology to many other fields of study. Q methodology presents as a means for engaging participants in the development of data collection tools and a way to better understand those perspectives from a range of subjective viewpoints.

Q methodology presents many strengths, as it provides a multidimensional perspective on human experiences and engages users both directly and indirectly in the process of inquiry. For the researchers, the steps are easy to follow, although sometimes time intensive. Reducing the logistical requirements for time is supported by software which assists with data collection and analysis so

Table 2 Pros and cons of Q methodology for the example studies

Example studies	Pros	Cons
Challenges in the provision of SRH care study (Mengesha et al. 2017)	Q methodology has demonstrated suitable to generate diverse accounts of a cross-cultural nature that are difficult to capture using other pattern analysis approaches	Sorting using Q-Assessor has been confusing for some participants
	The use of Q-Assessor has simplified data collection, analysis, and interpretation stages	Time-consuming (inception to publication ~ 2 years)
	Allows researchers to compare the subjective viewpoints of the HCPs and the extracted factors	Generalizations could not be made due to the gender and age composition of the participants
	The purposeful nature of the sampling strategy helped to achieve a comprehensive sample that ensured the perspectives identified reflected the experiences of providing SRH care to refugee and migrant women in a broad sense	
Constructions of SRH from the perspective of 1.5 generation migrant study (Dune et al. 2017)	The ability to explore perspectives in relation to individuals (vs the other way around)	Generalizations cannot be made as the sample was limited to 42 migrants who lived in metropolitan Sydney and near university campuses
	Apply principles of quantitative variance to concepts derived from subjective viewpoints	Requires some upskilling and some mentorship to adequately complete a Q study
	Clarity around variables that are impactful yet not accounted for in the Q set provides clear avenues for further research	
	A research strategy that is multidimensional and easy to do provided adequate mentorship is provided	
	Larger sample compared to other traditional qualitative research methods	Forced choice might have slightly influenced the findings
	Results from Q methodology could be triangulated with other research methods, such as open-ended questions and validated questionnaires	Some participants may have experienced difficulties in understanding the process of completing the study

(continued)

Table 2 (continued)

Example studies	Pros	Cons
	Participants could participate from home using Q-Assessor	Findings cannot be generalizable and require further research to understand whether reducing stress can help to better cope with chronic low back pain (LBP)
	Participation was not time-consuming (around 40 min)	
	Findings provided new insights for i) clinical practice on the importance to screen and then intervene to reduce stress in people with chronic LBP; ii) further research to find the most effective strategies and interventions for stress management	

that interpretation can be expedited. As with most software, the price for such a service can be prohibitive to some researchers especially students, early career researchers, independent researchers, or smaller organizations with limited research capacity. While there is a range of software available for free or at a minimal cost, time is required to ensure that the researcher is well versed in using the software and procedures for accurate input and analysis of the data. Given that Q methodology is research strategy being rediscovered by researchers across a range of fields outside of psychology and marketing, it is likely that software packages will continue to develop and become more affordable. In doing so, more researchers can gain information, access, and engagement with this innovative research strategy.

References

Brown SR. Political subjectivity: applications of Q methodology in political science. New Haven: Yale University Press; 1980.

Brown SR. A primer on Q-methodology. Operant Subjectivity. 1993;16:91–138.

Brown RD, Pirtle T. Beliefs of professional and family caregivers about the sexuality of individuals with intellectual disabilities: examining beliefs using a Q-methodology approach. Sex Educ. 2008;8(1):59–75. https://doi.org/10.1080/14681810701811829.

Buscemi V, Dune T, Liston MB, Schabrun SM. How do people with chronic low back pain perceive everyday stress? A Q-study. Unpublished paper in preparation for submission. 2018.

Churruca K, Perz J, Ussher JM. Uncontrollable behavior or mental illness? Exploring constructions of bulimia using Q methodology. J Eat Disord. 2014;2(1):22. https://doi.org/10.1186/s40337-014-0022-2.

Coogan J, Herrington N. Q methodology: an overview. Res Second Teach Educ. 2011;1(2):24–8.

Corr S. Exploring perceptions about services using Q methodology. In: Research in occupational therapy: methods of injury for enhancing practice. Philidelphia: FA Davis Company; 2006.

Creswell JW, Plano Clark VL. Designing and conducting mixed methods research. 3rd ed. Thousand Oaks: Sage; 2018.

Creswell JW, Klassen AC, Plano Clark VL, Smith KC. Best practices for mixed methods research in the health sciences. Bethesda: National Institutes of Health, Office of Behavioral and Social Sciences Research; 2011.

Cross RM. Exploring attitudes: the case for Q methodology. Health Educ Res. 2005;20(2):206–13.

Dune T, Perz J, Mengesha Z, Ayika D. Culture clash? Investigating constructions of sexual and reproductive health from the perspective of 1.5 generation migrants in Australia using Q methodology. Reprod Health. 2017;14:50. https://doi.org/10.1186/s12978-017-0310-9.

Dziopa F, Ahern K. A systematic literature review of the applications of Q-technique and its methodology. Methodology. 2011;7(2):39–55.

Ellingsen IT, Størksen I, Stephens P. Q methodology in social work research. Int J Soc Res Methodol. 2010;13(5):395–409.

Fairweather J. Reliability and validity of Q-method results: some empirical evidence. Operant Subjectivity. 1981;5(1):2–16.

Fetters MD, Curry LA, Creswell JW. Achieving integration in mixed methods designs–principles and practices. Health Serv Res. 2013;48(6 pt 2):2134–56.

Herron-Marx S, Williams A, Hicks C. A Q methodology study of women's experience of enduring postnatal perineal and pelvic floor morbidity. Midwifery. 2007;23(3):322–34.

Lazard L, Capdevila R, Roberts A. Methodological pluralism in theory and in practice: the case for Q in the community. Qual Res Psychol. 2011;8(2):140–50.

Liamputtong P. Qualitative research methods. 4th ed. Melbourne: Oxford University Press; 2013.

McKeown B, Thomas B. Q-methodology. Newbury Park: Sage; 1988.

Mengesha ZB, Perz J, Dune T, Ussher J. Challenges in the provision of sexual and reproductive health care to refugee and migrant women: a Q methodological study of health professional perspectives. J Immigr Minor Health. 2017. https://doi.org/10.1007/s10903-017-0611-7.

Montgomery CM, Gafos M, Lees S, Morar NS, Mweemba O, Ssali A, Pool R. Re-framing microbicide acceptability: findings from the MDP301 trial. Cult Health Sex. 2010;12(6): 649–62. https://doi.org/10.1080/13691051003736261.

Omeri A, Lennings C, Raymond L. Beyond asylum: implications for nursing and health care delivery for Afghan refugees in Australia. J Transcult Nurs. 2006;17(1):30–9.

Paige JB. Making sense of methods and measurement: Q-methodology – Part I – Philosophical background. Clin Simul Nurs. 2014;10(12):639–40. https://doi.org/10.1016/j.ecns.2014.09.008.

Paige JB. Making sense of methods and measurement: Q-methodology – Part II – Methodological procedures. Clin Simul Nurs. 2015;11(1):75–7. https://doi.org/10.1016/j.ecns.2014.10.004.

Perz J, Ussher JM, Gilbert E. Constructions of sex and intimacy after cancer: Q methodology study of people with cancer, their partners, and health professionals. BMC Cancer. 2013;13:270. https://doi.org/10.1186/1471-2407-13-270.

Reber B, Kaufman S. Q-Assessor: developing and testing an online solution to Q method data gathering and processing. World Association for Public Opinion Research, Amsterdam, Netherlands. 2011. Retrieved from www.file://Users/karenkavanaugh/Downloads/wapor-2011_Reber_Kaufman, 20(1).

Reber BH, Kaufman SE, Cropp F. Assessing Q-assessor: a validation study of computer-based Q sorts versus paper sorts. Operant Subjectivity. 2000;23(4):192–209.

Robbins P, Krueger R. Beyond bias? The promise and limits of Q method in human geography. Prof Geogr. 2000;52(4):636–48.

Rogers SR. Q methodology. In: Smith JA, Harre R, Van Langenhove L, editors. Rethinking methods in psychology. Thousand Oaks: Sage; 1995.

Stenner P, Dancey C, Watts S. The understanding of their illness amongst people with irritable bowel syndrome: a Q methodological study. Soc Sci Med. 2000;51(3):439–52.

Stoller EP, Webster NJ, Blixen CE, McCormick RA, Hund AJ, Perzynski AT, ... Dawson NV. Alcohol consumption decisions among nonabusing drinkers diagnosed with hepatitis C: an exploratory sequential mixed methods study. J Mixed Methods Res. 2009;3(1):65–86. https://doi.org/10.1177/1558689808326119.

Tariq S, Woodman J. Using mixed methods in health research. JRSM Short Rep. 2013;4(6): 2042533313479197.

Tariq S, Elford J, Cortina-Borja M, Tookey PA. The association between ethnicity and late presentation to antenatal care among pregnant women living with HIV in the UK and Ireland. AIDS Care. 2012;24(8):978–85. https://doi.org/10.1080/09540121.2012.668284.

Tashakkori A, Teddlie C. Sage handbook of mixed methods in social & behavioral research. Thousand Oaks: Sage; 2010.

Tubergen NV, Olins RA. Mail vs. personal interview administration for Q sorts: a comparative study. Operant Subjectivity. 1978;2(2):51–9.

Van Exel J, de Graaf G. Q methodology: a sneak preview. 2005. Retrieved from http://www.qmethodology.net/PDF/Q-methodology.

Watts S, Stenner P. Doing Q methodology: theory, method and interpretation. Qual Res Psychol. 2005;2(1):67–91. https://doi.org/10.1191/1478088705qp022oa.

Watts S, Stenner P. Doing Q methodological research: theory, method and interpretation. London: Sage; 2012.

Social Network Research

Janet C. Long and Simon Bishop

Contents

1	Introduction	770
2	Network Analysis in the Social Sciences: A Brief History	771
3	Social Network Concepts	771
4	Structure Versus Agency	772
5	Methods	774
6	Key Players in Collaborative Networks	775
7	Social Network Analysis and Healthcare Research	776
8	Conclusion and Future Directions	780
References		780

Abstract

Analysis of networks is increasingly seen as important for understanding the patterns, processes, and consequences of social relationships in healthcare. Networks can be formal, mandated structures (e.g., a clinical network), can emerge from sharing a common passion, or can be from routine exchanges such as referrals. Braithwaite and colleagues (2009) call for the fostering of naturally emerging networks suggesting these underpin the delivery of healthcare and play an important role in driving quality and safety. Social network analysis (SNA) emphasizes patterns of relationships and interactions between network members (actors) rather than individual attributes/behaviors or abstract social structures. SNA conceptualizes networks as composed of nodes (the actors in the group) and

J. C. Long (✉)
Australian Institute of Health Innovation, Macquarie University, Sydney, NSW, Australia
e-mail: janet.long@mq.edu.au

S. Bishop
Centre for Health Innovation, Leadership and Learning, Nottingham University Business School, Nottingham, UK
e-mail: Simon.Bishop@nottingham.ac.uk

ties (the relationship between the actors). Ties form the structure of the network, and the nodes occupy positions within that structure. This proves a basis to investigate a wide range of issues, including communication pathways between actors (including gaps, bottlenecks, or opportunities to increase connectivity), the presence of "tribes" or silos, key players, networks of social support, and patterns of social influences on behaviors. This also allows researchers to investigate relationships between network structures (e.g., communication flows) and important outcomes (e.g., rapid dissemination of ideas). In this chapter, we will introduce readers to key debates, concepts, methods, and applications of SNA, drawing on the authors' own studies and the growing body of healthcare literature adopting this approach. This demonstrates the contribution of SNA to understanding different types of networks, including at the individual, group, and organizational level.

Keywords

Interprofessional relationships · Collaboration · Connectivity · Brokerage · Knowledge exchange

1 Introduction

Analysis of networks is increasingly seen as important for understanding the patterns, processes, and consequences of collaborative relationships in healthcare. Networks can give a more holistic picture of the complex interactions which define the health system. Networks can be formal, mandated structures (e.g., a clinical network Haines et al. 2012), can emerge from sharing a common passion (e.g., a special interest group or community of practice Wenger et al. 2002), or can be from routine exchanges (e.g., referrals Fuller et al. 2007). Braithwaite et al. (2009) call for the fostering of naturally emerging, bottom-up networks, suggesting these underpin the delivery of healthcare and play an important role in driving quality and safety.

A network is any group of people or objects that can be said to interact or have some kind of relationship between them. Network theory provides a powerful lens through which to understand how the elements within such a group are organized, following a set of principles. The study of networks led to the realization that there are similarities between very diverse types of networks such as the neural networks of nematodes (Morita et al. 2001), power grids (Nasiruzzaman 2013), and the Internet (Carmi et al. 2007). In the social sciences, network theory is used to explain interpersonal relationships at various scales: from whole of communities (Putnam 1995) to a few clinicians exchanging information about a patient (Benham-Hutchins and Effken 2010). It provides insight into such phenomena as the influence of opinion leaders, why some companies have a competitive edge, and how effective teams work.

This chapter starts with a brief history of social network studies, followed by an introduction to basic network concepts and methods. We then describe studies which have used social network methodology to study aspects of health service delivery.

2 Network Analysis in the Social Sciences: A Brief History

The study of patterns of social relationships has been an enduring aspect of social science (Durkheim 1895; Simmel 1950). Here, we focus on social network analysis (SNA) as a distinct methodology, emerging in the mid-1930s in the social and behavioral sciences and advancing slowly but constantly over the next 60 years by a small core of researchers at Harvard. As Wasserman et al. (2005, p. 1) put it: "It was easy to trace the evolution of network theories and ideas from professors to student, from one generation to the next."

The psychiatrist, Jacob Levy Moreno (1889–1974), is often cited as the father of network analysis although Freeman (1989) argues that the structure of networks was recognized long before this in the kinship structures such as descendant lists in the Old Testament (e.g., Genesis 5). The first use of the term "network" as it is understood today (Freeman 2004, p.35) was in Moreno's seminal study on Hudson School for Girls and Sing-Sing Prison (Moreno and Jennings 1934). Moreno stated that the schoolgirls' action of running away was influenced more by their position within their social network than with a conscious, independent decision. Moreno used the term "sociometry" to describe "the mathematical study of psychological properties of populations ... methods which inquire into the evolution and organisation of groups and the position of individuals within them" (p.10). In other words, it is a method for eliciting and mapping the subjective feelings of individuals toward each other (Borgatti et al. 2009), focusing analytic attention on patterns of social relationships.

During the 1940s and 1950s, social network research developed through matrix algebra and graph theory, allowing the groups to be objectively identified within networks (Luce and Perry 1949). This led to work exploring concepts such as leadership, group cohesiveness, group productivity, cooperation, competition, communication and problem solving, and the spread of influence within groups (Borgatti et al. 2009; Freeman 2004). Around 1990, there was a massive rise of interest in networks, as other disciplines outside of sociology saw their potential, disciplines as diverse as physics and epidemiology (Wasserman and Faust 1994). A major contribution to network analysis was the characterization and modeling of small-world networks (Travers and Milgram 1969; Watts and Strogatz 1998). Small-world networks have been found in many settings including brain networks (Zhang et al. 2016) and food webs (Montoya and Solé 2002). Small-world networks display properties that transcend the characteristics of the individuals within it.

3 Social Network Concepts

SNA emphasizes patterns of relationships and interactions between network members (actors) rather than individual attributes. Actors can be individuals or entities such as departments or whole organizations, while relationships, which must be tightly defined, can be things such as collaboration, friendship, information exchange, or attendance at a particular event. While attribute data (e.g., gender, age, job position, seniority) is usually also collected, the focus is on this relational data that defines the network structure (Scott 2000). Different types of relational tie can lead to very different network structures; for example, a network of friendship ties between actors may be different from the same actors' network of reporting ties.

Ties can be directional (e.g., providing information to, seeking advice from) or nondirectional (e.g., works in the same building, attend the same meeting). Ties can be recorded as simply present or absent or weighted to signify the weakness or strength of a relationship. This can be based on emotional intensity, level of reciprocity, or more usually frequency of contact (Granovetter 1973).

Relational tie data can be collected in different ways depending on the nature of the interaction. Face-to-face communication patterns may be directly observed (e.g., Obstfeld 2005). Referral patterns, email communications, or collaboration may be gathered using a self-report survey (Bishop and Waring 2012; Chan et al. 2016; Long et al. 2016) or documentary evidence (Fattore et al. 2009; Zheng et al. 2010).

SNA conceptualizes networks as composed of nodes (the actors in the group) and ties (the relationship between the actors) to generate sociograms. The ties form the structure of the network, and the nodes occupy positions within that structure. This proves a basis to investigate a wide range of issues, including communication pathways between actors (including gaps, bottlenecks, or opportunities to increase connectivity), the presence of "tribes" or silos, identification of key players, defining networks of social support, and revealing patterns of social influences on behavior. This also allows researchers to investigate relationships between network structures (e.g., communication flows) and important outcomes (e.g., rapid dissemination of ideas). Table 1 summarises key terms in social network analysis.

Social network theory has been used to understand processes and phenomena across a range of different industries and settings including market competition (Burt 1992; Uzzi 1997), generation of innovative ideas (Bercovitz and Feldman 2011; Hargadon and Sutton 1997), influence and leadership (Lambright et al. 2010; Long et al. 2013b; Valente and Pumpuang 2007), and group dynamics (Balkundi et al. 2009; Susskind et al. 2011).

Within healthcare, social network theory and analysis have been used to look at coordination and integration of health services (e.g., Ayyalasomayajula et al. 2011; Khosla et al. 2016; Lower et al. 2010; Ryan et al. 2013), interprofessional communication and practice (e.g., Benham-Hutchins and Effken 2010; Chan et al. 2016; Creswick et al. 2009), strategies for translational research (e.g., Long et al. 2016; Rycroft-Malone et al. 2011), influence and leadership (e.g., Grimshaw et al. 2006; Kravitz et al. 2003), and quality and safety (e.g., Cunningham et al. 2012; Meltzer et al. 2010).

4   Structure Versus Agency

A debate within SNA research is the difference between two conceptualizations, usually referred to as structure and agency to explain human behavior and social networks. A structuralist view focuses on the recurring patterns of social interactions

Table 1 Some social network terms and their definitions

Term	Definition
Actor	A member of a network
Broker	An actor in a network that acts as an intermediary between two unlinked actors and clusters of actors
Brokerage	A strategy described by Burt (2005) of maximizing opportunities by increasing variation in the network through weak, bridging links to multiple, nonredundant contacts outside the group. This strategy contrasts with closure
Central actor	The actor who is nominated most often or who interacts with the most other members of a network
Centrality	A measure of which actor or actors are the most connected or who interact with the most other actors
Closure	A strategy described by Burt (2005) of increasing cohesion by reducing variation within a group by forming strong links to members of the network. This strategy contrasts with brokerage
Cluster	A subgroup of a network in which the local density of ties is higher than across the whole network
Contagion	The process of spreading disease (in epidemiology), ideas, knowledge, or uptake of new technology through direct contact or social influence in social networks
Degree	The number of ties that actors have to other actors
Density	The ratio of the number of ties present in a network divided by the number of possible ties
Directed tie	A tie that contains information about who initiated the tie and who receives it (e.g., information given by Actor A and received by Actor B)
Node	Element of interest in a network. In a social network, it may be an individual or organization. In nonsocial networks, it may be an object, e.g., a station in a railway network
Edge (or tie)	A link or relationship between actors in a network shown on sociograms as a line
Ego	Focal actor in a network
Egonet	Social network of a single focal actor
Homophily	Defined by Rogers (2003) as the extent to which linked actors share similar attributes such as education, gender, or social status
Reciprocity	A tie is said to be reciprocated when both actors acknowledge the tie
Social capital	A measure of the advantage that comes through social ties. May refer to the advantage held by an individual through their egonet (Burt 1992) or may refer to the quality of an entire group, e.g., an entire community (Putnam 1995)
Strength of tie	A measure of emotional intensity, level of reciprocity, or frequency of interaction associated with a tie
Strength of weak ties	A phenomenon described by Granovetter (1973) to describe the often advantageous, novel information that comes from weak links from outside of one's closely tied network (who all tend to know the same information)
Tie (or edge)	A link or relationship between actors in a network shown on sociograms as a line
Undirected tie	A tie that does not require information about who initiated the tie or who received it (e.g., two actors on the same board, kinship ties)
Whole network survey	A survey that aims to elicit data from every member of the network, rather than a sample of members

that appear to provide opportunities to an individual or constrain their behavior (Ansell et al. 2009). Agency, on the other hand, refers to an individual's power to act and purposefully change their world (Apelrouth and Edles 2008).

A structuralist perspective of networks takes the view that a certain individual's position in a network influences their actions (and consequences) as network positions afford certain opportunities. An actor in a central position in a network might be expected to have the same opportunities and constraints as another central actor in a different network. This approach focuses on the presence or absence of ties and tends to ignore the actual content of the ties ("ties conceptualised as girders" (Borgatti and Foster 2003, p.1003)). An example of this approach is a study of hospital facility managers (Heng et al. 2005) in which they illustrated through a sociogram that managers were situated centrally in the overall network between departments. This meant that they were able to act as coordinators and brokers between the many departments with which they linked.

An agency perspective perceives the actor taking a greater role and using the resources of the network to his or her own end. Agency-focused studies of networks try to understand how the individual's actions and behavior have shaped their environment. This approach focuses on the nature of the ties, more specifically, on the resources that are delivered in the ties ("ties conceptualised as pipes" Borgatti and Foster 2003, p.1003). A small study by Kalish (2008) considered the personality traits of students in brokerage positions in a multicultural class to understand the nature of personal agency in defining their network position.

Networks are not static structures, so some studies have used both agency and structural perspectives in the same study. For example, Johnson et al. (2003) described the relationships between crew members at an Antarctic science base over three successive winters. As well as network structural data ("who hung out with who"), they observed the social roles that people took within the networks ("clown," "leader who got things done"). By combining the data, they were able to describe the emergence and evolution of the network. Both viewpoints have merit and are inherently interesting to explore. Borgatti and Foster (2003), in their review of network research, however note that the vast majority of SNA studies take a structuralist perspective.

5 Methods

Social network data can be collected through self-report surveys, observation, or use of documentary data (e.g., emails, minutes of meetings). Before starting to collect data, the most important step is to define the relationship of interest. Referral or specific advice relationships may be straightforward, but for self-report surveys especially, the tie needs to be understood in the same way by all participants. Long et al. (2016), for example, used the following explanation of collaborative ties since collaboration is a multifaceted concept that had the potential to be understood in a number of different ways: "By 'collaboration' we mean either formally (e.g., on a funded project) or informally (e.g., have discussed aspects of research, supplied expertise, advice or equipment to others) ... Please select those people with whom you are currently collaborating on a network activity, event or project ..." (p. 6). This allowed the researchers to capture informal collaborative ties as well as the formal.

Two main methods of eliciting relationship data in the self-report survey method are roster style and name generator. If the boundaries of the network are known (e.g., people signed up to an online community of practice, staff on a ward, members of a committee), a roster of names may be used (pending ethical and governance approval). In the roster style survey, the members of the network are listed, and the respondent is asked to consider each person as a potential tie. In the name generator style of survey, the respondent is asked to write down the names of the people with whom they consider they have the defined tie without any prompting. This is useful if the membership of the network is not known (e.g., social support networks). The following resources provide detailed discussion of SNA methods and the various advantages and limitations associated with them (Borgatti et al. 2013; Scott 2000; Wasserman and Faust 1994).

6 Key Players in Collaborative Networks

Highly influential actors can be identified within networks, defined by their position in the overall structure. These actors are often called key players (Borgatti 2006), but there are a range of terms used in literature to describe them. Highly connected actors that occupy central positions in the network are termed opinion leaders (Gifford et al. 1999; Valente 2006; Valente and Pumpuang 2007), hubs (Buchanan 2003; Watts and Strogatz 1998), or connectors (Gladwell 2000 p.38). Moreno used the term communication "stars," referring to actors who are chosen as friends by the most people (Moreno and Jennings 1934 p.72), and Allen used this to refer to actors who are approached most often for advice in a work setting (Allen 1970). Central actors appear to sit at the center of a star when ties are graphed (see Fig. 1a).

Brokers are actors that link together individuals or groups of individuals (see Fig. 1b). They have been identified using a range of terms, the most common being bridges (Burt 1992; Valente and Fujimoto 2010), brokers (Cross and Prusak 2002; Gould and Fernandez 1989; Shi et al. 2009), and boundary spanners (Howse 2005; Tushman 1977). The broker is considered a key player as their position is inherently powerful; they may be the sole link between two noncommunicating groups. This can be used for a competitive advantage in business (e.g., having information from group A that group B does not, means the broker has a competitive edge) or to cause

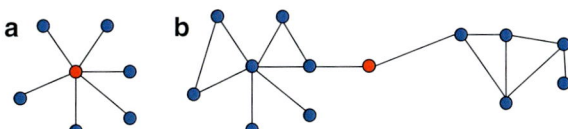

Fig. 1 (**a**) Star-shaped graph: the central actor is colored red. (**b**) Broker (in red) bridges two separate clusters of actors

mischief (e.g., hoarding relevant information and not passing it along; acting as a gatekeeper and not allowing access to resources held by the other group). More positively, in collaborative networks, they can broker beneficial introductions, mediate between parties that are at odds, or provide a service of some kind to both parties (e.g., an interpreter, an expert).

Both key player roles have costs associated with them as well as advantages (Long et al. 2013c). Maintaining ties is a time-consuming exercise and beyond a certain number is unfeasible (Burt 1992, 2002).

7 Social Network Analysis and Healthcare Research

Social network analysis is a powerful approach to apply to healthcare settings. It can provide a framework to examine information flows, social and professional influence, and the phenomenon of siloed thinking and action (Long et al. 2016). While SNA has been well noted for its potential to map epidemiological phenomenon (e.g., the spread of HIV (Lin et al. 2012) or SARs (Chen et al. 2011)), over the past 10 years, it has also been increasingly taken up in research on healthcare organizations and systems. A number of reasons for this interest can be suggested. The increasing focus on the shape of social networks can be seen to follow from a concern with network forms of governance and policy attempts to engage with, and harness, embedded professional networks. Rather than an integrated hierarchy, it has increasingly been recognized that multiple "decentered" professional and organizational networks are involved in shaping and controlling health systems; SNA offers an approach to study such network forms.

A related concern of healthcare researchers is the nature of relationships between heterogeneous professional and occupational groups, how work is divided, and the implications for the coordination of care and fostering of collaboration. Rather than focusing on the aggregate relations, as has been common in perspectives such as sociology of the professions, SNA allow empirical investigation of patterns of relationships at the individual and subgroup level.

Third, an increasing concern of healthcare researchers over the past 15 years has been how knowledge, particularly new knowledge from research evidence and innovation, is translated and diffused into practice. SNA has also been used to examine the strategy of using translational research networks to bridge the "valley of death" (Butler 2008) between basic science and bedside, "real-life" practice. Again, SNA has shed light the patterns of relationships that underpin this process and how knowledge translation and improvement efforts can be supported. Two examples of author projects demonstrate recent applications.

Example 1 SNA of translational research strategies
Translational research undertakes the crucial role of moving biomedical discoveries out of the highly controlled laboratory environment and applying it in the complexity of patient and service delivery realities (Goldblatt and Lee 2010; Woolf 2008). Expertise and understanding through collaboration between both fields are necessary

to achieve this, yet the gaps between research and clinical domains are widening through increased specialization and complexity (Schwartz and Vilquin 2003; Zerhouni 2005). Translational research networks are a strategy to facilitate collaboration by establishing a clear, joint vision and setting up an administrative structure to provide funding for joint projects, project officers, and shared resources as well as a social structure to maximize opportunities for collaboration, innovation, and knowledge exchange. While potential partners in such networks may abound, clusters within disciplines, professions, or geographic sites and the gaps between them may hinder their initiation. This study used SNA at baseline and three further points in time to examine changes in collaborative ties between members with reference to these clusters (Long et al. 2012, 2013a, b, 2014, 2016).

The translational research network of interest was established in late 2011, and initial membership was 68 cancer clinicians and researchers drawn from 6 hospital and university sites in New South Wales. An online, whole network survey was administered to all registered members of the network in early 2012, in 2013, and again in 2015. Membership changed in that time from 68 to 263 to 244 (respectively) as people joined or left. SNA showed that at baseline, ties of the original members were reflective of long-standing teaching and research arrangements and clustered by field (clinician or researcher) and by geographic proximity. Over the next 4 years, collaborative ties were shown to be bridging the field gap and including consumers in both research- and clinically based projects, although geographic proximity remained a feature. Key player analysis showed that the network manager was enacting a significant brokerage role in bringing new collaborative partners together, a quantitative finding that was confirmed through interviews (Long et al. 2013b).

In a similar project (unpublished Long and McDermott 2017), SNA was used to examine the growth of collaborative ties within a translational research network in the field of dementia. The network was shown, by the second year of operation, to have successfully brokered collaborations across formerly siloed sectors of academia, industry (largely staff in residential care facilities), consumers (people living with dementia and representatives from consumer advocacy groups), and government (policy-makers, regulators, and accreditation purveyors). Sociograms from the first survey at baseline (Fig. 2) and after 2 years of operation (Fig. 3) show this growth of intersectoral collaboration. External/internal (E/I) index analyses at the two points in time showed that at baseline, members from each sector were more likely to collaborate with people within their sector than with people in another sector, while after 2 years, members were more likely to collaborate with members outside their sector. In the last survey, there were 857 new ties ($n = 121$) described as "I have only worked with this person since joining the network." Again, key player analysis showed both the centrality of the network manager and director and their brokerage roles.

Example 2 Mixed methods SNA: relations between health and social care
The second example focuses on a study of knowledge sharing on issues of patient safety within a UK NHS hospital day surgery department. In light of well-recognized professional silos within health organizations (Waring 2004; Currie et al. 2008), this

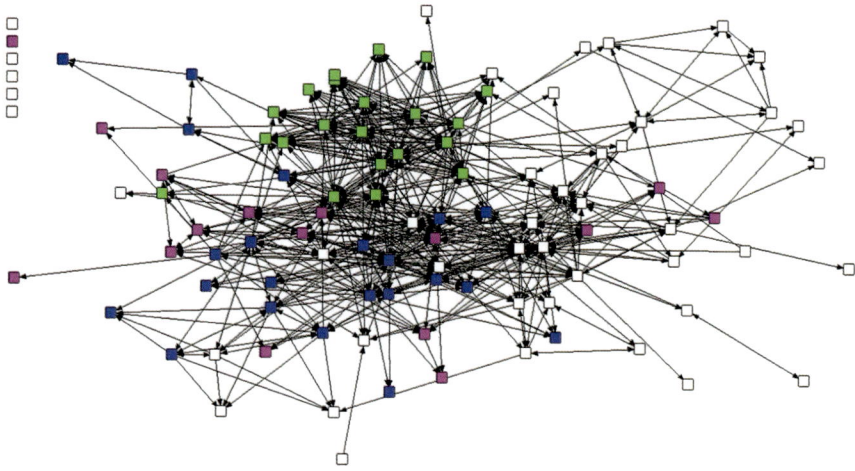

Fig. 2 Baseline collaboration in a dementia translational research network ($n = 104$). The four sectors are shown by color: green = consumers, blue = academics, white = industry, pink = government. Gray nodes indicate missing sector data

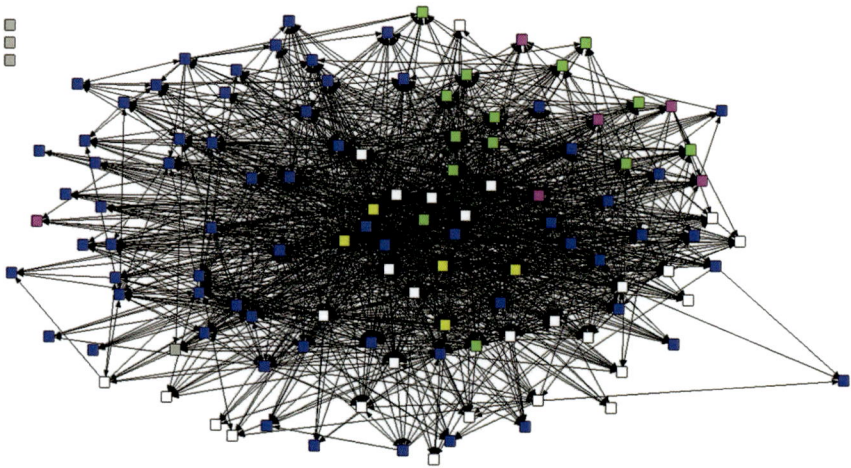

Fig. 3 Collaboration ties after 2 years of operation of a dementia translational research network ($n = 121$). While academics outnumber other sectors, cross-sectoral collaboration was demonstrated quantitatively and is now more evident visually (Legend as for Fig. 2)

study aimed to investigate the patterns of knowledge sharing within and between professional groups. The methodology involved both a quantitative SNA survey and a period of ethnographic observations. The quantitative SNA survey was designed to elicit respondents' close advice-giving contacts, asking respondents to provide named individuals within the department from whom they most commonly sought knowledge around patient safety, as well as the frequency of advice. Demographic

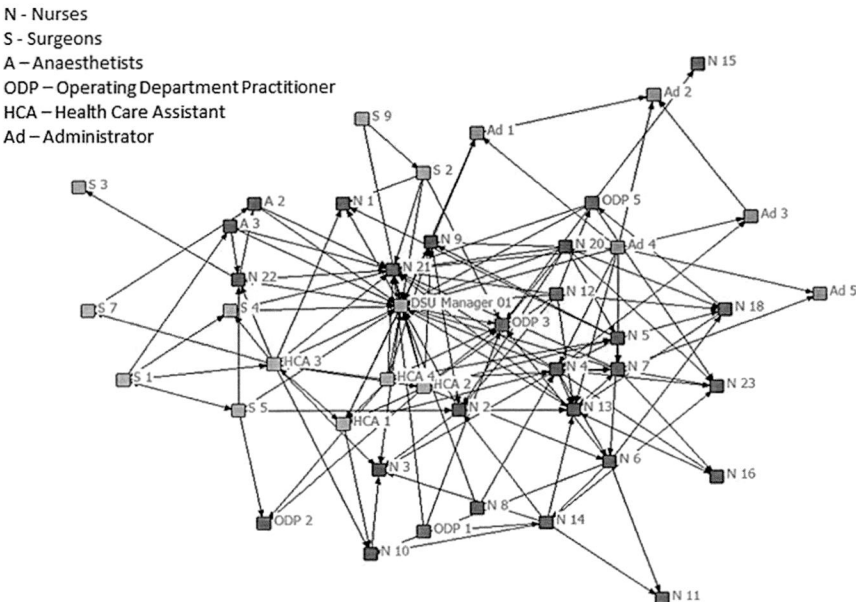

Fig. 4 Network of advice-seeking ties within a UK NHS hospital day surgery department

data was also collected on the professional background, tenure, and work role of the respondent. Full network data was sought from all members of the department, identified both through an initial staff list and through following up new individuals identified in the name generator of respondents ($n = 47$, 85% response rate). Alongside this, 250 hours of ethnographic observations were undertaken, focusing on working practices and communication across settings within the department, as well as 40 qualitative interviews (see Bishop and Waring 2012).

Results from the study brought to light a number of key issues surrounding knowledge sharing within the department (see Fig. 4). Quantitative SNA results illustrated the complex web of intra- and interprofessional knowledge-sharing relationships in the department and highlighted (1) medics' position toward the periphery of the network, (2) the central role of senior nurses in the advice network of the department, (3) the peripheral role of part time and temporary staff, and (4) that a higher number of advice-seeking ties were held within professional groups than between groups. These findings supported prior theorizing in relation to trust and knowledge sharing within professional groups (Chan et al. 2016; Creswick et al. 2009). They also appeared to reflect aspects of service organization, for example, the central administrative role played by senior nurses within the department and medics attached to external departments of their clinical specialisms.

Alongside the quantitative SNA findings, the qualitative component of the study allowed further exploration of the patterns of advice giving within the quantitative

SNA and provided insight into the meaning of the identified relationships. This work included examination of how work practice shaped the opportunities for interaction and hence knowledge sharing within and between groups. It also explored important factors shaping how individuals sought to negotiate relationships within the department while responding to conflicting demands. Bringing together quantitative SNA and qualitative research methods could, therefore, help to develop both an understanding of the structure of social relationship and the way these relationships are formed and maintained within the everyday practice of health organizations.

8 Conclusion and Future Directions

Researchers of health systems are increasingly recognizing that the socio-professional relationships are an essential component of quality, safety, and efficient delivery of care. SNA is a valuable tool to quantify these relationships at both an individual and organizational level. Patterns of collaboration, referral, and knowledge exchange are revealed by SNA and in combination with complementary qualitative methods such as ethnographic observation or interviews, fleshed out to give insight into social processes in healthcare. In addition, SNA is an important methodology for understanding emergent networks which have been shown to drive safety initiatives (Braithwaite et al. 2009).

SNA is an important methodology to analyze new social structures to drive policy and reform, cross-sectoral collaboration, integration of services, and dissemination of best practice. The use of SNA to reveal the utility of translational research networks as a strategy to create a common vision and broker-bridging relationships has been shown. SNA is also an important methodology for examining managed network structures as mechanisms of policy and reform. As public policy emphasizes dispersed leadership and accountability within networks, an understanding of the strength of relationships and how network roles such as brokerage are enacted is important. Further theory around network development and durability of relationships is another avenue for future research.

References

Allen TJ. Communication networks in R & D labs. R & D Manag. 1970;1:14–21.

Ansell C, Reckhow S, Kelly A. How to reform a reform coalition: outreach, agenda expansion, and brokerage in Urban School Reform. Policy Stud J. 2009;37(4):717–43. https://doi.org/10.1111/j.1541-0072.2009.00332.x.

Apelrouth S, Edles L. Classical and contemporary sociological theory: text and readings. Thousand Oaks: Pine Forge Press; 2008.

Ayyalasomayajula B, Wiebe N, Hemmelgarn BR, Bello A, Manns B, Klarenbach S, Tonelli M. A novel technique to optimize facility locations of new nephrology services for remote areas. Clin J Am Soc Nephrol. 2011;6(9):2157–64. https://doi.org/10.2215/CJN.01820211.

Balkundi P, Barsness Z, Michael JH. Unlocking the influence of leadership network structures on team conflict and viability. Small Group Res. 2009;40(3):301–22.

Benham-Hutchins MM, Effken JA. Multiprofessional patterns and methods of communication during patient handoffs. Int J Med Inform. 2010;79(4):252–67.

Bercovitz J, Feldman M. The mechanisms of collaboration in inventive teams: composition, social networks, and geography. Res Policy. 2011;40(1):81–93. https://doi.org/10.1016/j.respol.2010.09.008.

Bishop S, Waring J. Discovering healthcare professional-practice networks: the added value of qualitative SNA. Qual Res Organ Manag Int J. 2012;7(3):308–22. https://doi.org/10.1108/17465641211279770.

Borgatti SP, Everett MG, Johnson JC. Analyzing social networks. Thousand Oaks: SAGE; 2013.

Borgatti SP, Foster PC. The network paradigm in organizational research: a review and typology. J Manag. 2003;29(6):991–1013. https://doi.org/10.1016/s0149-2063_03_00087-4.

Borgatti SP, Mehra A, Brass DJ, Labianca G. Network analysis in the social sciences. Science. 2009;323(5916):892–5.

Borgatti SP. Identifying sets of key players in a social network. Comput Math Organ Theory. 2006;12:21.

Braithwaite J, Runciman WB, Merry AF. Towards safer, better healthcare: harnessing the natural properties of complex sociotechnical systems. Qual Saf Health Care. 2009;18(1):37–41.

Buchanan M. Nexus: small worlds and the groundbreaking science of networks. New York: WW Norton; 2003.

Burt RS. Bridge decay. Soc Networks. 2002;24(4):333–63.

Burt RS. Brokerage and closure: an introduction to social capital. New York: Oxford University Press; 2005.

Burt RS. Structural holes: the social structure of competition. Cambridge, MA: Harvard University Press; 1992.

Butler D. Crossing the valley of death. Nature. 2008;453:840–2.

Carmi S, Havlin S, Kirkpatrick S, Shavitt Y, Shir E. A model of Internet topology using k-shell decomposition. Proceedings of the National Academy of Sciences. 2007;104(27):11150–4.

Chan B, Reeve E, Matthews S, Carroll P, Long JC, Held F, ⋯ Hilmer SN. Medicine information exchange networks among health care professionals and prescribing in geriatric medicine wards. Br J Clin Pharmacol. 2016;83(6):1185–96. https://doi.org/10.1111/bcp.13222.

Chen Y-D, Chen H, King C-C. Social network analysis for contact tracing. In: Castillo-Chavez C, Chen H, Lober WB, Thurmond M, Zeng D, editors. Infectious disease informatics and biosurveillance: research, systems and case studies. Boston: Springer US; 2011. p. 339–58.

Creswick N, Westbrook JI, Braithwaite J. Understanding communication networks in the emergency department. BMC Health Serv Res. 2009;9:247.

Cross R, Prusak L. The people who make organizations go – or stop. Harv Bus Rev. 2002;80:105–12.

Cunningham FC, Ranmuthugala G, Plumb J, Georgiou A, Westbrook JI, Braithwaite J. Health professional networks as a vector for improving healthcare quality and safety: a systematic review. BMJ Qual Saf. 2012;21(3):239–49. https://doi.org/10.1136/bmjqs-2011-000187.

Currie G, Waring J, Finn R. The limits of knowledge management for UK Public Services modernization: the case of patient safety and service quality. Public Administration. 2008;86(2):363–85.

Durkheim E. Les Règles de la Méthode Sociologique. Paris: Revue philosophique; 1895.

Fattore G, Frosini F, Salvatore D, Tozzi V. Social network analysis in primary care: the impact of interactions on prescribing behaviour. Health Policy. 2009;92(2–3):141–8.

Freeman LC. Network representations. In: Freeman LC, White DR, Romney AK, editors. Research methods in social network analysis. Fairfax, Virginia: George Mason University; 1989.

Freeman LC. The development of social network analysis: a study in the sociology of science. Vancouver: Empirical Press; 2004.

Fuller J, Kelly B, Sartore G, Fragar L, Tonna A, Pollard G, Hazell T. Use of social network analysis to describe service links for farmers' mental health. Aust J Rural Health. 2007;15(2):99–106. https://doi.org/10.1111/j.1440-1584.2007.00861.x.

Gifford D, Holloway R, Frankel M, Albright C, Meyerson R, Griggs R, et al. Improving adherence to dementia guidelines through education and opinion leaders. Ann Intern Med. 1999;131:237–46.

Gladwell M. The tipping point: how little thing can make a big difference. New York: Back Bay Books/Little, Brown and Company; 2000.

Goldblatt EM, Lee W-H. From bench to bedside: the growing use of translational research in cancer medicine. Am J Transl Res. 2010;2(1):1–18.

Gould RV, Fernandez RM. Structures of mediation: a formal approach to brokerage in transaction networks. Sociol Methodol. 1989;19:89–126.

Granovetter M. The strength of weak ties. Am J Sociol. 1973;78:1360–80.

Grimshaw J, Eccles M, Greener J, Maclennan G, Ibbotson T, Kahan J, Sullivan F. Is the involvement of opinion leaders in the implementation of research findings a feasible strategy? Implement Sci. 2006;1:3.

Haines M, Brown B, Craig J, D'Este C, Elliott E, Klineberg E, ... Research Group, C. N. Determinants of successful clinical networks: the conceptual framework and study protocol. Implement Sci. 2012;7(1):16.

Hargadon A, Sutton RI. Technology brokering and innovation in a product development firm. Adm Sci Q. 1997;42(4):716–49.

Heng HKS, McGeorge WD, Loosemore M. Beyond strategy: exploring the brokerage role of facilities manager in hospitals. J Health Organ Manag. 2005;19(1):16–31.

Howse EL. Factors that motivate hospital nurse middle managers to share knowledge related to boundary spanning roles. Ph.D., University of Toronto (Canada). 2005. Retrieved from http://search.ebscohost.com/login.aspx?direct=true&db=cin20&AN=2009283278&site=ehost-live.

Johnson JC, Boster JS, Palinkas LA. Social roles and the evolution of networks in extreme and isolated environments. J Math Sociol. 2003;27(2–3):89–121. https://doi.org/10.1080/00222500305890.

Kalish Y. Bridging in social networks: who are the people in structural holes and why are they there? Asian J Soc Psychol. 2008;11(1):53–66.

Khosla N, Marsteller JA, Hsu YJ, Elliott DL. Analysing collaboration among HIV agencies through combining network theory and relational coordination. Soc Sci Med. 2016;150:85–94. https://doi.org/10.1016/j.socscimed.2015.12.006.

Kravitz RL, Krackhardt D, Melnikow J, Franz CE, Gilbert WM, Zach A, ... Romano PS. Networked for change? Identifying obstetric opinion leaders and assessing their opinions on caesarean delivery. Soc Sci Med. 2003;57(12):2423–34.

Lambright KT, Mischen PA, Laramee CB. Building trust in public and nonprofit networks: personal, dyadic, and third-party influences. Am Rev Public Adm. 2010;40(1):64–82. https://doi.org/10.1177/0275074008329426.

Lin H, He N, Ding Y, Qiu D, Zhu W, Liu X, ... Detels R. Tracing sexual contacts of HIV-infected individuals in a rural prefecture, Eastern China. BMC Public Health. 2012;12(1):533. https://doi.org/10.1186/1471-2458-12-533.

Long JC, Cunningham FC, Braithwaite J. Network structure and the role of key players in a translational cancer research network: a study protocol. BMJ Open. 2012;2(3):e001434. https://doi.org/10.1136/bmjopen-2012-001434.

Long JC, Cunningham FC, Carswell P, Braithwaite J. Who are the key players in a new translational research network? BMC Health Serv Res. 2013a;13:338. https://doi.org/10.1186/10.1186/1472-6963-13-338.

Long JC, Cunningham FC, Wiley J, Carswell P, Braithwaite J. Leadership in complex networks: the importance of network position and strategic action in a translational cancer research network. Implement Sci. 2013b;8:122. https://doi.org/10.1186/1748-5908-8-122.

Long LC, Cunningham FC, Braithwaite J. Bridges, brokers and boundary spanners in collaborative networks: a systematic review. BMC Health Serv Res. 2013c;13(1).

Long JC, Cunningham FC, Carswell P, Braithwaite J. Patterns of collaboration in complex networks: the example of a translational research network. BMC Health Serv Res. 2014;14(1):225. https://doi.org/10.1186/1472-6963-14-225.

Long JC, Hibbert P, Braithwaite J. Structuring successful collaboration: a longitudinal social network analysis of a translational research network. Implement Sci. 2016;11:19. https://doi.org/10.1186/s13012-016-0381-y.

Long JC, McDermott S. Social Network analysis of a Dementia translational research network. 2017; unpublished data

Lower T, Fragar L, Depcynzksi J, Fuller J, Challinor K, Williams W. Social network analysis for farmers' hearing services in a rural community. Aust J Prim Health. 2010;13(1):47–51.

Luce RD, Perry A. A method of matrix analysis of group structure. Psychometrika. 1949; 14(2):95–116.

Meltzer D, Chung J, Khalili P, Marlow E, Arora V, Schumock G, Burt R. Exploring the use of social network methods in designing healthcare quality improvement teams. Soc Sci Med. 2010; 71(6):1119–30.

Montoya JM, Solé RV. Small world patterns in food webs. J Theor Biol. 2002;214(3):405–12. https://doi.org/10.1006/jtbi.2001.2460.

Moreno JL, Jennings HH. Who shall survive? A new approach to the problem of human interrelations. Washington, DC: Nervous and Mental Disease Publishing Co; 1934.

Morita S, Oshio KI, Osana Y, Funabashi Y, Oka K, Kawamura K. Geometrical structure of the neuronal network of Caenorhabditis elegans. Physica A: Statistical Mechanics and its Applications 2001;298(3–4):553–561.

Nasiruzzaman A. Complex network framework based comparative study of power grid centrality measures. Int J Electr Comput Eng. 2013;3(4):543.

Obstfeld D. Social networks, the *tertius iungens* orientation, and involvement in innovation. Adm Sci Q. 2005;50:100–30.

Putnam R. Bowling alone: America's declining social capital. J Democr. 1995;6(1):65–78.

Rogers E. Diffusion of innovations. 4th ed. New York: Free Press; 2003.

Ryan DP, Puri M, Liu BA. Comparing patient and provider perceptions of home- and community-based services: social network analysis as a service integration metric. Home Health Care Serv Q. 2013;32(2):92–105. https://doi.org/10.1080/01621424.2013.779352.

Rycroft-Malone J, Wilkinson J, Burton C, Andrews G, Ariss S, Baker R, ⋯ Thompson C. Implementing health research through academic and clinical partnerships: a realistic evaluation of the Collaborations for Leadership in Applied Health Research and Care (CLAHRC). Implement Sci. 2011;6(1):74.

Schwartz K, Vilquin J-T. Building the translational highway: toward new partnerships between academia and the private sector. Nat Med. 2003;9(5):493–5.

Scott J. Social network analysis: a handbook. 2nd ed. London: Sage; 2000.

Shi W, Markoczy L, Dess GG. The role of middle management in the strategy process: group affiliation, structural holes, and tertius iungens. J Manag. 2009;35(6):1453–80. https://doi.org/10.1177/0149206309346338.

Simmel G. The sociology of Georg Simmel (trans: Wolff KH). New York: Free Press; 1950.

Susskind A, Odom-Reed P, Viccari A. Team leaders and team members in interorganizational networks: an examination of structural holes and performance. Commun Res. 2011;38(5): 613–33. https://doi.org/10.1177/0093650210380867.

Travers J, Milgram S. An experimental study of a small world problem. Sociometry. 1969;32(4): 425–43.

Tushman ML. Special boundary roles in the innovation process. Adm Sci Q. 1977;22(4):587–605.

Uzzi B. Social structure and competition in interfirm networks: the paradox of embeddedness. Adm Sci Q. 1997;42(1):35–67.

Valente T, Fujimoto K. Bridging: locating critical connectors in a network. Soc Networks. 2010;23:212–20.

Valente T, Pumpuang P. Identifying opinion leaders to promote behavior change. Health Educ Behav. 2007;34(6):881–96. https://doi.org/10.1177/1090198106297855.

Valente T. Opinion leader interventions in social networks. Br Med J. 2006;333(7578):1082–3. https://doi.org/10.1136/bmj.39042.435984.43.

Waring JJ. A qualitative study of the intra-hospital variations in incident reporting. International J Quality in Health Care. 2004;16(5):347–352.

Wasserman S, Faust K. Social network analysis. Cambridge: Cambridge University Press; 1994.

Wasserman S, Scott J, Carrington PJ. Introduction. In: Carrington PJ, Scott J, Wasserman S, editors. Models and methods in social network analysis. Cambridge, England: Cambridge University Press; 2005

Watts DJ, Strogatz SH. Collective dynamics of 'small-world' networks. Nature. 1998;393(6684):440–2.

Wenger E, McDermott R, Snyder WM. Cultivating communities of practice. Boston: Harvard Business School Press; 2002.

Woolf SH. The meaning of translational research and why it matters. JAMA. 2008;299(2):211–3. https://doi.org/10.1001/jama.2007.26.

Zerhouni EA. Translational and clinical science: time for a new vision. N Engl J Med. 2005;353(15):1621–3. https://doi.org/10.1056/NEJMsb053723.

Zhang J, Lin X, Fu G, Sai L, Chen H, Yang J, ⋯ Yuan Z. Mapping the small-world properties of brain networks in deception with functional near-infrared spectroscopy. 2016;6:25297. https://doi.org/10.1038/srep25297.

Zheng K, Padman R, Krackhardt D, Johnson MP, Diamond HS. Social networks and physician adoption of electronic health records: insights from an empirical study. J Am Med Inform Assoc. 2010;17(3):328–36. https://doi.org/10.1136/jamia.2009.000877.

Meta-synthesis of Qualitative Research

45

Angela J. Dawson

Contents

1 Introduction	786
2 Meta-synthesis: A Historical Overview	787
3 Approaches to the Meta-synthesis of Qualitative Data	788
3.1 Purpose	788
3.2 Position	789
3.3 Paradigm	790
4 Practical Approaches to Undertaking a Meta-synthesis	793
4.1 Defining the Scope of the Study	793
4.2 Question Design	793
4.3 Developing an Inclusion Criterion and Forming Keywords	795
4.4 Selecting a Method or Approach to the Synthesis of Qualitative Data	796
4.5 Developing a Protocol	797
4.6 Technology to Support Meta-synthesis	797
4.7 Searching and Selecting/Sampling Studies	798
4.8 Appraising Qualitative Studies in the Review	798
4.9 Extraction and Synthesis Processes	799
5 Conclusion and Future Directions	800
References	802

Abstract

A meta-synthesis of qualitative health research is a structured approach to analyzing primary data across the findings sections of published peer-reviewed papers reporting qualitative research. A meta-synthesis of qualitative research provides evidence for health care and service decision-making to inform improvements in both policy and practice. This chapter will provide an outline

A. J. Dawson (✉)
Australian Centre for Public and Population Health Research, University of Technology Sydney, Sydney, NSW, Australia
e-mail: angela.dawson@uts.edu.au

© Springer Nature Singapore Pte Ltd. 2019
P. Liamputtong (ed.), *Handbook of Research Methods in Health Social Sciences*,
https://doi.org/10.1007/978-981-10-5251-4_112

of the purpose of the meta-synthesis of qualitative health research, a historical overview, and insights into the value of knowledge generated from this approach. Reflective activities and references to examples from the literature will enable readers to:

- Summarize methodological approaches that can be applied to the analysis of qualitative research.
- Define the scope of and review question for a meta-synthesis of qualitative research.
- Undertake a systematic literature search using standard tools and frameworks.
- Examine critical appraisal tools for assessing the quality of research papers.

Keywords

Meta-synthesis · Qualitative research synthesis · Systematic review · Meta-ethnography · Meta-summary · Narrative synthesis

1 Introduction

A meta-synthesis of qualitative health research is a structured approach to analyzing primary data across the findings sections of published peer-reviewed papers reporting qualitative research. The analysis of this evidence involves qualitative and sometime quantitative methods with the aim of improving health outcomes, research, services, or policy. Meta-synthesis involves a process that enables researchers to examine a phenomenon such as health-care experiences or understandings of health or well-being, as well as environmental, organizational, and individual factors that affect the implementation or effectiveness of a health-care service or clinical intervention and the complex interaction of these. The approach involves searching for, selecting, appraising, summarizing, and combining qualitative evidence to answer a specific research question. A meta-synthesis of qualitative evidence involves interpreting the interpretations made by the authors of the studies included in the meta-synthesis to provide insights for practice or policy. These insights are theoretical and involve methodological development combined with critical reflexivity, thereby adding to, refining, and extending the meaning of original qualitative studies.

Meta-synthesis, therefore, involves the study of the underlying assumptions of various qualitative findings in the included studies, comparing different types of data according to their quality and utility and synthesizing and interpreting the findings of research studies that relate to the same phenomenon. The approach has been described as "the bringing together and breaking down of findings, examining them, discovering the essential features, and, in some way, combining phenomena into a transformed whole" (Schreiber et al. 1997). This approach differs from other qualitative reviews of literature including the integrative literature review both in the explicit, systematic, and replicable steps taken to searching for, screening, and identifying literature and in the approach to the analysis of data contained within the findings of the included research papers. The approach taken to the analysis is empirical rather than a discursive critique of academic discourse that involves a study of studies rather than an overt examination of their findings (Thorne et al. 2004, p. 1360).

2 Meta-synthesis: A Historical Overview

There is increasing recognition of the important role qualitative research can play in medical research. The movement toward patient-centered care has stimulated much of this recognition alongside public skepticism of medical expertise and scientific knowledge and the change in understanding the patient as a client and consumer of health care. While this has done much to raise the voices of patients/clients/consumers in decisions concerning health care and engage communities in the co-production of health services, there has been an acknowledgment of the important role of attitudes and beliefs in health-related behaviors and beyond the social determinants of health. This understanding has recognized that to improve health, an examination of sociocultural values is needed that necessitates qualitative data and research methods. Despite many researchers rejecting the British Medical Journal's suggestion that qualitative research, lacking in conclusive findings or clinical implications, should be published in specialist journals (Greenhalgh et al. 2016), there is a growing acknowledgment that experimental designs are not suited to answering many health research questions (see also ▶ Chap. 7, "Social Constructionism").

In addition to these developments, there has been an increasing focus on systematic reviews to support evidence-based health care including reviews of qualitative studies and qualitative approaches to reviewing qualitative and quantitative research. This change is visible in the actions of the Cochrane Library whose traditional statistical methodological expertise in meta-analyses has expanded to incorporate qualitative evidence in syntheses. In 2006, the Cochrane Qualitative and Implementation Methods Group (CQIMG) was established, and the latest guidance on qualitative reviews is now included in the Cochrane Handbook (Noyes et al. 2015). The meta-syntheses of qualitative studies have an important role to play in health-care decision-making by either augmenting quantitative meta-analysis through investigative questions linked to effectiveness or complementing meta-analysis to address questions on aspects other than effectiveness. In addition, meta-synthesis of qualitative evidence can help to frame research questions and identify gaps where further qualitative and quantitative research and/or systematic reviews should be undertaken. The CQIMG has recently published guidance on methods for integrating qualitative and implementation evidence within intervention effectiveness reviews (Harden et al. 2017). The group published a paper to illustrate how qualitative evidence can be used to compliment quantitative evidence to provide a more nuanced understanding of service delivery in the case of tuberculosis (Noyes and Popay 2007). However, it was not until 2013 for the first review of qualitative evidence to be published in the Cochrane Library (Glenton et al. 2013).

The meta-synthesis of qualitative research has its roots in the social sciences and early efforts such as those by Noblit and Hare who developed a procedure in the 1980s to work out "how to 'pull together' written interpretative accounts" (1988, p. 7). An increased trend in the number of published meta-synthesis of qualitative evidence has been noted (Hannes and Macaitis 2012; Tong et al. 2012), indicating a growing interest in the field and rapid development and refinement of approaches and methods.

3 Approaches to the Meta-synthesis of Qualitative Data

There are approximately 20 methods to synthesize qualitative evidence (Hannes and Lockwood 2012, p. 5). This variation in review methods highlights the nature of the evolving field of meta-synthesis but also the different understandings of knowledge, worldviews, and aims of researchers. This involves the researchers articulating the purpose of their review, the epistemological position, and the paradigm aligned with the choice of methodology.

3.1 Purpose

Researchers have set out to achieve several objectives in their meta-synthesis of qualitative research that represent a range of motivations that can be understood across a continuum (see Fig. 1). Those wishing to deliver practical insights for health policy and practice decision-making are more likely to adopt an approach that is highly structured and involves systematic sampling of research reports and use an a priori conceptual framework derived from the literature to help direct and define the study and triangulate the findings. The approach to the analysis of data will, therefore, involve summation, enumeration, or aggregation in a directed or deductive

Purpose		
Utilitarian/ practical		Theoretical
Test	Explore	Generate
Examine apriori values		conceive values a posteriori
Aggregate		Interpret
Reason deductively		Reason inductively
Generalizability of findings		Transferability of findings
Context specific insights		Multi-context insights
Synthesize quantitative or quantitative and qualitative evidence		Synthesize qualitative evidence

Fig. 1 A continuum of purposes dictating the approach to the synthesis of data

manner. Quantitative approaches to the analysis of qualitative evidence could be used for this purpose such as Bayesian methods for synthesis. This can involve converting qualitative data into numerical data so that it can be used together with quantitative data to calculate the likelihood of a hypothesis being invalid. For example, in Voils et al.' study (2009), qualitative data is quantified and analyzed to test if decreased adherence to antiretroviral medication would be associated with more complex drug regimes. The purpose is to draw conclusions that could be applied or generalized across different populations. This approach is aligned with a meta-analysis rather than a meta-synthesis as the focus is on statistical methods to improve estimates of the size of the effect of a given intervention.

Structured approaches employed to synthesize both qualitative and quantitative evidence from research reports that do not employ inferential statistical methods include narrative synthesis and critical interpretive synthesis. If the purpose is to synthesize qualitative evidence only, then a qualitative meta-summary (Martsolf et al. 2010) or ecological triangulation (Banning 2005) may be better suited. These approaches are more exploratory and less on the testing extreme of the continuum.

On the other hand, researchers may be aiming to generate theory such as in the case of Walder and Molineux (2017) who sought to develop a framework to understand occupational adaptation and identity for those affected by chronic disease and serious illness. The authors did not employ a priori organizing categories for data analysis. The context of the data is particularly valuable in this analysis to provide insight into how the framework can be applied to understand the experiences, perceptions, and needs of other populations.

While researchers may conduct a review with a particular selection of studies to gain insight into one specific context to provide targeted insights, reviews involving multiple contexts will require detailed information about cultural and organizational factors so that this can be considered for relevancy (Hannes and Harden 2011). This highlights the importance of a research team that is diverse in culture, age, gender, and disciplines so that new theoretical insights can be jointly created to address and explore a common issue.

3.2 Position

As meta-synthesis is an interpretive syntheses of qualitative data, it involves drawing upon methodological insights from phenomenology, ethnography, grounded theory, and other theories to integrate and make comprehensible descriptions or accounts of phenomena, cases, or events (Sandelowski and Barroso 2006). The choice of methodology, therefore, depends upon the question, the understanding of knowledge, and study lens or theoretical perspective. This involves adopting an epistemological position that underpins the approach to the review (see also ▶ Chap. 6, "Ontology and Epistemology"). Barnett-Page and Thomas (2007) have described approaches to the synthesis of qualitative research along epistemological dimensions where, at one end, there is subjective idealism, a theory that espouses that knowledge cannot be known and that truth is subjective. As such, there can be no shared reality

Position		
Realist		Idealist
Objectivist	Constructivist	Subjectivist

Fig. 2 Epistemological dimensions

that is independent from the many human constructions of knowledge. At the other end of this epistemological dimension is naïve realism where knowledge can be known but exists independently from human construction as objective truth. Crotty (1998) has described qualitative research across three epistemological positions: objectivist where meaning and reality exist separate from human consciousness; constructivist, where meaning originates from our interactions with the world; and subjectivist, where meaning is imposed on an object by a human (see Fig. 2).

At the scientific realist, objective end of the dimension, the review by Seymour et al. (2010) is focused on delivering robust findings for health-care decision-making that involves defining a clear question that remains constant throughout the study and a structured process of analysis (framework analysis) and interpretation to answer the review question. Here, knowledge is understood as representing an external reality that can be extracted and used to improve health professional communication with family members following genetic testing for cancer risk. An example of a review positioned toward the idealist constructivist end of the dimension is a review by Franzen et al. (2017) that employed a meta-narrative methodology to explore different approaches to health research capacity development in low- and middle-income countries over time. In this study, all approaches are different and alternative ways of understanding capacity building, and as such, the authors embrace heterogeneous evidence from multiple contexts.

3.3 Paradigm

The epistemological position and the methodological premise of a meta-analysis can be understood according to the paradigm it is aligned with (see Fig. 3). The associated paradigm provides a set of guiding principles that dictates how researchers approach the study of phenomena and how the results can be interpreted. This "set of beliefs" (Guba 1990, p. 17) includes post-positivism, constructivist/interpretive, critical theory, subjectivism, and pragmatism (Denzin and Lincoln 2011). These paradigms, however, are not discrete categories, but rather a continuum of theoretical assumptions underpinning reviews that are not always clear cut (Gough et al. 2012).

While post-positivists agree that reality can be measured, observation is imperfect, and as a result errors are common, and, therefore, theory is revisable. Bayesian meta-analysis involving the quantification of qualitative data to determine variables and calculate effect sizes (Crandell et al. 2011) is aligned with a post-positivist paradigm. Realist qualitative meta-synthesis reviews approach the study of qualitative evidence (Sager and Andereggen 2012) with this worldview to determine "what

Paradigm dimension	Methodology/approach	Associated theory	Epistemological focus
Post-positivist	Bayesian meta-analysis Bayesian meta-synthesis Meta-summary Realist synthesis Ecological triangulation Case survey Content analysis Narrative synthesis	Logical positivism	Reality can be measured, and the focus is on reliable and valid tools to achieve this
Constructivist/interpretive	Framework synthesis Narrative synthesis Meta-ethnography Mata-narrative Critical interpretative synthesis Meta-study Thematic analysis Formal grounded theory Heuristic inquiry	Symbolic interactionism Hermeneutics Phenomenology Critical inquiry Feminism	Phenomena must be interpreted to discover the many alternate underlying meanings
Critical	Meta-ethnography Grounded theory Discourse analysis Action/ participatory research	Marxism Critical social theory Feminism Queer theory	Knowledge is not neutral, but socially constructed and the product of power relations in society
Subjectivism	Discourse analysis Meta-ethnography	Post-modernism Structuralism Post structuralism	Knowledge and reality is a matter of perspective
Pragmatism	Meta-aggregative, use of multiple methodologies	Deweyan	The focus is on solving the problem or issues using the most appropriate means

Fig. 3 Research paradigms, meta-synthesis methods, theories, and epistemological foci

works for whom and in what circumstances?" (Pawson and Tilley 1997, p. xvi). Qualitative evidence from evaluations of health interventions can be synthesized to determine different context-mechanism-outcome configurations. One example is the examination of community engagement in health programs using qualitative comparative analysis by Thomas et al. (2014) where theory is refined according to its application to context.

The meta-synthesis of qualitative data involving meta-ethnography, grounded theory, critical interpretive synthesis, and phenomenological analysis are inquiries that can be applied within the constructivist/interpretive paradigm to focus on social processes and the ways in which individuals and communities enact their realities and give them meaning (see also ▶ Chap. 7, "Social Constructionism"). Meta-syntheses within a critical theory paradigm include the use of critical or feminist theory to analyze and interpret the qualitative data retrieved from studies in ways that involve a reflective and critical assessment of power relations in culture and society. Authors have suggested that grounded theory may be a useful methodology here (Thorne et al. 2004) where reflexivity can be employed to describe the study purpose and interactions with the data and researchers. Discourse analysis and semiotics may also be used to analyze text in qualitative research synthesis (Onwuegbuzie et al. 2012) to reveal institutionalized patterns of knowledge and power (Elfenbein 2016). Transformative frameworks have been applied in meta-synthesis to generate theory from participatory action or practice to help people address change in the interest of social justice (see ▶ Chap. 17, "Community-Based Participatory Action Research"). Oliver et al. (2015) have trialed the use of participatory methods to involve young people in the analysis and interpretation of qualitative data as part of a systematic review to ensure rigor as well as the relevance and feasibility of the findings to reduce childhood obesity. Patients have reported improved confidence and skills as a result of their involvement in such processes (Bayliss et al. 2016). The application of postmodern and poststructuralist methodologies such as deconstruction and genealogical inquiry that are aligned with a paradigm of subjectivism has been little explored in qualitative meta-synthesis.

A pragmatic theoretical perspective underpins a meta-aggregative approach to synthesis to deliver practical and applied insights (Hannes and Lockwood 2011). Value is assigned according to what is useful and the contribution the findings make to the research question. The meta-synthesis is, therefore, the result of questions that have arisen from current situations, needs, and consequent phenomena, not antecedent phenomena (Cherryholmes 1992). Fegran et al. (2014) review of adolescents' and young adults' transition experiences when transferring from pediatric to adult care uses multiple methods in line with a pragmatic approach. A meta-summary is used to quantitatively aggregate qualitative findings followed by a phenomenological-hermeneutic analysis. However, other pragmatic decisions may be made in the case of meta-methods where researchers analyze and interpret the interpretations of various methodological applications across many qualitative research studies. Meta-theory involves the analysis and interpretation of different theoretical and philosophical perspectives, sources and assumptions, and contexts across many qualitative studies. This contrasts with a formal grounded theory where the researchers use grounded theory studies to synthesize a new theory.

The choice of method or approach for a meta-synthesis is informed by the purpose and epistemological position but also the nature of the evidence in the field. If the evidence is mainly descriptive and there are no established theoretical models for understanding the phenomena, an inductive approach such as critical interpretative synthesis may be useful. In cases where the evidence is extremely theorized or conceptual, framework synthesis could be employed. Studies selected for synthesis that are heterogeneous may not allow an understanding to be built of one study in the light of another, a process known as reciprocal translation. A thematic synthesis, therefore, may be more appropriate than meta-ethnography (see ▶ Chap. 48, "Thematic Analysis"). If the research reports selected for synthesis do not provide "thick" data, then a meta-summary may be more appropriate than grounded theory. The choice of method will also depend on the expertise of the research team, the available financial resources, and time.

4 Practical Approaches to Undertaking a Meta-synthesis

There are many key steps (see Fig. 4) involved in undertaking a meta-synthesis that are not discrete but rather are iterative and may be constantly revised.

4.1 Defining the Scope of the Study

Scoping at the early stages of a meta-synthesis is necessary to determine the feasibility of undertaking a synthesis, the nature of the literature, and the resource implications. This process in turn will help to define the review question, the method of synthesis, and search terms or keywords, identify relevant databases or other repositories of research, and develop the inclusion/exclusion criteria for the screening and selection of research reports. Scoping involves preliminary searches of literature, reading, and discussions with experts to clarify research questions and identify papers that can provide theoretical insights, a priori frameworks, or conceptual leads that can provide a useful starting point for analysis. Ongoing discussion is required with the research team to make decisions about how the study should be designed and undertaken so that a protocol for the study can be developed.

4.2 Question Design

There are various approaches and tools that may be useful to guide question formulation whether it be an effectiveness question, an exploratory one, or a question designed to generate theoretical insights (Harris et al. 2017). Different approaches include the PICOS (Population, Intervention/phenomena of interest, Control, Outcome, and study design) model (Hannes and Pearson 2012; Dawson et al. 2013), the SPIDER (Sample, Phenomenon of Interest, Design, Evaluation, Research type) tool (Cooke et al. 2012), the SPICE (Setting, Perspective, Intervention/Exposure/Interest, Comparison, Evaluation) tool, or the CHIP (Context, How, Issues, Population) tool (Shaw 2010). An example of the use of the PICOS approach

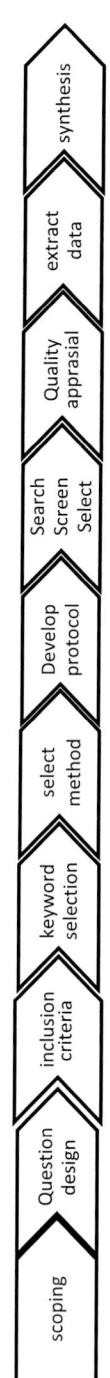

Fig. 4 Key steps

can be seen in a narrative synthesis on task shifting in reproductive health (Dawson et al. 2013). The question was defined by identifying the following key areas:

- Population: medical doctors, doctor assistants, nurses, midwives, auxiliary nurses, auxiliary midwives, and community health workers and lay health workers in low- and middle-income countries
- Intervention: worker substitution, delegation, and task sharing across teams
- Outcome: clinical performance, patient outcomes, training outcomes, provider needs and experiences, and cost-effectiveness
- Study design: qualitative studies, observational studies, quasi experimental and nonexperimental descriptive studies

The final review question therefore was: "For health-care providers in low- and middle-income countries, how have task-shifting interventions influenced the provision of reproductive health care and the capacity and needs of providers?"

The use of the SPIDER mnemonic to guide question development and the inclusion criteria can be seen in the narrative synthesis by Oishi and Murtagh (2014) that sought to identify the challenges of uncertainty and interprofessional collaboration in palliative care for non-cancer patients in the community. The question was defined by clarifying the following:

- Sample: patients with life-limiting diseases other than cancer, carers, health-care professionals
- Phenomenon of interest: primary palliative care for non-cancer patients at home
- Design: any design
- Evaluation: the views of patients
- Research type: any design

4.3 Developing an Inclusion Criterion and Forming Keywords

Clarifying the review question is important to identifying both the criteria that will be used to screen the papers and the keywords that will be used to direct the search for the literature. Table 1 describes the inclusion/exclusion criteria that was developed and applied to the screening of papers for a content analysis of qualitative and quantitative evidence (Dawson et al. 2015). This study aimed to identify approaches to improving the contribution of the nursing and midwifery workforce to increasing universal access to primary health care for vulnerable populations. For example, papers were included in the initial screening if the studies they reported were focused at the primary health-care level and the interventions were led by nurses and/or midwives.

Keywords can be identified from both the question definition and the inclusion criteria. A research librarian can be helpful to assist the study team to identify standard keywords used in specific databases and the Boolean operators (and, or, not) that can be used to connect and define the relationship between your search terms. Using the relevant MESH (Medical Subject Headings) as keywords will also help to optimize the search to ensure the retrieval of relevant literature. MESH is a comprehensive standard

Table 1 The inclusion/exclusion criteria applied to the screening of papers

Included	Excluded
Primary health care	Hospital-based care
Nurse-/midwifery-led health delivery	Care delivered by doctors, community, or lay health workers
Study participants: vulnerable population groups	Study participants: general population, high socioeconomic index
Interventions included nurse/midwife education/training and/or increase in supply and/or human resource management (HRM) strategy and/or policy/practice change and/or collaborative partnership arrangements	Interventions did *not* include nurse/midwife education/training and/or increase in supply and/or HRM strategy and/or policy/practice change and/or collaborative partnership arrangements
Outcomes included improvement in acceptability/satisfaction/uptake of services and/or service quality and/or health outcomes and/or nurse/midwife capacity to promote, care, and manage health issues	Outcomes did *not* include improvement in acceptability/satisfaction/uptake of services and/or service quality and/or health outcomes and/or nurse/midwife capacity to promote, care, and manage health issues
Primary research	Discursive or descriptive outlines of projects
English	Non-English
>2005	<2005

vocabulary for indexing journal articles that was established by the National Library of Medicine in the United States. They are constantly revised and updated. MESH, therefore, serves as a thesaurus that facilitates searching databases such as PubMed®. The keywords at Table 2 were used to retrieve literature from eight bibliographic databases for a content analysis discussed above (Dawson et al. 2015). This includes examples of MESH keywords and nonstandard keywords that were combined with the Boolean operator "and" to retrieve 111 records that were screened using the inclusion criteria at Table 1 to identify 115 records.

4.4 Selecting a Method or Approach to the Synthesis of Qualitative Data

As previously discussed, the approach for the meta-synthesis is dependent upon the purpose of the review, the position and paradigm, the nature of the current evidence, the expertise of the team, and available resources. The choice of content analysis to identify approaches to improving the contribution of the nursing and midwifery workforce to increasing universal access to primary health care for vulnerable populations in the study by Dawson et al. (2015) was based upon initial scoping and the purpose of the paper that was undertaken in partnership with the World Health Organization. A directed approach was deemed the most useful as the findings were to be employed to develop a plan of action at a Global Forum for Government Chief Nursing and Midwifery Officers. A conceptual framework was, therefore, developed to clearly articulate the workforce and leadership interventions that research had identified as contributing to increasing access to universal health care.

Table 2 Search terms applied to the database search for literature

	Search terms
	Nursing, nurse
and	Midwifery, midwife
and	Leadership
and	Policy
and	Universal access, effective coverage, equitable access, health-care access, appropriate health care, acceptable health care, health equity
and	Quality, availability, accessibility, acceptability
and	MESH heading "manpower" "Nursing" "Primary Care Nursing" "Maternal-Child Nursing" "Public Health Nursing" OR "Primary Nursing Obstetrical Nursing"[Mesh] OR "Nursing, Team" "Nursing Staff" "Midwifery" "Nurse Midwives" "Policy" "Universal Coverage"
	Primary health care, community health
Study type	
Date of publication	2003–2015

The data was, therefore, extracted, categorized, enumerated, and synthesized according to the elements of this framework. This approach enabled the findings to be immediately utilized for decision-making in multiple contexts. However, theorizing is largely absent and limited by the lack of rich detail of the original study contexts.

4.5 Developing a Protocol

A protocol is necessary to articulate clearly the study objectives and methods that serve as a point of reference to guide the implementation of the review, which ensures rigor and quality control and can be registered to promote transparency, avoid the duplication of reviews, and assist others to locate the research easily. Meta-synthesis registration can occur with organizations such as the Cochrane Collaboration, the Campbell Collaboration, or PROSPERO, an international database of prospectively registered systematic reviews where the outcome concerns health.

4.6 Technology to Support Meta-synthesis

Technology is useful to support the design of meta-synthesis, database searching, screening, appraisal, and analysis. This includes tools produced by the Cochrane Collaboration (RevMan and Covidence), QARI (Qualitative Assessment and Review) software developed by the Joanna Briggs Institute (JBI), and EPPI-Centre (Evidence for Policy and Practice Information and Co-ordinating Centre) Reviewer 4: software for research synthesis. There are also a number of tools for qualitative data analysis such as text mining tools (Thomas et al. 2011) including those to categorize research into themes (Stansfield et al. 2013) and other standard qualitative data management softwares QSR NVivo and ATLAS.ti (see also ▶ Chap. 52, "Using Qualitative Data Analysis Software (QDAS) to Assist Data Analyses").

4.7 Searching and Selecting/Sampling Studies

Decisions will need to be made concerning the approach to the selection of evidence in terms of either a comprehensive, purposeful, theoretical, or random sample of studies to be included in the synthesis. This decision will determine the number of studies included and depend on the purpose of the synthesis, the epistemology, and the available evidence clarified in the inclusion criteria. As noted above, if the evidence is mostly descriptive and the aim is to aggregate all available data, then a comprehensive approach may be appropriate involving many studies. However, if the aim is specific, then techniques such as stratified, snowball, or criterion sampling may be undertaken. For example, if human-lived experience is sought with a focus on personal consciousness and research reports are available, then a purposive selection of phenomenological studies may be relevant, such as the synthesis by Röing et al. (2017). Searches undertaken with the aim of theoretical inquiry are based on key concepts that are relevant to researchers, and attention is focused on inconsistent findings or similar study populations that are of conceptual relevance. Theoretical selection is iterative; the aim is to draw upon what is previously known to identify and examine the complexity of different conceptualizations. Selection is constantly refined until no new data is revealed. Whatever the approach, it must be demonstrated that the findings of the synthesis are well supported by adequate raw data such as participant quotes, observations, and so on.

Some researchers may make the decision to only search for peer-reviewed primary research papers in electronic databases using standard keywords with added hand or manual searching of reports based on references identified in selected research papers. Other researchers may extend their search to gray literature if the field is an emerging one and insights are to be gained from unpublished research reports. This can include documents from websites of organizations, online libraries such as the World Health Organization's Reproductive Health Library, meta-indexes, or a locally hosted website where there are collections of Internet resources on specific topics, electronic gateways (i.e., https://knowledge-gateway.org/), or data obtained from contacting the authors of research. Researchers may also seek data from these sources using nonstandard keywords. There is no consensus on the best data sources for meta-synthesis, and the decision should be well justified and information retrieval methods clearly articulated and documented using a diagram known as PRISMA (Preferred Reporting Items for Systematic reviews and Meta-Analyses) guidelines (Moher et al. 2009).

4.8 Appraising Qualitative Studies in the Review

Assessing the methodological quality of studies is one consideration in the process of deciding whether a study should be included or excluded from the synthesis.

Excluding methodological "weak" studies have not demonstrated to affect the results of a synthesis; however removing papers based on their methodological limitations can result in the loss of contextual or deviant data that might be useful to the synthesis findings. Therefore, the final decision regarding a paper's inclusion/exclusion must involve an explicit sensitivity analysis and consensus among the research team. Standard tools for appraisal include the United Kingdom's National Health Service Critical Appraisal Skills Programme (CASP) qualitative checklist and the JBI QARI Critical Appraisal Checklist (see also ▶ Chap. 58, "Appraisal of Qualitative Studies"). However, the latest Cochrane advice recommends the use of tools that focus on an assessment of methodological strengths and limitations of the research report (Noyes et al. 2017). This includes an assessment of the available detail on the data sampling, collection, and analysis processes and the coherence of the paradigm underpinning the study involving the fit between data gathered and the conceptual work of analysis and interpretation.

4.9 Extraction and Synthesis Processes

Once papers have undergone quality appraisal, data for the included studies will need to be systematically extracted for analysis, and the approach to this will again depend on the purpose, epistemological position, and paradigmatic worldview chosen by the researchers. First, the researcher will need to decide what constitutes qualitative data to be extracted and from what sections of the research report this data is to be extracted. The qualitative data may only include data pertaining to the findings sections of the papers such as qualitative quotations of participants, data from tables, photographs, diagrams, or quotations from researcher's field notes or recorded observations or the themes, sub-themes, and associated explanations. However, other researchers such as those undertaking a meta-ethnography may see data from the whole text itself including the title, abstract, introduction, and discussion. Supporting materials may also be included such as correspondence, interviews, or discussions with researchers about the study via email, video, or mobile phone (Onwuegbuzie et al. 2012). The rationale for the inclusion of multiple data may be to ensure adequate meaning is generated and, therefore, contributing to the quality and rigor of the synthesis. Despite the approach, research must articulate how and why data was extracted and integrated. These processes must be transparent, and an audit trail should be available documenting the logic of the analysis so that decision-making is traceable. This can be achieved through the use of standard forms and templates and frameworks developed from conceptual/theoretical models, logic models, or iteratively using qualitative data analysis tools such as NVivo (Noyes et al. 2017; see also ▶ Chap. 52, "Using Qualitative Data Analysis Software (QDAS) to Assist Data Analyses").

A meta-synthesis will involve two types of data analysis within-study analysis to help contextualize the findings across study analysis and synthesis. One of the key processes involved is reciprocal translation where researchers read within studies and across them to understand how studies relate, build upon, or differ from each other and knit this together in a critical manner. The approach to this can be linear and aggregative in line with meta-summary or iterative in the case of meta-ethnography. Melendez-Torres et al. (2015) have identified four approaches to reciprocal translation: visual representation, key paper integration, data reduction and thematic extraction, and line-by-line coding. Furthermore methods and tools to integrate qualitative and process evaluation evidence within intervention effectiveness reviews have recently been identified by CQIMG (Harden et al. 2017). Harden et al. have noted that in these pragmatic syntheses, researchers can either undertake a sequential or separate analysis of qualitative evidence and quantitative evidence and then integrate the data or take a convergent approach where themes and outcomes are integrated from the start using a common framework.

5 Conclusion and Future Directions

Over time, the rigor of reviews has been found to increase with search procedures, quality appraisal, and synthesis procedures more clearly articulated (Hannes and Macaitis 2012). Guidance on these processes is also more available such as those provided by the CQIMG and the increasing number of papers and handbooks demonstrating a variety of approaches that serve as points of reference for other researchers. Standards for the reporting of syntheses have emerged such as the generic tool for reporting qualitative meta-synthesis (Tong et al. 2012) and specific methodological guidance for the reporting of realist and meta-narrative reviews (RAMESES) (Wong et al. 2014). New tools are in development for meta-ethnographies (France et al. 2015), and it may only be a matter of time before these become available for other approaches such as formal grounded theory and critical interpretative synthesis. However, a balance must be made between the soundness of reporting and the innovation and creativity of approach that demands that these standards remain flexible.

A necessary element of conducting a meta-synthesis is updating the review to keep the findings current to health-care policy and practice. There is a lack of guidance on updating the meta-synthesis of qualitative evidence, a requirement if registered with Cochrane or PROSPERO. France and colleagues (2016) outline a structured approach to updating a meta-ethnography both in terms of deciding when an update is necessary and how this should be undertaken. The authors identify the characteristics of various approaches including reviews that focus on revising the original, undertaking a separate and independent review that can serve as a comparison or declare the original defunct and redo the review. In addition, France et al. provide some guidance for those appraising the update both in terms of the rationale for undertaking an update and the process and methods applied. While other authors have trialed these guidelines (Rodríguez-Prat et al. 2017), gaps remain in terms of updating other methods from different paradigms.

While tools have been developed to guide within-study quality appraisal and insights to support across-study quality meta-synthesis and the reporting of these, an approach to assessing the confidence in the findings of meta-syntheses is relatively recent. The GRADE-CERQual approach (Grading of Recommendation, Assessment,

Development, and Evaluation-Confidence in the Evidence from Qualitative Reviews) has been developed to ensure that evidence syntheses focused on providing evidence to inform health service or care decision-making processes are conducted and reported in a transparent manner. This tool enables an assessment of a meta-synthesis based upon:

> the methodological limitations of the qualitative studies contributing to a review finding, the relevance to the review question of the studies contributing to a review finding, the coherence of the review finding, and the adequacy of data supporting a review finding. (Lewin et al. 2015, p. 1)

The CERQual approach is the qualitative equivalent of the Grading of Recommendations, Assessment, Development, and Evaluation (GRADE) approach for evidence of effectiveness for quantitative meta-analysis. The tool provides a structured means to interrogate how issues in the design and conduct of the primary studies in a synthesis impacted upon the findings and the relevance of these study findings are to the context specified in the review question. In addition, the tool also assists with an investigation of the extent to which the synthesis is based upon the findings of the primary studies that are convincingly and adequately represented in the synthesis findings. Lewin et al. (2015) provide an example of how the tool can be used to generate a narrative to provide an overall CERQual assessment of confidence and an explanation of this judgment. However, this work is still in development along with the Cochrane qualitative Methodological Limitations Tool (CAMELOT) for use with the first component of CERQual. While CERQual has been tested largely on synthesis using framework analysis or narrative synthesis approaches (Chatfield et al. 2017), its use with synthesis that employ theory generation or interpretative approaches is yet to be seen.

Another area under investigation is the mega-synthesis of meta-syntheses. Tomlin and Borgetto (2011) developed a three-dimensional model to describe different levels of evidence to inform occupational health practice. Mega-synthesis has been argued to be the highest level of evidence in Tomlin's model. This, according to Tomlin et al., is achieved when there is a convergence of meta-synthesis and meta-analysis or when high-level/internally and externally valid qualitative, experimental, and outcome evidence meet. Hannes et al. (2015) have instead proposed a four-dimensional model with mega-synthesis in the center that promotes a question-led approach to synthesizing evidence from both external evidence (qualitative and quantitative scientific) and internal evidence from the experience and needs of clinicians, patients/consumers. They argue that review questions should be driven by what matters rather than what works. This highlights the importance of an inclusive reasoning process, not tiers of evidence informed by study methods such as randomized control trials to achieve treatment choices and inform shared decision-making. Mega synthesis, therefore, is a process where the internal evidence base is the starting point from where the available and relevant external, research-driven evidence base can be used to inform coherent decision-making. This approach to mega-synthesis then takes us back to the principles of the patient-centered movement, to ensure that evidence and decision-making process are directed by the health needs of the community together with health professionals, rather than by the medical establishment.

References

Banning JH. Ecological triangulation: an approach for qualitative meta-synthesis. US Department of Education, School of Education, Colorado State University, Colorado; 2005.

Barnett-Page E, Thomas J. Methods for the synthesis of qualitative research: a critical review, Evidence for Policy and Practice Information and Co-ordinating (EPPI) Centre, Social Science Research Unit Institute of Education, London 01/09; 2007.

Bayliss K, Starling B, Raza K, Johansson EC, Zabalan C, Moore S, Skingle D, Jasinski T, Thomas S, Stack R. Patient involvement in a qualitative meta-synthesis: lessons learnt. Res Involv Engagem. 2016;2:18. https://doi.org/10.1186/s40900-016-0032-0.

Chatfield SL, DeBois K, Nolan R, Crawford H, Hallam JS. Hand hygiene among healthcare workers: a qualitative meta summary using the GRADE-CERQual process. J Infect Prev. 2017;18(3):104–20.

Cherryholmes CH. Notes on pragmatism and scientific realism. Educ Res. 1992;21(6):13–7.

Cooke A, Smith D, Booth A. Beyond PICO: the SPIDER tool for qualitative evidence synthesis. Qual Health Res. 2012;22(10):1435–43.

Crandell JL, Voils CI, Chang Y, Sandelowski M. Bayesian data augmentation methods for the synthesis of qualitative and quantitative research findings. Qual Quant. 2011;45(3):653–69.

Crotty M. The foundations of social research: meaning and perspective in the research process. St Leonards: Sage; 1998.

Dawson AJ, Buchan J, Duffield C, Homer CS, Wijewardena K. Task shifting and sharing in maternal and reproductive health in low-income countries: a narrative synthesis of current evidence. Health Policy Plan. 2013;29(3):396–408.

Dawson A, Nkowane A, Whelan A. Approaches to improving the contribution of the nursing and midwifery workforce to increasing universal access to primary health care for vulnerable populations: a systematic review. Hum Resour Health. 2015;13:97. https://doi.org/10.1186/s12960-015-0096-1.

Denzin NK, Lincoln YS. The Sage handbook of qualitative research. Thousand Oaks: Sage; 2011.

Elfenbein DM. Confidence crisis among general surgery residents: a systematic review and qualitative discourse analysis. JAMA Surg. 2016;151(12):1166–75.

Fegran L, Hall EO, Uhrenfeldt L, Aagaard H, Ludvigsen MS. Adolescents' and young adults' transition experiences when transferring from paediatric to adult care: a qualitative meta-synthesis. Int J Nurs Stud. 2014;51(1):123–35.

France E, Ring N, Noyes J, Maxwell M, Jepson R, Duncan E, Turley R, Jones D, Uny I. Protocol-developing meta-ethnography reporting guidelines (eMERGe). BMC Med Res Methodol. 2015;15:103. https://doi.org/10.1186/s12874-015-0068-0.

France EF, Wells M, Lang H, Williams B. Why, when and how to update a meta-ethnography qualitative synthesis. Systematic reviews. 2016;5(44). https://doi.org/10.1186/s13643-13016-10218-13644. https://doi.org/10.1186/s13643-016-0218-4

Franzen SR, Chandler C, Lang T. Health research capacity development in low and middle income countries: reality or rhetoric? A systematic meta-narrative review of the qualitative literature. BMJ Open. 2017;7(1):e012332. https://doi.org/10.1136/bmjopen-2016-012332.

Glenton C, Colvin CJ, Carlsen B, Swartz A, Lewin S, Noyes J, Rashidian A. Barriers and facilitators to the implementation of lay health worker programmes to improve access to maternal and child health: qualitative evidence synthesis. Cochrane Database of Syst Rev. 2013;10. https://doi.org/10.1002/14651858.CD010414.pub2.

Gough D, Thomas J, Oliver S. Clarifying differences between review designs and methods. Syst Rev. 2012;1:28. https://doi.org/10.1186/2046-4053-1-28.

Greenhalgh T, Annandale E, Ashcroft R, Barlow J, Black N, Bleakley A, Boaden R, Braithwaite J, Britten N, Carnevale F. An open letter to The BMJ editors on qualitative research. BMJ. 2016;352:i563. https://doi.org/10.1136/bmj.i563.

Guba EG. The paradigm dialog. Newberry Park: Sage; 1990.

Hannes K, Harden A. Multi-context versus context-specific qualitative evidence syntheses: combining the best of both. Res Synth Methods. 2011;2(4):271–8.

Hannes K, Lockwood C. Pragmatism as the philosophical foundation for the Joanna Briggs meta-aggregative approach to qualitative evidence synthesis. J Adv Nurs. 2011;67(7):1632–42.

Hannes K, Lockwood C. Qualitative evidence synthesis: choosing the right approach. West Sussex: Wiley-Blackwell; 2012.

Hannes K, Macaitis K. A move to more systematic and transparent approaches in qualitative evidence synthesis: update on a review of published papers. Qual Res. 2012;12(4):402–42.

Hannes K, Pearson A. Obstacles to the implementation of evidence-based practice in Belgium: a worked example of meta-aggregation. In: Synthesizing qualitative research: choosing the right approach. Chichester: Wiley; 2012. p. 21–39.

Hannes K, Behrens J, Bath-Hextall F. There is no such thing as a one dimensional hierarchy of evidence: a critique and a perspective. Vienna, Austria: Paper presented at the Cochrane Colloquium; 2015.

Harden A, Thomas J, Cargo M, Harris J, Pantoja T, Flemming K, Booth A, Garside R, Hannes K, Noyes J. Cochrane Qualitative and Implementation Methods Group guidance paper 4: methods for integrating qualitative and implementation evidence within intervention effectiveness reviews. J Clin Epidemiol. 2017;In Press, Accepted Manuscript. https://doi.org/10.1016/j.jclinepi.2017.11.029.

Harris JL, Booth A, Cargo M, Hannes K, Harden A, Flemming K, Garside R, Pantoja T, Thomas J, Noyes J. Cochrane Qualitative and Implementation Methods Group guidance series-paper 6: methods for question formulation, searching and protocol development for qualitative evidence synthesis. J Clin Epidemiol. 2017;In Press, Corrected Proof. https://doi.org/10.1016/j.jclinepi.2017.10.023.

Lewin S, Glenton C, Munthe-Kaas H, Carlsen B, Colvin CJ, Gülmezoglu M, Noyes J, Booth A, Garside R, Rashidian A. Using qualitative evidence in decision making for health and social interventions: an approach to assess confidence in findings from qualitative evidence syntheses (GRADE-CERQual). PLoS Med. 2015;12(10):e1001895. https://doi.org/10.1371/journal.pmed.

Martsolf DS, Draucker CB, Cook CB, Ross R, Stidham AW, Mweemba P. A meta-summary of qualitative findings about professional services for survivors of sexual violence. Qual Rep. 2010;15(3):489–506.

Melendez-Torres GJ, Grant S, Bonell C. A systematic review and critical appraisal of qualitative metasynthetic practice in public health to develop a taxonomy of operations of reciprocal translation. Res Synth Methods. 2015;6(4):357–71.

Moher D, Liberati A, Tetzlaff J, Altman DG, The PRISMA Group. Preferred reporting items for systematic reviews and meta-analyses: the PRISMA statement. PLoS Med. 2009;6(7): e1000097. https://doi.org/10.1371/journal.pmed.

Noblit GW, Hare RD. Meta-ethnography: synthesizing qualitative studies. Newbury Park: Sage; 1988.

Noyes J, Popay J. Directly observed therapy and tuberculosis: how can a systematic review of qualitative research contribute to improving services? A qualitative meta-synthesis. J Adv Nurs. 2007;57(3):227–43.

Noyes J, Hannes K, Booth A, Harris J, Harden A, Popay J, Pearson A, Cargo M, Pantoja T. Qualitative research and cochrane reviews. In: Higgins J, Green S, editors. Cochrane handbook for systematic reviews of interventions version 5.3.0. The Cochrane Collaboration. 2015. http://qim.cochrane.org/supplemental-handbook-guidance

Noyes J, Booth A, Flemming K, Garside R, Harden A, Lewin S, Pantoja T, Hannes K, Cargo M, Thomas J. Cochrane Qualitative and Implementation Methods Group guidance paper 2: methods for assessing methodological limitations, data extraction and synthesis, and confidence in synthesized qualitative findings. J Clin Epidemiol. 2017;In Press, Corrected Proof. https://doi.org/10.1016/j.jclinepi.2017.06.020.

Oishi A, Murtagh FE. The challenges of uncertainty and interprofessional collaboration in palliative care for non-cancer patients in the community: a systematic review of views from patients, carers and health-care professionals. Palliat Med. 2014;28(9):1081–98.

Oliver K, Rees R, Brady L-M, Kavanagh J, Oliver S, Thomas J. Broadening public participation in systematic reviews: a case example involving young people in two configurative reviews. Research Synthesis Methods. 2015;6(2):206–217. https://doi.org/10.1002/jrsm.1145

Onwuegbuzie AJ, Leech NL, Collins KM. Qualitative analysis techniques for the review of the literature. Qual Rep. 2012;17(28):1–28.

Pawson R, Tilley N. Realistic evaluation. London: Sage; 1997.

Rodríguez-Prat A, Balaguer A, Booth A, Monforte-Royo C. Understanding patients' experiences of the wish to hasten death: an updated and expanded systematic review and meta-ethnography. BMJ Open. 2017;7(9):e016659. https://doi.org/10.1136/bmjopen-2017-016528.

Röing M, Holmström IK, Larsson J. A metasynthesis of phenomenographic articles on understandings of work among healthcare professionals. Qual Health Res. 2017;28(2):273–91. https://doi.org/10.1177/1049732317719433.

Sager F, Andereggen C. Dealing with complex causality in realist synthesis: the promise of qualitative comparative analysis. Am J Eval. 2012;33(1):60–78.

Sandelowski M, Barroso J. Handbook for synthesizing qualitative research. New York: Springer; 2006.

Schreiber R, Crooks D, Stern PN. Qualitative meta-analysis. In: Morse M, editor. Completing a qualitative project: details and dialogue. Thousand Oaks: Sage; 1997. p. 311–26.

Seymour KC, Addington-Hall J, Lucassen AM, Foster CL. What facilitates or impedes family communication following genetic testing for cancer risk? A systematic review and meta-synthesis of primary qualitative research. J Genet Couns. 2010;19(4):330–42.

Shaw R. Conducting literature reviews. In: Forreste MA, editor. Doing qualitative research in psychology: a practical guide. London: Sage; 2010. p. 39–56.

Stansfield C, Thomas J, Kavanagh J. 'Clustering' documents automatically to support scoping reviews of research: a case study. Res Synth Methods. 2013;4(3):230–41.

Thomas J, McNaught J, Ananiadou S. Applications of text mining within systematic reviews. Res Synth Methods. 2011;2(1):1–14.

Thomas J, O'Mara-Eves A, Brunton G. Using qualitative comparative analysis (QCA) in systematic reviews of complex interventions: a worked example. Syst Rev. 2014;3:67. https://doi.org/10.1186/2046-4053-3-67.

Thorne S, Jensen L, Kearney MH, Noblit G, Sandelowski M. Qualitative metasynthesis: reflections on methodological orientation and ideological agenda. Qual Health Res. 2004;14(10):1342–65.

Tomlin G, Borgetto B. Research pyramid: a new evidence-based practice model for occupational therapy. Am J Occup Ther. 2011;65(2):189–96.

Tong A, Flemming K, McInnes E, Oliver S, Craig J. Enhancing transparency in reporting the synthesis of qualitative research: ENTREQ. BMC Med Res Methodol. 2012;12:181. https://doi.org/10.1186/471-2288-12-181.

Voils C, Hassselblad V, Crandell J, Chang Y, Lee E, Sandelowski M. A Bayesian method for the synthesis of evidence from qualitative and quantitative reports: the example of antiretroviral medication adherence. J Health Serv Res Policy. 2009;14(4):226–33.

Walder K, Molineux M. Occupational adaptation and identity reconstruction: a grounded theory synthesis of qualitative studies exploring adults' experiences of adjustment to chronic disease, major illness or injury. J Occup Sci. 2017;24(2):225. https://doi.org/10.1080/14427591.2016.1269240.

Wong G, Greenhalgh T, Westhorp G, Pawson R. Development of methodological guidance, publication standards and training materials for realist and meta-narrative reviews: the RAMESES (Realist and Meta-narrative Evidence Syntheses–Evolving Standards) project. Health Serv Deliv Res. 2014;2(30):1. https://doi.org/10.3310/hsdr02300.

Conducting a Systematic Review: A Practical Guide

46

Freya MacMillan, Kate A. McBride, Emma S. George, and Genevieve Z. Steiner

Contents

1	Introduction	806
2	Stages of a Systematic Review	808
3	Constructing the Research Question	808
4	Conducting a Scoping Search	809
5	Developing a Systematic Review Protocol	811
6	Searching for Relevant Literature	812
7	Managing Citations	813
8	Documenting the Characteristics of Included Studies and Summary of Findings	814
	8.1 Participants	815
	8.2 Methods	816
	8.3 Intervention Characteristics	816
	8.4 Outcomes	816
	8.5 Additional Characteristics for Qualitative Studies	817
9	Data Extraction	817
	9.1 What to Extract	817
	9.2 Data Extraction Forms and Databases	817

F. MacMillan (✉)
School of Science and Health and Translational Health Research Institute (THRI), Western Sydney University, Penrith, NSW, Australia
e-mail: F.Macmillan@westernsydney.edu.au

K. A. McBride
School of Medicine and Translational Health Research Institute, Western Sydney University, Sydney, NSW, Australia
e-mail: K.Mcbride@westernsydney.edu.au

E. S. George
School of Science and Health, Western Sydney University, Sydney, NSW, Australia
e-mail: E.George@westernsydney.edu.au

G. Z. Steiner
NICM and Translational Health Research Institute (THRI), Western Sydney University, Penrith, NSW, Australia
e-mail: G.Steiner@westernsydney.edu.au

© Springer Nature Singapore Pte Ltd. 2019
P. Liamputtong (ed.), *Handbook of Research Methods in Health Social Sciences*,
https://doi.org/10.1007/978-981-10-5251-4_113

9.3 Software 818
10 Methods for Assessing Risk of Bias and Considering Heterogeneity 818
  10.1 Measuring Study Quality 818
  10.2 Heterogeneity 819
  10.3 Risk of Bias Tools 819
  10.4 When Risk of Bias Measurement May Need to Be More Flexible 820
11 Meta Analyses 820
  11.1 What Is Meta-analysis? 820
12 Creating a Narrative and Interpreting Findings 822
13 Conclusion and Future Directions 824
References 825

Abstract

It can be challenging to conduct a systematic review with limited experience and skills in undertaking such a task. This chapter provides a practical guide to undertaking a systematic review, providing step-by-step instructions to guide the individual through the process from start to finish. The chapter begins with defining what a systematic review is, reviewing its various components, turning a research question into a search strategy, developing a systematic review protocol, followed by searching for relevant literature and managing citations. Next, the chapter focuses on documenting the characteristics of included studies and summarizing findings, extracting data, methods for assessing risk of bias and considering heterogeneity, and undertaking meta-analyses. Last, the chapter explores creating a narrative and interpreting findings. Practical tips and examples from existing literature are utilized throughout the chapter to assist readers in their learning. By the end of this chapter, the reader will have the knowledge to conduct their own systematic review.

Keywords

Systematic review · Search strategy · Risk of bias · Heterogeneity · Meta-analysis · Forest plot · Funnel plot · Meta-synthesis

1 Introduction

One of the key principles in evidence-based practice is the synthesis of all available evidence on a particular research topic. As there has been exponential growth in the volume of scientific research over the last few decades, a method to synthesize this evidence has become necessary. For some time, literature or narrative reviews have been used to give a broad overview of a particular research topic. Typically, these types of literature or narrative reviews refer to a number of articles published within a research topic area and can give a reasonable description of an issue. Literature or narrative reviews are generally considered to be opinion pieces as they usually do not systematically search the literature and instead often focus on a small group of relevant studies in a chosen area, based on author selection (Uman 2011). Following this process, selection bias can be introduced into this type of review. This is what makes systematic reviews distinct from narrative and literature reviews as (using multiple investigators to ensure rigor) systematic reviews use explicit, structured, predefined methods to identify all relevant literature and to minimize risk of bias (Koelemay and Vermeulen 2016).

Systematic reviews should, therefore, have a clear definition of inclusion and exclusion criteria, feature a wide-ranging search which identifies all relevant literature, use explicit and reproducible selection criteria for included studies, have a rigorous appraisal of potential biases in the included studies and systematically synthesizes results of the included studies (Cook et al. 1997). This is why systematic reviews have been adopted as a more trustworthy and robust means of synthesizing the available evidence on a particular research topic (Mulrow et al. 1997).

When well conducted to ensure high quality, systematic reviews are considered to be one of the highest forms of scientific evidence to inform recommendations for health promotion, intervention design, policy development, and best practice approaches in health (Moher et al. 2015). Systematic reviews can be complicated to conduct however, with quality largely dependent on the type of studies available on a particular topic, research methodology, the outcomes measured, and the quality of reporting. The best quality systematic reviews summarize results of intervention studies (e.g., randomized controlled trials), but sometimes due to a lack of evidence in a particular field, a systematic review may also include other study designs. For example, a systematic review could collate evidence on the effectiveness of diagnostic tools, on the findings of observational studies, or on qualitative studies. Historically, systematic reviews have been conducted on quantitative data. More recently, systematic reviews on qualitative data have been increasing in popularity; the process by which qualitative data are synthesized is called a meta-synthesis (see also ▶ Chap. 45, "Meta-synthesis of Qualitative Research").

When conducting a systematic review, authors develop detailed search strategies that are carried out in several discipline-specific databases. Potentially relevant articles are rigorously screened against a set of inclusion and exclusion criteria. Data are then extracted from all included research articles and study quality is assessed by exploring potential risk of bias in relation to study design. The authors will then make use of this evidence to make an overall judgment about the effectiveness of the findings of the included studies weighted by study quality. Sometimes, effect sizes from two or more included studies can be combined quantitatively to give an overall combined measure of effect. This pooling of statistical data is called a meta-analysis (Cook et al. 1997). A meta-analysis can usually give a more accurate idea of the overall effect of the combined studies. However, often it is not possible to quantitatively combine results of included studies due to variance in the reporting of outcomes (e.g., missing information from papers required for pooling data), or incomparable tools used to measure efficacy of an intervention across papers (e.g., some self-report vs. objective measurements). Quantitative systematic reviews can, therefore, also be narrative, where the results of the combined studies are synthesized descriptively. Both of these methods alone or in combination are able to summarize existing research on a particular topic.

Systematic reviews can improve not only dissemination of evidence but may also help in examining heterogeneity in results of different studies, make overall findings more generalizable, improve understanding of a particular research issue, and guide practice and policy decision-making (Cook et al. 1997; Greenhalgh 2010).

2 Stages of a Systematic Review

A methodologically sound systematic review consists of several stages:

1. Constructing the research question
2. Scoping search
3. Protocol development
4. Comprehensive and systematic search
5. Selection of studies against eligibility criteria
6. Data extraction
7. Appraisal of studies using a quality checklist
8. Analysis of results
9. Interpretation of findings
10. Dissemination

For the remainder of this chapter, each of these stages will be discussed.

3 Constructing the Research Question

Before beginning a systematic review, it is important to have a clear focus as this will guide decisions on search terms, the databases to be used, and the main types of research studies to be included. Once this has been established, it is necessary to formulate clear, unambiguous, and structured questions. A PICO structure is the most useful method of creating an appropriate and specific review question. PICO is defined as

- Population, participants, patient or problem
- Intervention(s), therapy, treatment
- Comparison (other intervention or treatment, control group)
- Outcome(s)

For example, if you were interested in a review on the impact of breast screening on the early detection of breast cancer in women, your PICO could look like this:

P – women
I – breast screening
C – no intervention
O – early detection of breast cancer

In some cases, a PICO question may also have the addition of the study design, making it a PICOS question.

4 Conducting a Scoping Search

Once a clear focus and PICO question have been developed, the next step is to conduct an initial scoping search of the literature. A scoping search is a basic, fairly brief search of the existing literature that will give an idea of the breadth of studies on a chosen topic, and whether a systematic review has already been conducted on the same topic. If another systematic review has already been published recently, you may choose to look at the literature from a different perspective, for example, you may focus on a different patient group. A scoping search will help shape the final research question, identify relevant search terms (by checking the titles and abstracts of the papers found in your scoping search), and develop inclusion/exclusion criteria for the systematic review. Additionally, the search will help to identify the quantity of primary research on the topic to inform the scale of the review. Last, a scoping search can help in identifying key papers on a chosen topic so that once the main search begins, it is possible to check the effectiveness of the search strategy in finding those key papers.

To conduct a scoping search, a search strategy that includes all potential synonyms for key search terms needs to be developed. For example, if a search was being conducted to identify breast screening programs, authors may use alternate terms such as "mammograms" or "cancer screening" instead of "breast screening." These additional search terms should be included the search strategy (see Table 1).

The next step is to select resources relevant to the research topic area that should be searched. These resources should provide access to all types of literature including systematic reviews, clinical guidelines, and primary research. For health-based systematic reviews, using databases such as the Cochrane Library (to search for systematic reviews), National Institute for Health and Care Excellence (systematic reviews, clinical trials, clinical guidelines), Turning Research Into Practice (systematic reviews, clinical trials, clinical guidelines, primary research), and MEDLINE (systematic reviews, clinical trials, clinical guidelines, primary research) is recommended.

Each database is different, so it is important to be familiar with the platform selected before searching is started. The search may need to be adjusted slightly with each database to ensure the most relevant results are yielded. A minimum of two to three databases should be selected to ensure access to a wide range of resources. In each database, it is also important to be familiar with the correct use of truncation, Medical Subject Headings (known as MeSH headings), wildcard features and Boolean operators (see Table 2), which will make searching more efficient and

Table 1 Example synonyms for a PICO search relating to breast cancer screening in women

	Search terms
P	Middle-aged women, women, females, postmenopausal women
I	Breast screening, mammogram, cancer screening
C	No mammogram, no intervention
O	Early detection, breast cancer, early diagnosis

Table 2 Using truncation, MeSH, wildcards, and Boolean operators in a search strategy

Truncation	Truncation is a technique to broaden a search to include various word endings and spellings. To use **truncation**, enter the root of a word and put the **truncation** symbol at the end. The database will return results that include any ending of that root word. Commonly used truncation symbols are * and $, but these vary between databases
MeSH headings	MeSH headings are a comprehensive controlled vocabulary used to index journal articles and books in the life sciences and can also be used as a thesaurus when creating alternate keywords. Different databases may have different terminology for their medical headings (e.g., in the Current Nursing and Allied Health Literature (CINAHL), they are called CINAHL headings)
Wildcard	Wildcards are a method of searching for alternative spellings of the same words. For instance, there are a number of words that are spelt differently in the United States versus the United Kingdom, such as organization (organisation in the UK) and pediatric (paediatric in the UK). To use the wildcard feature, simply substitute the wildcard symbol, which is often "?", to replace a missing letter. For example, you could search for "organi?ation" to capture both "organization" and "organisation"
Boolean operators	After identifying all relevant keywords, synonyms, and phrases within a search, Boolean operators need to be used to combine topic areas together. The Boolean operators are "AND", "OR", and "NOT": Using OR combines all the individual synonym terms together into one search and broadens your results by including references that have ANY ONE of the search terms within it Using AND focusses the search by combining keywords to find references that contain ALL of the keywords, to narrow your search Using NOT will eliminate items and limit a search further

thorough. Information on how to apply these can usually be found in the help sections of each database. It is also important to create an account for each database as this will allow searches to be saved and the creation of alerts based on the search strategy to be set up.

The scoping search should then be run in each database. It is important to note each of the search terms used and the total number of hits for each search. Saving a search to an account is a good way to keep track of searches. Once this is done, a scan through the titles of the papers identified in the scoping search should be undertaken, and citations downloaded for relevant papers. It is also useful to save the pdf of each paper in a designated "scoping search" folder or reference manager library, such as in EndNote. Reading the abstracts and papers is not too important at this stage – only complete a quick scan of the abstract if there is uncertainty about the eligibility of a particular paper based on the title. The scoping search will provide you with an estimate of the number of papers available on a particular topic and will allow you to make modifications to your search strategy. Irrelevant papers may also have been found that are not necessarily relevant to the current PICO question. It is usual to have captured some irrelevant papers, but if these make up the majority of your returned hits, then there are several ways to deal with this:

- Revision of the synonyms used – the search may have been too broad. Remove one search term at a time and record the number of hits. There may be specific terms that are expanding your search considerably and away from your PICO question.
- Use the "NOT" Boolean operator to exclude a specific search term from the search.
- Create a scoping summary table. This table does not require as much detail as the table to be created from the formal systematic review search. In the table, enter the following information from the papers found:
  – Authors, year of publication, title of paper, trial/program name (in case multiple papers report on the same trial), aim of the study.
  – This information can be inserted directly from the abstract as the purpose of this phase is to provide scope of the papers that should be identified in the formal search.

Use this table to explore patterns across studies and how you might refine the search if you are returning too many hits.

Following the above steps, elements of the PICO(S) question will then be refined to determine the specific inclusion and exclusion criteria that will be used to select studies for inclusion in the review.

5   Developing a Systematic Review Protocol

Systematic review protocols help assure that decisions made during the review process are not arbitrary and that decisions to include or exclude studies are not made with knowledge of individual study results. A systematic review protocol should describe the rationale, hypothesis, and planned methods of the review, including the research question. A protocol should always be prepared before the review is started and be used as a guide throughout the review. Several guidelines for preparing a systematic review protocol are available including the Preferred Reporting Items for Systematic review and Meta-analysis extension for Protocols (PRIMSA-P) (Moher et al. 2015) and The Cochrane Handbook of Systematic Reviews (Higgins and Green 2011).

Following PRISMA-P guidelines (Moher et al. 2015), the protocol should include the following:

1. Introduction
    (a) **Rationale** described for the review in the context of what is already known about the topic
    (b) An explicit statement of the **research question** including PICO(S) terms
2. Methods
    (a) **Eligibility criteria** should be specified for report characteristics (language, publication status) as well as study characteristics. These should be based on your PICO question, considering all aspects of the topic including age groups, geographical areas, study designs, illness stage (if applicable), and any outcome measures. Clear eligibility criteria make it easier to identify

relevant articles at the screening stage. Be wary of date range limits; these must be justifiable.
   (b) All **information sources** should be described (databases to be used, personal communications, use of trial registers, grey literature sources) with anticipated dates of coverage.
   (c) A draft of the **search strategy** that will be used in at least one database should be presented, including limiters, so that the search can be repeated.
   (d) A description of how data and records will be **data managed** throughout the review.
   (e) The **selection process** for study inclusion and exclusion (e.g., two independent reviewers) for each part of the review must be described (screening, eligibility, and inclusion in meta-analysis).
   (f) The planned **data collection process** method must describe data extraction methods (e.g., was this done independently, and were data extraction forms used to ensure consistency) as well as processes for obtaining and/or confirming data with study authors.
   (g) Any pre-planned **data assumptions or simplifications** should be listed and defined.
3. **Outcomes and prioritization** must be listed including a definition and list of all outcomes for which data will be sought and prioritization of primary and secondary outcomes detailed.
4. The proposed methods should be detailed for assessing **risk of bias** and how the information will be used at the data synthesis stage.
5. **Data synthesis** methods need to be described, including the following:
   (a) Criteria for which studies will be **quantitatively synthesized.**
   (b) Where data are appropriate for quantitative synthesis, planned summary measures, data handling and combination methods as well as heterogeneity exploration methods.
   (c) Any proposed **additional analyses**, such as sensitivity or subgroup analyses
   (d) Where quantitative analysis is not feasible, the type of summary planned.
6. Any **planned assessment of meta-biases**, for example, publication bias, should be described.
7. Confidence in cumulative evidence should be outlined with a description of how the body of evidence will be assessed.

The systematic review protocol should be adhered to throughout all stages of the review with any amendments to a protocol tracked and dated.

6 Searching for Relevant Literature

Once a search strategy has been generated from the research question, a scoping search has been undertaken, and eligibility criteria (inclusion/exclusion criteria) have been developed and refined, the formal systematic review search can then be conducted. The advanced search strategy for the systematic review will need to be

Table 3 Example of a search conducted in the MEDLINE database

P	AB Mid* age* wom#n OR AB wom#n (S1)
I	AB Physical activity* OR AB sport* OR AB fitness OR AB walk* OR AB exercise (S2)
C	AB control OR AB usual care OR AB no intervention
O	AB Type 2 diabetes OR AB diabetes OR AB T2DM OR AB diabetes mellitus OR diabetes management (S3)
Final Search	S1 AND S2 AND S3

Note: AB, Abstract; *, Truncation; #, Wildcard; S, Search; P, Population; I, Intervention; C, Comparison; O, Outcome

finalized for each database, using appropriate MeSH terms, wildcards, truncation, and Boolean operators. An example of an advanced search strategy for MEDLINE (looking for the selected words in abstracts) is detailed in Table 3 for the research question: Is physical activity effective for the management of type 2 diabetes in middle-aged women compared to usual care?

As with the scoping search, logging into each database before conducting your systematic searches allows for a record of the search to be saved, and for alerts to be configured. As systematic reviews can take several months to prepare, it is important for database alerts to be enabled so that authors can be notified when any new studies matching their search strategy are added to a database. Again, as with the scoping search, limiters may also be applied before searching (e.g., publications in English only, publication dates, full text availability, searching abstracts only, and so on). Note that in some instances, you may not need to include comparison or outcome terms – removing these terms will widen your search. All databases should be searched within the same week to ensure consistency. A summary table is useful for keeping track of the search strategy (and slight variations between databases), including synonyms and MeSH terms, limiters, dates of searches, and number of hits for all databases that you have searched.

Librarians are a useful resource to advise on the selection of databases and should be consulted in this process. There is no consensus on the number of databases that should be searched, but rather the final selection should allow for a wide net to be cast to pick up the majority of studies on the chosen area. Citation and reference list searches can also be conducted on identified eligible studies and experts in the area can be consulted, to help find any additional papers not captured by the initial database searching. These additional steps will help to minimize the possibility of published work being overlooked.

7 Managing Citations

Conducting scoping and systematic searches can produce a large number of references that can be difficult to manage and categorize. Having a systematic approach to managing citations can not only make the sorting process less logistically challenging, but it can also improve the rigor and reproducibility of the results. There are a range of different citation management (bibliographic) software tools available that can facilitate this process, for example, EndNote, Mendeley, Refworks, and Review Manager. Once the search results have been downloaded from each database, citation management tools can be used to automatically remove duplicates and to sort and categorize results during the screening process. Note that citation management tools often do not capture all duplicates for removal, but this automatic process can help to speed up the process of duplicate removal considerably.

The PRISMA statement (Moher et al. 2009) provides a similar set of guidelines to the PRISMA-P statement (Moher et al. 2015) described above. The PRISMA statement focuses on the reporting of systematic reviews rather than protocol development, and recommends including a flow diagram detailing the study selection processes through each stage of the review (e.g., number of hits identified from databases, number of papers meeting eligibility criteria, selection of studies for final inclusion, and so on). Citation management software is, therefore, a useful and powerful tool in the organization and screening of large numbers of references, and ensuring that data are reported correctly for the PRISMA flow diagram (King et al. 2011). Tools in EndNote such as Groups, Smart Groups, Group Sets, and Labels can be particularly useful for categorizing references. For example, Groups may include all publications within a certain year range (e.g., 2005–2010 and 2011–2015); these can be combined using a Publication Date Group Set (Peters 2017). Groups are useful to file citations into include, exclude, and uncertain categories, for transparency of the review flow with the wider review team. The review process to adhere to the PRISMA flow diagram requirements, as comprehensively detailed in Peters (2017), typically includes identification of studies, title and abstract screening of studies, eligibility of studies for inclusion (based on inclusion/exclusion criteria), eligibility for inclusion in meta-analysis based on methodological quality, and coding of included references (e.g., this could be done based on patterns of results: beneficial/detrimental findings from RCTs). Because citation management software such as Endnote also details the number of references in each Group, it simplifies the process of populating the PRISMA flow diagram and ensures the process is transparent and systematic.

8 Documenting the Characteristics of Included Studies and Summary of Findings

Once you have finalized your database searches and identified the articles that meet inclusion criteria for the review, you will need to document the characteristics of the included studies. The characteristics you document and report on should be relevant to your research question and the intended users of your systematic review (Higgins and Green 2011). Although the studies you deem eligible will typically focus on a similar population group, intervention, or outcome (depending on your research question and the focus of your review), not all studies follow appropriate recommended guidelines for reporting (e.g., the CONSORT statement for randomized trials (Moher et al. 2001; see also ▶ Chap. 56, "Writing Quantitative Research Studies"). It is, therefore,

Table 4 Information to be included in the *Characteristics of Included Studies* table

	Characteristics
Publication details	Author details Year of publication
Participants	Study setting (e.g., workplace) and location (country and city/town in which the study was conducted) Participant demographic profile, including age, sex, and any other specific information relevant to your review (e.g., body mass index, health status, presence or absence of disease, ethnicity)
Methods	Study design (e.g., randomized controlled trial), including details of group allocation and if and how the study differs from a standard parallel group design (e.g., cluster randomized trial) Study duration
Intervention characteristics (for intervention studies)	The name of the study or intervention (if applicable, to ensure you are reporting on unique interventions) Details of all intervention control groups For drug trials, include details of the drug, mode of administration, dose, frequency, and any other relevant information For intervention trials, include details of the mode and frequency of delivery, content of the intervention, and any other relevant information on intervention elements and materials
Outcomes	List all outcomes that are relevant to the review at each time-point and the tools used to measure outcomes

imperative that you develop a structured approach to collate and report on the characteristics of the included studies to ensure consistency in reporting. Such tables will aid you in writing the narrative section of your results. The following information in Table 4 should be (as a minimum) included in the "Characteristics of included studies" table:

8.1   Participants

When documenting the participant characteristics, it is important to include as much detail as possible to allow users of your review to determine the applicability of the study to the population group targeted. This information is also an important factor when comparing findings across multiple studies and unique population groups (Higgins and Green 2011), as heterogeneity in samples may explain some of the differences in findings. In this section, you should report details of the study setting and location, and any characteristics of the sample that are applicable to your review and the interpretation of the study results. For example, if you are conducting a review on the effectiveness of weight loss programs in culturally and linguistically diverse women, it may be important to report on characteristics including country of birth, languages spoken at home, and year of arrival in the host country, if reported, as these characteristics may be associated with the outcome of interest.

8.2 Methods

To accurately report on the methods for each of your included studies, you must identify and report on the study design. If you are only including one specific study design in your review, for example, randomized controlled trials, it may not be necessary to include this information unless there are distinctions in design between the trials included. If you are including any studies with a pre- and post-test assessment, however, you may wish to include a column to detail type of study in your characteristics table (e.g., randomized controlled trials, nonrandomized experimental trials, or pseudorandomized controlled trials). In this case, it is important to specify between study designs as the users of your review will need to consider the rigor of the studies and hence quality of the results in relation to their research and population group.

8.3 Intervention Characteristics

When reporting on intervention characteristics, you should make note of the name of the intervention or program to help identify studies that have reported on the same study in multiple papers. This will prevent doubling-up on data extraction. Each individual study should provide enough detail on each intervention or treatment so that the study could be replicated (Higgins and Green 2011); you should be able to clearly describe the study and the characteristics of each group or treatment. For intervention trials, you should identify and report on the mode and frequency of intervention delivery (e.g., a 30-min face-to-face intervention delivered once per week for 12 weeks), details of facilitators (e.g., delivered by a trained exercise physiologist, general practitioners, community volunteers, or researchers), and details of intervention elements or materials (e.g., access to a mobile app or lifestyle peer support groups). In this particular phase, the focus is on providing the readers of your review with context. You will examine the quality of reporting for each study during the risk of bias phase, which is discussed in detail later in this chapter.

8.4 Outcomes

You may decide to only report on the outcomes that are relevant to your research question and review (Higgins and Green 2011). For example, if your review is focused on the effectiveness of workplace-based physical activity interventions and one of your included studies assessed physical activity levels, blood pressure, and psychological distress, you may decide to only include the results for physical activity as this will be most applicable to your review focus. You may also wish to consider only reporting on a specific time-point that is most relevant to the review, but the findings of each study would need to be interpreted in light of the time-points reported (e.g., did findings differ over the long-term compared to shorter interventions?).

8.5 Additional Characteristics for Qualitative Studies

Systematic reviews of qualitative studies will use a similar approach to that of quantitative studies, but should also include information relating to methodological underpinnings, data analysis techniques and approaches, and a more in-depth summary of the results including themes, quotes, and author interpretations (Butler et al. 2016). More detailed information on conducting systematic reviews using qualitative studies exists (see Dixon-Woods et al. 2006; Butler et al. 2016; see also ▶ Chap. 45, "Meta-Synthesis of Qualitative Research").

9 Data Extraction

9.1 What to Extract

Prior to commencing data extraction, it is important to plan out carefully exactly what you want to gain from your review (this means thinking back to your research question). This will allow you to come up with a list of outcomes, intervention components, and data that you will identify from each included article. As well as the common descriptive characteristics detailed earlier, if you aim to undertake a meta-analysis, then extraction of outcome data will also be required. For qualitative reviews, extraction of themes relevant to your topic is necessary. Remember to check if authors have published associated protocol papers or supplementary online materials that might incorporate the information you are looking for. Authors can also be contacted to ask for any data not reported in a publication or to even provide a full data set.

9.2 Data Extraction Forms and Databases

Creating a standardized data extraction form is important to ensure that only data relevant to your review is extracted and that extraction is completed in a consistent manner across included papers and the research team. Prior to undertaking the full data extraction process, it is recommended that you trial your data extraction forms by having members of your research team extract data from one or two included studies. This will ensure that all relevant data is captured by your data extraction form, and will avoid having to re-visit papers multiple times to extract additional information. It is worth setting up an organized database or spreadsheet to store the information that you extract for your review. Although this might take some time to set up initially, it can save time in the long run by removing the need to refer back to individual papers and manually tallying up the number of papers reporting on a specific outcome. It will also make the process of reporting on extracted information narratively in your write-up, and the construction of summary information tables, quicker, easier, and more transparent for your co-authors. Existing software packages can be useful and are worth considering (see Sect. 9.3 below) or you might set up a simple Microsoft Excel spreadsheet, such as that pictured in Fig. 1. In this

	A	B	C	D	E	F	G
1	Lead Author & publication date	Population	Age (yrs)	Intervention duration (wks)	Follow up Duration (wks)	Num of intervention groups	Sample size (baseline)
2	MacMillan, 2016	Men	18-65	24	24	2	90
3	McBride, 2012 & 2014	Women	30-50	12	12	3	38
4	Steiner, 2012	Older adults	>65	12	24	2	1234
5	George, 2014	Children	5-12	24	104	2	408

Fig. 1 Example Microsoft Excel data extraction spreadsheet

figure, column headings represent study characteristics and outcome data to be extracted, with each row representing a different study included in the review.

Where you have identified more than one paper for the same study, it is important to clearly describe how you have dealt with the data extracted. For example, have you only included one row in your extraction sheet for all the papers identified related to that study or multiple rows, such as the McBride example above (there is a 2012 and 2014 paper associated with the same study)? Spreadsheets and databases can also be set up to assist you in reporting on study quality. For example, you may use a risk of bias tool to rate each study on a number of study quality areas. The questions used to decide on risk of bias ratings can be used as column headings to assist reviewers in the extraction of this information.

9.3 Software

The Systematic Review Tool box (http://www.systematicreviewtools.com) web-based catalogue includes links to tools that support all stages of the systematic review process. Literature reviewing softwares such as Covidence, GRADEpro GDT, Review Manager, and LitAssist include templates available for data extraction and can thus help speed up the data extraction process. For qualitative data extraction, software packages such as those used for individual qualitative studies (e. g., NVivo and Quirkos) can also be used to code data that will be summarized narratively in the results section of a systematic review. Other tools, such as EPPI-Reviewer, can be used for either quantitative or qualitative data synthesis. TaskExchange is software that bridges connections between those requiring assistance in completing a systematic review with experts in the conduction of reviews.

10 Methods for Assessing Risk of Bias and Considering Heterogeneity

10.1 Measuring Study Quality

Attempts to measure the quality of studies included in a review are necessary to interpret findings in respect to the strength of evidence (e.g., better designed, high-quality studies provide stronger evidence than poorly designed low-quality studies). It is often the case that studies with the poorest quality (poor methodology and study design) and highest risk of internal validity overestimate treatment effect size. It is, therefore, important to critically appraise studies (e.g., measure study quality) and interpret findings in the weight of their risk of bias. If risk of bias is not taken into consideration, then the overall findings of a review are likely to be biased and may report that an intervention is effective when it is not, or ineffective when in fact it is.

This stage of a review involves selection, adaptation, or development of an appropriate tool to assess the risk of bias of the included studies. It is recommended that risk of bias ratings be completed by at least two raters independently to ensure a consistent and reliable process is followed. Ideally, this process should be undertaken following a blinded process so that raters are unaware of the authors of each included study that they assess. When assessing study quality, it is important to consider various sources of heterogeneity.

10.2 Heterogeneity

Differences in findings across studies with the same outcome measure can be due to a number of reasons. First, if a review is focused on determining effectiveness of a particular intervention, there may be differences in the delivery of the intervention that account for differences in the outcome achieved (e.g., cooking classes may be group-based, 30 min in duration and delivered by a dietitian twice weekly in one study vs. one-to-one, 60 min in duration and delivered by a community volunteer once weekly in another study). Second, demographic characteristics of samples of participants and baseline outcome values may differ across studies. Third, although the same outcome may be measured, the methods and tools to measure that outcome may be quite different across studies, with varying levels of validity and reliability in the target group (e.g., physical activity can be measured using objective measurement tools, such as accelerometers, and also using subjective tools, such as surveys and questionnaires). Fourth, setting may have an impact on outcomes achieved (e.g., a hospital-based yoga intervention vs. a community-based yoga intervention). Finally, the statistical analyses selected may impact on the conclusions drawn from the data. It is, therefore, important in any systematic review to consider the potential sources of heterogeneity and the impact heterogeneity may play on overall conclusions.

10.3 Risk of Bias Tools

There are many tools available to assess study quality. Some will produce scores, whilst others are checklists. Cochrane do not advise utilizing quality scales for the appraisal of clinical trials, but rather suggest the focus should be on methodological domains, which should be considered and tailored to the review (e.g., inclusion or exclusion, or more or less weight given to a particular domain based on the relevance of that domain to the review topic). Note that validity of a given tool will be affected by making amendments to it, but adjustments may be necessary to produce a more relevant tool for the studies included in the review. For example, several risk of bias tools include a measure of the quality of blinding participants to treatment group. This might be very applicable to some studies (e.g., clinical trials of a new drug therapy), but in behavioral interventions, it is impossible to blind participants to treatment group (e.g., participants will know if they are in an exercise intervention group or not). Different tools will also be required based on the design of studies included in your review (e.g., the Risk of Bias in Non-Randomized Studies of Interventions, ROBINS-I tool (Sterne et al. 2016), and the Cochrane Collaboration's tool for assessing risk of bias in randomized trials (Higgins et al. 2011) are specifically designed to assess study quality of nonrandomized and randomized trials, respectively). Guidelines also exist for critical appraisal and quality publishing of qualitative research (Barbour 2001; Tong et al. 2007, 2016; Hannes et al. 2010; see also ▶ Chap. 59, "Critical Appraisal of Quantitative Research"). Systematic reviews may not necessarily be undertaken to evaluate effectiveness, but rather explore areas of process; therefore, other frameworks may be useful in assessing study quality (e.g., a review exploring implementation of interventions may use the RE-AIM framework, such as in a systematic review of physical activity interventions in practice for adults with type 2 diabetes (Matthews et al. 2014)).

10.4 When Risk of Bias Measurement May Need to Be More Flexible

There are some cases where rigid risk of bias tools can be too strict and result in down-rating of evidence from particular types of studies, such as natural experiments (Humphreys et al. 2017) and "real-world" community-based interventions. For example, in the case of community-based interventions, flexibility in delivery of interventions may be necessary to suit a particular site compared to another (e.g., what might work in one school may not work in another school due to differences in structures in place at each location). To ensure success, there may need to be some flexibility in how an intervention is delivered and strict measures of intervention fidelity will rate such studies including flexibility as low quality. In this instance, it may be appropriate to adjust an existing risk of bias tool or create an appropriate indicator of quality applicable to this type of study, bearing in mind, as mentioned earlier, that the validity of tools will be affected if they are adjusted. With natural experiments, researchers are not in control of the intervention itself and steps such as blinding and allocation concealment may therefore be impossible (Humphreys et al. 2017). It is, therefore, recommended that authors interpret the ratings from risk of bias tools cautiously.

11 Meta Analyses

11.1 What Is Meta-analysis?

A meta-analysis is when similar outcome data are pooled statistically from across more than one study in an attempt to provide a more precise, larger, effect estimate than relying on the findings of a single study. The benefit of using such an analysis is that a single study may report nonsignificant findings, but when data are pooled across several studies, statistical power to detect small effects is increased and overall significant effects may be identified. Therefore, there is lower probability of missing small effects than if exploring one or only a small number of studies. Subanalyses can also be undertaken to explore heterogeneity and variability in study results in more depth. For example, it might be of interest to explore differences in effect sizes based on total intervention length or the intensity of an intervention (frequency of delivery per week and individual session duration), such as in a review of physical activity and sedentary behavior interventions in young people with type 1 diabetes (MacMillan et al. 2014). Subanalyses can also be useful as sensitivity analyses by exploring differences in findings when data are analyzed based on study quality (e.g., comparing effect sizes in studies with low or high overall risk of bias ratings).

A meta-analysis involves calculating the effect size for individual studies, by utilizing data such as mean and standard deviation (to calculate the standardized mean difference), correlation coefficients, and risk or odds ratios for each group, and then calculating an overall summary effect size by pooling effect size data from across studies in the review. Calculations may be necessary to transform data into a common metric when the same type of data is not consistently reported across studies. By using standardized outcome data, if the same outcome has been measured in different ways (e.g., different questionnaires or scales are used to measure an outcome), these data can then be pooled. In some instances, it may not be appropriate to combine data on a particular outcome. For example, consider physical activity. There are many types (e.g., aerobic activities, muscular strength activities, flexibility), dimensions (e.g., frequency of sessions/activities, intensity, duration of sessions/activities), and resulting outcomes that could be measured and reported in an intervention study (e.g., step counts per day, maximal oxygen consumption, one repetition maximum during a deadlift, upper body flexibility). Conceptually and methodologically, it would not make sense to pool data for some of these outcomes together (e.g., step counts are very different to upper body flexibility). Therefore, the researcher needs to exercise judgment on what makes conceptual and methodological sense to include in a meta-analysis.

Homogeneity of outcome data should be explored to guide the type of model to use in a meta-analysis. Chi-square tests can be used to statistically explore homogeneity. If there is no heterogeneity across the results from studies (other than that from sampling), then a fixed effects model should be used to combine effect sizes. If heterogeneity is present, then a random effects model should be utilized.

Forest plots can be used to graphically display effect sizes across studies in a meta-analysis. An example of a forest plot produced in Cochrane RevMan Software is provided in Fig. 2.

Data for each study included in the meta-analysis are included alongside a square to indicate the effect size and a confidence interval (a line) to display the level of uncertainty around that effect (in this case in a random-effects model). The overall effect size for pooled data is shown by the diamond displayed underneath the squares for each study (in the above example the overall effect size is -2.21

(95% CI, −2.84 to −1.59)). In Fig. 2, because the confidence interval for the overall effect does not cross the line of no effect, this indicates that, overall, there is a significant effect in favor of the intervention. The test for homogeneity suggests that a random effects model was suitable because heterogeneity was significant, ($p = 0.040$, $I^2 = 63\%$).

Funnel plots can be used to assist in identifying the chances of publication bias and can also be created in software such as RevMan. Funnel plots are scatter plots of study sample size (as an indicator of study precision) against effect size. The underlying premise of a funnel plot is that as sample size increases, variation in effect sizes will decrease and, therefore, effect estimates are more precise, forming an inverted funnel shape. If effect sizes do not form a typical funnel shape, then this would suggest publication bias, and potentially other biases too. Publication bias is when studies finding negative results (nonsignificant effects) remain unpublished due to authors not submitting their work to journals, or journals not publishing such findings. Publication bias can lead to an overestimation of overall intervention effectiveness. To avoid publication bias, researchers can search for relevant articles in nonpeer reviewed sources, including searching through trial and other study registers, and include non-English language studies. Additionally, there are statistical methods that can be used to explore and adjust for publication bias, but these tests are low-powered.

12 Creating a Narrative and Interpreting Findings

Most reviews will incorporate either a mixed narrative synthesis with a meta-analysis or, where a meta-analysis is not appropriate, solely a narrative descriptive synthesis. A narrative synthesis is a useful way of qualitatively summarizing the results of a quantitative systematic review when the studies included in the review are sufficiently heterogeneous, such that a meta-analysis is not possible. As mentioned earlier, when the evidence base is broad and studies are diverse (in their designs, outcome measures, and populations), a narrative synthesis can be a more appropriate way of summarizing patterns in the data, and providing an overall picture (or story) of the results.

Where a meta-analysis usually follows a strict framework, a narrative synthesis can be more subjective, which can be a disadvantage. One way to conduct a narrative synthesis is to summarize patterns of information using the number of studies reporting particular trends. For example, you may report that 10 studies showed a beneficial effect of an intervention, whilst 4 studies reported detrimental results. However, this method does not take into account the methodological quality of the studies and may lead to an inaccurate representation of the overall results due to bias. An alternative method involves weighting higher quality studies that have a low risk of bias in a qualitative synthesis. This might involve including a separate paragraph synthesizing the high-quality studies, which can then be used to make recommendations (see Steiner et al. 2017 for an exemplar) or reporting the findings in the context of the risk of bias (see MacMillan et al. 2017 for an exemplar). The FORM

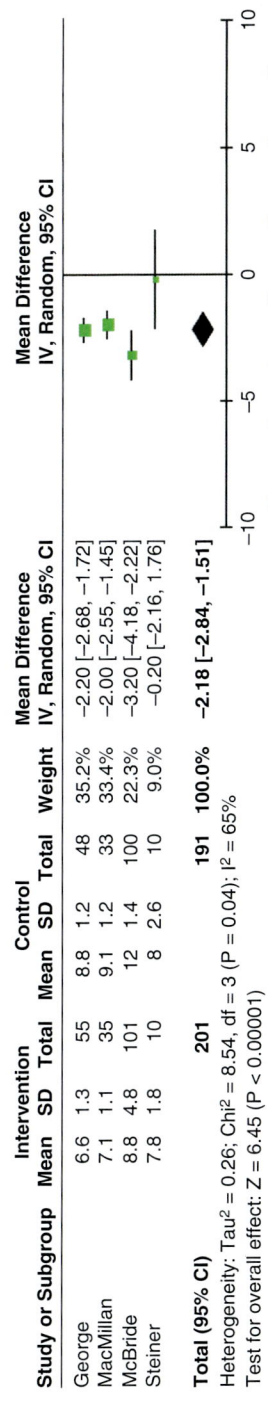

Fig. 2 Example of a forest plot

framework is a guide for formulating and grading recommendations for clinical guidelines, and can be applied to structure narrative syntheses and recommendations in systematic reviews (Hillier et al. 2011). The FORM framework comprizes five components:

1. *Evidence base*, which covers the quality and quantity of evidence
2. *Consistency* of results across studies
3. *Clinical impact* of the work on the target population
4. *Generalizability* of the findings to the research question being asked
5. *Applicability* of the pattern of results to the population

By applying the FORM framework to a narrative synthesis, it is possible to make a clear recommendation about the pattern of results from a systematic review (e.g., that the intervention should be adopted, or results are mixed and further research is required). (For more information on reporting quantitative research findings, see ▶ Chap. 56, "Writing Quantitative Research Studies").

Similar to narrative syntheses, there is no consensus on one standardized method for undertaking a meta-synthesis of qualitative studies (Lucas et al. 2007), although several frameworks have been proposed (Dixon-Woods et al. 2006). For example, a thematic approach groups or clusters data into themes, where a textual narrative synthesis involves describing and summarizing the major aspects of the study (e.g., study characteristics, results, similarities and differences, and study quality) (Lucas et al. 2007). (For more information on conducting and reporting a meta-synthesis, see ▶ Chap. 45, "Meta-Synthesis of Qualitative Research," and Writing/Reporting Qualitative Research Studies".)

13 Conclusion and Future Directions

To summarize, systematic reviews can provide high-level evidence to inform decision making in policy and practice, particularly when conducted rigorously. Systematic reviews synthesize evidence, both quantitatively and qualitatively, across several studies focused on the same research topic. The steps within a systematic review are the development of a protocol including a detailed search strategy, database searching to identify citations meeting review inclusion criteria, screening of titles and abstracts and then full articles against inclusion and exclusion criteria, assessment of study quality using a risk of bias tool, pooling of quantitative outcome data in a meta-analysis where relevant and possible, and summarizing and interpretation of findings across included studies narratively.

With the ever-increasing quantity of health and social research being published, and the drive for evidence-based practice, special attention should be paid to the reporting of literature summaries. Rigorously conducted systematic reviews are, therefore, a crucial approach to ensuring that future policy and practice are guided by the most recent, high- quality evidence available.

References

Barbour RS. Checklists for improving rigour in qualitative research: a case of the tail wagging the dog? BMJ. 2001;322(7294):1115–7.

Butler A, Hall H, Copnell B. A guide to writing a qualitative systematic review protocol to enhance evidence-based practice in nursing and health care. Worldviews Evid-Based Nurs. 2016; 13(3):241–9.

Cook DJ, Mulrow CD, Haynes RB. Systematic reviews: synthesis of best evidence for clinical decisions. Ann Intern Med. 1997;126(5):376–80.

Dixon-Woods M, Bonas S, Booth A, Jones DR, Miller T, Sutton AJ, ... Young B. How can systematic reviews incorporate qualitative research? A critical perspective. Qual Res. 2006; 6(1):27–44. https://doi.org/10.1177/1468794106058867.

Greenhalgh T. How to read a paper: the basics of evidence-based medicine. 4th ed. Chichester/Hoboken: Wiley-Blackwell; 2010.

Hannes K, Lockwood C, Pearson A. A comparative analysis of three online appraisal instruments' ability to assess validity in qualitative research. Qual Health Res. 2010;20(12):1736–43. https://doi.org/10.1177/1049732310378656.

Higgins JPT, Green S. Cochrane handbook for systematic reviews of interventions (Version 5.1.0 [updated March 2011]). The Cochrane Collaboration; 2011. http://handbook-5-1.cochrane.org/

Higgins JPT, Altman DG, Gøtzsche PC, Jüni P, Moher D, Oxman AD, ... Sterne JAC. The Cochrane Collaboration's tool for assessing risk of bias in randomised trials. BMJ. 2011;343. https://doi.org/10.1136/bmj.d5928.

Hillier S, Grimmer-Somers K, Merlin T, Middleton P, Salisbury J, Tooher R, Weston A. FORM: an Australian method for formulating and grading recommendations in evidence-based clinical guidelines. BMC Med Res Methodol. 2011;11:23. https://doi.org/10.1186/1471-2288-11-23.

Humphreys DK, Panter J, Ogilvie D. Questioning the application of risk of bias tools in appraising evidence from natural experimental studies: critical reflections on Benton et al., IJBNPA 2016. Int J Behav Nutr Phys Act. 2017;*14*(1):49. https://doi.org/10.1186/s12966-017-0500-4.

King R, Hooper B, Wood W. Using bibliographic software to appraise and code data in educational systematic review research. Med Teach. 2011;33(9):719–23. https://doi.org/10.3109/0142159x.2011.558138.

Koelemay MJ, Vermeulen H. Quick guide to systematic reviews and meta-analysis. Eur J Vasc Endovasc Surg. 2016;51(2):309. https://doi.org/10.1016/j.ejvs.2015.11.010.

Lucas PJ, Baird J, Arai L, Law C, Roberts HM. Worked examples of alternative methods for the synthesis of qualitative and quantitative research in systematic reviews. BMC Med Res Methodol. 2007;7:4–4. https://doi.org/10.1186/1471-2288-7-4.

MacMillan F, Kirk A, Mutrie N, Matthews L, Robertson K, Saunders DH. A systematic review of physical activity and sedentary behavior intervention studies in youth with type 1 diabetes: study characteristics, intervention design, and efficacy. Pediatr Diabetes. 2014;15(3):175–89. https://doi.org/10.1111/pedi.12060.

MacMillan F, Karamacoska D, El Masri A, McBride KA, Steiner GZ, Cook A, ... George ES. A systematic review of health promotion intervention studies in the police force: study characteristics, intervention design and impacts on health. Occup Environ Med. 2017. https://doi.org/10.1136/oemed-2017-104430.

Matthews L, Kirk A, MacMillan F, Mutrie N. Can physical activity interventions for adults with type 2 diabetes be translated into practice settings? A systematic review using the RE-AIM framework. Transl Behav Med. 2014;4(1):60–78. https://doi.org/10.1007/s13142-013-0235-y.

Moher D, Schulz KF, Altman DG. The CONSORT statement: revised recommendations for improving the quality of reports of parallel group randomized trials. BMC Med Res Methodol. 2001;1:2. https://doi.org/10.1186/1471-2288-1-2.

Moher D, Liberati A, Tetzlaff J, Altman DG. Preferred reporting items for systematic reviews and meta-analyses: the PRISMA statement. PLoS Med. 2009;6(7):e1000097. https://doi.org/10.1371/journal.pmed.1000097.

Moher D, Shamseer L, Clarke M, Ghersi D, Liberati A, Petticrew M, et al. Preferred reporting items for systematic review and meta-analysis protocols (PRISMA-P) 2015 statement. Syst Rev. 2015;4:1. https://doi.org/10.1186/2046-4053-4-1.

Mulrow CD, Cook DJ, Davidoff F. Systematic reviews: critical links in the great chain of evidence. Ann Intern Med. 1997;126(5):389–91.

Peters MDJ. Managing and coding references for systematic reviews and scoping reviews in EndNote. Med Ref Serv Q. 2017;36(1):19–31. https://doi.org/10.1080/02763869.2017.1259891.

Steiner GZ, Mathersul DC, MacMillan F, Camfield DA, Klupp NL, Seto SW, ... Chang DH. A systematic review of intervention studies examining nutritional and herbal therapies for mild cognitive impairment and dementia using neuroimaging methods: study characteristics and intervention efficacy. Evid Based Complement Alternat Med. 2017;2017:21. https://doi.org/10.1155/2017/6083629.

Sterne JA, Hernán MA, Reeves BC, Savović J, Berkman ND, Viswanathan M, ... Higgins JP. ROBINS-I: a tool for assessing risk of bias in non-randomised studies of interventions. BMJ. 2016;355. https://doi.org/10.1136/bmj.i4919.

Tong A, Sainsbury P, Craig J. Consolidated criteria for reporting qualitative research (COREQ): a 32-item checklist for interviews and focus groups. Int J Qual Health Care. 2007;19(6):349–57. https://doi.org/10.1093/intqhc/mzm042.

Tong A, Palmer S, Craig JC, Strippoli GFM. A guide to reading and using systematic reviews of qualitative research. Nephrol Dial Transplant. 2016;31(6):897–903. https://doi.org/10.1093/ndt/gfu354.

Uman LS. Systematic reviews and meta-analyses. J Can Acad Child Adolesc Psychiatry. 2011;20(1):57–9.

ns to # Content Analysis: Using Critical Realism to Extend Its Utility

47

Doris Y. Leung and Betty P. M. Chung

Contents

1	Introduction	828
2	What Is Content Analysis?	828
3	Types of Content Analysis	829
4	Historical Overview of Content Analysis	830
5	The Philosophy of Critical Realism	832
6	The Roots of Critical Realism	833
7	Critical Realism as a Stance in Content Analysis	836
8	Example of the Use of Critical Realism	837
9	Conclusion and Future Directions	839
References		840

Abstract

Content analysis (CA) has become one of the most common forms of data analysis, but it is often criticized for a lack of rigor and limited utility of its findings. We define CA and describe its general procedures and the three most frequently used forms of CA. Next, we review the history of CA leading up to its current popularity within diverse disciplines, including social science and healthcare disciplines. Its origins highlight concerns about researchers' motivations underlying their interpretations of communications. In response, improved transparency and the application of CA in understanding underlying connotation in communications have furthered its evolution. CA can now be located on a

D. Y. Leung (✉)
School of Nursing, The Hong Kong Polytechnic University, Hong Kong, SAR, China

The Lawrence S. Bloomberg Faculty of Nursing, University of Toronto, Toronto, ON, Canada
e-mail: doris.leung@polyu.edu.hk

B. P. M. Chung
School of Nursing, The Hong Kong Polytechnic University, Hong Kong, SAR, China
e-mail: betty.chung@polyu.edu.hk

© Springer Nature Singapore Pte Ltd. 2019
P. Liamputtong (ed.), *Handbook of Research Methods in Health Social Sciences*,
https://doi.org/10.1007/978-981-10-5251-4_102

continuum representing depth of interpretation, from surface description of phenomena to the uncovering of deeper meanings. We explore how CA may be used to uncover deeper underlying meanings and answer questions concerning how social relations, in connection with their context, affect outcomes such as individual behavior. By investigating deeper meanings, researchers can explore the core of the phenomenon and posit explanations of why the phenomenon is as it is. Finally, we argue that adopting an explicit philosophical orientation for inquiry will improve rigor and enhance practical utility of findings. We examine the philosophical and theoretical position of critical realism with CA. We then provide an example to illustrate the use of critical realism and outline the position's key aspects to consider in future directions of CA.

Keywords

Content analysis · History · Qualitative description · Critical realism · Rigor · Qualitative research utility

1 Introduction

This chapter focuses on a comprehensive understanding of content analysis (CA) and the ways in which CA can be used to investigate phenomena. The chapter opens with a definition of CA and a description of its process and then examines different forms of CA that reflect a continuum from shallow to deep interpretation. We review CA's history up to its current popularity as an analytic method across a variety of disciplines. To enhance its potential, we propose that the use of critical realism in CA can extend interpretation beyond surface description to a deeper understanding of how and why a phenomenon occurs. This approach will be illustrated using data gathered from a meta-synthesis focused on identifying influences on decision-making concerning transitioning patients to end-of-life care in intensive care units.

2 What Is Content Analysis?

We adopt a broad definition of CA that is applicable to both qualitative and quantitative research in areas ranging from social psychology to healthcare. According to Hsieh and Shannon (2005), CA is the process in which a researcher interprets the meaning or usage of written or visual data. These interpretations are organized into categories or themes using everyday language (Sandelowski 2010). This technique can be used to answer superficial questions of perception of a phenomenon, such as *What is it*? But it can also address more critical questions concerning the phenomenon's meaning: *What is it about? How is it happening? Where is it happening?* or *What may be its consequences?* (Sandelowski 2000). Addressing questions of meaning provides a richer picture of the phenomenon. With answers to these questions, the researcher can compare the data to existing evidence on the phenomenon, which will further shape the researcher's initial interpretations (Hsieh and Shannon 2005).

CA uses a systematic set of analytical steps. First, researchers examine the data about the phenomenon of interest in small chunks or phrases with meaning. Researchers give each chunk of data a label, or code, that reflects its meaning. They compare and contrast coded chunks against each other and group similar chunks to form conceptual categories. These categories reflect meanings and inferences about different aspects of the phenomenon (Hsieh and Shannon 2005). The derived categories are then compared to theories about the phenomenon in existing literature from the same discipline or across disciplines. Finally, a summary of the findings and their implications for practice, education, and future research are presented (Hsieh and Shannon 2005).

The way in which researchers apply the analytical steps of CA depends on previous evidence and theories. If existing literature about the phenomenon is scarce or inconsistent, then researchers may use an inductive process, whereby findings are derived from the collected data and then combined to form broader conceptual meanings or general statements (Elo and Kyngäs 2008, p. 109). Alternatively, if researchers have a theory or conceptual framework in mind, they may apply a deductive process, whereby the existing theoretical constructs are compared to the data observed from a group of participants from a specific context; this process validates the credibility of the theory (Elo and Kyngäs 2008, p. 109).

In social inquiry, researchers use CA to achieve one of two main purposes: (a) to uncover how a phenomenon is socially arranged and how it implicitly or explicitly works or (b) to describe how participants explain their behavior, to generate a cumulative understanding of the phenomenon (Miles et al. 2014). However, the degree to which the analysis accomplishes this goal varies widely depending on the research aim and questions. The purposes of CA are diverse, and while this can create confusion in how it should be conducted, its flexibility makes it a valuable method with which to get at multiple realities (White and Marsh 2006).

3   Types of Content Analysis

The researchers' research purpose or research question determines the use of one of three approaches to CA: summative, directed, or conventional (Hsieh and Shannon 2005). These approaches differ substantively in the depth of interpretation they offer and, thus, can be distinguished along a continuum of interpretation from shallow description to multifaceted construction of the phenomenon's meaning, processes, and mechanisms. We provide a brief overview of each approach below.

The first approach is summative CA, which draws more from quantitative than qualitative techniques. The purpose is to explore the contextual use of certain words in the text (Hsieh and Shannon 2005, p. 1283). The analysis, often referred to as *manifest* interpretation because it focuses on readily apparent meaning, is conducted by counting the frequency of particular words or phrases used by speakers in specific contexts. Further identification of the contexts in which the key terms are used – for example, whether use of particular terms for the same concept differs between patients versus healthcare providers or across age groups – informs the usage of these terms (Hsieh and Shannon 2005). These findings may be reported using descriptive statistics (e.g., proportions, means). In addition, additional interpretation may involve *latent* or underlying meanings of key words or phrases. The findings can offer basic insights into the range of meanings conveyed about the topic and its key messages (Hsieh and Shannon 2005).

The second approach is directed CA (Hsieh and Shannon 2005) or conceptual analysis (Wilson 2016). Its purpose is to add credibility to or conceptually extend a theoretical framework or theory (Hsieh and Shannon 2005, p. 1281). Initial codes and categorizations are derived from existing literature or a theoretical framework and applied deductively to code the data. Data that do not reflect the concepts within the a priori coding scheme are inductively assigned a new code through manifest or latent interpretations. Researchers, thus, report on categories that validate the theoretical framework but also on new concepts that further expand or refine the theoretical framework or phenomenon (Hsieh and Shannon 2005).

The third approach is conventional CA, which is a strictly descriptive approach using both manifest and latent inferences to understand a phenomenon. This is the most inductive type of CA and is used when limited research or theory exists about the phenomenon. Researchers assign codes to label key thoughts or ideas. Related codes are then grouped into meaningful categories, and initial categories help organize subsequent codes into meaningful clusters (Hsieh and Shannon 2005). This type of CA provides a much more comprehensive picture of the phenomenon, through new insights exclusively grounded in the data, than the other two types (Hsieh and Shannon 2005).

Next, we provide an overview of the history of CA, which originated from concerns about motivations underlying public communications but also motivations underlying people's interpretations of these communications. Improvements in transparency and the application of CA to understanding underlying meanings and motivations in communications have led to progress in the evolution of CA, and it has become a popular methodology across disciplines. Despite such progress, criticism is still frequently leveled at CA as an atheoretical and often poorly conducted methodology (Sandelowski 2010, pp. 79–80).

4 Historical Overview of Content Analysis

CA originated in the eighteenth century in Sweden, from political debate about meaning in journalistic writing that went beyond straight reporting of facts. Beginning early in the twentieth century, the wide circulation of newspapers led to the first kind of CA, known as quantitative newspaper analysis. Journalists analyzed how the news was reported to support their arguments and provided commentary on public sentiment about an issue. It was then that Schools of Journalism studied how public sentiment could be quantified to reveal what was important to the public (Krippendorff 2013).

However, as journalists became adept at reporting public sentiment, it became apparent that the analytical processes they used were not transparent, and people began to question how representative journalistic commentaries were of the general population's opinions (Krippendorff 2013). Questions about the journalists' attitudes and biases arose, which others felt reflected hidden agendas, such as motives of profit. In response, psychologists investigated the psychological relation between newspapers and their readers to form a systematic assessment of biases in newspapers (Allport and Faden 1940). This was published in the Public Opinion Quarterly, in a document entitled *The Psychology of Newspapers: Five Tentative Laws* (Allport and Faden 1940). These "laws" were psychological principles aimed at assisting journalists in creating guidelines to provide fair and objective communication of information to readers (Allport and Faden 1940). This was the first attempt to create transparency and standards in how CA was conducted. However, social scientists continued to question the representativeness of findings in CAs, raising such issues as the following: How were constructs theoretically motivated? How were ideas operationally defined? How did stereotypes occur? What social values did news ideas represent? (Krippendorff 2013).

In 1941, CA came to be known as the systematic analysis of mass communications such as propaganda. In particular, analysis of propaganda during World War II led to the awareness that underlying meanings, or latent content, communicated motivations and political interests (George 1959). Indeed, analysis of speeches given by the Nazi propagandist Joseph Goebbels enabled the United States' Federal Communication Commission to predict major Nazi military and political campaigns (Krippendorff 2013). Thus, people began to realize how words not only informed but also emotionally aroused and led to predictable responses in individuals. Moreover, the understanding of legitimate meaning required more than quantitative analysis, such as counting the frequency of words, but also qualitative understanding of the context connected to communicators' motivations (Krippendorff 2013). Hence, a shift from quantitative to qualitative CA began to occur.

From the 1960s to 1990s, CA was taken up by various disciplines – sociology, psychology, political science, literature, anthropology, library and information studies, linguistics, medicine, and nursing, to name a few – for intensive studies of individuals in naturalistic settings (Elo and Kyngäs 2008). According to Miles et al. (2014), literature from these disciplines followed a trend from description of manifest content to include: (1) inferring a phenomenon's antecedent conditions, (2) identifying symbolic or latent meaning (Krippendorff 2013), and (3) predicting stable social patterns, such as behaviors and relationships, from in-depth knowledge of the phenomenon (Miles et al. 2014). During the 1990s, CA focused on how messages were communicated, such as in the use of figurative language or illustrations. From patterns in communications, researchers inferred communicators' motives and their effects on others (White and Marsh 2006). In this way, researchers using CA could explore the social world not only to understand what was being said but also to interpret the *how* and explain the *why* of social phenomena (White and Marsh 2006; Sandelowski 2010).

In 2000, Sandelowski energized healthcare researchers, particularly nurses, to consider the utility of CA as basic or fundamental to descriptive research. She asserted that, although it was less interpretive compared to other methods such as interpretive description, CA was useful in that it could adopt any theoretical basis, such as the disciplinary or professional understanding of a phenomenon (Sandelowski 2010). In this way, CA was pragmatic and flexible and could enable inference of meaning in the most basic sense, "close to their data and to the surface of words and events" (Sandelowski 2000, p. 334), but also from experience, previous research, or existing theories (White and Marsh 2006). Further, CA allowed triangulation of qualitative and quantitative data to enhance validity and reliability of the social patterns being posited (White and Marsh 2006).

Overall, the current popularity of CA across disciplines is attributable to its capacity not only to produce findings close to the data as given but also to allow for inferences about how and why events or experiences happen. Thus, unlike earlier applications of CA that focused on the superficial purpose of quantifying qualitative data (e.g., frequency of words), current trends extend CA to examining context-dependent factors that may explain patterns of the phenomenon. This aligns with the original objectives for CA, to explain the nature of the social world, and how it came to be that way.

Although the methodology of CA continues to improve, the problem of representation of phenomena exists. What theoretical basis guides interpretation? How can researchers' assumptions of the phenomenon be made more explicit? Even more worrisome is Sandelowski's (2010) suggestion that interpretation could be reduced to "no interpretation" other than a "presentation of the facts of the case" (p. 79). Consequently, researchers may skim the surface in their interpretations and see phenomena as much less complex than it is.

We suggest that the lack of in-depth interpretation may be attributed to insufficient attention to adopting a theoretical position in CA. Representation of any phenomena requires a theoretical orientation to inquiry. A theoretical orientation positions researchers to know how to identify data that are relevant to the phenomenon of interest and to differentiate the relationships among different data elements. This is especially important when CA is used to understand complex social and healthcare problems, which requires looking at these particular phenomena from many different perspectives. To present one way to meet this goal and enhance the practical utility of descriptive research, we recommend conducting CA from the philosophical position of critical realism.

5 The Philosophy of Critical Realism

Critical realism is a philosophical theory that has been applied within social science, the systematic study of society and social relationships. The theory essentially suggests that, within both the natural and the social world, there are unobservable objects and events that have causal properties and, thus, cause observable events (Bhaskar 1998). Thus, the realist stance is that such entities may exist independent of our awareness of them. How reality comes to be understood epistemologically is relative to how we subjectively perceive it. Critical realists believe that perceptions of reality can change depending on unobservable subjective factors such as our individual ideology and value commitments, mental state, and situation (Maxwell and Mittapalli 2010).

Critical realism in the social realm focuses on the qualitative nature of social objects and their relations with social outcomes of interest. In particular, it seeks to understand the underlying unobservable mechanisms that may or may not cause certain social outcomes (Elder-Vass 2010). Mechanisms refer to processes of interactions among social objects, structures, events, and so on that result in observed outcomes. They can be naturalistic or programmatic and work in different ways under different circumstances to generate diverse outcomes (Maxwell and Mittapalli 2010). The nature of mechanisms, as well as the objects from which they arise, is thought to be more dynamic in the social world than in the natural world (Elder-Vass 2010).

One significant application of critical realism has been to understand the influence of social structures, which refer to interactions within social groups (e.g., organizations, families) that possess historical or social properties or relations that govern the behavior of their members (Elder-Vass 2010; Cruickshank 2012). Thus, defined groups of people have causal properties (e.g., social norms), and these, as unobservable mechanisms, may condition individuals' thoughts and behaviors, though not necessarily determine them (Elder-Vass 2010; Cruickshank 2012). Knowledge production also occurs in a contextual process (Nairn 2012). Mechanisms in this case can induce normative or ritualistic ways of knowing and being in their local contexts. Social norms tend to be relatively stable for a period, so we can make assumptions about how they may influence events within that period and context (Elder-Vass 2010). An example is how we understand the way to behave in elevators: facing forward to the door. If someone stood facing the opposite way in an elevator, towards the back, this behavior would reflect resistance to conformity to social norms and elicit discomfort in others in the elevator. Thus, research questions applying critical realism attempt to uncover the underlying mechanisms that cause the phenomenon of interest to be as it is observed locally: *How, why, for whom, and under what circumstances does the phenomenon occur?*

6 The Roots of Critical Realism

The philosophy of critical realism originated in 1996 with Roy Bhaskar (b.1944-), a British philosopher who was concerned with the practice and philosophy of social theory. Since then, an increasing number of disciplines have applied critical realism to their research (Easton 2002).

Bhaskar (1998) viewed reality across three strata: (1) the empirical, (2) the actual, and (3) the real. The closest stratum to our experience is the empirical, referring to what we observe about the event and, most often, what we tell each other about the event. The actual is what objectively occurs, regardless of our perception of it and our knowledge of its occurrence; these are the events that are influenced by unobservable mechanisms. Finally, the real refers to the mechanisms that are involved in causing changes in actual events and that are the focus of understanding in critical realism.

We illustrate this stratified reality using the following example. Suppose you see a man having difficulty breathing. He tells you that he started to feel light-headed and experienced pain in his chest while having lunch. He also tells you that he feels nauseous and is tasting vomit. In this example, what you observe (i.e., a man having shortness of breath) and what the man tells you about his symptoms, including what he was doing when these symptoms began, represent the empirical reality, or the storied construction of experience. The actual reality comprises the material things or events that actually exist or happen – in this case, likely some kind of medical event. Since no measureable medical data have been gathered from the man, such as an electrocardiogram, most individuals would not perceive this actual reality until such evidence is presented. Lastly, the real encompasses the possible causal mechanisms of the event, such as the man's unhealthy lifestyle habits or genetic predisposition to cardiac events. We cannot observe those mechanisms that contributed to the medical event because they are part of a complex process that operates outside of our awareness (Elder-Vass 2010).

The three strata of reality described in critical realism are partially compatible with three existing philosophies and theoretical positions about knowledge: positivism, constructionism, and critical theory (Elder-Vass 2010). However, critical realism also represents a movement away from all three philosophies and, thus, differs from them on significant elements. Each will be described below with reference to our medical-event example.

Positivism reflects the perspective of basic science, in that it views reality as objective and as that which is directly observable with the senses or with instruments; nothing that is not directly observable exists. Positivists focus on identifying consistent empirical relationships, typically through controlled experiments; such relationships are then assumed to represent law-like causal relationships. From this perspective, knowledge must be empirically derived (Philosophical Foundations 2006; see also ▶ Chap. 9, "Positivism and Realism"). However, critical realists oppose regularity in causation but rather suggest that patterns may be produced from unobservable mechanisms, such as mental states or group attributes (Maxwell and Mittapalli 2010). In the above example, we suspect that the man is having a heart attack. We infer its presence from empirical data gathered (e.g., what the man is telling you). We could further determine if it is a heart attack by conducting empirical investigations, such as conducting electrocardiograms or blood tests. This type of knowledge reflects positivism. However, we could also infer underlying explanations that are not objectively knowable, such as whether he has a hidden motive to approach you. So, while critical realism supports material or measurable events as elements of the nature of knowledge (ontology), its fundamental difference to positivism lies in its epistemologically relativistic viewpoint (Maxwell and Mittapalli 2010), which we discuss momentarily.

Constructivism is a philosophy of social science that opposes the positivist philosophy in that it views all reality as subjective. The perspective focuses on understanding and interpreting the meanings of human behavior and experience that have developed socially and that are dependent on time and context. Like critical realists, constructivists take a skeptical attitude towards reality; they believe that reality is socially constructed rather than objectively observed and examine how narratives are constructed within a relational context (Cruickshank 2012). In understanding complex social situations, a philosophy of constructivism is helpful because, like critical realism, it espouses the epistemologically relativistic viewpoint that the reality of a situation is relative to the person's motives and social norms within the local context (Cruickshank 2012). Hence, the person and reality are intertwined, and what is happening is connected to the context (see also ▶ Chap. 7, "Social Constructionism"). In the example of the man's medical event, when a physician intervenes as we are helping the man, we may defer to the physician because we, as a society, agree that physicians have the knowledge and authority to best assess and treat medical events. This belief is socially constructed, and it influences our actions because the practice of medicine is sanctioned by the medical system. Further, social structures, such as the medical system, and social norms possess power because we repeatedly reinforce and reproduce norms. However, critical realism differs in that it posits that we are affected by our social environment but can act to resist or transform it. Thus, we are not fully determined by social conditioning, as constructivism at a macro-level suggests, but still possess some autonomy over our agency or actions (Elder-Vass 2010; Cruickshank 2012).

Critical theory seeks to analyze what is in the interests of society and the societal assumptions about what is normal (i.e., ideology and cultural norms) that exist independently of individuals (Kellner and Roderick 1981). The theory recognizes that meanings inherent in the language we use shape groups of people within a specific time, culture, and societal structure. In our medical example, when the physician intervenes, we tend to talk and behave in a way that acknowledges his/her authority, and the physician tends to accept this responsibility and act on it. This is because we have been socialized within our society to the role that physicians play and its significance, as well as to the appropriate ways to relate to each other in these roles (e.g., deferring medical decision-making to the physician's expertise). This type of knowledge is the focus of critical theory (see also ▶ Chap. 8, "Critical Theory: Epistemological Content and Method").

Within social inquiry, both critical realism and critical theory, we (1) try to understand naturally occurring behavioral patterns that are created by social structures and mechanisms and that promote change in social systems (Elder-Vass 2010), (2) believe that language has pre-existing origins and implicit meaning, and (3) consider the broader social consequences of processes such as language and social relations. Language possesses causative properties, and both critical realism and critical theory are concerned with how language reinforces social structures or triggers other social structures. Unlike critical theorists, critical realists assume that social structures and mechanisms change over time and contexts and that these changes can lead to different outcomes (Elder-Vass 2010). In addition, critical realists take into account how difficult it can be for changes to occur in the face of shifting social forces in those who are ingrained in their ways of being; critical theorists do not consider this issue (Elder-Vass 2010).

Our understanding of how mechanisms trigger social structures to change outcomes provides us with a fuller understanding of the structures themselves (Elder-Vass 2010). Sometimes, opposing forces can work persistently within a structure. For example, findings from a literature review exploring decision-making in intensive care units indicated that conditions supported both resistance to and initiation of palliative end-of-life care for patients with a high risk of mortality post-ICU (Leung et al. 2016). We discuss this study later to illustrate the use of critical realism.

7   Critical Realism as a Stance in Content Analysis

Researchers may use CA following critical realism if they are interested in understanding how emergent causal properties occur in a social situation. Of particular interest for critical realists is how different patterns within social structures facilitate or block changes or outcomes in a naturalistic system, with broad social consequences (Pawson et al. 2005; Elder-Vass 2010). In addition, critical realism allows for different types of data to be used in the same study because this perspective accepts multiple strata of reality – the empirical, the actual, and the real (Nairn 2012; Porter and O'Halloran 2012). Critical realism also does not center on specific texts or numbers but rather can utilize quantitative and qualitative data to produce a unified interpretation. It can, therefore be used when different research approaches are concurrently employed, as in mixed-methods studies (Maxwell and Mittapalli 2010).

Commonly, CA is applied in qualitative descriptive studies. Qualitative description is a distinctive qualitative methodology suitable for variants of CA (Sandelowski 2010). This is because qualitative description reflects a theoretical position, known as the factist perspective, that limits the degree of interpretation to addressing material or measurable constructs that have predictive power (Sandelowski 2010). However, we agree with Sandelowski (2010) when she states that qualitative description should not be viewed as a "fixed way of ordering the world of inquiry" (p. 80). Rather, the boundaries of what we believe about reality are permeable. This means that qualitative description does not need to remain static. It may act as a foundation from which to extend the degree of meaning in interpretation further along the interpretive continuum, from shallow to deep interpretation. We suggest that critical realism can provide much-needed flexibility in understanding other types of realities by permitting us to access different types of knowledge. Its flexibility in CA draws not only from manifest meaning but also from the latent meaning of the context of the situation (Mayan 2009). The key is to look for and explore "what is into, between, over, and beyond text" (Sandelowski 2010, p. 78).

In summary, we propose that qualitative descriptive research applying critical realism to direct CA provides several advantages in getting at causal explanations about phenomena. Critical realism allows researchers to (1) explore a phenomenon from multiple realities with data collected using different methods; (2) study emergent causal properties in open, naturalistic contexts; (3) investigate explanations of consequences and outcomes that occurred; (4) acknowledge the ways individuals can actively resist or defy social structures, such as by engaging in protests or avoiding responses to requests; and (5) acknowledge opposing forces between individuals and societal structures (Elder-Vass 2010). We now turn to an example of the use of critical realism in CA.

8 Example of the Use of Critical Realism

To illustrate how researchers can use critical realism with methods of CA, we describe the methodology from a published paper, *Transitions to end-of-life care for patients with chronic critical illness: A meta-synthesis*, by the first author of this chapter (Leung et al. 2016). While a detailed account of the study can be found in the publication, we provide a brief background and synopsis of the purpose and methods of the study.

Despite the high rates of morbidity and mortality post-discharge from intensive care units (ICUs) in patients with chronic critical illness, palliation is often not augmented during their treatment (Camhi et al. 2009). We conducted a critical literature review with the following purpose: "to identify social structures that contribute to timely, context-dependent decisions for transitions from acute care to end-of-life care for patients with chronic critical illness, their families, or close friends, and/or healthcare providers in an ICU environment" (Leung et al. 2016, p. 729). This purpose fits the aim of a particular type of literature review called a metasynthesis, which involves a systematic process of critically evaluating studies with qualitative data from different contexts (see ▶ Chap. 45, "Meta-synthesis of Qualitative Research"). The aim of a metasynthesis is "to develop a better conceptual or theoretical understanding or a different perspective of the phenomenon" (Tong et al. 2012, p. 181). The conclusions highlight knowledge gaps and provide direction in developing future research and practice interventions about the phenomenon of interest (Tong et al. 2012).

Critical realism fits well with the study's purpose because treatment decision-making near the end of life is a complex, context-dependent process. In particular, society's death-denying culture forms a social structure. Competing with this structure are mechanisms that reinforce society's value of health as a right, the belief that we ought to benefit from access to all service interventions when making decisions about our health, and the desire to die well at the end of life. These structures and mechanisms dominate ICU conversations, but there is little recognition of how they may affect the outcomes of treatment decision-making near the end of life (Boniatti et al. 2011).

Leung et al. (2016) utilized the principles of critical realism to guide analyses that moved from theory construction to theory refinement. The focus of the research questions was to describe specific mechanisms of healthcare providers' end-of-life decision-making and initiation of decision-making discussions with patients and family members: (1) How were decisions made? (2) Who communicated decisions to patients and families, and how were decisions communicated? (3) What resources and strategies helped in decision-making? and (4) What problems were resolved or persisted during decision-making? The researchers searched the literature for relevant qualitative and mixed-method studies. Inclusion criteria were as follows: studies, written in English, reporting on (1) adults with chronic critical illness, their families or close friends, and/or ICU healthcare providers, and (2) processes of decision-making on end-of-life care for patients with chronic critical illness (Leung et al. 2016, p. 730).

Leung et al. (2016) found five relevant qualitative articles reporting interview and observational data. After reviewing the articles for overall comprehension of the data, the authors grouped data with unique meaning and labeled them with descriptive codes. These codes captured both their manifest and latent meanings. Manifest codes reflected empirically based interpretations that remained close to the overt, literal meanings of the data, whereas latent codes reflected possible underlying mechanisms, such as behavioral dispositions, habits, and communication skills, connected to the ICU context in which they occurred (Leung et al. 2016).

To illustrate, we use a chunk of data from one of the reviewed papers, by Sinuff et al. (2009), that Leung et al. (2016) included in their metasynthesis. Sinuff et al. (2009, p. 155) reported the following:

> The central theme of family members' experience with mechanical ventilation was "living with dying." Families felt perpetually challenged to understand the patient's state while ventilated: is my loved one alive or dead, recovering or dying? The ventilator symbolized efforts to keep the patient alive but simultaneously indicated the patient's proximity to death.

Sinuff et al. (2009, p. 155) presented a quote from one interview to support their interpretation: "It makes you think of death...If they take him off he's going to die...you're living with death."

Recall Bhaskar's (1998) three strata of reality, theorized as causal processes by which objects and social relations constituting social structures may or may not generate predictable consequences. Leung et al. (2016) conceptualized aspects of the families' experiences of a threat to mortality along each of the three strata of reality. The participant's statement, "It makes you think of death..." was considered representative of the empirical. The contiguous context, such as the *sporadic, unplanned, decision-making in response to the deteriorating state of patients on mechanical ventilation*, was representative of the actual. Finally, what emerged in Leung et al.'s (2016) analysis was evidence of an underlying mechanism, they labelled as a process of *questioning of patient' quality of life*, and *the healthcare providers' dispositions to collaborate with families,* when families struggled to understand the patient's condition. This was one important mechanism that was considered representative of the real. Thus, all three strata of reality were identified as present in the one chunk of data from Sinuff et al. (2009) (Fig. 1).

Leung et al. (2016) posited that structures, such as the culture of cure within the ICU, and mechanisms, such as healthcare providers' work to suppress resistance by patients and families' recognition of dying (e.g., when families were not allowed to see the patients struggling to breathe when they were turned), could *lock in place routinized work of life preservation*, a derived subtheme. On the other hand, healthcare providers' disposition to collaborate with patients and families to

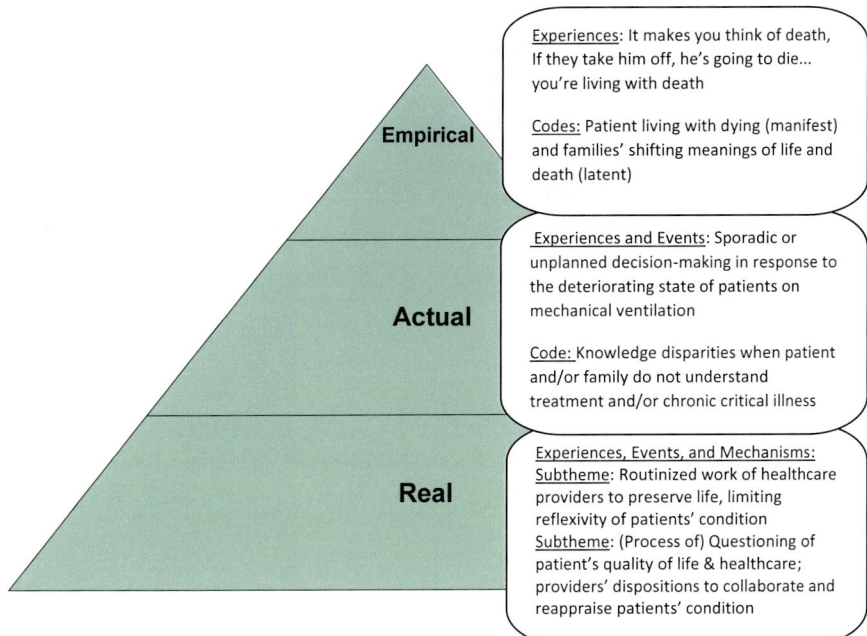

Fig. 1 Illustration of data in a stratified reality using critical realism

reappraise patients' condition, as well as families' questioning of patients' quality of life, could *open up access to augmenting end-of-life care*, another subtheme that conflicts with the preceding subtheme. Often, mechanisms worked together in the real strata, to trigger sporadic or unplanned decision-making and reveal the patient and/or families' knowledge disparities when it was clear that they did not understand treatment and/or the condition of chronic critical illness. This explanatory formula was prospectively tested against other data from selected studies. When the authors found it to be a concurrent pattern in all studies, they posited a main theme of *morally ambiguous expectations of ICU treatment*. This theme became the main process theorized to possess strong causal powers in social interactions among patients, families, and the healthcare providers. These social relations constituted emergent but transient social structures in the ICU. The theory was refined to improve the theoretical understanding of how patients with chronic critical illness did or did not transition from acute to end-of-life care in ICUs.

9   Conclusion and Future Directions

Content analysis (CA) has become the most common form of data analysis, despite criticisms about its lack of rigor and limited utility of its findings. The three most frequently used forms of CA and their capacity to represent interpretation along a continuum lend to the appeal of CA: researchers may access surface description or deeper meanings of representation. We suggest for future directions that CA can posit explanations of why the phenomenon is as it is by applying critical realism to identify social structures and important mechanisms. Without the clarity of a theoretical orientation for inquiry, researchers using CA may not possess a clear understanding of what they are seeking to identify. This minimizes the symbolic and conceptual value of findings from CA.

We examined the philosophical and theoretical position and assumptions of critical realism as well suited to extending the instrumental effectiveness of CA. Future directions using critical realism with CA involve connecting social outcomes with processes that include both measures of observable facts and socially influenced factors (Maxwell and Mittapalli 2010). Critical realism places more weight on causal explanations than on the causal description of consistent relationships between phenomena that experimental research emphasizes (Elder-Vass 2010); the researcher applying critical realism emphasizes the processes that lead to the occurrence of certain phenomena under certain conditions. Hence, critical realism uncovers context-dependent mechanisms that provide causal explanations for phenomena (Maxwell and Mittapalli 2010).

We conclude by suggesting that researchers pay attention to diversity not as something extraneous but as something essential to understanding the processes of a phenomenon (Maxwell and Mittapalli 2010). Some social structures may emerge and others may not because of the diversity of individuals. Indeed, understanding particular cultural and social norms are critical to understanding when certain social outcomes or consequences may or may not emerge. As a result, CA using critical realism may generate more than one final, often contradictory, theme, and the validity of its findings may be constrained to local or similar situations, not to all situations (Elder-Vass 2010).

We do not suggest that CA using critical realism is the best approach; rather, critical realism provides flexibility for researchers to explore deeper latent meanings and mechanisms underlying the phenomenon and to posit explanations of how and why the phenomenon is as it is. This advantage can empower researchers conducting qualitative descriptive research to make data from less visible psychological, social, and cultural domains more tangible and findings more instrumentally useful, as is necessary in this age of evidence-based practice (Sandelowski 2004), and, indeed, essential to the evaluation of complex service interventions (Pawson et al. 2005).

Acknowledgement Editorial services of Research Maven Consulting Services (http://www.researchmaven.ca) were enlisted to support substantive and copyediting of the manuscript.

References

Allport GW, Faden JM. The psychology of newspapers: five tentative laws. Public Opin Q. 1940;4:687–703.

Bhaskar R. Chapter 2: Philosophy and scientific realism. In: Archer M, Bhaskar R, Collier A, Lawson T, Norrie A, editors. Critical realism: essential readings. New York: Routledge; 1998. p. 16–47.

Boniatti MM, Freidman G, Castilho RK, Vieira SRR, Fialkow L. Characteristics of chronically critically ill patients: comparing two definitions. Clinics. 2011;66:701–4.

Camhi SL, Mercado AF, Morrison RS, Platt DM, August GI, Nelson JE. Deciding in the dark: advance directives and continuation of treatment in chronic critical illness. Crit Care Med. 2009;37:919–25.

Cruickshank J. Positioning positivism, critical realism, and social constructionism in the health sciences: a philosophical orientation. Nurs Inq. 2012;19:71–82.

Easton G. Marketing: a critical realist approach. J Bus Res. 2002;55:103–9.

Elder-Vass D. The causal powers of social structures: emergence, structure, and agency. Cambridge, UK: Cambridge University Press; 2010.

Elo S, Kyngäs H. The qualitative content analysis process. J Adv Nurs. 2008;62:107–15.

George AL. Propaganda analysis: a study of inferences made from Nazi propaganda in World War II. Evanston: Row, Peterson; 1959.

Hsieh HF, Shannon SE. Three approaches to qualitative content analysis. Qual Health Res. 2005;15:1277–88.

Kellner D, Roderick R. Recent literature on critical theory. New German Crit. 1981;23:141–77.

Krippendorff K. Content analysis: an introduction to its methodology. 3rd ed. Thousand Oaks: Sage; 2013.

Leung D, Angus JE, Sinuff T, Bavly S, Rose L. Transitions to end-of-life care for patients with chronic critical illness: a meta-synthesis. Am J Hosp Palliat Med. 2016;34:729–36.

Maxwell JA, Mittapalli K. Realism as a stance for mixed method research. In: Tashakkori A, Teddlie C, editors. Handbook of mixed methods in social & behavioral research. 2nd ed. Thousand Oaks: Sage; 2010. p. 145–68.

Mayan MJ. Essentials of qualitative inquiry. Walnut Creek: Left Coast Press; 2009.

Miles MB, Huberman AM, Saldana J. Qualitative data analysis: a methods sourcebook. 3rd ed. Thousand Oaks: Sage; 2014.

Nairn S. A critical realist approach to knowledge: implications for evidence-based practice in and beyond nursing. Nurs Inq. 2012;19:6–17.

Pawson R, Greenhalgh T, Harvey G, Walshe K. Realist review-a new method of systematic review designed for complex policy interventions. J Health Serv Res Policy. 2005;10(Suppl 1):21–34.

Philosophical foundations: critical realism. In: Mingers J, editor. Realising systems: knowledge and action in management science. Boston: Springer; 2006. pp. 11–31. https://doi.org/10.1007/0-387-29841-X_2.

Porter S, O'Halloran P. The use and limitation of realistic evaluation as a tool for evidence- based practice: a critical realist perspective. Nurs Inq. 2012;19:18–28.

Sandelowski M. Whatever happened to qualitative description. Res Nurs Health. 2000;23:334–40.

Sandelowski M. Using qualitative research. Qual Health Res. 2004;14:1366–86.

Sandelowski M. What's in a name? Qualitative description revisited. Res Nurs Health. 2010;33:77–84.

Sinuff T, Giacomini M, Shaw R, Swinton M, Cook DJ. "Living with dying": the evolution of family members' experience of mechanical ventilation. Crit Care Med. 2009;37:154–8.

Tong A, Fleming K, McInnes E, Oliver S, Craig J. Enhancing transparency in reporting the synthesis of qualitative research: ENTREQ. BMC Med Res Methodol. 2012;12:181. Retrieved from https://bmcmedresmethodol.biomedcentral.com/track/pdf/10.1186/1471-2288-12-181?site=bmcmedresmethodol.biomedcentral.com.

White MD, Marsh EE. Content analysis: a flexible methodology. Libr Trends. 2006;55:22–45.

Wilson V. Research methods: content analysis. Evid Based Libr Inf Pract. 2016;6:177–9. Retrieved from https://journals.library.ualberta.ca/eblip/index.php/EBLIP/article/view/12180/13124.

Thematic Analysis

48

Virginia Braun, Victoria Clarke, Nikki Hayfield, and Gareth Terry

Contents

1	Introduction	844
2	Thematic Analysis: A Brief History	844
3	Mapping the Terrain of Thematic Analysis: What Is a Theme?	845
4	Mapping the Terrain of Thematic Analysis: Different Schools of TA	847
5	Some Design Considerations for (Reflexive) Thematic Analysis	849
6	Six Phases of Reflexive Thematic Analysis	852
7	Conclusion and Future Directions	857
References		858

Abstract

This chapter maps the terrain of thematic analysis (TA), a method for capturing patterns ("themes") across qualitative datasets. We identify key concepts and different orientations and practices, illustrating why TA is often better understood as an umbrella term, used for sometimes quite different approaches, than a single qualitative analytic approach. Under the umbrella, three broad approaches can be identified: a "coding reliability" approach, a "codebook" approach, and a "reflexive" approach. These are often characterized by distinctive – sometimes radically

V. Braun (✉)
School of Psychology, The University of Auckland, Auckland, New Zealand
e-mail: v.braun@auckland.ac.nz

V. Clarke · N. Hayfield
Department of Health and Social Sciences, Faculty of Health and Applied Sciences, University of the West of England (UWE), Bristol, UK
e-mail: victoria.clarke@uwe.ac.uk; nikki2.hayfield@uwe.ac.uk

G. Terry
Centre for Person Centred Research, School of Clinical Sciences, Auckland University of Technology, Auckland, New Zealand
e-mail: gareth.terry@aut.ac.nz

© Springer Nature Singapore Pte Ltd. 2019
P. Liamputtong (ed.), *Handbook of Research Methods in Health Social Sciences*,
https://doi.org/10.1007/978-981-10-5251-4_103

different – conceptualizations of what a theme is, as well as methods for theme identification and development, and indeed coding. We then provide *practical* guidance on completing TA within our popular (reflexive) approach to TA, discussing each phase of the six-phase approach we have developed in relation to a project on men, rehabilitation, and embodiment. We conclude with a discussion of key concerns related to ensuring the TA you do – within whatever approach – is of the highest quality.

> **Keywords**
> Code · Codebook · Coding reliability · Epistemology · Latent · Reflexive thematic analysis · Semantic · Thematic map · Theme

1 Introduction

Thematic analysis (TA) is often misconceptualized as a single qualitative analytic approach. It is better understood as an umbrella term, designating sometimes quite different approaches aimed at identifying patterns ("themes") across qualitative datasets. In this chapter, we first define key concepts and map the terrain of TA; we identify three distinct "schools" of TA, highlighting differences between these schools, particularly in relation to underlying philosophy and approach to data analysis. We then provide practical guidance on completing TA, focused on one of the most popular approaches – developed by two of the authors of this chapter (Braun and Clarke 2006, 2012, 2013).

2 Thematic Analysis: A Brief History

Philosopher of science Gerard Houlton has been credited with inventing TA in the 1970s, in his work on "themata" in scientific thought (Holton 1973; see also Joffe 2011). But the term was in use well before then: musicologists in the 1930s described the analysis of musical scores as TA (e.g., Kinsky and Strunk 1933); sociologists in the 1940s used the term to describe a method for analyzing mass propaganda (e.g., Lazarsfeld and Merton 1944); psychoanalysts in the 1950s used it to refer to techniques for analyzing the results of projective tests (e.g., Winder and Hersko 1958). The conceptualization of TA as an approach for analyzing patterns of meaning *may* reflect a methodological evolution from (quantitative) content analysis. The terms "TA" and "content analysis" have often been used interchangeably, and the hybrid term "thematic content analysis" is also common (e.g., Brewster et al. 2014). Regardless of its developmental origins, TA clearly has shared history with content analysis (see ▶ Chap. 47, "Content Analysis: Using Critical Realism to Extend Its Utility").

In the 1980s and 1990s – around the time there was a general explosion of interest in qualitative research – TA started to appear as a particular approach for analyzing *qualitative data* in the health and social sciences (e.g., Dapkus 1985; Aronson 1994). But it was the countless published papers that described some version of "themes emerging" from data, without reference to an established methodological approach to TA, that led us to describe TA as "a poorly demarcated and rarely acknowledged, yet widely used qualitative analytic method" (Braun and Clarke 2006, p. 77). That was 2006; just over a decade later, how things have changed! TA is now increasingly recognized as an approach to analysis in its own right (there is still some debate around this; see (Willig 2013)), and there are *many* different approaches to TA. The shared name "TA" obscures divergence, both in terms of procedures, and, more importantly, in underlying philosophy and the conceptualization of key elements of the method (e.g., a theme, a code, or coding). It is not uncommon to see researchers cite sources on, and sometimes follow procedures for, approaches to TA that do not align, conceptually or in practice. Not grasping these distinctions can result in published papers where the approach to TA used is unclear, procedures and assumptions are misattributed or mixed up, and underlying conceptual clashes between different approaches are not recognized. This does a disservice to TA. Avoiding such errors requires understanding of the conceptual and procedural differences within the terrain of TA. To aid clarity, we will define some key concepts in TA and consider the distinctive features of three "schools" of TA – which we refer to as "coding reliability," "codebook," and "reflexive TA."

3 Mapping the Terrain of Thematic Analysis: What Is a Theme?

First up, it is vital to understand how "a theme" is conceptualized, as there are two competing ideas in TA research: domain summaries versus shared meaning-based patterns. We – and many others – view themes as reflecting a *pattern* of shared meaning, organized around a core concept or idea, a central organizing concept (see Braun et al. 2014). In this conceptualization, themes capture the essence and spread of meaning; they unite data that might otherwise appear disparate, or meaning that occurs in multiple and varied contexts; they (often) explain large portions of a dataset; they are often abstract entities or ideas, capturing implicit ideas "beneath the surface" of the data, but can also capture more explicit and concrete meaning; and they are built from smaller meaning units (codes) (DeSantis and Ugarriza 2000). An example of this type of theme comes from our research on meaning around male body hair (Terry and Braun 2016). A theme "men's hair as natural" captured the way body hair was often described as natural for men and "a dominant expression of masculine embodiment" (p. 17). As well as reporting participants' overt statements about the naturalness of male body hair (a "surface" level of meaning) – the theme explored more nuanced manifestations of this idea – gendered assumptions that men *should* be hairy and women hairless and that men's embodiment is biological (natural) while women's is socially produced (worked upon), constructions which were naturalized and essentialized in the dataset.

In contrast to our conceptualization, a theme in a "domain summary" conceptualization *summarizes* what participants said in relation to a topic or issue, typically at the *semantic* or surface level of meaning, and usually reports multiple or even contradictory meaning-content. The ("theme") issues are often based around data collection tools, such as responses to a particular interview question. Take the first theme in Roditis and Felsher's (2015) research on adolescents' perceptions of the risks and benefits of conventional cigarettes, e-cigarettes, and marijuana. The title – "perceived risks and benefits of conventional cigarettes compared to marijuana" – indicates a theme-as-domain-summary conceptualization, not least because it combines risks *and* benefits. And indeed, that is what is reported. Overview-type statements – "Youth either stated there was nothing good about using conventional cigarettes or stated that using cigarettes could help someone relax. Students easily recited a long list of negative consequences related to conventional cigarette use such as …." (p. 182) – highlight that the theme is a summary of youth perceptions in relation to a particular topic area.

Although some see domain summaries as a meaningful and useful conceptualization of a theme, others (e.g., Sandelowski and Leeman 2012; Connelly and Peltzer 2016) characterize them as at best underdeveloped or not fully realized themes and, at worst, misconceptualized. Some TA reports *do* read as if the analysis is only partly developed. For example, in Weatherhead and Daiches's (2010) paper on Muslim views on mental health and psychotherapy, the seven themes – "causes," "problem management," "relevance of services," "barriers," "service delivery," "therapy content," and "therapist characteristics" – were effectively domain summaries. Discussing this paper in class, one of our students evocatively dubbed them "bucket themes": you collect all the information gathered about X in one place, without considering shared meaning or difference. The theme "causes," for example, described participants' attributions for their mental health problems – explanations as diverse as reactions to life events and secular or religious notions that "life is a test." Yet Weatherhead and Daiches' discussion explored the "continued interweaving of religious and secular influences in participants' account of mental distress and well-being" (p. 85), hinting at the potential for themes as shared-meanings, where the analysis is developed further and deeper. A domain-summary approach risks conceptualizing TA as simply a data reduction activity, where the purpose of analysis is to succinctly summarize the diversity of responses across the scope of a project. However, this *can* sometimes simply be an issue of ensuring shared-meaning themes are well-named (see Braun and Clarke 2013); we recommend avoiding one-word theme names to avoid this.

Approaches to TA also vary on whether themes are conceptualized as analytic *inputs* – patterns identified and developed at the *start* of the analytic process (usually following some data familiarization) which guide the data coding process – or as analytic *outputs*, patterns identified and developed *later* in the analytic process, building on, and representing the *outcome* of, coding. To some extent, these conceptualizations align with the two different ideas about what a theme is.

4 Mapping the Terrain of Thematic Analysis: Different Schools of TA

We refer to *schools* of TA when we describe three broad "types" of TA, because there is not just *one* approach associated with each type. The names we use for the schools – coding reliability, codebook, and reflexive – emphasize the key distinctive element of each approach.

Coding reliability approaches – associated with authors like Boyatzis (1998), Guest et al. (2012), and Joffe (2011) – represent what we characterize as a *partially* qualitative approach to TA. Qualitative data are collected and analyzed using qualitative techniques of coding and theme development; the data are reported qualitatively as themes, typically illustrated by extracts of data. However, the underlying logic of these processes is firmly (post-)positivist, and some characterize these coding reliability approaches as "bridging the divide" between qualitative and quantitative methods. According to Boyatzis (1998, p. vii), (coding reliability) TA "is a translator of those speaking the language of qualitative analysis and those speaking the language of quantitative analysis." Coding reliability TA researchers share values with quantitative researchers' – for example, the importance of the reliability and replicability of observation – values at odds with (fully) qualitative paradigms.

In coding reliability TA, themes are often conceptualized as domain summaries (often derived from data collection questions), and as analytic inputs, as well as outputs – they *drive* the coding process and are the *output of* the coding process. The coding *process* is designed to prioritize "reliable" data coding, by which they mean identification of "accurate" codes/themes within data, usually based on agreement among multiple coders. Coding is guided by a codebook/coding frame, which typically contains a list of codes/themes – each has a label/name, a definition, information on how to identify the code/theme, a description of any exclusions or qualifications to identifying the code/theme, and data examples (Boyatzis 1998). This is designed to allow the researcher to categorize the data into (predetermined) themes. Despite a sometimes interchangeable use of the terms code and theme, *coding* is essentially conceptualized as a *process* (for identification of theme-relevant data and thus themes). Ideally, the codebook is applied to the data by more than one coder, each working independently; for some, the ideal coder has no prior experience with or knowledge of the topic of concern and comes to the coding process "cold." After coding, the level of "agreement" between the coders is calculated (using Cohen's kappa). A score of 0.80 and above is generally thought to signal accurate or reliable coding; lower scores are problematic, and lack of agreement needs to be resolved.

This coding approach can be understood as *consensus* coding – because it builds toward a singular, shared, and "correct" analysis of the data. The process has strong echoes of "the scientific method" – the researcher develops a hypothesis (themes), tests these (searches for evidence of the themes using the codebook), is concerned about, and seeks to control for, "researcher bias" or "influence", and, if the right procedures are followed, claims reliable and potentially replicable results. It reflects what Kidder and Fine (1987) described as "small q" qualitative research – qualitative research conceptualized as tools and techniques, not as a paradigm or underlying philosophy for research. We consequently characterize this school of TA as "partially" qualitative. To us, the idea that (such) TA can "bridge a qualitative-quantitative divide" is problematic, because it requires discarding what we see as central to good qualitative research practice – depth of engagement ("commitment and rigor" in (Yardley's 2015) open-ended and flexible principles for qualitative research quality), an open and exploratory design and analytic process, and a prioritization of researcher subjectivity and reflexivity (Finlay and Gough 2003; Gough and Madill 2012). The *reflexive* school of TA emphasizes these elements.

Reflexive TA approaches include our (e.g., Braun and Clarke 2006) popular version of TA, as well as others (e.g., Langdridge 2004). In these, TA is conceptualized as a *fully* qualitative approach – with data collection and analysis techniques underpinned by a qualitative philosophy or paradigm – a "Big Q" approach (Kidder and Fine 1987). Although there is no widely agreed definition of a qualitative paradigm or, indeed, whether there is just one qualitative paradigm (Madill 2015), a qualitative orientation usually emphasizes meaning as contextual or situated, reality or reali*ties* as multiple, and researcher subjectivity as not just valid but a *resource* (Braun and Clarke 2013). We characterize this school as *reflexive* TA to emphasize the active role of the researcher in the knowledge production process.

In reflexive TA, themes are conceptualized as meaning-based patterns, evident in explicit (semantic) or conceptual (latent) ways, and as the *output* of coding – themes result *from* considerable analytic work on the part of the researcher to explore and develop an understanding of patterned meaning across the dataset. Coding is an organic and open iterative *process*; it is not "fixed" at the start of the process (e.g., through the use of a codebook or coding frame). Codes – the product of coding – can evolve throughout the coding process. An initial code might be "split" into two or more different codes, renamed, or combined with other codes. The aim of such changes during coding is to better capture the researcher's developing conceptualization of the data. It is relatively easy to determine themes as domain summaries at the start of the analytic process; it is difficult to determine themes as conceptually founded patterns at the start, because it requires depth of (close and critical) engagement to move beyond the surface or obvious content of the data and to identify implicitly or unexpected unifying patterns of meaning. The aim of coding and theme development in reflexive TA is not to "accurately" summarize the data, nor to minimize the influence of researcher subjectivity on the analytic process, because neither is seen as possible nor indeed desirable. The aim is to provide a coherent and compelling *interpretation* of the data, grounded in the data. The researcher is a *storyteller*, actively engaged in interpreting data through the lens of their own cultural membership and social positionings, their theoretical assumptions and ideological commitments, as well as their scholarly knowledge. This subjective, even political, take on research is very different to a positivist-empiricist model of the researcher. Many reflexive TA researchers do indeed have some kind of social justice motivation – be it "giving voice" to a socially marginalized group, or a group rarely allowed to speak or be heard in a particular context, or a more radical agenda of social critique or change.

We use the term *codebook TA* to describe a third "school" of TA – although many of these, which include framework (e.g., Ritchie and Spencer 1994; Ritchie and Lewis 2003; Smith and Firth 2011), template (King 2014; Brooks et al. 2015) and matrix analysis (Miles and Huberman 1994; Nadin and Cassell 2014), among others, do not use the actual term TA. This school of TA sits somewhere *between* "coding reliability" and "reflexive" TA, sharing the structured approach to coding with coding reliability TA (though often *without* the use of coding reliability measures) with the broadly qualitative underlying philosophy of reflexive TA. In codebook TA, some if not all themes are determined in advance of full analysis, and themes are typically conceptualized as domain summaries.

Some TA researchers, including template analysis proponents Brooks et al. (2015), have argued that researchers should not be precious about their way of working with TA. Although some friends and colleagues have (jokingly) suggested Virginia and Victoria are the "TA police," issuing edicts about how TA should be done and rigorously punishing crimes against TA that do not follow our guidelines, we actually *somewhat* agree with Brooks et al.'s sentiment. Overall, what is important is that researchers use the approach to TA that is most appropriate for their research, they use it in a "knowing" way, they aim to be thoughtful in their data collection and analytic processes and practices, and they produce an overall coherent piece of work. Yet, we do *advocate* certain practices. From our perspective, the use of a structured codebook, determining themes in advance of analysis or following only data familiarization (using themes as analytic inputs) and conceptualizing themes as domain summaries, delimits the depth of engagement and flexibility central to *qualitative* research practice. There are, however, clear pragmatic advantages to codebook approaches – the coding framework allows teams of researchers to more easily work *together* on data analysis, facilitates a relatively quick analytic process, and provides some structure for qualitative novices. Taken out of a "consensus" and reliability framework, this has potential to produce rich nuanced analysis. But pragmatic factors should not (always) be the sole determinant of method.

5 Some Design Considerations for (Reflexive) Thematic Analysis

TA offers researchers great flexibility, meaning it can be used to do lots of the things that qualitative researchers are interested in. This flexibility stems from TA's status as an analytic method, rather than a methodology, the latter referring to a theoretically-informed *framework* for research. Although the school of TA chosen delimits a broad paradigm ([post]positivist or qualitative), beyond that there is scope to design and locate the method – and indeed, a requirement to do so, for reflexive TA.

Locating your overall theoretical and interpretative frameworks is important. Some treat TA as particularly compatible with phenomenological approaches (e.g., Joffe 2011; Guest et al. 2012), and it is indeed often used to describe or summarize participants' experiences, rather than to do more interpretative or conceptual work. But *why* TA should be limited to such an interpretative framework is unclear, and we think treating TA as a descriptive approach focused on experience underappreciates its flexibility and full potential. Indeed, it works well with many different interpretative frameworks, ranging from phenomenological ones to critical constructionist interrogations of meaning. And it, therefore, has the potential to answer different research questions. TA *can* address questions about, and be used to describe, the "lived experiences" of particular social groups (e.g., sex workers (Mellor and Lovell 2012), people with Parkinson's disease (Redmond et al. 2012), Asian migrants (Terry et al. 2011), adults with gay, lesbian, or bisexual parents (Titlestad and Pooley 2014)) or about particular aspects of their lives (e.g., the experience of freezing for people with Parkinson's disease (Redmond et al. 2012) or the health needs of street-based sex workers (Mellor and Lovell 2012)). It can also examine the "factors" that influence, underpin, or contextualize particular processes or phenomena (such as the factors that shape nurses' values in relation to compassionate care (McSherry et al. 2017)), identify views about particular phenomena (such as contested views about who is best placed to provide expertise in legal proceedings related to children's care (Hill et al. 2017)), or interrogate dominant patterns of meaning surrounding particular phenomena (such as the discourses underpinning the normalization of female genital cosmetic surgery on a cosmetic surgery website (Moran and Lee 2013)). Research questions for TA need to be aligned with the theoretical orientation of your TA.

TA also offers flexibility around data collection: interviews are common; focus groups are popular; diaries, visual methods, participatory methods, surveys, a wide range of secondary sources – such as online forums, blogs, websites, magazines, newspaper articles, and police reports – and many other methods have been used in TA research. As TA is generally a method for *across* dataset analysis (although it has been also used in case study research; e.g., Cedervall and Åberg 2010; Manago 2013), what is an adequate sample size? How many interviews should be conducted? How many participants should be recruited? How many hours of data should be recorded? These questions are perhaps some of the thorniest for qualitative research, although some concerns around sample size justification perhaps hark back to broader positivist-empiricist concerns with representation and generalizability, now connected to power analyses in statistical research.

Perhaps the most commonly used criterion for determining sample size in TA is "saturation" – such as claims that participants "were recruited until saturation was reached" (Gershgoren et al. 2016, p. 130; see also ▶ Chaps. 55, "Reporting of Qualitative Health Research," and ▶ 58, "Appraisal of Qualitative Studies"). Here, "saturation" typically refers to information redundancy, or collecting data until no new information is generated (there are other definitions), and some TA researchers have suggested saturation can be achieved in as few as 6–12 interviews (Guest et al. 2006; Ando et al. 2014) or 5 focus groups (Namey et al. 2016). Such papers are often cited to provide justification for relatively small sample sizes in TA research, but bold claims about saturation warrant interrogation. One problem with an information redundancy conceptualization is that it relies on an understanding of meaning as transparent and obvious *prior to* analysis. As TA (often) involves identifying new patterns of meaning, and this usually happens *after* data collection, *analysis* is necessary to judge whether the information generated by participants offers something new or not. Researchers who claim saturation, then, seem to rely on potentially superficial impressions made of data during data collection. This approach is more compatible with coding-reliability versions of TA. Where saturation has (attempted to be) operationalized, it is often within an implicit coding-reliability approach to TA (e.g., Guest et al. 2006). With such often surface-meaning-based and early-conceptualized analysis, it is easy to see how the appearance of "saturation" might be achieved in relatively few interviews. A recent paper suggested that if coding in TA moves beyond the surface level, larger samples are needed to achieve saturation (Hennink et al. 2016), and reported more conceptual codes achieved "saturation" with 16–24 interviews or not at all.

More problematic for using saturation as the rationale for sample size relates to the underlying philosophy of the research – in qualitative approaches that emphasize the partial, multiple, and contextual nature of meaning, and view knowledge as the actively *created* product of the interpretive efforts of a particular researcher (or researchers), combined with the dataset, the concept of saturation stops making sense (Malterud et al. 2016). We remain skeptical of the usefulness of the saturation concept, particularly when conceptualized as information redundancy, for determining sample size in TA research, and do not think it is useful for much "Big Q" TA. Researchers who use saturation need to do so from a position of theoretical "knowingness," understanding the assumptions embedded in (their particular iteration of) this concept, and whether those are compatible with the underlying philosophy of their research.

What does this mean for sample sizes in TA research? Unfortunately, there are *no magic formulas* for determining sample size in TA research! We urge readers to be skeptical of anyone proffering simple formulas (e.g., Fugard and Potts 2015), as they always contain inbuilt assumptions (see Braun and Clarke 2016). Sample size is most often informed by various contextual and pragmatic considerations, some of which cannot be (wholly) determined in advance of data collection. Imagine a PhD student conducting TA research – their sample size could be informed by local "norms" around the appropriate scope of doctoral research, what is considered an acceptable sample size in journals the student hopes to publish their research in, and other such pragmatic "rules of thumb," as well as more contextual considerations such as the breadth of their research question, the diversity within the population of study, and the amount and richness of data collected from each participant/case. Our pragmatic "rule of thumb" is *at least* five or six interviews for a (*very*) small project, assuming the data are rich, the sample relatively homogenous, the research question focused, and the output an unpublished dissertation (for more "rules of thumb" advice on sample size, see Braun and Clarke 2013). It is also important to reflect on the sorts of claims made about themes developed, in light of the sample size.

6   Six Phases of Reflexive Thematic Analysis

Having discussed some conceptual and design issues, we now provide research-illustrated (see Box 1) discussion around the phases of *doing* (reflexive) TA, aligned to the "six-phase" approach we have developed, noting this as a reflexive and recursive, rather than strictly linear, process. For more practical step-by-step guidelines, see Braun and Clarke (2006, 2012, 2013) and Terry et al. (2017).

> **Box 1 The Men's Embodiment in Rehabilitation Study**
> Despite three decades of calls for more research into men's health, it continues to be underresearched, in general and in rehabilitation studies in Aotearoa/New Zealand. This project, theoretically located at the intersection of critical health psychology, critical rehabilitation studies, and critical disability studies, was designed to explore men's experiences, practices, and sensemaking regarding male bodies undergoing rehabilitation for illness or impairment. Data were generated through one-to-one qualitative interviews with 20 men in various states of health and fitness, who had experienced recent (and extensive) rehabilitative treatment of some kind. Gareth Terry was the primary investigator and David Anstiss a postdoc researcher on the project. Rehabilitation was a new area of research focus for both.

Familiarization, which requires the researcher to shift focus from data generation (including transcription) to analysis, is fundamentally about appreciating the data *as* data. The process involves becoming "immersed" in the data and connecting with them in different ways: engaged, but also relaxed; making *casual* notes, but being thoughtful and curious about what you are reading. It is not about attaching formal labels – that comes later – but about looking for what is interesting about the data and what you notice about possibilities, connections (between participants, data, and existing literature), and quirks, which may add depth and nuance to your later coding. It can be one of the most enjoyable phases of the analytic process, and by providing a solid foundation of interrogating and thus "knowing" your data, it certainly makes the rest of the analysis *much* more enjoyable.

Practically, familiarization includes listening to audio data, watching video data, and/or reading and rereading textual data, "noticing" interesting features, and making notes about individual data items, as well as the whole dataset. These notes should be shaped by your research question(s), as well as broader questions about what is going on in the data. For instance, in the men and embodiment project (Box 1), some key questions that Gareth and David asked of the data related to how "typical" understandings of men's health might intersect with newly-experienced impairments and disabling environments – both social and physical. As researchers are new to the topic, familiarization was a crucial "entry point" into the data, providing them with an opportunity to closely read and thoroughly engage with the data, and giving room for reflexivity – asking questions of themselves and how they responded to the data. Consequently, much of the familiarization that occurred in the project involved making sense of ideas in the data that were new to them, which correspondingly made them aware of their own *abled* experiences and assumptions. Gareth and David engaged in the familiarization process concurrently, meeting several times to discuss their "noticings" and notes in detail. This process was not intended to produce any "consensus," but rather to gain greater initial insight through sharing each other's perspective on the data.

Generating codes moves to more detailed and systematic engagement with the data. We sometimes suggest familiarization could be done with a glass of wine, but coding needs coffee (or a good cup of tea). The coding phase in (our) TA is about focused attention, to systematically and rigorously make sense of data. If the familiarization phase could be considered a somewhat "loose" route into engaging with the data, the coding phase is about succinctly and systematically identifying meaning throughout the dataset. Data are organized around similar meanings and the content reduced into collated chunks of text. As a process, coding involves attaching pithy, clear labels (codes) to "chunks" of data, to help you organize the data around meaning-patterns (developed in later phases).

There are two broad orientations to coding: an *inductive* orientation, where the researcher starts the analytic process *from* the data, working "bottom-up" to identify meaning without importing ideas, and a *deductive* orientation, where the researcher approaches the data with various ideas, concepts, and theories, or even potential codes based on such, which are then explored and tagged within the dataset. In practice, any researcher *will* approach the data with preconceived ideas based on their existing knowledge and viewpoints. Coding inductively does not mean that we assume the researcher is a "blank state," but, instead, that the starting point of the analysis is with the data, rather than existing concepts or theories (Terry et al. 2017).

Another consideration is the level at which "meaning" is identified and coded for – something partly informed by the epistemological approach of a project. *Semantic* codes stay at the "surface" of the data, capturing explicit meaning, close to participant language. *Latent* codes focus on a deeper, more *implicit* or conceptual level of meaning, sometimes quite abstracted from the explicit content of the data. The boundaries between these types of codes in practice are not always distinct; these codes represent ends of a continuum of ways of looking at data, rather than a binary. Initial coding for most TA projects is often semantic, and it can be hard to move beyond this level, to start to see the meaning *beyond* the obvious. As researchers become more experienced, or an analysis develops, latent-level meaning can be easier to "see" – but whether latent meaning is included can depend on the aims of the project.

Table 1 Example of coding, P1 ("Derek") from the men's embodiment and rehabilitation study

Data	Code
GT: So has that (pause) has that (pause) situation resolved itself a little bit	
P1: Nah I didn't see my father at all (inbreath). I was really hopeless, lots of things happened that (pause) he was not supportive at all	"Harden up" mentality
	Relational breakdown with father
	Lack of affective recognition = unsupportive
GT: Yep	A long time to begin accepting
P1: He was a real cuppa concrete guy you know which is just not how it works (pause, inbreath); you know I was a lot worse then coz that was sort of two years in	Statute of limitations on acceptance
	Importance of supportive partner
	Importance of ACC support
	Financial means to get best treatment
GT: Yep	Recovery as relational
P1: Sort of accept it and move on you know. I was (pause) I'm not giving up (inbreath); there have been a few people that (pause) I mean I was fortunate my wife was very into research and stuff and she found (inbreath) my [private rehab clinic], and ACC paid for it; I think about 500 hours of rehab	

Transcription notation: *underline*, participant emphasis; (pause), pause in speech; (laughs), laughter from speaker

Gareth and David coded both semantically and latently, first working independently in the early part of this phase, then more collaboratively. Earlier familiarization discussions informed coding, but their emphasis was on generating a wide variety of codes to discuss and refine (see Table 1 for a brief example). Understanding that researchers look at data through their own lenses, and make interpretative choices throughout the analytic process (see Braun and Clarke 2016), they aimed to develop a diverse range of codes to build themes from, rather than trying to reach a consensus. Their practice demonstrates how more than one coder can work effectively with reflexive approaches to TA.

Constructing themes continues the active process of the previous phases. Themes are built, molded, and given meaning at the intersection of data, researcher experience and subjectivity, and research question(s). Because themes *do not emerge* fully-formed from the data, the process of constructing them is akin to processes of engineering or design. Prototypes (or *candidate themes*) are developed from the analytic work of the earlier phases, and "tested out" in relation to the research question/dataset overall. Knowing that not all candidate themes will necessarily survive this early development process is vital to not getting too attached. Good themes are those that tell a coherent, insightful story about the data in relation to the research question.

There are two key ways to develop codes into candidate themes. The first involves using codes as building blocks – similar codes are collated, together with their associated data, into coherent clusters of meaning that tell a story about a particular aspect of the dataset. This approach is most commonly how researchers move from codes to constructing (candidate) themes. However, sometimes a code may be "substantial" enough to be "promoted" to a theme – if it contains a central organizing idea that captures a *meaningful* pattern across the dataset, as well as different manifestations of that pattern. A common pitfall in (reflexive TA) theme development is identifying a feature of the data, rather than meaning-based patterns – features are somewhat akin to the idea of themes as domain summaries. For instance, men in the embodiment and rehabilitation project would often use humor to deflect questions about emotion and rehabilitation. This is potentially important information, but "humor" in and of itself is a *feature* of the dataset, not a meaning-based pattern. If the researchers could identify a conceptual meaning related to the use of humor, it might work as a theme, but alone, it does not.

Gareth and David again worked independently and collaboratively in the early stages of theme construction, meeting regularly to discuss candidate themes. Their meetings took the form of a kind of "theme off": each presented their candidate themes, including preliminary theme names and definitions (discussed soon); they then "tussled" with each theme, and the collection of themes, to identify the most meaningful potential themes, the ones that collectively told the best story of the data. Thematic mapping – a process of visually exploring potential themes and subthemes, and connections between them (Braun and Clarke 2006) – was useful. Figure 1 maps the six candidate themes produced through this process, with all being relevant to the research question. Three of the initial themes Gareth and David constructed independently were similar enough to collapse into the single "bodies about more than roles and functions" theme – Gareth had identified a theme called "multiple embodiments"; David had two called "demanding embodiment" and "knowledge about bodies." Their process demonstrates a way of working together, analytically, outside a consensus-building model.

The phases of *revising* and *defining themes* are particularly important, precisely because candidate themes *are* effectively prototypes. Sometimes they do not work! It can be difficult to "let go" of our early ideas, but holding too tightly to a candidate theme can potentially result in analytic "thinness" or conceptual overlap. The story being told about the data risks being diminished in richness, or conceptually confused, by inclusion of weak or overlapping themes. Having clear definitions of each theme – a paragraph delineating the theme's boundaries and central organizing concept (for examples, see Terry et al. 2017) – helps clarify the essence and scope of each theme. Indeed, it was such descriptions of Gareth and David's three candidate themes that highlighted their similarity and led to combination as a single theme (note that typically, not every facet survives this review process!).

Key to reviewing and defining is compiling all coded data for each of the candidate themes and reviewing them to ensure that the data relate to a central organizing concept; another stage of review involves checking the themes against the whole dataset. It is also important to develop a clear sense of how each theme

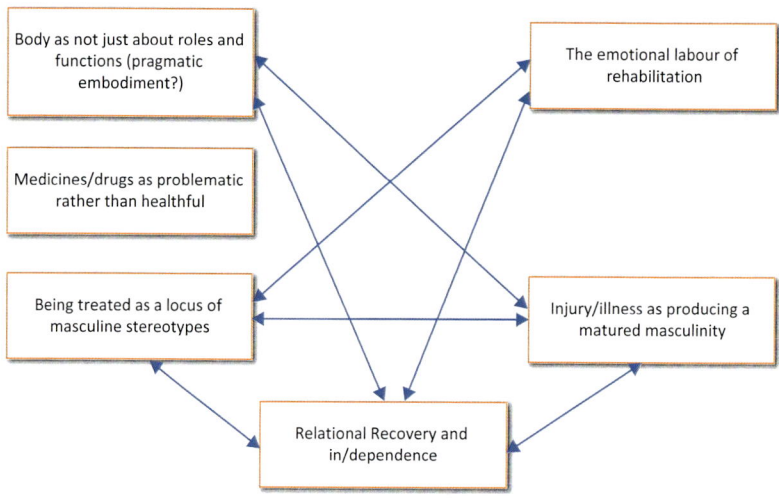

Fig. 1 Map of candidate themes from the men's embodiment and rehabilitation study

relates to the others. Thematic maps can be useful to visualize how the themes fit together and tell the *overall story* of your data – and to check that themes do not overlap. We often move from "early" maps for candidate themes, through to "final" maps when the revising phase is complete. For the men's embodiment project, it became clear when developing the (early) thematic map (Fig. 1), and comparing the definitions of each candidate theme, that some of the relationships between themes were stronger than others. The "resistance to medication" theme appeared quite independent (no connecting arrows); the other five themes all spoke to "relational outworking" of rehabilitation - how the men related to their bodies, and to others, and how they were related to *by* others. Thinking more deeply about the central organizing concept for each theme led Gareth and David to conclude that a notion of *recovery as relational* acted as an overarching theme (see Braun and Clarke 2013), an "umbrella" that contained three, related, themes: "bodies about more than roles and functions;" "being treated as a locus of masculine stereotypes;" and "the emotional labor of rehabilitation." As analysis is a task of telling a compelling story about (aspects of) the data, they set aside one strong theme, "injury/illness producing a matured masculinity," for future analysis oriented to *identity*. Through this process of revising themes, you aim for an in-depth and nuanced understanding of the central organizing concept and boundaries of each theme, including any subthemes (and overarching themes), and the overall theme story. Tables and similar tools can also facilitate in clearly identifying boundaries and structures of themes, in these phases.

Defining themes often leads to tighter/clearer theme names, which signal the scope and "core" of each theme. When you arrive at this point in the process, theme names will likely be somewhat makeshift – perhaps lengthy, or alternatively just one word – and only *provisionally* capture the content and scope of each theme. Final theme names should succinctly cue the reader in to what they can expect to read about in the theme, *and* draw them into wanting to read the analysis!

The revising and defining phases seek to ensure that themes, *and* theme names, clearly, comprehensively *and* concisely capture what is meaningful about the data, related to the research question, getting you close to a "completed" analysis. The final phase, *producing the report*, is not, however, purely a writing-up exercise. Producing the report often serves as a final test of how well the themes work, individually in relation to the dataset, and overall. Revisiting the research question, your notes from the earlier phases of familiarization and coding, your lists of codes, and theme definitions can be useful to ensure that the final themes remain close to the data *and* answer the research question well (your research question can be "tweaked" for better fit at this point). The scholarly process of making connections to existing research and literature on the topic of interest, and weaving this in to the written results and discussion, may offer final moments of inspiration and a deeper insight into the analysis. Therefore, we urge researchers to view this phase as the final stage of analysis, and be open to making further revisions to the themes' content, structure, and names. It may be that when you start to write your analytic narrative around the data extracts, you decide that some participant quotations do not as clearly demonstrate the point as other quotations do – your analytic claims may shift to reflect this. Or, as you write up your themes, you might notice that a change to the order in which they are presented would help with the flow of the story of your data, and so on. It can be useful to draw on our 15-point checklist (see Braun and Clarke 2006; Terry et al. 2017) to check the strength of your analysis and consistency across the report. We emphasize, again, that the epistemological position you have claimed will inform the terminology you use and the way you treat the data.

7 Conclusion and Future Directions

Despite TA providing an accessible method for (novice) qualitative researchers, there *are* plenty of potential traps and ways you can go wrong. Having read this chapter, you will be well-equipped to avoid many of these! In our view, good quality TA requires a degree of "theoretical knowingness'" – an understanding of the philosophical basis of enquiry. This means, for instance, understanding the assumptions underpinning coding reliability or consensus coding practices, and understanding why these are *not* compatible with Big Q qualitative enquiry. Viewing theory as something that we *do*, rather than an abstract consideration divorced from the practical processes of conducting research, helps realize how essential this is. Imagine a supervisor telling a student "stop overthinking things and just get on with it." Such advice suggests theory is *separate from* "getting on with it," but theory is meshed into everything. We *do* theory all the time, in how we relate to participants, in our interviewing "style," in how exactly we transcribe our interviews... All of these, and *many* other practical elements of research, reflect theoretical assumptions (implicitly made *or* actively chosen) about the nature of enquiry and what counts as meaningful knowledge.

Understanding theoretical bases also means you can avoid inconsistencies which mar too much published TA work. Too often we read research that contains statements like "following the procedures outlined by Braun and Clarke..." and then a description of analytic procedures that have little or no relation to those we outline. Sometimes coding reliability and reflexive approaches to TA are both cited, or the procedures combined, without any acknowledgment or justification of merging two philosophically divergent approaches. Sometimes researchers attribute analytic processes and concepts associated with grounded theory to TA (such as constant comparison, line-by-line coding, open coding, categories and subcategories, saturation; see Braun and Clarke 2013). Sometimes researchers claim domain summaries as conceptually founded patterns... Sometimes we really are left wondering whether the authors have actually *read* Braun and Clarke (2006)!

Is this us being picky, or does this really matter? Are we succumbing to methodolatry – the prioritizing of procedure above all else – something that "method-obsessed" psychologists, such as ourselves, are thought to be particularly vulnerable to? Does it matter if the end product ("the results" section) is good? We think it does matter! To us, these method(ological) "choices" – for example, combining consensus coding with Braun and Clarke (2006) – seem rarely to be made knowingly or reflexively. They seem instead to reflect a lack of understanding of, or caring about, the philosophical underpinnings of (Big Q) qualitative research or, perhaps, an (knowing or unknowing) acquiescence to the notion that (post)positivism is the only valid philosophy for research. Vitally, in a context in which there is not only much confusion about qualitative research, and its philosophical underpinnings, but much critique, these practices serve to create further confusion and even give the critiques some validity!

Furthermore, when published work is internally incoherent, or does not follow best- or even good-practice guidelines for any "school" of TA, this can be confusing, particularly for qualitatively inexperienced or student readers, who may assume publication is a guarantee of quality! We have often led a critical discussion of a published paper that ends with a student asking some variety of the question: "how does this stuff get published?" For the future of TA, all of us – scholars and researchers doing TA, reviewers, and editors – need to work hard to ensure consistency and quality in published TA.

References

Ando H, Cousins R, Young C. Achieving saturation in thematic analysis: development and refinement of a codebook. Compr Psychol. 2014;3:4.

Aronson J. A pragmatic view of thematic analysis. Qual Rep. 1994;2(1):1–3. http://www.nova.edu/ssss/QR/BackIssues/QR2-1/aronson.html.

Boyatzis RE. Transforming qualitative information: thematic analysis and code development. Thousand Oaks: Sage; 1998.

Braun V, Clarke V. Using thematic analysis in psychology. Qual Res Psychol. 2006;3(2):77–101.

Braun V, Clarke V. Thematic analysis. In: Cooper H, editor. APA handbook of research methods in psychology, Research designs, vol. 2. Washington, DC: APA books; 2012. p. 57–71.

Braun V, Clarke V. Successful qualitative research: a practical guide for beginners. London: Sage; 2013.

Braun V, Clarke V. (Mis)conceptualising themes, thematic analysis, and other problems with Fugard and Potts' (2015) sample-size tool for thematic analysis. Int J Soc Res Methodol. 2016;19(6):739–43.

Braun V, Clarke V, Rance N. How to use thematic analysis with interview data. In: Vossler A, Moller N, editors. The counselling and psychotherapy research handbook. London: Sage; 2014. p. 183–97.

Brewster ME, Velez BL, Mennicke A, Tebbe E. Voices from beyond: a thematic content analysis of transgender employees' workplace experiences. Psychol Sex Orient Gend Divers. 2014;1(2):159–69.

Brooks J, McCluskey S, Turley E, King N. The utility of template analysis in qualitative psychological research. Qual Res Psychol. 2015;12(2):202–22.

Cedervall Y, Åberg AC. Physical activity and implications on well-being in mild Alzheimer's disease: a qualitative case study on two men with dementia and their spouses. Physiother Theory Pract. 2010;26(4):226–39.

Connelly LM, Peltzer JN. Underdeveloped themes in qualitative research: relationships with interviews and analysis. Clin Nurse Spec. 2016;30:51–7.

Dapkus MA. A thematic analysis of the experience of time. Personal Process Individ Differ. 1985;49(2):408–19.

DeSantis L, Ugarriza DN. The concept of theme as used in qualitative nursing research. West J Nurs Res. 2000;22(3):351–72.

Finlay L, Gough B, editors. Reflexivity: a practical guide for researchers in health and social sciences. Oxford: Blackwell Science; 2003.

Fugard AJB, Potts HWW. Supporting thinking on sample sizes for thematic analysis: a quantitative tool. Int J Soc Res Methodol. 2015;18(6):669–84.

Gershgoren L, Basevitch I, Filho E, Gershgoren A, Brill YS, Shink RJ, Tenebaum G. Expertise in soccer teams: a thematic enquiry into the role of shared mental models within team chemistry. Psychol Sport Exerc. 2016;24:128–39.

Gough B, Madill A. Subjectivity in psychological research: from problem to prospect. Psychol Methods. 2012;17(3):374–84.

Guest G, Bunce A, Johnson L. How many interviews are enough? An experiment with data saturation and variability. Field Methods. 2006;18(1):59–82.

Guest G, MacQueen KM, Namey EE. Applied thematic analysis. Thousand Oaks: Sage; 2012.

Hennink MM, Kaiser BN, Marconi VC. Code saturation versus meaning saturation: how many interviews are enough? Qual Health Res. 2016;27(4):591–608.

Hill M, Welch V, Gadda A. Contested views of expertise in children's care and permanence proceedings. J Soc Welf Fam Law. 2017;39(1):42–66.

Holton GJ. Thematic origins of scientific thought: Kepler to Einstein. Cambridge, MA: Harvard University Press; 1973.

Joffe H. Thematic analysis. In: Harper D, Thompson AR, editors. Qualitative methods in mental health and psychotherapy: a guide for students and practitioners. Chichester: Wiley; 2011. p. 209–23.

Kidder LH, Fine M. Qualitative and quantitative methods: when stories converge. In: Mark MM, Shotland L, editors. New directions in program evaluation. San Francisco: Jossey-Bass; 1987. p. 57–75.

King N. Using templates in the thematic analysis of text. In: Cassell C, Symon G, editors. Essential guide to qualitative methods in organisational research. London: Sage; 2014. p. 256–70.

Kinsky G, Strunk WO. Was Mendelssohn indebted to weber? An attempted solution of an old controversy. Music Q. 1933;19(2):178–86.

Langdridge D. Introduction to research methods and data analysis in psychology. Harlow: Pearson; 2004.

Lazarsfeld PF, Merton RK. The psychological analysis of propaganda. In: Proceedings, Writers' Congress: Berkeley and Los Angeles: University of California Press; 1944.

Madill AL. Qualitative research is not a paradigm. Qual Psychol. 2015;2(2):214–20.

Malterud K, Siersma VK, Guassora AD. Sample size in qualitative interview studies: guided by information power. Qual Health Res. 2016;26(13):1753–60.

Manago AM. Negotiating a sexy masculinity on social networking sites. Fem Psychol. 2013;23(4):478–97.

McSherry W, Bloomfield S, Thompson R, Nixon VA, Birch C, Griffiths N, Fisher S, Boughey AJ. A cross-sectional analysis of the factors that shape adult nursing students' values, attitudes and perceptions of compassionate care. J Res Nurs. 2017;22(1–2):25–39.

Mellor R, Lovell A. The lived experience of UK street-based sex workers and the health consequences: an exploratory study. Health Promot Int. 2012;27(3):311–22.

Miles MB, Huberman AM. Qualitative data analysis: an expanded sourcebook. 2nd ed. Thousand Oaks: Sage; 1994.

Moran C, Lee C. Selling genital cosmetic surgery to healthy women: a multimodal discourse analysis of Australian surgical websites. Crit Discourse Stud. 2013;10(4):373–91.

Nadin S, Cassell C. Using data matrices. In: Cassell C, Symon G, editors. Essential guide to qualitative methods in organisational research. London: Sage; 2014. p. 271–87.

Namey E, Guest G, McKenna K, Chen M. Evaluating bang for the buck: a cost-effectiveness comparison between individual interviews and focus groups based on thematic saturation levels. Am J Eval. 2016;37(3):425–40.

Redmond L, Suddick K, Earhart GM. The lived experience of freezing in people with Parkinson's: an interpretive phenomenological approach. Int J Ther Rehabil. 2012;19(3):169–77.

Ritchie J, Lewis J. Qualitative research practice: a guide for social science students and researchers. London: Sage; 2003.

Ritchie J, Spencer L. Qualitative data analysis for applied policy research. In: Bryman A, Burgess RG, editors. Analysing qualitative data. London: Taylor & Francis; 1994. p. 173–94.

Roditis ML, Halpern-Felsher B. Adolescents' perceptions of risks and benefits of conventional cigarettes, e-cigarettes, and marijuana: a qualitative analysis. J Adolesc Health. 2015;57:179–85.

Sandelowski M, Leeman J. Writing usable qualitative health research findings. Qual Health Res. 2012;22(10):1404–13.

Smith J, Firth J. Qualitative data approaches: the framework approach. Nurs Res. 2011;18(2):52–62.

Terry D, Ali M, Lê Q. Asian migrants' lived experience and acculturation to Western health care in rural Tasmania. Aust J Rural Health. 2011;19(6):318–23.

Terry G, Braun V. "I think gorilla-like back effusions of hair are rather a turn-off": 'excessive hair' and male body hair (removal) discourse. Body Image. 2016;17:14–24.

Terry G, Hayfield N, Clarke V, Braun V. Thematic analysis. In: Willig C, Stainton Rogers W, editors. The sage handbook of qualitative research in psychology. 2nd ed. London: Sage; 2017. p. 17–37.

Titlestad A, Pooley JA. Resilience in same-sex-parented families: the lived experience of adults with gay, lesbian, or bisexual parents. J GLBT Fam Stud. 2014;10(4):329–53.

Weatherhead S, Daiches A. Muslim views on mental health and psychotherapy. Psychology and Psychotherapy: Theory, Research and Practice. 2010;83(1):75–89.

Willig C. Introducing qualitative research in psychology: Adventures in theory and method. 3rd ed. Maidenhead, Berkshire: McGraw-Hill Education (UK); 2013.

Winder AE, Hersko M. A thematic analysis of an outpatient psychotherapy group. Int J Group Psychother. 1958;8(3):293–300.

Yardley L. Demonstrating validity in qualitative psychology. In: Smith JA, editor. Qualitative psychology: a practical guide to research methods. 3rd ed. London: Sage; 2015. p. 257–72.

Narrative Analysis

49

Nicole L. Sharp, Rosalind A. Bye, and Anne Cusick

Contents

1	Introduction	862
2	What Is a Narrative?	864
3	Narrative Inquiry in Health and Social Sciences	865
4	Introduction to the Research Exemplars	866
	4.1 Exploring the Experience of Emerging Adults with Cerebral Palsy	866
	4.2 Exploring Family Life Following the Acquired Brain Injury of an Adolescent Child	867
5	Introduction to Two Approaches: Narrative Analysis and Analysis of Narratives	868
	5.1 Narrative Cognition and Narrative Analysis	868
	5.2 Paradigmatic Cognition and Analysis of Narratives	869
	5.3 Complementary Use of Narrative and Paradigmatic Approaches to Narrative Inquiry	869
6	Narrative Analysis Techniques	871
7	Analysis of Narrative Techniques	872
	7.1 Inductive Paradigmatic Analysis	873
	7.2 Deductive Paradigmatic Analysis with a Theoretical Focus	874
8	Examples of the Outcomes of Narrative Analysis and Analysis of Narratives	875
9	Conclusion and Future Directions	877
References		878

Abstract

Narrative inquiry methods have much to offer within health and social research. They have the capacity to reveal the complexity of human experience and to understand how people make sense of their lives within social, cultural, and

---

N. L. Sharp (✉) · R. A. Bye
School of Science and Health, Western Sydney University, Campbelltown, NSW, Australia
e-mail: n.sharp@westernsydney.edu.au; r.bye@westernsydney.edu.au

A. Cusick
Faculty of Health Sciences, Sydney University, Sydney, Australia
e-mail: anne.cusick@sydney.edu.au

© Springer Nature Singapore Pte Ltd. 2019
P. Liamputtong (ed.), *Handbook of Research Methods in Health Social Sciences*,
https://doi.org/10.1007/978-981-10-5251-4_106

historical contexts. There is no set approach to undertaking a narrative inquiry, and a number of scholars have offered interpretations of narrative inquiry approaches. Various combinations have also been employed successfully in the literature. There are, however, limited detailed accounts of the actual techniques and processes undertaken during the analysis phase of narrative inquiry. This can make it difficult for researchers to know where to start (and stop) when they come to do narrative analysis. This chapter describes in detail the practical steps that can be undertaken within narrative analysis. Drawing on the work of Polkinghorne (Int J Qual Stud Educ. 8(1):5–23, 1995), both narrative analysis and paradigmatic analysis of narrative techniques are explored, as they offer equally useful insights for different purposes. Narrative analysis procedures reveal the constructed story of an individual participant, while paradigmatic analysis of narratives uses both inductive and deductive means to identify common and contrasting themes between stories. These analysis methods can be used separately, or in combination, depending on the aims of the research. Details from narrative inquiries conducted by the authors to reveal the stories of emerging adults with cerebral palsy, and families of adolescents with acquired brain inquiry, are used throughout the chapter to provide practical examples of narrative analysis techniques.

Keywords

Narrative analysis · Social constructionism · Analysis procedures · Paradigmatic analysis of narratives · Story analyst · Storytellers · Narrative inquiry

1 Introduction

Narrative inquiry has been defined as a methodology "in which stories are used to describe human action" (Polkinghorne 1995, p. 5). Schwandt (2007) further explains that stories are central to all aspects of narrative inquiry. Narrative inquiry includes not only generating data in the form of stories but is also a means of analyzing stories about life experiences and a method of representing and reporting the findings of that analysis (Schwandt 2007). Narrative inquiry is grounded in the assumption that stories "constitute a fundamental form of human understanding, through which individuals make sense of themselves and of their lives" (Ylijoki 2001, p. 22). A life narrative allows people to interpret who they are and where their life is headed (Polkinghorne 1988). Even as children, stories are fundamental to learning to understand the world around us and our place within that social world. Stories continue to absorb us throughout our lives and become the "vehicle through which the reality of life is made manifest...we live by stories – both in the telling and the realising of the self" (Gergen 1994, p. 186). Through listening to stories, understanding is gained about the way people make meaning of their everyday lives within historical, social, and cultural contexts. Shared beliefs are understood, and values are conveyed (Polkinghorne 1988; Bruner 1990; Gergen 1994; Kramp 2004). Stories "assist humans to make life experiences meaningful. Stories preserve our memories, prompt our reflections, connect us with our past and present, and assist us to envision our future" (Kramp 2004, p. 107).

Polkinghorne (1988) argues that adopting a narrative approach to research is critical to understanding the meaning of life events and many others agree (e.g., Bruner 1990; Gergen 1994; Mishler 1986, 1995; Reissman 1993, 1997; Ennals and Howie 2017). Bruner (1990) suggests that humans have an innate readiness or predisposition that leads to using the narrative form to organize experiences into stories that have a temporal sequence and an unfolding plot structure. Gergen (1994) adds that the tendency to tell stories is also culturally influenced. Family stories are told to conserve memory across generations and pass on family values and beliefs. In the home and school settings, young people are encouraged to tell their own stories about day-to-day life to evaluate the day's events and their place in them (Gergen 1994). Narrative inquiry research is based on this same premise and assumes that by listening to people's stories, understanding can be gained about the way they make meaning of their everyday lives within a broader social context (see also ▶ Chap. 24, "Narrative Research").

This overarching view of the positive products of storytelling, based on a social constructionist perspective, is the foundation of narrative inquiry and narrative analysis. Narrative inquiry refers to an overall methodological approach to research, in a similar way that terms like phenomenology and ethnography refer to overarching qualitative research methodologies. According to Chase (2011, p. 421), narrative inquiry "revolves around an interest in life experiences as narrated by those who live them." Within this broad agenda, however, narrative researchers' interests, research processes, and outcomes differ significantly. There is a huge variety in what constitutes narrative inquiry. Some narrative researchers emphasize during data analysis the *whats* of narratives, that is, the content, plot, characters, and themes. Others focus more on the *hows* of narrative production, for example, the discourse used, semantics, structure, and form of language (Holstein and Gubrium 2012a). This chapter does not seek to provide an exhaustive account of all possible approaches to narrative inquiry; rather it introduces some key approaches with potential relevance to the health and social sciences. For interested readers, Chase (2011), Elliott (2005), Holstein and Gubrium (2012b), Mishler (1995), Reissman (2008), and Ennals and Howie (2017) all offer useful discussions of the broad varieties of narrative inquiry (see also ▶ Chap. 24, "Narrative Research").

Narrative analysis refers to specific procedures undertaken during the data analysis stage of narrative inquiry. Narrative analysis is the primary focus of this chapter. To set the scene, the chapter commences with a brief discussion of narratives, narrative inquiry, and the complexity and variety of methods deemed within the literature to constitute narrative inquiry. The core of the chapter focuses on exploring data analysis procedures which can usefully be employed during narrative inquiries. The narrative analysis conducted by the authors in conducting two major narrative inquiries is outlined in detail, with examples from the research findings included to highlight the outcomes of narrative analysis. Finally, the chapter concludes with an overview of future directions in narrative analysis.

2 What Is a Narrative?

Narratives involve the integration of events and human actions into a goal-directed story which is organized temporally (Polkinghorne 1988, 1995). Narratives consider the influence of the past, present, and future and are presented within a specific place or sequences of places (Clandinin and Connelly 2000; Kramp 2004; Ennals and Howie 2017). Integration occurs when a thematic thread, or plot, is employed to allow individual events and happenings to take on a "narrative meaning" (Polkinghorne 1995, p. 5). Events and happenings are then understood in relation to their contribution and impact on a specific outcome (Polkinghorne 1995). A plot weaves together "a complex of events to make a single story" (Polkinghorne 1988, p. 19). This story, or narrative whole, reveals greater meaning than when individual events are viewed in isolation. An example is helpful to demonstrate this. The first author conducted a doctoral research project which aimed to understand the stories of emerging adults aged 18–25 years with cerebral palsy from their own perspective. The following life events were shared by a participant during an in-depth interview: "I left school in year 10," "I started an apprenticeship," and "I took some time off." Viewed individually, these statements do not provide the complexity of understanding that is revealed through the narrative excerpt:

> I left High School in Year 10 and started a hair dressing apprenticeship. That went down the gurgler. I hated it. Went to TAFE, started a course, hated that, it wasn't what I wanted. Then I had a bit of a mental health crisis, took some time off, went back and did [another course] at TAFE, which was brilliant… I didn't really know what I wanted to do…I kept going from one thing to another.

In accord with other authors in the area of narrative inquiry (e.g., Kramp 2004; Polkinghorne 1995; Reissman 2008; Liamputtong 2013; Ennals and Howie 2017), we believe that the terms story and narrative can be used interchangeably, and this convention is adopted throughout this chapter. It is important to note that some researchers in the field of narrative inquiry, particularly in social linguistics, hold a different view and define story and narrative as different but related concepts. Other authors previously held this view (e.g., Reissman 1993), but have more recently come to adopt "contemporary conventions" of using story and narrative interchangeably (Reissman 2008, p. 7).

Narratives are not only conceived differently by scholars, but they are also defined in various ways across different disciplines; there is no single meaning. For example, in social linguistics, narrative relates to a discrete unit of discourse, while in social history and anthropology, narrative can refer to an entire life story. Narrative in psychology and sociology generally refers to extended accounts of lives within a clear context. These accounts may develop over a single interview or multiple interviews. In the human sciences, the definition of narrative is even broader, referring to both the stories told by research participants during interviews and the interpretive accounts developed by the researcher as a result of data analysis (Reissman 2008). Reissman (2008, p. 6) labels this a "story about stories." Reissman's broad definition of narrative has been adopted in this chapter.

3   Narrative Inquiry in Health and Social Sciences

The social constructivist underpinning of narrative inquiry can be particularly useful in understanding the lives of those whose experiences depart from "normative" expectations, such as those with disability or whose lives have taken an unexpected turn (Bruner 1990; Reissman 1993). For this reason, narrative inquiry has much to offer the health and social sciences. Narrative inquiry can be an approach that is most relevant to understanding the experiences, actions, motivations, and life journeys of people who are challenged by health, disability, trauma, change, adaptation, loss, or other significant life challenge. Narrative inquiry can contribute significantly to knowledge within the health and social sciences through its capacity to give voice to people whose voices have so often been discounted (Clandinin and Raymond 2006; Holloway 2007; Ennals and Howie 2017).

Narrative researchers, like other qualitative researchers, see their work as having potential to create positive change and address injustices. Many collect, present, and disseminate stories as a way of encouraging change (Chase 2011). For this reason, narrative inquiries are particularly relevant within the health and social sciences, where they have the potential to inform positive changes to practice, policy, education, and theory. Recommendations for change can be made on the basis of life experiences of people directly impacted by practices and policies. The potential of narrative inquiry to create positive social change has led to its popularity in research with marginalized and vulnerable groups (Liamputtong 2007). Chase highlights that "the urgency of speaking and being heard drives the ongoing collection and publication of narratives about many forms of social injustice" (Chase 2011, p. 428).

In the area of disability studies, for example, where our research has been focused, narrative inquiry puts the perspectives of people with disability, their families, and people who care for or work with them, "in the foreground of debates about care and constructions of impairment and disability" (Goodley and Tregaskis 2006, p. 632). Telling stories of disability has the potential to teach people with disability, and others, including service providers and policy-makers, about the complexity and diversity of experience. Smith and Sparkes (2008, p. 19) highlight that telling stories of disability has the capacity to "displace the tragedy story, challenge and resist social oppression."

Outside of health and disability, narrative inquiry has potential to contribute greatly to wide ranging social research. Just some of the areas noted to use narrative as a methodological tool include studies of human behavior, criminology, family and relationships, sexuality, and the sociology of education (Elliott 2005). This is by no means an exhaustive list, and the potential of narrative inquiry is likely to be recognized in increasingly diverse areas, and by researchers with increasingly varied backgrounds, in the future.

The diversity of narrative inquiry also extends beyond the field of research to the types of data utilized in telling and analyzing stories. In-depth interviews remain a predominant data collection tool used by many narrative researchers, including the authors of this chapter (see ▶ Chap. 23, "Qualitative Interviewing"). However, the data used in the field of narrative inquiry has broadened significantly in recent years. The addition of data including ethnographic observations, photographs, autobiographical writings, diaries, websites, and documents adds to the complexity and diversity of narrative inquiries now present within published literature (see ▶ Chaps. 26, "Ethnographic Method," ▶ 65, "Understanding Health Through a Different Lens: Photovoice Method," ▶ 30, "Autoethnography," and ▶ 29, "Unobtrusive Methods"). Data gained through rapidly developing technological sources including social media, blogs, video logs, and podcasts is likely to continue adding to the complexity and diversity in the future (see ▶ Chap. 77, "Blogs in Social Research").

Similarly, much variation exists in published narrative inquiries in relation to data analysis processes. Unlike other qualitative research methods, such as grounded theory, there is no standard analytical procedure for narrative inquiry research (Reissman 1993; Edwards 2016). It is impossible within this chapter to cover all possibilities. We will focus on the two distinct approaches to analysis in narrative inquiry detailed by Polkinghorne (1995). These are paradigmatic analysis of narrative data in the form of stories to produce categories, themes, or typologies and narrative analysis of data in the form of actions, events, and happenings to produce a narrative or explanatory story. As mentioned, these are not the only approaches to data analysis in narrative inquiry. Nor are we suggesting that they are superior to other approaches. Rather, this chapter aims to introduce the reader to two commonly used approaches, which individually, or in combination, may be useful for answering a broad range of research questions in the health and social sciences.

4 Introduction to the Research Exemplars

The following sections will utilize two key pieces of health research as exemplars to highlight techniques that can usefully be employed during narrative analysis. As mentioned, this chapter does not aim to be an authoritative source on the whole range of techniques that may constitute narrative analysis, as it is a hugely varied method open to significant interpretation. Rather, it aims to provide clear and replicable examples of those particular narrative analysis techniques and processes which have usefully been employed by the authors in health research and which may be useful to readers with similar worldviews and research aims. This section will briefly introduce the two research studies to set the context for the discussions that follow.

4.1 Exploring the Experience of Emerging Adults with Cerebral Palsy

Emerging adulthood has been identified as a unique developmental stage within developed countries (Arnett 2000). Generally associated with ages 18–25, emerging adulthood is seen as a time of identity exploration, instability, self-focus, a sense of feeling in-between, and possibilities. It is a time of profound change when people develop, review, and update their plan for the future (Arnett 2014). While emerging adulthood is a time of significant change and transition for all young people, it may be even more challenging for young people with a lifelong disability such as cerebral palsy (CP). The first author's doctoral research aimed to address a significant gap in knowledge around the experiences of emerging adults with CP from their own perspective. It also aimed to reflect on the usefulness of both developmental (the theory of emerging adulthood) and disability (the International Classification of Functioning, Disability and Health (ICF) [World Health Organisation 2001]) frameworks in understanding the experience of emerging adults with CP. Specifically, this study aimed to answer the following research questions:

1. What are the stories of emerging adults aged 18–25 years with CP?
2. What key themes are evident within the stories of emerging adults aged 18–25 years with CP?
3. How do these stories and themes inform and/or extend the theory of emerging adulthood and the ICF, and are these two conceptual models useful in understanding the experiences of these young people?

Eighteen emerging adults with CP participated in multiple unstructured in-depth interviews, guided by Holstein and Gubrium's (1995) active interview approach. Active interviews allow for in-depth responses and do not dictate the course of the interview or use preset questions. Rather, they "loosely direct" the interview by setting certain parameters and introducing certain topics during the interview to elicit responses that are relevant to the aims of the research (Holstein and Gubrium 1995, p. 29). Active interviews invite participants to talk about their experiences narratively and are therefore highly relevant to narrative inquiries.

4.2 Exploring Family Life Following the Acquired Brain Injury of an Adolescent Child

In Australia, adolescents aged 15–19 years are the most likely group to sustain a brain injury (Australian Institute of Health and Welfare 2003). The purpose of the second author's doctoral research study was to explore family life after the acquired brain injury (ABI) of an adolescent child. The specific aim of the study was to understand how parental caregivers shape everyday family life during the first 2 years post-injury to accommodate the disrupted development of their adolescent child.

This investigation of family adaptation was underpinned by ecocultural theory (Gallimore et al. 1999). Ecocultural theory builds upon systems approaches to extend the understanding of how family ecology (e.g., income, housing, transportation, neighborhood) and family culture (e.g., family values, goals, needs) shape family life. Ecocultural theory applies a social constructivist perspective to family life. It recognizes the capacity of families to "organize, understand and give meaning to their everyday lives" and thus create a unique family "ecocultural niche" through the daily routine (Bernheimer et al. 1990, p. 223). Ecocultural theory holds that parents construct daily life to meet the developmental needs of the children and family. Weisner (2002) refers to this process as the family's adaptive project.

The specific research questions addressed in this study were:

1. What is the adaptive project families engage in following the ABI of an adolescent child?
2. Do families engage in this process in different ways, and if so, how?
3. What implications do family ways of engaging in this adaptive project have for theory, practice, policy, education, and research?

Parents from 12 families where an adolescent aged 15–19 years had sustained a severe brain injury were interviewed using a narrative approach. Interviews took place over the first 2 years post-injury to capture the process of adaptation during the critical initial years of recovery (Khan et al. 2003).

The narrative analysis techniques utilized in the exemplar research studies will be explored in depth in the following sections.

5 Introduction to Two Approaches: Narrative Analysis and Analysis of Narratives

As previously mentioned, Polkinghorne (1995) made an important distinction between two types of narrative inquiry and, therefore, two types of narrative analysis. He based the two types on the two different, but complementary and equally valid, ways of understanding the world identified by Bruner (1985), narrative cognition and paradigmatic cognition. Discussion of how these two approaches can usefully be used, either separately or in combination, will be a focus of this chapter.

5.1 Narrative Cognition and Narrative Analysis

Narrative cognition is designed to understand the outcome of the interaction between a person's previous learning and experiences, their present situation, and their future goals and purposes. Narrative inquiry based on narrative cognition is labeled by Polkinghorne as *narrative analysis* (Polkinghorne 1995). Understanding is expressed by way of a story, with a plot that retains the complexity of the situation under exploration, and the emotions and motivations attached to it. Stories defined as using narrative cognition are complex accounts with a beginning, middle, and end, as distinct from a simple listing of a series of events. The plot, or point, of the story functions to enable researchers to select, from the multitude of data collected, those descriptions of happenings, events, and actions that relate to each other and are directly relevant to the story (Polkinghorne 1995). The result of narrative analysis is "an explanation that is retrospective, having linked past events together to account for how a final outcome might have come about" (Polkinghorne 1995, p. 16). The events are recounted in a time-ordered way that makes it clear how they contributed to the overall "point" (Gergen 1994, 1999). Researchers who conduct narrative analysis have been termed *storytellers*, as the outcome of data analysis is itself a story (Smith and Sparkes 2009).

5.2 Paradigmatic Cognition and Analysis of Narratives

Paradigmatic cognition, in contrast, refers to methods that classify instances into categories and subcategories based on common attributes. Paradigmatic reasoning is common to both qualitative and quantitative research design; however, it is used in different ways. In quantitative research, the categories, or units of measurement, are usually identified before data collection. In qualitative research, conceptualization may take two forms. It includes the inductive discovery of categories or themes from within the data, and may also utilize deductive processes to explore how well data fits with predetermined concepts, usually those reflected in an existing theoretical framework or frameworks (Berg 2007; Polkinghorne 1995). Narrative inquiry based on paradigmatic cognition is referred to by Polkinghorne (1995) as *analysis of narratives*. It requires the collection of stories as data, followed by paradigmatic analysis that results in "descriptions of themes that hold across the stories" (Polkinghorne 1995, p. 12). Researchers who conduct analysis of narratives have been termed *story analysts*. They go beyond the collection and construction of narratives, which is the focus of *storytellers*. Story analysts further analyze stories in order to extrapolate categories, themes, or other theoretical propositions (Smith and Sparkes 2009).

The key benefit of analysis based on paradigmatic cognition is the ability to "bring order to experience by seeing individual things as belonging to a category" (Polkinghorne 1995, p. 10). It enables general knowledge about a collection of stories to be gained. It does however by necessity "underplay the unique and particular aspects of each story" (Polkinghorne 1995, p. 15), which are maintained in the alternative narrative analysis. In order to achieve the benefits of both approaches to narrative inquiry, and depending on the aims of the research, they can be utilized in combination.

5.3 Complementary Use of Narrative and Paradigmatic Approaches to Narrative Inquiry

Polkinghorne (1995) reinforced that both types of narrative inquiry can make important contributions to knowledge and produce valuable results. Within the literature, a range of different approaches are described, and various combinations of approaches have been successfully utilized. The different approaches are not mutually exclusive (Smith and Sparkes 2008). For example, Ylijoki (2001) interviewed 72 students and used a narrative analysis approach to construct four different core narratives around the experience of writing a master's research thesis. McCance et al. (2001) identified common themes using paradigmatic analysis of narratives and also constructed six storied case studies using narrative analysis in their research exploring caring in nursing practices. Cussen et al. (2012) used narrative analysis to construct individual stories for all of their adolescent participants and then derived common themes about aspirations for the future using paradigmatic analysis of narratives. Other researchers have also previously employed Polkinghorne's dual techniques successfully (e.g., Bailey and Jackson 2003; Kramp 2004; McCormack 2004). Bleakley (2005) goes as far as describing narrative analysis and analysis of narratives as being as complementary as a "lock and key: approaches of analysis and synthesis look different apart but constitute a unit together" (Bleakley 2005, p. 537).

This dual approach has been utilized by the authors in their research studies. Each type of analysis was employed for a different purpose. For example, in the first authors' research with emerging adults with CP, each participant's story was initially told individually, through the outcomes of narrative analysis. Each participant had an important story to tell, and the researcher wanted their story (as constructed by the researcher) to be included as a whole, not only parts of it in a fractured way. Narrative analysis provided the answer to the research question: what are the stories of emerging adults aged 18–25 years with CP?

Additionally, the research aimed to understand themes evident across stories and to understand how the stories inform the theory of emerging adulthood and the ICF. For this reason, a paradigmatic analysis of narratives approach was subsequently utilized to identify the common and contrasting themes present across stories. These themes were identified using both an inductive approach, directly from the data, and a deductive approach, whereby key concepts within the theory of emerging adulthood and the ICF were considered in relation to the data. Both processes will be outlined in the sections to follow.

Similarly, the second authors' research with families of adolescents with acquired brain injury initially presented a constructed story, developed through narrative analysis, describing each family's individual experience during the first 2 years post-injury. Analysis of narratives followed in order to explore and describe the adaptive family projects of the participating families. An initial common experience during the acute phase of recovery was identified and categorized thematically across all stories, along with three different adaptive family projects, or ways of living, during the post-acute phase of recovery over 2 years post-injury. Each adaptive family project was representative of several families' new way of living post-injury. Similar to the first exemplar, both inductive and deductive paradigmatic analysis of narratives techniques were employed, with the deductive analysis in this case associated with how families' adaptive projects related to key concepts from the guiding ecocultural theory.

The following section will explain in detail the processes for both narrative analysis and analysis of narratives used by the authors, to provide examples to guide the reader in planning their own data analysis procedures. For ease of explanation, procedures used during narrative analysis and analysis of narratives are described in this chapter separately. In reality, both forms of analysis can be utilized concurrently or at the very least considered throughout the analysis process. For example, while a researcher may focus initially on narrative analysis, early identification of common and contrasting themes may occur during this process. These can be recorded and then further explored during more targeted analysis of narratives.

6 Narrative Analysis Techniques

The aim of narrative analysis is to construct a narrative which details a participant's experience in relation to the research question(s). Polkinghorne (1995) provides a useful description of criteria, first articulated by Dollard in 1935, which constructed narratives should aim to address. These criteria include:

(a) Descriptions are included of the cultural and social context of the story.
(b) Information is provided about the subject of the story (the research participant), for example, their age, developmental stage, and other information relevant to the aims of the research.
(c) Explanations are included of the relationships between the participant and other significant people in their life.
(d) The story concentrates on the goals, choices, interests, plans, purposes, and actions of the participant, on their meanings, and on their vision of the world.
(e) Recognition is given to historical experiences and events that have influenced the participant's life story.
(f) The story is bound by time; it has a beginning, middle, and end.
(g) The narrative offers a meaningful explanation of the participant's experiences and actions, drawing together separate data elements in a credible and understandable way (Polkinghorne 1995).

Polkinghorne (1995) offers guidance around the actual process of constructing narratives. The researcher is encouraged to first consider the story's ending or outcome. This provides a lens through which parts of the data that are relevant to, or contribute to, that outcome can be identified. These data elements can then be arranged chronologically, and the connections between events and happenings can be articulated. Numerous direct quotes from the participant should be included to demonstrate key points using the voice of the participant. This ensures the narrative is grounded in data and authentic in tone. The overall aim of this process is to construct a narrative in which "the range of disconnected data elements are made to cohere in an interesting and explanatory way" (Polkinghorne 1995, p. 20). Developing narratives involves a recursive process of movement between the participant's data and the emerging story. Early attempts at writing the narrative should be tested against the data, and if events or actions are identified that contradict the emerging plot, further development and refinement is required (Polkinghorne 1995).

It is important to note the importance of the term "constructing narratives." The narratives discussed above are the researcher's construction of the participant's experiences. Narratives in research findings are not neutral depictions of the participants' life stories. The construction is influenced by the researcher's own experiences, views, and priorities (Polkinghorne 1995). It is also influenced by the data produced through the dynamic and collaborative interactions between researcher and participants during in-depth interviews and by the narrative terrain of the research (Holstein and Gubrium 1995). The stories told through narrative analysis are not the same stories the participants would tell if they were asked to write their own story. They are unlikely to be the same stories another researcher may construct after undertaking similar research. But they offer an understandable and credible explanation of the participants' experiences, including many examples in their own voices. The stories constructed through narrative analysis should thus aim to "fit the data while at the same time bringing an order and meaningfulness that is not apparent in the data" (Polkinghorne 1995, p. 16).

7 Analysis of Narrative Techniques

In contrast to narrative analysis techniques described above, analysis of narratives is designed to compare and contrast various narratives, identify key themes, and/or explore narratives through a theoretical lens. While the analysis of narrative form of narrative inquiry is different to other types of qualitative research in that its data is in the form of stories, it is similar in many ways to other types of qualitative research in its analysis methods. Some researchers may find qualitative data analysis software helpful to facilitate the process of analysis of narratives, for example, NVivo, ATLAS, or Quirkos (see also ▶ Chap. 52, "Using Qualitative Data Analysis Software (QDAS) to Assist Data Analyses"). Interested readers are directed to Liamputtong (2013) and Serry and Liamputtong (2017) for a discussion of computer-assisted qualitative data analysis, its key functions, benefits, cautions, and considerations. However, we conducted our coding and analysis using more traditional, low-tech strategies during the exemplar research. These strategies will be outlined below.

As briefly outlined earlier in the chapter, analysis of narratives is based on a paradigmatic view of the world and is, therefore, designed to identify categories, themes, or typologies based on attributes common to stories (Polkinghorne 1995). There are two approaches to undertaking a paradigmatic analysis of narratives, inductive and deductive. Inductive analysis refers to deriving categories and themes directly from the data using such processes as constant comparison, made explicit initially by Glaser and Strauss (1967) and later by Strauss and Corbin (1990). Deductive analysis of narratives refers to analysis which is focused on exploring data for examples of theoretical concepts or pre-existing knowledge relevant to the research aims. Berg (2007) suggests that while inductive analysis allows the researcher to ground categories within the data, and therefore, most directly present the perceptions of participants, it is also reasonable that researchers should draw on their own experience with the study phenomena during analysis. Understandings gained from theoretical perspectives, existing scholarly literature and research undertaken in the field, can be drawn on and further clarified or explored through deductive analysis methods (Berg 2007). Researchers can thus explore whether data fits with a predetermined theory or pre-existing knowledge in the field (Polkinghorne 1995). Techniques that can be usefully employed during both approaches will now be explored.

7.1 Inductive Paradigmatic Analysis

Polkinghorne (1995) describes the data analysis methods proposed for grounded theory by Glaser and Strauss (1967), and later Strauss and Corbin (1998), as being appropriate tools for inductive analysis of narratives. Two key grounded theory techniques are now introduced: asking questions and making comparisons (Strauss and Corbin 1998; Corbin and Strauss 2014). These analytical tools can be employed to assist the researcher to grasp the meaning of events, to sensitize the researcher to undiscovered properties and dimensions within the data, and to facilitate the linking of categories.

Asking questions enables a researcher to focus in on what the data is indicating, to understand structure and process, to identify connections between concepts, to stimulate thinking, and to increase sensitivity to what to look for in future data (Strauss and Corbin 1998; Corbin and Strauss 2014). Key questions that can be asked of the data include "Who? When? Why? Where? What? How? How much? With what results?" (Strauss and Corbin 1998, p. 89; Corbin and Strauss 2014). For example, "what is going on here?", "why is this important?", "how did this experience make them feel?", and "how did they respond?".

Making comparisons allows incidents within the data to be compared to other incidents in order to group them into categories according to similarities and differences (Strauss and Corbin 1998; Corbin and Strauss 2014). Grouped data is then coded by giving it a name or conceptual label that best summarizes the data (Strauss and Corbin 1990; Corbin and Strauss 2014). Questions the researcher can ask of the data to help guide comparisons include "did other participants have a similar experience?", "did others report different experiences?", and "what seems to be influencing the differences in experience?".

Other paradigmatic qualitative data analysis techniques could also provide useful insights into techniques that could effectively be employed during analysis of narratives. For example, thematic analysis technique as described by Braun and Clarke (2006, 2013) is another example of an approach that can be used to derive themes inductively from data (see also ▶ Chap. 48, "Thematic Analysis"). Overcash (2003) provides a discussion of the relevance of thematic analysis in making sense of narrative data. Carson et al. (2017), in their study of the birth stories of young mothers, provide a useful example of a narrative analysis which utilizes thematic analysis procedures.

In the first author's research with emerging adults with CP, inductive coding was initially recorded in the margins of hard copy transcripts, and various color markers were utilized to highlight related passages of text. For example, codes were identified during this process including bullying, transport and driving, mental health problems, disclosure, and giving back. As analysis continued, the researcher developed knowledge of codes which had previously, or were commonly, appearing in the data; this "pool of concepts" was used to help guide further analysis (Strauss and Corbin 1998, p. 114; Corbin and Strauss 2014). Early codes were updated and developed as greater understanding of the nature of the concepts was gained (Strauss and Corbin 1998; Corbin and Strauss 2014). Codes and illustrative direct quotes from participants were recorded electronically within Word documents as data analysis progressed. Related codes were grouped into themes, and these were reviewed against new data for consistency and revised as necessary prior to the final compilation of results.

7.2 Deductive Paradigmatic Analysis with a Theoretical Focus

As introduced above, deductive analysis of narratives refers to analysis which is focused on exploring data to identify examples of theoretical concepts or pre-existing knowledge that is of relevance to the aims of the research (Polkinghorne 1995). This approach to paradigmatic analysis can sensitize the researcher to seeing concepts in the data that may not be evident without knowledge of relevant conceptual frameworks or theories. For example, the data can be examined to see whether key concepts that are widely understood and accepted in research literature or theory are evident in the unique experiences of participants. The theory is not used as a set of categories to fit the data into, but rather as a conceptual framework to examine the data in relation to, thus being able to possibly extend, refute, or support concepts and theories in relation to the participants' experiences being investigated.

For example, in the first author's research, deductive coding was completed concurrently with inductive coding and according to key concepts within the guiding theories. Data related to the five features of emerging adulthood (identity explorations, instability, self-focus, feeling in-between, and possibility) were recorded when they appeared in the narratives. Concepts related to the ICF were also coded: interpersonal interactions and relationships; community, social, and civic life; education, work, and economic life; attitudes; support; services and systems; and experiences of accessing the community. For example, where a participant was talking about how a particular experience had influenced the way he or she viewed him or herself, this was labeled "identity exploration." When a participant spoke of the influence of others' attitudes on their experiences, it was labeled "attitudes."

In the second author's narrative inquiry, ecocultural theory was used to guide deductive analysis of narratives. This involved coding data relevant to the adaptive family project post-adolescent ABI, in particular how parental primary caregivers shaped the daily routine to undertake this adaptive project. Weisner (2002) states that the adaptive family project can be analyzed in terms of its meaningfulness, ecological fit, and congruence. These three concepts are the summary dimensions of family life according to ecocultural theory, and therefore, coding was also undertaken to identify data related to these three theoretical concepts. The types of data included under the adaptive family project heading included how the daily routine was structured and the aims that drove the construction of this way of life. Data pertaining to the meaning behind the construction of this project was then categorized under the heading of meaningfulness. The ecological fit category contained data relating to the resources families drew upon to continue with this routine and whether these resources were adequate to meet the aims of the adaptive family project. The congruence category contained data relating to how this adaptive project impacted on family members, for example, how parents juggled the demands of raising other children at the same time as helping their adolescent with ABI.

The following section will provide examples of the outcomes of narrative analysis and analysis of narratives. Depending on the aims of research, the structure of research results of narrative inquiries can vary significantly. Two examples are provided in detail from the authors' inquiries, and a range of alternatives are also highlighted from the literature, which may offer additional guidance and insight to the reader as they consider the most appropriate narrative inquiry approach to answer their research question or questions.

8 Examples of the Outcomes of Narrative Analysis and Analysis of Narratives

In the first author's research with emerging adults with CP, implementing a narrative analysis approach allowed the synthesis of participants' descriptions and stories into an individualized narrative of each emerging adult's experience, relevant to the research aims, which was organized by time. These constructed narratives were reflective of the data, but at the same time offered a new level of order and meaningfulness (Polkinghorne 1995). Each story included numerous direct quotes in the participants' own voices, offering a depth that is unequaled by other forms of data analysis. In the presentation of results, each of the 18 narratives was initially told as a stand-alone story. The outcome of inductive and deductive analysis of narratives was then presented. Fourteen themes were identified during analysis of narratives as important across the 18 participants' stories. These were presented and discussed, with multiple direct quotes from participants used to highlight the themes and to compare and contrast participant experiences. Examples of themes identified included: "The journey to find myself: Identity exploration," "Getting back on the road: resilient versus resigned," and "A bumpy road: Finding the right job."

In the second author's research, data analysis revealed an initial crisis period post-injury common to all parents. This was termed "willing survival and recovery" and detailed parents efforts to organize daily life to spend time at the hospital so they could focus on the survival of the adolescent. As recovery took place, and the adolescent returned home, parents assessed how the injury had altered their adolescent's development and how family life would need to change to accommodate the altered child. Three factors influenced parents' construction of daily life: the developmental discrepancy between their adolescent's abilities pre- and post-injury, the developmental pressure to resolve this discrepancy during the initial 2 years of recovery, and a sense of developmental uncertainty about whether their adolescent would return to the previous developmental level given unknown recovery following ABI.

Three divergent ways of living emerged to manage this developmental uncertainty, discrepancy, and pressure. Some parents constructed "a rehabilitative life." These parents worked hard to rehabilitate their child to facilitate the best recovery possible, altering family routines to concentrate on both formal and home-based rehabilitation. Others constructed a daily routine that focused around "the search for a settled life." These families faced many disruptions due to their adolescents' behavioral changes post-injury, and parents aimed to achieve a stable family routine. Another group of parents constructed "a cautious life." The adolescents in these families had good recoveries post-injury and were able to gradually resume normal life. However, the trauma of the brain injury of their child led these parents to be more cautious, ensuring their child remained safe from harm during the recovery period.

In the presentation of results of this research, the initial common crisis was outlined, with multiple direct quotes from participants utilized to highlight the challenges of this crisis for families. The three divergent ways of living that emerged after the initial crisis period were then presented in different sections. These three ways of living were derived from the experiences of the 12 families, and each of the 12 families was categorized under the way of living that suited their new life post ABI. Initially, a constructed narrative for each participant within the group was presented, the result of narrative analysis. The overall adaptive family project of the group was then presented, the result of deductive analysis of narratives. Each adaptive family project was structured around the group's experiences in relation to the ecocultural theory concepts of meaningfulness, ecological fit, and congruence. The meaning of everyday life, its ecological fit, and the congruence of the daily routine were differentiated within each adaptive family project with their own unique category names which reflected the dominant theme within the narrative data of this group. For example, within the "rehabilitative life" adaptive family project, meaningfulness was summed up and described by the category "hope and hard work toward a more normative life," ecological fit was described under the category "rehabilitation resources: parents and professionals," and congruence was described under the category "we all have to pull together 'otherwise everything falls apart'."

Within the literature, the reader will find a myriad of similar, and disparate, ways of presenting the findings of narrative analysis. Readers are encouraged to review the literature to locate exemplar research that is in line with their own research aims and worldview. In doing so, researchers can take guidance on approaches to narrative analysis that may be most appropriate to their own research. Just a few examples from the health and social sciences are presented here as a starting point for the reader's further exploration.

Brown and Addington-Hall (2008) analyzed longitudinal narrative interviews with 13 people with motor neuron disease in order to explore how participants talked about living and coping with their illness. Four types of narrative storylines were identified within the stories of participants. The first was a "sustaining storyline," about living life as well as possible and remaining engaged. The second was the "enduring storyline," which concerned feelings of disempowerment that resulted from living with a life-limiting illness. Survival was the essence of the third storyline, termed the "preserving storyline." The final storyline concerned loss and fear of the future and was labeled the "fracturing storyline." Storylines were not mutually exclusive, with more than one, or even all four, being present within some participant's stories (Brown and Addington-Hall 2008).

In another valuable example, Davies et al. (2016) utilized narrative analysis in their study of older widows' experiences of loneliness. Data analysis resulted in the identification of three themes: "experiencing the absence," "loss of routine connection," and "establishing new routines." These themes were then reviewed and an overarching narrative, or collective story, of participants was identified:

> The overarching narrative followed a trajectory, from the onset of loneliness associated with widowhood through a process of transition from marriage to widowhood. The loss of the spouse was experienced as a sense of absence that resulted in a loss of usual routines which had enabled a sense of connection with others. The participants had transitioned through establishing new routines thus creating new connections with others. (Davies et al. 2016, p. 535)

Lastly, Papathomas et al. (2015) explored the experiences of a 21-year-old elite triathlete with an eating disorder and the experiences of her parents. Open-ended, participant-led interviews were conducted over 1 year in order to provide a longitudinal perspective. Narrative analysis considered both the content and structure of the data and was guided by Frank's (1995) typology of illness narratives. Analysis identified that the stories of the participants reflected a restitution narrative: "yesterday I was a healthy athlete, today I have an eating disorder, tomorrow I'll be a healthy athlete again" (Papathomas et al. 2015, p. 317), with a turn to a quest narrative in the final interviews: "I'm just trying to make the most of it" (Papathomas et al. 2015, p. 322).

The examples presented within this section, both from our own research and published literature, provide the reader with a snapshot of the broad range of possible outcomes of narrative analysis and analysis of narratives. Many more possibilities are available to researchers who choose a narrative inquiry approach. This chapter concludes with a discussion of some of those possibilities and future directions for narrative analysis.

9 Conclusion and Future Directions

Stories and storytelling are inherent to human experience, help us make sense of our lives, and find meaning from events. Similarly, researchers can adopt narrative inquiry as an approach that is well suited to understanding people's experiences related to a research phenomenon. In the health and social sciences, narrative inquiry is particularly relevant to understanding challenges people face and how they cope and adapt throughout their lives to significant turns in life: change to health, adaptation to or living with disability, trauma, nonnormative events, aging, and so on. Narrative inquiry provides rich data from which researchers can better understand the details of individuals' lives and from that point understand the implications of research findings in relation to necessary changes to policy, practice, education, and theory.

The purpose of this chapter was to provide readers with a background briefing on narrative inquiry and illustrate key approaches to narrative analysis and analysis of narratives, using examples from research. This chapter did not aim to be, and does not pretend to be, an exhaustive or prescriptive source of information about narrative analysis. Instead, it aimed to provide practical guidance on those approaches the authors' have effectively utilized and which may be relevant for other research in the health and social sciences. In any narrative inquiry, the researcher needs to make choices about the best approach to data analysis and provide a clear rationale for these choices. There is, therefore, a need for researchers to articulate their worldview and the influence on the narrative analysis procedures they employ, as there is a huge variety in scope and actual data analysis techniques. Researchers are advised to carefully consider and justify the reasons guiding a decision to employ narrative analysis, analysis of narratives, a combination, or indeed other types of analysis not discussed in depth within this chapter.

Readers are also encouraged to share their own experiences and insights gained through conducting narrative inquiries. This will continue the important dialogue about narrative inquiry and narrative data analysis and support the further development and advancement of the methodology. In the words of Overcash (2003, p. 183), "the beauty of narrative methods is in the diversity and malleability of the methodology in capturing the human experience." Through sharing this diversity, including practical discussions of the *how* of data analysis, new ways of capturing and understanding human experience will evolve. We look forward to continued discussion, debate, and development of narrative inquiry and analysis.

References

Arnett JJ. Emerging adulthood: a theory of development from the late teens through the twenties. Am Psychol. 2000;55(5):469–80.
Arnett JJ. Presidential address: the emergence of emerging adulthood: a personal history. Emerg Adulthood. 2014;2(3):155–62. https://doi.org/10.1177/2167696814541096.
Australian Institute of Health and Welfare. Disability prevalence and trends. Canberra: Author; 2003.
Bailey DM, Jackson JM. Qualitative data analysis: challenges and dilemmas related to theory and method. Am J Occup Ther. 2003;57(1):57–65.
Berg BL. Qualitative research methods for the social sciences. 6th ed. Boston: Pearson Education; 2007.
Bernheimer LP, Gallimore R, Weisner TS. Ecocultural theory as a context for the individual family service plan. J Early Interv. 1990;14(3):219–33.
Bleakley A. Stories as data, data as stories: making sense of narrative inquiry in clinical education. Med Educ. 2005;39(5):534–40. https://doi.org/10.1111/j.1365-2929.2005.02126.x.

Braun V, Clarke V. Using thematic analysis in psychology. Qual Res Psychol. 2006;3(2):77–101. https://doi.org/10.1191/1478088706qp063oa.

Braun V, Clarke V. Successful qualitative research: a practical guide for beginners. London: Sage; 2013.

Brown J, Addington-Hall J. How people with motor neurone disease talk about living with their illness: a narrative study. J Adv Nurs. 2008;62(2):200–8. https://doi.org/10.1111/j.1365-2648.2007.04588.x.

Bruner J. Actual minds, possible worlds. Cambridge: Harvard University Press; 1985.

Bruner J. Acts of meaning. Cambridge: Harvard University Press; 1990.

Carson A, Chabot C, Greyson D, Shannon K, Duff P, Shoveller J. A narrative analysis of the birth stories of early-age mothers. Sociol Health Illn. 2017;39(6):816–31. https://doi.org/10.1111/1467-9566.12518.

Chase SE. Narrative inquiry: still a field in the making. In: Denzin NK, Lincoln YS, editors. The sage handbook of qualitative research. 4th ed. Thousand Oaks: Sage; 2011. p. 421–34.

Clandinin DJ, Connelly FM. Narrative inquiry: experience and story in qualitative research. San Francisco: Jossey-Bass; 2000.

Clandinin DJ, Raymond H. Note on narrating disability. Equity Excell Educ. 2006;39(2):101–4. https://doi.org/10.1080/10665680500541176.

Corbin J, Strauss A. Basics of qualitative research: techniques and procedures for developing grounded theory. 4th ed. Thousand Oaks: Sage; 2014.

Cussen A, Howie L, Imms C. Looking to the future: adolescents with cerebral palsy talk about their aspirations – a narrative study. Disabil Rehabil. 2012;34(24):2103–10. https://doi.org/10.3109/09638288.2012.672540.

Davies N, Crowe M, Whitehead L. Establishing routines to cope with the loneliness associated with widowhood: a narrative analysis. J Psychiatr Ment Health Nurs. 2016;23(8):532–9. https://doi.org/10.1111/jpm.12339.

Edwards SL. Narrative analysis: how students learn from stories of practice. Nurse Res. 2016;23(3):18–25.

Elliott J. Using narrative in social research: qualitative and quantitative approaches. London: Sage; 2005.

Ennals P, Howie L. Narrative inquiry and health research. In: Liamputtong P, editor. Research methods in health: foundations for evidence-based practice. 3rd ed. South Melbourne: Oxford University Press; 2017. p. 105–20.

Frank AW. The wounded storyteller: body, illness and ethics. Chicago: University of Chicago Press; 1995.

Gallimore R, Bernheimer LP, Weisner TS. Family life is more than managing crisis: broadening the agenda of research on families adapting to childhood disability. In: Gallimore R, Bernheimer LP, MacMillan DL, Speece DL, Vaughn S, editors. Developmental perspectives on children with high-incidence disabilities. Mahwah: Lawrence Erlbaum Associates; 1999. p. 55–80.

Gergen KJ. Realities and relationships: soundings in social construction. Cambridge: Harvard University Press; 1994.

Gergen KJ. An invitation to social construction. London: Sage; 1999.

Glaser BG, Strauss AL. The discovery of grounded theory: strategies for qualitative research. New York: Aldine de Gruyter; 1967.

Goodley D, Tregaskis C. Storying disability and impairment: retrospective accounts of disabled family life. Qual Health Res. 2006;16(5):630–46. https://doi.org/10.1177/1049732305285840.

Holloway I. Vulnerable story telling: narrative research in nursing. J Res Nurs. 2007;12(6):703–11. https://doi.org/10.1177/1744987107084669.

Holstein JA, Gubrium JF. The active interview. Thousand Oaks: Sage; 1995.

Holstein JA, Gubrium JF. Introduction: establishing a balance. In: Holstein JA, Gubrium JF, editors. Varieties of narrative analysis. Thousand Oaks: Sage; 2012a. p. 1–12.

Holstein JA, Gubrium JF. Varieties of narrative analysis. Thousand Oaks: Sage; 2012b.

Khan F, Baguley IJ, Cameron ID. Rehabilitation after traumatic brain injury. Med J Aust. 2003;178:290–5.

Kramp MK. Exploring life and experience through narrative inquiry. In: deMarrais KB, Lapan SD, editors. Foundations of research: methods of inquiry in education and the social sciences. Mahwah: Lawrence Erlbaum Associates; 2004. p. 103–22.

Liamputtong, P. Researching the vulnerable: a guide to sensitive research methods. London: Sage; 2007.

Liamputtong P. Qualitative research methods. 4th ed. South Melbourne: Oxford University Press; 2013.

McCance TV, McKenna HP, Boore JRP. Exploring caring using narrative methodology: an analysis of the approach. J Adv Nurs. 2001;33(3):350–6. https://doi.org/10.1046/j.1365-2648.2001.01671.x.

McCormack C. Storying stories: a narrative approach to in-depth interview conversations. Int J Soc Res Methodol. 2004;7(3):219–36. https://doi.org/10.1080/13645570210166382.

Mishler EG. Research interviewing: context and narrative. Cambridge: Harvard University Press; 1986.

Mishler EG. Models of narrative analysis: a typology. J Narrat Life Hist. 1995;5(2):87–123. https://doi.org/10.1075/jnlh.5.2.01mod.

Overcash JA. Narrative research: a review of methodology and relevance to clinical practice. Crit Rev Oncol Hematol. 2003;48:179–84.

Papathomas A, Smith B, Lavallee D. Family experiences of living with an eating disorder: a narrative analysis. J Health Psychol. 2015;20(3):313–25. https://doi.org/10.1177/1359105314566608.

Polkinghorne DE. Narrative knowing and the human sciences. New York: State University of New York Press; 1988.

Polkinghorne DE. Narrative configuration in qualitative analysis. Int J Qual Stud Educ. 1995;8(1):5–23. https://doi.org/10.1080/0951839950080103.

Reissman CK. Narrative analysis. Newbury Park: Sage; 1993.

Reissman CK. A short story about long stories. J Narrat Life Hist. 1997;7(1–4):155–8.

Reissman CK. Narrative methods for the human sciences. Thousand Oaks: Sage; 2008.

Schwandt T. The Sage dictionary of qualitative inquiry. 3rd ed. Thousand Oaks: Sage; 2007.

Serry T, Liamputtong P. Computer-assisted qualitative data analysis (CAQDAS). In: Liamputtong P, editor. Research methods in health: foundations for evidence-based practice. 3rd ed. South Melbourne: Oxford University Press; 2017. p. 437–50.

Smith B, Sparkes AC. Narrative and its potential contribution to disability studies. Disabil Soc. 2008;23(1):17–28. https://doi.org/10.1080/09687590701725542.

Smith B, Sparkes AC. Narrative analysis and sport and exercise psychology: understanding lives in diverse ways. Psychol Sport Exerc. 2009;10(2):279–88. https://doi.org/10.1016/j.psychsport.2008.07.012.

Strauss AL, Corbin J. Basics of qualitative research: grounded theory procedures and techniques. Newbury Park: Sage; 1990.

Strauss AL, Corbin J. Basics of qualitative research: techniques and procedures for developing grounded theory. 2nd ed. Thousand Oaks: Sage; 1998.

Weisner TS. Ecocultural understanding of children's developmental pathways. Hum Dev. 2002;45(4):275–81.

World Health Organisation. International classification of functioning, disability and health: ICF. Geneva: WHO; 2001.

Ylijoki O-H. Master's thesis writing from a narrative approach. Stud High Educ. 2001;26(1):21–34. https://doi.org/10.1080/03075070020030698.

Critical Discourse/Discourse Analysis

50

Jane M. Ussher and Janette Perz

Contents

1	Introduction	882
2	Discourse Analysis and Discursive Psychology	882
3	Foucauldian Discourse Analysis	884
4	Synthesizing DA and FDA: A Combined Approach	886
5	Defining Discourse Analytic Methods	886
6	Analyzing Premenstrual Distress: A Feminist Discourse Analysis	887
	6.1 Study Aim and Method	887
	6.2 Women's Descriptions of Premenstrual Change	888
	6.3 Premenstrual Change as Pathology: Woman as Victim or Monster	889
	6.4 PMS as an Understandable Reaction: Facilitating Agency and Self-Care	892
7	Conclusion and Future Directions	893
References		894

Abstract

Discourse analysis (DA) conceptualizes language as performative and productive, central to the construction of social reality and subjectivity. This chapter examines two identifiable, but overlapping, schools of DA, discursive psychology (DP) and Foucauldian discourse analysis (FDA). DP draws on the practices of ethnomethodology and conversations analysis and focuses on the *action orientation* of talk and text in social practice: what is the text *doing,* rather than what does the text mean, or "what is the text saying?" Analysis focuses on "interpretive repertoires" or "discourses": sets of statements that reflect shared patterns of meaning. Foucauldian discourse analysis (FDA) originates within poststructuralist theory, influenced by the philosophical work of Michel Foucault. Within FDA, language is deemed to be constitutive of social life,

---

J. M. Ussher (✉) · J. Perz
Translational Health Research Institute, School of Medicine, Western Sydney University, Sydney, NSW, Australia
e-mail: j.ussher@westernsydney.edu.au; j.perz@westernsydney.edu.au

© Springer Nature Singapore Pte Ltd. 2019
P. Liamputtong (ed.), *Handbook of Research Methods in Health Social Sciences*,
https://doi.org/10.1007/978-981-10-5251-4_105

making available certain subject positions, which influence and regulate subjectivity and experience – the way we think or feel, our sense of self, and the practices in which we engage. FDA is thus concerned with identifying discourses, the subject positions they open up (or disallow), and the implications of such positioning for subjectivity and social practice, rather than the form or structure of interaction within talk or text. Following discussion of a range of DP and FDA research studies, a detailed example of feminist FDA is provided, including steps of analysis, based on a study of women's accounts of PMS (premenstrual syndrome). It is concluded that there is no one correct method of DA, as multiple methods have been identified, and practitioners interpret and present analyses in a range of different ways.

Keywords

Foucauldian discourse analysis · Discursive psychology · Feminist discourse analysis · Premenstrual syndrome (PMS) · Discursive practices

1 Introduction

Discourse analysis (DA) is a form of qualitative analysis that followed the "turn to language" in the social sciences in the second half of the twentieth century. This "turn" marked a shift from the conceptualization of language as a reflection of thought or experience, to seeing language as both performative and productive, central to the construction of social reality and subjectivity. Discourse analysis has, therefore, always been more than a method – it is part of a broader critique of positivism-realism within social science disciplines (see Potter 2012; see also ▶ Chap. 9, "Positivism and Realism").

DA evolved within a range of diverse disciplinary contexts, including literary criticism, linguistics, psychology, philosophy, and sociology. The influence of these disciplines shapes the nature and emphasis of the various strands of DA practiced today. For example, within linguistics, the focus is on fine grained examination of sentence structure and utterances; within cognitive psychology, attention is paid to the role of mental schemas and scripts in the comprehension of language (see Potter 2004). Within the Social Sciences, there is much rich and vigorous debate about the nature and function of discourse analysis, leading to a range of interpretations and developments (Hollway 1989; Wetherell 1998; Billig 2012; Parker 2012). This has crystallized into two identifiable, but overlapping, schools of thought, described as discursive psychology (DP) and Foucauldian discourse analysis (FDA). In this chapter, we will examine the theoretical and conceptual framework of both of these approaches, their commonalities and differences, and their applicability for the Health Social Sciences.

2 Discourse Analysis and Discursive Psychology

Initially described as "discourse analysis" (Potter and Wetherell 1987), and more recently as "discursive psychology" (DP) (Edwards and Potter 1992), this methodology is influenced by poststructuralist literary theory, as well as the sociology of scientific knowledge (Potter 2012). DP draws on the practices of ethnomethodology and conversations analysis, and focuses on the *action orientation* of talk and text in social practice: what is the text *doing,* rather than what does the text mean, or "what is the text saying?" (Willig 2008, p. 98; see also ▶ Chaps. 16, "Ethnomethodology," and ▶ 28, "Conversation Analysis: An Introduction to Methodology, Data Collection, and Analysis"). Analysis focuses on "interpretive repertoires" or "discourses": sets of statements that reflect shared patterns of meaning. Topics recognized in mainstream health research, such as "attitudes," "causal attribution," "script," and "knowledge," are reframed as "discourse practices" (Edwards and Potter 2005, p. 241). Thus, for example, the study of attitudes is replaced by analysis of argumentative practices in discourse (Potter and Wetherell 1987).

Individuals conducting DP are concerned with the management of issues of "stake or interest" within talk (Potter 2004). For example, an individual may say "I've got nothing against gay people, but I don't agree with gay marriage," as a way of disclaiming a homophobic identity, then legitimate their position by appealing to a higher authority: "scientific research makes it clear that children need to be brought up by a man and a woman." Within DP, attention is also paid to the *negotiation* of meaning within language and to the interaction between speakers in everyday situations. The *form* of language is also of interest, in order to examine "what people do" with language (Potter 2004, p. 203), and how they manage social interactions. Emphasis is thus placed on the rhetorical or argumentative nature of talk and texts, metaphors and analogies, extreme case examples, graphic descriptions, and consensus formulations (see Edwards and Potter 1992). The association of gay marriage with bestiality and polygamy within political and media debate in both the UK and Australia would provide ideal fodder for discursive analysis of extreme cases. Analysis is often focused on the often contradictory interpretative repertoires individuals draw on in their accounts (Wetherell 1998) and the rhetorical context within which such repertoires are deployed.

While many health social scientists conduct interviews as a means of analyzing discourse, naturally occurring talk is often the focus of analysis, as research questions center on how people account for themselves and interact in everyday life, with the intention of achieving personal objectives (Willig 2008). For example, Curl and Drew (2008) examined language use in phone calls to emergency services, and Rowe and colleagues (2003) investigated representations of "depression" in the print media in Australia. Lay accounts of health issues are a focus of attention, exemplified by studies men's accounts of PMS in on-line chat rooms (King et al. 2014), and young women's accounts of smoking (Triandafilidis et al. 2017a).

In a health or clinical context, interactions between patients and health professionals, therapists and clients, or lay people discussing health issues on the radio or TV, also provide appropriate material for investigation. For example, Rapley examined the social construction of intellectual disability, identifying the ways in which talk and text actively constitute the truths about being "mental" as well as "disabled" (Rapley 2004, p. 10), through the analysis of "official" texts, and the interactions of health professionals with people described as intellectually disabled. In a further study, McHoul and Rapley (2005) examined the transcript of a diagnostic session involving a young boy, his parents and a pediatrician, to contest the diagnosis of attention deficit disorder (ADHD). In the extract below (p. 425), the mother's account of how she came to bring her son to see the pediatrician is analyzed:

Mo: >I j's think < (I was) just picking up (.) things along the way
Dr: Mm hm
Mo: We (.) just basic'ly decided to eliminate
Dr: Mm hm
Mo: the possibility
Dr: Mm hm (0.7)
Dr: So you really hadn't got- had great problems until he'd got to schoo:l (mm) °is'at right°

The interaction within this account is interpreted as representing the mother adopting a scientific process of deduction through collecting data and evidence "just picking things up along the way," in order to refute the conjecture that she was overly concerned. "Problems" are identified, and tied to school, but not, as yet, identified as ADHD. In combination with analysis of the remainder of the interview, this local instance of talk-in-interaction is used to examine how routine and mundane it is for children to be positively diagnosed and medicated merely on presentation with the possibility of ADHD, even when parents are manifestly skeptical about the diagnosis (McHoul and Rapley 2005).

3   Foucauldian Discourse Analysis

Foucauldian discourse analysis (FDA) originates within poststructuralist theory, influenced by the philosophical work of Michel Foucault (1972). Within FDA, discourses are described as "sets of statements that constitute objects and an array of subject positions" (Parker 1994, p. 245) that are "a product of social factors, of powers and processes, rather than an individual's set of ideas" (Hollway 1983, p. 231). Language is deemed to be constitutive of social life, making available certain subject positions, which influence and regulate subjectivity and experience – the way we think or feel, our sense of self, and the practices in which we engage (Gavey 1989). FDA is, thus, concerned with identifying discourses, the subject positions they open up (or disallow), and the implications of such positioning for subjectivity and social practice, rather than the form or structure of interaction within talk or text. This includes analysis of expert discourse and institutional practice, talk generated through interviews, diaries or group discussions, and broader cultural representations.

In contrast to the DP focus on interpersonal communication, FDA centers on the examination of the relationship *between* discourse, subjectivity, practice, and the material conditions within which experience takes place (Willig 2008). This leads to attention being paid to wider social processes and power and how social order and the political realm is produced and reproduced through discourse. For example, the biomedical discourse, which positions health professionals (in particular doctors) as all-knowing and powerful, and health problems as pathologies to be eradicated through expert intervention, has been identified as one of the most powerful discourses in the field of health and illness (Foucault 1987). Legitimating the subject positions "expert" and "patient," it leads to a focus on somatic or psychological "symptoms," which are deemed to be located within the individual, and conceptualized in a realist manner as existing outside of language or cultural interpretation (Fee 2000). Thus, individuals who experience changes in mood, sleep patterns, or energy, and who report such changes to a medical practitioner, are positioned as "depressive," and their future experiences interpreted through a medical lens, when they may have previously normalized or accepted such changes (LaFrance 2007). This positioning of emotional and behavioral changes as psychiatric illness, or madness, can be identified as serving to maintain the boundaries of normality, leading to self-policing on the part of the individual, in order to avoid diagnosis (Ussher 2011), or to return to "normality" after diagnosis is given.

In this vein, FDA has been used to examine expert accounts of diagnosis or treatment (Larsson et al. 2012), broader cultural representations (Bilić and Georgaca 2007), and lay experiences of psychological distress or diagnosis (LaFrance 2007). In each example, FDA is used to identify the multiple and sometimes contradictory subject positions adopted, the implications of these subject positions for subjectivity, and their association with broader social discourse.

Physical health and health behavior has also been subjected to FDA, including a deconstruction of the meaning of being an un/healthy fat woman (Tischner and Malson 2012), negotiating sexual changes after cancer (Ussher et al. 2013; Parton et al. 2017), and representations of menopause in medical textbooks (Niland and Lyons 2011).

As part of a FDA analysis, attention is often paid to the relationship between discourse and institutions, or institutional practice, described as "ways of organising, regulating and administering social life" (Willig 2008, p. 113). For example, the proliferation of psychiatric diagnosis through the Diagnostic and Statistical Manual of the American Psychiatric Association (DSM) serves to maintain the authority of the "psy-professions" – psychiatry and psychology – the experts who are legally empowered to execute psychiatric judgment and administer "treatment." The development of new diagnostic categories in every edition of DSM also serves to shore up the power and profits of BigPharma, the drug companies that sponsor the experts, who create the diagnostic categories, which new drugs are then developed to treat.

Within FDA, attention is also paid to the ways in which discourses change over time, the *genealogy* of discourse and discursive practice. For example, the development of the modern clinical diagnosis "anorexia nervosa" can be traced to historical accounts of "fasting girls" (Malson 1998), and the diagnostic category premenstrual dysphoric disorder (PMDD) traced to diagnoses of hysteria and neurasthenia, as well as to nineteenth-century pronouncements on the vagaries of menstruation (Ussher 2006). This genealogical analysis identifies continuities in both discourse and discursive practice – in both of the examples cited above, these practices center upon the pathologization and regulation of the female body, and the maintenance of the boundaries of acceptable femininity. FDA has been influential in feminist health research (Gavey 1989; Hollway 1989) because of its ability to be used in the analysis of the gendered construction of subjectivity, power relations, and social practice. However, feminist health researchers have also adopted DP in analyzing expert and lay accounts (Sheriff and Weatherall 2009; Lafrance et al. 2017), as discursive practice is also often gendered, and thus research arising from both DP and FDA traditions is described as "feminist discourse analysis" (Gavey 2011).

4 Synthesizing DA and FDA: A Combined Approach

Whilst many reviewers and discourse analysts distinguish between discursive psychology and FDA (e.g., Burr 2003; Potter 2004; Willig 2008), it has also been argued that the two strands should not be differentiated so sharply and that the analysis of discursive practices and resources should be combined (Rapley 2004; Sims-Schouten et al. 2007). For example, Margaret Wetherell (1998) has argued for a synthesis of the two approaches, "which reads one in terms of the other" (p. 388). Sims-Schouten and colleagues (2007) have combined discursive psychology and FDA in their description of a critical realist discourse analysis, and Rapley (2004) used a combination of discursive psychology and Foucauldian theory in analyzing constructions of disability. These approaches examine both discursive practices, the performative qualities of discourse, and the role of discourse in the constitution of subjectivity, self-hood, and power relations (Willig 2000).

At the same time, across both DP and FDA, differences can be identified between those who conduct fine grained analysis of selected sections of talk or text, using either DP or FDA (for example, Potter 2004; McHoul and Rapley 2005), and those who combine elements of thematic analysis with discourse analysis (see also ▶ Chap. 48, "Thematic Analysis"). The latter focuses on identifying discursive themes across accounts through thematic decomposition (Stenner 1993; Parton et al. 2016; Churruca et al. 2017), or within individual cases, described as thematic composition (Watts et al. 2009).

5 Defining Discourse Analytic Methods

This leads to the question: how do you do discourse analysis, and how does it differ across the different strands? There is no simple recipe, which can make discourse analysis seem daunting to the novice researcher. Parker (1992) provides a detailed 20 step guide for conducting FDA, which Willig (2008) has simplified into a six stage guide, as well as providing three steps for DP analysis. Potter and Wetherell (1987) provide details of 10 stages of discourse analysis, which later came to be known as DP, but also argue that it is reliant upon "intuition," craft skills and tacit knowledge and that there is "no analytic method" (p. 169). Similarly, Hollway (1989) emphasizes her own intuitive feelings about the identification of discourse, and Billig (1997) argues that discourse analysis cannot be simply learnt as a procedure, separate from its wider theoretical critique of psychology, and that methodological guidelines should not be followed too rigidly. It has also been argued that the process of analysis and writing are not separate, as analysis will be refined and clarified throughout the writing process (Potter and Wetherell 1987).

With these caveats in mind, in Table 1 below we provide a general summary of steps in the process of conducting discourse analysis, drawing on the guidelines provided by Willig (2008) and Potter and Wetherell (1987), with distinctions between DP and FDA indicated in the analysis section. We then provide a

Table 1 Components of discourse analysis

Reading: Read through transcripts, and listen to interview recordings, to gain an overview of the data, and what the text is doing. For textual analysis, read through texts or representations.
Coding: Select the material for analysis, using the research questions as the basis for selection. Develop a coding frame, based on reading and re-reading of the data. Highlight and select relevant text and file it under the coding frame. Computer software, such as NVivo, can be used to manage the organization of coded data (see also ▶ Chap. 52, "Using Qualitative Data Analysis Software (QDAS) to Assist Data Analyses").
Analysis: Read through the coded data, paying attention to the functional aspects of discourse: how does the text construct subjects and objects? What is the discursive context within which the account is produced? Are there contradictions or variability in the accounts? Can particular discursive themes, or interpretative repertoires, be identified?
Discursive Psychology: *Focus of analysis*: How are particular versions of reality manufactured, negotiated and deployed in conversation? *Steps:* What terminology is used; what are the stylistic and grammatical features, the preferred metaphors and figures of speech? What is the action orientation of the account – what people do with language? How is meaning negotiated in local interaction?
Foucauldian Discourse Analysis: *Focus of analysis*: Examine the social, psychological, and physical effects of discourse; the availability of discourses within a culture and the implications for those within. *Steps: Discourse and discursive constructions*: Locate the various discursive constructions of the object, and identify their association with wider cultural discourses. If conducting a genealogy, examine the historical development of such discourses and discursive practices. *Function*: What is the function of such constructions? What is gained by constructing the object in this way? *Positioning*: what subject positions are offered by the text? *Practice*: how does discourse open up or close down opportunities for action? What are the implications in terms of power relations? *Subjectivity*: what are the consequences of taking up, or resisting, subject positions made available? What can be thought, felt, or experienced from within various discourses?
Writing: Contextualize your research in the context of other DA studies using a similar method. Provide details of theory and method adopted: how you did the analysis. Depending on the research question, focus the analysis on the identification of discourses (or interpretative repertoires) in talk or representation; on the discursive construction of objects (such as menopause) and/or subject positions (such as "aging woman"); or on discursive strategies and their consequences. Illustrate each with examples from the data. Analysis and discussion sections are often combined, with a separate shorter conclusion section drawing out the wider theoretical implications and suggestions for future research.

description of a Foucauldian discourse analysis of women's negotiation of negative premenstrual change, as an illustration of how this analysis can function in practice.

6 Analyzing Premenstrual Distress: A Feminist Discourse Analysis

6.1 Study Aim and Method

The aim of this study is to examine women's construction and negotiation of negative premenstrual change. The research involved 60 heterosexual and lesbian women who were taking part in a mixed method study examining the construction and experience of premenstrual change in self-diagnosed Premenstrual Syndrome (PMS) sufferers (Ussher and Perz 2013). We conducted one-to-one semi-structured interviews, which lasted between 45 and 90 mins. We kept the interview questions open and general to avoid being leading and conducted the interview as a discussion between interviewer and participant. The interviewer began by asking women to describe a typical experience of PMS and how this varied across relational contexts, and then explored strategies of coping. In the analysis, we adopted a feminist FDA approach, which examines the role of discourse in the constitution of subjectivity and social practice, while also acknowledging the material conditions which influence such experiences, and the role of discourse in wider social processes of legitimation and power (Gavey 1989, 2011).

6.1.1 Reading and Coding

All of the interviews were transcribed verbatim. A subset of the interviews were then read and reread by both authors and a research assistant to identify first order codes, such as "embodied changes," "emotional distress," "relational issues," "PMS at work," "coping," and "triggers for premenstrual distress." The entire data set was then coded using NVivo, a computer package that facilitates organization of coded qualitative data.

6.1.2 Analysis

All of the coded data was then read through independently by both authors, making detailed notes of patterns, commonalities, variability across the data, and uniqueness within cases. This was a reflexive process that allowed us to interpret participant accounts from our different perspectives: as a woman who experienced premenstrual change (Ussher) and one who did not (Perz). Through a process of discussion, we then identified the discursive constructions of PMS and premenstrual change in the context of broader cultural discourse. These constructions included PMS as illness, as sign of weakness, as natural, and as a relational experience. The function these discourses served for women and for their partners was identified, and attention paid to the subject positions made available through various discursive constructions of premenstrual change. The implications of discursive constructions for practice, in particular for styles of coping, and power relations in both a relational and broader social context were also examined. Finally, the consequences of taking up or resisting subject positions, in terms of women's subjectivity, were attended to: women's accounts of thoughts or feelings, and of self-hood, from various discursive positions. The genealogy of discursive constructions of PMS is examined elsewhere (Ussher 2006).

6.2 Women's Descriptions of Premenstrual Change

All of the women interviewed described premenstrual change as "PMS," characterized by heightened premenstrual irritability, intolerance of others, and oversensitivity, using terms such as "irritable," "cranky," "short-tempered," "snappy,"

Table 2 A feminist Foucauldian discourse analysis of PMS

Discursive construction	PMS as pathology	PMS as an understandable reaction
Discourses	Biomedical	Feminist psycho-social
Function	PMS as a thing that causes distress Locates PMS within the body Exonerates women from unacceptable anger	Locates PMS within relationships or life stress Legitimates expression of premenstrual needs and emotions Tolerance and normalization of premenstrual change
Positioning	Woman as out of control Victim of hormonal imbalances Woman as monster; abject	Woman as agentic and rebellious Woman as sensitive or vulnerable premenstrually
Practice	Pharmaceutical interventions Avoidance to protect others Premenstrual self-control and self-silencing	Legitimating time-out and avoidance for self-care Attention to situational issues associated with distress Rejection of acquiescent femininity
Subjectivity	Self-blame and guilt Shame	Acceptance of premenstrual emotions Catharsis

"confrontational," having a "short fuse," "bitey," "impatient," "grumpy," "stroppy," "frustrated," "stressed" "annoyed," or "teary." However, how women discursively positioned these changes, and the implications for subjectivity, self-positioning and practice, varied. Two discursive constructions of PMS identified in women's accounts are presented below: PMS as pathology, drawing on biomedical discourse, and leading to a victim or monster subject position; and PMS as an understandable reaction, drawing on a feminist life-stress discourse, which facilitated tolerance and self-care, as summarized in Table 2.

6.3 Premenstrual Change as Pathology: Woman as Victim or Monster

> When you're being taxed physically by the PMS, because it does something, I don't know what it does, whether it... I know it depletes certain vitamins and you've got low magnesium and this and this and this and that, and your hormones can be off, and not where they should be. And I think just that being taxed with those things helps you, or makes you not be able to handle other things, where if you're not taxed with all that, you can handle other difficulties that you'd normally be able to handle.

6.3.1 Discourse and Discursive Constructions

The construction of premenstrual change as pathology, drawing on a biomedical discourse, is illustrated in the above account. PMS is described as a thing ("the PMS") that results in the woman's inability to handle daily stresses that she is "taxed" with, and which she would "normally" be able to handle. PMS is blamed for depleting "certain vitamins" and for causing hormones to be "off." This "thingifying" of PMS has been previously identified in published expert accounts of premenstrual change (Ussher 2003), demonstrating that women are drawing on wider cultural discourses in adopting this construction of premenstrual change. This serves to position PMS as the *cause* of women's premenstrual distress, rather than a label which is given to an array of premenstrual changes that women may report.

6.3.2 Function and Positioning

We asked what is the function of such constructions? and what is gained, or lost, by constructing premenstrual change is such a manner? One of the implications of the adoption of a biomedical discourse is the self-positioning of the premenstrual woman as "out of control" and a victim of her hormonal "imbalances," as is evident in the account below:

> The imbalances were happening in the body and all that sort of stuff that I had absolutely no control over. I mean, sure, I had other issues I had to contend with, but I was dealing with that so, to me, this was something that was so out of my control that I felt like I was being blamed for actually knowing how to deal with it, and it's like, 'Well, I don't.'

Through reading the coded data, we identified a process of splitting, wherein women discursively separate their normal sane selves from the abhorrent nature of premenstrual emotion, thus exonerating themselves for what they construe to be mad or bad behavior. Extreme descriptions of the premenstrual self were used to illustrate the dramatic nature of this change, including "crazy," "mad," a "nut case," "absolute psycho," "Schizo," "out of my mind," or "a complete loony." Positioning the premenstrual self as mad serves to reinforce the notion of premenstrual changes as outside of the woman's agency, and as someone whose behavior is a sign of pathology:

> You're like a person that's probably, you know a crazy drunk that's had a lot of alcohol or on pot, really high on drugs and they just snap and you can't control their anger. I would um react like that, without having had anything.

However, the expression of premenstrual irritation or anger was invariably followed by reports of guilt and self-criticism, suggesting that self-positioning as afflicted by hormones premenstrually is not effectively serving to exonerate women from "bad" behavior, as previously suggested. Thus, one woman told us, "(Y)ou feel horrible about it the next week... it makes you feel sick," and another said she feels "really upset" and "angry" with herself. This is associated with shame, and with a sense of the need to apologize, which involves woman taking up a position of abjection:

> There's fair bit of violence coming through in all of this isn't there. Throwing things, um, venting a frustration and trying to learn to control these things so that people don't look at

you funny and you don't make a fool of yourself which you later regret. When you come back to normality, you think, um, I've got to go around and apologise to all these people I've just bitten their heads off over absolutely nothing. No, it's, in the back of your head you know what's normal.

6.3.3 Practice

We asked what are the implications of this positioning in terms of women's coping strategies in the face of negative premenstrual change? What opportunities for action are made available by the subject positions women adopt?

Biomedical constructions of PMS leave women with one obvious opportunity for action: pharmaceutical interventions. In this vein, we identified that women who adopted a biomedical discourse talked of taking self-prescribed herbs (such as St John's Wort, Vitex, or evening primrose oil), and vitamins, or minerals (in particular B complex and magnesium), as part of their premenstrual self-care. A small minority of women had also been prescribed the contraceptive pill, with one participant prescribed antidepressants, to reduce premenstrual distress. The majority reported satisfaction with such remedies, reinforcing the positioning of PMS as a bio-medical phenomenon which can be treated: "I know that I wouldn't cope too well if I wasn't on St Johns Wort which has helped me 100%"; "I think if I remember to take them, like, as I should, I mostly get away without almost without any symptoms."

6.3.4 Implications for Power Relations

These strategies served to position women as dependent upon medical advice and intervention, as "patients" whose bodies are the focus of intervention. However, all of these dietary supplement or pharmaceutical coping strategies were also accompanied by psychological or behavioral strategies, demonstrating that the adoption of a bio-medical discourse does not inevitably preclude women's agency. This is analogous to the "tight-rope talk" identified by Sue McKenzie-Mohr and Michelle Lafrance (2011), wherein women construct themselves as both "agents and patients: both active and acted upon" (p. 64), enabling credit for agency in coping and deflection of blame for "having" PMS.

The coping strategy most commonly reported by participants was avoidance of people or situations that had the potential to provoke anger and irritation. For example, one woman told us, "the kids ... I try not to get into conflict, into confrontation with them." Anticipation or avoidance of stressful situations is not always possible, however, and many participants gave accounts of experiencing unexpected situations that elicited premenstrual anger or irritation. As a result, they described coping with occurrences of negative premenstrual emotion through exertion of self-control, leaving a situation when they had become angry, in order to avoid escalation of conflict.

> I usually feel stressed in the lead up. If it gets to the point where I actually need to say that, I know the pressure cooker, little thing on the top bouncing up and down, you know, um at that

point and it's almost like a last resort for me. If I know I'm going to explode, I try to train myself to step back and chill.

In some accounts, self-control was described as necessary for the protection of others, reinforcing the positioning of the woman as monstrous and out of control. For example, one woman described "hibernating," because of a fear of not being able to "rein yourself in," and wanting to avoid "hurting people with words," because "it's not their fault."

6.3.5 Subjectivity

The consequence of constructing premenstrual change as pathology, and adopting a biomedical discourse, was that women reported shame and dislike of the self during the premenstrual phase of the cycle. The body was a focus of negative emotion, being blamed for women's distress, and disconnected from the self: "I feel quite odd and almost out of body"; "you feel like you're a blimp um even though you feel ugly and this is this is all PMS."

6.4 PMS as an Understandable Reaction: Facilitating Agency and Self-Care

P. I get very snappy and short with people I know... you know... um... yeah, things that normally don't bother me, bother me. It tends to be about housework, and my role as a woman, and why do I have to do all of this and... yeah. Things that I kind of repress... come out to the surface.
I. Okay.
P. That's what I've found.
I. Okay. So they're not things that aren't there the rest of the time and suddenly they appear, they're things that you think are...?
P. They're underlying. They're underlying issues and I just kind of tick along nicely and think, "Well, you know, it's okay, I can deal with it," and then, PMS comes and I can't deal with it. Yeah.

6.4.1 Discourse and Discursive Constructions

In contrast, in the account above, premenstrual emotion is described as a reflection of a woman's anger or frustration with "underlying issues" associated with domestic concerns and her "role as a woman." This reflects the adoption of a feminist psychosocial discourse, wherein gendered inequalities and over-responsibility are positioned as a cause of women's distress, and anger or frustration is deemed to be a legitimate reaction. A similar construction of PMS was adopted in accounts of intolerance or anger towards male partners premenstrually. For example, one woman described her husband as "a bit of a hoarder and a collector, and 3 weeks of the month that does not bother me." However, when premenstrual, she said, "it bothers me a lot and I want to throw everything out, to put everything into plastic bags and dump it on his desk [laughs]."

6.4.2 Function

One of the primary functions of adoption of a feminist psycho-social discourse was to position premenstrual emotions as understandable and reasonable, resisting the discursive positioning of the premenstrual woman as mad, bad, or dangerous (Chrisler and Caplan 2002). Thus, one woman described awareness of premenstrual sensitivity as "a weight off my mind. 'Cause at first I used to think I was going a little crazy... it's helped me deal with, 'those are PMS feelings'." Similarly, another woman tells herself, "Oh, OK, I know now, you're not actually the wicked witch." This construction of PMS was also associated with tolerance of negative premenstrual change, which could then be normalized, or embraced as an "opportunity" to "be emotional":

> I think I embrace it as an opportunity to, um, just rest and, um, be more in tune with myself, and be emotional. I think I was probably fighting being emotional in the past, and that's what caused so much discomfort and stress.

Recognition and acceptance of negative premenstrual change can also function to give women permission to engage in coping strategies to avoid or reduce premenstrual distress – without a sense of guilt. Thus, one woman told us: "I'll actually give myself permission to actually go and lie down for half an hour. Even half an hour will make substantial amount of difference."

6.4.3 Positioning and Subjectivity

In these accounts, women not only avoid taking up a subject position of victim or patient, wherein their bodies (or minds) warrant medical treatment, they also subvert the self-surveillance which they adopt for the remainder of the month (see Ussher and Perz 2010). Implicit in these accounts is the transgression of self-silencing, which is broken when PMS "comes" along.

6.4.4 Implications for Power Relations

Premenstrual emotional expression is associated with a sense of catharsis. In this way, "PMS" signifies rebellion and resistance, rather than weakness and pathology, and breaks in self-silencing are a sign of women's agency.

This case example illustrates the ways in which discourse analysis can be used to understand the construction and experience of a specific health issue, as well as the role of language in the course of distress, and in facilitating coping.

7 Conclusion and Future Directions

Discourse analysis has a relatively short history in the Health Social Sciences, but has had a significant impact upon the conceptualization of the sick or healthy subject and on the conduct of qualitative research. Discourse analysis offers the opportunity to explore the construction of health and illness within a range of cultural representations and expert accounts, the meaning and experience of health and illness for laypeople, and the implications of discursive constructions of health and illness for subjectivity. There is no one correct method of discourse analysis, as multiple methods have been identified (Wetherell 2001), and practitioners interpret and present analyses in a range of different ways. This flexibility is its strength in the Health Social Sciences, given the uniqueness of health problems. This chapter has summarized two identifiable strands of DA, discursive psychology and Foucauldian discourse analysis, and presented one interpretation of feminist FDA as an example. However, the specific form of DA adopted will always be determined by the research questions that drive a project, with the mode of analysis and presentation influenced by the theoretical orientation, skills, and creativity of the researcher.

Future directions include the integration of a range of nontextual methodologies alongside or instead of analysis of interview or written text, including photography (Triandafilidis et al. 2017b), body-mapping, and other arts-based methods (Boydell et al. 2016). The move towards an intersectional analysis (Hankivsky and Cormier 2009) in feminist research has resulted in discourse analysis, with its attention to complexity of subjectivity and experience, the negotiation of multiple and potentially contradictory subject positions, and the importance of power relationships, becoming the qualitative research method of choice. As critical Health Social Scientists also embrace intersectionality, discourse analysis may also become the preferred method of qualitative analysis in the Health Social Sciences.

Acknowledgements Sections of this chapter draw on an earlier publication by the authors: Ussher, JM & Perz, J. (2014) Discourse Analysis. In: Poul Rohleder and Antonia Lyons (eds) Qualitative Research in Clinical and Health Psychology. London: Palgrave MacMillan (p. 218–237)

References

Bilić B, Georgaca E. Representations of 'mental illness' in Serbian newspapers: a critical discourse analysis. Qual Res Psychol. 2007;4(1–2):167–86. https://doi.org/10.1080/14780880701473573.

Billig M. Rhetorical and discursive analysis: how families talk about the royal family. In: Hayes N, editor. Doing qualitative analysis in psychology. Hove: Psychology press; 1997.

Billig M. Undisciplined beginnings, academic success, and discursive psychology. Br J Soc Psychol. 2012;51(3):413–24. https://doi.org/10.1111/j.2044-8309.2011.02086.x.

Boydell KM, Hodgins M, Gladstone BM, Stasiulis E, Belliveau G, Cheu H, ... Parsons J. Arts-based health research and academic legitimacy: transcending hegemonic conventions. Qual Res. 2016;16(6):681–700. https://doi.org/10.1177/1468794116630040

Burr V. Social constructionism. 2nd ed. London: Routledge; 2003.

Chrisler JC, Caplan P. The strange case of Dr. Jekyll and Ms Hyde: how PMS became a cultural phenomenon and a psychiatric disorder. Annu Rev Sex Res. 2002;13:274–306.

Churruca K, Ussher JM, Perz J. Just desserts? Exploring constructions of food in women's experiences of bulimia. Qual Health Res. 2017;27(10):1491–506. https://doi.org/10.1177/1049732316672644.

Curl TS, Drew P. Contingency and action: a comparison of two forms of requesting. Res Lang Soc Interact. 2008;41(2):129–53. https://doi.org/10.1080/08351810802028613.

Edwards D, Potter J. Discursive psychology. London: Sage; 1992.

Edwards D, Potter J. Discursive psychology, mental states and descriptions. In: Molder H, Potter J, editors. Conversation and cognition. Cambridge: Cambridge University Press; 2005. p. 241–59.

Fee D. Pathology and the postmodern: mental illness as discourse and experience. London: Sage; 2000.

Foucault M. The archeology of knowledge and the discourse on language. New York: Pantheon Books; 1972.

Foucault M. Mental illness and psychology. Berkley: University of California Press (first published 1976); 1987.

Gavey N. Feminist poststructuralism and discourse analysis: contributions to feminist psychology. Psychol Women Q. 1989;13(1):459–75. https://doi.org/10.1177/0361684310395916.

Gavey N. Feminist poststructuralism and discourse analysis revisited. Psychol Women Q. 2011;35(1):183–8. https://doi.org/10.1177/0361684310395916.

Hankivsky O, Cormier R. Intersectionality: moving women's health research and policy forward. Vancouver: Women's Health Research Network; 2009. http://www.whrn.ca/intersectionality-download.php.

Hollway W. Heterosexual sex: power and desire for the other. In: Cartledge S, Ryan J, editors. Sex and love: new thoughts on old contradictions. London: Women's Press; 1983. p. 124–40.

Hollway W. Subjectivity and method in psychology: gender, meaning and science. London: Sage; 1989.

King M, Ussher JM, Perz J. Representations of PMS and premenstrual women in men's accounts: an analysis of online posts from PMSBuddy.com. Womens Reprod Health. 2014;1(1):3–20. https://doi.org/10.1080/23293691.2014.901796.

LaFrance MN. A bitter pill. A discursive analysis of women's medicalized accounts of depression. J Health Psychol. 2007;12(1):127–40.

Lafrance MN, Stelzl M, Bullock K. "I'm not Gonna fake it": University women's accounts of resisting the normative practice of faking orgasm. Psychol Women Q. 2017;41(2):210–22. https://doi.org/10.1177/0361684316683520.

Larsson P, Loewenthal D, Brooks O. Counselling psychology and schizophrenia: a critical discursive account. Couns Psychol Q. 2012;25(1):31–47. https://doi.org/10.1080/09515070.2012.662785.

Malson H. The thin woman: feminism, post-structuralism and the social psychology of anorexia nervosa. London: Routledge; 1998.

McHoul A, Rapley M. A case of attention-deficit/hyperactivity disorder diagnosis: Sir Karl and Francis B. Slug it out on the consulting room floor. Discourse Soc. 2005;16(3):419–49. https://doi.org/10.1177/0957926505051173.

McKenzie-Mohr S, LaFrance M. Telling stories without the words: 'Tight-rope talk' in women's accounts of coming to live well after rape or depression. Fem Psychol. 2011;21(1):49–73.

Niland P, Lyons AC. Uncertainty in medicine: meanings of menopause and hormone replacement therapy in medical textbooks. Soc Sci Med. 2011;73(8):1238–45. https://doi.org/10.1016/j.socscimed.2011.07.024.

Parker I. Discourse dynamics: critical analysis for social and individual psychology. London: Sage; 1992.

Parker I. Reflexive research and the grounding of analysis: social psychology and the psy-complex. J Community Appl Soc Psychol. 1994;4(4):239–52.

Parker I. Discursive social psychology now. Br J Soc Psychol. 2012;51(3):471–7. https://doi.org/10.1111/j.2044-8309.2011.02046.x.

Parton CM, Ussher JM, Perz J. Women's construction of embodiment and the abject sexual body after cancer. Qual Health Res. 2016;26(4):490–503. https://doi.org/10.1177/1049732315570130.

Parton C, Ussher JM, Perz J. The medical body: women's experiences of sexual embodiment across the cancer illness trajectory. Women's Reprod Health. 2017;4(1):46–60. https://doi.org/10.1080/23293691.2017.1276370.

Potter J. Discourse analysis as a way of analysing naturally occuring talk. In: Silverman D, editor. Qualitative research: theory, method, practice. London: Sage; 2004. p. 200–21.

Potter J. Re-reading discourse and social psychology: transforming social psychology. Br J Soc Psychol. 2012;51(3):436–55. https://doi.org/10.1111/j.2044-8309.2011.02085.x.

Potter J, Wetherell M. Discourse and social psychology: beyond attitudes and behaviour. London: Sage; 1987.

Rapley M. The social construction of intellectual disability. Cambridge: Cambridge University Press; 2004.

Rowe R, Tilbury F, Rapley M, O'Ferrall I. 'About a year before the breakdown I was having symptoms': sadness, pathology and the Australian newspaper media. Sociol Health Illn. 2003;25(6):680–96. https://doi.org/10.1111/1467-9566.00365.

Sheriff M, Weatherall A. A feminist discourse analysis of popular-press accounts of postmaternity. Fem Psychol. 2009;19(1):89–108. https://doi.org/10.1177/0959353508098621.

Sims-Schouten W, Riley SCE, Willig C. Critical realism in discourse analysis: a presentation of a systematic method of analysis using women's talk of motherhood, childcare and female employment as an example. Theory Psychol. 2007;17(1):101–24. https://doi.org/10.1177/0959354307073153.

Stenner P. Discoursing jealousy. In: Burman E, Parker I, editors. Discourse analytic research. London: Routledge; 1993. p. 114–34.

Tischner I, Malson H. Deconstructing health and the un/healthy fat woman. J Community Appl Soc Psychol. 2012;22(1):50–62. https://doi.org/10.1002/casp.1096.

Triandafilidis Z, Ussher JM, Perz J, Huppatz K. Doing and undoing femininities: an intersectional analysis of young women's smoking. Fem Psychol. 2017a. https://doi.org/10.1177/0959353517693030.

Triandafilidis Z, Ussher JM, Perz J, Huppatz K. 'It's one of those "It'll never happen to me" things': young women's constructions of smoking and risk. Health Risk Soc. 2017b;19:1–24. https://doi.org/10.1080/13698575.2017.1384801.

Ussher JM. The role of premenstrual dysphoric disorder in the subjectification of women. J Med Humanit. 2003;24(1/2):131–46.

Ussher JM. Managing the monstrous feminine: regulating the reproductive body. London: Routledge; 2006.

Ussher JM. The madness of women: myth and experience. London: Routledge; 2011.

Ussher JM, Perz J. Disruption of the silenced-self: the case of pre-menstrual syndrome. In: Jack DC, Ali A, editors. The depression epidemic: international perspectives on women's self-silencing and psychological distress. Oxford: Oxford University Press; 2010. p. 435–58.

Ussher JM, Perz J. PMS as a gendered illness linked to the construction and relational experience of hetero-femininity. Sex Roles. 2013;68(1–2):132–50.

Ussher JM, Perz J, Gilbert E, Wong WKT, Hobbs K. Renegotiating sex after cancer: resisting the coital imperative. Cancer Nurs. 2013;36(6):454–62. https://doi.org/10.1097/NCC.0b013e3182759e21.

Watts S, O'Hara L, Trigg R. Living with type 1 diabetes: a by-person qualitative exploration. Psychol Health. 2009;25(4):491–506. https://doi.org/10.1080/08870440802688588.

Wetherell M. Positioning and interpretative repertoires: conversation analysis and post-structuralism in dialogue. Discourse Soc. 1998;9(3):387–412. https://doi.org/10.1177/0957926598009003005.

Wetherell M. Debates in discourse research. In: Wetherell M, Taylor S, Yates SJ, editors. Discourse theory and practice: a reader. London: Sage; 2001.

Willig C. A discourse-dynamic approach to the study of subjectivity in health psychology. Theory Psychol. 2000;10(4):547–70.

Willig C. Introducing qualitative methods in psychology: adventures in theory and method. Maidenhead: McGraw Hill; 2008.

Schema Analysis of Qualitative Data: A Team-Based Approach

51

Frances Rapport, Patti Shih, Mia Bierbaum, and Anne Hogden

Contents

1	Introduction	898
2	What Is Schema Analysis?	900
3	The Schema Analysis Process	901
	3.1 Stage 1: Constructing Individual Schemas	901
	3.2 Stage 2: Group Work and the Development of Group or "Meta-schemas"	903
	3.3 Stage 3: Interpretation of Meta-schemas	904
4	An Exemplar: The Role of Multidisciplinary Teams in Breast Cancer Risk Communication	905
	4.1 Stage 1: Individual Schemas	905
	4.2 Stage 2: The Group or Meta-schema	908
	4.3 Stage 3: Interpretation of Meta-schema	911
5	Validity, Rigor, and Trustworthiness of Data	912
6	Lessons Learned: Enriching Qualitative Analytic Practices	912
7	Conclusion and Future Directions	913
References		914

Abstract

Schema analysis is a novel, qualitative data analysis technique that uses a summative approach to make sense of complex, nuanced, textual data. It aims to ensure that key features of a text, or "essential elements," are revealed *before* any interpretation of

---

F. Rapport (✉) · M. Bierbaum · A. Hogden
Centre for Healthcare Resilience and Implementation Science, Australian Institute of Health Innovation (AIHI), Macquarie University, Sydney, NSW, Australia
e-mail: frances.rapport@mq.edu.au; mia.bierbaum@mq.edu.au; anne.hogden@mq.edu.au

P. Shih
Centre for Healthcare Resilience and Implementation Science, Australian Institute of Health and Innovation (AIHI), Macquarie University, Sydney, NSW, Australia
e-mail: patti.shih@mq.edu.au

© Springer Nature Singapore Pte Ltd. 2019
P. Liamputtong (ed.), *Handbook of Research Methods in Health Social Sciences*,
https://doi.org/10.1007/978-981-10-5251-4_104

those key elements takes place, based on the assertion that data should be handled in a principled, informed, and strategic manner to achieve phenomenal clarity. Teamwork is a central element of schema analysis, enabling researchers to effectively co-create meaning across disciplinary boundaries through consensus-driven strategies. The research team's shared accountability for interpretive decisions is clearly linked to a study's original questions and the research team's desire to be rigorous in their collaborative stance and equally vocal. Schema analysis can be used in the context of a wide range of qualitative studies or can sit alongside outputs from other mixed-methods studies to add substance to their findings. The technique has been successfully developed and refined by the lead author, to suit a wide variety of healthcare scenarios, evidenced by published research projects from UK and Australian contexts. In this chapter, the method is presented in detail, step by step. An example of the use of schema analysis in practice is offered up from a recent Australian study that examined multidisciplinary team-working practices in oncology. Study data from an interview with an oncology psychologist working as part of an oncological multidisciplinary team is considered, to reveal how healthcare professionals present information about risk to women with breast cancer. The chapter considers methodological implications for achieving validity and rigor, upholding trustworthiness in data, and creating data that are transferable to different settings. To conclude, the chapter reflects on future opportunities for the method's use in qualitative research.

Keywords
Schema analysis · Team-working · Qualitative data · Consensus-driven approach

1 Introduction

Schema analysis is a novel, qualitative data analysis technique that uses a summative approach to make sense of complex, nuanced, qualitative data. It concentrates on qualitative data in textual form only and aims to ensure that the key features of a text, or "essential elements," are revealed *before* any interpretation of those key elements takes place. Schema analysis is dependent on accurate descriptions of data content that can be used in the context of wider qualitative studies or that can sit alongside outputs from mixed-methods studies.

Schema analysis derives from the notion that data should be handled in a principled, informed, and strategic manner (Rapport 2010). Accountability, regarding the decisions that are made, lies with all the researchers involved. As this is a group-working method, people's roles are equivocal. The group comes together out of a shared need to understand a text or a number of texts, in their fullest expression, while wishing to derive an essential understanding of each piece of text, to achieve "phenomenal clarity." Thus, being accountable for decisions made is clearly linked, not only to the original study questions but to the researchers' desire to be rigorous and equally vocal, gained by abiding to methodological principles of how to create schema.

Schema analysis was developed by Frances Rapport (first author) and helped to refine her summative analysis method (Rapport 2010). It was the result of the need to handle and manage difficult and complex datasets that had been collected from disenfranchised and often dispersed population groups (Rapport 2010). It was developed, first, as a group activity to manage qualitative data from extended focus groups with women with breast cancer who were exploring the value of decision aids to support complex treatment choices (Rapport et al. 2006; Iredale et al. 2008) and was adapted, over the next 10 years, to fit various summative analytic styles. The development of schema analysis also influenced co-joint working practices, such as those undertaken by Rapport and Sparkes (2009) and Rapport and Hartill (2010, 2016) to examine survivor's life and health trajectories following the trauma of overcoming an extraordinary event such as the Holocaust. Finally, it took account of others' writings on, for example, thematic analysis (Ryan and Bernard 2000), the use of oral and written testimony in research (see, e.g., "talk and text" (Silverman 2000)), and the research participant's role in research (see, e.g., Fleming and Ward 2017). More latterly, schema analysis has been adapted as a result of Rapport's involvement in Nominal Group Work (NGW) activity, through multiple stakeholder group consultations (Rapport et al. 2010, 2015).

Teamwork is a central element of schema analysis, enabling researchers to effectively co-create meaning across disciplinary boundaries (what this authorship group calls "meaning-making in the round"). This leads to narratively rich, impactful, and highly original outputs. Furthermore, by working collaboratively and on a basis of equality, the final research product (irrespective of form) will enhance the meaning-making activity.

In order to achieve an inclusive research process, in a multidisciplinary context, through clear research collaboration, it is important for the method to be adaptable to different group dynamics and inclusive of all involved. Schema analysis achieves this through its versatility, accommodating those working in both public and professional domains. The method is clearly defined for all involved and purposefully inclusive of people who only have a general interest in qualitative research methods or very little qualitative analytic experience. Consequently, irrespective of one's knowledge base or expertise, the process is clearly defined for all to contribute.

A clinician or consumer can fully immerse themselves in team-working, while other researchers can learn more about their own topic of interest from laypeople or healthcare professionals, using what is learned to inform their own research. Thus, the method is adaptable and impactful and should be widely available to academics, researchers, and students who may wish to apply it within their own work or educators and policy-makers to explore educational opportunities or societal effects.

Schema analysis has developed around the principles of equality, availability, and malleability. The method has to be appropriate across a variety of disciplinary fields, such as the sciences, social sciences, and humanities. In order to ensure that this is the case, it foregrounds the need for clearly labeled decision trails, transferable and practical outcomes, and transparent, interdisciplinary working methods and backgrounds hierarchical working methods.

In this chapter, we present the method in detail, step by step. We reveal the implications for research validity, rigor, and the trustworthiness of data, and we highlight how the method can be used. We present an example of the method in use from a recent study examining multidisciplinary team-working practices and presentations of risk to women with breast cancer in an Australian healthcare context. To conclude, we consider opportunities for the use of this method in qualitative research.

2 What Is Schema Analysis?

Schema analysis is a way of summarizing, and then offering a clear and succinct presentation, of the essential elements within an original text. It uses group-working activities with groups of researchers revealing essential textual elements in data before interpreting what data means. It avoids preemptive, superimposed meaning-making, refraining from either breaking text down into small chunks (as with thematic analysis) or presenting text through authorial presence (as with realist tales). By so doing, it captures essential elements of a text richly and fluidly, without losing any of the original text's nuance or ambiguity. It uses group-working activities to derive succinct, collaboratively created textual overviews that, once created, can then be examined for their underlying meaning. These are known as group schemas (or "meta-schemas"), and once created, they can be fully interpreted, to inform study reports or publications.

Analysis is applicable to any raw textual material but, most commonly, works best with interview transcripts, biographies, autobiographies, or focus group reports. Group schemas are co-created by members of a research team who work together towards well-crafted and concise summaries, paying particular attention to the inclusion of person-focused, topic-sensitive, verbatim quotations. Schema analysis is underpinned by ongoing teamwork meetings (also known as workshops), so sense-making is built over time, iteratively and through ongoing debate. During workshops, team members must agree on which aspects of the raw material to retain and which to lose. Views are presented, discussed, and negotiated. Ideas become refined, as raw materials are honed down into their core components, and are narratively presentative. At the same time, schema analysis offers a way of handling large quantities of qualitative data succinctly, ensuring they can be reduced to easy-to-follow half-page presentations that encapsulate a holistic understanding of each full piece of text while revealing the key features it holds.

Schema analysis is wholly dependent on the ethos of group-working. Researchers move quickly from an individual consideration of data to a shared understanding. The wider team (optimally no more than six people) come together to read one another's work. This is followed by an assessment of how the group compliments one another or how views are contrasted. Qualitative researchers are encouraged to not only discuss the raw material with one another but also their own personal views – the impact of the "self" on data, in a desire to withhold personal preconceptions or assumptions. Consequently, schema analysis depends on honesty and openness from the whole team, often with a senior academic overseeing this aspect of group work. At the same time, individuals are encouraged to keep a study diary noting personal presuppositions as they work, changes to these over time, and their views of others' opinions. This aids individual awareness, helps the senior academic to manage the group-working activity most effectively, and encourages team members to move smoothly from an individual narrative to a meta-schema.

This approach significantly differs from many other qualitative methods, such as the creation of a standard or "realist" tale. The realist tale, as Andrew Sparkes (2002) reminds us, when well-constructed, can provide a useful, compelling, and detailed depiction of one's own or another's social world (see also ▶ Chap. 47, "Content Analysis: Using Critical Realism to Extend Its Utility"). Realist tales are powerful ways of presenting qualitative data but are, nevertheless, predicated on the presentation of a single viewpoint. They are wholly dependent on experiential authority and the interpretive omnipotence of a single researcher. Unlike the realist tale, neither summative analysis (Rapport 2010) nor schema analysis (its most recent derivation) accepts the authorial presence of interpreter omnipotence, nor do they abide by the backgrounding of others' intentions, to allow for an unchallenged authority on data or its interpretative aspirations (Van Manen 1998; Sparkes 2002). On the contrary, schema analysis moves away from the notion of being overly deterministic, to provide opportunities for a group's shared creativity.

Schema analysis also differs from thematic analysis, which concentrates on codifying data into themes and their concomitant categories. Thematic analysis works by breaking data down into small chunks, or "meaning units," and by applying techniques such as line-by-line coding, and code-nesting, which classifies meaning units under thematic headings. This tends to place themes in clear silos, each separated and demarcated (see also ▶ Chap. 48, "Thematic Analysis"). Thematic analysis can, as a result, run the danger of decontextualizing data, often failing to resituate data in their wider context, and missing out on "whole-body" presentations of meaning.

Schema analysis avoids the unnecessary dissolution and discombobulation of text by emphasizing the need to continue to review narratives holistically. It is worth noting, however, that schema analysis can be an excellent way of complementing a thematic analysis or a realist tale, with outputs open to comparison across methods.

3 The Schema Analysis Process

There are three stages to schema analysis: Stage 1, the development of individual summative schemas; Stage 2, the development of meta-schemas from the individual schemas; and Stage 3, the development of interpretive accounts from the meta-schemas. Figure 1 illustrates the three stages involved.

3.1 Stage 1: Constructing Individual Schemas

In Stage 1, individual schemas are developed by individual analysts. These are people who have been brought together by a senior academic, or team leader, identified according to their interest in the topic, what they hold in common, or the different qualities they can bring to the group. Groups of co-researchers may be

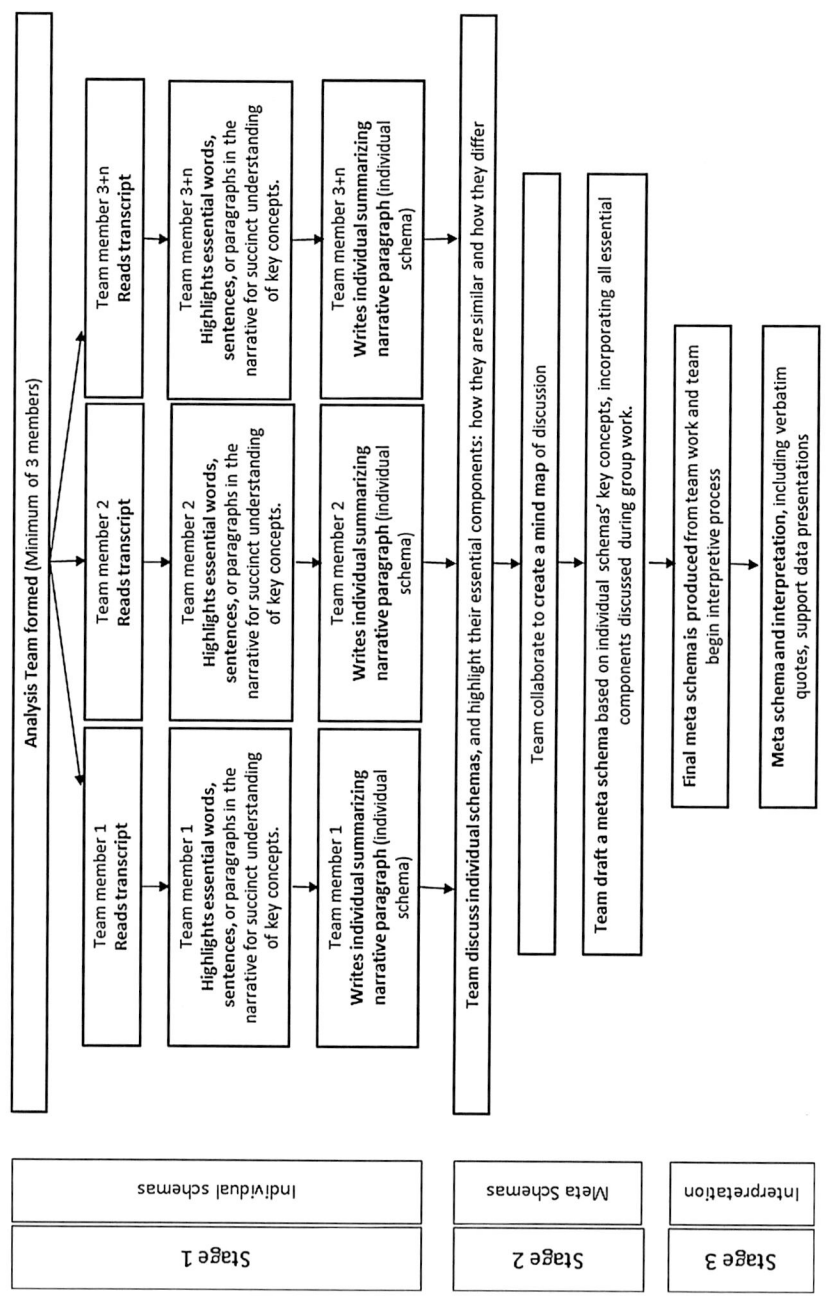

Fig. 1 Simplified presentation of schema analysis, an in-depth teamwork analysis process (Adapted from Rapport et al. under review)

selected for a variety of reasons, such as (a) their homogeneity (familiarity with the topic, research question, or people involved), (b) their impartiality (coming from very different backgrounds, with different expertise and interests), or (c) their lack of knowledge of the topic. The choice of co-researchers, as with other group-working methods, is dependent on the individual study and the needs of the primary researcher or senior academic overseeing the work.

Each individual reviews the research data separately and, following their review, identifies the essential elements belying each piece of text. The essential elements (also known as a text's key features or critical moments) are those aspects that provide a point of entry into an understanding of the whole text. Rather than taking account of the individual analyst's personal view of the data, however, individual schemas should be written as an ongoing, free-flowing narrative that tells the story of the text, by weaving the key features together fluidly, and which presents the research participant's perspective in a vastly reduced form. The individual analyst's schema should concentrate on the participant's use of words and phrasing (their "ordinary language"), while schemas show interweave their own writing with quotations, to add richness and individuality to each piece.

Essential elements of a text do not purport to a single, critical "truth" but together offer a composite picture that contains multiple truths, all of which go equally toward an understanding of the whole. They must be considered in their own right, as critical component parts, which once revealed, can be fully accounted for through the group-working activity to follow. Thus, it could be said that the essential elements are those without which the topic (the phenomenon) would lack its full coherence and cannot be comprehensively defined or revealed in all of its complexity and uniqueness.

From individual schemas, succinct group schemas (or meta-schemas) are crafted (see Stage 2 below), and individuals create one schema for every piece of text examined, each approximately half an A4 page in length, ready for sharing with others (Stage 2).

3.2 Stage 2: Group Work and the Development of Group or "Meta-schemas"

Team-working, in Stage 2, allows multiple perspectives to come to the fore (Mauthner and Doucet 2008). Teamwork takes place over an extended period of time. The team meets through multiple workshops to take part in consensus-building discussions. From individual schemas, meta-schemas must include the original text's essential elements with full agreement from all present. Stage 2 begins with discussions of each individual's schema, while each individual's view is compared and contrasted with the views of every other person in the group. The group must agree on which aspects of a text to hold dear and which to let go of. They must discuss which are essential, and which are incidental, which are outliers and which are critical experience, as well as which quotes support or undermine the essential elements in their holistic presentation. Agreeing on this takes time and is a consequence the desire to achieve consensus opinion must be paramount. Team members learn to recognize, and take advantage of, the scope and value of all present and to enjoy the mutually beneficial, multidisciplinary, and collaborative process while discerning the impact of group understanding – more valuable than any one individual's perspective alone.

Data analysis in qualitative studies can often be a lonely and isolating experience, demanding long hours of solitary working. Individual researchers may occasionally seek help in making sense of the "bigger picture" (the broader understanding of their data), but this is rare. Similarly, only occasionally will a data analyst consult on the "minutiae" of data (a text's nuances, ambiguity, or unexpected revelations) or turn to a colleague or supervisor for assistance. More often than not, peer review is piecemeal, and while help can be forthcoming, others' input is temporary, leading to gains that are short-lived. The infrequent involvement of others can lead to an introspective activity that adds little to a researcher's sense of shared responsibility for data management and handling.

This is not the case in schema analysis. In creating and discussing individual and group schemas, through multiple workshops that take place over time, under the clear steerage of a senior academic with some expertise and yet no dominant worldview, everyone involved is fully and equally invested in the outcome. Researchers must, as a consequence, completely commit to the process, from the beginning to the end, and take equal responsibility for the end product, while sharing in the rewards.

3.3 Stage 3: Interpretation of Meta-schemas

The number of group schemas produced depends on the number of pieces of text being analyzed at any given time. The primary researcher or senior academic's role, on account of their knowledge of schema analysis, can, if necessary, be to manage the team, as long as members adhere closely to these three stages through self-direction and self-determination. This is wholly dependent on the makeup and needs of the team. Furthermore, team members can work on a sub-sample or the complete dataset, depending on the needs of the study.

Meta-schemas stand as a record of all the work that has gone before, and from their creation, interpretation of the phenomenon can be finally derived. Interpretation is avoided until Stage 3, where a final analysis of all meta-schemas through the triangulation of data takes place. From the group schemas, a comprehensive interpretable presentation of the whole dataset being studied is written in the form of a final report or publication that contains all group schemas and their interpretations. During Stage 3, time must be spent discussing meta-schemas in terms of the final study product, aiming to highlight the team's shared understandings and shared meanings derived across datasets. The interpretative stage, Stage 3, illustrates what the data has to offer, and indicates how study outputs may hold value for wider audiences, with potentially far-reaching consequences. At this stage, aspects of context can be brought to bear, with descriptions, in whatever format, including key quotations and revelations of setting and environment.

4 An Exemplar: The Role of Multidisciplinary Teams in Breast Cancer Risk Communication

In this section, we present an exemplar of the method in use. The study in question, approved by the Macquarie University's Human Research Ethics Committee (Reference Number 5201600446), was undertaken at the Australian Institute of Health Innovation (AIHI), Macquarie University, Sydney, to examine the role of multidisciplinary teams in breast cancer risk communication. It took place between 2016 and 2017, and in this exemplar we examine the views of one study participant, an oncology psychologist (who will be known under the pseudonym "Chloe"), who took part in a semi-structured, face-to-face interview with one of the study researchers (PS). The full team included four staff members from AIHI, a social scientist, an ethnographer, a biomedical scientist, and a clinically trained mixed-methods researcher (see Fig. 2 for detail).

The transcript of the interview described the role of the psychologist in a wider multidisciplinary team-working context, her responsibility to the breast cancer program, and how she understood and presented patients with the concept of "risk" as well as how she communicated with other colleagues. Within the transcript, the oncology psychologist also discussed her role in the multidisciplinary team and members' relationships with one another and with patients.

4.1 Stage 1: Individual Schemas

The team of researchers, with varied degree of qualitative research experience (see Fig. 2), developed the meta-schema from individual schemas and the interview transcript. The primary researcher accountable for the study (FR) oversaw the process, creating her own succinct overview of the data source (see Example 1). Concentrating on the perspective of the research participant, the individual schema

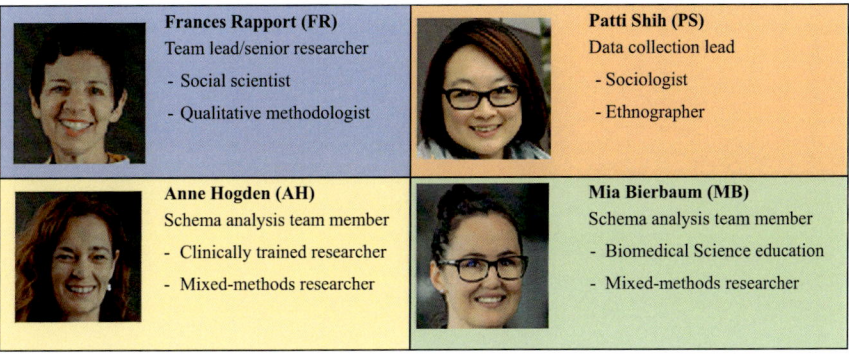

Fig. 2 Roles and disciplinary backgrounds of the research team from Macquarie University responsible for the case study

offered a brief vignette of the text's content. It took the form of a coherent, free-flowing narrative of approximately a paragraph in length (half a page), and the approach was then taken up by each of the other researchers in the team who created their own individual schemas (see Example 2 for an example of one of the group member's schemas). Once these were created, they were circulated among the group, and an initial workshop was convened to discuss similarities and differences in individual's schemas' content and style and to compare and contrast the views of the group on the written exercise. Individual schemas were collated and distributed before the first workshop took place to give team members a chance to read each other's work and consider their content. The initial workshop led to extensive discussion of essential elements that differed in individual presentations.

Example 1 of an Individual Schema
The psychologist's primary role is to offer emotional support to patients and help them decipher what doctors have recommended, while clarifying underlying meanings expressed. Patients are appealing to the psychologist for: "insider knowledge" or an "informal opinion," hoping to gain clues about the minutiae of conversations with oncology specialists, while looking for translations of nonverbal inference. For the psychologist, discussing risk with patients is not about deciphering quantification. Rather, it is about assessing what patients are scared about, what they need to be reassured about, and what their tolerance level is for more information. The psychologist must assess how to bridge the gap between clinical team members' presentations, their actions and reactions to patients, and patients' expectations and understandings. The psychologist is the arbiter of the peace – there to direct patients along a steadier route than might otherwise have been the case. They are well aware that the patient is afraid of dying. Patients are often "scrambled and distressed," and as a result have poor recall of what is being said within any given consultation. The psychologist wants to support them as best she can, to take a part in the "play" that is the dynamic of clinic life. She uses this analogy to encourage patients to address risk appropriately, to use the expertise of all the players involved, and to consider who might be the most relevant team member, at any given time, to support them. "Different people take different parts on stage at different times." The psychologist has to manage patient anxiety, while helping patients contain their emotions and overcome their worst fears. She is there to uphold confidences, and above all else, not let the patient get too disillusioned with their treatment, or the team's response to treatment outcomes. "We all want certainty in an uncertain world," but information needs to be provided sparingly, in order for patients to realign their goals to reality. "We have a gentler way of doing risk assessment," in order to: "talk about risk of values not being met." This is at the crux of the psychologist's role, to steer the patient, carefully and empathically, through a maze of

(continued)

emotions and moods, and subtly help facilitate discussions with oncologists that are going to encourage patients to be realistic in the face of a very daunting disease.

Example 2 of an Individual Schema
Chloe saw herself as an intermediary between patients and the medical system. She interprets doctors' messages for patients: not just medical information on tumor and their treatment progress, but the elephant in the room about the key question of short- and long-term survival. "Whether they're going to die" is an often unspoken, yet most important question for patients in the complex and traumatic journey of cancer treatment. Chloe works closely with patients to help them navigate the medical complexities of treatment, communication challenges with oncologists, and the healthcare system. Chloe recognizes that the cancer care system is responsible for some uncertainty and inconsistency in patient care. Firstly, there are variations between healthcare professionals' practices. Different doctors talk about, and emphasize, different types of risks: "surgeons are more likely to use numbers and talk about risks of surgery, complications and side effects; palliative care doctors focus on goals of care," and the way they communicate to patients is different as they: "vary in skills, vary between teams, and in seniority." Secondly, the treatment journey is full of different types of risks: tumor risk, treatment risk, as well as emotional, financial, and psychosocial risk. Even in posttreatment survivorship, patients are faced with the risk of recurrence, posttreatment medication side effects, and posttreatment trauma. Thirdly, each patient will vary in their emotional response to cancer diagnosis and treatment; where some are more emotionally vulnerable, others are less so.

Chloe's role is to assist patients in managing the range of anxieties and fear that may arise as they face these complex uncertainties. Chloe saw her role as helping patients gain: "certainty in an uncertain world", so that they can cope with the grueling process of treatment and enable the medical team to do their job: "my whole role is to make sure [patients] keep confidence in the medical team." She recognizes that face-value reassurance does not reflect the reality of dealing with the trauma of disease. For example, doctors are hesitant to be frank to a patient who is emotionally vulnerable: "there is an agreement within the team to keep information limited, because if we convey exact risk, she won't cope." Clearly, for Chloe, psychosocial risk will impact on medical risk, and thus her role as an intermediary between doctors and patients is to manage a balance between these two. As a psychologist, Chloe's expertise is to help patients bring medical facts into line with their personal values: "we talk about risks of values not being met" and her role is to help patients: "get a more realistic idea about their risk."

4.2 Stage 2: The Group or Meta-schema

Following the initial workshop, Stage 2 was planned. The team met on an ongoing basis to agree on the most essential elements of the individual schemas, acting as an interpretive community.

During Stage 2, an important aspect of the group work was to meld understanding. For the purpose of this study, understanding revolved around how Chloe described her consultations with breast cancer patients, her experience of these consultations, and her presentations of "risk" to patients. The team also examined how the implications of being part of a multidisciplinary team in this context impacted on patient-professional interactions.

As mentioned earlier, essential elements drive greater clarity through the schematic process. They highlight integral pathways through conversations and biographic insights, and they lead to a greater team awareness and commitment. They suggest the participant's point of view, but by so doing, draw the co-researchers together, to take shared ownership of re-presenting that point of view. They indicate a story yet at the same time offer vital insights into self-expression and all from the perspective of the research participant.

> **Example 3 of a Group Schema**
> As a psychologist, Chloe has an intermediary role in interpreting medical information provided by the clinician and sharing it with the patient, while guiding patients along an appropriate care pathway. She is there to offer patients a route through the medical system while helping them gain more certainty in their lives. Patients' personalities, their circumstances, and life experiences are so different, and so are their responses to a cancer diagnosis. One of Chloe's key roles is to help patients manage the range and fluctuation of their fears and other emotions associated with cancer diagnosis and treatment. By "providing certainty in an uncertain world," Chloe can help patients maintain confidence in the medical team. This is crucial for successful treatment as well as for the ability to manage patients' anxieties. Chloe's conversations with patients are gentler in approach than those of her peers. Her discussions add nuance to medical representations of risk in a way that is more aligned with patients' own personal goals and values. This can encourage patients to be more realistic in their understanding of risk when in discussion about treatment with oncologists. As a psychologist, Chloe angles her conversations with patients towards three things: (1) the patient's prognosis, (2) patient drivers, e.g., their fears, values, and goals, and (3) doctors' drivers, e.g., what the doctors would like the psychologist to help them achieve, e.g., Psychological readiness for treatment and for receiving challenging information.

Coming to recognize the essential elements of any text is a road to team discovery, and following listing and labeling (including key quotations presented verbatim) in the case of the exemplar study, the group were able to redefine their

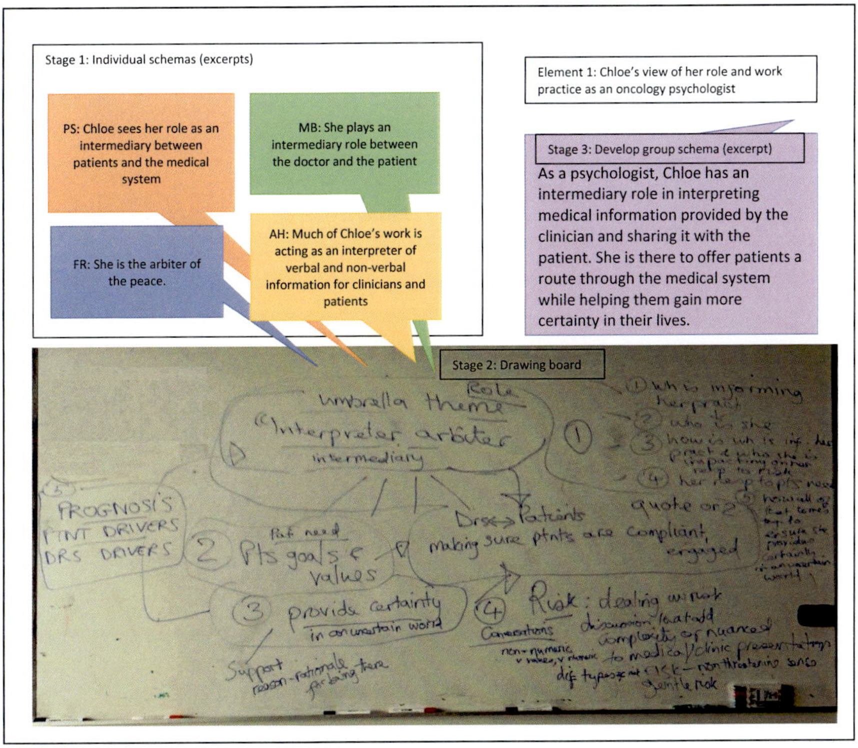

Fig. 3 Developing a part of the group schema for Element 1 (Chloe's view of her role and work in oncology psychologist)

individual schemas as one meta-schema that retold the story, succinctly and clearly (see Example 3 for detail of this group's example).

The Teamwork Drawing Board in the center of Fig. 3 demonstrates the development of five core elements that needed to be presented in the meta-schema, derived, in this instance, to explicate the interviewee's view of her role and working practices as a psychologist within an oncology multidisciplinary team, and the elements were:

Element 1: Psychologist as interpreter, intermediary, and arbitrator between clinicians and patients.
Element 2: Psychologist as identifier with patients' goals and values.
Element 3: Psychologist as provider of a sense of certainty in an uncertain world.
Element 4: Psychologist as expert, presenting "risk" to patients in a nuanced and person-oriented way.
Element 5: Work practices are grounded in the patients' prognoses, patient drivers (e.g., their fears, values, goals), and clinicians' drivers (e.g., ensuring patients are psychologically prepared for certain treatment decisions or for receiving challenging information).

These five elements were developed by using, as a core to their meta-presentation, the ideas generated from each person's individual schema. The following section

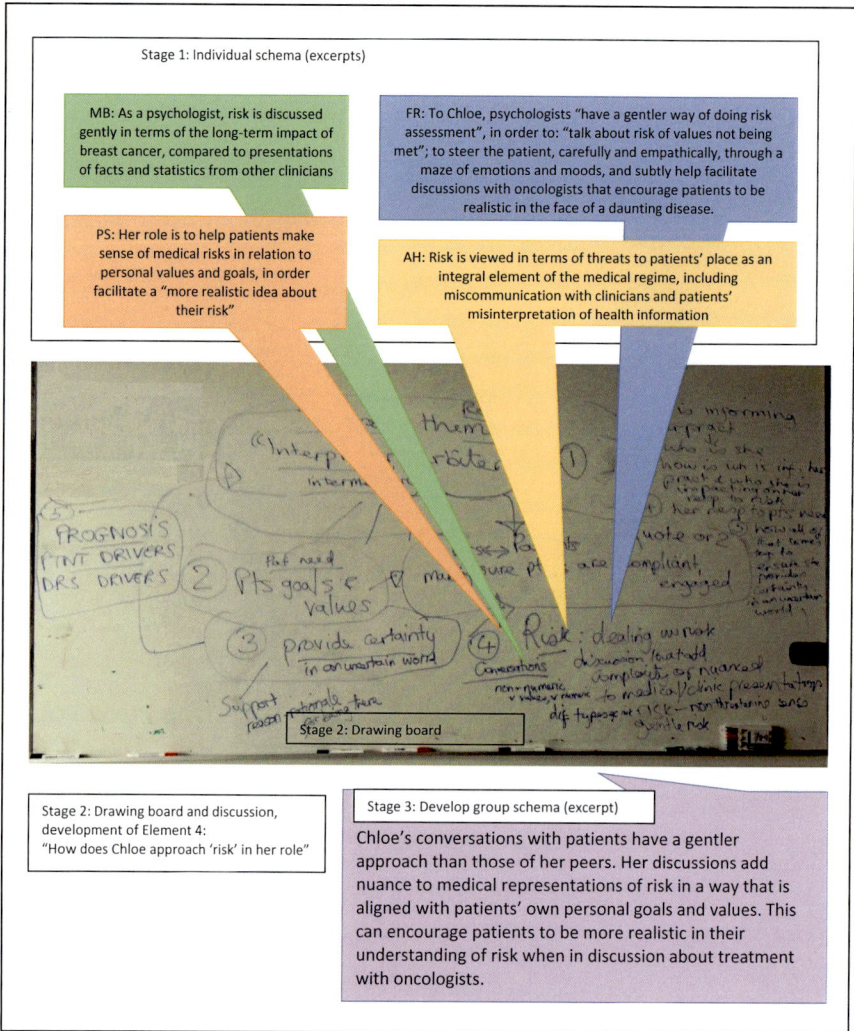

Fig. 4 Developing a part of the group schema relating to Element 4 ("How does Chloe approach 'risk' in her role?")

explains how some of the elements were analyzed. For example, Element 1 came from the general consensus (Fig. 3) that as the interviewee was acting as an intermediary between patients and clinicians, she was frequently presenting each party's goals and objectives to the other party.

Element 4, on the other hand, was developed from a more nuanced discussion about what "risk" might mean to the psychologist working with patients in these circumstances, and the different types of risk the psychologist might need to deal with, as well as how this has impacted on her descriptions of her work and approach to patients (Fig. 4). Through discussion and some ongoing debate, the group schema

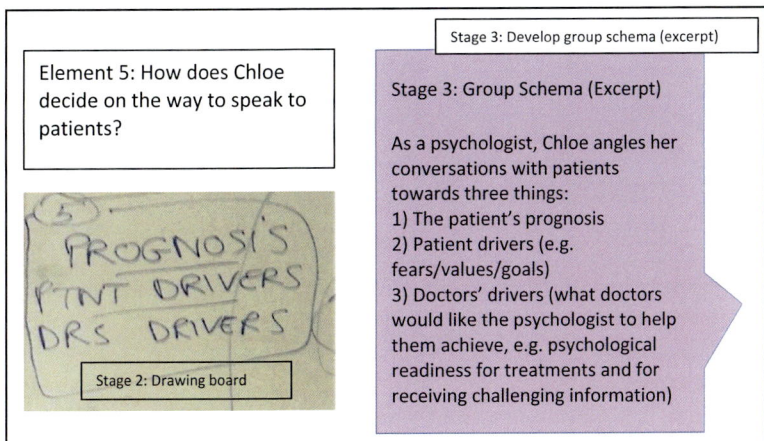

Fig. 5 Developing a group schema to include Element 5 ("How does Chloe decide on the way to speak to patients")

included Chloe's understanding of risk from the point of view of her role as a psychologist and the effect this has on the way she approached the topic of "risk" with her patients.

Element 5 arose from group discussions of the elements embedded in the individual schemas and the group's subsequent understanding of the way Chloe used her own response to risk, and her role as a psychologist, to inform her professional practice (Fig. 5).

4.3 Stage 3: Interpretation of Meta-schema

Once Stage 2 was completed, the team felt at liberty, for the first time in the analytic process, to consider the meaning underlying the meta-schema and to discuss what the meta-schema led the group to conclude. The interpretative stage, or Stage 3, enabled the researchers to examine why the situation was as it was and why a view was expressed in a certain way or a relationship created with others. Stage 3 unfolded in line with the literature on the topic and others' appreciation of the topic and its current relevance. Interpretations were revealed once the meta-schema was crafted, and it was kept intact and shared alongside the interpretation of its contents, line by line. This encouraged the whole team to reflect back upon the raw material and ensure nothing of importance had been left out. Verbatim quotations embedded in the meta-schema gave the piece its flow and direction, but also helped give weight to the research team's views of the most important points spoken by the participant, with the participant's voice clearly present. Understanding was embellished by consideration of other participant transcripts and observational field notes. On this occasion, field notes were used corroboratively, to highlight dynamic moments in patients' consultations, the way people interacted in the consultation, and to add to the veracity of the schema analysis method, by providing additional and contextual information to support schema findings.

5 Validity, Rigor, and Trustworthiness of Data

Validity of data, in this case, refers to the ability of data to be seen as sound, reasonable, and credible and presented in such a way that accurately represents Chloe's views. Validity also suggests that the study's conclusions link back to the raw material and thus convincingly achieve meaningful findings.

The core techniques of schema analysis, group-working, checking and rechecking of data, and the development of individual and meta-schemas, lend itself to an accurate and thorough representation of the raw material through consensus-building activities (Thomas and Magilvy 2011). This adds to the credibility of data and the knowledge that the process can be replicated by another research group to generate similar results (Thomas and Magilvy 2011). The group work discussion adds rigor to the working methods (Long and Johnson 2000), and the engagement of multiple researchers counteracts any one individual having excessive influence over the production of study results.

Trustworthiness of data consists of four components: credibility, transferability dependability, and confirmability (Lincoln and Guba 1986). In schema analysis, credibility is assured when multiple researchers agree that the data analysis aligns with the reality of the original texts. Using group consensus, when the group arrive at a position that the emergent schemas are uncontested, and study findings are representative of the participants' views, it can be said that data are credible. Transferability is assured during the interpretation phase, when the contextualization of data reveals how meta-schema can be applied to other situations and settings. The dependability of data can be assured when others can repeat the data analysis process, if desired, with a different group of researchers, arriving at the same or similar conclusion, regardless of who is involved. Confirmability ensures that the study's findings are robust. The teamwork aspect of schema analysis enables key findings within meta-schemas to be conferred across analysts (see also ▶ Chap. 63, "Mind Maps in Qualitative Research").

The final stage of the process, the interpretation stage, ensures that no one component of data is prioritized, before wholistic descriptions of all data are produced (Lincoln and Guba 1986).

6 Lessons Learned: Enriching Qualitative Analytic Practices

The degree of difference between schema analysis and other qualitative methods becomes apparent as researchers immerse themselves in this new approach, and there are distinct methodological and practical strengths and weaknesses in re-presenting and interpreting complex datasets in this way.

Strengths: Firstly, in terms of the method's strengths, schema analysis arises from a *process* of representation, collation and interpretation, and extensive researcher collaboration. Secondly, individual aspects of the method ensure a deep grounding in the data without the influence of others: ("This is my understanding of what X is saying about herself") before group understanding is sought. Thirdly, each researcher is encouraged to "get inside the participant's head" and engender, in their individual schema, a strong representation of the person's voice, lending itself to a person-centered approach to analysis. Fourthly, once individual schemas are considered by the team, similarities and differences between them become the salient points for discussion, and learning occurs as a team. The collaboration needed to refine individual representations into a group representation, which embodies a shared understanding of the data and a collaborative representation of the participant voice, ensures clear agreement across group members, adding veracity to the value of group-working. Fifthly, a mentoring process can be built into the schema analysis, with more junior researchers or those less familiar with qualitative methods learning the techniques from the active modeling of others. The value of this is that ongoing reflective discourse between researchers results in a more cohesive and rigorous analysis, with equal buy-in from all researchers. Finally, as each member of the team has the same amount of work, tasks cannot be delegated or parceled out to others, suggesting equal commitment to the process at hand.

Weaknesses: Firstly, if the precise steps of the process are not followed, the schema produced will be inadequate. Secondly, researchers must resist the temptation of interpreting data preemptively. Thirdly, removing one's researcher perspective or clinical self from the representation can be difficult, but failure to do so can result in a flawed representation of the participant's point of view. Fourthly, team members must refrain from sharing their own opinions too readily, but equally must be sure not to omit their opinions at the interpretation stage: ("It's not about me").

Combined strengths and weaknesses: Generating schemas takes time. Group schemas rely on the availability and cooperation of all workshop members, and as meta-schemas are based on more than one representation, the process can be disadvantageous to a researcher wishing to work alone. While schemas can be developed and then analyzed by individual researchers, there is no in-built mechanism to check the quality of representation of the participant voice and no capacity to develop a collaborative group schema. While the team approach to analysis is a resounding strength, methodologically, it is also a weakness of practicability.

7 Conclusion and Future Directions

Since its first development, schema analysis has been used extensively by the authors to make sense of complex data across health service and medical research contexts. For example, it has been successfully applied to assess breast cancer patients' responses to decision aids for treatment and care (Rapport et al. 2006; Iredale et al. 2008). It has been used to clarify multidisciplinary team practices for assessment of breast cancer risk (Shih et al. under review), and it has helped examine differences between communication styles in individual breast cancer consultations and multidisciplinary care (Rapport 2010; Rapport et al. 2015). Schema analysis is currently being used to underpin a study assessing cochlear implants for use by older Australians to improve hearing health and reduce listening effort (Rapport et al. under review).

Schema analysis has the potential for broader application than this and may be highly applicable to research that requires an in-depth understanding of a particular client group. For example, it could be useful in diverse fields such as business, education, and the environmental sciences where a strong interest has grown in the needs of stakeholders to drive and give meaning to a system, in order to deepen understanding of clients' needs and experiences and ensure improvement in these fields is in keeping with client expectation. With versatility and flexibility built into the method, and a growing interest in it as a very practical approach to team-building and team awareness, the potential of this method, as it continues to be refined, is extensive. We urge others to try it out for themselves, to assess its value in their own research studies, and to experiment with its application across human and health science domains and topic areas.

References

Fleming J, Ward D. Self-directed groupwork–social justice through social action and empowerment. Crit Radic Soc Work. 2017;5(1):75–91.

Iredale R, Rapport F, Sivell S, Jones W, Edwards A, Gray J, Elwyn G. Exploring the requirements for a decision aid on familial breast cancer in the UK context: a qualitative study with patients referred to a cancer genetics service. J Eval Clin Pract. 2008;14(1):110–5.

Lincoln YS, Guba EG. But is it rigorous? Trustworthiness and authenticity in naturalistic evaluation. N Dir Eval. 1986;1986(30):73–84.

Long T, Johnson M. Rigour, reliability and validity in qualitative research. Clin Eff Nurs. 2000;4(1):30–7.

Mauthner NS, Doucet A. 'Knowledge once divided can be hard to put together again': an epistemological critique of collaborative and team-based research practices. Sociology. 2008;42(5):971–85.

Rapport F. Summative analysis: a qualitative method for social science and health research. Int J Qual Methods. 2010;9(3):270–90.

Rapport F, Hartill G. Poetics of memory: in defence of literary experimentation with holocaust survivor testimony. Anthropol Humanism. 2010;35(1):20–37.

Rapport F, Hartill G. Making the case for poetic inquiry in health services research poetic inquiry II–seeing, caring, understanding. Rotterdam, Springer; 2016. p. 211–26.

Rapport F, Iredale R, Jones W, Sivell S, Edwards A, Gray J, Elwyn G. Decision aids for familial breast cancer: exploring women's views using focus groups. Health Expect. 2006;9(3):232–44.

Rapport F, Jerzembek G, Seagrove A, Hutchings H, Russell I, Cheung W-Y. Evaluating new initiatives in the delivery and organization of gastrointestinal endoscopy services (the ENIGMA study): focus groups in Wales and England. Qual Health Res. 2010;20:922–30. https://doi.org/10.1177/1049732309354282.

Rapport F, Clement C, Doel MA, Hutchings HA. Qualitative research and its methods in epilepsy: contributing to an understanding of patients' lived experiences of the disease. Epilepsy Behav. 2015;45:94–100.

Rapport F, Bierbaum M, Hughes S, Lau A, Boisvert I, Braithwaite J, McMahon C. Behavioural and attitudinal responses to cochlear implantation in Australia and the UK: a study protocol. under review.

Rapport F, Sparkes AC. Narrating the holocaust: In pursuit of poetic representations of health. Medical Humanities. 2009;35(1):27–34.

Ryan GW, Bernard HR. Data management and analysis methods. In: Denzin NK, Lincoln Y, editors. Handbook of qualitative research. 2nd ed. Thousand Oaks: Sage; 2000. p. 769–802.

Shih P, Rapport F, Hogden A, Bierbaum M, Hsu J, Boyages J, Braithwaite J. Relational autonomy in breast diseases care: a qualitative study of contextual and social influences on patients' capacity for decision-making. under review.

Silverman D. Analyzing talk and text. In: Denzin NK, Lincoln Y, editors. Handbook of qualitative research, vol. 2. 2nd ed. Thousand Oaks: Sage; 2000. p. 821–34.

Sparkes A. Telling tales in sport and physical activity: a qualitative journey. Champaign: Human Kinetics Publishers; 2002.

Thomas E, Magilvy JK. Qualitative rigor or research validity in qualitative research. J Spec Pediatr Nurs. 2011;16(2):151–5.

Van Manen M. Modalities of body experience in illness and health. Qual Health Res. 1998;8(1):7–24.

Using Qualitative Data Analysis Software (QDAS) to Assist Data Analyses

52

Pat Bazeley

Contents

1 Introduction 918
2 Establishing the Goals of Analysis 919
3 Selecting Data for Analysis 921
4 Managing Data for Analysis 922
5 Reviewing and Exploring Data 924
6 Sorting and Coding Data: What Are My Data About? 926
7 Investigate and Interrogate 931
8 Conclusion and Future Directions 933
References 934

Abstract

Qualitative data analysis software (QDAS) has much to offer the health researcher. Software facilitates efficient management of qualitative and mixed methods data through a variety of tools to organize and keep track of multiple data sources and types and of the ideas flowing from those data. Coding tools provide structure to the categories and themes evidenced in the data, allowing for rapid retrieval of information. Increased depth and rigor of analysis are facilitated through capacity to search and interrogate the data sources using a combination of coding and other data management tools. Questions can be asked about how often and how different categories or themes are expressed by different groups within a sample or within different contexts or times. Similarly, experiential data might be compared for those with different measures on health-related variables. Relationships between different aspects of experience (or attitudes or feelings, etc.) can be explored and/or verified using coding queries, through a range of visual displays,

P. Bazeley (✉)
Translational Research and Social Innovation group, Western Sydney University, Liverpool, NSW, Australia
e-mail: pat@researchsupport.com.au

or through statistical analyses using exported coding information. Such queries can be limited to one type of data, or multiple types of sources can be imported, coded, and analyzed together. Linking tools are used throughout to connect reflective thoughts to the data that prompted them or interim results to the evidence that supports them. Explanations of these processes are illustrated by figures and examples.

Keywords

Qualitative · Software · Analysis · Mixed methods · Coding · Visualizing · Theory building

1 Introduction

Various forms of software to support qualitative data analysis were developed for public use in the 1980s and 1990s. These software developments followed closely on the rapid developments that occurred in the use of qualitative approaches to research that began a decade earlier and were supported by the revolution occurring in computer science with the shift from mainframe to personal computers. While statistical software to analyze numeric data was widely adopted without question in the quantitatively dominated research climate that prevailed at the beginning of the digital era, adoption of qualitative software as a legitimate tool to assist analysis of textual data has been a much slower process – and continues to be resisted in some quarters. The majority of early programs focused on "code and retrieve" as a primary tool to assist analysis, because this is what computers were good at doing. The capacity to query connections in coding was also there as an extension to code retrieval but was often ignored. This was partly because querying required a higher level of technical skill in the user, but also for some, the use of logic-based queries was (and is) seen as a poor substitute for the "intuitive" connections arising from deep immersion in data, manifested in the work of skilled qualitative researchers. Qualitative software, in its current form, is a product of the developments that have taken place in both computer technology and methodology in the twenty-first century; it provides a much wider and ever-expanding range of tools to emulate the tasks involved in qualitative analysis, yet its heritage in those early 1980s developments, with their emphasis on coding as a primary analytical task, is still very much evident.

One of the issues faced by developers of software for qualitative analysis was, and is, the diversity of qualitative methodologies and the inherent complexity of the analytic and interpretive processes associated with those methodologies – processes that are not readily reduced to linear algorithms, such as occurred with statistical analyses. This complexity requires flexibility in the way software tools function, which in turn engenders complexity in software design – and for the user, a relatively steep "learning curve" to achieve effective use of the software for her particular purposes (Gilbert 2002). This further inhibits adoption by some researchers using qualitative methods. Nevertheless, adoption is now widespread, and the use of software for analysis of text and visual data is becoming more or less expected in many fields.

Just as the qualitative "revolution" challenged and changed foundations for research from the 1970s, so also a mixed methods revolution has been changing research practices, primarily since the 2000s (Mertens et al. 2016). There has been some adaptation in statistical software, but it is qualitative software that is really taking up the challenge of developing ways to merge quantitative and qualitative data and analyses. Further challenges in a digital era of big data and social media are being met by both statistical and qualitative analysis programs. Software of both types is constantly evolving, as developers work to meet the rapidly evolving landscape of research methods in a digital age.

Although technology plays an increasingly important role in the health sciences, health is essentially about human populations and the individuals that make up those populations, with all of their physical, mental, emotional, and spiritual qualities (Huber et al. 2011; Liamputtong in press). Used wisely and with sensitivity, qualitative data analysis software (QDAS) can assist health researchers in their work, insofar as it is focused on the human interface of health care. It does so by offering a flexible suite of tools to use in managing and working through qualitative (and mixed methods) data, aiding analysis and interpretation in the process. Users of QDAS find coding, memoing, linking, querying, and visualization tools that help them to work with, reflect on, and see their data in new ways. It can assist, but not substitute, for deep immersion in data and the skilled recognition and interpretation of linkages and patterns across data by the researcher that are the hallmarks of ground-breaking qualitative analysis. Additionally, the use of software ensures more systematic and transparent analytic processes, with conclusions not only supported by evidence but by evidence and a chain of thinking that can be traced. Ultimately, however, the analytic interpretation remains in the researcher's skilled hands.

Software is often dismissed as something that is useful only at the latter stages of a project, when analysis is in full swing. I propose in this chapter to take the reader through the main stages of conducting a research project and to demonstrate that QDAS has a role to play from start to finish.

2 Establishing the Goals of Analysis

If we are to consider how software can assist health researchers achieve their goals for analysis, a first step must be to establish what those goals might be. And immediately one is faced with the diversity of health research, just as health care and health services are diverse – a reflection of the breadth of what health itself entails. Research goals can range from the conceptual to the practical, from measurement-oriented assessment of the benefit of a drug or surgical intervention through to exploring the phenomenological nature of particular health-related experiences (which might, coincidentally, be of drug or surgical intervention), and from a focus on policies, organizations, and services to a focus on ordinary people and their daily lives.

The role (and value) of software for analysis has its starting point right here, at the point of establishing goals for analysis. What will this project be designed to achieve? What, and whose, purposes will it serve? QDAS contributes in three ways at this pre-analytical point: through recording reflections, mapping preliminary ideas, and reviewing and analyzing literature.

Keeping a record within the software of one's musings over purposes and goals is helpful insofar as the act of writing assists in building clarification. Having a written journal recording ideas and steps taken additionally provides the basis for an audit trail relating to project development, something that becomes useful when justifying choices made and explaining directions taken, during a later stage of reporting. Applying some codes to this document as it is being written assists during analysis and writing with locating particular ideas, and it generates starter codes to apply to literature or participant data. For those new to QDAS, this provides a gentle introduction to using software.

Modeling (mind-mapping) ideas about the topic being studied also helps in clarifying those ideas, and with project planning, by pointing to those things for which data will be needed (see also ▶ Chap. 63, "Mind Maps in Qualitative Research"). Items entered in a mind map or conceptual model become represented as codes that will capture data and links drawn between items in an early conceptual model become assumptions to test (Fig. 1). Ideas sparked through mapping processes are recorded in the project journal, again assisting clarification and for later reference.

Most researchers now use bibliographic software to manage the literature relating to their research. That literature can be selectively imported from the bibliographic tool into the qualitative software, with abstracts, notes, metadata, and, where available, the original .pdf sources. Once imported, those materials can be coded interactively or searched using selected terms to identify relevant passages for consideration (rather than a search simply identifying whole references, as occurs within the bibliographic software). Alternatively, the content of articles might be rapidly scanned, aided by predictive auto coding tools that are available to varying degrees in different qualitative programs. Working through the literature on a subject is, of course, a key tool for refining objectives for the research and for identifying concepts and frameworks that, in turn, become tools to guide both design and analysis. When literature is incorporated into a project file, passages within it can be linked to new data, contributing to building a connected web of knowledge. Alternatively, queries can be used to compare what is being revealed through this project with established understanding as expressed in the literature. For example, professional understanding of caregiving for an aged person as a set of instrumental tasks is inconsistent with families' perceptions of caregiving as something more akin to psychosocial care; similarly discharge planners' views of home care needs can be very different from patients' perceptions of home care needs, with such inconsistencies leading to problems of noncompliance (Bowers 1989). (Coding and querying processes will be explained below.)

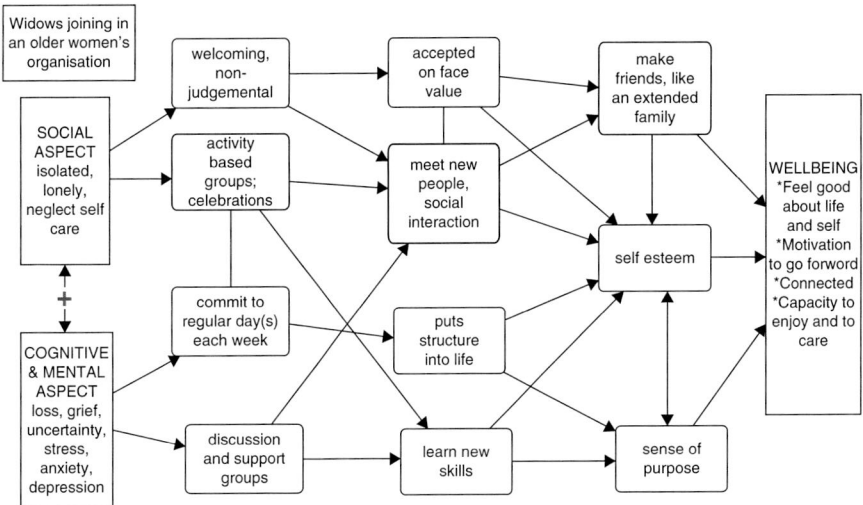

Fig. 1 Conceptual model: how older women's groups benefit the social and mental well-being of widows

3 Selecting Data for Analysis

Traditionally, qualitative health researchers have relied heavily on interviews or, to a lesser extent, focus groups to provide research data. Diaries, journals, and observations also provide data for some, while visual data (video, images, drawings) and documents (e.g., medical records, nursing notes) offer less frequently accessed forms of qualitative data. Researchers are also exploring the potential of network, geographic, and social media sources (see chapters in the "Innovative Research Methods in Health Social Sciences" section). Health services and evaluation researchers taking a more quantitative or mixed orientation have relied largely on surveys and questionnaires, sometimes including some open-ended questions within those (see ▶ Chaps. 20, "Evaluation Research in Public Health," and ▶ 32, "Traditional Survey and Questionnaire Platforms"). Increasingly, health researchers are broadening their traditional methodological approaches to combine these multiple sources and types of data, in what has become known as multimethod or mixed methods research (see chapters in the ▶ Chaps. 39, "Integrated Methods in Research," ▶ 40, "The Use of Mixed Methods in Research," ▶ 41, "The Delphi Technique," ▶ 42, "Consensus Methods: Nominal Group Technique," ▶ 43, "Jumping the Methodological Fence: Q Methodology," ▶ 44, "Social Network Research," ▶ 45, "Meta-Synthesis of Qualitative Research," and ▶ 46, "Conducting a Systematic Review: A Practical Guide" section).

Relatively recent developments in QDAS ensure that almost any type of data can find a place in a project managed using software. As well as text data, most programs allow for importing video and image data in a variety of formats, several allow for recording geocodes (e.g., manually, or from GPS) and for displaying these in map format, some for incorporating geographic images as a base on which to "map" other visually displayed qualitative data, and some now facilitate importing information directly from websites and/or social media platforms (including associated metadata). Additionally, survey data that includes open-ended responses can be imported directly from either Excel or online databases such as SurveyMonkey or Qualtrics. Basic demographic data for participants has always been considered useful in qualitative projects; in mixed methods projects, this is likely to be extended to include categorical responses and scaled data. QDAS usually deals with this demographic information and also any other quantitative variable data relating to participants or cases in the research by treating it as variable or attribute data which is then associated with related non-numeric data for those participants or cases. Finally, for those using ethnographic or other methods where data are continually added to ongoing field notes, records can be created and maintained within the software.

4 Managing Data for Analysis

Data management is rarely discussed in texts, and yet it is a crucial element in analysis, especially for the complex kinds of data that might be part of a qualitative project or a mixed methods project. Miles and Huberman (1994, p. 43) observed that "qualitative studies, especially those done by the lone researcher or the novice graduate student, are notorious for their vulnerability to poor study management." Approaches to and requirements for data management should be considered *before* data are gathered for a study, because these can impact on what can be done with those data during analysis – how they can be sorted, compared, and queried (Bazeley 2013). In particular, some forms of data (e.g., focus group data, survey data) require special preparation to access the features of software that facilitate linking demographic, categorical, and scaled data with qualitative case data (Bazeley and Jackson 2013).

Being able to trace and identify the source of evidence used, as well as the context for that evidence, is important for the transparency and transferability of results. Keeping track of sources, and of segments of data from those sources, during the process of coding and analysis is facilitated in QDAS through the use of folders, descriptions, and labeling (Fig. 2). Different kinds of sources are placed into folders or document groups, so that it becomes easier to see what data are available within the project. Contextual and reference information about specific sources can be recorded in descriptions, linked memos (a good place for associated field notes), or in associated variable data. And, whenever a particular segment of data is retrieved, the software indicates the source of that segment and provides the option to link directly from the segment back into its original context to assist in its interpretation (Fig. 2).

Managing sources in a way that allows for their analysis both as separate entities *and* as a common body of knowledge has always been a problem for qualitative researchers dealing with voluminous and messy data. QDAS assists this process by

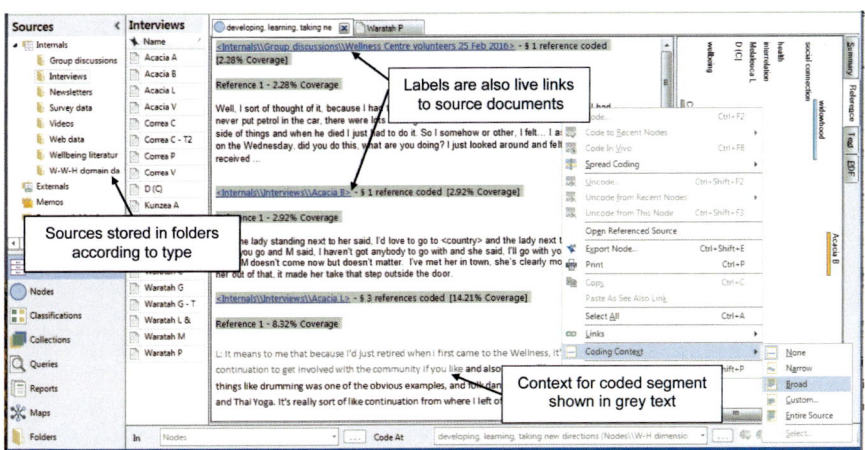

Fig. 2 Keeping track of sources and context for coded segments

providing multiple systems for sorting and "cutting" data, in ways that facilitate both targeted and comparative analyses.

Folders used for sorting and viewing sources become useful at the analysis stage as well. Some types of analyses are more or less appropriate with specific kinds of data, and so folders allow for scoping (focusing) a query to a specific group of sources.

Qualitative and mixed research projects are often structured around *cases* – the participants or other entities that exemplify the phenomenon being studied (Bazeley 2013; Yin 2014). NVivo facilitates the development of a dedicated structure for holding together diverse sources of data about each of the cases – the units of analysis – in a project, thus facilitating case-based analyses. Data for each case are *coded* to specifically designated categories, which means that data relating to a particular case can include one or multiple sources of the same or different types, and/or parts of sources. Thus, a participant's case data might include a combination of interviews along with their contribution to a focus group and their contributions to a meeting as discerned from the minutes of the meeting. In addition, attribute data (demographic, categorical, or scaled variables) relating to the case would be entered or imported and attached to that case's code, which means it is automatically applied to all data coded to that case, even if added later. Attribute values can then be used as a basis for comparative analyses (as described in a later section). The alternative, for QDAS that do not offer a specialized case structure option, is to use the general coding structure to create a parent code for cases with a set of subcodes. Each subcode would then represent a case to which all relevant material is coded. In this situation, however, attribute (variable) data can be attached only to whole sources.

Sets are created in QDAS to hold together aliases for collections of sources, or of codes (or in NVivo, also a combination of sources and codes). Aliases update as the

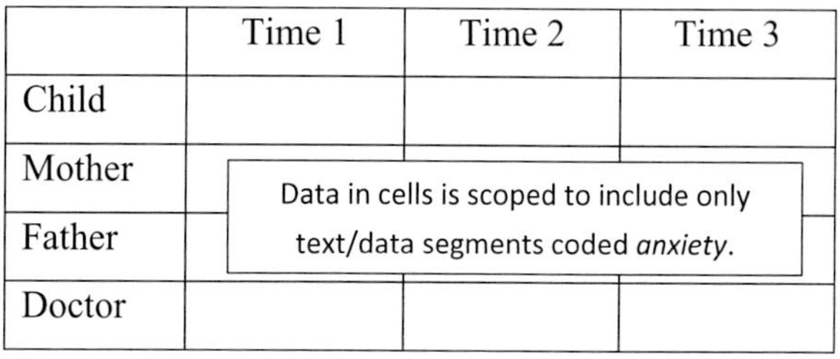

Fig. 3 Using data management tools to support a complex comparative query

	Time 1	Time 2	Time 3
Child			
Mother			
Father	Data in cells is scoped to include only text/data segments coded *anxiety*.		
Doctor			

material they represent is updated. Sets are useful, for example, for separating sources created in different phases of a longitudinal data collection ($Time_1 \ldots Time_n$), or for designating different categories of contributors to a set of cases (e.g., when interviews are held with mothers, fathers, doctors, and the target child for each child [the case] with juvenile diabetes). Making sets of codes can be a useful strategy for identifying concepts that "hang together," perhaps making for a more comprehensive or abstract category. The particular value of sets in analyses is that the software will treat the contents of a set as a single item, facilitating their use in comparative queries, or again, for scoping queries to a particular set of data.

Using these structural tools in combination will allow, for example, for a comparison of what different people [identified by sets] had to say about the anxiety [achieved by scoping the query to a particular code] over time [identified by sets] for children with juvenile diabetes [the cases] (Fig. 3). Or, the anxiety [code used for scoping the query] of specific children [cases] or groups of children [identified using attribute values] might be compared over time (sets), based on the combined input of those who are connected to each case.

While folders, cases, and sets can be set up at any stage during a project, it is nevertheless important to have considered what they might look like at the start of data collection, especially where these impinge on the way data are prepared (what formatting is required) and how they are labeled. Good management facilitates good analysis; it also prompts the qualitative researcher toward clarity and transparency, with potential impacts for ongoing design as well as analysis.

5 Reviewing and Exploring Data

Simply reviewing and/or exploring each data item as it is gathered is traditionally recommended as a first step in qualitative analysis (Bazeley 2013; Liamputtong and Serry 2017). In the early stages of a project, it can be found necessary to make adjustments to data collection procedures, based on these early reviews. Reviewing the whole of a source again, before detailed work is begun, helps to ensure specific elements within the source will be seen in the context of that whole. Many researchers prefer to read through and annotate a paper copy at this stage, although using software to add notes on or linked to specific points while reading through is not only possible but has the advantage that those initial impressions are not lost to later analysis. Some also find it helpful to summarize the key points learned from a

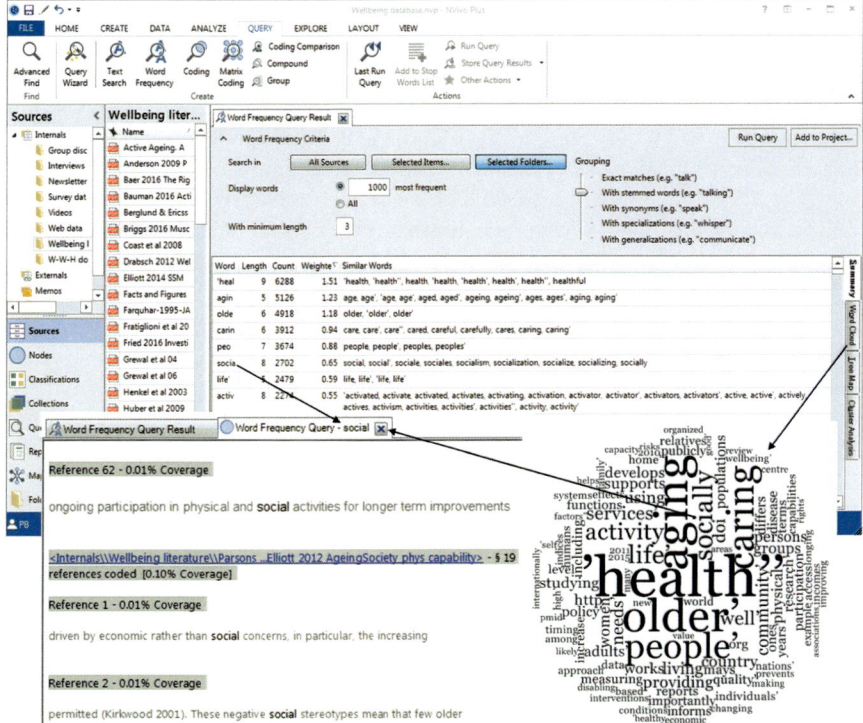

Fig. 4 Word frequency query with links to associated text

new data source; these can be recorded in a memo linked to that source. Another alternative is to use the mapping tools in software to create a visual diagram representing the narrative or perhaps the interconnected ideas being gleaned from the data source during this exploratory phase.

Of course, not all projects have data coming in one piece at a time, to allow the kind of work just described. Qualitative data can come in larger volumes, especially when sources other than interviews are being used. It is still valuable, however, to obtain an overview of what these new data are saying, before delving into detail. For this task, QDAS offers the options of either a global word frequency count or targeted word searches. A word frequency count will identify the most commonly occurring words – the overview – and present these as a summary or in a visual display. Each line in the summary then provides a direct link from the listed word to the passages where it was found, and from there to the broader context of that find (e. g., the paragraph in which it appeared), or to the source documents, allowing for a deeper exploration of any words of interest, and the concepts they represent. These strategies are useful, also, for scanning the literature, as shown in Fig. 4.

Recording ideas as they occur and reflecting on what is being learned are useful at any stage of the analysis process, and this time of initial review and exploration is a good time to establish lasting habits for doing so. Ideas and reflections can be recorded either in a general journal for the project or in memos linked to specific sources – the choice depends on what is being recorded about what (e.g., insights about the project as a whole, or its methodology, or reflections about the particular source). In the process of writing these reflections (of either kind), it is useful to link them to the source text or items that prompted them. In NVivo, a specific segment of data (or literature) that prompts an analytic or reflective thought is copied and pasted *as a link* within the journal or memo document at the point where the reflection is written. When the link is activated, it takes the researcher back to the copied evidence, showing it in its original context. In MAXQDA, separate memos for each passage are attached directly to the paragraph containing the passage that prompts the reflection; thus both programs achieve the same goal of connecting reflective thoughts to the evidence that supports them. Similarly, connections can be recorded linking from text (or other detail) in one item to text (or other detail) in another (e.g., from a data item to relevant literature; from one reference to another; from part of a picture or video to a discussion about that component or event).

Journal or memo entries are optionally coded to categories used also for coding data. This has the benefit of allowing the researcher to easily retrieve not only data on a particular topic but also reflections on that topic, potentially with links back to source data that provided the basis for the reflections. It overcomes any problems of trying to remember whether that note was made in the project journal or a particular memo.

Decisions made about the way the research is being conducted and about the direction of the analysis are also usefully recorded throughout the project (applying time stamps to these entries is a good idea). Some researchers choose to record these methodological steps in a separate journal; others use the general project journal. This builds an "audit trail" for the project – a record that explains what was done and why it was done that way (Carcary 2009; Richards 2014; see also ▶ Chap. 63, "Mind Maps in Qualitative Research"). This will help the researcher show the pathway taken in reaching the project's results and conclusions and allows for a more transparent write-up, for the benefit of the eventual readers.

6 Sorting and Coding Data: What Are My Data About?

As much as making connections between data items, the ideas in those data items are an important aspect of qualitative analysis, and there is no question that coding is the basic tool that underpins most forms of qualitative analysis (Strauss 1987). Coding is essentially a form of indexing that allows the researcher to find all the passages in their sources that relate to a particular topic, with the topic being represented by a label (the name for the code; see ▶ Chap. 48, "Thematic Analysis"). Coding, therefore, is an analytic task that connects data to ideas; it requires a level of interpretation by the person applying the codes (Bazeley 2013). During the coding process, the researcher is also making connections between those ideas (these should be noted in the journal, to be further explored). Codes might be topic-based, contextual, or more abstract and conceptual. One of the simplest ways to move from descriptive coding to more abstract coding is to keep asking "Why am I interested in that?" (Richards 2014). The codes developed in this way have significance beyond the immediate source, making them more analytically useful, although more descriptive codes retain their value also in recontextualizing more abstract concepts.

When coding was done manually, researchers would either write labels in the margin next to relevant passages, make copies of their documents, and then cut them into sections that were then sorted into piles for each code or create index cards that listed where relevant passages could be found, with a summary of what was in those passages. All of these methods imposed limitations on the researcher, either in terms of retrieving all that could be known about a particular topic or, more particularly, in sorting and connecting passages to explore or demonstrate a relationship between codes (e.g., a patient's mood when at work, compared with his mood at home; or, to identify the nature of a child's response to bullying at school, and then the impact of this on their sibling relationships). Because manual coding systems tend to limit the level of detail one can build into a coding system, they foster the use of more global codes in order to capture all of what is happening in a passage in one code, with a consequent tendency to rely on simple thematic analysis.

Coding with QDAS involves selecting passages and identifying what that passage is about. Each element of what is "going on" in a passage is captured with separate codes, and each code represents a single concept. Thus, responses or actions that signify improvement in, say, a medical condition might be coded for the type of response and/or action, descriptively for the particular condition being referred to and for the circumstances where the improvement became evident, as well as for the more abstract category of "improvement in condition," with each of these aspects being picked up by different codes and applied to the same passage of text. This has two consequences: firstly, all material relating to any particular code can be readily retrieved (with the source identified and accessible for each retrieved passage), allowing the researcher to explore in some depth what that concept is about – to see data in terms of the category rather than the original sources; and secondly, connections between codes, in terms of passages that are coded by particular combinations of codes, can be readily found using the query functions in the software. This means, for example, that the connection between a "what" and a "how" can always be reestablished, such as how and where improvement were demonstrated, with relevant passages retrieved. More than that, however, connections between the same "what" and different "hows" can also be explored. Thus, using the examples given above, if codes for where a mood is being experienced are applied at the same time as codes for the type of mood being experienced, the software can sort out text to illustrate and compare which moods are being experienced where. Or, text about a child's response to a bullying experience can be associated with that child's way of interacting with siblings, and perhaps also with responses, say, for children living in different family circumstances being compared.

The way in which a QDAS coding system is structured has implications for how well a researcher can use the codes to explore and interrogate her data. The most effective coding systems, in terms of facilitating querying, are those that categorize

Fig. 5 Sample coding structures for two projects

codes by what kind of thing they are about, in a taxonomic system (rather like the way plants are categorized), that is, its branching structure is not theoretically determined. Thus, for example, different events will be listed as subcodes within one "tree," emotions will be listed in another, as will times, places, or who is involved in particular events or emotions, or whatever else is being coded (Fig. 5 shows partially expanded examples for two projects). In an analytical sense, it means that any code can be associated with any other code, without limitations, with queries being able to find particular links as required, or patterns across groups of codes. In a purely practical sense, this means that a coding system is kept manageable (subcodes are not repeated across different trees; each can be a subcategory of only one "kind of thing" and therefore has its own place). This type of structure helps the researcher to "see" what her data are about – what kinds of concepts are involved. An added benefit for a supervisor of students engaged in qualitative research is that the structure and content of the coding system efficiently conveys the clarity and depth of a student's ongoing analysis.

As codes are being moved in the process of structuring a coding system, increasingly attention is turned from the individual sources to retrieving, reviewing, and rethinking the concepts being worked with in a project. Descriptions will have been provided for codes when they were first created, or perhaps as further data are being added to them and further clarification is needed, or they are provided now as they are being moved around. During this structuring process, further memos about the coding process are likely to be added to existing journals or memos, or new memos are created and linked to specific codes, with each holding information or thoughts about that particular category, its nature, and what it might be associated with (to guide future queries). Consideration is given to what "job" each code is doing and how it might be used in later analyses. Codes that have only a few passages coded to them can be revisited and compared with others representing similar concepts, to see if they might be combined (if so, this warrants a note in the description and/or in the methodological journal). Others might benefit from being split – coded on to different categories instead of, or in addition to, the original categories. As with text from sources, any thought-provoking segment found in the coded data can be linked to a note in a memo. When the linked segment is retrieved, it will be shown in its original (document or image) context.

All of this activity (and interpretive thinking) is building an ever-deepening understanding of the data in the researcher's mind. As the coding, connecting, and memoing processes are proceeding, the researcher will sometimes use query tools offered by the software to check on aspects of coding. For example, has text coded for emotions also been coded for what gave rise to those emotions? This can be checked simply by retrieving the codes for various emotions and reviewing coding stripes. Was coding applied to all the times when patients expressed satisfaction with a service? A text search will locate all the times satisf* or related words (e.g., good OR lik* OR happ*) occurred in the text and retrieve these showing the surrounding paragraph, so that coding can be checked. Even better, the search can be set to exclude passages already coded with satisfaction by using a query that combines a text search with a coding query.

Visual charting and mapping tools also continue to have a role as coding proceeds. Charts provide an overview of the coding for particular sources; coding stripes show the dominance and combinations of codes used in coding a source; coding stripes also show those codes that intersect with a code being reviewed; and simple conceptual maps created by the researcher serve as a visual aid in capturing ideas about connections in data, including literature. Visual tools based on coding are useful in an exploratory sense. For example, a coding comparison diagram shows the cases (or sources) coded at one or another or both of two concepts, prompting thoughts about the relationship between these concepts. Thus, Fig. 6 suggests the possibility of a strong relationship between social connection and engaging in physical activity in that most participants attending "Wellness Centers" for older women spoke of both, and no one spoke of activity without also speaking of social connection. Alternatively, a cluster analysis of codes, based on commonality of words used, suggests relationships between pairs of codes that might warrant further investigation, as well as, in this case, pointing to the possible dimensionality of the concept of well-being for older women, based on the grouping of codes (indicators) used to describe it by those attending Wellness Centers (Fig. 7).

A final step, before interrogation of data starts in earnest, is to review the content coded at each coding category, firstly just to check that all the content coded there "belongs" (checking consistency is a way of assessing coding reliability, without confounding the situation by having someone with a different perspective try to duplicate the coding process). More usefully, from the point of view of analysis, a review and summing up of each code clarifies both focus and boundaries for that

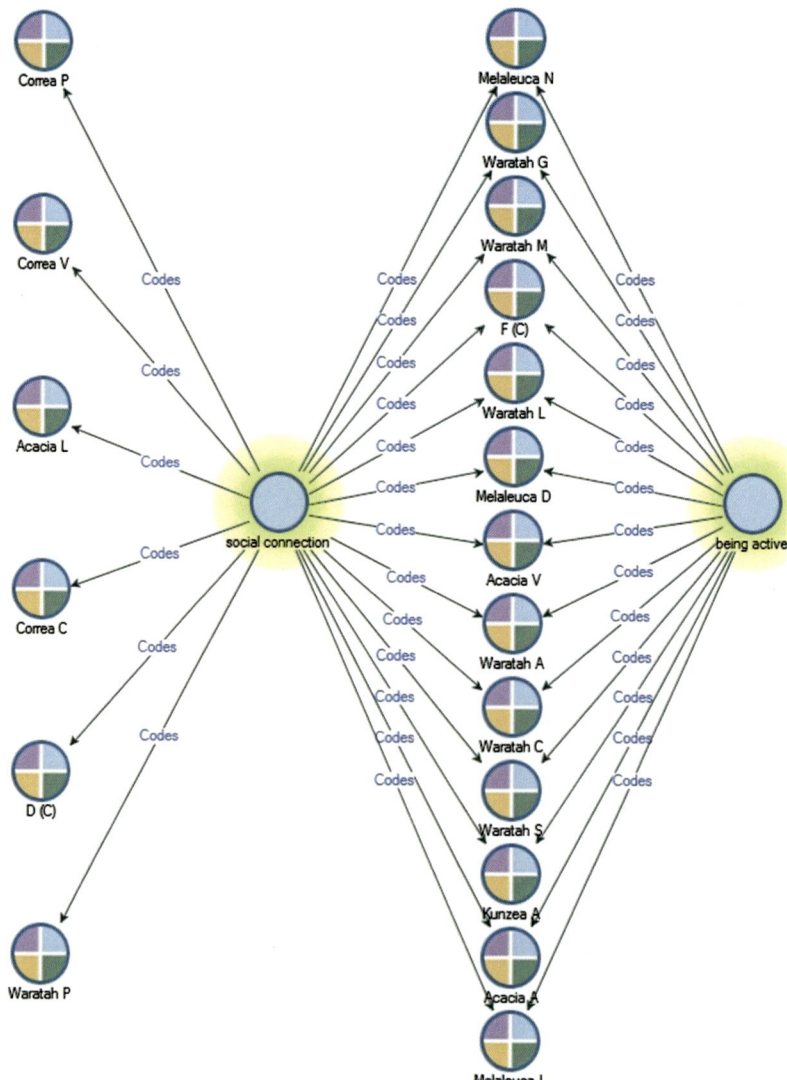

Fig. 6 Coding comparison diagram showing cases coded at social connection and being active

concept, provides the descriptive beginnings of a report of results (helping to solve writer's block), and brings to light issues for further exploration – lines of investigation that will take the analysis beyond description of "themes." These summaries might be recorded in memos linked to the codes, with the issues for further investigation noted in the project journal. Alternatively, record the summaries directly into Word, using headings to identify each so that the navigation pane (document map) can be used to quickly locate them when further information becomes available.

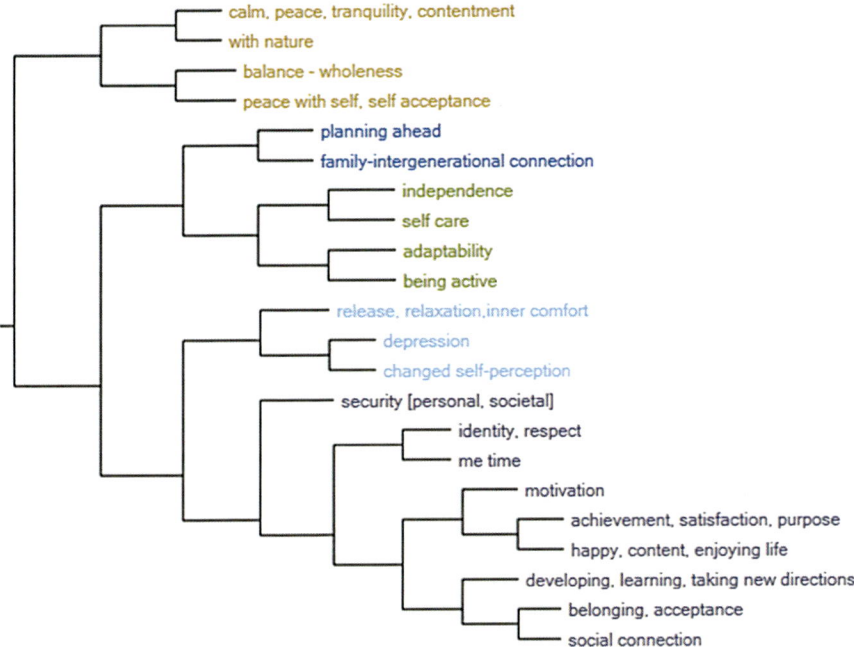

Fig. 7 Clustering of codes used to describe well-being by older women

7 Investigate and Interrogate

The qualitative researcher is developing analytic ideas throughout coding and related processes, particularly as a review of coded data is undertaken. A point is reached, however, where the focus shifts to a more deeply interpretive phase, where the researcher seeks to not just see but to understand the data and what can be learned from them, to answer the research questions, and to meet the purpose of the study. Actually, reviewing those research questions is advised at this stage because questions often shift to some degree through the course of a project. Once reviewed and realigned, the questions provide a guide to the direction of these further investigations, although some flexibility is advised as interesting leads can appear "late in the day" (although perhaps these should be set aside for later investigation, especially when a deadline is involved).

Some incidental queries using the software might have been conducted along the way, as coding and structuring of the coding system was proceeding. At this stage, however, a process of deliberate and systematic querying is more likely, as patterns and connections in the data are investigated in the search for answers. Comparative analyses are good ways to start interrogating the data. Qualitative text (or other data) is compared for different subgroups of the sample, defined by values of demographic or other quantitative variables that have been stored as attributes attached to cases.

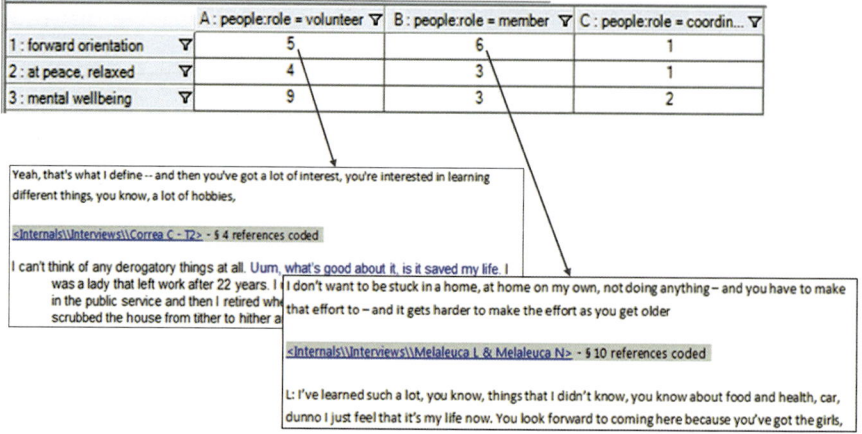

Fig. 8 Comparison of some ways in which well-being is described, for people playing different roles in Wellness Centers

In NVivo, a matrix coding query is used for this purpose (other QDAS offer equivalent tools). This provides the analyst with two kinds of data – the frequency with which different subgroups discuss the topics or experiences or issues being considered and what is actually said about each of those by each subgroup. For example, Fig. 8 compares what is said about some aspects of well-being for those who have different roles at Wellness Centers. Comparative analyses such as these are directly (descriptively) useful where comparative research questions were asked; analytically they are of value in prompting further investigation, such as why it is that this group had a different pattern of responses from that one – was it because…? Comparative analyses sometimes also reveal (sub)dimensions within concepts, when sampled groups have different ways of talking about the same concept. For example, communication patterns attributed to doctors might be compared according to whether they were described by nurses who do or who do not rate their approachability as an important issue, to reveal tacit differences in the way communication is perceived (e.g., two-way versus one-way) by these groups of nurses. These kinds of comparative queries are, therefore, useful for both exploratory analysis and for more directed analyses. Additionally, they can be extended to answer more complex three-way questions combining information from cases, sets, and codes, as described earlier.

The matrix coding query is useful also for investigating patterns of relationships between groups of codes. For example, different (reported or observed) patient in-hospital experiences might be examined in relation to emotional responses of those patients, or perhaps in relation to patterns of adjustment or recovery once discharged. For a project evaluating strategies used to build adherence to guidelines for healthy eating (or exercise or medication use), the way in which specific strategies were implemented can be reviewed in relation to whether they were considered helpful or not in terms of observable outcomes, as well as to how patients responded to each, and under what circumstances they were helpful – or indeed, some combination of these.

The way a coding system was structured is critical if it is to allow for asking a variety of questions, such as these – in the latter evaluation example, segments of texts describing responses to strategies would have been coded for the strategies being referred to, and in separate codes, whether they were observed to be useful, how the patient responded, and what the circumstances were at the time, with each of these codes being grouped in different "trees." This might appear, at the time of coding, to be unnecessarily fracturing the data, but the consequence is that it allows for flexibility in asking questions of the data when it comes time for analysis. Had strategies been listed in just two trees, as being helpful or not, then it would be difficult to ask further questions about other responses or influences.

Particular and/or more complex combinations of specific codes (or codes and attribute values) suggested either in earlier memos based on impressions when working with the data or through the kinds of visual and pattern analyses just described, might be assessed using regular coding queries, which will return all text satisfying the criteria set for the query. As for results of matrix queries, it is then up to the researcher to interpret patterns within the data that are revealed through the query.

Reporting from a qualitative or mixed methods study is best developed as the study proceeds, initially using the in-project document/memo system and the program's visual tools for mapping concepts and relationships, but then using a word processor alongside the qualitative software to record descriptions, insights, quotes, and helpful visual displays. The software thus provides an "evidentiary database" (Yin 2014) to be drawn on to support claims in the report.

8 Conclusion and Future Directions

What should be evident from the description given of these various processes for working with data when using software is the intensity of the way in which the researcher interacts with her data, putting to rest any possible claim or thought that using a computer might create distance between researcher and data (Jackson in press). Qualitative researchers using software find it contributes to the rigor and transparency of their research processes and especially to the depth of analysis they are able to achieve. The software does not provide neat answers to the research questions; rather it is a tool that facilitates working with the data in ways that will provide the evidence needed for the researcher to make judgments and reach conclusions that are supported by data. Ultimately, of course, the responsibility for the depth and quality of the analysis and interpretation of results lies with the researcher.

In an increasingly digital world, qualitative and mixed methods researchers will also increasingly adopt digital technologies to aid their research work. And the imperative of business survival means that software developers will continue to develop new tools to assist that work. Qualitative software has moved progressively into managing an ever-expanding range of data types. There has been progressive development also in the kinds of analysis strategies that are supported, with the focus shifting most recently to automation of coding and now to mixed methods analysis, including basic statistical operations. Programs are handling ever-increasing volumes of data, which in itself requires new strategies for analysis. These trends in development will bring new opportunities but also pose (or repose) threats to those who see such moves as eroding the "real," "intimate" character of qualitative research. Software *will* continue to support small-scale, intensive research; at the same time, it will increasingly develop to capture new, volume-oriented markets. The future scenario will offer choice, but that choice will not necessarily be "either-or." Predictive coding strategies incorporated into software are improving in their capacity to apply machine learning, but perhaps of more interest are the moves afoot to develop and capitalize on "citizen science" and crowdsourcing strategies for hand-coding large volumes of data (Williams and Burnap 2015; Adams 2016) that are then fed into software used to assist with the (researcher-directed) analysis of those data.

References

Adams NA. A crowd content analysis assembly line: scaling up hand-coding with text units of analysis. Social Science Research Network (SSRN). 2016. Retrieved from http://ssrn.com/abstract=2808731.
Bazeley P. Qualitative data analysis: practical strategies. London: Sage; 2013.
Bazeley P, Jackson K. Qualitative data analysis with NVivo. 2nd ed. London: Sage; 2013.
Bowers BJ. Grounded theory. In: Sarter B, editor. Paths to knowledge: innovative research methods for nursing, vol. 15–2233. New York: National League for Nursing; 1989. p. 33–59.
Carcary M. The research audit trail – enhancing trustworthiness in qualitative inquiry. Electron J Bus Res Methods. 2009;7(1):11–24.
Gilbert LS. Going the distance: 'closeness' in qualitative data analysis software. Int J Soc Res Methodol. 2002;5(3):215–28.
Huber M, Knottnerus JA, Green L, van der Horst H, Jadad AR, Kromhout D, ... Smid H. How should we define health? Br Med J. 2011;343(d4163):1–3. https://doi.org/10.1136/bmj.d4163.
Jackson K. Where qualitative researchers and technologies meet: lessons from interactive digital art. Qual Inq. in press. https://doi.org/10.1177/1077800417731086.
Liamputtong P. Health, illness and well-being: an introduction to social determinants of health. In: Liamputtong P, editor. Social determinants of health: individuals, society and healthcare. Melbourne: Oxford University Press; in press.
Liamputtong P, Serry T. Making sense of qualitative data. In: Liamputtong P, editor. Research methods in health: foundations for evidence-based practice. 3rd ed. Melbourne: Oxford University Press; 2017. p. 421–36.
Mertens DM, Bazeley P, Bowleg L, Fielding N, Maxwell J, Molina-Azorín JF, Niglas K. The future of mixed methods: a five year projection to 2020. Mixed Methods International Research Association. 2016. Retrieved from www.mmira.org.
Miles MB, Huberman AM. Qualitative data analysis: an expanded sourcebook. 2nd ed. Thousand Oaks: Sage; 1994.
Richards L. Handling qualitative data. 3rd ed. London: Sage; 2014.
Strauss AL. Qualitative analysis for social scientists. Cambridge: Cambridge University Press; 1987.
Williams ML, Burnap P. Crime sensing with social media. Methods news: newsletter from the National Centre for Research Methods. 2015.
Yin RK. Case study research: design and methods. 5th ed. Thousand Oaks: Sage; 2014.

Sequence Analysis of Life History Data

Bram Vanhoutte, Morten Wahrendorf, and Jennifer Prattley

Contents

1 Introduction 936
2 Why Use Sequence Analysis? 936
3 What Are Sequence Data? 939
4 Sequence Analysis Toolbox 941
    4.1 Descriptive Graphical Visualizations 942
    4.2 Numerically Descriptive Tools 945
    4.3 How to Compare and Group Sequences? On Costs and Distances 948
5 Conclusion and Future Directions 950
References 951

Abstract

This chapter is an entry-level introduction to sequence analysis, which is a set of techniques for exploring sequential quantitative data such as those contained in life histories. We illustrate the benefits of the approach, discuss its links with the life course perspective, and underline its importance for studying personal histories and trajectories, instead of single events. We explain what sequential data are and define the core concepts used to describe sequences. We give an overview of tools that sequence analysis offers, distinguishing between visually descriptive,

---

B. Vanhoutte (✉)
Department of Sociology, University of Manchester, Manchester, UK
e-mail: bram.vanhoutte@manchester.ac.uk

M. Wahrendorf
Institute of Medical Sociology, Centre of Health and Society (CHS), Heinrich-Heine-University Düsseldorf, Medical Faculty, Düsseldorf, Germany
e-mail: wahrendorf@uni-duesseldorf.de

J. Prattley
Department of Social Statistics, University of Manchester, Manchester, UK
e-mail: jennifer.prattley@manchester.ac.uk

numerically descriptive, and more analytical techniques, and illustrate concepts with examples using life history data. Graphical methods such as index plots, chronograms, and modal plots give us an intuitive overview of sequences. Numerically descriptive tools including the cumulative duration, number and duration of spells, as well as sequence complexity give a more statistical and quantitative grasp on the key differences between sequences. Comparing how similar sequences are, by calculating distances, either between sequences or to an ideal type, allows grouping sequences for more analytical research purposes. We conclude with a discussion on the possibilities in terms of hypothesis testing of this mainly explorative analytical technique.

Keywords

Sequence analysis · Sequence data · Life course · Life history data · Retrospective data · Visualization · Optimal matching analysis

1 Introduction

Our lives are like threads unwinding from a spool: the longer we have lived, the more thread lies behind us. As time progresses from birth onward, biographical life histories unfold for all of us over different aspects of our lives: where we have lived, what jobs we have had, and what partners we lived with. These personal histories contain valuable information to understand the situation of a person today. Just like a patient's medical history is essential for a doctor making a diagnosis, personal histories can be a valuable part of quantitative research. Sequential data, or ordered successions of stages and states, can be statistically investigated with a tailored set of techniques: sequence analysis. Rather than designating a specific method, sequence analysis refers to an approach that comprises a group of analytic methods. The adaptation of this technique, from its origins in bioinformatics to study DNA and RNA to social processes, began with the work of Abbott (1995) and has recently gained impetus by a range of new methodological developments in response to life course applications (Aisenbrey and Fasang 2010). This chapter provides a gentle introduction to the main concepts and ideas of sequence analysis and illustrates some applications in the field of epidemiology and sociology from a life course perspective, without going into too much technical detail.

More in-depth information, as well as a broader discussion of the key concepts of sequence analysis, can be found in Blanchard et al. (2014) and Cornwall (2014).

2 Why Use Sequence Analysis?

Most aspects of our life can be imagined in a dynamic way, as a series of transitions from one stage to another, reflecting both developmental and social changes we go through as we age. This personal trajectory through life, and its relation with the specific historical context in which each phase or transition takes place, should be taken into account to better understand someone's current situation (Abbott 1995).

A first reason to use sequence analysis is to bring *personal history to the foreground* of the analysis. We usually implicate a person's background by using synchronous measures and assume they refer in the same way to the past and social origins of all respondents.

As an example, let's discuss the social gradient of health in later life. Health inequalities have shown to be strongly related to social class during working life (Marmot et al. 1988). Taking a fundamental cause approach (Link and Phelan 1995), many researchers use educational level or occupational class to highlight these disparities (see, e.g., Arber and Ginn 1993), although the graduation ceremony lies decades in the past and older people are no longer on the job market. Implicitly, these models assume that the knowledge acquired at school or the past relation to the means of production plays a role in understanding health. Nevertheless, the actual meaning of education in terms of job opportunities differs greatly between generations (Roberts 2009), and equally when and how someone retires has shown to matter greatly for mental and physical health aspects (Marshall and Nazroo 2016; Wetzel et al. 2016). These commonly ignored historical and biographical contexts nevertheless strongly influence health in later life: actual trajectories of employment over the life course offer additional explanations of how well-being differs between people, over and above both parental and own occupational class, as well as current income (Wahrendorf 2015). In conjunction with work histories, family histories have equally shown to be strongly predictive of later life disability and mortality (Benson et al. 2017). More important than the empirical fact that these trajectories matter statistically is that they offer a more complete, and compelling, narrative about how specific patterns in a person's life lead to a certain outcome than a combination of isolated pieces of information, such as educational level, occupational class, and material circumstances, will ever do.

Second, sequence analysis is about *focusing on holistic trajectories instead of single events* (Aisenbrey and Fasang 2010). Investigating single events, such as transitions from one state to another, can help to uncover the processes that lead to the event occurring but by definition can only focus on an isolated event. As such, it importantly leaves those who never experience the event out of the picture and narrows the focus by implicitly seeing event occurrence as the end of a trajectory, while many events are reversible. Take as an example buying a house, a crucial event that is closely related to family formation (Clark et al. 1994), as well as the most common form of capital accumulation. A recent investigation of housing careers of the English population aged 50 or older (Vanhoutte et al. 2017) illustrated that a large share tend to live in owned housing, always rented, or moved from owned to rented accommodation but never acquired a house of their own (see Fig. 1). Studying only the transition to first ownership would ignore about 40% of this sample. It would equally ignore the substantial group of people that made two separate transitions from renting to ownership at different points in their lives. Investigating the trajectory in its entirety on the other hand allows not only to investigate differences between different timings of this transition but equally to consider which people are not at risk of experiencing this event.

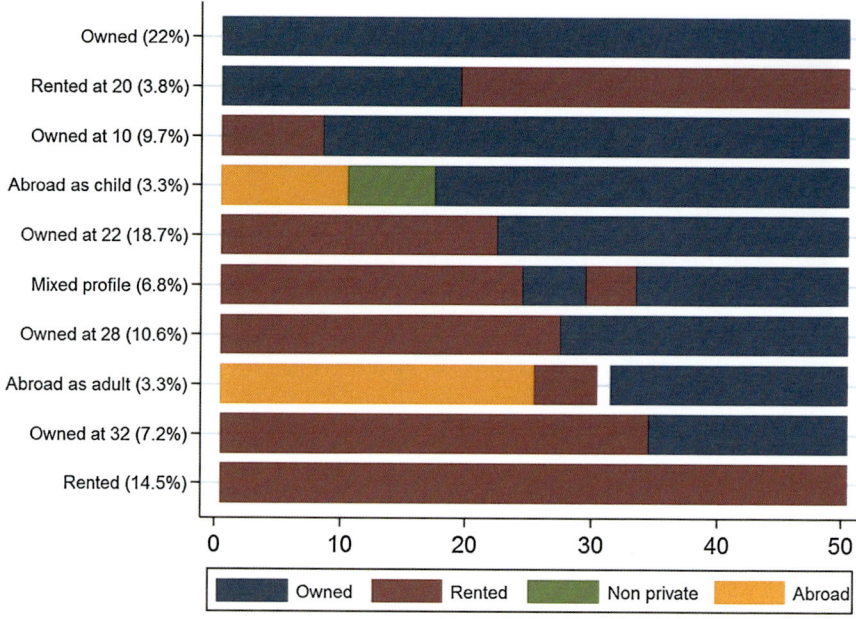

Fig. 1 Modal plot of the ten housing careers (proportion of respondents) (N = 7505) from (Vanhoutte et al. 2017). (The figure is created with Stata, using the sqmodalplot command from the SQ package. A detailed description of this package can be found in Brzinsky-Fay et al. (2006). For more detailed information on this type of graph, representing the modal sequence of which individual trajectories within clusters can differ, please see Sect. 3.1.3.)

Third, sequence analysis enables *translating concepts from the life course perspective* in relatively straightforward ways. Life course studies emphasize that each generation develops under unique historical circumstances that affect this development profoundly (Elder et al. 2004). Empirically, distinguishing the influence of different conceptualizations of time, such as duration, timing, and order, is crucial to use insights from the life course perspective in research (Vanhoutte and Nazroo 2016; Wahrendorf and Chandola 2016; Vanhoutte et al. 2017). Three key life course mechanisms, aligned with these three ways of looking at time, outline how exposures over the life course influence health and well-being: accumulation, critical period, and social mobility (Kuh et al. 2003; Niedzwiedz et al. 2012).

Accumulation (or cumulative duration) relates to the idea that inequalities are expressed over time and not instantaneously (Dannefer 1987; Willson et al. 2007). Small differences in an initial observation can develop into large disparities in health, wealth, and well-being, as exposures to stressors and health insults are unequally distributed over different social locations and have a strong compounded impact on health outcomes (Singh-Manoux et al. 2004; Lee et al. 2016). Accurately capturing the duration of exposure to these potentially damaging environments, such as inferior accommodation, precarious employment, or intimate relationships, in that way is an essential step to understanding the limits of human resilience to adverse circumstances. It allows investigating to what extent a dose-response type relation exists between the stressor over the life course and the outcome under study. Therefore, the cumulative duration of time spent in a certain state can be seen as a novel and different way of looking at how (dis)advantages are embedded in the life course.

Critical period (or timing) lies behind most research on scarring early life effects (Ben-shlomo and Kuh 2002) and means that when an event happens can be crucial for the impact it has. Alongside the occurrence of the event itself, unfavorable timing can set in motion a whole cascade of knock-on effects that negatively influence the outcome. Giving birth to a child as a teenager in itself is not detrimental to health and well-being, but it can disrupt educational and occupational trajectories, having negative effects further down the road. The timing of a single transition can adequately be studied using event history analysis, but sequence analysis explicitly analyzes multiple types of transitions (between similar and different states), as well as how often they occur in specific life phases.

Social mobility (or order) is the idea that many people experience trajectories that can be characterized in terms of (social) improvement and/or deterioration. Using holistic trajectories, as sequence analysis does, allows us to establish what a typical order of succession of states is, what is not typical, and how common these specific normative or deviating trajectories are (Elzinga and Liefbroer 2007).

Aspects of duration and timing relate to descriptive aspects of sequences, and will be treated in more detail in Sects. 4.1 and 4.2, while ordering in whole sequences relates to the investigation of how entire histories unfold and are described in Sect. 4.3.

In sum, there are strong reasons to use sequence analysis when examining the social aspects of health that concern the importance of personal history for the present, the power of holistic trajectories to uncover social processes, and the possibilities to translate life course concepts.

3 What Are Sequence Data?

In contrast to other statistical techniques, the aim of sequence analyses is not to merely link different variables to each other but to enable an in-depth analysis of entire life courses, by visualizing, summarizing, and grouping sequences. The unit of analyses is the sequence itself, and it contains detailed information on individual life courses for an extended time frame.

Now what exactly is a sequence? A life history sequence represents a series of successive experienced states. Sequences are characterized by the nature of these states and the structure of the timeline along which they are positioned. The finite list of states that can occur at each time point is given by the sequence *alphabet*. Each state in the alphabet has an associated symbol that is usually reflective of its meaning (Cornwell 2015). A *spell* is a set of adjacent identical states, and a *transition* occurs when an individual moves between two different states.

Sequences depicting life course phenomena can be positioned along an age, calendar year, or some other appropriate timelines, including wave of data collection. The most suitable metric for time depends on the outcome and aims of the research project. The chosen unit and scale should reflect the rate at which the process under study is expected to develop and unfold throughout the period of interest (Singer and Willet 2003).

As such, sequence data require the descriptions of a specific state (e.g., whether the respondent rents or owns his accommodation as above) available for a chronological order (e.g., the age of the individual). While such data used to be rare, their availability has recently grown quickly, for at least three reasons: first, the number of prospective studies is rising together with longer observation periods. In some cases, this provides sequence data that can last over 20 years. However, the richness of this data does also depend on the number of measurement points within the study. For example, if data collection lasts 20 years and occurs every 5 years, then the sequence is likely to be restricted to four measurement points only (even without attrition). Second, the use of administrative data has increased substantially. Examples are administrative records of employment histories from national pension insurances, with details on different jobs (including start and end). Restrictions of use of administrative data, though, exist in terms of data protection regulations and the fact that data is not collected for research purposes. Administrative data, nonetheless, are of high accuracy and – if accessible – an efficient and cost-effective way to gain data, as it does not require personal interviews. A third, important reason for the increasing availability of sequence data is the methodological developments in the field. An increasing number of studies collect information on individual life courses retrospectively, using an event history calendar. Hereby, data collection does not occur in a conventional face-to-face interview where people just give answers to various questions. Rather, the collection of data occurs on the basis of a graphical representation of a life course (or a "calendar") that is filled out during the interview (Belli 1998). This calendar usually contains different life domains (e.g., work, partnership, accommodation, and children histories). Studies show that calendar interviews improve the accuracy of retrospective information, as they help to memorize previous life events (Belli et al. 2007; Drasch and Matthes 2013). Furthermore, the life grid approach allows for comparable information (referring to different time points) to be collected, without producing missing data due to panel attrition in a prospective survey. Calendar interviews have now been used in several studies, including the English Longitudinal Study of Ageing (ELSA); the Survey of Health, Ageing and Retirement in Europe (SHARE) (Schröder 2011); the Australian Life Histories and Health (LHH) study (Kendig et al. 2014); and the US Health and Retirement Study (HRS). Access to harmonized key sequence data in ELSA, SHARE, and HRS is forthcoming through the website www.g2aging.org. (See also ▶ Chaps. 68, "Semistructured Life History Calendar Method," and ▶ 69, "Calendar and Time Diary Methods.")

In all three cases (prospective, administrative, and retrospective data), the resulting sequence data can be presented an alphabetical string, where each letter represents a specific state at a given time. As an example, the table below presents

Table 1 Examples of employment sequences and brief description

Id	Sequence 25–45	Brief description
1	WWWWWhhhhwwwwwwwwwwwww	5 years of full-time work, followed by an episode of home or family work and part-time employment thereafter
2	WWWhhhhwwwwwwwwwwwwwww	3 years of full-time work, followed by an episode of home or family work and part-time employment thereafter
3	WWuuuuWWWWWWWWWWWWWWWW	Full-time work with an early 4-year episode of unemployment
4	WWuWWWWWWWuWWWuWWWuWWW	Full-time work with repeated episodes of unemployment

Note: "*W*" full-time work, "*w*" part-time work, "*u*" unemployment, "*h*" home or family work

four possible employment sequences. We distinguish between four occupational situations ("W" = full-time work; "w" = part-time work; "u" = unemployment; "h" = home or family work) and focus on histories from age 25 to 45, thus covering 21 years of the working career.

The sequences in Table 1 differ with respect to the set of constituent states, the age when each state occurs, and the order in which they are experienced. The duration of individual spells, and the total time spent in a state, can also vary across sequences (Studer and Ritschard 2016). The first sequence represents a person that worked full-time at the beginning, then had a 4-year episode of home or family work (possibly maternity or paternity leave), and thereafter started to work part-time until age 45. Similarly, the second person also started with full-time work and had a 4-year episode of home or family work before reentering the labor market as part-time worker. However, compared with the first person, the episode of home or family work occurred 2 years earlier. The first two sequences consist of the same set of states but differ in the timing of home or family work and the duration of full-time and part-time work spells. The third and fourth sequences in Table 1 contain a different set of states than the first two and differ from each other with respect to the timing and duration of the constituent unemployment and work spells. While both individuals spend a total of 4 years in unemployment, person three experiences these consecutively, following 2 years of full-time work and with further uninterrupted full-time employment thereafter. The fourth person experiences repeated episodes of unemployment of a 1-year duration.

4   Sequence Analysis Toolbox

The sequence analysis toolkit includes both visual and numeric methods for describing life history data. Essential definitions and key concepts that underpin these methods are defined in this section, and an overview of index plots and chronograms is also given. These are two graphical approaches that give insight into the composition and structure of life history data. Brzinsky-Fay (2014) has comprehensive advice on presenting and interpreting the type of graphs introduced here.

The examples used below are based on complete data, in that there are no missing values in the sequences. Every individual in the sample has a known state at each time point. However, in more complex cases, sequences may be incomplete with missing values either at the left or right end of the timeline and/or at any point within the period of interest. For the purposes of this exposition, sequences have the same beginning time, the same end point, and no gaps in between. Cornwell (2015) and Halpin (2016) discuss methods for addressing missing values in sequence data.

4.1 Descriptive Graphical Visualizations

4.1.1 Index Plot

An index plot uses line segments to graph each individual sequence in the sample. Plots are read from left to right and show how individuals move between states over time. A change in color denotes a change in state or transition. Fig. 2 plots family history using data collected for 2547 women, from the life history module of the English Longitudinal Study of Ageing (ELSA). For each year of age between 15 and 60 inclusive, a woman's family status is recorded as either single with no children, married with no children, single with children, or married with children. In this figure, a single childless woman who marries, has children, and then becomes single again is represented by a horizontal line that changes from green to red, to blue, and finally to yellow.

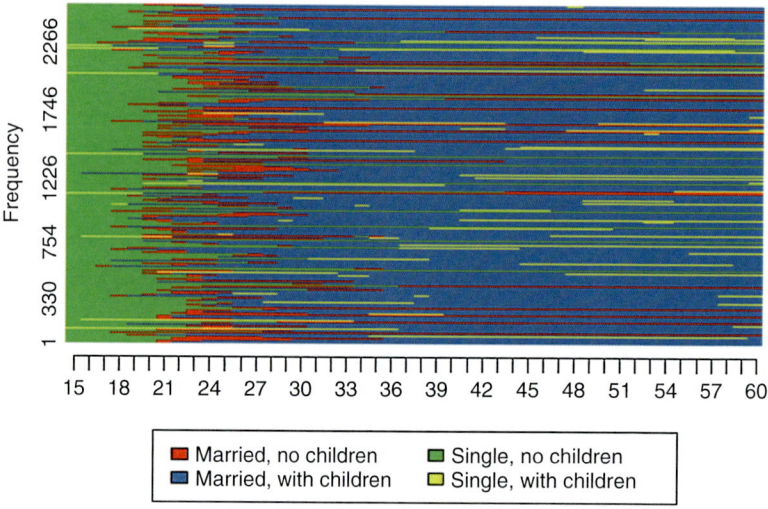

Fig. 2 Index plot of women's family history (N = 2547). (The index plots in this chapter are created using the Traminer package in Gabadinho et al. (2011), offering a comprehensive tool to conduct sequence analyses. See also http://traminer.unige.ch/ for more information.)

53 Sequence Analysis of Life History Data

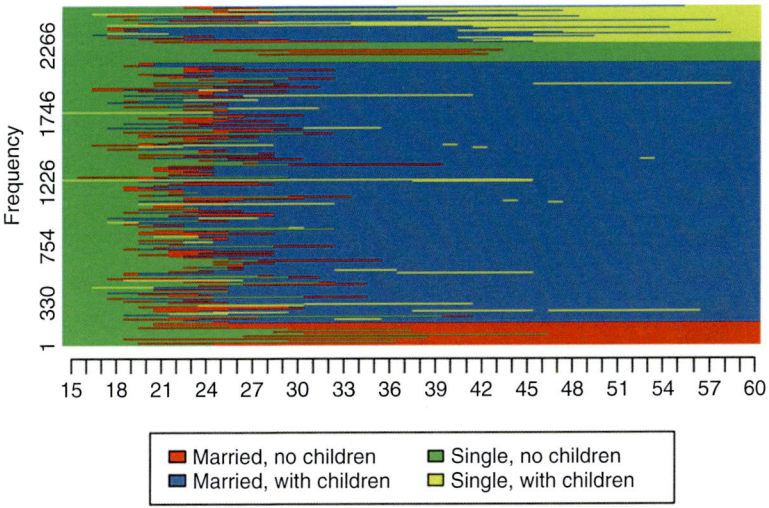

Fig. 3 Index plot of women's family history, sorted by state at age 60 (N = 2547)

The readability of an index plot, and its utility for detecting patterns, can be improved by sorting the plotted sequences according to either the first or last recorded state or value of some other attributes of interest. Fig. 3 plots the same sequences as in Fig. 2 but organized according to the marital and family status of each woman at age 60. This gives four distinct groups. Reading from bottom to top of the graph, women who are married with no children at age 60 are plotted first. Typically, these individuals married in their 20s and remained childless throughout their life course. The next group of women forms the majority of the sample. They differ from the previous group in that they married but had children. The third set of sequences is comprised primarily of women who remained single for the duration of the studied time. A minority of this group married in their late 20s but were single again by their mid-40s. The final set of sequences, at the top of the plot, represent women who were also single at 60 but had children. Most of these individuals were married during their midlife and became single by age 55. The duration of their marriage spell varies as shown by the blue portion of these trajectories.

Figure 4 shows the women's family history sequences grouped according to level of education. The top left figure is an index plot for 480 women with high levels of qualification; the bottom left plots 585 sequences for women with a mid-level of education, and top right is an index plot for 1456 women educated to a low level. Twenty-six women are removed from the sample due to missing education data. Within each cluster, sequences are sorted by work status at age 15.

Grouping index plots in the way described allows cross-sectional comparisons of the proportion of women in each state at each time point. Longitudinal sequence features, including the duration and timing of spells, can also be compared. Fig. 4 shows that, compared to the low and medium education groups, the high attainment sample contains a higher proportion of women who are single and childless at

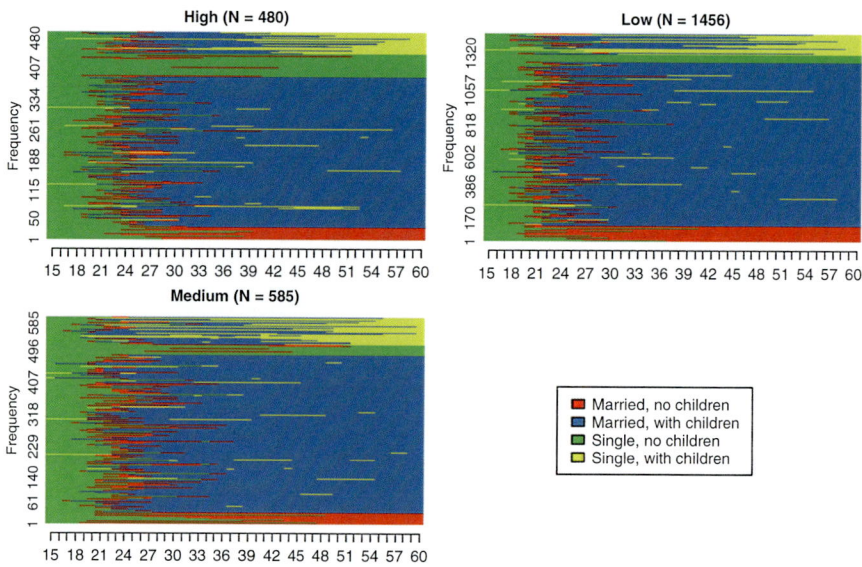

Fig. 4 Index plot of women's family history, by level of education

age 60, and there is some variation in the timing and duration of marriage spells among women who are single with children at 60. Those with a high level of education tend to have shorter marriage spells, and spend a correspondingly longer time single, than less well-qualified women. The proportion of 60-year-old women that are married without children is similar irrespective of education level, but the age of marriage tends to be younger among women with low education than those more highly qualified.

4.1.2 State Distribution Plot/Chronogram

A state distribution plot shows the proportion of observations in each state for every time point. This graph, alternatively known as a chronogram or state proportion plot, contains no information on individual sequences; rather, it is useful for describing changes in the composition of a sample or group over time. Time is positioned on the horizontal axis, the proportion or percentage of observations in each state is given by the vertical axis, and the graph is read vertically at each time point.

Figure 5 is a chronogram of women's family sequences from ELSA, as graphed in the index plots in Figs. 2 and 3 above. At 15 years old, 99% of women are single with no children and 1% single but with children. By age 23, there is more diversity in the sample. The proportion of single women with no children has fallen to one third; 29% are married, but childless, and 35% are married with children. The proportion of single women with children increases over time, to a maximum of 11% at age 60. At age 60, 77% of women are married with children, 7% are married without children, and 5% are single and childless. As with sequence index plots, separate

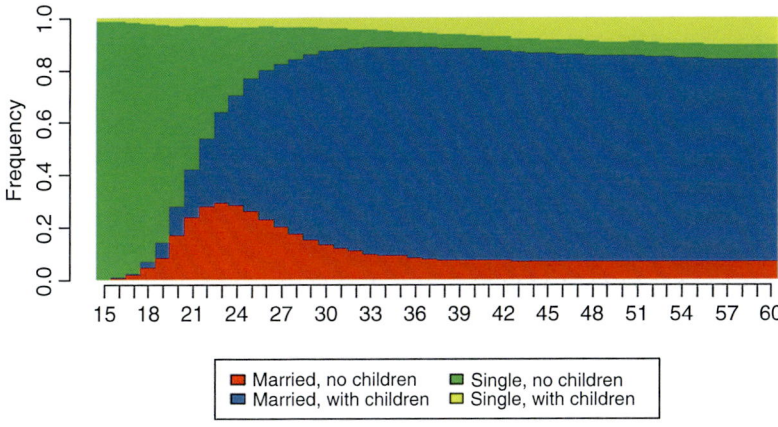

Fig. 5 Density plot of women's family history (N = 2547)

chronograms can be drawn for subsets of a sample, allowing state distributions to be compared across groups differentiated by an attribute or characteristic of interest.

4.1.3 Modal Plots

A modal plot shows a representative sequence for a cluster or group (Aassve et al. 2007). The most common state at each time point is plotted, although the resulting sequence may be artificial and not present in the studied dataset. Fig. 1 above is an example of a modal plot, which shows representative sequences for each of ten clusters, formed from plotting the most common housing status at each year of age.

4.2 Numerically Descriptive Tools

While some life courses are quite simple and stable, others are more complex and varied. As illustrated in the previous paragraph, visual inspection can give us a feel of how sequences differ from each other. Quantifying their properties allows an even more precise and in-depth comparison, as well as enabling the use of descriptive summary statistics of sequences in further analysis. Examining these properties of sequences by different genders, social groups, or cohorts can help to answer a number of research questions, such as those concerned with the increasing complexity within and between life courses, named, respectively, the differentiation and destandardization of the life course (Aisenbrey and Fasang 2010).

As said earlier, sequences can differ from each other in five aspects: experienced states, state distribution (or total time in each state), timing (or time point at which a state appears), duration (or length of a spell), and sequencing (order of states) (Studer and Ritschard 2016). In terms of descriptive metrics of sequences, three main properties related to these aspects emerge: cumulative duration; timing, number, and duration of spells; and complexity (Utilities to calculate these measures are available in the SADI package in Stata (Halpin 2017).).

4.2.1 Cumulative Duration

The most evident descriptive statistic of a sequence is the cumulative duration spent in a certain state. This metric reflects both the experienced states and the state distribution of a sequence. Comparison of the average total time spent in a state by cohort is a straightforward way to understand demographic changes (Billari 2001). Cumulative duration, as a total sum or count of the amount of time spent in a certain state, can equally function as a measure of exposure to circumstances associated with this specific state, in a logic of cumulative (dis)advantage (Wahrendorf and Chandola 2016; Vanhoutte et al. 2017). Cumulative duration might help to understand what type of relationship exists between exposure and outcome. Is there a dose-response relationship between number of years out of work and health in later life, or is it simply about passing a certain threshold? A crucial benefit is that using sequence data to grasp the length of exposure gives a more reliable answer than the straightforward question how long a person was unemployed. While duration is straightforward to calculate, it does have a number of drawbacks which are intuitively understood when examining our example sequences.

Consider Table 2, which contains three different, straightforward employment sequences of ten observed time points (years), with two possible states U and W, designating unemployment and employment, respectively.

Person A has a cumulative duration of 10 years in an employed state and a duration of zero in an unemployed state. Both persons B and C spent 5 years in each state, although their sequences are very different from each other. Note that person A has a different set of experienced states (only W) than persons B and C (U and W). A second point is that all three people might have been working in several workplaces during their years of employment. As such, duration as a metric does not say anything about the number of transitions, either within or between states. The maximum cumulative duration is the total time observed, while the minimum cumulative duration is the unit in which time is measured.

4.2.2 Timing, Number, and Duration of Spells

A second set of sequence descriptors relates to the timing, number, and duration of spells. A spell is defined as a period of continuous time spent in one particular state. Spells can be of short or long duration, and there can be multiple spells of the same state within a sequence, with each starting at a different point in time. The minimum

Table 2 Examples of 10-year employment sequences

Person	Sequence 25–35
A	WWWWWWWWWW
B	UUUUUWWWWW
C	UWUWUWUWUW

Note: "*W*" full-time work, "*U*" unemployment

number of spells is one, and the maximum possible is the total number of observations that each person has. Comparing the number, duration, and starting time of spells across sequences allows us to compare the timing and continuity of peoples experiences, and distinguish between those who have experienced the same states, and the same amount of cumulative time in each state, but who differ in terms of how stable their life histories have been.

Comparing spell attributes is meaningful when looking at life histories of a career type, like occupational or housing. Consider the difference between person D with sequence U U U W W W W W W and person E with a sequence UWUWUWUWWW. In both cases, they have spent a total of 4 years in unemployment and 6 years in work. However the spells within each sequence differ in terms of number and duration; person D has two spells, one of unemployment and one of work. They are 4 and 6 years long, respectively. Person E has a more disrupted work history, with eight spells in their sequence; seven are 1-year long, and one is 3 years. The time at which each person settled into longer-term employment differs, with person D entering more stable work in their fifth year, but person E not doing so until the eighth year. This example illustrates the importance of concurrently examining both the number and duration of spells, as well as timing, to maximally distinguish between sequences. While it is important to study transitions between different states, for example, the transition from unemployment to employment, it is equally important to look at how time within one state is experienced over the life course, and these differences can only be investigated when sequences are examined in terms of the features described here.

4.2.3 Complexity

A third set of descriptive statistics summarizes how complex sequences are internally. One often used measure is Shannon's entropy index, which originated in information theory and reflects how predictable a sequence is (Billari 2001; Widmer and Ritschard 2009; Elzinga 2010). Shannon's entropy is calculated based on the cumulative duration of each observed state in a sequence. The index is 1 when all possible states have the same duration and 0 when only one state is observed. As stated before, cumulative duration in itself does not reflect number of spells, so some authors use an adapted index multiplied by the number of spells over the maximum possible number of spells (Vanhoutte et al. 2017).

A second measure of sequence complexity is turbulence, named after the term from hydrodynamics that designates an unstable speed and direction of flow (Elzinga 2003; Elzinga and Liefbroer 2007). Turbulence is based on how many different subsequences we can detect in each sequence and reflects differences in order (Elzinga 2010). Subsequences are the unique building blocks of sequences, and the more different building blocks can be distinguished in a sequence, the more complex it is. For example, while the sequence "aaa" counts three subsequences (a, aa, and aaa), sequence "aba" counts five subsequences (a, b, ab, ba, and aba) and hence is more complex. The measure takes into account duration in a state, and is the logarithm of the weighted count of subsequences, which can be normalized.

The following example highlights the difference between the entropy and turbulence measures. Consider the sequence abcabcabc and a sequence aaabbbccc; these have the same total time spent in each state, and as such the Shannon's entropy index will be the same for both sequences (0.333). However, the second sequence is more stable and less complex than the first one and will hence have a lower adapted Shannon's entropy index. The adapted index, accounting for the number of spells or transitions, is higher for the first sequence, as it contains 9 transitions over 9 time points, and hence is weighted as 1 = 0.333. The second one contains 3 transitions over 9 time points and is weighted by a third = 0.111. For the first sequence, turbulence is 8.35 (or 1 normalized) and for the second it is 7.04 (or 0.70 normalized).

4.3 How to Compare and Group Sequences? On Costs and Distances

The examples in this chapter illustrate how sequence analysis provides a set of different numeric measures (e.g., cumulative duration in a state) and techniques of visualization (e.g., index plot with a line for each sequence in the data) to describe entire sequences. Beyond that, sequence analysis allows us to explore if the observed sequences in the data share similar patterns and regroup sequences with similar patterns into meaningful clusters. This helps to reduce the complexity and variety of individual trajectories and to identify specific types of sequences (e.g., types of employment histories).

An important step toward this aim is to quantify how similar or different an individual sequence is relative to another sequence and to calculate a dissimilarity measure. One commonly used approach is to ask how many changes are needed to turn one sequence into another (for a more comprehensive discussion of dissimilarity measures and their approaches, see Studer and Ritschard 2016). Most simply, we could count the number of necessary substitutions to make one sequence equal to the other. In the case of the two sequences ABCABC and CABCAB, for example, this would mean that we compare sequences element by element and in this case need to substitute each element to turn the first sequence into the second. To quantify this transformation (and to obtain a dissimilarity measure), we need to define the substitution cost. Substitution costs are the value assigned to each performed substitution and are commonly defined by the researcher (for other possibilities, including "data-driven" costs, see Studer and Ritschard 2016). For example, if substitution costs are set to one in our example, then the transformation would cost six points (which is the maximum distance for a sequence of this length). In the literature, this strategy is usually referred to as "naïve distance" or "traditional Hamming distance" with consistent substitution costs set to 1 (Hamming 1950; Eerola and Helske 2016).

This strategy has one important limitation: similarities at different time points in the sequence are not recognized. Specifically, in our example (ABCABC and CABCAB), the subsequence ABCAB is part of both sequences, only shifted by one place. This similarity is not recognized in case of the naïve distance and the maximum distance is assigned. Optimal matching (OM), a specific form of calculating distances within

Table 3 Examples of employment sequences and dissimilarity measures

Id	Sequence 25–45	
1	WWWWWhhhhhwwwwwwwwwwww	
2	WWWhhhhhwwwwwwwwwwwwww	
Naïve distance (hamming distance)	(subst = 1)	4
Levenshtein distance (OM distance)	(subst = 1; indel = 0.5)	2

Note: "*W*" full-time work, "*w*" part-time work, "*u*" unemployment, "*h*" home or family work

sequence analysis, resolves this problem by using a different strategy to calculate the dissimilarity, using the so-called "Levenshtein distance" (Levenshtein 1966; Eerola and Helske 2016). Hereby, two additional possible operations exist to turn one sequence into another: to insert a specific state into a sequence ("insertion") and to delete specific states from a sequence ("deletion"). Turning back to our example (ABCABC and CABCAB), that would mean that we could insert a state at the beginning of the first sequence (insertion of C) and delete the last state (deletion of C). In that case two operations of one insertion and one deletion rather than six substitutions would be necessary to make both sequences equal. These "indel" costs (insertion and deletion) are usually set to half of the substitution cost.

To summarize these two different strategies (naïve Hamming distance and Levenshtein distance), the following table returns to the two first sequences from Table 3, that is, the two sequences that were characterized by a 4-year episode of home or family work that occurred at different time points. In case of the naïve distance, the calculated distances would be 4, while 2 would be the resulting Levenshtein distance.

Each strategy has its advantages. The OM distance is possibly more appropriate in case a researcher is interested in detecting similar patterns, while the Hamming distance puts a strong emphasis on the timing of an episode that may also be of interest in some cases.

As stated above, both strategies require costs to be defined and set. Up to this point, we have assumed that substitution costs are 1 for each possible replacement. As an example, we assumed that replacing "full-time work" (W) with "part-time work" (w) would be the same (or cost as much) as replacing "full-time work" (W) with "unemployment" (u). Depending on the goals of the analyses, this may be appropriate. But for theoretical reasons, we may want to treat a substitution of full-time work with part-time work as less expensive, and hence more alike, than a substitution of full-time work with unemployment. In that case, for example, we could specify that replacing W with w would cost 0.5 (and not 1 as before). The researcher can specify costs to vary according to replacements that occur in substitution operations. Importantly, this type of variable costing requires a clear theoretical rationale; however, there is often no reason to assume different substitution costs, and a standard unique cost structure is applied with substitution costs set to 1 and indel costs to 0.5 (as in Table 2) (Abbott and Angela 2000).

The conducted comparisons allow us to quantify how similar (or dissimilar) sequences are. When applying these comparisons in the frame of a study and based on large datasets, two scenarios or approaches of comparisons are possible. First, we can predefine an ideal-typical reference sequence and calculate the distance of each person in the data relative to this reference sequence. For example, a researcher may be interested to know how different a person's employment history is compared with a prototypical history of continued full-time work, that is, a sequence uniquely marked by W in our example. As a result, we would quantify to what extent the employment history differs from a given typology. One such example is a recent study investigating links between work-family life courses between age 16 and 42 years and health during midlife, based on the National Child Development Study (NCDS) (Lacey et al. 2015). In that case, the authors predefined 12 ideal types based upon previous knowledge (e.g., continuous full-time work without family or early family without work) and calculated the distance from each individuals' work family sequence to this set of 12 ideal types, so that they could identify the closest ideal types. This strategy, though, has the disadvantage that it merely informs the researcher about the degree to which a respondent differs from a self-chosen prototype. It does not allow the observed sequences to be grouped into empirically derived types and thus identify sequences with similar patterns and complexity in the data.

A second approach is to compare every sequence to every other sequence in the data (without defining a reference sequence) and to use this information as a basis to group similar sequences into empirically distinct clusters. In that case, for every pair of sequences, a distance is calculated. If our sample consists of 1000 people, comparisons would produce a large distance matrix that quantifies the distance between each pair of individuals in the sample (i.e., a 1000 × 1000 matrix). For the purpose of grouping similar sequences into types, this information is crucial as individuals who have similar sequences will have low distances to each other, and dissimilar sequences will have high dissimilarity measures to each other. However, the resulting information is vast and in itself impossible to digest. Therefore, the standard practice is to complement the calculation of the dissimilarities with a cluster analysis that uses the matrix as basis and moves from distance matrices to typologies of sequences. The goal of the cluster analysis is to organize the sequences into groups in a way that the similarity is maximized for the sequences within a group and minimized between groups (for an overview of different clustering techniques as applied to distance matrices, see Studer 2013).

5 Conclusion and Future Directions

This chapter aims to be a gentle introduction to sequence analysis, focusing on its application using life history data to uncover the social determinants of health. We show how there are both theoretical and empirical parallels between the life course approach and sequence analysis, introducing methods for providing empirical evidence of the influence long-term processes can have on current health and well-being. As sequence analysis is a technique under continuous development, many currently topical issues did not receive attention, so as to avoid overburdening the reader. The most debated issues focus around the calculation on costs and distances between sequences (for a recent overview of possible approaches, see Studer and Ritschard 2016), how to investigate multiple sequences (e.g., from different life domains) at the same time (see, e.g., Gauthier et al. 2010), as well as how to use sequence analysis as a predictive and explanatory tool instead of a descriptive set of techniques. While the first two issues quickly become technical and complicated, and fall outside the scope of this chapter, the third is worthwhile considering. The most commonly used sequence analysis techniques up until now, such as optimal matching, have exploited the algorithmic, pattern-seeking nature inherent in grouping sequences according to their relative distance. The resulting outcome of this process, a categorical classification of sequences that differ minimally within groups and maximally between groups, can then be used as an independent (or dependent) variable within other methods. For example, we could relate resulting clusters of employment histories in a regression to predict health in later life. While this is a justifiable approach, it remains difficult to ascertain causality in strict statistical terms between exposure and outcome, as well as to tackle shared explanatory power between these sequence categorizations and common predictors such as gender, social position, and cohort. A different approach to this issue that is currently gaining ground is the use of numerical descriptors of sequences, such as the complexity measures we described, to explain differences between social groups in a regression. These sequence metrics in a sense present a middle road between the broad all-encompassing categorical classifications, and the minute detail of the actual sequence, and can be tailored to fit specific conceptual frameworks (see, e.g., Brzinsky-Fay 2007). As such, the way forward lies in alternating the explorative and hypothesis testing possibilities of sequence analysis in combination, to achieve more insight into how our personal histories structure our present.

References

Aassve A, Billari FC, Piccarreta R. Strings of adulthood: a sequence analysis of young British Women's work-family trajectories. Eur J Popul/Revue européenne de Démographie. 2007;23(3–4):369–88.
Abbott A. Sequence analysis: new methods for old ideas. Annu Rev Sociol. 1995;21(1):93–113.
Abbott A, Tsay A. Sequence analysis and optimal matching methods in sociology. Sociol Methods Res. 2000;29(1):3–33.
Aisenbrey S, Fasang E. New life for old ideas: the 'second wave' of sequence analysis bringing the 'course' back into the life course. Sociol Methods Res. 2010;38(3):420–62.
Arber S, Ginn J. Gender and inequalities in health in later life. Soc Sci Med. 1993;36(1):33. http://www.sciencedirect.com/science/article/pii/027795369390303L. 26 Aug 2014.
Belli RF. The structure of autobiographical memory and the event history calendar: potential improvements in the quality of retrospective reports in surveys. Memory. 1998;6(4):383–406.
Belli RF, Smith LM, Andreski PM, Agrawal S. Methodological comparisons between CATI event history calendar and standardized conventional questionnaire instruments. Public Opin Q. 2007;71(4):603–22.
Ben-shlomo Y, Kuh D. A life course approach to chronic disease epidemiology: conceptual models empirical challenges and interdisciplinary perspectives. International Journal of Epidemiology. 2002;31(6):285–293.

Benson R, Glaser K, Corna LM, Platts LG, Di Gessa G, Worts D, Price D, Mcdonough P, Sacker A. Do work and family care histories predict health in older women? Eur J Pub Health. 2017;27:1–6. https://doi.org/10.1093/eurpub/ckx128.

Billari FC. The analysis of early life courses: complex description of the transition to adulthood. J Popul Res. 2001;18(2):119–42.

Blanchard P, Bühlmann F, Gauthier J-A. 2 Advances in Sequence Analysis: Theory, Method, Applications. 2014. http://link.springer.com/10.1007/978-3-319-04969-4_7%5Cn; http://ebooks.cambridge.org/ref/id/CBO9781107415324A009%5Cn; http://link.springer.com/10.1007/978-3-319-04969-4.

Brzinsky-Fay C. Lost in transition ? Labour market entry sequences of school leavers in Europe. Eur Sociol Rev. 2007;23(4):409–22.

Brzinsky-Fay C. Graphical representation of transitions and sequences. In: Blanchard P, Bühlmann F, Gauthier J-A, editors. Advances in sequence analysis: theory, method, applications. Cham: Springer International Publishing; 2014. p. 265–84.

Brzinsky-Fay C, Kohler U, Luniak M. Sequence analysis with Stata. Stata J. 2006;6(4):435–60.

Clark WAV, Deurloo MC, Dieleman FM. Tenure changes in the context of micro-level family and macro-level economic shifts. Urban Stud. 1994;31(1):137–54.

Cornwall B. Social sequence analysis : methods and applications. Cambridge: Cambridge University Press; 2014.

Cornwell B. Social sequence analysis: methods and applications. Cambridge: Cambridge University Press; 2015.

Dannefer D. Aging as Intracohort differentiation: accentuation, the Matthew effect, and the life course. Sociol Forum. 1987;2(2):211–36. http://link.springer.com/article/10.1007/BF01124164. 27 Nov 2014.

Drasch K, Matthes B. Improving retrospective life course data by combining modularized self-reports and event history calendars: experiences from a large scale survey. Qual Quant. 2013;47(2):817–38.

Eerola M, Helske S. Statistical analysis of life history calendar data. Stat Methods Med Res. 2016. https://doi.org/10.1177/0962280212461205.

Elder GH, Johnson MK, Crosnoe R. The emergence and development of life course theory. In: Mortimer JT, Shanahan MJ, editors. Handbook of the life course handbook of the life course. New York: Springer; 2004. p. 3–19.

Elzinga CH. Sequence Similarity. Sociol Methods Res. 2003;32(1):3–29. https://doi.org/10.1177/0049124103253373.

Elzinga CH. Complexity of categorical time series. Sociol Methods Res. 2010;38(3):463–81. https://doi.org/10.1177/0049124109357535.

Elzinga CH, Liefbroer AC. De-standardization of family-life trajectories of young adults: a cross-national comparison using sequence analysis. Eur J Popul. 2007;23(3–4):225–50.

Gabadinho A, Ritschard G, Müller NS, Studer M. Analyzing and visualizing state sequences in R with **TraMineR**. J Stat Softw. 2011;40(4):1–37. http://www.jstatsoft.org/v40/i04/.

Gauthier J-A, Eric DW, Bucher P, Notredame C. Multichannel sequence analysis applied to social science data. Sociol Methodol. 2010;40(1):1–38. http://onlinelibrary.wiley.com/doi/10.1111/j.1467-9531.2010.01227.x/abstract\n; http://onlinelibrary.wiley.com/store/10.1111/j.1467-9531.2010.01227.x/asset/j.1467-9531.2010.01227.x.pdf?v=1&t=hz9y8pje&s=10fad405bf31f699c7c8813c5c88242667be8ef9.

Halpin B. Multiple Imputation for Categorical Time-series. Stata Journal. 2016;16(3):590–612.

Halpin B. SADI: sequence analysis tools for Stata. Stata J. 2017;17(3):546–72.

Hamming RW. Error detecting and error correcting codes. Bell Syst Tech J. 1950. https://doi.org/10.1002/j.1538-7305.1950.tb00463.x.

Kendig HL, Byles JE, Loughlin KO, Nazroo JY, Mishra G, Noone J, Loh V, Forder PM. Adapting data collection methods in the Australian life histories and health survey: a retrospective life course study. BMJ Open. 2014;4:e004476. https://doi.org/10.1136/bmjopen-2013-004476.

Kuh D, Ben-Shlomo Y, Lynch JW, Hallqvist J, Power C. Life course epidemiology. J Epidemiol Community Health. 2003;57(10):778–83. http://www.ncbi.nlm.nih.gov/pmc/articles/PMC173 2305/. 15 Jan 2013.

Lacey RE, Sacker A, Kumari M, Worts D, McDonough P, Booker C, McMunn A. Work-family life courses and markers of stress and inflammation in mid-life: evidence from the National Child Development Study. Int J Epidemiol. 2015;45(4):1247–59. https://academic.oup.com/ije/article-lookup/doi/10.1093/ije/dyv205.

Lee DM, Vanhoutte B, Nazroo J, Pendleton N. Sexual health and positive subjective Well-being in partnered older men and women. J Gerontol B Psychol Sci Soc Sci. 2016;71(4):698–710.

Levenshtein VI. Binary codes capable of correcting deletions, insertions, and reversals. Sov Phys Dokl. 1966;10:707. https://doi.org/citeulike-article-id:311174.

Link BG, Phelan JC. Social conditions as fundamental causes of disease. J Health Soc Behav. 1995; (Extra Issue: Forty Years of Medical Sociology: The State of the Art and Directions for the Future):80–94.

Marmot M, Smith GD, Stansfeld S, Patel C, North F, Head J, White I, Brunner E, Feeney A. Health inequalities among British civil servants: the Whitehall II study. Lancet. 1988;337:1387–93.

Marshall A, Nazroo J. Trajectories in the prevalence of self-reported illness around retirement. Popul Ageing. 2016;9(11):11–48.

Niedzwiedz CL, Katikireddi AV, Pell JP, Mitchell R. Life course socio-economic position and quality of life in adulthood: a systematic review of life course models. BMC Public Health. 2012;12(1):628. http://www.pubmedcentral.nih.gov/articlerender.fcgi?artid=3490823&tool=pmcentrez&rendertype=abstract. 27 Nov 2012.

Roberts K. Opportunity structures then and now opportunity structures then and now. J Educ Work. 2009;22(5):355–69.

Schröder M. Retrospective data collection in the survey of health, ageing and retirement in Europe. SHARELIFE methodology. Mannheim: Mannheim Research Institute for the Economics of Ageing (MEA); 2011.

Singer JD, Willet JB. Applied longitudinal data analysis. Modelling change and event occurrence. Oxford: Oxford University Press; 2003.

Singh-Manoux A, Ferrie JE, Chandola T, Marmot M. Socioeconomic trajectories across the life course and health outcomes in midlife: evidence for the accumulation hypothesis? Int J Epidemiol. 2004;33(5):1072–9. http://www.ncbi.nlm.nih.gov/pubmed/15256527. 12 Nov 2012.

Studer M. Weighted cluster library manual: a practical guide to creating typologies of trajectories in the social sciences with R. LIVES working papers. 2013;24. https://doi.org/10.12682/lives.22 96-1658.2013.24.

Studer M, Ritschard G. What matters in differences between life trajectories: a comparative review of sequence dissimilarity measures. J R Stat Soc Ser A Stat Soc. 2016;179(2):481–511.

Vanhoutte B, Nazroo J. Perspective: life-history data. Public Health Res Pract. 2016;26(3):4–7.

Vanhoutte B, Wahrendorf M, Nazroo J. Duration, timing and order: how housing histories relate to later life wellbeing. Longitud Life Course Stud. 2017;8(3):227–43.

Wahrendorf M. Previous employment histories and quality of life in older ages: sequence analyses using SHARELIFE. Ageing Soc/First View Article. 2015;9:1928–59. http://journals.cambridge.org/ASO.

Wahrendorf M, Chandola T. A life course perspective on work stress and health. In: Siegrist J, Wahrendorf M, editors. Work stress and health in a globalized economy, the model of effort reward imbalance. Cham: Springer; 2016. p. 43–66.

Wetzel M, Huxhold O, Tesch-Roemer C. Transition into retirement affects life satisfaction : short- and long-term development depends on last labor market status and education. Soc Indic Res. 2016;125:991–1009.

Widmer ED, Ritschard G. The de-standardization of the life course: are men and women equal? Adv Life Course Res. 2009;14(1–2):28–39.

Willson AE, Shuey KM, Elder GH. Cumulative advantage processes as mechanisms of inequality in life course health. Am J Sociol. 2007;112(6):1886–924.

Data Analysis in Quantitative Research

54

Yong Moon Jung

Contents

1 Introduction 956
2 Nature of Data for Quantitative Data Analysis 956
   2.1 Significance of Understating of Levels of Measurement 956
   2.2 Four Levels of Measurement 957
3 Types of Analysis Models 958
   3.1 Types of Research Questions 960
   3.2 Different Types of Variate Analysis 961
   3.3 Types of Analysis by Purpose 962
4 Conducting Data Analysis 965
   4.1 Choice of a Suitable Analysis Model 965
   4.2 Practice of Data Analysis in Quantitative Research 965
5 Conclusion and Future Directions 968
References 968

Abstract

Quantitative data analysis serves as part of an essential process of evidence-making in health and social sciences. It is adopted for any types of research question and design whether it is descriptive, explanatory, or causal. However, compared with qualitative counterpart, quantitative data analysis has less flexibility. Conducting quantitative data analysis requires a prerequisite understanding of the statistical knowledge and skills. It also requires rigor in the choice of appropriate analysis model and the interpretation of the analysis outcomes. Basically, the choice of appropriate analysis techniques is determined by the type of research question and the nature of the data. In addition, different analysis techniques require different assumptions of data. This chapter provides

---

Y. M. Jung (✉)
Centre for Business and Social Innovation, University of Technology Sydney, Ultimo, NSW, Australia
e-mail: yong.jung@uts.edu.au

© Springer Nature Singapore Pte Ltd. 2019
P. Liamputtong (ed.), *Handbook of Research Methods in Health Social Sciences*,
https://doi.org/10.1007/978-981-10-5251-4_109

introductory guides for readers to assist them with their informed decision-making in choosing the correct analysis models. To this end, it begins with discussion of the levels of measure: nominal, ordinal, and scale. Some commonly used analysis techniques in univariate, bivariate, and multivariate data analysis are presented for practical examples. Example analysis outcomes are produced by the use of SPSS (Statistical Package for Social Sciences).

Keywords

Quantitative data analysis · Levels of measurement · Choice of analysis model · SPSS

1 Introduction

Quantitative data analysis is an essential process that supports decision-making and evidence-based research in health and social sciences. Compared with qualitative counterpart, quantitative data analysis has less flexibility (see ▶ Chaps. 48, "Thematic Analysis," ▶ 49, "Narrative Analysis," ▶ 28, "Conversation Analysis: An Introduction to Methodology, Data Collection, and Analysis," and ▶ 50, "Critical Discourse/Discourse Analysis"). Conducting quantitative data analysis requires a prerequisite understanding of the statistical knowledge and skills. It also requires rigor in the choice of appropriate analysis model and in the interpretation of the analysis outcomes. In addition, different analysis techniques require different assumptions of data. When these conditions are not fully satisfied, the analysis is regarded as inappropriate and misleading.

This chapter provides introductory guides for readers to assist them with their informed decision-making in choosing the correct analysis models. The chapter begins with discussion of the levels of measure: nominal, ordinal, and scale. Some commonly used analysis techniques in univariate, bivariate, and multivariate data analysis are presented for practical examples. Example analysis outcomes are produced by the use of SPSS (Statistical Package for Social Sciences).

2 Nature of Data for Quantitative Data Analysis

2.1 Significance of Understating of Levels of Measurement

For proper quantitative analysis, information needs to be expressed in numerical formats. In order for the numbers to be the basic components of the dataset in quantitative analysis, every attributes of the variables need to be converted into numbers. This process is call coding, where numbers are assigned to each attribute of a variable. For instance, the attribute of male in *sex* variable is given the value of 1, and the attribute of female is given the value of 2. In this case, the numbers represent the attributes expressed in lengthier text terms for the purpose of quantitative analysis.

While the coding process is applied to replace any non-numerical information such as letters and symbols, the numbers do not represent attributes in the same way. This means that the nature and value of the same number can be different. For instance, the value of number 1 in the above sex variable does not have any numerical meaning but is just a shorter placeholder for male (Trochim and Donnelly 2007). However, the value of number 1 in *income* variable represents the actual value of the number. In these two cases, different meanings are attached to the same value of number 1. Put another way, the number represents different levels depending on the nature of the variable.

Level of measurement has critical implications for data analysis. Firstly, an understanding of level of measurement helps with evaluating the measurement. Each variable must be measured in such way that its attributes are clearly represented in the measurement, and the maximum amount of information is collected. Measurement evaluation can judge if the most appropriate level of measurement was selected for the collection of the best information of a variable. This will be further explained in the following section. Secondly, understanding of the level of measurement ensures the correct interpretation of the analysis outcomes. A clear understanding of what the value of the number exactly represents and what the distance among the values means guides proper interpretation. Lastly but most importantly, the level of measurement of a variable determines the choice of the analysis model. For example, *mean* value is not produced for sex variable, and subsequently an analysis model for mean comparison for sex variable simply does not make sense. The choice of inappropriate analysis model has a possibility of misdealing the discussions and conclusions.

2.2   Four Levels of Measurement

The level of measurement defines the nature and the relationship of the values assigned to the attributes of a variable (Trochim and Donnelly 2007). That is to say that the relationship between the values of 1 and 2 in sex variable is different from that in income variable. Unlike the latter case, 2 is neither greater than 1 nor double the magnitude or quantity of 1. A variety of measurements is in use in quantitative research. Most of the texts for quantitative research present four levels of measurement (Brockopp and Hastings-Tolsma 2003; Trochim and Donnelly 2007; Babbie 2016).

The first or lowest level of measurement is *nominal*. The nominal measurement is characterized by variables that are discrete and noncontinuous (Brockopp and Hastings-Tolsma 2003). The values are assigned arbitrarily, and thus it is not assumed that higher values mean more of something. The nominal level is appropriate for categorical variables such as sex (male, female), marital status (married, unmarried), and religion (Christianity, non-Christianity).

The second level of measurement is *ordinal*, where the attributes are rank-ordered along the continuum of the characteristics. The ordinal measurement is more than classifying information, and higher values signify more of things in this measurement. However, distances between values do not represent the numerical differences. Educational attainment (high school or less, undergraduate, postgraduate), level of socioeconomic status (high, medium, low), and degree of agreement (agree, neutral, disagree) can be measured by the ordinal measurement.

The third level of measurement is *interval*. While the values do not have numerical meanings in the previous two measurements, the distances between the values are interpretable in this measurement. In this measurement, computing average makes sense. However, the interval measurement does not have an absolute zero, and ratios do not make sense in this measurement. Temperature is a typical example of the interval measurement, where 0° does not mean there is no temperature and 40° is not twice as hot as 20°.

The last or highest measurement is *ratio*. This measurement is characterized by variables that are assessed incrementally with equal distances and has a meaningful absolute zero. Height, weight, number of clients, and annual income can be measured by the ratio measurement and meaningful ratios or fraction can be calculated in this measurement. Although interval and ratio measurement are distinctive in their concepts, it is noted that they are not strictly distinguished in the data analysis in social and behavioral sciences. For instance, SPSS combines these two measurements into one measurement.

As was indicated, four levels of measurement form a hierarchy by the amount of information. Also the nature of the values in each measurement defines appropriate statistics that can be produced (McHugh 2007). Higher levels of measurement have capacity to produce more statistics. Table 1 summarizes the key features of the levels of measurement and the producible statistics.

It should be noted here that variables that can be measured by the interval and ratio measurements can also be measured by lower measurements. For instance, *income* variable can be measured by nominal (yes, no), ordinal (low, medium, high), and ratio (actual amount of income). If income is measured by the ratio measurement, it can be later reduced to ordinal or nominal measurements. However, variables measured by the lower measurements cannot be converted into higher ones (Babbie 2016). The ability to manipulate higher measurements means that a wider variety of statistics can be used to test and describe variables at that level (McHugh 2007). Therefore, it is suggested that a variable should be measured at the highest level possible (Brockopp and Hastings-Tolsma 2003; Babbie 2016). If lower measurements are used when higher measurements are applicable, it causes loss of information and decreases the variety and the power of statistics.

3 Types of Analysis Models

There is a range of analysis models available, and each model has different requirements to be satisfied. Choosing an appropriate analysis model requires a decision-making process (Pallant 2016). There are a number of factors to be considered. Basically, the choice of analysis model in quantitative data analysis is determined by

Table 1 Levels of measurement and appropriate descriptive statistics

Nominal		Ordinal		Interval		Ratio	
Features	Statistics	Features	Statistics	Features	Statistics	Features	Statistics
Discrete categories	Mode	Discrete categories	Mode	Discrete categories	Mode	Discrete categories	Mode
	Percentage		Percentage		Percentage		Percentage
		Ordered categories	Median	Ordered categories	Median	Ordered categories	Median
			Range		Range		Range
				Equal distances between the categories	Mean	Equal distances between the categories	Mean
					Standard deviation		Standard deviation
						Meaningful absolute zero	Ratio

(1) the nature of the variable or the level of measurement of the variable to be analyzed, (2) the types of research question, and (3) the types of analysis.

The importance of understanding of different levels of measurements was already discussed in the previous section. Lower levels of measurements can produce only limited statistics, and this defines analysis models that can be employed. For example, when income variable is measured in a nominal way (yes, no), only limited option is available in choosing an analysis model. However, when it is measured in a ratio way, a full range of statistics is available, and researchers are given broadened options.

This section will outline the other two considerations: types of research question and types of analysis. This will be followed by the analysis models that suit different types of research question and analysis. It is noted that each analysis model assumes certain characteristics of the data, which is also known as assumptions. Violation of these assumptions can mislead the conclusion (Wells and Hin 2007). It will assist researchers with informed decision-making in choosing appropriate analysis model.

3.1 Types of Research Questions

Research question is a question that the study intends to address and defines the purpose of the study (Creswell 2014). Research question also defines the research method and the analysis plan. There can be a range of classification of research question, but most research questions can be divided into three different categories: exploratory, relational, and causal.

Exploratory research questions seek answers to what is it or how it does. They seek to "describe or classify specific dimensions or characteristics of individuals, groups, situations, or events by summarizing the commonalities found in discrete observations" (Fawcett 1999, p. 15) by exploring the characteristics of phenomenon, the prevalence of phenomenon, and the process by which the phenomenon is experienced. The percentage or the proportion of people on various opinions and the average of any variable are primarily exploratory information in nature.

While exploratory questions typically deal with a single variable, relational research questions are interested in the connection between two or more variables. Relational research raises the following types of questions: is one variable related to the other variable? or to what extent do two (or more) variables tend to occur together (Fawcett 1999)? In other words, they seek to explore the existence of the relationship between variables (yes, no), the direction of the relationship (positive, negative), and the strength of the relationship (weak, medium, strong) (Polit and Beck 2004) (Table 2).

Causal research questions are part of the relational ones but further explore the nature of the relationship to predict the causative relationship between variables. They assume that natural or social phenomena have antecedent factors or causes (Polit and Beck 2004). Therefore, causal questions are raised after the non-causal relationships between variables are formulated. In quantitative data analysis, causal research sets dependent and independent variables

Table 2 Type of research questions

Type	Features	Example question
Exploratory (descriptive)	Categorization	What is the percentage of people who are obese?
	Identification of commonalities	
	Analyzing single variable	
Relational	Connection between two or more variables	Is gender related to obesity?
	Analyzing multiple variables	
Causal (explanatory)	Prediction of causative relationships	Does exercise make a difference in the occurrence of obesity?
	Dependent and independent variables	

to identify the causative relationships. Causal questions are also named as explanatory questions.

3.2 Different Types of Variate Analysis

The term of variate is widely used in statistical texts, but it is difficult to locate statistical literature that provides a clearly workable definition of the term. Not surprisingly "the term is not used consistently in the literature" (Hair et al. 2006, p. 4). More often than not, the term of variate is used interchangeably with the variable. For instance, some literature defines multivariate analysis as simply involving multiple number of variables in the analysis. However, the variate is strictly a different concept from the variable.

The variate is broadly an object of statistical analysis as is expressed in numbers. In this regard, it refers to the values or data. However, the use of the variate with nominal or categorical measurements is not appropriate. This is because it assumes the variance, a statistic that describes the variability in the data for a variable and is "the sum of the squared deviations from the mean divided by the number of values" (Trochim and Donnelly 2007, p. 267). Strictly, the variate is "a linear combination of variables" (Hair et al. 2006, p. 8) and, thus, requires at least two continuous variables. While keeping the conceptual difference between variable and variate, this section outlines statistical analyses in line with the convention of statistical literature that does not strictly distinguish them from each other.

When a single variable is involved in analysis, it is called univariate data analysis (when the variable is nominal, the appropriate name of the analysis is univariable analysis). Univariate analysis examines one variable at a time without associating other variables. Frequency analysis or percent distribution that describes the number of occurrences of the values is a typical form of univariate data analysis. Univariate analysis is also referred to as descriptive analysis that deals with central tendency and dispersion of variables. z-test is also can be categorized as univariate analysis. Descriptive analysis will be further explained in the next section.

Table 3 Type of analysis by the number of variables

Type	Features	Example analysis models
Univariate analysis	Examination of single variable	Descriptive analysis
		z-test/one sample t-test
Bivariate analysis	Examination of relationship between two variables	Cross-tabulation
		Bivariate correlation
		T-test/ANOVA
		Simple regression
Multivariate analysis	Examination of more than two variables	Multiple regression
		MANOVA
		Factor analysis
		Discriminant analysis

When more than one variables are simultaneously included in analysis, it is called multivariate analysis (again multivariable analysis is an appropriate naming when multiple variables are included in the analysis regardless of the level of measurement of the variables (Katz 2006)). However, statistical literature distinguishes the analysis that involves exactly two variables from multivariate data analysis and calls it bivariate analysis. Bivariate analysis usually aims to examine the empirical relationship between two variables. Cross-tabulation and correlation analysis are the examples of bivariate analysis (cross-tabulation can be appropriately called bivariable analysis as the variables tested are nominal). Analyses with a purpose of subgroup comparison such as t-test, analysis of variance (ANOVA), and simple regression can also fall under this category (Babbie 2016) (Table 3).

Multivariate data analysis simultaneously involves multiple measurements and usually more than two variables just to distinguish it from bivariate analysis. The techniques of multivariate analysis are mostly the extension of univariate and bivariate analyses (Hair et al. 2006). For example, simple regression is extended to multivariate analyses by including multiple independent variables. In a similar way, ANOVA can be extended to multivariate analysis of variance (MANOVA). However, the design of some multivariate analysis such as factor analysis is not based on univariate or bivariate analysis, and they are designed based on completely different principles and assumptions.

3.3 Types of Analysis by Purpose

Quantitative data analysis can also be categorized into descriptive and inferential statistics by the purpose of analysis. Descriptive analysis simply describes the variables in the sample. Descriptive analysis reduces the large amount of data into a simpler summary. The outcomes of descriptive analysis vary depending on the level of measurement of the variable. If the variable is nominal, a frequency table or a percentage distribution is a typical outcome. If the variable is a continuous

Table 4 Examples of descriptive statistics (single variable (left) and two variables (right))

Statistics		
Overall satisfaction with life		
N	Valid	151
	Missing	2
Mean		3.77
Median		4.00
Mode		4
Std. deviation		0.716
Variance		0.513
Skewness		−0.509
Std. error of skewness		0.197
Kurtosis		0.376
Std. error of kurtosis		0.392
Range		3
Minimum		2
Maximum		5

Descriptives				Statistic	Std. Error
Gender					
Overall satisfaction with life	Male	Mean		3.81	0.101
		95% confidence interval for mean	Lower bound	3.61	
			Upper bound	4.02	
		5% trimmed mean		3.83	
		Median		4.00	
		Variance		0.603	
		Std. deviation		0.776	
		Minimum		2	
		Maximum		5	
		Range		3	
		Interquartile range		1	
		Skewness		−0.117	0.311
		Kurtosis		−0.457	0.613
	Female	Mean		3.74	0.071
		95% confidence interval for mean	Lower bound	3.60	
			Upper bound	3.88	
		5% trimmed mean		3.77	
		Median		4.00	
		Variance		0.459	
		Std. reviation		0.677	
		Minimum		2	
		Maximum		5	
		Range		3	
		Interquartile range		1	
		Skewness		−0.929	0.251
		Kurtosis		1.123	0.498

*The above tables were produced by the use of a dataset for my research on migrants' social inclusion

measurement, a variety of statistics of central tendency and dispersion are producible to describe the distribution of the sample.

The central tendency is "an estimate of the centre of a distribution of values" (Trochim and Donnelly 2007, p. 266), and there are three types of central tendency: mean, median, and mode. Dispersion or variability refers to the spread of the values around the central tendency, and the common statistics of dispersion include the range, variance (standard deviation), minimum, maximum, and quartiles. Checking the shape of distribution through skewness (the degree of symmetry of the distribution) and kurtosis (the degree of pointiness of the distribution) is also part of descriptive analysis.

Although descriptive analysis generally examines single variable, it can also involve two variables to explore their relationship. For instance, the left table in Table 4 is an outcome of univariate descriptive analysis of life satisfaction, whereas the right table shows a relationship between gender and the life satisfaction. Cross-tabulation that explores the relationship between two categorical variables is also a type of descriptive analysis.

While descriptive analysis seeks to simply describe the sample, inferential analysis aims to reach conclusions that extend beyond the description of the sample (Trochim and Donnelly 2007). Literally, inferential analysis infers from the sample data of the population, the entire pool from which a statistical sample is drawn. Usually, quantitative research deals with the sample data except for the Census and

Fig. 1 Flow chart of inferential analysis

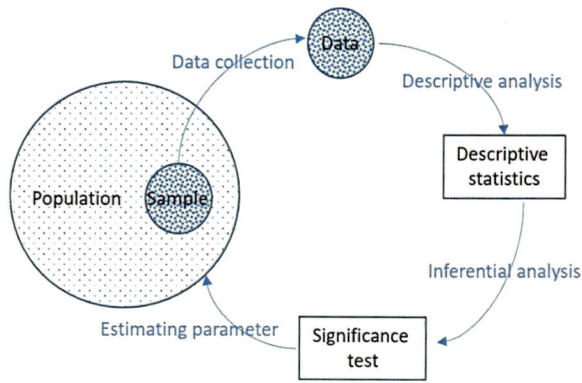

tries to estimate the parameters, a measurable characteristic of a population, from the sample statistics. Thus, the purpose of inferential statistical analysis is to generalize the findings from the sample data into the wider target population (Babbie 2016). Inferential analysis is usually the final phase of data analysis (Grove et al. 2015).

The process of generalization in inferential analysis requires a test of significance. Significance test tells the researcher the likelihood that the sample statistics can be attributed to sampling error. In other words, significance test enables informed judgment about if the observation found in the sample data occurred by chance and how confident the researcher can be in generalizing the sample outcomes. It should be noted that no sample data perfectly represents the population and guarantees accurate estimates for generalization. Significance test defines the confidence level of the estimates to the researcher.

In statistics, the level of confidence is expressed in probability such as 95%, 99%, or 99.9%. They can alternatively be expressed in probability values such as 0.05, 0.01, or 0.001. In social and behavioral sciences, 95% confidence level is commonly applied. If a significance test satisfies the criterion of 95% of confidence, it is regarded that the researcher can get the same sample statistics from 95 times of repeated sample surveys out of 100 times. In univariate statistics, the accuracy of the sample statistics is expressed in range. For instance, when the sample size is 1,000 and the 95% confidence level is applied, the population parameter is approximate estimated to be within the range of plus or minus sampling statistics (Babbie 2016) (Fig. 1).

Despite some possible flaws and criticism (Armstrong 2007), inferential data analysis relies on the custom of significance test for generalization. Test of statistical significance is a "class of statistical computations that indicate the likelihood that the relationship observed between variables in a sample can be attributed to sampling error only" (Babbie 2016, p. 461). Significance test starts with setting up a null hypothesis and an alternative hypothesis. They are also known as a statistical hypothesis and a research hypothesis (Grove et al. 2015). Null hypothesis predicts there is no relationship between variables tested in the analysis (Grove et al. 2015). It assumes that there is no predicted effect of the experimental manipulation (Field 2013). In univariate statistics, it suggests that the sample statistics is the same as the population parameter. Alternative hypothesis is contrary to null hypothesis and assumes that certain variables in the analysis will relate to each other. Significance test enables a judgment if the null hypothesis is rejected or not.

In the decision-making of adopting or ejecting the null hypothesis, the significance value, also known as *p*-value, is used (*p* represents probability). Every inferential analysis produces significance values, and the research interprets the outcomes against the α*(alpha)*-level. α-Level is a cutoff criterion for statistical significance and usually sets at 0.05 (95% significance level) or 0.01 (99% significance level). If the *p*-value is greater than the α-level, the outcome is regarded as "statistically not significant," and the researcher rejects the alternative hypothesis and adopts the null hypothesis. In other words, "significant at the 0.05 level ($p \leq 0.05$)" means that the probability that a relationship observed in the sample analysis occurs by the sampling error in no more than 5 in 100 (Babbie 2016).

4 Conducting Data Analysis

4.1 Choice of a Suitable Analysis Model

Each analysis model has been designed to serve different type of research questions, analysis purposes, and the level of measurement of the variables included in the analysis. Table 5 summarizes the appropriate analysis model by the nature of research question and the variable requirements. It should be noted that the analysis models presented in the table are only the samples of all the different models. It is suggested that researchers should refer to the manual texts to choose the most suitable model. More detailed information for decision-making trees are available in the following references (Hair et al. 2006; Grove et al. 2015; Pallant 2016).

There is a range of software available for statistical analysis. While some software is designed for specialized purposes such as LISREL or AMOS for structural equation modeling, one of the most commonly used software in social and behavioral sciences is Statistical Package for Social Sciences (SPSS) and Statistical Software Analysis (SAS). Excel also has capacity for statistical analysis, but it requires additional processes for inferential statistics compared with statistical software.

4.2 Practice of Data Analysis in Quantitative Research

Although each analysis model produces different statistics, they generally share the structure of outcomes. That is, statistical analysis usually presents descriptive statistics first and then proceeds to the outcomes of significance test. This section will demonstrate an example of an inferential data analysis using the ANOVA model, which explores the mean difference between more than two groups. The ANOVA requires one nominal variable as a grouping variable and one test variable measured at a scale level that can produce the mean. The following examples are the products

Table 5 Statistical analysis and the requirements

Research question	No. of variable	Level of measurement	Analysis model	Analysis purpose
Exploratory	Single	Nominal	Frequency percentage	Descriptive
		Scale	Central tendency, dispersion, distribution	Descriptive
			z-test/one sample t-test	Inferential
Relational (general)	Two	Nominal	Cross-tabulation	Descriptive
			χ^2 (chi-square) goodness of fit	Inferential
		Ordinal[a]	Spearman's ρ (roh)	Inferential
		Scale	Bivariate correlation (Pearson's correlation)	Inferential
	Multiple	Scale	Factor analysis	Inferential
			Structural equation model	
Relational (group comparison)	Two	Nominal (grouping variable)/scale (test variable)	Independent samples t-test	Inferential
			One-way analysis of variance (ANOVA)	
	Multiple		Multivariate analysis of variance (MANOVA)	Inferential
Causal	Two	Nominal (grouping variable)/scale (test variable)	Repeated measures ANOVA	Inferential
		Scale	Simple linear regression	Inferential
	Multiple	Scale	Multiple regression	Inferential
			Path analysis	

[a]Ordinal variables are sometimes difficult to deal with in statistical analysis because of its in-between nature. They have been sometimes accepted to the data analysis for scale measurements in social sciences. However, analysis models have been developed to more properly deal with them, and now nonparametric alternative models for ordinal variables are available for most cases (Pallant 2016)

of SPSS. The dataset was from my recent pilot study of social inclusion of migrants in Australia.

The research question to be examined in this analysis is if perceived life satisfaction of migrants is related to their visa status. The null hypothesis of this analysis is that all of the groups' population means are equal. The alternative hypothesis is that the mean is not the same for all groups or there is at least one group whose mean differs from all of the others.

The first sub-table in Table 6 provides descriptive statistics of each groups and the whole sample. From this table, an overview idea about the mean difference by different groups is obtained. Obviously, the sample statistics show that citizenship holders present the highest level of life satisfaction and the temporary visa holders

Table 6 Outcomes of ANOVA

Descriptives

Overall satisfaction with life

	N	Mean	Std. Deviation	Std. Error	95% confidence interval for mean		Minimum	Maximum
					Lower bound	Upper bound		
Temporary	12	3.50	0.798	0.230	2.99	4.01	2	5
Permanent	62	3.60	0.664	0.084	3.43	3.77	2	5
Citizen	77	3.95	0.705	0.080	3.79	4.11	2	5
Total	151	3.77	0.716	0.058	3.65	3.88	2	5

Test of homogeneity of variances

Overall satisfaction with life

Significance test for equal variance across groups

Levene statistic	Df1	Df2	Sig.
1.998	2	148	0.139

Anova

Overall satisfaction with life

Significance test for mean difference between all groups

	Sum of squares	Df	Mean square	F	Sig.
Between groups	5.176	2	2.588	5.341	0.006
Within groups	71.712	148	0.485		
Total	76.887	150			

Multiple comparisons

Dependent variable: overall satisfaction with life
tukey hsd

Significance test for mean difference within each pair of

(I) visa status group	(J) visa status group	Mean difference (i-j)	Std. error	Sig.	95% confidence interval	
					Lower bound	Upper bound
Temporary	Permanent	−0.097	0.220	0.899	−0.62	0.42
	Citizen	−0.448	0.216	0.099	−0.96	0.06
Permanent	Temporary	0.097	0.220	0.899	−0.42	0.62
	Citizen	−0.351*	0.119	0.010	−0.63	−0.07
Citizen	Temporary	0.448	0.216	0.099	−0.06	0.96
	Permanent	0.351*	0.119	0.010	0.07	0.63

*. The mean difference is significant at the 0.05 level.

have the lowest mean. However, it is a wrong interpretation if it is concluded that life satisfaction is actually different by visa status among all migrants in Australia. This is because the descriptive outcomes are immediate statistics of the sample data.

The second sub-table is an outcome of a test for equal variance between groups. The group comparison models assume that the variances are the same across the groups. This is because if the variances are unequal, it can increase the possibility for the incorrect rejection of a true null hypothesis (Type I error) (there are two types of error involved in decision-making in significance test. Type I error occurs when the null hypothesis is rejected when it is true. On the contrary, Type II error occurs when the null hypothesis is adopted when it is false. The risk of errors is indicated by the level of significance. That is, there is a greater risk of a Type I error with a 95% significance ($\alpha = 0.05$) level than with a 99% significance level ($\alpha = 0.01$). Conversely, the risk of a Type II error is greater increases when the significance level is 99% than when it is 95% (Grove et al. 2015)). If the Sig. value (p-value) is greater than 0.05, the assumption of homogeneity of variances is not violated (the null hypothesis for equal variance is accepted). As the p-value is greater than 0.05 in this case, it is safe to move on to the next table.

The last sub-table presents significance test of the mean difference. It is by this table that a conclusion is made about the mean difference of the population. The F-value means variance of the group means divided by the mean of the within group variances. The detailed logic and the equations for between groups and within groups sums of squares and the calculation of the F-value can be found in (Tabachnick and Fidell 2013) and many other texts. As the p-value of the ANOVA is less than 0.05, the alternative hypothesis is adopted, which means that at least one group has a statistically significantly different mean.

Although statistically significant mean difference was observed in the ANOVA table, it is still not certain about which group has a significantly different mean. The last sub-table of post hoc test shows the results of mean comparisons of each combination, through which the pair where the significant mean difference occurred is identified. According to the outcomes, Citizen group has a significantly higher mean than Permanent group ($p \leq 0.05$), whereas mean difference between Temporary and Permanent groups is not significant ($p \geq 0.05$).

5 Conclusion and Future Directions

This chapter was designed to provide introductory understandings of quantitative data analysis. A special focus was given to the considerations in the choice of appropriate data analysis model and the process of quantitative data analysis. It is admitted that this chapter was unable to cover diverse range of analysis models. However, this chapter provided understandings of key concepts that underpin across quantitative data analysis.

In consideration of the statistical understandings of intended readers at beginner or intermediate levels, this chapter took a conceptual approach rather than a formulaic approach, avoiding explaining by the involvement of the numerical equations. Thus, it is suggested that readers who intend to verify the conceptual understandings through mathematical formula refer to other statistical manuals. Despite limitations, it is hoped that this chapter provided a useful guide for conducting quantitative analysis.

References

Armstrong JS. Significance tests harm progress in forecasting. Int J Forecast. 2007;23(2):321–7.
Babbie E. The practice of social research. 14th ed. Belmont: Cengage Learning; 2016.

Brockopp DY, Hastings-Tolsma MT. Fundamentals of nursing research. Boston: Jones & Bartlett; 2003.

Creswell JW. Research design: qualitative, quantitative, and mixed methods approaches. Thousand Oaks: Sage; 2014.

Fawcett J. The relationship of theory and research. Philadelphia: F. A. Davis; 1999.

Field A. Discovering statistics using IBM SPSS statistics. London: Sage; 2013.

Grove SK, Gray JR, Burns N. Understanding nursing research: building an evidence-based practice. 6th ed. St. Louis: Elsevier Saunders; 2015.

Hair JF, Black WC, Babin BJ, Anderson RE, Tatham RD. Multivariate data analysis. Upper Saddle River: Pearson Prentice Hall; 2006.

Katz MH. Multivariable analysis: a practical guide for clinicians. Cambridge: Cambridge University Press; 2006.

McHugh ML. Scientific inquiry. J Specialists Pediatr Nurs. 2007;8(1):35–7. Volume 8, Issue 1, Version of Record online: 22 FEB 2007

Pallant J. SPSS survival manual: a step by step guide to data analysis using IBM SPSS. Sydney: Allen & Unwin; 2016.

Polit DF, Beck CT. Nursing research: principles and methods. Philadelphia: Lippincott Williams & Wilkins; 2004.

Trochim WMK, Donnelly JP. Research methods knowledge base. 3rd ed. Mason: Thomson Custom Publishing; 2007.

Tabachnick, B. G., & Fidell, L. S. (2013). Using multivariate statistics. Boston: Pearson Education.

Wells CS, Hin JM. Dealing with assumptions underlying statistical tests. Psychol Sch. 2007;44(5):495–502.

Reporting of Qualitative Health Research

55

Allison Tong and Jonathan C. Craig

Contents

1 Introduction	972
2 The Problems in the Reporting of Qualitative Health Research	973
3 The Challenges of a Standardized Approach to Reporting Qualitative Research	974
4 Reporting Guidelines for Qualitative Research	975
5 Reporting Qualitative Health Research	977
5.1 Research Team and Reflexivity	978
5.2 Methodology	978
5.3 Participant Selection and Description of the Sample	978
5.4 Data Collection	979
5.5 Data Analysis	980
5.6 Study Findings	981
6 Conclusion and Future Directions	982
References	982

Abstract

Transparent and comprehensive reporting can improve the reliability and value of research. Reporting guidelines have been developed for different quantitative research designs including CONSORT for randomized controlled trials, STROBE for observational studies, and PRISMA for systematic reviews. Only a few reporting guidelines are available for qualitative studies – such as the

---

A. Tong (✉)
Sydney School of Public Health, The University of Sydney, Sydney, NSW, Australia

Centre for Kidney Research, The Children's Hospital at Westmead, Westmead, Australia
e-mail: allison.tong@sydney.edu.au

J. C. Craig
Sydney School of Public Health, The University of Sydney, Sydney, NSW, Australia

Centre for Kidney Research, The Children's Hospital at Westmead, Sydney, NSW, Australia
e-mail: jonathan.craig@sydney.edu.au

© Springer Nature Singapore Pte Ltd. 2019
P. Liamputtong (ed.), *Handbook of Research Methods in Health Social Sciences*,
https://doi.org/10.1007/978-981-10-5251-4_116

Consolidated Criteria for Reporting Qualitative Health Research (COREQ), which includes reporting items that address the research team and reflexivity, methodological framework, data collection, data analysis, and presentation of the findings. This chapter will address the current problems in reporting qualitative research, discuss the challenges of a standardized approach to reporting qualitative research, provide an overview of current reporting guidelines, propose principles for reporting the methods and findings of qualitative studies, and discuss strategies to improve the quality of reporting of qualitative health research.

Keywords

Qualitative research · Reporting guidelines · Quality · Publishing · Interviews · Focus groups

1 Introduction

Health and medical research aims to generate knowledge to improve healthcare and outcomes for people. However, the lack of transparency of reporting research studies can diminish the value and reliability of research because readers and potential users cannot assess the validity and relevance of the study (Simera et al. 2010). Consequently, this leads to "research waste" (Glasziou et al. 2014). The problem of poor reporting of research is pervasive and long-standing and perhaps has been made more evident with the increasing number of systematic reviews (Simera et al. 2010; Moher et al. 2014; Altman and Simera 2016). Systematic reviews of quantitative and qualitative studies consistently show incomplete, variable, and generally poor reporting across the primary studies included in the review (Altman and Simera 2016).

The International Committee of Medical Journal Editors has emphasized that "the research enterprise has an obligation to conduct research ethically and to report it honestly" (DeAngelis et al. 2004, p. 606). Reporting guidelines for research help to improve the clarity about the study design, methods, and process so readers and potential users of the research can understand, replicate, appraise, translate, and implement the findings. In 2006, the Enhancing the Quality and Transparency of Health Research (EQUATOR) Network was established to "improve the reliability and value of published health research literature by promoting transparent and accurate reporting and wider use of robust reporting guidelines" (Altman and Simera 2016, p. 2). There are approximately 400 reporting guidelines listed in the EQUATOR Network Library (http://www.equator-network.org/). The past decade has also seen an increase in the number of journals that endorse or mandate the use of reporting guidelines (Stevens et al. 2014).

Many reporting guidelines have been produced for different types of quantitative studies including the Consolidated Standards for Reporting Trials (CONSORT) statement for randomized trials, the Strengthening the Reporting of Observational Studies in Epidemiology (STROBE) statement for observational studies, the Preferred Reporting Items for Systematic Reviews and Meta-Analyses (PRISMA) statement for systematic reviews, and the Standards for Reporting

Diagnostic Accuracy Studies (STARD) for diagnostic accuracy studies (see ▶ Chaps. 56, "Writing Quantitative Research Studies," and ▶ 59, "Critical Appraisal of Quantitative Research"). For qualitative health research, reporting guidelines also exist. The Consolidated Criteria for Reporting Qualitative Health Research (COREQ) was developed for interview and focus group studies (Tong et al. 2007), the Standards for Reporting Qualitative Research: A Synthesis of Recommendations (SRQR) was developed for primary qualitative studies (O'Brien et al. 2014), and the Enhancing Transparency in Reporting the Synthesis of Qualitative Research (ENTREQ) statement was developed for reporting the systematic review and synthesis of qualitative health research (Tong et al. 2012). These guidelines for reporting qualitative health research will be covered in more detail later in this chapter.

This chapter will cover reporting of qualitative health research, with a particular focus on reporting in peer-reviewed journal articles as this is perhaps the most common and traditional approach to disseminating health research. Specifically, this chapter will address the current problems in reporting qualitative research, discuss the challenges of a standardized approach to reporting qualitative research, provide an overview of current reporting guidelines, propose principles for reporting the methods and findings of qualitative studies, and discuss strategies to improve the quality of reporting of qualitative health.

2 The Problems in the Reporting of Qualitative Health Research

The problems in reporting the methods and findings are most evident in systematic reviews of qualitative studies, which have consistently shown variability in the quality of reporting of the included studies with many details about the research process lacking (see also ▶ Chap. 45, "Meta-synthesis of Qualitative Research").

A systematic review and synthesis of 16 studies on the experiences of parents caring for children with chronic kidney disease found that important details were not provided in most studies (Tong et al. 2008). None of the studies provided the interview guide or questions, and across all studies, sparse details were provided on the methods used for data analysis. In a thematic synthesis of qualitative studies that included 30 studies involving 1552 bone marrow and peripheral blood stem cell donors on their motivations, experiences, and perspectives of donation, 19 (63%) studies specified the participant selection strategy, 2 (7%) reported the use of software to facilitate data analysis, and 20 (67%) provided participant quotations to support the findings (Garcia et al. 2013). Similarly, a recent thematic synthesis of 26 studies on patients' perspectives and experience of living with systematic sclerosis found that only half of the studies described the participant selection strategy, methods of recording data (audio or visual recording) were reported in 18 (70%) of studies, and quotations were provided in 20 (77%) of the studies (Sumpton et al. 2017).

3 The Challenges of a Standardized Approach to Reporting Qualitative Research

The lack of transparency in published qualitative studies highlights the need to improve reporting, which may be achieved by establishing guidance or standards for reporting qualitative health research. While establishing standards for rigor remains highly contentious (Mays and Pope 2000; Yardley 2000; Barbour 2001; Dixon-Woods et al. 2004; Cohen and Crabtree 2008; Dalton et al. 2017), there is growing recognition and agreement of the need for clarity and completeness of reporting so that readers can appraise the study and assess the transferability of the findings to their setting (Tong et al. 2007; Cohen and Crabtree 2008; Dunt and McKenzie 2012; O'Brien et al. 2014). In fact, proposed quality criteria or characteristics of good qualitative research usually include transparency of the methods and findings (Yardley 2000; Cohen and Crabtree 2008). Despite this, there has not yet been any consensus on reporting items for qualitative health research (Moher et al. 2014).

There are some potential conceptual and practical challenges to acknowledge in establishing or following standards of reporting of qualitative research (Moher et al. 2014). Qualitative research is an umbrella term for a wide array of methodologies and methods, and so developing a single set of criteria that considers all the different approaches and findings may be difficult to achieve (Yardley 2000). There are also concerns that a prescriptive framework may inadvertently "reduce qualitative research to a list of technical procedures and result in 'the tail wagging the dog'" (Barbour 2001). Also, qualitative data (e.g., transcription of interviews, videos, images, field notes, artifacts) are diverse, detailed, and nuanced, and may be presented in different ways depending on the type and nature of the data collection, and also the target audience. This can also add complexity in reporting qualitative research. Also, in view of the increasing number of qualitative articles published in biomedical journals, the limited word count also pose a challenge given the thick and detailed description of the methods and findings to be described in an article reporting qualitative research.

With these things considered, there is still a need for qualitative research to be systematically, completely, and carefully documented so that readers can assess the trustworthiness and rigor of the findings. Of particular relevance, *dependability* is the extent to which the research process is logical, auditable, and transparent and refers to coherence across the methodology, methods, data, and findings, which reiterates the case for clear and transparent reporting (Liamputtong 2013).

Reporting guidelines can improve the clarity, accuracy, and transparency of reporting and should be used as appropriate such that it does not constrain researchers in conveying the richness, meaning, and nuances in the findings from qualitative studies. It should be clarified that reporting guidelines for not purport that a specific reporting item, e.g., "use of software" improves the quality of the study given the lack of empiric basis to make this claim, but instead the reporting item is intended to make the research process more transparent and "auditable."

4 Reporting Guidelines for Qualitative Research

While there are many resources and guidelines available for presenting qualitative research (Wolcott 2009) and assessing the conduct of qualitative studies, guidelines have also been developed specifically for the reporting of qualitative health research. Three reporting guidelines for qualitative health research are recommended for use by the EQUATOR Network (Tong et al. 2007, 2012; O'Brien et al. 2014), which will be discussed in this section.

The COREQ checklist is a reporting guideline to support explicit and comprehensive reporting of interviews and focus group studies (Tong et al. 2007). The reporting items were generated based on a comprehensive review of 22 different checklists that were identified in systematic reviews of qualitative studies, author and reviewer guidelines of medical journals, and existing checklists used to appraise qualitative studies. The items from these checklists were extracted and compiled into a complete list. Items that were duplicative, ambiguous, and impractical to assess were removed. Two new items considered relevant for reporting qualitative research that were not explicitly addressed in previous checklists were added; these were identifying the authors involved in data collection and reporting the presence of nonparticipants during data collection, for example, if others were present at the interview or focus groups. The final COREQ checklist comprises 32 criteria grouped into 3 domains: research team and reflexivity, study design, and data analysis and reporting. The COREQ reporting items for participant selection, with descriptors, are shown in Table 1.

So far, COREQ is endorsed and recommended for use in authorship policies of many general and discipline-specific journals including *BMJ Open*, *PLOS One*, *BMC*, *Palliative Care*, *Scandinavian Journal of Work, Environment and Health*, *Transplantation*, *The American Journal of Kidney Disease*, *Peritoneal Dialysis International*, *Journal of Graduate Medical Education*, *Physiotherapy*, *Journal Psychiatric and Mental Health Nursing*, *American Journal of Occupational Therapy*, *Journal of Emergency Nursing*, *British Journal of General Practice*, and the *Journal of the Academy of Nutrition and Dietetics*, among others.

The Standards for Reporting Qualitative Research (SRQR) was also developed to define standards for the broad spectrum of primary qualitative research (O'Brien

Table 1 COREQ checklist – an example reporting items for participant selection

No	Item	Guide questions/description
Participant selection		
10	Sampling	How were participants selected? For example, purposive, convenience, consecutive, snowball
11	Method of approach	How were participants approached? For example, face-to-face, telephone, mail, email
12	Sample size	How many participants were in the study?
13	Nonparticipation	How many people refused to participate or dropped out? Reasons?

Reference for the full guideline: (Tong et al. 2007)

Table 2 SRQR checklist – an example reporting item for participant selection

No	Item	Guide questions/description
S9	Sampling strategy	How and why research participants, documents, or events were selected; criteria for deciding when no further sampling was necessary (e.g., sampling saturation); rationale[a]

Reference for the full guideline: (O'Brien et al. 2014)
[a]The rationale should briefly discuss the justification for choosing that theory, approach, method, or technique rather than other options available, the assumptions and limitations implicit in those choices, and how those choices influence study conclusions and transferability. As appropriate, the rationale for several items might be discussed together

et al. 2014). Similarly, the authors searched for guidelines, reporting standards, and criteria for critical appraisal of qualitative studies, to generate and refine a set of items deemed important to include in guideline. The SRQR includes 21 items that cover title and abstract, introduction, methods, results/findings, and discussion. The SRQR reporting item for sampling strategy is shown in Table 2.

The SRQR has also been recommended by journals including the *British Journal of Dermatology* and *Archives of Physical Medicine and Rehabilitation*.

ENTREQ is a reporting guideline for the synthesis of multiple primary qualitative studies (Tong et al. 2012). The synthesis of findings from qualitative studies brings together evidence from different populations and healthcare contexts to provide more comprehensive information to generate new theoretical or conceptual models, to identify knowledge gaps, inform the design or primary studies, and to inform practice and policy (Thomas and Harden 2008; Ring et al. 2011). There are a range of methods to synthesize the findings, with the more common methods being thematic synthesis, meta-ethnography, critical interpretive synthesis, and narrative synthesis (Ring et al. 2011; Tong et al. 2012). Although this is an emerging type of research synthesis, they are increasingly being published and regarded as important evidence to support practice and policy (see also ▶ Chap. 45, "Meta-synthesis of Qualitative Research"). To develop the ENTREQ checklist, guidance and reviews relevant to the synthesis of qualitative research, methodology articles, and published syntheses of qualitative studies were identified. Reporting items were generated inductively. The preliminary ENTREQ framework was piloted against 40 published systematic reviews and/or syntheses of qualitative studies that spanned a range of time of publication, health topics, and methodologist, to ensure that it could be broadly applied. The ENTREQ checklist comprises 21 items classified into five domains: introduction, methods and methodology, literature search and selection, appraisal, and synthesis of the findings. As an example, the ENTREQ reporting items for the search, screening, and selection of primary qualitative studies are provided in Table 3.

Thus far, the ENTREQ checklist has also been included in authorship policies of peer-reviewed journals including *Australian Critical Care, British Journal of Dermatology, Journal of the Academy of Nutrition and Dietetics, International Journal*

Table 3 ENTREQ checklist – an example reporting items for the search, screening, and selection of primary qualitative studies

No	Item	Guide questions/description
3	Approach to screening	Indicate whether the search was preplanned (comprehensive search strategies to seek all available studies) or iterative (to seek all available concepts until the theoretical saturation is achieved)
4	Inclusion criteria	Specify the inclusion/exclusion criteria (e.g., in terms of population, language, year limits, type of publication, study type)
5	Data sources	Describe the information sources used (e.g., electronic databases (MEDLINE, EMBASE, CINAHL, psycINFO, Econlit), gray literature databases (digital thesis, policy reports), relevant organizational websites, experts, information specialists, generic web searches (Google Scholar) hand searching, reference lists) and when the searches were conducted; provide the rationale for using the data sources
6	Electronic search strategy	Describe the literature search (e.g., provide electronic search strategies with population terms, clinical or health topic terms, experiential or social phenomena-related terms, filters for qualitative research, and search limits)
7	Study screening methods	Describe the process of study screening and sifting (e.g., title, abstract and full text review, number of independent reviewers who screened studies)

Reference for the full guideline: (Tong et al. 2012)

of Surgery, European Journal of Oncology Pharmacy, Palliative Medicine, American Journal of Kidney Disease, and *Transplantation.*

In 1999, Elliot and Fischer proposed guidelines for the publication of qualitative research studies in psychology and related fields. In addition to providing guidelines applicable to both quantitative and qualitative approaches, they specified seven reporting items particularly pertinent to qualitative research: owning one's perspective, situating the sample, grounding in examples, providing the credibility checks, coherence, accomplishing general versus specific research tasks, and resonating with readers. Each of these is accompanied by a discussion and examples of poor practice and good practice. Overall, the principles put forward by the authors are covered in the reporting guidelines recommended by the EQUATOR Network.

5 Reporting Qualitative Health Research

This section will broadly outline some principles and suggestions for reporting key aspects and stages of qualitative research including research team and reflexivity, methodology, participation selection, data collection, data analysis, and study findings. The section will largely draw from existing reporting guideline outlines in the previous section of this chapter, with examples provided from a range of published qualitative studies.

5.1 Research Team and Reflexivity

As qualitative researchers are intrinsic to the research process and inevitably bear influence on all aspects of the study including data collection and analysis, there is a need to report details about the research team. Reflexivity involves reflection and documentation of how the background, motivations, assumptions, actions, or intentions of the researchers may have impacted the product of the research investigation (Yardley 2000). Any relevant preestablished relationship between the researchers and participants may be important to acknowledge if it influences rapport with participants. For example, a physician interviewing their own patients may inhibit honest and open responses.

5.2 Methodology

The methodology guides the approach to data collection and analysis, which should be stated if a specific methodology was used in the study. The examples of qualitative methodologies listed in COREQ include grounded theory, discourse analysis, ethnography, and phenomenology (Tong et al. 2007). However, there are other methodological frameworks that qualitative researchers may situate their research within. These include feminist methodology, symbolic interactionism, postmodernism, and participatory research (Liamputtong 2013).

5.3 Participant Selection and Description of the Sample

The selection of participants should be described in terms of the strategy and rationale as this can allow readers to ascertain the appropriateness of the sample to the research question, the range of perspectives obtained, and the extent to which the data elicited from the sample is transferable to other settings (Kuper et al. 2008; Anderson 2010; Moher et al. 2014). Sampling strategies may be determined based on the specific methodology used, but strategies that are commonly reported in qualitative research publications include:

- Purposive sampling, to select participants who can provide a rich and meaningful data pertinent to the research questions (key informants) and may involve various approaches including typical case sampling to select the usual cases of a phenomenon; maximum-variation sampling to obtain the widest range of perspectives as possible, i.e., based on demographic or clinical characteristics, in order to obtain a broad range of data; and deviant case sampling to select the most extreme case of a phenomenon (Giacomini and Cook 2000; Kuper et al. 2008; Patton 2015). Also, theoretical sampling is a type of purposive sampling used in grounded theory whereby participants are chosen based on the theories that emerge from the concurrent data analysis (Bryant and Charmaz 2010).

- Convenience sampling, to select participants who are easily accessible (Liamputtong 2013).
- Snowball sampling, to ask participants to suggest other potential participants who can offer a different and relevant perspective, or to access hard to reach populations (Liamputtong 2007, 2013).

The sample size and characteristics of the participants should also be reported. For focus groups, specifying the number of focus groups and the number participants in each group may allow readers to determine whether "enough" focus groups were convened for each stratum of investigation (if any) and the extent to which participants were given opportunity to express their opinion.

Also, the number of those who did not participate in the study and the reasons for nonparticipation should be stated, as this may have relevant implications for the findings that readers may need to consider in assessing the transferability of the data to their context and population (Tong et al. 2007). Examples of reporting items for participant selection are shown in Tables 1 and 2 in the previous section.

5.4 Data Collection

The setting, mode, and techniques of data collection can have an impact on the findings, and it is important that these are carefully described in a qualitative study (Graffigna and Bosio 2006; Kuper et al. 2008; Irvine 2011). These aspects are addressed in the COREQ (Tong et al. 2007) and SRQR (O'Brien et al. 2014) reporting guidelines and discussed in the following section.

In terms of the setting, data collection may occur in the participants' home, workplace, or in the clinic. However, patients may feel intimidated, disempowered, and less inclined to express their views honestly and freely if the research interviews are conducted in a clinical setting.

There may also be differences in the data collected through face-to-face, telephone, or online interviews, and thus the mode of data collection should be stated. There are various approaches to collecting qualitative data including unstructured or semi-structured interviews, focus groups, documents (journals, online, and digital media), and observations (see ▶ Chaps. 23, "Qualitative Interviewing," ▶ 24, "Narrative Research," ▶ 25, "The Life History Interview," ▶ 26, "Ethnographic Method," ▶ 27, "Institutional Ethnography," ▶ 28, "Conversation Analysis: An Introduction to Methodology, Data Collection, and Analysis," ▶ 29, "Unobtrusive Methods," ▶ 30, "Autoethnography," ▶ 31, "Memory Work," ▶ 32, "Traditional Survey and Questionnaire Platforms," ▶ 33, "Epidemiology," ▶ 34, "Single-Case Designs," ▶ 35, "Longitudinal Study Designs," ▶ 36, "Eliciting Preferences from Choices: Discrete Choice Experiments," ▶ 37, "Randomized Controlled Trials," and ▶ 38, "Measurement Issues in Quantitative Research"). The specific method (or methods) of data collection used in the study should be reported.

For interviews and focus groups, the question guide and prompts should be provided so readers can assess whether the focus and scope of the questions align with the research aims and whether they were appropriate in allowing participants to convey their opinions on their own terms (see ▶ Chap. 23, "Qualitative Interviewing"). For journal publications, this may be published as an online supplementary file.

There may also be other relevant aspects to report depending on the population. For example, studies involving children may need to report how the methods were age-appropriate and considered their developmental and communication needs (Ireland and Holloway 1996; Punch 2002; see also ▶ Chap. 115, "Researching with Children"). It may also be relevant to report the language in which the interview was conducted if participants were non-English speaking and the study is reported in English language as linguistic nuances and cultural meaning may be potentially lost or diluted if the data were not collected in the primary language of the participants (Twinn 1997; Esposito 2001; Liamputtong 2010; see also ▶ Chaps. 94, "Finding Meaning: A Cross-Language Mixed-Methods Research Strategy," and ▶ 95, "An Approach to Conducting Cross-Language Qualitative Research with People from Multiple Language Groups").

The method for recording the participant's perspectives should also be detailed, with audio-recording and transcription generally regarded as the "best practice" approach to accurately capture verbal data rather than relying on the researchers' notes or memory. Otherwise, reasons for not recording the collection of data should be provided. For example, participants who are asked to give opinions on very sensitive topics may not give permission to be recorded, and thus, it would not be ethical to audiotape them. It is also relevant to state if data collection, usually interviews, was conducted prospectively as this may contribute to rapport and richness of the data collected. Whether field notes were used to capture nonverbal communication and the contextual details surrounding data collection should also be reported (Popay et al. 1998).

The duration of data collection (e.g., length of the interview, or time frame of the observation, and length of the recorded clinical consultation) allows readers to gauge the extent of data collection.

Data saturation occurs when little or no new information or concepts are being identified in subsequent data collection (Mays and Pope 2006; Kuper et al. 2008; Liamputtong 2013). The intention, definition, and extent to which this was achieved should be stated or discussed so readers can consider whether it was adequate to enable comparisons of shared and divergent perspectives, experiences, and meanings related to the topic of inquiry.

5.5 Data Analysis

There are many accepted approaches to analyzing qualitative data, which should be clearly detailed (Bradley et al. 2007; Kuper et al. 2008). The number of investigators and their involvement in the different stages of analysis (e.g., independent coding, feedback on the preliminary analysis) should be reported. For example, it may be more defensible to report the use of investigator triangulation to reassure readers that the complete range and depth of the data collected have been captured in the findings and are not just a reflection of the agenda or assumptions of a single researcher. Researchers should describe the process for identifying, defining, and interpreting the data, including how concepts were identified and coded, how concepts were grouped together, the development of the coding tree, and the derivation of the themes (if applicable) (see ▶ Chaps. 48, "Thematic Analysis," ▶ 49, "Narrative Analysis," and ▶ 50, "Critical Discourse/Discourse Analysis"). Any software packages to manage retrieve and store qualitative data should be reported, as the use of software may facilitate a systematic and traceable coding process (Anderson 2010; see ▶ Chap. 52, "Using Qualitative Data Analysis Software (QDAS) to Assist Data Analyses"). Researchers should also state whether member checking was carried out, in which the preliminary findings are sent to the participants for further feedback, comment, and integrated into the final analysis (Liamputtong 2013).

The COREQ reporting items for data analysis include number of data coders, description of the coding tree, derivation of themes, use of software, and participant (member) checking (Tong et al. 2007). Similarly, the reporting items in the SRQR include data processing (methods for processing the data before and during the analysis, i.e., transcription, data management and security, verification of data entry, coding), data analysis (process by which inferences, themes, and so on were identified and developed, including the researchers involved in the data analysis, reference to a specific methodology if relevant, and the rationale), and techniques to enhance trustworthiness (e.g., member checking, audit trail, triangulation) and the rationale (O'Brien et al. 2014).

5.6 Study Findings

Researchers conducting qualitative studies have a responsibility to "construct a [convincing] version of reality" that readers can recognize to be meaningful to them (Yardley 2000). The findings of the qualitative study need to be described in a comprehensive and compelling way such that it captures the richness, depth, and context of the data to provide insights pertinent to the research question. There is also a need to demonstrate, to some degree, the transferability of the findings to other populations and settings.

The synthesis and interpretation may be presented in different ways, for example, as a set of themes or a theory or model (Bradley et al. 2007). The presentation of the findings or output depends on the methodology or methods used. For example, a qualitative study using thematic analysis should report themes; or a grounded theory study would be expected to describe theory grounded in the participants' perspectives.

The COREQ (Tong et al. 2007) and SRQR (O'Brien et al. 2014) guidelines both recommend providing a link between the findings and empirical data. Providing raw data, i.e., quotations and images, can strengthen the confirmability of the research findings as this can allow readers to evaluate the consistency between the data presented and the findings of the study (Lincoln and Guba 1985). Providing quotations from a range of participants is recommended so researchers can enhance the transparency and trustworthiness of their findings and interpretations of the data (Tong et al. 2007; O'Brien et al. 2014). Due to word count restrictions in biomedical journals, it may be necessary to present quotations in tables or as supplementary files.

6 Conclusion and Future Directions

Qualitative research encompasses a suite of different theoretical and methodological frameworks and methods for collecting and analyzing the data, and there are multiple ways in which the findings can be presented. Transparent and complete reporting of qualitative health research can help readers assess the rigor of the methods and trustworthiness of the findings and thus support the translation and implementation of findings for patients, caregivers, and health professionals, to help improve the care and health outcomes for people. Reporting guidelines can serve to help legitimize qualitative research (Elliot et al. 1999). There are reporting guidelines available for primary qualitative studies and synthesis of multiple qualitative studies, which have increasingly been endorsed and adopted by biomedical journals in an effort to improve the quality of reporting of qualitative research.

Obtaining broader consensus on reporting items to inform revisions of existing reporting frameworks, and further efforts to develop and evaluate guidelines for specific qualitative methodologies and methods, possibly as extensions to current guidelines, may help to contribute to the broader efforts in improving the quality of reporting health research.

Qualitative research can generate evidence about patients' experiences, values, and goals, to inform strategies for patient-centered healthcare. Researchers conducting qualitative studies need to report their rationale, methods, and findings in a clear, transparent, and compelling way and thereby maximize the potential uptake of the findings into practice and policy to improve quality of care and patient outcomes.

References

Altman D, Simera I. A history of the evolution of guidelines for reporting medical research: the long road to the EQUATOR network. J R Soc Med. 2016;109(2):67–77.

Anderson C. Presenting and evaluating qualitative research. Am J Pharm Educ. 2010;74(8):141.

Barbour RS. Checklists for improving rigour in qualitative research: a case of the tail wagging the dog? Br Med J. 2001;322(7294):1115–7.

Bradley E, Curry LA, Devers KJ. Qualitative data analysis for health services research: developing taxonomy, themes, and theory. Health Serv Res. 2007;42(4):1758–72.

Bryant A, Charmaz K. The Sage handbook of grounded theory. Thousand Oaks: Sage; 2010.

Cohen DJ, Crabtree BF. Evaluative criteria for qualitative research in health care: controversies and recommendations. Ann Fam Med. 2008;6(4):331–9.

Dalton J, Booth A, Noyes J, Sowden AJ. Potential value of systematic reviews of qualitative evidence in informing user-centered health and social care: findings from a descriptive overview. J Clin Epidemiol. 2017;88:37–46.

DeAngelis CD, Drazen JM, Frizelle FA, Haug C, Hoey J, Horton R,. .. Van Der Weyden MB. Clinical trial registration: a statement from the International Committee of Medical Journal Editors. J Am Med Assoc. 2004;292(11):1363–4.

Dixon-Woods M, Shaw RL, Agarwal S, Smith JA. The problem of appraising qualitative research. Qual Saf Health Care. 2004;13:223–5.

Dunt D, McKenzie R. Improving the quality of qualitative studies: do reporting guidelines have a place? Fam Pract. 2012;29(4):367–9.

Elliot R, Fischer CT, Rennie DL. Evolving guidelines for publication of qualitative research studies in psychology and related fields. Br J Clin Psychol. 1999;38:215–29.

Esposito N. From meaning to meaning: the influence of translation techniques on non-English focus group research. Qual Health Res. 2001;11(4):568–79.

Garcia MC, Chapman JR, Shaw PJ, Gottlieb DJ, Ralph A, Craig JC, Tong A. Motivations, experiences, and perspectives of bone marrow and peripheral blood stem cell donors: thematic synthesis of qualitative studies. Biol Blood Marrow Transplant. 2013;19(7):1046–58.

Giacomini MK, Cook DJ. Users' guides to the medical literature XXIII. Qualitative research in health care. A. Are the results of the study valid? J Am Med Assoc. 2000;284:357–62.

Glasziou P, Altman D, Bossuyt P, Bouton I, Clarke M, Julious S, ... Wagner E. Research: increasing value, reducing waste. Reducing waste from incomplete or unusable reports of biomedical research. Lancet. 2014;383(9913):267–76.

Graffigna G, Bosio AC. The influence of setting on findings produced in qualitative healht research: a comparison between face-to-face and online discussion groups about HIV/AIDS. Int J Qual Methods. 2006;5(3):55–76.

Ireland L, Holloway I. Qualitative health research with children. Child Soc. 1996;10(2):155–64.

Irvine A. Duration, dominance and depth in telephone and face-to-face interviews: a comparative exploration. Int J Qual Methods. 2011;10(3):202–20.

Kuper A, Lingard L, Levinson W. Critically appraising qualitative research. Br Med J. 2008;337: a1035.

Liamputtong P. Researching the vulnerable: a guide to sensitive research methods. London: Sage; 2007.

Liamputtong P. Performing qualitative cross-cultural research. Cambridge: Cambridge University Press; 2010.

Liamputtong P. Qualitative research methods. 4th ed. Melbourne: Oxford University Press; 2013.

Lincoln YS, Guba EG. Naturalistic inquiry. Newbury Park: Sage; 1985.

Mays N, Pope C. Assessing quality in qualitative research. Br Med J. 2000;320(7226):50–2.

Mays N, Pope C. Quality in qualitative health research. In: Pope C, Mays N, editors. Qualitative research in health care. UK: Blackwell; 2006.

Moher D, Altman DG, Schulz KF, Simera I, Wager E. Guidelines for reporting health research. West Sussex: Wiley; 2014.

O'Brien BC, Harris IB, Beckman TJ, Reed DA, Cook DA. Standards for reporting qualitative research: a synthesis of recommendations. Acad Med. 2014;89(9):1245–51.

Patton MQ. Qualitative research and evaluation methods. 4th ed. Thousand Oaks: Sage; 2015.

Popay J, Rogers A, Williams G. Rationale and standards for the systematic review of qualitative literature in health services research. Qual Health Res. 1998;8:341–51.

Punch S. Research with children. Childhood. 2002;9(3):321–41.

Ring N, Ritchie K, Mandava L, Jepson R. A guide to synthesising qualitative research for researchers undertaking health technology assessments and systematic reviews. Scotland: NHS Quality Improvement Scotland; 2011.

Simera I, Moher D, Hoey J, Schulz KF, Altman DG. A catalogue of reporting guidelines for health research. Eur J Clin Investig. 2010;40(1):35–53.

Stevens A, Shamseer L, Weinstein E, Yazdi F, Turner L, Thielman J, et al. Relation of completeness of reporting of health research to journals' endorsement of reporting guidelines: systematic review. Br Med J. 2014;348:g3804.

Sumpton D, Thakkar V, O'Neill S, Singh-Grewal D, Craig JC, Tong A. "It's not me, it's not really me". Insights from patients on living with systematic sclerosis: an interview study. Arthirtis Care Res. 2017;69:1733–42.

Thomas J, Harden A. Methods for the thematic synthesis of qualitative research in systematic reviews. BMC Med Res Methodol. 2008;8:45.

Tong A, Sainsbury P, Craig JC. Consolidated criteria for reporting qualitative research (COREQ): a 32-item checklist for interviews and focus groups. Int J Qual Health Care. 2007;19(6):349–57.

Tong A, Lowe A, Sainsbury P, Craig JC. Experiences of parents who have children with chronic kidney disease: a systematic review of qualitative studies. Pediatrics. 2008;121(2):349–60.

Tong A, Flemming K, McInnes E, Oliver S, Craig JC. Enhancing transparency in reporting the synthesis of qualitative research: ENTREQ. BMC Med Res Methodol. 2012;12:181.

Twinn S. An exploratory study examining the influence of translation on the validity and reliability of qualitative data in nursing research. J Adv Nurs. 1997;26(2):418–23.

Wolcott HF. Writing up qualitative research. 3rd ed. Thousand Oaks: Sage; 2009.

Yardley L. Dilemmas in qualitative health research. Psychol Health. 2000;15(2):215–28.

Writing Quantitative Research Studies

Ankur Singh, Adyya Gupta, and Karen G. Peres

Contents

1 Introduction	986
2 Reporting Guidelines and Checklists	987
2.1 Title and Abstract: Attract Reader's Attention	987
2.2 Background/Introduction Section: Generate Interest about the Study	988
2.3 Methods/Methodology Section: Describe the Process Involved	989
2.4 Results/Findings Section: Reveal the Findings	992
2.5 Discussion Section: Relate to the Bigger Picture	993
2.6 Conclusion: Leave the Reader Enlightened	995
3 Conclusion and Future Directions	995
References	996

Abstract

Summarizing quantitative data and its effective presentation and discussion can be challenging for students and researchers. This chapter provides a framework for adequately reporting findings from quantitative analysis in a research study for those contemplating to write a research paper. The rationale underpinning the reporting methods to maintain the credibility and integrity of quantitative studies

---

A. Singh (✉)
Centre for Health Equity, Melbourne School of Population and Global Health, The University of Melbourne, Melbourne, VIC, Australia
e-mail: ankur.singh@unimelb.edu.au

A. Gupta
School of Public Health, The University of Adelaide, Adelaide, SA, Australia
e-mail: adyya.gupta@adelaide.edu.au

K. G. Peres
Australian Research Centre for Population Oral Health (ARCPOH), Adelaide Dental School, The University of Adelaide, Adelaide, SA, Australia
e-mail: karen.peres@adelaide.edu.au

© Springer Nature Singapore Pte Ltd. 2019
P. Liamputtong (ed.), *Handbook of Research Methods in Health Social Sciences*,
https://doi.org/10.1007/978-981-10-5251-4_117

is outlined. Commonly used terminologies in empirical studies are defined and discussed with suitable examples. Key elements that build consistency between different sections (background, methods, results, and the discussion) of a research study using quantitative methods in a journal article are explicated. Specifically, recommended standard guidelines for randomized controlled trials and observational studies for reporting and discussion of findings from quantitative studies are elaborated. Key aspects of methodology that include describing the study population, sampling strategy, data collection methods, measurements/variables, and statistical analysis which informs the quality of a study from the reviewer's perspective are described. Effective use of references in the methods section to strengthen the rationale behind specific statistical techniques and choice of measures has been highlighted with examples. Identifying ways in which data can be most succinctly and effectively summarized in tables and graphs according to their suitability and purpose of information is also detailed in this chapter. Strategies to present and discuss the quantitative findings in a structured discussion section are also provided. Overall, the chapter provides the readers with a comprehensive set of tools to identify key strategies to be considered when reporting quantitative research.

Keywords

Quantitative analysis · Reporting · Research methodology · Writing strategies · Empirical studies

1 Introduction

Research and scientific enquiry forms the basis for generating evidence pertaining to a specific question of interest. Quantitative data collection and analysis guided by statistical principles are common methods of choice to address key questions related to studying social and health phenomenon at the population level. Mimicking the population phenomenon within the sample through application of statistical models carefully allows researchers to generalize findings obtained from a sample to a larger population of interest.

The consistency of statistical analysis with the research question and the nature of data, along with the quality of design as well as the conduct of the study (for example: compliance with protocol), deeply impact the extent to which findings from a study can be trusted. These aspects to quantitative research are the building blocks or a basic recipe for a good study. In addition to fulfilling the fundamental requirements of appropriate scientific methods and rigor, due attention must be given to presentation of study findings. When presented inadequately, a study conducted with appropriate scientific methods and rigor can considerably limit the interest of journal editors, peer reviewers, and ultimately the readers. Given the low acceptance rates in high-quality academic journals, and more and more articles being rejected editorially even before peer review, appropriate presentation of quantitative research can substantially benefit (Szklo 2006; Kool et al. 2016). Therefore, adequate presentation of quantitative research is central to successful publication and dissemination of study findings.

This chapter aims to provide a general and preliminary framework for consideration for both students and researchers in public health discipline, for appropriately presenting quantitative results when writing journal articles. We would caution the readers to use this chapter as a general framework for preparing quantitative research studies and also alert the readers to more closely follow specific journal and disciplinary rules/guidelines suited to their purpose as they often vary a lot in their approach and styles.

2 Reporting Guidelines and Checklists

Depending on the nature of the quantitative research (interventional or observational), reporting guidelines and checklists exist that can help researchers organize and present their information in an accurate and consistent manner. Two widely applied checklists for reporting include "STrengthening the Reporting of OBservational studies in Epidemiology" (STROBE) statement and "Consolidated Standards of Reporting Trials" (CONSORT) statements (Vandenbroucke et al. 2007, Mannocci et al. 2015). STROBE statement is applied to observational study designs including cohort, case-control, and cross-sectional studies. CONSORT statement is applied to interventional studies, mainly randomized trials (Schulz et al. 2010). Completion of these checklists during submission of articles for peer review is a requirement in many journals. These checklists can also be useful tools for critical appraisal of quantitative studies (see also ▶ Chap. 59, "Critical Appraisal of Quantitative Research"). Reviewers seek a structural presentation of quantitative studies to maintain a logical flow of content for readers. Most journals seek submissions that report quantitative research with relevant information structured in the form of title and abstract, keywords, introduction/background, methods, results, discussion, references, acknowledgments, and tables and figures. These sections are elaborated below to capture the minimum information they should aim to capture.

2.1 Title and Abstract: Attract Reader's Attention

The title of the research paper is the first and foremost point of contact that the authors make with their reader. A succinct, attractive, and descriptive title brings the reader's attention to a paper. A well-informed title also assists in proper indexation in electronic databases so that it is identified with ease by researchers (Dickersin et al. 2002). While this applies to both qualitative studies and reviews, this is critical for quantitative studies. The title of a quantitative study should ideally inform the population that it addresses the type of research design/methodology and the key question that the study is answering. These are useful aspects that inform regarding the generalizability of the study, novelty in research idea or methodology, and the relevance of the topic for the readers. Thus, the authors must explicitly state the important identifiable aspects of their research paper. In addition to the title of the study, listing relevant *keywords* for the work presented can also assist in the indexation purposes. For example (Singh et al. 2016):

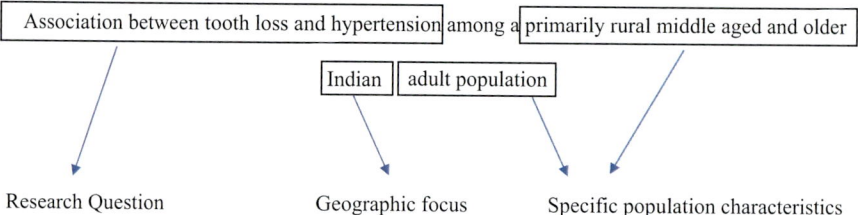

An abstract is a further elaboration of the title of the research paper, with a length of about 200–300 words, depending on the journal guidelines. Majority of the times in one paragraph, the abstract summarizes the major aspects of the paper in the following sequence: Introduction (purpose leading to aims and objectives), Methods (the study design and key information about the sample, setting, and other methods used to conduct the study), Results (the main findings of the study in the form of mean, frequency, prevalence, standard deviations, absolute and relative measures such as ratios, rates, proportions, percentages, trends, etc.), and Discussion/Conclusion (relating the results back to the aims and objectives in a summary form). Some journals request for an additional component such as implications of the research that requires the authors to state future implications of the work presented. Often, some journals prefer a nonstructured abstract where these headings are not supposed to be identified. Mostly, the abstract is the point where the reader decides whether to read the entire paper or not. Thus, it should include enough key details to enable the reader to make an informed decision (Szklo 2006).

Titles and abstracts are also critical, as they form the first criteria for selection of studies within systematic reviews (Bhaumik et al. 2015). Sometimes researchers may also cite a research paper based on the appropriateness of the abstract. Though the abstract is an opening section of any research paper, it is advised that the abstract be written at the very last to ensure that the information presented in the abstract completely aligns with the paper.

2.2 Background/Introduction Section: Generate Interest about the Study

The key purpose of the background/introduction (BI) section is to establish the context of the study being presented. Some key questions that the authors are expected to answer in the introduction section of a research paper are as follows: What is the problem? Why is it a problem? What is the strength and magnitude of the problem? Why is it important? What is already known about the topic? What are the gaps in knowledge? and how will the study contribute toward filling the gap? Summarizing the existing evidence (published literature) on the research topic of

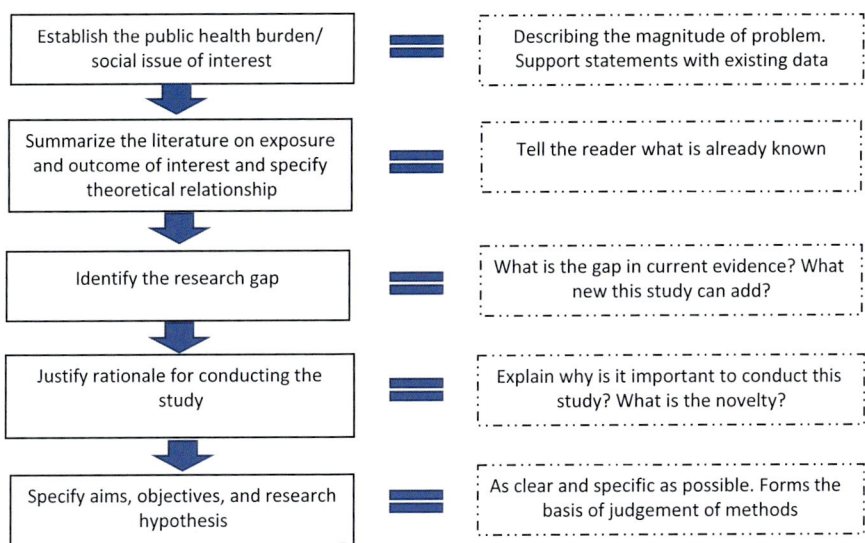

Fig. 1 Key aspects of the structure for BI

interest and providing an understanding of the magnitude of the problem being investigated can achieve this. It is the most important section to establish the plausibility of the hypothesis being tested within the study (Szklo 2006).

Writing the BI section for quantitative study can benefit by following a structure. As an example, for a journal that has a word limit of 2500–3000 words, the BI needs to be very concise so that the authors can accordingly allocate more space to the methods and discussion section. In such cases, using a structure within the BI section can enhance the logical flow while keeping the text concise and succinct (see Fig. 1).

It is advised that the BI section of the manuscript reporting quantitative findings follows an inverted funnel approach where authors start from presenting the research context from a broad overview and arrive to narrowly defined specific research objectives. However, this approach comes with a risk. By the time the readers reach specific aims of the study, the presentation of multiple related ideas to paint a general overview of research may lead to confusion among readers regarding researchers' intended research question. Therefore, consistency of material presented within the literature section to the specific research aims and objectives is paramount. This section should ideally be limited within 500–1000 words and presentation of any superfluous or irrelevant information must be avoided.

2.3 Methods/Methodology Section: Describe the Process Involved

Methods section is one of the most important components of the research paper as it enables the reader to develop confidence in the work being conducted through transparent and reproducible methods (Szklo 2006). Consequently, authors may choose to present this section with subheadings that may include study population, study period and settings, sample selection and size and data collection, ethics consideration, study measures (outcomes, explanatory variable/s, and covariates), and statistical analysis. Reporting of sufficient and relevant information under these headings is critical. STROBE and CONSORT guidelines/checklists can be very useful to guide this process (Vandenbroucke et al. 2007, Mannocci et al. 2015). Methodological detail of a study is often absent and insufficiently described for the reviewers to judge methods for robustness and appropriateness (Kool et al. 2016). Carefully adopting the recommendations within these checklists for reporting of study design, information on participants, variables, data sources/measurement, bias, study size, and handling and categorization of quantitative variables and statistical methods can substantially make the methods section of the studies transparent and clear.

Study population and settings should identify the characteristics of the population of interest such as age group, nationality, survey purpose, and methodology if the study is a secondary analysis of already collected data. Clarity on *study settings and locations* where the data was collected is highly relevant for quantitative studies. This information on the settings and locations is crucial to judge the applicability and generalizability of the study findings. This could include the country, city if applicable, and the study environment (e.g., community, hospital, research organization). If more than one setting were used, the authors must mention each of them as appropriately as possible. Reporting other aspects of the setting (including the social, economic, and cultural environment and the climate) is critical for the readers to be able to extrapolate the results of the study to their own settings (external validity or generalizability). This is also important for guiding policy and practice (Rothwell 2005; Weiss et al. 2008).

Next, *study design* should be explicitly recognized. In quantitative research, the choice of study designs is often made at the start of the study when developing a research question and hypothesis. Study designs are broadly divided into two categories: descriptive and analytical (see Fig. 2). Descriptive studies are conducted with the aim to study the amount and distribution of the disease within a population. If the descriptive study includes individual's cases, it can be specified as a case report or a case series. Descriptive studies carried out at a population level aim to describe and summarize health and disease characteristics at the population level. The other broad category, analytical study design, is employed to study the determinants of the disease. This can be further divided into observational study and experimental study. Observation studies are further subdivided into cross-sectional studies (analytical), cohort studies, and case-control studies, while the experimental studies can be subdivided into randomized controlled trials (RCTs) and non-randomized controlled trials (non-RCTs) (see also ▶ Chaps. 35, "Longitudinal Study Designs," and ▶ 59, "Critical Appraisal of Quantitative Research"). Hence, it is very important to identify and state study design for clarity along with the rationale for choosing a particular study design over others and how best will it be able to answer the research question.

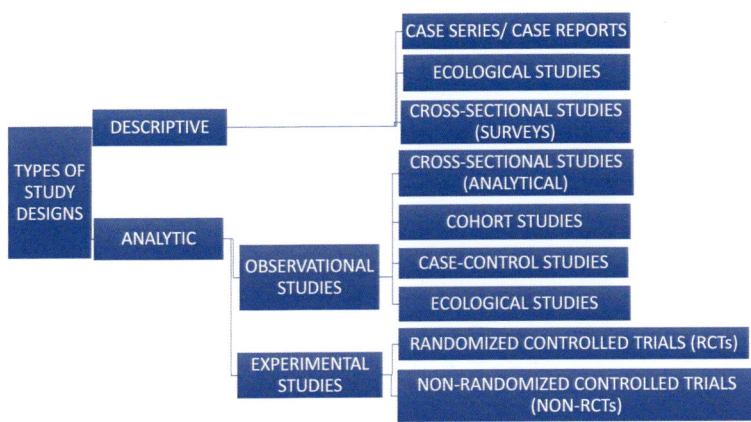

Fig. 2 Types of study designs

Data collection method describes the procedures for collecting the data for the study in sufficient detail such that the work could be replicated with minimal error if required. It is vital that the method of data collection be chosen and analyzed such that it is consistent to the study design and is best suited to answer the research question. There are a variety of techniques that can be used to collect data in a quantitative research study. These include observations, interviews, questionnaires, and so on. Information collected through these mediums are further defined as the key measures/variables for the quantitative analysis. The data can be sourced either from primary data (collected first hand) or secondary data (previously collected data). However, the types of quantitative variables that may comprise within the data could be of various types. These may include qualitative variables (binary, ordinal, nominal) and quantitative variables (discrete and continuous). These definitions should be clearly identified when describing the variables within the methods section as the choice of statistical modeling is also dependent on the types of variables. Another important aspect to be included in data collection section is to report on the explanatory and outcome variables, sample size, eligibility criteria for participants, and the description of the intervention used in the study. Participant eligibility criteria may refer to demographics, clinical diagnosis, and comorbid conditions and many others factors that make them eligible for the inclusion in the study. It is desirable to report any protocol developed prior to conducting the study that provides more detailed information on how the data was collected and any ethics approval that was sort or exempted to obtain the data for the purpose of the study.

Statistical analysis must be detailed appropriately as its consistency with the research question and the nature of the data is the keystone for any quantitative study. Statistical methods for both descriptive and inferential analysis should be informed, and the choice of statistical methods be justified by relevant text or existing references. Often, well-known statistical techniques may not require a description within the paper; instead, an appropriate reference citation may be sufficient. Choice of summary measures for describing the data, and associations, needs to be consistent with statistical analysis and can be explained at this stage. Consequently, the readers will be acquainted with the measures that they will be interpreting in the following results section. Statistical software with the versions that has been used for the analysis should also be mentioned (see also ▶ Chap. 54, "Data Analysis in Quantitative Research").

A priori presentation of the theoretical model that the authors aim to follow must be explicitly declared at this stage. This is paramount as the readers gain confidence that authors are not chasing "significant" findings to publish and have a clear theoretical concept that they aim to test empirically using population-based data. A useful tool to explicitly declare an author's position on presumed theoretical relationship between outcome, exposure, and covariates is a "directed acyclic graph" (DAG) (Greenland et al. 1999).

2.4 Results/Findings Section: Reveal the Findings

In the results section, it is important for authors to concisely and objectively present the key results, in a logical sequence using both text and illustrative materials (tables and figures) (see Fig. 3). The results section can be organized in a sequential format to present the key findings in a logical order. The text of the results section can be crafted in response to the questions/hypotheses set out in the beginning. In addition to positively toned results (significant findings), it is equally important to report the negative results. Providing a description of the missing data can help readers make cautious interpretations of the findings. Often, authors write the text of the results section based upon the sequence of tables and figures to provide clarity and flow in

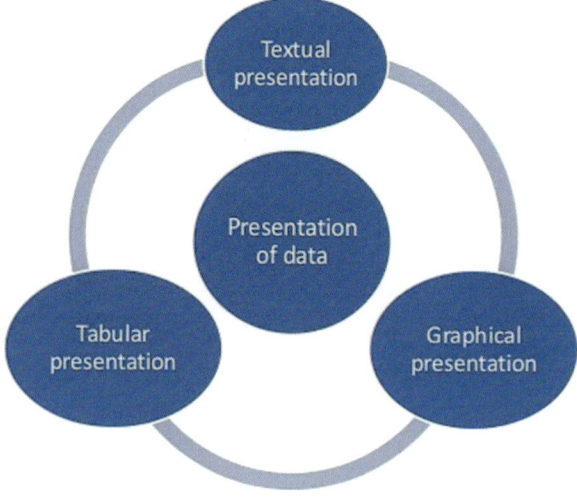

Fig. 3 Different types of presenting numerical data

understanding. Most commonly, the results begin with describing the basic characteristics of the study leading to simple (crude) statistical analysis results and finally into the more complex statistical analysis results. Regardless of the complexity of the analysis employed, authors can ensure that the results are written in a simple and intelligible manner. This section includes reporting of the frequency/occurrence of various variables and their distribution in the sample, called as *frequency distribution*, *means*, *medians*, and *percentages* (Szklo 2006). Following this, further effect estimates in the form of differences, directionality, and magnitude between variables are reported to provide an understanding to the reader about the nature of differences or relationships observed. Few points to consider when writing the results section – avoid repetition of the values from the figures and tables in the text; only the key result or trends must be stated. It can make the tables redundant and a waste of space and energy. When reporting results in tables, authors can be more author friendly by (Weiss et al. 2008):

1. Presenting most understandable measures.
2. Appropriately presenting results of interactions, if tested.
3. Titles of tables and figures should be informative so that the readers do not need to look at corresponding text in results section.
4. Specifying units for variables.
5. Avoiding redundancy (preferably report confidence intervals than p-values, no need to present both).
6. Present results without discussing them.
7. Include specific results in the text; which are not included in tables (if this is necessary).

Text of the results section should be a summary of the key results, while tables and figures should serve to illustrate, explain, and justify this information (Kool et al. 2016). A general advice would be to ensure that the results are presented in way that interests the readers.

2.5 Discussion Section: Relate to the Bigger Picture

Discussion section is the place where the creativity and the capacity of the authors to discuss the results leading to implications are established. This is the hardest section to be written as here the authors need to convince the editor and the reader that their study is relevant and important. It ties the entire research together and offers an explanation for the bigger picture. The purpose of the discussion section is to interpret the results in the light of existing evidence on the subject. The discussion section seeks to explain a new understanding of the problem under investigation based on the obtained results. The discussion must be connected to the introduction section by way of the question(s) or hypotheses stated before and the literature cited to support that. It is not a repetition or a rearrangement of the introduction section but instead provides a more comprehensive understanding to the problem. It is advised that the discussion section follows the following structure (Docherty and Smith 1999):

1. Statement of principal findings
2. Strengths and weaknesses of the study
3. Strengths and weaknesses in relation to existing literature. Similarities and dissimilarities with previous findings
4. A summary of the results: its plausibility, possible mechanisms explaining associations and implications
5. Unanswered questions and future research

Structuring the discussion section in this form is beneficial as it reduces overall length, prevents unjustified extrapolation and selective repetition, reduces reporting bias, and improves the overall quality of reporting (Docherty and Smith 1999). The discussion section is largely the author's playground where the author has the freedom to provide an interpretation of what the results mean in the larger context of the problem (Szklo 2006). However, speculations should be minimal and based on the findings from the study (Docherty and Smith 1999).

Strengths and limitations of the study provide the authors to display transparency and highlighted the strengths of the study alongside acknowledging any potential bias or limitation that may have affected the study findings (such as generalizability, potential bias). Strengths of the study could reflect the potential to make an important contribution in the existing pool of evidence through its rigorous data collection strategies or use of robust analytical techniques. Not every study is perfect, and there would, therefore, be some pitfalls or limitations of the study. It is necessary for the authors to be truthful and clear about the potential biases such as small sample size, low participation rate, presence of unmeasured confounders that could have affected the associations being tested, and many more depending on the context of the study. Authors should also discuss any imprecision of the results that may have arisen as a result of measurement of a primary outcome or diagnostic tool used. Perhaps the scale used was validated on an adult population but used in a pediatric one, or the assessor was not trained in how to administer the instrument.

Caution must be made to not repeat the results; instead sentences interpreting the results in a broader context are encouraged. Additionally, no new results must be presented in the discussion section; instead it must be in line with the results reported in the results section only. Uses of appropriate references to support the interpretations are another important feature of a good discussion section. If helpful, subheadings may be used to organize the presentation of the discussion. When comparing and contrasting the findings with other published literature, the authors must ensure that they discuss reasons for similarities and differences between their and others' findings. This will allow the authors to draw a stronger argument of how their findings could be combined with existing evidence to better substantiate an understanding of the problem. The authors must ensure to cite references additional to that used in the introduction section to display the use of a wide range of evidence.

Depending on the relevance of the study, implications of the findings for research, policy, clinical care, or any other must be provided by the authors. This allows the opportunity for translating research into practice. As part of reflecting on what the findings mean, drawing out the implications of the findings for the field itself and/or societies is crucial. Contextualizing the findings within previous research helps readers to grasp the significance of the research and its contribution to knowledge. As a researcher, it is the purpose of any research conducted to offer any future implications or future recommendations (e.g., policy implications or research implications) that may benefit the society as a whole or help improvise future research.

For illustration, we use example of a published study. A paper entitled "Association between tooth loss and hypertension among a primarily rural middle aged and older Indian adult population" by Singh et al. (2016), who conducted a secondary analysis of a cross-sectional data sourced from the Longitudinal Aging Study of India. In the discussion section, the authors began by reporting the key finding of the study in line with the purpose of the study as mentioned in the last paragraph of the introduction section. The authors described the results in terms of both the direction and the magnitude of associations. This was followed by a discussion of their findings with the existing evidence highlighting both similarities and dissimilarities. Possible potential explanations were provided to support the findings that were contrary to the existing evidence. This is an important step to identify what new this study adds to the literature. The authors then highlighted the strengths of their study followed by the study's limitations. This subsection enables the reader to view the study in confidence. After feeding the reader with all the necessary information about the study's results, the author then drew a closure to the paper by summarizing the entire study in the conclusion section (Singh et al. 2016).

2.6 Conclusion: Leave the Reader Enlightened

In simple words, a conclusion is, in some ways, like the introduction where the author restates the research questions and summarizes key findings in line with questions. It is usually written in a single paragraph. It captures the essence of the findings of the study and provides the significance of the study findings. It is the place where the authors tie all the ends together and complete a full circle by integrating the key idea from the introduction to the key take away message from the study. The authors could end the conclusion by either making a suggestion or a recommendation or posing a question calling for an action through future investigation.

3 Conclusion and Future Directions

A scientific research paper is a piece of academic writing based on the author's original research on a particular topic and the analysis and interpretation of the research findings. How well a research paper is written entirely depends upon its structure, format, content, and style of writing. This is an iterative process, and it may take several drafts and even rewritings of sections to achieve coherence and consistency in the presentation of the idea, all under the word limit as instructed by the journal. Although difficult, this process has considerable value as scientific articles can lay foundations for policies and evidence-based practices and guidelines. The push on evidence-based medicine and public health practices in the last few decades has reinforced the value in generating and publishing good research.

There are profound benefits of publications in academics' research impact within the current pressures of publish or perish. Scientific publications in most academic contexts continue to play an important role in judging an individual researcher's productivity, more so at the earlier stages of academic career when applying for competitive scholarships, awards, fellowships, and jobs. Well-written scientific papers convey the capacity of researchers to deliver research independently and collaboratively to its intended audiences. Authorship statements and authorship criteria explicitly outline the independent contribution of each researcher to the complete package. It must be noted that scientific publications often take a long time from its inception to the publication stage and finally citations. Readers may find the basic principles of writing quantitative research highlighted in this chapter helpful at any stage of publication.

Ultimately, benefits of well-written scientific papers are not only limited to researchers and research teams but extend to the research community and the overall society. As authors of scientific papers, one has a key role to play in this process. The value of scientific paper writing can be best summarized by following quote from Richard Horton on the rhetoric of research (Horton 1995):

> The text of a scientific paper is not an atlas that offers readers several equally appealing routes through terrain mapped out by the authors. Rather, the text describes a specific path, carefully carved by the authors, through a complex undergrowth of competing arguments. By examining this path more closely, we come to see the authors' intention and the means by which they convey this intention.

References

Bhaumik S, Arora M, Singh A, Sargent JD. Impact of entertainment media smoking on adolescent smoking behaviours. Cochrane Database Syst Rev. 2015;6:1–12. https://doi.org/10.1002/14651858.CD011720.

Dickersin K, Manheimer E, Wieland S, Robinson KA, Lefebvre C, McDonald S. Development of the Cochrane Collaboration's CENTRAL register of controlled clinical trials. Eval Health Prof. 2002;25(1):38–64.

Docherty M, Smith R. The case for structuring the discussion of scientific papers: much the same as that for structuring abstracts. Br Med J. 1999;318(7193):1224–5.

Greenland S, Pearl J, Robins JM. Causal diagrams for epidemiologic research. Epidemiology. 1999;10(1):37–48.

Horton R. The rhetoric of research. Br Med J. 1995;310(6985):985–7.

Kool B, Ziersch A, Robinson P, Wolfenden L, Lowe JB. The 'Seven deadly sins' of rejected papers. Aust N Z J Public Health. 2016;40(1):3–4.

Mannocci A, Saulle R, Colamesta V, D'Aguanno S, Giraldi G, Maffongelli E, et al. What is the impact of reporting guidelines on public health journals in Europe? The case of STROBE, CONSORT and PRISMA. J Public Health. 2015;37(4):737–40.

Rothwell PM. External validity of randomised controlled trials: "to whom do the results of this trial apply?". Lancet. 2005;365(9453):82–93.

Schulz KF, Altman DG, Moher D. CONSORT 2010 statement: updated guidelines for reporting parallel group randomised trials. PLoS Med. 2010;7(3):e1000251.

Szklo M. Quality of scientific articles. Rev Saude Publica. 2006;40 Spec no:30–5.

Vandenbroucke JP, von Elm E, Altman DG, Gotzsche PC, Mulrow CD, Pocock SJ, et al. Strengthening the reporting of observational studies in epidemiology (STROBE): explanation and elaboration. PLoS Med. 2007;4(10):e297.

Weiss NS, Koepsell TD, Psaty BM. Generalizability of the results of randomized trials. Arch Intern Med. 2008;168(2):133–5.

Singh A, Gupta A, Peres MA, Watt RG, Tsakos G, Mathur MR. Association between tooth loss and hypertension among a primarily rural middle aged and older Indian adult population. J Public Health Dent. 2016;76:198–205.

Traditional Academic Presentation of Research Findings and Public Policies

Graciela Tonon

Contents

1 Introduction 1000
2 Power Mechanisms in the Scientific Field and Research Freedom 1001
3 Transmission of Research Results in the Context of Public Policy 1002
   3.1 Public Policies 1002
   3.2 The Relation Between Public Policies and Research Knowledge 1003
   3.3 Relationship Between Researchers and Policy-Makers 1006
4 Conclusions and Future Directions 1009
References 1010

Abstract

The transmission process of research results in the field of public policy reveals various peculiarities. In order to approach its study, it ought first to be pointed out that, in the academic field, power mechanisms have been observed, which have traditionally institutionalized certain topics and certain actors, making others invisible during the process. Furthermore, the last decades have revealed the role of politics in research production, which may be understood as a chance that the research results might be taken into account and exert an influence on public policy decisions oriented toward the people's life improvement – though this does not always happen, in practice. The aim of this chapter is to review the subject of power in the scientific field (Bourdieu, Sociología y cultura. México: Grijalbo,

---

G. Tonon (✉)
Master Program in Social Sciences and CICS-UP, Universidad de Palermo, Buenos Aires, Argentina

UNICOM- Universidad Nacional de Lomas de Zamora, Buenos Aires, Argentina
e-mail: gtonon1@palermo.edu

© Springer Nature Singapore Pte Ltd. 2019
P. Liamputtong (ed.), *Handbook of Research Methods in Health Social Sciences*,
https://doi.org/10.1007/978-981-10-5251-4_145

1984) and the researchers' freedom to make further progress in the definition of the concept of public policy and, on that basis, study the possible ways in which research results may be reflected in the arena of political decisions by identifying its obstacles and facilitators and proposing options to bridge the gap.

Keywords
Academic presentation · Research findings · Public policies · Knowledge

1 Introduction

The notion of universality of social knowledge greatly depends on its contextualization in time and space. Therefore, in order to be considered as scientific knowledge, its concepts must be able to build a coherent understanding of some kind of reality. However, in order to be regarded as pertinent, it ought to be attached to the reality contexts in which specific and unrepeatable events take place (Sotolongo Codina and Delgado Diaz 2006, p. 90). This necessarily gives rise to the question regarding the *forms of knowledge production* – both in the fields of strictly scientific and academic knowledge and in the participation of social and political actors, in the aforementioned production (Carrizo 2004, p. 73). Crewe and Young (2002, p. v) suggest that "research and policy defy neat separation but can be conceptually distinguished by their goals and methods: research produces knowledge, policy aims for continuity or change of a practice."

Caplan (1979) marks a difference between conceptual use and instrumental use of knowledge. Caplan considers instrumental use to be associated with microlevel decisions, that is to say, with the day-to-day policy issues, involving administrative policy issues pertaining to bureaucratic management and efficiency, rather than substantive public policy issues. Regarding the conceptual use, Caplan (1979) describes the gradual shifts in terms of policy-makers' awareness and reorientation of their basic perspectives, which involve important policy matters and are mostly associated with macrolevel decisions. On the other hand, Weiss (1991) argues that conceptual use is more common and deems as central to her enlightenment research model (Nielson 2001).

The difference between the cultural level of the professionals in government agencies and that of university researchers leads to a lack of communication between them which eventually results in low knowledge utilization (Landry et al. 2003). Thus, in order to enhance research utilization in the orbit of public policy, it is necessary to achieve a two-way exchange between both professional groups in order to integrate research findings with the policy-making process. Sharing the results of research findings with policy-makers has become a worldwide challenge (Uzochukwu et al. 2016, p. 12).

This chapter is organized by subthemes. In the first place, it will deal with the subject of power in the scientific field and freedom of research. It will later dwell on the review of the concept of public policy, on the analysis of the transmission of research results in the context of public policies – taking into account the different characteristics of knowledge and the way in which researchers and policy-makers interrelate in this field. It will finally identify and comment on the handicaps which have historically turned up in the development of the process and the possibility of strengthening and improving it in the future.

2 Power Mechanisms in the Scientific Field and Research Freedom

The university structure has been historically based on scientific knowledge, and the science system is formulated through basic research which allows the development of knowledge while highlighting the utility and social relevance of science (Plascencia Castellanos 2006, p. 31). Yet, "the institutionalization of knowledge has brought about power mechanisms privileging certain types of research and theories which disseminate notions of truth that we finally learn, naturalize and reproduce" (Sotolongo Codina and Delgado Diaz 2006, p. 225).

Bourdieu (2000) recognizes two forms of power in the scientific field: an institutionalized political power exerted by those who actually hold decision-making positions and a specific power based on personal prestige, with a lower degree of institutionalization, exerted by invisible colleges of scientists united by mutual recognition. Bourdieu (2000) further mentions that the former type of power is achieved through political strategies which respond to the rules of bureaucratic capital, while the latter is achieved by purely accumulating science capital, solely generated by personal effort and scientific work – which explains why, in general lines, those who are most prestigious are not the most powerful.

Taking into consideration the two abovementioned situations, I stop to wonder about researchers' actual possibilities and, most importantly, the degree of freedom they may enjoy when deciding upon their research topics (Tonon 2010a). My starting point is the idea that the only possible access to knowledge is through freedom, either built within a sociocultural research context or on the basis of each researcher's individual achievements (Sotolongo Codina and Delgado Diaz 2006). When referring to freedom, I follow the lines of Sen (2000) who proposes that freedom should be focused on people's possibility to become creative actors and agents of their own development. Plascencia Castellanos (2006, p. 25) writes: "The university is the place par excellence where unconditional free speech and questioning shall be guaranteed and exercised."

In the case of social researchers, when referring to freedom and contextualization, it ought to be borne in mind that there are different forms of power – formal and informal – which they must reckon with in order to be able to comply with their research work. Furthermore, the struggle in the scientific field described by Bourdieu (1984, p. 135) becomes more complex according to each context, for it leads to a conflict of interest among the protagonists, since both evaluators and subjects of assessment turn out to be colleagues and competitors; if this situation should become more complex still, the research field may become an "apparatus," thus giving way to pathological situations (Bourdieu 1984, p. 158).

Bourdieu (1984, pp. 157–158) expresses the difference between research field and research apparatus by stating that the research field is the scenario of the struggle between agents and institutions, with different forces, and in accordance with the constitutive rules of this space. And while those who exert power also have the means to do so, they are also resisted by the rest. On the other hand, when those who dominate the field possess the means to counteract resistance, the field becomes an "apparatus."

It is, therefore, necessary to elucidate the power mechanisms derived from established knowledge, and delve into the institutional network of this research work, thus discovering the existence of the so-called institution science (Bourdieu 2008) which tends to establish, as a model of scientific activity, and became the routine practice in which more scientifically decisive operations may be carried out without reflection or critical control under the impeccable appearance of visible processes.

In this light, Carrizo (2003) suggests conducting a political analysis of institutional relations, i.e., of the distribution of power, which turns out to be important in order to unveil certain factors that constitute limitations to the outlooks and practices of researchers as well as of others with whom, and for whom, they work. And, even though the transmission of research results in the scientific field has been traditionally conducted by the researchers in the presence of their peers and/or their students and postgraduates, the twenty-first century calls for a further step in this line of work in order to secure the transference of those results to the field of public policy.

3 Transmission of Research Results in the Context of Public Policy

3.1 Public Policies

In a broader sense, public policies may be defined as the response governments give to arising problems. This leads us to evoke the historical existence of two opposite models which presuppose different ways of engaging in politics, i.e., public policy formulation and public policy implementation (Tonon 2010b). In that respect, Oszlak (1980, p. 4) sustains that, in their formulation, public policies are the genuine expression of public interest, and their legitimacy derives from a democratic legislative process or from the act of applying technically rational criteria and knowledge to the solution of social problems. However, implementation, which occurs in the bureaucratic context of the state, is related (in the popular imaginary) to routine, inefficiency, and corruption. Oszlak (1980, p. 11) goes on to say that the action of state institutions ought not to be merely regarded as the implementation of a set of norms but also as an attempt to achieve compatibility between their clientele's interests and their own, i.e., the interests upheld in their political projects by regimes that succeed one another in office.

In a previous research study, Oszlak and O'Donnell (1976, p. 21) had already defined public policies as "a set of actions-omissions which reveal a certain form of state intervention regarding an issue that may claim the attention, interest, or mobilization of other civil society actors." They further made it plain that this process included successive simultaneous decisions emanated from several organisms of the state, thus ensuring that the position adopted would not necessarily be univocal, homogeneous, or permanent. Later on, Regonini (1989) identified five analytical categories to explain the public policy formulation and implementation processes, namely, characteristics of the most influential and recurrent actors, style of decision-making processes, dynamics of the stages of the public policy's life cycle, structure of the policy problem, and the rules of the game. Thus, an analysis of public policy ought to consider, within a space-time dimension, the conditions of its emergence, the dynamics of its development, and the tendencies and contradictions of its political- institutional unfolding (Fleury 1997, p. 172).

Although traditional public policies have been designed in terms of the satisfaction of social or collective rights, as an external activity and provided by the state, a change of focus has recently been promoted. This shifts from the traditional one to a focus on human rights, characterized by an attempt to construct a reflective capacity oriented to the development of a type of citizenship not merely involved in political-state recognition but also in a sociocultural kind. In other words, the tendency is to promote a process of public policy formation characterized by stage interaction and the possibility of a permanent actor-decision adjustment with the object of enhancing results (INDES 2006).

3.2 The Relation Between Public Policies and Research Knowledge

Landri, Lamari, and Amara (2003, p. 194) explored the studies related to knowledge utilization and discovered that the pioneering studies in this field paid attention to variables related to the characteristics of the research products (see Caplan 1975; Weiss 1981). Another group or scholars were dedicated to the importance of policy contextual factors (Sabatier 1978; Webber 1984; Lester 1993). More recently, a group was devoted to the study of the importance of other explanatory factors, such as dissemination and links and exchanges between researchers and the use of research (Huberman 1994; Lomas 1997, 2000).

In 1979, Weiss published her emblematic article *The Many Meanings of Research Utilization* in which she expresses the following:

> This is a time when more and more social scientists are becoming concerned about making their research useful for public policy-makers, and policy-makers are displaying spurts of well publicized concern about the usefulness of the social science research that government funds support. There is mutual interest in fathoming whether or not social science research aiming to influence policy is actually being used; but before that important issue can be properly addressed, it is essential to understand what using research actually means. (p. 426)

Weiss sums up the seven meanings she proposes for research utilization (pp. 427–430):

- Knowledge-driven model: it is the scheme that derives from natural science, composed of successive stages, basic research, applied research, development, and application.

- Problem-solving model: it implies the direct utilization of knowledge generated in the scientific field, with the aim of solving a pending problem in the social and political arena. Two cases may arise: (a) that the knowledge may already exist, in which case policy-makers will utilize it in order to make a decision, and (b) that the policy-makers may have a notion of what is needed and thus require that the researchers should work in that direction in order to generate the abovementioned knowledge.
- Interactive model: in this case, the policy-makers utilize information, not only produced by researchers but also by other social actors (citizens, planners, journalists, other politicians, administrators, groups of interest, NGOs, etc.).
- Political model: at times, a nucleus of interest regarding a certain political topic may bias the position of policy-makers, thus rendering them non-receptive to new knowledge generated in the research field.
- Tactical model: it is used by policy-makers in certain situations in which research is considered to be proof of their responsibility in the issue or when they wish to delay a decision, arguing that it is being developed in the research field, or to shirk their responsibilities in unpopular decisions.
- Enlightenment model: enlightenment is the most frequent way in which research may relate to decision-making.
- Research as part of the intellectual enterprise of the society: research utilization looks upon social science research as one of the intellectual pursuits of a society.

Weiss concludes that the time has come when importance must be given, not only to an increase in the utilization of research results applied to political decision-making but also to the increase and improvement of research contributions with political knowledge.

In that same year, the OECD Committee for Scientific and Technological Policy published the report *Social Science in Policy Making,* which also constituted an important document for the analysis of the relation between social sciences and the generation of public policies. James Mullin, coordinator of the abovementioned committee, made special emphasis on the fact that one of the major responsibilities of political decision-makers and researchers is the development of a mutual understanding of each other's tasks. This shows that, back then, there was a concern about progress and the most effective utilization of scientific knowledge for the resolution of social problems (Carrizo 2004).

In 1996, Davis and Howden-Chapman posed the question regarding the ways in which research work was reflected in public policies, and they arrived at the conclusion that "it was by some process of natural diffusion; or should it be conceived more actively as a matter of dissemination supplemented by a conscious program of implementation?" (p. 867).

The interaction between science and public policy, which was initially conceived through engineering models that presupposed a utilitarian or linear incorporation of scientific results to public policies, has nowadays become the object of interactive, reflective, or critical studies, which are not merely based on a practical utilization of knowledge (Estébanez 2004). At this point, we ought to identify the difference between transference of research to policy and the term "knowledge transfer" which describes the activities conducted during the process of generating knowledge based on user needs, disseminating it, building capacity for its uptake by decision-makers, and finally tracking its application in specific contexts (Almeida and Báscolo 2006).

Several decades later, Crewe and Young (2002) carried out a review of the current literature on the subject and concluded that there are three major dimensions to consider regarding the impact of research on public policy:

- *Context: politics and institutions, which includes* the interests of key policy-makers and researchers, structures, and ideologies they were limited by; whether the policy changes were reformist or radical, how organizational pressures operated, and to what extent policies were adapted, developed, or distorted when put into practice.
- *Evidence: credibility and communication, which implies* investigating the impact of research findings and raising questions about the credibility of the research that has made some impact and the way in which it was communicated to policy-makers.
- *Links: influence and legitimacy* are important to find out the identity of the key actors, the roles they have played, the links between them, and the extent to which the research methodology has given legitimacy to the findings (Crewe and Young 2002, p. 5).

In 2004, Carrizo developed a research study in which he identified authors who pointed out the existence of different obstacles in the transformation of political culture – i.e., in the ranking of social research and its products as improvement factors in decision-making. Carrizo (2004, pp. 77–79) identified some of them. Tactic obstacles are derived from the possibility that the research might provide knowledge which may promote decisions considered to be outside the electoral acceptance zone. Temporary obstacles reveal how the offer of an answer to urgent issues is regarded as being menaced by scientific culture, since the latter requires longer processes and high academic standards. Communicational obstacles suggest that the characteristics of scientific discourse might discourage its usage by political actors. Epistemic-praxis obstacles point to the fact that the need to simplify increasingly complex realities in decision-making does not articulate easily with the complexity of social research. Historical-political obstacles suggest that in circumstances when political parties have a central role, they tend to hinder any relevant incorporation of technical and scientific logics in the process of decision-making. Philosophical obstacles tend to warn political actors against the risks of technocratic elitism which may not contribute to the construction of democracy.

Scientific and technological advice is a process which links at least two well-defined sectors: producers of scientific and technological knowledge acting as advisors and the state, as knowledge user in decision-making (Estébanez 2004). This link may derive from advisory boards composed of scientists acting in governmental institutions; consulting processes with scientists, on the face of a crisis or problem; legislative science and technology advisors; and the presence of scientists in technical functions or government policies (Estébanez 2004).

According to the traditional view, scientific knowledge is regarded as an *accumulable product* which decision-makers can resort to according to their needs. This conception is generally allied with a simplified view of the decision-making process (Pellegrini 2000), under the assumption that policy formulation and implementation are a linear process comprising a chain of rational decisions made by privileged actors. In this respect, the problem seems to lie in making the right information available to decision-makers, at the right moment (Almeida and Báscolo 2006).

Evidence can occupy an important place in the policy-making process if it is available when needed, if it is communicated in terms that fit with policy direction, and if it points to practical action (Nutbeam and Boxall 2008). Orton et al. (2011) identified in their study of barriers that hinder the use of research evidence, decision-makers' perceptions of research evidence, the gulf between researchers and decision-makers, the culture in which decision-makers operate, competing influences on decision-making, and practical constraints.

Bowen and Zwi (2005, p. 600) propose an "evidence-informed policy and practice pathway to help researchers and policy actors navigate the use of evidence." The pathway involves three active stages of progression, influenced by the policy context: sourcing, using, and implementing the evidence. It also involves decision-making factors and a process they call *adopt, adapt, and act*. At the same time, Orton et al. (2011) also suggested ways of overcoming these barriers by proposing that research should meet the needs of decision-makers, research clearly highlighting key messages and capacity building.

Nielson (2001) summarized the different theories to explain the under-/non-utilization of knowledge or research by policy-makers for decision-making purposes. Caplan (1979) proposes his theory of the "two communities," arguing that the limited use of research by policy-makers is, in part, due to the fact that researchers and policy-makers have different worldviews. Later explanations based on the writings of Weiss (1977), Webber (1991), Sabatier and Jenkins-Smith (1993), and others include the idea that the research-policy link is not a direct one, particularly regarding data and information sources. These writings support the position that research is only one of many sources of information for policy-makers and that it is not a simple dichotomy between "use" and "non-use" but rather that knowledge/research utilization is built on a gradual shift in conceptual thinking over time. This represents what Weiss coined as the "enlightenment function" of research.

3.3 Relationship Between Researchers and Policy-Makers

Regarding the relationship between researchers and policy-makers, Crewe and Young (2002, p. 14) stated that when both have close personal links, with appropriate chains of legitimacy to those whom they represent, researchers should have a higher influence, and policy-makers could make better use of research. Crewe and Young (2002, p. v) suggest that "policy makers and researchers cut across categories but their position of power and the aims of the organizations they work for, can be identified."

Policy decision-makers as well as researchers and academics need to see eye to eye and take into consideration the social macro- and micro-dimensions in their daily tasks (Tonon 2015). Thus, Torres Carrillo (2006, p. 94) comes to mind when he points out that subjectivity crisscrosses social life and is present in all the social dynamics of daily life, both in micro-social and macro-social spaces and in daily intersubjective experience and in the institutional structure of the time.

Interrelations between researchers and decision-makers have been considered a prime factor in analyzing knowledge transfer processes. Analysis of the models of interrelations between researchers and decision-makers is relevant when one realizes that the use of scientific knowledge depends largely on certain characteristics of the actors, that is to say, the researchers' behavior and the decision-makers' receptiveness (Almeida and Báscolo 2006).

At this point, it is vital to reflect upon the characteristics of both researchers and policy-makers. Social researchers are individuals who related to other individuals; thus the products generated by those research processes are both socially and historically constructed (Tonon 2015). From the point of view of the social researcher, "his objects are not only objects for his observation but also beings that possess their own pre-interpreted world and carry out their own observation; they are fellow creatures inserted in a social reality" (Natanson 1974, p. 23). Brown (1991) argues that quantitative and qualitative databases and statistical indicators cannot be considered research results per se but are the raw material on which research is shaped and without which it cannot be conducted. He also emphasizes that "documentation is not always a step towards action; sometimes it stultifies it" (p. 28).

In the case of policy decision-makers, they require quantitative and qualitative indicators to obtain the information they supply – if they are to generate public policies that not only cater for the subjects' external living conditions but also to their quality of life in multidimensional terms (Tonon 2015). In this matter, we coincide with Veenhoven (2000) who expressed that public policies are not merely limited to material issues but are also extensive to affairs related to people's mentality.

There is a considerable difference between what scientists and policy-makers consider as knowledge, as well as the difference between how that knowledge has been developed or obtained. Social scientists generally regard knowledge as something that is theoretically and methodologically sound and/or defensible. Policy-makers see knowledge as the result of experience (Nielson 2001, p. 6).

Research quality often determines the credibility of the organization that either conducts or financially supports the research and, as such, may also determine the credibility and/or integrity of the research field itself as a source of useable knowledge. In relation to the concept of research quality, Seck and Phillips (2001) regard rigorousness as a primary quality or characteristic which may help to determine the quality of the research – a rigorous research being the one that is free of fault in design, method, and interpretation. They further point out that it is not synonymous with academic or pathbreaking theoretical research. The authors also explain that another important characteristic is completeness, which is a concept related to the exploration of all potential options, as well as making available all relevant facts and figures that research can uncover in the search for intrinsically good policy options (Seck and Phillips 2001, p. 4).

Nielson (2001, p. 11, quoting Weiss 1991) recognizes three models of "research" used by policy-makers: research as data, research as ideas, and research as argumentation. The first model is more mechanistic in terms of its application to the problem at hand. It assumes that the data or sets of findings obtained meet the users' needs and that there is no conflict in terms of what solution, or goal, is desired or required in order to resolve the problem. The second one is perceived to be more general in nature, applied to situations in which problems are regarded as complex in nature, i.e., when uncertainty is high and ideas are in demand. Finally, research as argumentation is used when a decision has already been made and policy-makers and/or interest groups draw on research to take an advocacy position.

Reimers and McGinn (1997, p. 22) propose four types of research. First is academic research which is the research where systems of explanation are composed by theories, models, and conceptual frameworks. Second, planning research is a kind of research that uses statistical analysis to generate patterns of relationships among variables. Third is instrumentation research where repeated trial and error methods are used in the instrumentation process of preparing a new curriculum. Fourth is action research which has its focus on the outcomes themselves rather than in the knowledge of how to achieve them.

Uzochukwu et al. (2016) point out that the failure of the utilization of high-quality research evidence by decision-makers has been singled out as the gap between research and policy. The authors quote Lomas (1997), identifying four misunderstandings between the evidence production and the policy-making effort. The first point is that researchers and policy-makers consider each other's activity as the act of generating products rather than engaging in processes. Secondly, scientific research attempts to focus on the question, so that a clear and crisp answer may be provided, whereas policy-making considers other variables such as interests, ideology, values, or opinions. Thirdly, decision-makers are not sensitive to the incentives that drive researchers, such as attracting grant money and publishing in peer-reviewed journals, causing them to be reluctant to respond to issues that are politically current for government policy-makers. In the fourth place, researchers rarely take into account their potential research audiences (Lomas 1997).

Various authors have examined and promoted different ways of improving interrelations between researchers and decision-makers, such as collaborative or "allied research" (Pittman 2004), constructivist approaches or evaluative research (Furtado 2001), including strategies to improve the knowledge output for decision-making.

Campbell et al. (2009) developed a study which expresses that better communication is often suggested as fundamental to increasing the use of research evidence in policy, but little is known about how researchers and policy-makers work together or about the barriers they are faced with. This study has explored the views and practice of policy-makers and researchers regarding the use of evidence in public policy, including current use of research evidence to inform policy, dissemination of and access to research findings for policy, communication and exchange between researchers and policy-makers, and incentives for an increase in the use of research in policy-making (Campbell et al. 2009, p. 1). The results of the study show that policy-makers and researchers acknowledge the potential of research as a contribution to policy and are making significant attempts to integrate research into the policy-making process, although only half of the researchers believe their research to have been used to get issues on the policy agenda, or to select preferred policy options, in the past 2 years. These findings suggest four strategies to assist in increasing the use of research in policy-making: making research findings more accessible to policy-makers; increasing opportunities for interaction between policy-makers and researchers; addressing structural barriers, such as research receptivity in policy agencies and a lack of incentives to lure academics to link with policy; and highlighting the relevance of research to policy-making (Campbell et al. 2009, p. 1).

4 Conclusions and Future Directions

I propose reconsidering social science research in the light of the challenges of the twenty-first century. González Perdomo (2006, p. 28) suggests that "social research must be conceived as a kind of experience with life, in which, through an exhortation to dare to think, such experience can become significant and transforming." About the use of knowledge, Nielson (2001, p. 7) quotes Webber (1991, pp. 5–6) who states that use is understood to mean consideration, while the exact process of use has been given different interpretations – and little effort has been made to compare approaches to measuring knowledge use in the same sample of policy-makers.

It ought to be borne in mind that scientific arguments are important in some areas of policy-making and that policy-makers could make more constructive use of research, while researchers could communicate their findings more effectively in order to influence policy-making. Crewe and Young (2002, p. 1) contend that "if more were understood about the context within which researchers, policy makers, and stakeholders are working, the links between them would be enhanced, and good quality research would be disseminated more effectively, thus, better policy making might ensue."

In order to increase the relevance of research, policy-makers need be able to clearly identify and acquaint researchers with gaps in knowledge and policy priorities that require research. A greater understanding of the policy context by researchers could increase relevance by focusing the research on more useful questions, collecting information instrumental to policy decisions, and improving the description of the research results and their implications (Campbell et al. 2009). With the right level of interaction between researchers and decision-makers, the translation of research findings into actionable policy and programmatic guidance is an achievable goal (Uzochukwu et al. 2016).

References

Almeida C, Báscolo E. Use of research results in policy decision-making formulation, and implementation: a review of the literature, vol. 22. Rio de Janeiro: Cad Saude Publica; 2006. p. 7–33.

Bourdieu P. Sociología y cultura. México: Grijalbo; 1984.

Bourdieu P. Los usos sociales de la ciencia. Buenos Aires: Ediciones Nueva Visión; 2000.

Bourdieu P. Homo Academicus. Mexico: Siglo XXI Editores; 2008.

Bowen S, Zwi AB. Pathways to "evidence-informed" policy and practice: a framework for action. PLoS Med. 2005;2:600–5.

Brown L. Knowledge and power: health services research as a political resource. In: Ginzberg E, editor. Health services research: key to health policy. Cambridge: Harvard University Press; 1991. p. 20–45.

Campbell D, Redman S, Jorm L, Cooke M, Zwi AB, Rychetnik L. Increasing the use of evidence in health policy: practice and views of policy makers and researchers. Aust N Z Health Policy. 2009;6(21):1–11. Retrieved from https://anzhealthpolicy.biomedcentral.com/track/pdf/10.1186/1743-8462-6-21?site=anzhealthpolicy.biomedcentral.com. 7 Sept 2017.

Caplan N. The use of social science information by Federal Executives. In: Lyons G, editor. Social science and public policies. Hannover: Dartmouth College Public Affairs Center; 1975. p. 47–67.

Caplan N. The two-communities theory and knowledge utilization. Am Behav Sci. 1979;22(3): 459–70.

Carrizo L. El investigador y la actitud transdisciplinaria. Condiciones, implicancias y limitaciones. Documento de debate. Programa MOST-UNESCO. Washington, DC; 2003. p. 58–78.

Carrizo L. Producción de conocimiento y políticas públicas. Desafíos de la universidad para la gobernanza democrática. Cuadernos del CLAEH n° 89, Montevideo 2° serie, año 27-2; 2004. p. 69–84.

Crewe E, Young J. Bridging research and policy: context, evidence and links. WP 173. London: ODI; 2002.

Davis P, Howden-Chapman P. Translating research findings into health policy. Soc Sci Med. 1996;43(5):865–72. Great Britain. Elsevier.

Estébanez ME. Conocimiento científico y políticas públicas: un análisis de la utilidad social de las investigaciones científicas en el campo social. Espacio Abierto, vol. 13, núm. 1, enero-marzo. Maracaibo: Universidad del Zulia; 2004. p. 7–37.

Fleury S. Estado sin ciudadanos. Buenos Aires: Lugar Editorial; 1997.

Furtado JP. Um método construtivista para a avaliação em saúde. Ciênc Saúde Coletiva. 2001;6:165–81.

González Perdomo A. La interdisciplinariedad en la formación del pensamiento y el espíritu crítico que lo guía. In: Cuadernos de Sociología, vol. 40. Bogotá: Universidad Santo Tomás; 2006. p. 17–34.

Huberman M. Research utilization: the state of the art. Knowl Policy. 1994;7(4):13–33.

INDES. Documento de trabajo Medición del desarrollo y políticas públicas. Washington, DC: BID; 2006.

Landry R, Lamari M, Amara N. The extent and determinants of the utilization of university research in government agencies. Public Adm Rev. 2003;63(2):195–205.

Lester J. The utilization of policy analysis by state agency officials. Knowl Creat Diff Util. 1993;14(3):267–90.

Lomas J. Improving research dissemination and uptake in the health sector: beyond the sound of one hand clapping. Hamilton: Centre for Health Economics and Policy Analysis; 1997.

Lomas J. Using linkage and exchange to move research into policy at a Canadian foundation. Health Aff (Millwood). 2000;19:236–40.

Natanson M. Introducción. In: Schutz A, editor. El problema de la realidad social. Amorrortu: Barcelona; 1974. p. 15–32.

Nielson S. Knowledge utilization and public policy processes: a literature review. Evaluation UNIT IDRC. 2001. Retrieved from https://idl-bnc-idrc.dspacedirect.org/bitstream/handle/10625/31356/117145.pdf?sequence=1. 3 Sept 2017.

Nutbeam D, Boxall A. What influences the transfer of research into health policy and practice? Observations from England and Australia. Public Health. 2008;122:747–53.

Orton L, Lloyd-Williams F, Taylor-Robinson D, O'Flaherty M, Capewell S. The use of research evidence in public health decision making processes: systematic review. PLoS One. 2011;6(7): e21704. https://doi.org/10.1371/journal.pone.0021704.

Oszlak O. Políticas Públicas y Regímenes Políticos: Reflexiones a partir de algunas experiencias Latinoamericanas. Documento de Estudios CEDES Vol. 3 N° 2. Buenos Aires; 1980.

Oszlak O, O'Donnell G. Estado y políticas estatales en América Latina.: hacia una estrategia de investigación. Documento CEDES/G. E. CLACSO 4. Buenos Aires; 1976.

Pellegrini FA. Ciencia en pro de la salud. Notas sobre la organización de la actividad científica para el desarrollo de la salud en América Latina y el Caribe. Washington DC: Organización Panamericana de la Salud; Publicación Científica y Técnica; 2000. p. 578.

Pittman P. Allied research: experimenting with structures and processes to increase the use of research in health policy. In: Global Forum for Health Research – final documents [CD-ROM]. Mexico: Global Forum for Health Research; 2004.

Plascencia Castellanos G. Palabra libre. Condición de la Universidad. México: Universidad Iberoamericana; 2006.

Regonini G. El estudio de las políticas públicas. In: Panebianco A, editor. El análisis de la política. Bologna: Il Mulino; 1989.

Reimers F, McGinn N. Informed dialogue: using research to shape education policy around the world. Connecticut: Praeger; 1997.

Sabatier P. The acquisition and utilization of technical information by administrative agencies. Adm Sci Q. 1978;23(3):396–417.

Sabatier P, Jenkins-Smith H, editors. Policy change and learning: an advocacy coalition approach. Boulder: Westview Press; 1993.

Seck D, Phillips LC. Adjusting structural adjustment: the research-policy Nexus: conceptual and historical perspectives. In: Adjusting structural adjustment: best practices in policy research in Africa (Draft Manuscript); 2001.

Sen A. Desarrollo y Libertad. Bogota: Planeta; 2000.

Sotolongo Codina P, Delgado Diaz C. La revolución contemporánea del saber y la complejidad social. Buenos Aires: CLACSO Libros; 2006.

Tonon G. La propuesta teórica de la calidad de vida como escenario facilitador de construcción de redes de investigación. Hologramática – Año. 2010a;VI(7):15–21. Fac. C. Soc. UNLZ.

Tonon G. La utilización de indicadores de calidad de vida para la decisión de políticas públicas. Número 26 *Revista Polis. Universidad Bolivariana.* Santiago de Chile agosto; 2010b.

Tonon G. Los sujetos como protagonistas de las políticas de bienestar: una reflexión desde la calidad de vida y las human capablities. In: Gómez Álvarez D, Ortiz Ortega V, (comp.). El bienestar subjetivo en América Latina. Guadalajara: Universidad de Guadalajara; 2015. p. 75–87.

Torres Carrillo A. Subjetividad y sujeto: perspectivas para abordar lo social y lo educativo. In: Revista Colombiana de Educación N° 50. Primer semestre 2006: Universidad Pedagógica Nacional; 2006. p. 87–103.

Uzochukwu B, Onwujekwe O, Mbachu C, Okwuosa C, Etiaba E, Nyström ME, Gilson L. The challenge of bridging the gap between researchers and policy makers: experiences of a Health Policy Research Group in engaging policy makers to support evidence informed policy making in Nigeria. Glob Health. 2016;2016:12–67.

Veenhoven R. Why social policy needs subjective indicators? In: Casas F, Saurina C, editors. Proceedings of the Third Conference of the ISQOLS: Universidad de Girona; 2000. p. 807–17.

Webber D. Political conditions motivating Legislator's use of policy information. Policy Stud Rev. 1984;4(1):110–8.

Webber DJ. The distribution and use of policy knowledge in the policy process. Knowl Policy. 1991;4(4):6–36.

Weiss C. Research for policy's sake: the enlightenment function of social science research. Policy Anal. 1977;3(4):531–45.

Weiss C. The many meanings of research utilization. Public Adm Rev. 1979;39(5 (sept.–Oct.)):426–31.

Weiss. Measuring the use of evaluation. In: Ciarlo J, editor. Utilizing evaluation. Concepts and measuring techniques. Beverly Hills: Sage; 1981. p. 17–33.

Weiss C. Policy research as advocacy: pro and con. Knowl Policy. 1991;4(1/2):37–56.

Appraisal of Qualitative Studies

Camilla S. Hanson, Angela Ju, and Allison Tong

Contents

1 Introduction	1014
2 Debates and Challenges of Appraising Qualitative Research	1015
3 Conceptualizing Rigor in Qualitative Research	1016
4 Techniques to Enhance Rigor	1017
4.1 Participant Selection and Recruitment	1018
4.2 Data Collection	1018
4.3 Data Analysis	1019
4.4 Interpretations and Conclusions	1020
5 Guidelines for Appraising Qualitative Research	1021
6 Proposed Strategy for the Appraisal of Qualitative Health Research	1024
7 Conclusion and Future Directions	1025
References	1025

Abstract

The appraisal of health research is an essential skill required of readers in order to determine the extent to which the findings may inform evidence-based policy and practice. The appraisal of qualitative research remains highly contentious, and there is a lack of consensus regarding a standard approach to appraising qualitative studies. Different guides and tools are available for the critical appraisal of qualitative research. While these guides propose different criteria for assessment,

C. S. Hanson (✉) · A. Ju
Sydney School of Public Health, The University of Sydney, Sydney, NSW, Australia

Centre for Kidney Research, The Children's Hospital at Westmead, Westmead, NSW, Australia
e-mail: Camilla.hanson@sydney.edu.au; angela.ju@sydney.edu.au

A. Tong
Sydney School of Public Health, The University of Sydney, Sydney, NSW, Australia

Centre for Kidney Research, The Children's Hospital at Westmead, Westmead, Australia
e-mail: Allison.tong@sydney.edu.au

overarching principles of rigor have been widely adopted, and these include credibility, dependability, confirmability, transferability, and reflexivity. This chapter will discuss the importance of appraising qualitative research, the principles and techniques for establishing rigor, and future directions regarding the use of guidelines to appraise qualitative research.

> **Keywords**
> Appraisal · Quality criteria · Qualitative research · Rigor · Trustworthiness

1 Introduction

Qualitative research, like all research, must be read and judged with a critical eye before accepting their findings as trustworthy and relevant to practice. Critical appraisal is, therefore, an essential process in the translation of research into health policy and practice (Hill and Spittlehouse 2003). Significant responsibility falls onto the researchers, who must report their methods in sufficient detail and highlight the limitations that may impact the validity and relevance of their findings (Patton 1999). Health research has a broad audience, including stakeholders with limited expertise in research methods. Standard checklists and tools for evaluating different types of quantitative studies have been widely adopted in order to equip readers with the skills to systematically examine research evidence, for example, the Cochrane collaborations' risk of bias tool for clinical trials (Higgins et al. 2011). Various approaches have been developed to appraise qualitative research, but there is little agreement among researchers as to standard criteria or a gold-standard approach to conducting qualitative research. The diverse approaches used in qualitative studies, for example, phenomenology, ethnography, and grounded theory, are founded on varying theoretical frameworks and philosophical assumptions about the nature of reality and knowledge (Patton 2015). It is unlikely that one standard set of criteria for qualitative research will be developed that will be applicable to the whole range of qualitative approaches (Barbour 2001).

There has been greater demand and interest in qualitative research in health and also notably in medicine. This may reflect the paradigm shift toward patient-centered care, which requires an understanding of the values, goals, beliefs, and attitudes of patients. The British Medical Journal and the Medical Journal of Australia have published educational articles providing criteria for the appraisal of qualitative research (Kitto et al. 2008; Kuper et al. 2008; Mays and Pope 2000). However, the audiences of biomedical journals are largely trained in a positivist tradition of science, which seeks to uncover an objective reality through scientific methods (Meyrick 2006). Qualitative research is interpretive and naturalistic, using distinct methods to explain and describe a phenomenon (Liamputtong 2013; Patton 2015). Readers who are unfamiliar with qualitative research are typically focused on aspects of rigor that are associated with the quantitative methods (Meyrick 2006). Qualitative research is frequently judged by inappropriate standards and criteria (Tong and Dew 2016). With uncertainty as to how to interpret and assess the findings of qualitative studies, it is no surprise that their methods are viewed with some skepticism and even as "second class" by funding agencies, clinicians, and policy makers (Tong et al. 2007). The appraisal of qualitative studies is, therefore, particularly important for qualitative research to be accepted and recognized for its strengths and their findings to produce detailed and nuanced evidence that can inform ways to improve health (Patton 2015). Many of the tools that are currently available require considerable understanding and experience in qualitative research methods and principles (Mays and Pope 2007).

This chapter will discuss the importance of the critical appraisal of qualitative health research, with particular focus on the principles of rigor including credibility, dependability, confirmability, transferability, and reflexivity. Specifically, this chapter will address the debates and challenges that surround the appraisal of qualitative research, discuss techniques to enhance rigor, outline examples of available guidelines for the appraisal of qualitative research, and discuss future directions in the appraisal of qualitative research.

2 Debates and Challenges of Appraising Qualitative Research

There is much contention about appraising qualitative research, which will be briefly summarized in this section. The extreme relativist perspective argues against the use of criteria to judge quality, rejecting the existence of a single truth independent of the researcher and study context (Mays and Pope 2000; Meyrick 2006). Among those who support the appraisal of qualitative research, there is debate as to how quality should be judged. Anti-realists argue that the distinct qualitative research paradigm requires a set of criteria to assess quality that is unique from the quantitative criteria (i.e., validity and reliability) (Mays and Pope 2000; Meyrick 2006). This approach emphasizes that multiple accounts can be produced of a phenomenon because of the influence of the researcher and the methods on the knowledge generated (Dixon-Woods et al. 2004). Others believe that the principles of validity, reliability, and objectivity can be broadly applied to qualitative research, with some adjustment in the criteria to take into account the unique goals and methods of qualitative research (Mays and Pope 2000; Noyes et al. 2008). This is based on the subtle realist argument that an underlying reality, rather than single truth, can be obtained from qualitative research (Mays and Pope 2000). Core concepts of rigor have been developed specifically for qualitative research, with some overlap with the concepts used in the quantitative research paradigm. These are discussed in the following section.

The appraisal of qualitative research is not straightforward and requires an understanding of diverse qualitative principles, methodologies, and methods. Readers are required to judge whether the study design is appropriate to the research question and whether the methods were rigorous such that the findings can be trusted (Giacomini et al. 2000). Unlike quantitative research, the appraisal of qualitative research cannot be reduced to black and white questions, for example, is the sample size sufficient for statistical power? (see ▶ Chap. 59, "Critical Appraisal of Quantitative Research"). Rather, the appraisal requires the reader to make a judgment regarding the appropriateness of the methods and the adequacy of the sample, data collection, and analysis (O'reilly and Parker 2013). Similarly, there is no agreement regarding the gold-standard approach to conducting qualitative research (O'reilly and Parker 2013). For example, interviews are not considered superior to focus groups or field observation. The appropriateness of the methods depends on the research question and the methodological framework guiding the study. The appraisal process is, therefore, necessarily interpretive and can be quite challenging for researchers who are unfamiliar with the qualitative research paradigm and principles of rigor.

3 Conceptualizing Rigor in Qualitative Research

The traditional criteria for rigor in the quantitative paradigm are well known. These include trustworthiness (internal validity), generalizability (external validity), consistency (reliability), and objectivity (Mays and Pope 2007). Different terminology and criteria for rigor been proposed for qualitative research, which have used these same core principles (Mays and Pope 2007). Lincoln and Guba (1986) have proposed four constructs to appraise rigor in qualitative research that align with these principles from the conventional paradigm: credibility, dependability, transferability, and confirmability (see Table 1).

Credibility aligns with the principle of internal validity and addresses whether the findings and judgments made by the researchers can be trusted and the extent to which they provide comprehensive and sensible interpretations of the data (Lincoln and Guba 1986). The question guide used in interviews or focus groups could be examined to determine whether the questions that were asked were appropriate and enabled participants the opportunity to provide in-depth responses relevant to the research question. Authors should provide thick description, which means to describe the data in detail and provide contextual information (Liamputtong 2013). Member checking, which involves obtaining feedback from the participants on the preliminary analysis to ensure that it reflects their perspectives, and investigator triangulation can ensure that the interpretations capture the range and depth of the data that was generated (Liamputtong 2013).

Confirmability parallels the principle of objectivity, such that the authors attempt to demonstrate that the findings and interpretations are linked to the data, and reflects the perspectives of participants (Lincoln and Guba 1986). This can convince the audience that the results and conclusions are not unduly influenced by the biases, assumptions, and experiences of the researcher. Qualitative research is inherently interpretive, but involving multiple investigators, providing quotations that support the findings, and

Table 1 Criteria for rigor in qualitative and quantitative research (Lincoln and Guba 1986; Kitto et al. 2008)

Common principle	Qualitative terminology	Quantitative terminology
Truth/reality	Credibility	Internal validity
Applicability/relevance	Transferability	Generalizability or external validity
Consistency	Dependability	Reliability
Neutrality	Confirmability	Objectivity

having participants check the interpretations can ensure that the findings reflect the full depth and scope of the participants' experiences and perspectives.

Dependability is similar to the criteria of reliability (Lincoln and Guba 1986). While it is not feasible to reproduce a qualitative study, the research process should be logical and transparent, such that the process and procedures can be auditable and traced. The research process and analytical decisions should be recorded (Liamputtong 2013). There should be coherence across the methods and findings. These aspects can be demonstrated by recording and transcribing the data and using software to facilitate coding and to store, manage, and retrieve data.

Generalizability, or external validity, is not directly applicable to qualitative research. Qualitative studies often include a small number of participants who can provide rich information on a topic. The criteria of transferability, which refer to the relevance and potential applicability of findings to other settings and contexts, are applied in qualitative research (Lincoln and Guba 1986). The reader looks for resonance between the study setting and sample and their own context (Kuper et al. 2008) and makes a judgment about whether the findings are relevant to them. This requires researchers to report their study with sufficient detail so that the reader understands the context of the study and the findings. Transferability can also be demonstrated by comparing the results with other health-care contexts and populations or existing theoretical models.

Some additional criteria for rigor have been proposed. Theoretical or conceptual rigor establishes that the research design is appropriate to the research question and the aims of the study (Liamputtong 2013). This requires a clear research question that is articulated in the aims of the study (Kitto et al. 2008). The theoretical approach and methods should be clearly justified in terms of the relevance to the research question (Kitto et al. 2008). Many published qualitative studies do not specify the study design or theoretical perspective that has been used (Dixon-Woods et al. 2004). Of note, some studies may purport to use a specific methodological approach, e.g., phenomenology or grounded theory, but the process by which this was applied may not be aligned with the stated methodology (Barbour 2001). Interpretive rigor assesses the extent to which the researchers produce accurate, trustworthy interpretations of the findings, which are relevant to the intended audience, and produce a depth and breadth of understanding of the phenomenon (Liamputtong 2013). This covers credibility, confirmability, and transferability but also considers the impact of the interpretations that are produced. Some studies may have been rigorously conducted and reported yet provide little insight into the phenomenon being studied (Noyes et al. 2008).

4 Techniques to Enhance Rigor

A number of techniques can be used to enhance the confirmability, transferability, dependability, and credibility of qualitative research (Liamputtong 2013). However, the use of a single technique does not guarantee that a qualitative study is rigorous. In the following section, we have outlined different strategies for improving rigor, focusing on the different components of study design and process including participant selection and recruitment, data collection, data analysis, and the interpretations and conclusions.

4.1 Participant Selection and Recruitment

The appropriateness of the sample should be assessed in reference to the aims of the research study (Kuper et al. 2008). There are a range of strategies for selecting participants. Qualitative research generally seeks to include a broad range of information-rich participants to capture wide variation in the phenomenon and perspectives, and this can be achieved using purposive sampling (Sandelowski 1995; Patton 2015). Snowballing sampling strategy involves asking initial participants to identify other people how may fit the inclusion criteria and be willing to participate in the study. Snowballing is suitable for identifying participants from hidden populations who are difficult to approach directly, for example, drug users (Liamputtong 2007, 2013). Convenience sampling is not recommended for qualitative research as this is likely to lead to a homogenous sample with a single experience or perspective, rather than capturing a range of experiences (Liamputtong 2013). Nevertheless, it is commonly used as it is an efficient, cheap, and easy strategy to select participants. The potential biases and limitations of the sample should be discussed by the researchers.

The selection criteria may change during the analysis process, but the rationale for including participants should be clearly justified (Giacomini et al. 2000). Limitations in the sampling should be discussed, particularly regarding the absence of participants with important characteristics (Kuper et al. 2008; Patton 2015). Qualitative studies will usually not have a predetermined sample size, as the sampling is ceased when a thorough understanding of the phenomenon has been achieved and participants are no longer raising new insights (i.e., data saturation) (Kuper et al. 2008; Liamputtong 2013). Many researches do not justify their decision to stop recruitment (Giacomini et al. 2000). In reality, recruitment may need to cease due to resource constraints and feasibility before saturation has been reached (O'reilly and Parker 2013). This should be discussed in the paper. Alternatively, saturation is often cited, but it is unclear to readers how this was achieved, perhaps stemming from a lack of guidance and clarity among researchers the process of achieving saturation (O'reilly and Parker 2013). A recent study that interviewed first-time fathers' postnatal experiences and support needs in the early postpartum period reported that they reached after the 13th participant and conducted two additional interviews to confirm that they were not identifying new findings (Shorey et al. 2017). The reader should also examine the characteristics of the sample and the breadth and depth of the results to make a judgment as to whether the sample included varying cases and participants.

4.2 Data Collection

Readers should question whether the data were collected appropriately based on the research question (Kuper et al. 2008). For example, the selection of interviews, focus groups, or field observation as the method for collecting data will ultimately depend upon the aims of the study. The choice of methods should be clearly justified in the research report.

Field observation should consider whether the presence of a researcher may influence the behavior of participants rather than observing events as they naturally occur (Giacomini et al. 2000). Individual interviews are suitable for eliciting personal experiences and perspectives, particularly regarding personal and emotional topics. Focus groups are useful for generating data that capitalizes on group interactions and dynamics (Krueger 2014). While the group dynamic in focus groups may empower participants to speak about emotionally sensitive experiences among people with similar experiences, the inappropriate composition of the group can inhibit disclosure (Liamputtong 2011). For example, a focus group study is more suitable than interviews to study the barriers to medication adherence among young people living with HIV, as the rapport among participants may facilitate more open communication regarding this sensitive and stigmatized issue (Rao et al. 2007). Interview and focus group studies should minimize power imbalance between the researcher and participants, for example, where possible, patients should not be interviewed by a clinician involved in their care (Råheim et al. 2016). This should be mentioned in the report.

Triangulation using multiple methods for collecting data can help produce more comprehensive findings and compensate for any disadvantages associated with one particular method (Patton 2015). Different methods can be used to look for patterns of convergence or to help build the interpretations of the phenomenon. Triangulation commonly produces findings that are at odds with each other, due to producing different types of information under different settings (Patton 2015). Inconsistencies provide an opportunity to further understand the phenomenon and the relationship between data collection methods and the findings (Patton 2015). These issues should be explored in the research report.

It should also be demonstrated to readers that the data collection was systematic, organized, and comprehensive enough to produce robust and in-depth descriptions of the phenomenon (Giacomini et al. 2000; Kuper et al. 2008). This cannot be judged by the sample size alone, as a large sample may not extensively study each individual participant and produce findings that are shallower and less nuanced than a study of smaller sample with more in-depth analysis of the interactions with participants (Giacomini et al. 2000). In addition to sample size, readers should look at other indicators including the number of observations or interviews, duration of the data collection, diversity of data collection and data collection techniques, and the number of investigators involved in collecting the data (Giacomini et al. 2000).

4.3 Data Analysis

There are many different frameworks for qualitative analysis, thus the approach chosen should be justified and appropriate to the objectives of the study (Kuper et al. 2008). Unlike statistical analysis which follows formulas and rules and can be performed using statistical analysis software, qualitative researchers are the "instrument" in the analysis, as their interpretations and expertise are required to analyze the data (Patton 2015). It is, therefore, critical that qualitative analysis is performed systematically and clearly described (Patton 2015). Qualitative data collection and analysis therefore also require training and guidance by an experienced researcher (Patton 2015). Important skills that require practice and experience include asking questions, prompting for further information, building rapport with participants, recording detailed field notes, and distinguishing important and relevant data in the analysis (Patton 2015). Software can be used to perform the analysis systematically; however, it does not automatically ensure that a high-quality analysis is achieved (Giacomini et al. 2000).

The goal of qualitative analysis is to develop a conceptual framework that provides in-depth and broad understanding of the phenomenon. Researcher triangulation and respondent validation can also be associated with improved rigor of the analysis. Involving multiple investigators allow for discussion to further develop the analytical framework, ensuring all the data is captured and preventing the biases of a single researcher from unduly influencing the interpretations (Giacomini et al. 2000; Kitto et al. 2008). Respondent validation (or member checking) involves seeking feedback on the interpretations from the participants, ensuring it is meaningful to those who participated (Giacomini et al. 2000; Liamputtong 2013). This has been suggested to be the strongest marker of credibility of the analysis (Lincoln and Guba 1986). Their feedback generates new data that requires further analysis (Mays and Pope 2007). Researchers should also test for alternative explanations and consider negative cases that do not fit with their preliminary conceptualizations of the data, to refine and develop the analysis (Mays and Pope 2000; Patton 2015). Fair dealing is another technique to ensure that the perspectives of the participants are represented fairly. The analysis should capture a range of perspectives and avoid presenting the viewpoint of one group as the majority opinion or experience (Mays and Pope 2000).

4.4 Interpretations and Conclusions

The interpretive rigor of a study should be assessed in terms of whether accurate, trustworthy, relevant, and in-depth understanding of the phenomenon is presented (Liamputtong 2013). Given the direct involvement of the qualitative researcher in analyzing and interpreting the findings, a criterion for rigor in qualitative research involves the demonstration of reflective thought (i.e., reflexivity) to minimize researcher bias and ensure the findings capture participant's perspectives (Giacomini et al. 2000; Kuper et al. 2008; Liamputtong 2013). Reflexivity requires the researcher to acknowledge and address the influence they may have on the data collection and study results and assess the potential sociocultural, political, or ethical influences on the research (Liamputtong 2013). The researcher's background, gender, and professional role may influence their design of the study and shape the data that is collected from their interactions with participants and their interpretations of the data (Kuper et al. 2008; Patton 2015). For example, there may be a power imbalance between the researcher and the participants, an inclination to perform in the presence of an observer or provide desirable responses (Anyan 2013). The researcher's own attitudes and biases may change during data collection (Patton 2015). These changes should be systematically recorded and assessed in the research report. Readers should have sufficient information to make an assessment regarding the likely influence of the researcher on the study.

A number of techniques have been described in this section, which can be used to demonstrate the credibility, transferability, dependability, and confirmability of a qualitative study (see Fig. 1). Importantly, these techniques require detailed description, reporting, and justification to enable readers to make an assessment of the rigor of the study methods and interpretations (see also ▶ Chap. 55, "Reporting of Qualitative Health Research").

5 Guidelines for Appraising Qualitative Research

There is a lack of consensus among experts about the most appropriate criteria for assessing qualitative research and by what indicators these criteria should be assessed (Dixon-Woods et al. 2004). There have been over 100 proposed guidelines for assessing the conduct of qualitative studies, most of which are general criteria that can be applied across various qualitative methods (Dixon-Woods et al. 2004). The criteria used varies considerably across these frameworks; however most do address credibility to some degree (Noyes et al. 2008). Some argue that these criteria should be utilized as guides rather than rigid criteria presented as standard requirements for quality research (Kenwood and Pigeon 1992).

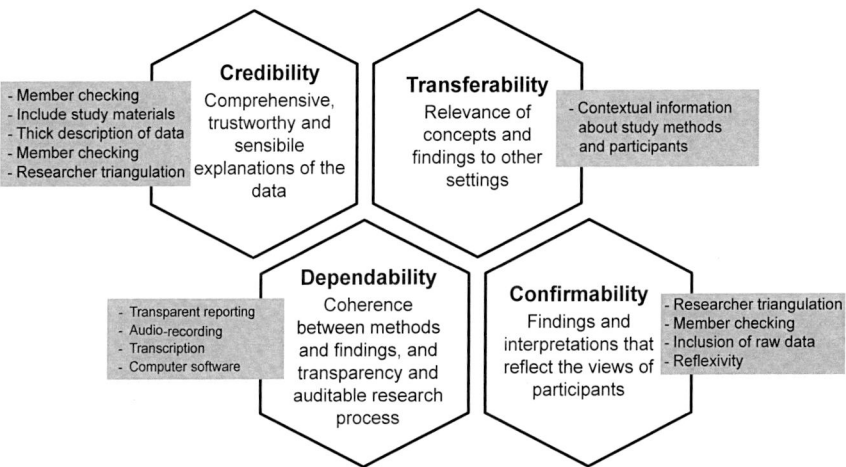

Fig. 1 Techniques to enhance the credibility, dependability, confirmability, and transferability of a study

Some guidelines provide a more structured procedure for evaluating rigor. Guidelines such as the Critical Appraisal Skills Programme (CASP 2017) and Spencer and colleagues' framework (Spencer et al. 2003) are widely used. However, neither of these tools proposes to be a strict set of standards or rules for qualitative research but rather provides an aid to inform judgments about the quality of a qualitative study. The CASP tool has been widely used in the synthesis of qualitative studies, to inform decisions about the exclusion of studies from a review (see ▶ Chap. 45, "Meta-synthesis of Qualitative Research"). It is a checklist consisting of ten questions that broadly cover methodological rigor, credibility, and relevance. Each item provides suggestions to consider when making an assessment. Despite offering a relatively structured process, agreement between researchers is not necessarily strong as the questions require subjective judgments (Dixon-Woods et al. 2007). An example of an appraisal item taken from CASP is shown in Table 2.

Spencer and colleague's framework (Spencer et al. 2003) has been described as one of the most comprehensive frameworks available (Pope and Mays 2006). The framework is underpinned by four central principles: contribution in advancing knowledge, the defensibility of the design, rigorous conduct, and credibility of the claims of the study. These principles were used to identify 18 questions that cover the study findings, design, sample, data collection, analysis, reporting, reflexivity, neutrality, ethics, and auditability. An advantage of this framework is the inclusion of comprehensive indicators to guide the readers in answering each question. This framework is also more general than other tools, including considerations for various types of data (e.g., document analysis) and methods (e.g., field observation). An example of the appraisal items related to sampling taken from this framework is shown in Table 3.

Kitto and colleagues (2008) developed criteria for authors and assessors of research articles submitted to the Medical Journal of Australia. These criteria cover domains including clarification and justification of the aims and methods, procedural rigor, representativeness of the sample, interpretive rigor, reflexivity and evaluative rigor, and transferability. Some of the items are straightforward as they are related to the reporting of aspects of the study design.

Many qualitative researchers have proposed general questions to guide the appraisal of qualitative studies (Popay et al. 1998; Mays and Pope 2000; Kuper

Table 2 CASP appraisal criteria related to the sampling (CASP 2017)

Item			Considerations
Was the recruitment strategy appropriate to the aims of the research?			If the researcher has explained how the participants were selected
☐ Yes	☐ Can't tell	☐ No	If they explained why the participants they selected were the most appropriate to provide access to the type of knowledge sought by the study
			If there are any discussions around recruitment (e.g., why some people chose not to take part)

Table 3 Spencer and colleague's appraisal criteria related to sampling (Spencer et al. 2003)

Appraisal questions	Quality indicators to consider
How well defended is the sample design/target selection of cases/documents?	Discussion of study locations/areas and how and why they were chosen
	Description of population of interest and how sample selection related to it (e.g., typical, extreme case, diverse constituencies, etc.)
	Rationale for basis of selection of target sample/settings/documents (e.g., characteristics/features of target sample/settings/documents, basis for inclusions and exclusions, discussion of sample size/number of cases/setting selected, etc.)
	Discussion of how sample/selections allowed required comparisons to be made
Sample composition/case inclusion – how well is the eventual coverage described?	Detailed profile of achieved sample/case coverage
	Maximum inclusion (e.g., language matching or translation, specialized recruitment, organized transport for group attendance)
	Discussion of any missing coverage in achieved samples/cases and implications for study evidence (e.g., through comparison of target and achieved samples, comparison with population, etc.)
	Documentation of reasons for nonparticipation among sample approached/noninclusion of selected cases/documents
	Discussion of access and methods of approach and how these might have affected participation/coverage

et al. 2008). Dixon-Woods et al. (2004) provide a small number of generic prompts to guide the appraisal of qualitative research, which they emphasize are not prescriptive criteria. These include items relating to both the conduct and reporting of the study (see Table 4). They intend for these prompts to be supplemented by items specific to different methods and theoretical frameworks.

Across all these guides for the appraisal of qualitative research, a large component of this process of assessing study conduct relies largely on subjective assessments. This is particularly evident in items related to judgments of the appropriateness of the study methods, the adequacy of the sample size, and the quality of the insights and interpretations produced. This can be particularly challenging for researchers who have limited experience in a range of qualitative methods and principles. Available tools typically focus on the technical aspects of a study (i.e., methodological rigor). Some have also recommended that an appraisal of theoretical rigor be conducted (Noyes et al. 2008). This would involve an assessment of the congruence between the theoretical framework, methodology, and methods chosen.

There are also guidelines available for the appraisal of the transparency of reporting of qualitative studies (Tong et al. 2007), which can inform judgments about the overall study conduct. Systematic reviews synthesizing the findings from

Table 4 Adapted from Dixon-Woods prompts for the appraisal of qualitative research (Dixon-Woods et al. 2004)

Topics	Items
Research question	Are the research questions clear?
	Are the research questions suited to qualitative inquiry?
Methods	Are the following clearly described? (sampling, data collection, analysis)
	Are the following appropriate to the research question? (sampling, data collection, analysis)
Claims	Are the claims made supported by sufficient evidence?
Overall	Are the data, interpretations, and conclusions clearly integrated?
	Does the paper make a useful contribution?

qualitative research frequently show that aspects of the study are infrequently reported (Tong et al. 2007; see also ▶ Chap. 55, "Reporting of Qualitative Health Research").

6 Proposed Strategy for the Appraisal of Qualitative Health Research

Given the vast differences in the appraisal guidelines that are available, and their generic nature, researchers should not rely on a single guideline or tool to appraise qualitative studies (Pope and Mays 2006). It is recommended that researchers read about rigor in the context of the relevant methods and methodological approach. For less experienced qualitative researchers, we suggest using guidelines that provide sufficient guidance regarding indicators to examine the criteria, for example, the CASP tool. It is recommended that the appraisal process is completed with the guidance of supervisor with experience in qualitative research methods and appraisal. Additional criteria could be added to available tools to ensure that components are assessed that are relevant to the methods of the chosen study (Dixon-Woods et al. 2004).

Some qualitative researchers have argued that the proliferation of standard criteria for quality in qualitative research has had a negative impact on study conduct and reporting. In particular, they attribute the prominent use of techniques such as "saturation," "purposive sampling," "respondent validation," and "researcher triangulation" to the attempts of researchers to conform to perceived "standards" of qualitative research, without actually achieving or explaining these processes in their study (Barbour 2001; Dixon-Woods et al. 2004; O'reilly and Parker 2013). For example, Barbour (2001) suggests that grounded theory methodology is frequently used as "an approving bumper sticker," while the data collection, analysis, and interpretations do not adhere to the assumptions and traditions that underpin this methodological approach. These issues are also symptomatic of the general problem with the limited transparency in the reporting of qualitative research methods, particularly the interpretive process.

Future work is needed to develop appraisal criteria that are specific to different methods and theoretical approaches in qualitative research (Dixon-Woods et al. 2004). This may help develop agreement on the "fatal flaws" in terms of limitations of a study that are critical or render one study better quality than another (Dixon-Woods et al. 2004; Noyes et al. 2008). This is particularly challenging in the context of appraising qualitative studies for inclusion in a systematic review. Currently, there is little guidance to assist with these decisions. The approach to appraising rigor using overarching principles rather than relying on the use of specific techniques, like the use of software, is therefore widely encouraged. It is therefore important for those appraising qualitative research to have a thorough understanding of the principles of rigor, in addition to using a guide or tool to assess quality in qualitative research.

7   Conclusion and Future Directions

The critical appraisal of qualitative research is essential to identify limitations in qualitative evidence and prevent readers from inappropriately transferring and applying their findings to decision-making, health care, and policy. Transparency in the reporting of qualitative research aims and methods is fundamental to enable readers to appraise various aspects of a qualitative study including the appropriateness of the methods and interpretations. Lincoln and Guba's principles of credibility, transferability, dependability, and confirmability are useful for conceptualizing rigor across qualitative research. Various techniques can be used to achieve these principles of rigor including saturation, member checking, triangulation, and the use of software, depending on the methodological approach and research question. It is unlikely that a standard set of criteria will ever be developed that is indicative of quality across all types of qualitative research. As we have discussed above, further work is needed to develop additional criteria that are specific to different methods and theoretical frameworks in qualitative research. There is a need to improve understanding of the core principles of rigor and to encourage the use of appraisal tools to assist with the critical appraisal of qualitative research. Increased understanding and skills in the critical appraisal of qualitative research are required to ensure the acceptance of qualitative studies and to ensure their impact on health care.

References

Anyan F. The influence of power shifts in data collection and analysis stages: a focus on qualitative research interview. Qual Rep. 2013;18(18):1.

Barbour RS. Checklists for improving rigour in qualitative research: a case of the tail wagging the dog? BMJ: Br Med J. 2001;322(7294):1115.

CASP. CASP qualitative research checklist. http://docs.wixstatic.com/ugd/dded87_25658615020 e427da194a325e7773d42.pdf. Critical Appraisal Skills Programme (2017).

Dixon-Woods M, Shaw RL, Agarwal S, Smith JA. The problem of appraising qualitative research. Qual Saf Health Care. 2004;13(3):223–5.

Dixon-Woods M, Sutton A, Shaw R, Miller T, Smith J, Young B, ... Jones D. Appraising qualitative research for inclusion in systematic reviews: a quantitative and qualitative comparison of three methods. J Health Serv Res Policy. 2007;12(1):42–7.

Giacomini MK, Cook DJ, Group, E.-B. M. W. Users' guides to the medical literature: XXIII. Qualitative research in health care A. Are the results of the study valid? JAMA. 2000;284(3):357–62.

Higgins JP, Altman DG, Gøtzsche PC, Jüni P, Moher D, Oxman AD, ... Sterne JA. The Cochrane collaboration's tool for assessing risk of bias in randomised trials. BMJ. 2011;343:d5928.

Hill A, Spittlehouse C. Evidence based medicine-what is critical appraisal. Newmarket: Heyward Med Commun; 2003.

Kenwood K, Pigeon N. Qualitative research and psychological theorising. Br J Psychol. 1992;83(1):97–112.

Kitto SC, Chesters J, Grbich C. Quality in qualitative research. Med J Aust. 2008;188(4):243.

Krueger RA. Focus groups: a practical guide for applied research. Singapore: Sage; 2014.

Kuper A, Lingard L, Levinson W. Critically appraising qualitative research. BMJ. 2008;337: a1035–a1035.

Liamputtong P. Researching the vulnerable: a guide to sensitive research methods. London: Sage; 2007.

Liamputtong, P. Focus group methodology: Principle and practice. Sage Publications. 2011.

Liamputtong P. Qualitative research methods. 4th ed. Melbourne: Oxford University Press; 2013.

Lincoln YS, Guba EG. But is it rigorous? Trustworthiness and authenticity in naturalistic evaluation. N Dir Eval. 1986;1986(30):73–84.

Mays N, Pope C. Assessing quality in qualitative research. BMJ. 2000;320(7226):50.

Mays N, Pope C. Quality in qualitative health research. In: Qualitative research in health care. 3rd ed. Blackwell Publishing Ltd, Oxford, UK. 2007. p. 82–101.

Meyrick J. What is good qualitative research? A first step towards a comprehensive approach to judging rigour/quality. J Health Psychol. 2006;11(5):799–808.

Noyes J, Popay J, Pearson A, Hannes K. 20 qualitative research and Cochrane reviews. Jackie Chandler (ed.) In: Cochrane handbook for systematic reviews of interventions. United Kingdom: The Cochrane Collaboration. 2008. p. 571.

O'reilly M, Parker N. 'Unsatisfactory saturation': a critical exploration of the notion of saturated sample sizes in qualitative research. Qual Res. 2013;13(2):190–7.

Patton MQ. Enhancing the quality and credibility of qualitative analysis. Health Serv Res. 1999;34(5 Pt 2):1189.

Patton MQ. Qualitative research and evaluation methods. 4th ed. Thousand Oaks: Sage; 2015.

Popay J, Rogers A, Williams G. Rationale and standards for the systematic review of qualitative literature in health services research. Qual Health Res. 1998;8(3):341–51.

Pope C, Mays N. Qualitative research in health care. 3rd ed. Malden: Blackwell; 2006.

Rao D, Kekwaletswe T, Hosek S, Martinez J, Rodriguez F. Stigma and social barriers to medication adherence with urban youth living with HIV. AIDS Care. 2007;19(1):28–33.

Råheim M, Magnussen LH, Sekse RJT, Lunde Å, Jacobsen T, Blystad A. Researcher–researched relationship in qualitative research: Shifts in positions and researcher vulnerability. International journal of qualitative studies on health and well-being. 2016;11(1):30996.

Sandelowski M. Sample size in qualitative research. Res Nurs Health. 1995;18(2):179–83.

Shorey S, Dennis CL, Bridge S, Chong YS, Holroyd E, He HG. First-time fathers' postnatal experiences and support needs: a descriptive qualitative study. J Adv Nurs. 2017;73:2987–96.

Spencer, L, Ritchie J, Lewis J, and Dillon L. Quality in Qualitative Evaluatoin: A framework for assessing research evidence, Government Chief Social Researcher's Office, London: Cabinet Office. 2003.

Tong A, Dew MA. Qualitative research in transplantation: ensuring relevance and rigor. Transplantation. 2016;100(4):710–2.

Tong A, Sainsbury P, Craig J. Consolidated criteria for reporting qualitative research (COREQ): a 32-item checklist for interviews and focus groups. Int J Qual Health C. 2007;19(6):349–57. https://doi.org/10.1093/intqhc/mzm042.

Critical Appraisal of Quantitative Research 59

Rocco Cavaleri, Sameer Bhole, and Amit Arora

Contents

1 Introduction	1028
2 Critically Appraising Systematic Reviews	1029
2.1 Tools for Appraising Systematic Reviews	1030
2.2 Important Questions to Ask when Appraising a Systematic Review	1030
2.3 Completeness and Quality of Reporting in Systematic Reviews	1034
3 Critically Appraising Experimental Studies (RCTs and Non-RCTs)	1035
3.1 Tools for Appraising Experimental Studies	1035
3.2 Important Questions to Ask when Appraising an Experimental Study	1036
3.3 Completeness and Quality of Reporting in Experimental Studies	1040

R. Cavaleri (✉)
School of Science and Health, Western Sydney University, Campbelltown, NSW, Australia
e-mail: R.Cavaleri@westernsydney.edu.au

S. Bhole
Sydney Dental School, Faculty of Medicine and Health, The University of Sydney, Surry Hills, NSW, Australia

Oral Health Services, Sydney Local Health District and Sydney Dental Hospital, NSW Health, Surry Hills, NSW, Australia
e-mail: Sameer.Bhole@health.nsw.gov.au

A. Arora
School of Science and Health, Western Sydney University, Sydney, NSW, Australia

Discipline of Paediatrics and Child Health, Sydney Medical School, Sydney, NSW, Australia

Oral Health Services, Sydney Local Health District and Sydney Dental Hospital, NSW Health, Sydney, NSW, Australia

COHORTE Research Group, Ingham Institute of Applied Medical Research, Liverpool, NSW, Australia
e-mail: a.arora@westernsydney.edu.au

© Springer Nature Singapore Pte Ltd. 2019
P. Liamputtong (ed.), *Handbook of Research Methods in Health Social Sciences*,
https://doi.org/10.1007/978-981-10-5251-4_120

4	Critically Appraising Observational Studies (Cohort, Case-control, and Cross-Sectional Studies)	1040
	4.1 Tools for Appraising Observational Studies	1041
	4.2 Important Questions to Ask when Appraising an Observational Study	1043
	4.3 Completeness and Quality of Reporting in Observational Studies	1046
5	Conclusion and Future Directions	1046
References		1047

Abstract

Critical appraisal skills are important for anyone wishing to make informed decisions or improve the quality of healthcare delivery. A good critical appraisal provides information regarding the believability and usefulness of a particular study. However, the appraisal process is often overlooked, and critically appraising quantitative research can be daunting for both researchers and clinicians. This chapter introduces the concept of critical appraisal and highlights its importance in evidence-based practice. Readers are then introduced to the most common quantitative study designs and key questions to ask when appraising each type of study. These studies include systematic reviews, experimental studies (randomized controlled trials and non-randomized controlled trials), and observational studies (cohort, case-control, and cross-sectional studies). This chapter also provides the tools most commonly used to appraise the methodological and reporting quality of quantitative studies. Overall, this chapter serves as a step-by-step guide to appraising quantitative research in healthcare settings.

Keywords

Critical appraisal · Quantitative research · Methodological quality · Reporting quality

1 Introduction

Critical appraisal describes the process of analyzing a study in a rigorous and methodical way. Often, this process involves working through a series of questions to assess the "quality" of a study by examining its strengths and limitations. A good critical appraisal should also explore whether or not a study's findings can be applied to your own patient or clinical context. In this chapter, we will explore the critical appraisal process and provide you with the foundation required to start appraising quantitative research.

Critical appraisal skills are important for anyone wishing to make informed decisions or improve the quality of healthcare delivery. Not all studies are carried out using rigorous methods. Studies may be influenced by many types of bias, irrespective of their level of evidence. Just because something has been published does not mean that you can automatically trust its findings. Further, even if you can trust the study's findings, it does not necessarily mean that they will be useful in your particular clinical situation. Critical appraisal skills are required to determine whether or not the study was well-conducted and if its findings are believable or useful.

Whenever a study is completed, there are three likely explanations for its findings (Mhaskar et al. 2009):

1. The study findings are correct and its conclusions are true. This is the ideal scenario but is not always the case.
2. The findings were due to random variation (chance). In any study, there is some degree of uncertainty, and we can never be absolutely sure that the results were not due to chance. Studies investigating larger groups of people are more likely to produce results that accurately reflect "reality," rather than random variation. To conceptualize this, imagine you have thrown a dart toward a target and immediately hit the bullseye. It is difficult to determine whether or not this reflects your true skill level or was simply a "fluke." If we were to record your performance following ten attempts, the average score across these attempts would give us a better idea of your overall skill level. There would be some random variation in these attempts, with you occasionally hitting above or below the bullseye. Increasing our number of recordings would give us an even more accurate reflection of your typical performance. Similarly, increasing the sample size of a study can "wash out" the influence of random variation.
3. The findings were affected by systematic error (bias). If the tail of your dart was bent, it may lead to you hitting well below the bullseye with every throw. In this case, we would be able to predict the direction of the result (above or below the bullseye) every time. This predictable, nonrandom error is known as bias. To an onlooker, a biased dart may make you appear to be a much better or worse dart player than you actually are. Increasing our number of recordings would not necessarily give us a better reflection of your true skill level. Likewise, increasing a study's sample size may not eliminate systematic errors.

Deviations from "reality" can be caused by a number of problems throughout the design and execution of a study. Bias can obscure up to 60% of the real effect of a healthcare intervention (Mhaskar et al. 2009). Increasing evidence indicates that "biased results from poorly designed and reported trials can mislead decision-making in healthcare at all levels" (Moher et al. 2001, p. 29). Media attention and false credibility can further increase the influence of poorly conducted studies. Critical appraisal skills are essential when it comes to identifying biases and making informed decisions. In the following sections, we are going to step you through the process of critically appraising common quantitative study designs. We will provide you with tools that can be used to assist with this process, as well as examples of important questions to ask when appraising each type of study.

2 Critically Appraising Systematic Reviews

Systematic reviews occupy the highest level on the Australian National Health and Medical Research Council (NHMRC) hierarchy of evidence (National Health and Medical Research Council 2009). These studies collect and analyze multiple studies in order to provide a complete, exhaustive summary of the literature currently available to answer a particular question. This literature should be obtained through transparent and reproducible database searches, based on predefined criteria (see also ▶ Chap. 46, "Conducting a Systematic Review: A Practical Guide"). However, as stated in the previous section, not all studies are conducted using rigorous methods. Although systematic reviews are considered to be a high level of evidence, they are not immune to biases and cannot be trusted by default. As with any study, there are three primary elements to consider when appraising a systematic review: the validity of the methodology, the size and precision of the effects, and the applicability of the findings to your specific client or population of interest (Hoffmann et al. 2013).

2.1 Tools for Appraising Systematic Reviews

There are a number of tools that can be used to assist with critically appraising systematic reviews. For systematic reviews of intervention studies, commonly employed appraisal tools include:

- A measurement tool to assess systematic reviews (AMSTAR 2) (Shea et al. 2017)
- The Critical Appraisal Skills Program (CASP) checklist for systematic reviews (Critical Appraisal Skills Program 2017)
- The Joanna Briggs Institute's critical appraisal tool for systematic reviews (Joanna Briggs Institute 2017)
- The Centre for Evidence-Based Medicine's systematic review appraisal tool (Centre for Evidence-based Medicine 2017)
- The criteria proposed by Greenhalgh and Donald (2000)

All of these tools provide checklists and guides designed to assist in critically appraising the methodological quality of systematic reviews and the extent to which they have addressed potential biases.

2.2 Important Questions to Ask when Appraising a Systematic Review

Table 1 presents a list of key questions to ask when appraising the methodological quality of a systematic review. These questions have been adapted from the AMSTAR 2 (Shea et al. 2017), CASP (Shea et al. 2017), and Greenhalgh and Donald (2000) checklists. In this section, we will explore each of these questions and provide examples to facilitate their application and interpretation. Note that the last two questions do not relate to methodological quality (which should always be considered first), and the results of a study should not influence your decision regarding the "best available" evidence. However, we have included these questions here in order to provide you with a holistic approach toward appraisal that will be most useful in clinical contexts.

Table 1 Important questions to ask when appraising a systematic review

1. Did the review address a clearly focused question?
2. Did the review have clearly defined eligibility criteria and select the most appropriate studies?
3. Does the review have a comprehensive literature search strategy?
4. Did the review include a risk of bias assessment?
5. Did the review perform a meta-analysis, and was it appropriate to do so?
6. What are the results of the review and how precise are they?
7. How relevant are the results to my patient or problem?

1. Did the Review Address a Clearly Focused Question?

Systematic reviews should clearly and concisely spell out their research question. A good research question will highlight the population of interest (P), the proposed intervention/issue (I), any comparison groups (C), as well as the outcomes being investigated (O). An example research question using this PICO format would be, "Is hand therapy (I) more effective than corticosteroid injections (C) in reducing pain levels (O) in people with de Quervain's disease (a repetitive strain injury in the hand) (P)" (Cavaleri et al. 2016). Including timeframes, such as "over a 6-month period," is also useful (see also ▶ Chap. 46, "Conducting a Systematic Review: A Practical Guide").

2. Did the Review Have Clearly Defined Eligibility Criteria and Select the Most Appropriate Studies?

Reviews should include studies that address their research question and are of a robust design. Randomized controlled trials are usually most appropriate for evaluating interventions but may not always be appropriate for ethical reasons (see also ▶ Chap. 37, "Randomized Controlled Trials"). For example, it would not be ethical to conduct a randomized controlled trial investigating the effects of a substance known to cause harm (e.g., illicit drugs). In these cases, authors should explain why the study designs they have included are most suitable. To answer "yes" to this question, systematic reviews should also elaborate on their PICO in a section that discusses their inclusion and exclusion criteria. This section should highlight study designs, populations, interventions, comparisons, and outcomes that were eligible to be included in the review and those that were not. To minimize bias, more than one independent reviewer should be involved in selecting studies based on these criteria (Critical Appraisal Skills Program 2017; Joanna Briggs Institute 2017).

3. Does the Review Have a Comprehensive Literature Search Strategy?

This question helps us determine whether or not important articles may have been missed during the search for relevant literature. A good search strategy must be reproducible. To answer "yes" to this question, a systematic review should:

- Search at least two databases that are relevant to the research question.
- Provide a search strategy that includes synonyms for all search terms.
- Justify any publication restrictions, such as language.
- Search the reference lists of included studies. This is called "pearling" and is useful for finding articles that may have been missed by database searches (Hoffmann et al. 2013).

Ideally, systematic reviews should also search for unpublished papers ("gray literature") to avoid publication bias. Publication bias occurs when journals selectively publish articles with significant findings. As studies with positive results are also more likely to be published in English, pooling data from English studies alone may lead to an overestimation of treatment effect (Higgins and Green 2009). Including non-English studies is, therefore, also beneficial in highlighting the true effect of a particular intervention.

4. Did the Review Include a Risk of Bias Assessment?

As with a good search strategy, a risk of bias assessment should be reproducible. Risk of bias assessments may incorporate the appraisal tools presented throughout this chapter. The risk of bias section is important in determining whether the articles included in the systematic review were of high quality. Note that you can answer "yes" to this question even if the studies included in the review were poorly designed, as long as the review has identified that this was the case. A systematic review can still have high methodological quality even if the only literature available for review was poor (i.e., we are critiquing the review itself and not necessarily the studies it found).

When reviewing randomized controlled trials, factors such as random allocation and blinding should be considered (see also ▶ Chap. 37, "Randomized Controlled Trials"). For all study designs, reproducible methods for assessing biases and confounding factors must be provided. A confounding factor refers to anything that may "mask" or confuse the true results of a study (Shea et al. 2017). For example, if you were investigating the effect of television viewing and eye damage, the association between these variables could be distorted if one group has a greater number of elderly individuals than the other. Television may appear to cause eye damage, but the confounding effect of age (which could also influence vision) makes it difficult to come to a definitive conclusion. More than one reviewer should perform risk of bias assessments according to a set of prespecified criteria (Critical Appraisal Skills Program 2017; Joanna Briggs Institute 2017). Situations where reviewers disagree should be resolved through discussion or consultation with an additional reviewer. Use of the Grading of Recommendations, Assessment, Development, and Evaluations (GRADE) guideline is now encouraged (Guyatt et al. 2011). This is a method for summarizing the overall quality of the reviewed evidence, with ratings of "high," "moderate," or "low" being used to indicate how confident we can be in the findings of the included studies.

5. Did the Review Perform a Meta-analysis, and Was It Appropriate to Do so?

To answer "yes" to this question, systematic reviews should justify combining (or not combining) the data from different studies in the form of a meta-analysis. Heterogeneity, or the differences between studies, should be examined and potential sources of heterogeneity investigated and discussed. In systematic reviews, heterogeneity may be tested statistically using the I^2 statistic. Ideally, reviews should also examine how overall results vary with the inclusion or exclusion of studies that were judged to be at a high risk of bias (Hoffmann et al. 2013).

6. What Are the Results of the Review and How Precise Are They?

While the results of a study should not influence our decision regarding its methodological quality, an understanding of a systematic review's results section is important when deciding to implement a particular intervention based on the best available evidence. Let us consider the forest plot and table below (see Fig. 1).

The first column from the left lists the studies (A, B, and C) that are included in the forest plot. The second and third columns present the number of people who recovered, as well as the total number of people included, in both the intervention and control groups for each study. Next is the risk ratio, which is a ratio describing the probability of an event (recovery) occurring in the intervention group versus the control group. Other ratios or mean differences between groups can also be presented in forest plots. While these values are useful in describing what happened in the actual study, they do not provide the whole picture in terms of the range of values that can be expected beyond the study setting. For this, we provide the 95% confidence interval (CI) in brackets next to each risk ratio. The confidence intervals tell us the range of results we could expect to see in the population, in this case with a 95% degree of certainty (precision). The weight column tells us how much "pull" or influence the study has over the combined results and is largely determined by its sample size.

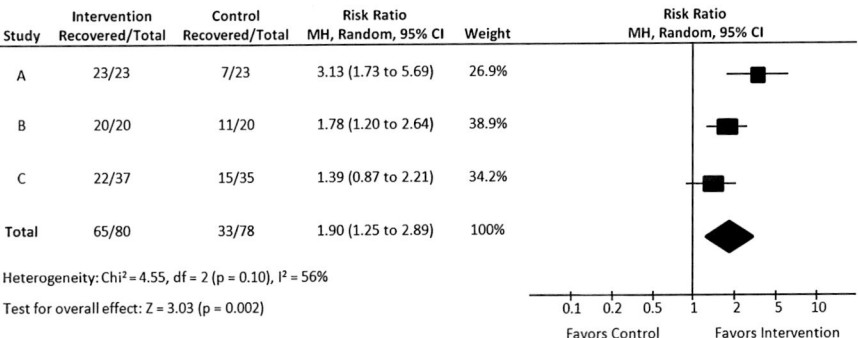

Fig. 1 Example forest plot

In the forest plot itself, each black square represents the intervention effect from the individual studies (A, B, and C). The horizontal lines either side of the black squares represent the 95% confidence intervals surrounding these results. As this plot is presenting ratio data, a value of 1 would indicate "no effect," represented by the vertical line in the middle of the plot. Results to the right of this line favor the intervention, and results to the left of this line favor the control group. The diamond at the bottom of the plot is the summary statistic, which is a value that represents the overall effect when all of the study results are combined. The center of the diamond indicates the risk ratio (or other study effect result), and the width of the diamond represents the overall confidence interval. In this plot, we can see that the overall results favor the intervention. Being able to interpret such data is important when making clinical decisions.

7. How Relevant Are the Results to My Patient or Problem?

Finally, you should consider whether the findings of the review are applicable to your patient or population of interest. Is the study population similar to your particular case in terms of age, diagnosis, chronicity, etc.? Do you have access to the proposed interventions? How do the results fit in with the patient's values and your own clinical experience? Another factor to consider is whether the results are strong enough to justify implementing the intervention. A good method of assessing this is by discussing a minimally important difference (MID) with your patient. The MID is the smallest improvement that your patient would consider worthwhile. If the systematic review indicates that, in the population, we can expect better improvements than that desired by your patient, we can say that the intervention is worth pursuing ("clinically significant"). If not, exploring other options may be worthwhile.

2.3 Completeness and Quality of Reporting in Systematic Reviews

There is an important distinction between methodological quality and reporting quality. Methodological quality is concerned with how well a study was designed and conducted (e.g., literature search, eligibility criteria, meta-analyses), while reporting quality describes how well these processes were described in the paper itself. Although a systematic review may have been conducted in a rigorous manner, it is not useful to the reader unless its methods and findings are sufficiently reported. There is a link between methodological quality and reporting quality in that research processes cannot be reported if they were not appropriately implemented in the first place. However, reporting quality also encompasses parts of the paper beyond the methods section. Thorough appraisals should, therefore, examine both methodological and reporting quality. Guidelines for assessing the reporting quality of systematic reviews have been developed to improve clarity and reduce research "waste" (Moher et al. 2009). The most popular tool used to assess reporting quality in systematic reviews is the preferred reporting items for systematic reviews and meta-analyses (PRISMA) statement (Moher et al. 2009). The PRISMA provides a minimum set of items for reporting in systematic reviews, particularly those evaluating the effects of interventions. These items cover the title, abstract, introduction, methods, results, and discussion sections of systematic reviews. The meta-analysis of observational studies in epidemiology (MOOSE) group also offers guidelines for assessing the reporting quality of systematic reviews based on observational studies (Stroup et al. 2000).

3 Critically Appraising Experimental Studies (RCTs and Non-RCTs)

Experimental studies involve actively altering, rather than just observing, exposures or interventions of interest and assessing their impact over time. That is, in experimental studies, the independent variable is under the control of the researcher. Experimental studies are less susceptible to confounding and bias than other study designs because researchers can take measures to control certain variables (Dawes et al. 2005). However, experimental studies are by no means perfect, and strong critical appraisals are required before you decide to implement their findings.

3.1 Tools for Appraising Experimental Studies

There is a wide array of tools available to assist with critically appraising experimental studies. Some of the most commonly used appraisal tools include:

Randomized controlled trials

– The Physiotherapy Evidence Database (PEDro) scale (2017)
– The CASP RCT appraisal tool (Critical Appraisal Skills Program 2017)
– The Joanna Briggs Institute's checklist for RCTs (Joanna Briggs Institute 2017)
– The Jadad scale (Clark et al. 1999)
– The Centre for Evidence-Based Medicine's RCT appraisal tool (Centre for Evidence-based Medicine 2017)

Non-randomized (quasi-randomized) controlled trials

– The Joanna Briggs Institute's appraisal checklist for non-randomized controlled trials (Joanna Briggs Institute 2017)
– The Cochrane risk of bias in non-randomized studies of interventions (ROBINS-I) tool (Sterne et al. 2016)

All of these tools assess the extent to which experimental studies are internally valid. External validity, or applicability to "real-world" settings, is also covered by some of these tools.

3.2 Important Questions to Ask when Appraising an Experimental Study

There are eight questions that are commonly asked when appraising the potential risk of bias in experimental studies. These questions are summarized in Table 2 and are based upon criteria presented in the PEDro (2017) and CASP (2017) RCT appraisal tools. As always, it is also important to consider the study in the context of your specific patient or population of interest.

1. **Was There a Control Group?**

To appreciate the true effect of an intervention, comparison to a control group is required. A control group is a group of participants who are as similar as possible to the intervention group, except that they do not receive the intervention or exposure being investigated (Portney and Watkins 2009). Participants in the control group may instead be given no intervention or a placebo intervention. If participants are followed over time with no comparison to a control group, it is impossible to tell how much of the observed change is due to the effect of the intervention itself and how much is due to some other explanation (e.g., the placebo effect). Further, while participants in an uncontrolled study may improve following the intervention, there is nothing to say that they would not still have improved without it. Comparison to a control group is required to detect whether the intervention causes effects that would not otherwise be expected over time (Portney and Watkins 2009).

2. **Was the Allocation of Participants to Groups Randomized?**

Randomized controlled trials and non-randomized controlled trials differ in one very important regard – randomization. This process ensures that participants have an equal chance of being allocated to any group, meaning that potential differences

Table 2 Important questions to ask when appraising an experimental study

1. Was there a control group?
2. Was the allocation of participants to groups randomized?
3. Was the group allocation sequence concealed?
4. Were the groups similar at baseline?
5. Were the participants and study personnel blinded?
6. Was there adequate follow-up of participants and a sufficient completion rate?
7. Was an intention-to-treat analysis performed?
8. Were between-group statistical comparisons performed?

between groups should be minimal and due to chance only (Altman and Bland 1999; see also ▶ Chap. 37, "Randomized Controlled Trials"). Studies that do not employ randomization are at risk of having systematic differences between groups. In such cases, we may not be able to determine if it is the intervention, or some other factor, that is influencing the results. For example, if we were to allocate every male to the intervention group, and every female to the control group, we would not be able to determine if differences between the groups were actually due to the intervention or simply due to gender differences. Randomly allocating people to groups ensures that participant characteristics (including gender, age, weight, height, occupation, recreational activities, belief in the effect of the intervention, and so on) are evenly distributed throughout the intervention and control groups (Hoffmann et al. 2013). This means that any differences between groups at the end of the study are most likely a result of the intervention and not some other factor.

Random allocation can be done by using a random number generator, flipping a coin, or pulling names out of a hat. The simplest randomization process involves randomly allocating individuals or clusters of individuals (such as clinics, schools, or teams) to each group. Randomization can also be stratified, meaning that participants are matched as well as randomly allocated to groups. This approach ensures that confounding factors are balanced between groups. For example, when investigating the influence of a training intervention on athletic performance, participants could be stratified according to their skill level, belonging to either "beginner," "intermediate," or "advanced" categories. When randomization occurs, the researchers then make sure that there are equal numbers of people from each category in the intervention and control groups (Portney and Watkins 2009).

3. **Was the Group Allocation Sequence Concealed?**

Group allocation describes the process of assigning participants to either the intervention or control group. Adhering to the allocation sequence is important when it comes to maintaining the benefits of randomization (Altman and Bland 1999). If researchers know the group to which a person will be allocated (that is, the sequence is not 'concealed'), they may think twice about including that person in the study or may interfere with the allocation process to ensure that a certain individual receives a particular intervention. This selective allocation of participants can introduce bias that affects the overall results (Moher et al. 2001). However, such problems can be avoided by concealing the allocation sequence. Allocation can be concealed by having the randomization sequence administered by someone who is not part of the study personnel (or is "off-site"). Opaque envelopes can also be used. In this case, the envelope, which has the participant's allocated group inside, is not opened until after the participant has been enrolled in the study (Hoffmann et al. 2013).

4. **Were the Groups Similar at Baseline?**

Authors should provide information regarding important baseline characteristics so that readers can determine whether or not the groups were truly similar at 'baseline', or prior to the introduction of the intervention. This is crucial, because we need to be sure that differences between groups at the end of the trial were due to the intervention alone and not simply the result of baseline differences. Consider, for example, a study investigating the effect of meditation on student stress levels. If one of the groups had higher stress levels at the start of the study, then differences between the groups at the end of the study may be due to this initial difference, rather than the intervention itself.

While randomization is an important step, it cannot absolutely guarantee that groups will be similar, particular when small samples are being investigated. Studies should therefore present key demographic data (e.g., age, gender, height, weight) and baseline scores on key outcome measures (Roberts and Torgerson 1999). As a critical appraiser, it is up to you to determine which of these factors may influence the results, and whether or not the groups are sufficiently similar at baseline. If baseline differences are present, researchers should account for these differences during their statistical analyses. If they do not, you will need to consider the baseline differences when interpreting the results.

5. Were the Participants and Study Personnel Blinded?

Blinding refers to the process of ensuring that participants and study personnel (researchers and health professionals) involved in a study do not know the groups to which participants were allocated (Hewitt and Torgerson 2006; see ▶ Chap. 37, "Randomized Controlled Trials"). Blinding and allocation concealment are often confused. Allocation concealment involves not disclosing the allocation sequence *before* the patient is enrolled into the study, while blinding ensures that participants and study personnel do not know the treatment allocation *after* enrolment. This is important because people involved in studies will often have preconceived ideas or expectations about the effects of particular interventions. These expectations can affect their behavior and, in turn, the results of the study (Hewitt and Torgerson 2006). For example, a participant may stop putting in effort during assessments if they discover that they are part of the control group. Similarly, a participant who knows they are in the intervention group is likely to experience exaggerated improvements in line with their expectations (Hewitt and Torgerson 2006). The intervention and control groups should, therefore, be indistinguishable, with there being no way of participants or study personnel discovering the groups to which people were allocated until the end of the study.

Assessors, or the people taking measurements during a study, can usually be blinded during objective assessments. Ideally, an "independent assessor" is used. This describes a person who is unaware of a participant's group allocation during baseline and follow-up assessments. However, health professionals and participants cannot always be blinded. Consider a study investigating the effects of aquatic physiotherapy on lower limb mobility in people with osteoarthritis. Both the participants and health professionals will know whether or not they are in the intervention (aquatic environment) or control (land-based environment) group. This factor would have to be taken into consideration when appraising the study and interpreting its results.

Experimental studies are often categorized according to the degree of blinding they employ. In a "single-blind" experiment, only the participants are blinded. During "double-blind" experiments, neither the participants nor the researchers know the groups to which participants belong. A "triple-blind" experiment involves the most rigorous form of blinding where none of the participants, researchers, or assessors know which participants are in the control group and which are in the intervention group.

6. Was There Adequate Follow-Up of Participants and a Sufficient Completion Rate?

It is not uncommon for participants to withdraw from studies or be unavailable for follow-up assessments. Such issues may introduce bias (Guyatt et al. 1993). For example, a situation in which ten participants were "lost to follow-up" due to changing occupations and moving away is vastly different to a situation in which ten participants were lost to follow-up because the intervention was making them ill. In the latter case, the overall results may favor the intervention simply because people who had negative side effects did not complete the study. This would mean that only people who improved or had no side effects were analyzed. Studies should, therefore, indicate clearly whether participants were properly accounted for, and why data may be missing for certain participants. As bias is increased when large numbers of participants withdraw from a study, various authors have suggested approaching studies with a completion rate of less than 85% with caution (Herbert et al. 2005; Dumville et al. 2006). Ideally, a diagram detailing the flow of participants through each stage of the study should be included alongside explanations for any participants lost to follow-up (Moher et al. 2001).

7. Was an Intention-to-Treat Analysis Performed?

Study participants may not always receive the intervention or control conditions as they were initially allocated. Imagine a study investigating the effectiveness of exercise in the treatment of knee pain. Some participants in the intervention group may not perform exercises as instructed due to a lack of motivation or excessive pain. In such cases, it may seem appropriate to analyze participants as if they were in the control group rather than the group to which they were actually allocated. However, doing so would increase the number of people in the control group who were unmotivated or had excessive pain (Hoffmann et al. 2013). This would undo the effects of randomization, leading to potentially unfair comparisons being made between groups. Likewise, we could not simply remove these participants from the analysis altogether as this would cause problems associated with incomplete follow-up. Instead, participants should be analyzed in the groups to which they were initially randomized, regardless of whether or not they remained in that group for the entire study. This principle is referred to as "intention-to-treat" analysis and is important in preserving the value of random allocation (Hoffmann et al. 2013).

8. **Were Between-Group Statistical Comparisons Performed?**

In an experimental study, the most valuable comparisons are between the intervention (or exposure) group and the control group. These between-group comparisons allow us to identify the extent to which the intervention has made a true impact. Consider a study investigating the effects of fruit consumption on the blood glucose levels of people with diabetes. If the intervention group's average blood glucose readings improved from baseline, it may be tempting to state that the intervention was successful based on this "within-group" information. However, by doing so, we would be ignoring the results of the control group entirely. Instead, our conclusions should be based on comparisons between the intervention and control groups at the end of the study, assuming that the groups were similar at baseline. Such information is far more valuable than performing within-group analyses, which are susceptible to many forms of bias.

3.3 Completeness and Quality of Reporting in Experimental Studies

The Consolidated Standards of Reporting Trials (CONSORT) statement provides a checklist for researchers and readers to assess the reporting quality of experimental studies (Moher et al. 2001). This statement does not assess the way in which the study was conducted but, rather, explores the quality of the reporting of these details within the paper itself. The template for intervention description and replication (TIDieR) checklist is a similar tool specifically designed for intervention studies (Hoffmann et al. 2014). The use of such tools is strongly recommended (and in most cases, compulsory) when submitting journal articles for publication. However, not all journals require authors to adhere to CONSORT or TIDieR recommendations, and many articles were published before these tools were created, so performing appraisals of reporting quality yourself is always a good idea.

4 Critically Appraising Observational Studies (Cohort, Case-control, and Cross-Sectional Studies)

During observational studies, the independent variable is not under the control of the researcher. These studies seek to observe the natural progression of a condition or response to treatment, without directly intervening or randomly allocating people to groups. As such, these studies are more susceptible to confounding and bias than experimental studies (Hoffmann et al. 2013). However, observational studies are particularly useful for situations in which experimental studies would not be ethical or practical, like when investigating potentially harmful exposures. To review, the main types of observational studies are:

- **Cohort studies**: Cohort studies are a type of longitudinal study in which participants are followed over time (see also ▶ Chap. 35, "Longitudinal Study Designs"). Participants with certain characteristics or exposures (e.g., risk factors, conditions, behaviors) are labeled as a "cohort." Following exposure, the cohort is followed for a period of time, outcomes are measured, and these outcomes are compared to the initial presentations of people in the cohort. Remember that exposure is not controlled by the researchers and would have occurred regardless of the study (unlike in experimental study designs). For example, a cohort study may involve identifying a group of mothers who recently starting breastfeeding and following them over time to observe if their children develop dental caries or tooth decay (Arora et al. 2011). Researchers would then identify differences between children who did or did not develop tooth decay in terms of factors such as parental health, feeding duration, and fluoride exposure.
- **Case-control studies**: Case-control studies are different to cohort studies in that they investigate participants who have experienced an outcome, "cases" (e.g., developing a disease, becoming injured), already. These participants are matched with similar participants, "controls" (for factors such as age, gender, and so on), who have not experienced that outcome. Case-control studies then compare the groups by looking-back and observing differences between them in terms of risk factors or exposures. For example, a case-control study may involve identifying people who already have lung cancer, matching them to people who do not, and then looking back to identify any differences in exposure to cigarette smoking. This design is not as robust as a cohort study as researchers are often not present when baseline data were collected, and the information regarding risk factors may be incomplete. Consider a study where the outcome was "death before the age of 50 years old" – researchers would need to rely on hospital records to identify risk factors, rather than collecting them from participants at baseline like in a cohort study.
- **Cross-sectional study**: Cross-sectional studies take a "snapshot" of a sample population at a single point in time and determine who has the outcome of interest. For example, a cross-sectional study may analyze a sample of current cigarette smokers and identify the proportion of people with early-stage lung cancer. The major advantage of this study design is that it is relatively inexpensive compared to cohort or case-control studies. However, cross-sectional studies require data on a particular subject to be readily available and routinely collected. The accuracy of this historical data cannot be guaranteed. Cross-sectional studies also provide no longitudinal information and so cannot produce claims as strong as other study types.

Figure 2 provides a summary of the main observation study types.

4.1 Tools for Appraising Observational Studies

Many tools are available to assist with the critical appraisal of observational studies. Some of these tools provide general guidelines, while others offer checklists specific to cohort, case-control, or cross-sectional studies. Many of the key appraisal items

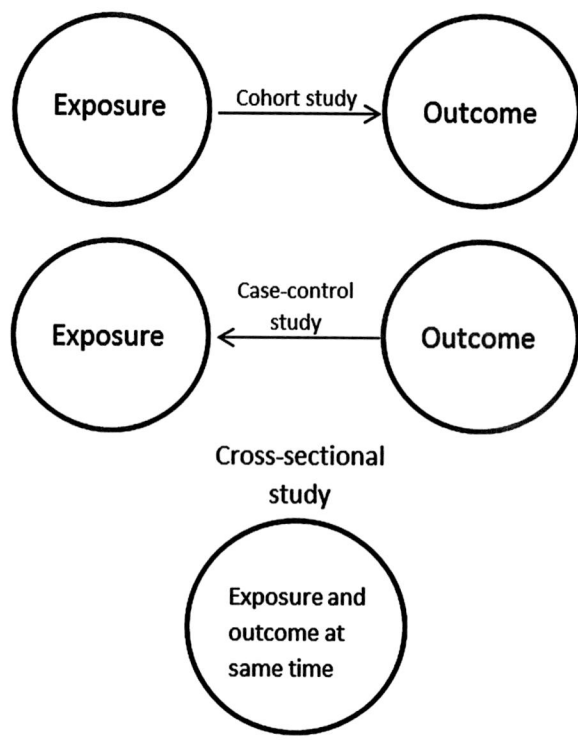

Fig. 2 Overview of observational studies

overlap across these study types. As always, you should consider methodological quality in conjunction with the applicability of the results to your individual patient or population of interest. Commonly employed appraisal tools include:

Tools that combine study types

- The quality assessment tools for observational studies from the National Heart, Lung, and Blood Institute (2017). This site provides a combined tool to appraise cohort and cross-sectional studies and a separate tool to appraise case-control studies.
- The Centre for Evidence-Based Medicine's prognostic and diagnostic study appraisal tools (Centre for Evidence-based Medicine 2017).

Cohort studies

- The CASP cohort study appraisal tool (Critical Appraisal Skills Program 2017)
- The Joanna Briggs Institute's appraisal checklist for cohort studies (Joanna Briggs Institute 2017)
- The cohort study appraisal tool from the Center for Evidence-based Management (2017)

Case-control studies

- The CASP case-control study appraisal tool (Critical Appraisal Skills Program 2017)
- The Joanna Briggs Institute's appraisal checklist for case-control studies (Joanna Briggs Institute 2017)
- The case-control study appraisal tool from the Center for Evidence-based Management (2017)

Cross-sectional studies

- The cross-sectional study appraisal tool from the Center for Evidence-based Management (2017)
- The Joanna Briggs Institute's appraisal checklist for cross-sectional studies (Joanna Briggs Institute 2017)

Tools such as the quality assessment of diagnostic accuracy studies (QUADAS-2) are also available for observational studies focusing on the evaluation of diagnostic tools and techniques (Whiting et al. 2011).

4.2 Important Questions to Ask when Appraising an Observational Study

There are six questions that are commonly asked when appraising the potential risk of bias in observational studies. These questions are summarized in Table 3 and are derived from criteria presented in the CASP (2017) and Joanna Briggs Institute's (2017) observational study appraisal tools.

1. Was a Clearly Focused Issue Addressed and Was an Appropriate Type of Study Used?

As with all studies, a good initial question to ask when appraising observational studies is whether a clearly focused issue was addressed. This is often considered a "screening question" along with asking if participants were recruited in an appropriate manner. Answering "no" to such questions indicates that a study is unlikely to be helpful and may not warrant further appraisal. Observational studies should use the PICO question format described in the systematic reviews section of this chapter (see also ▶ Chap. 46, "Conducting a Systematic Review: A Practical Guide").

It is also important to identify whether the type of study adopted is suitable to address the issue presented by the authors. Questions concerning natural history and risk factors are best answered by cohort studies, but costs or disease rarity may make case-control or cross-sectional studies more suitable. Observational studies are also most suitable when the outcome is harmful because RCTs in this case would be unethical. As outlined in the previous section, questions regarding

Table 3 Important questions to ask when appraising an observational study

1. Was a clearly focused issue addressed and was an appropriate type of study used?
2. Were participants recruited in an appropriate manner?
3. Was the exposure measured in a valid and reliable manner?
4. Were confounding factors identified and taken into account?
5. Were the outcomes measured in a valid and reliable manner?
6. Was follow-up complete and long enough for outcomes to be observed?

treatment effectiveness would likely be better suited to experimental study types (see ▶ Chap. 37, "Randomized Controlled Trials").

2. Were Participants Recruited in an Appropriate Manner?

This question explores whether or not a study's sample is representative of the larger population of interest. "Generalizability" or "representativeness" is important in determining if a study's findings may be applicable beyond the study setting and to our own patients (see also ▶ Chap. 38, "Measurement Issues in Quantitative Research"). Observational studies should provide sufficient detail on their sample to allow readers to identify if it is applicable to their individual case or population of interest. Clearly defined inclusion and exclusion criteria are useful in highlighting the study's target population.

Additional notes for cohort studies: In a cohort study, the odds of achieving a representative sample can be increased by recruiting all of the eligible patients ("consecutive cases") who present to the recruitment site. Recruiting all eligible patients prevents bias that could be introduced if eligible patients were avoided or missed (Joanna Briggs Institute 2017). For case-control and cross-sectional studies, achieving a representative sample can be more difficult as researchers are limited to the data already available. During cohort studies, it is also important that participants are recruited at a consistent and well-defined stage of a condition's progression. This is required because two people with the same condition can present with very different symptoms and experiences depending upon the length of time since their diagnosis (Hoffman et al. 2013). When participants are recruited at a consistent point early in their disease progression, the cohort is known as an "inception cohort." Many studies that recruit individuals upon diagnosis label their participants as an inception cohort. However, as highlighted by Hoffman et al. (2013, p. 173), certain people may receive their initial diagnosis further into their disease progression than others. You, as the critical appraiser, must therefore consider carefully whether a true inception cohort has been obtained.

Additional notes for case-control studies: In a case-control study, the case and control groups should be as similar as possible, apart from the presence of the disease or exposure of interest. This is usually done by individually matching controls to each case on the basis of certain characteristics.

3. Was the Exposure Measured in a Valid and Reliable Manner?

The validity of an exposure measurement relates to its ability to assess what it is intended to assess. This is usually determined by comparing the measure to a "gold standard" that is already known to be accurate. Reliability refers to the reproducibility of the measure's results over time and between assessors. To allow the reader to determine whether the exposure was measured appropriately, observational studies should clearly outline the method with which the exposure was measured (Joanna Briggs Institute 2017). For a study in which the exposure was "visiting a trained herbalist during a marathon preparation," it would be important to provide clear and reproducible definitions of "trained herbalist" and "marathon preparation." Wherever possible, objective criteria should be applied to minimize bias (see ▶ Chap. 38, "Measurement Issues in Quantitative Research").

4. Were Confounding Factors Identified and Taken into Account?

Confounding factors describe anything that can become confused with the outcome of interest and bias the results. Most often, this occurs when there are differences between comparison groups (apart from the exposure of interest). Common confounders include baseline characteristics, prognostic factors (e.g., severity of condition, age of diagnosis, comorbidities), or concomitant exposures (e.g., smoking) (Joanna Briggs Institute 2017). Further explanation of confounding factors can be found in Sect. 2.2 of this chapter. Observational studies should identify potential confounders and employ strategies to account for their influence. This is often done by reporting the results of various subgroups of participants (e.g., smokers, non-smokers). Statistical techniques can be employed to adjust for subgroup analyses (see also ▶ Chap. 54, "Data Analysis in Quantitative Research").

5. Were the Outcomes Measured in a Valid and Reliable Manner?

Although not always possible, outcomes are ideally assessed using objective criteria. As with exposure measurements, outcomes should be measured using previously validated definitions or diagnostic criteria. Once the objectivity, validity, and reliability of the outcome measure have been established, we should then explore the means by which measurements were conducted. We can look for information regarding the level of expertise of the assessors or the number of assessors involved in taking measurements (Joanna Briggs Institute 2017). A larger number of assessors may decrease the reliability of our recordings, particularly if they do not have similar levels of experience with the assessment tools (see ▶ Chap. 38, "Measurement Issues in Quantitative Research").

6. Was Follow-Up Complete and Long Enough for Outcomes to Be Observed?

It is important for observational studies to run long enough (or look back far enough) to demonstrate an association between the exposure and the outcome. You, as the critical appraiser, must use your knowledge of the condition of interest to determine whether or not the follow-up period was long enough to allow clinically meaningful changes to occur (see ▶ Chap. 35, "Longitudinal Study Designs").

Clinical practice guidelines, clinical research, and expert opinions can be useful in guiding your decision. Cross-sectional studies by design have no follow-up, so this question would not be considered when appraising that type of research.

4.3 Completeness and Quality of Reporting in Observational Studies

As outlined in the systematic review and experimental study sections of this chapter, a well-reported study can simplify the critical appraisal process. The reporting standard for observational studies is known as the STROBE (Strengthening the Reporting of Observational Studies in Epidemiology) statement (Von Elm et al. 2014). There are STROBE checklists available for observational studies in general, as well as statements designed specifically for cohort, case-control, and cross-sectional studies. The reporting quality of studies exploring diagnostic accuracy can be assessed using the Standards for Reporting Diagnostic Accuracy Studies (STARD) checklist (Bossuyt et al. 2003).

5 Conclusion and Future Directions

Not all studies are carried out using rigorous methods. Studies may be influenced by many types of bias, irrespective of their level of evidence. Critical appraisal skills are important for anyone wishing to make informed decisions or improve the quality of healthcare delivery. Important questions to ask during critical appraisals include:

For systematic reviews:

1. Did the review address a clearly focused question?
2. Did the review have clearly defined eligibility criteria and select the most appropriate studies?
3. Does the review have a comprehensive literature search strategy?
4. Did the review include a risk of bias assessment?
5. Did the review perform a meta-analysis, and was it appropriate to do so?
6. What are the results of the review and how precise are they?
7. How relevant are the results to my patient or problem?

For experimental studies:

1. Was there a control group?
2. Was the allocation of participants to groups randomized?
3. Was the group allocation sequence concealed?
4. Were the groups similar at baseline?
5. Were the participants and study personnel blinded?
6. Was there adequate follow-up of participants and a sufficient completion rate?

7. Was an intention-to-treat analysis performed?
8. Were between-group statistical comparisons performed?

For observational studies:

1. Was a clearly focused issue addressed and was an appropriate type of study used?
2. Were participants recruited in an appropriate manner?
3. Was the exposure measured in a valid and reliable manner?
4. Were confounding factors identified and taken into account?
5. Were the outcomes measured in a valid and reliable manner?
6. Was follow-up complete and long enough for outcomes to be observed?

A well-reported study can simplify the critical appraisal process, and there are also a number of tools available to assess reporting quality. This is distinct from methodological quality and focusses on the way in which study processes were described in the paper, rather than the actual study processes themselves. A good critical appraisal should assess a study's methodological and reporting quality, as well as the applicability of the results to one's own patient or clinical context.

Moving forward, make sure to use the skills you have acquired throughout this chapter in order to effectively appraise and implement the findings of quantitative research. Refining your critical appraisal skills will allow you to effectively identify relevant and rigorously conducted literature, saving you time and maximizing your treatment effectiveness. With the continuing acceleration of the healthcare research movement, employing strong critical appraisals will ensure that your patients can be confident that you are providing them with the best treatment options currently available.

References

Altman DG, Bland JM. Treatment allocation in controlled trials: why randomise? BMJ. 1999;318(7192):1209.

Arora A, Scott JA, Bhole S, Do L, Schwarz E, Blinkhorn AS. Early childhood feeding practices and dental caries in preschool children: a multi-centre birth cohort study. BMC Public Health. 2011;11(1):28.

Bossuyt PM, Reitsma JB, Bruns DE, Gatsonis CA, Glasziou PP, Irwig LM, ... Lijmer JG. The STARD statement for reporting studies of diagnostic accuracy: explanation and elaboration. Ann Intern Med. 2003;138(1):W1–12.

Cavaleri R, Schabrun S, Te M, Chipchase L. Hand therapy versus corticosteroid injections in the treatment of de quervain's disease: a systematic review and meta-analysis. J Hand Ther. 2016;29(1):3–11. https://doi.org/10.1016/j.jht.2015.10.004.

Centre for Evidence-based Management. Critical appraisal tools. 2017. Retrieved 20 Dec 2017, from https://www.cebma.org/resources-and-tools/what-is-critical-appraisal/.

Centre for Evidence-based Medicine. Critical appraisal worksheets. 2017. Retrieved 3 Dec 2017, from http://www.cebm.net/blog/2014/06/10/critical-appraisal/.

Clark HD, Wells GA, Huët C, McAlister FA, Salmi LR, Fergusson D, Laupacis A. Assessing the quality of randomized trials: reliability of the jadad scale. Control Clin Trials. 1999;20(5):448–52. https://doi.org/10.1016/S0197-2456(99)00026-4.

Critical Appraisal Skills Program. Casp checklists. 2017. Retrieved 5 Dec 2017, from http://www.casp-uk.net/casp-tools-checklists.

Dawes M, Davies P, Gray A, Mant J, Seers K, Snowball R. Evidence-based practice: a primer for health care professionals. London: Elsevier; 2005.

Dumville JC, Torgerson DJ, Hewitt CE. Research methods: reporting attrition in randomised controlled trials. BMJ. 2006;332(7547):969.

Greenhalgh T, Donald A. Evidence-based health care workbook: understanding research for individual and group learning. London: BMJ Publishing Group; 2000.

Guyatt GH, Sackett DL, Cook DJ, Guyatt G, Bass E, Brill-Edwards P, . . . Gerstein H. Users' guides to the medical literature: II. How to use an article about therapy or prevention. JAMA. 1993;270(21):2598–601.

Guyatt GH, Oxman AD, Akl EA, Kunz R, Vist G, Brozek J, . . . Jaeschke R. GRADE guidelines: 1. Introduction – GRADE evidence profiles and summary of findings tables. J Clin Epidemiol. 2011;64(4), 383–94.

Herbert R, Jamtvedt G, Mead J, Birger Hagen K. Practical evidence-based physiotherapy. London: Elsevier Health Sciences; 2005.

Hewitt CE, Torgerson DJ. Is restricted randomisation necessary? BMJ. 2006;332(7556):1506–8.

Higgins JPT, Green S. Cochrane handbook for systematic reviews of interventions version 5.0.2. The cochrane collaboration. 2009. Retrieved 3 Dec 2017, from http://www.cochrane-handbook.org.

Hoffmann T, Bennett S, Del Mar C. Evidence-based practice across the health professions. Chatswood: Elsevier Health Sciences; 2013.

Hoffmann T, Glasziou PP, Boutron I, Milne R, Perera R, Moher D, . . . Johnston M. Better reporting of interventions: template for intervention description and replication (TIDieR) checklist and guide. BMJ. 2014;348: g1687.

Joanna Briggs Institute. Critical appraisal tools. 2017. Retrieved 4 Dec 2017, from http://joannabriggs.org/research/critical-appraisal-tools.html.

Mhaskar R, Emmanuel P, Mishra S, Patel S, Naik E, Kumar A. Critical appraisal skills are essential to informed decision-making. Indian J Sex Transm Dis. 2009;30(2):112–9. https://doi.org/10.4103/0253-7184.62770.

Moher D, Schulz KF, Altman DG. The CONSORT statement: revised recommendations for improving the quality of reports of parallel group randomized trials. BMC Med Res Methodol. 2001;1(1):2. https://doi.org/10.1186/1471-2288-1-2.

Moher D, Liberati A, Tetzlaff J, Altman DG, Prisma Group. Preferred reporting items for systematic reviews and meta-analyses: the prisma statement. PLoS Med. 2009;6(7):e1000097.

National Health and Medical Research Council. NHMRC additional levels of evidence and grades for recommendations for developers of guidelines. Canberra: NHMRC; 2009. Retrieved from https://www.nhmrc.gov.au/_files_nhmrc/file/guidelines/developers/nhmrc_levels_grades_evidence_120423.pdf.

National Heart Lung and Blood Institute. Study quality assessment tools. 2017. Retrieved 17 Dec 2017, from https://www.nhlbi.nih.gov/health-topics/study-quality-assessment-tools.

Physiotherapy Evidence Database. PEDro scale. 2017. Retrieved 10 Dec 2017, from https://www.pedro.org.au/english/downloads/pedro-scale/.

Portney L, Watkins M. Foundations of clinical research: application to practice. 2nd ed. Upper Saddle River: F.A. Davis Company/Publishers; 2009.

Roberts C, Torgerson DJ. Understanding controlled trials: baseline imbalance in randomised controlled trials. BMJ. 1999;319(7203):185.

Shea BJ, Reeves BC, Wells G, Thuku M, Hamel C, Moran J, . . . Kristjansson E. AMSTAR 2: a critical appraisal tool for systematic reviews that include randomised or non-randomised studies of healthcare interventions, or both. BMJ. 2017;358:j4008. https://doi.org/10.1136/bmj.j4008.

Sterne JA, Hernán MA, Reeves BC, Savović J, Berkman ND, Viswanathan M, . . . Boutron I. ROBINS-I: a tool for assessing risk of bias in non-randomised studies of interventions. BMJ. 2016;355:i4919.

Stroup DF, Berlin JA, Morton SC, Olkin I, Williamson GD, Rennie D, ... Thacker SB. Meta-analysis of observational studies in epidemiology: a proposal for reporting. JAMA. 2000;283(15):2008–12.

Von Elm E, Altman DG, Egger M, Pocock SJ, Gøtzsche PC, Vandenbroucke JP, Initiative S. The strengthening the reporting of observational studies in epidemiology (strobe) statement: guidelines for reporting observational studies. Int J Surg. 2014;12(12):1495–9.

Whiting PF, Rutjes AW, Westwood ME, Mallett S, Deeks JJ, Reitsma JB, ... Bossuyt PM. QUADAS-2: a revised tool for the quality assessment of diagnostic accuracy studies. Ann Intern Med 2011;155(8):529–36.

Appraising Mixed Methods Research

60

Elizabeth J. Halcomb

Contents

1 Introduction	1052
2 Critically Appraising Mixed Methods Research	1052
2.1 Mixed Methods Appraisal Tool (MMAT)	1054
2.2 Bespoke Quality Framework for Mixed Methods Research	1057
2.3 Central Criteria for Appraisal of Mixed Methods Research	1062
3 Conclusion and Future Directions	1064
References	1065

Abstract

There is increasing interest in the use of mixed methods research approaches among health researchers. While mixed methods research has the potential to reveal rich data and deeper understandings of complex phenomena, it needs to be evaluated with the same level of critical appraisal as other methodologies. To date, however, much of the discourse around the critical appraisal of mixed methods research has discussed the challenges and considerations underlying critical appraisal. There has been limited agreement reached on optimal methods of evaluating this body of literature. This chapter will synthesize the literature on critically appraising mixed methods research and provide advice to those reviewing mixed methods papers around considerations in critical appraisal for this type of research.

Keywords

Mixed methods research · Critical appraisal · Methodological quality · Reporting quality

---

E. J. Halcomb (✉)
School of Nursing, University of Wollongong, Wollongong, NSW, Australia
e-mail: ehalcomb@uow.edu.au

© Springer Nature Singapore Pte Ltd. 2019
P. Liamputtong (ed.), *Handbook of Research Methods in Health Social Sciences*,
https://doi.org/10.1007/978-981-10-5251-4_121

1 Introduction

Critical appraisal of any research is important in order to evaluate the "quality" and rigor of the study (Heyvaert et al. 2013). Understanding the quality of research will inform decisions about utilizing its findings in policy, education, and practice. While some limitations will be acceptable or unavoidable in a particular study, given its design, others may impact the utility of the findings and their transferability into practice.

In recent years, a series of checklists and tools have proliferated to guide the researcher in critically appraising the various research methodologies. Most of these tools are specifically designed for the evaluation of research that uses a particular study design. For example, the Critical Skills Appraisal Program (CASP) (2017) comprises eight tools that guide the evaluation of systematic reviews, randomized controlled trials, cohort studies, case control studies, economic evaluations, diagnostic studies, qualitative studies, and clinical prediction rule. Additionally, other tools have also emerged to define the reporting criteria for different types of research (Collins et al. 2012). For example, the Preferred Reporting Items for Systematic Reviews and Meta-Analyses (PRISMA) (Moher et al. 2009) guidelines inform reporting of systematic reviews of randomised controlled trials, while the consolidated criteria for reporting qualitative research (COREQ) (Tong et al. 2007) have become the standard for reporting qualitative research. It is important to remember that these latter tools focus on the quality of information describing the study. So, while they also reflect the quality of the research, they predominately reflect the quality of the writing rather than the study conduct (Souto et al. 2015). It is important, therefore, that the user understands the purpose of the tool to ensure that they are using it for the purpose which it was intended.

The need to critically appraise mixed methods studies is as important as the appraisal of either purely qualitative or quantitative research. Just because a single study has used both qualitative and quantitative aspects, it is not necessarily robust or rigorous (Bryman 2006b; Lavelle et al. 2013). While the issue of quality in mixed methods research has been long debated, it remains a contentious issue (O'Cathain 2010; Heyvaert et al. 2013; Barnat et al. 2017; MacInnes 2009). The evaluation of mixed methods research faces a number of issues beyond those of pure qualitative or quantitative research (O'Cathain et al. 2008; Fàbregues et al. 2018; Collins et al. 2012; MacInnes 2009). These issues stem from the underlying paradigmatic tensions of mixing methods, as well as variations in methodology and methods (Sale and Brazil 2004; Bryman 2006b). This chapter will discuss the tools that have been developed to guide the appraisal of mixed methods research and explore the issues that need to be considered when evaluating mixed methods research.

2 Critically Appraising Mixed Methods Research

Three alternatives have been suggested for the critical appraisal of mixed methods research (Bryman 2006b; O'Cathain 2010; Collins et al. 2012; Heyvaert et al. 2013). Firstly, convergent criteria have been proposed, whereby the same broad criteria are used to evaluate both the quantitative and the qualitative components of the research. O'Cathain (2010) identified various simple and quick generic tools that can be used for critical appraisal regardless of study design. While these tools can be useful to provide a quick and broad evaluation, they are often too generalist to provide discrimination of quality and overlook the quality issues specific to mixed methods research.

Secondly, the individual component approach has been proposed, whereby the quantitative and the qualitative components are each evaluated in isolation based on their individual methodology or method (Dellinger and Leech 2007; O'Cathain 2010; Collins et al. 2012; Heyvaert et al. 2013) (see ▶ Chaps. 58, "Appraisal of Qualitative Studies," and ▶ 59, "Critical Appraisal of Quantitative Research"). An example of this kind of generic tool is the mixed methods appraisal tool (MMAT) discussed in detail below (Pace et al. 2012). Despite the logic that sound qualitative and quantitative components together strengthen the quality of a mixed methods study, the quality of one or other components may be adversely affected by its inclusion in a mixed methods investigation (O'Cathain 2010). In addition to the risk of underresourcing, underdeveloping, or underanalyzing one or other components, this method of appraisal ignores the fact that inferences are drawn from the whole study rather than just each component in isolation (O'Cathain 2010).

While both the convergent and individual component approaches do indeed provide critique of the qualitative and qualitative components that comprise the mixed methods study, they fail to capture the complexity of the integration that characterizes the mixed methods approach (Sale and Brazil 2004; Heyvaert et al. 2013; Barnat et al. 2017; Fàbregues et al. 2018). Therefore, in more recent years, attempts have been made to develop the bespoke tools, suggested as the third alternative for mixed methods quality appraisal (Bryman 2006b; O'Cathain 2010). These tools seek to capture both the individual considerations of the qualitative and qualitative methods and the considerations of mixing these in a single study. To date, however, consensus around a particular tool has not been reached. The most widely accepted tools, however, are discussed in more detail in subsequent sections (Sect. 2.1).

Fàbregues et al. (2018) have identified three key limitations in the current work around bespoke quality appraisal tools in mixed methods research. Firstly, most of the frameworks that have been developed have been based on the authors' personal views of what constitutes quality, rather than being empirically derived. Therefore, the personal preferences and ways of thinking of these individuals have likely shaped the tools that have been developed (Collins et al. 2012; Heyvaert et al. 2013). Secondly, the professional discipline of the researcher is likely to impact on their practices and thinking as well as their understanding of quality in research (O'Cathain 2010; Collins et al. 2012; Fàbregues et al. 2018). While "research quality is not a homogeneous concept" (Fàbregues et al. 2018, p. 3) and is perceived somewhat differently by researchers across disciplines, it does have a number of common properties (Sale and Brazil 2004; O'Cathain 2010). However, given the expansion of mixed methods across health services and related disciplines beyond those of the researchers who have developed the current quality criteria, it is important that efforts are made to ensure that criteria are relevant to mixed methods research across disciplines.

The final limitation is that the focus in the debate around quality in mixed methods research has been around the operationalization rather than conceptualization of quality criteria (Heyvaert et al. 2013; Fàbregues et al. 2018). So, while the focus has been on developing criteria to appraise mixed methods research, there has been little attention on researchers' conceptualization of quality in this methodology.

Despite calls to develop consensus in terms of quality appraisal criteria for mixed methods research (O'Cathain et al. 2008), others have identified that, given the heterogeneity of mixed methods research, a single set of quality criteria may not suit all studies (Bryman 2006b; O'Cathain 2010; Heyvaert et al. 2013). Proponents of this argument advocate that quality appraisal of mixed methods research should be guided by core criteria that reflect the most important aspects of mixed methods research and which can accommodate a range of different contexts, alternate designs, creativity, and various disciplinary foci (Sale and Brazil 2004; O'Cathain 2010; Bryman 2014; Fàbregues et al. 2018). Therefore, consideration of the basic principles of quality and the specific attributes of each study are required to ensure that the research is evaluated appropriately. In the absence of agreement, this chapter, I will present both a comprehensive bespoke quality framework (O'Cathain 2010) (see Sect. 2.2) and a set of broad bespoke quality criteria (Bryman 2014) (see Sect. 2.3) to allow the reader to see the various strategies for critical appraisal of mixed methods research. This will provide insight to inform the readers understanding of the complexity of critical appraisal in this methodology. It will also demonstrate that despite similarities, consensus has not been reached on a single quality evaluation approach that will suit all circumstances.

2.1 Mixed Methods Appraisal Tool (MMAT)

The MMAT was developed to facilitate the appraisal of mixed methods research within mixed studies reviews (Pluye et al. 2011; Pace et al. 2012; Souto et al. 2015) (see Table 1). The tool comprises five sets of criteria, namely:

(i) Qualitative set
(ii) Randomized controlled set
(iii) Non-randomized set
(iv) Observational descriptive set
(v) Mixed methods set

Each of the first four sets of criteria can be used for either a single study of that design or a specific component of a mixed methods study. The final, mixed methods, set of criteria focuses on mixed methods research only. So, for example, a mixed methods study that combines a randomized trial of an intervention and interviews with participants would utilize three sets of criteria, that is, the randomized controlled set, the qualitative set, and the mixed methods set. While the quality score

60 Appraising Mixed Methods Research

Table 1 Mixed methods appraisal tool (Pluye et al. 2011; Pace et al. 2012)

Types of mixed methods components or primary studies	Quality criteria	Responses			Comments
		Yes	No	Can't tell	
Screening questions	Are there clear qualitative and quantitative research questions (or objectives), or a clear mixed methods question (or objective)? Do the collected data allow to address the research question (objective), for example, consider whether the follow-up period is long enough for the outcome to occur (for longitudinal studies or study components)? Further quality appraisal may be not feasible when the answer is "No" or "Can't tell" to one or both screening questions.				
Qualitative	1.1 Are the sources of qualitative data (archives, documents, informants, observations) relevant to address the research question (objective)? 1.2 Is the process for analyzing qualitative data relevant to address the research question (objective)? 1.3 Is appropriate consideration given to how findings relate to the context, e.g., the setting, in which the data were collected? 1.4 Is appropriate consideration given to how findings relate to researchers' influence, e.g., through their interactions with participants?				
Randomized controlled	2.1 Is there a clear description of the randomization (or an appropriate sequence generation)? 2.2 Is there a clear description of the allocation concealment (or blinding when applicable)? 2.3 Are there complete outcome data (80% or above)? 2.4 Is there low withdrawal/dropout (below 20%)?				
Non-randomized	3.1 Are participants (organizations) recruited in a way that minimized selection bias? 3.2 Are measurements appropriate (clear origin, or validity known, or standard instrument, and absence of contamination between groups when appropriate) regarding the exposure/intervention and outcomes? 3.3 In the groups being compared (exposed vs. nonexposed; with intervention vs. without; cases vs. controls), are the participants comparable, or do researchers				

(*continued*)

Table 1 (continued)

Types of mixed methods components or primary studies	Quality criteria	Responses			Comments
		Yes	No	Can't tell	
	take into account (control for) the difference between these groups? 3.4 Are there complete outcome data (80% or above), and, when applicable, an acceptable response rate (60% or above), or an acceptable follow-up rate for cohort studies (depending on the duration of follow-up)?				
Observational descriptive	4.1 Is the sampling strategy relevant to address the quantitative research question (quantitative aspect of the mixed methods question)? 4.2 Is the sample representative of the population understudy? 4.3 Are measurements appropriate (clear origin, or validity known, or standard instrument)? 4.4 Is there an acceptable response rate (60% or above)?				
Mixed methods	5.1 Is the mixed methods research design relevant to address the qualitative and quantitative research questions (or objectives), or the qualitative and quantitative aspects of the mixed methods question (or objective)? 5.2 Is the integration of qualitative and quantitative data (or results) relevant to address the research question (objective)? 5.3 Is appropriate consideration given to the limitations associated with this integration, e.g., the divergence of qualitative and quantitative data (or results) in a triangulation design? Criteria for the qualitative component (1.1 to 1.4), and appropriate criteria for the quantitative component (2.1 to 2.4, or 3.1 to 3.4, or 4.1 to 4.4), must be also applied				

derived from implementing the MMAT is not considered informative (Pace et al. 2012), the description of study quality gained from utilizing the criteria is more useful to inform the reader about quality issues. If the quality score is used in mixed methods studies, Pace et al. (2012) advocate that the lowest score on the subscales that make up the study component should be considered the quality score. The rationale for this is that the overall study cannot be any stronger than its weakest component. While the MMAT has been fairly recently developed, reliability testing has demonstrated that this tool is efficient and shows promising reliability (Pluye et al. 2011; Pace et al. 2012; Souto et al. 2015).

2.2 Bespoke Quality Framework for Mixed Methods Research

Various groups have described different approaches for evaluating the quality of mixed methods research (Onwuegbuzie and Johnson 2006; O'Cathain et al. 2008; Tashakkori and Teddlie 2008; Teddlie and Tashakkori 2009; Creswell and Plano Clark 2018). However, O'Cathain (2010) synthesized this into a comprehensive 44 criteria of quality framework. An adapted version of this framework is presented in Fig. 1. This framework delineates eight domains, across four stages of the research process, which need to be considered when evaluating the quality of a mixed methods study. Each domain is composed of a number of specific items, each of which is discussed in detail below. While this framework is perhaps the best attempt to date to synthesize the literature around mixed methods quality appraisal into a coherent framework, the large number of items and degree of overlap between concepts impact its utility in critical appraisal (Heyvaert et al. 2013). It is presented here to provide the reader with an overview of the key concepts and complex interrelationship between quality factors inherent in mixed methods studies.

2.2.1 Planning Phase

This first domain encompasses how well the mixed methods study has been planned and focuses on considerations that should be demonstrated within the mixed methods research proposal. Firstly, the *foundational element* of a comprehensive, critical

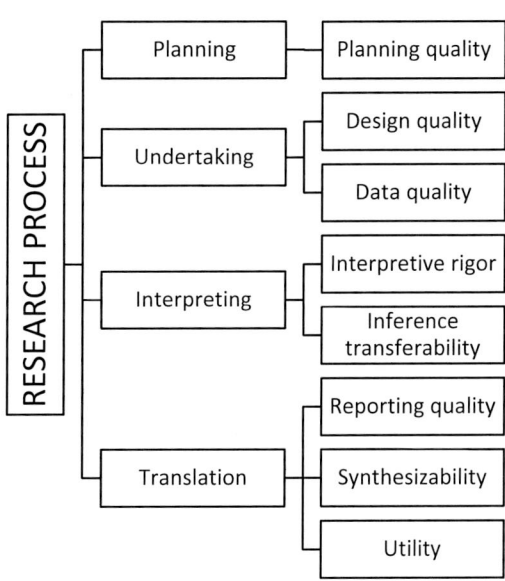

Fig. 1 Mixed methods quality framework. (Adapted from O'Cathain 2010)

literature review that situates the research question and study within a context is required to justify the choice of design (Dellinger and Leech 2007; O'Cathain 2010). Secondly, a clear justification of the choice of a mixed methods approach provides *rationale transparency*. This can highlight whether the approach has been selected for its intrinsic value or strategic purposes (e.g., attaining funding) (O'Cathain et al. 2007b). In cases where the justification is strategic, there is greater potential for neglect of one of the study components, lack of attention toward integration and lack of meta-inferences drawn from the whole study data (O'Cathain 2010).

Planning transparency refers to the clarity within the proposal around reporting the details of key aspects across research process. This includes aspects such as the paradigm, design, data collection plan, analysis methods, and integration approaches to be used.

Proposals of mixed methods research raise a number of issues around *feasibility*. Firstly, some mixed methods proposals demonstrate lack of researcher knowledge and experience by proposing components that are not feasible within the project time frame or resources (Halcomb and Andrew 2009). Additionally, in sequential studies, sufficient time may not be available or set aside within the proposed project time frame to complete all components (Halcomb and Andrew 2009). Limitations around feasibility have the potential to significantly impact on project quality.

2.2.2 Undertaking Phase

Within the conduct of the research, there are two domains of quality that should be considered. O'Cathain (2010) separates evaluation of the research design from that of data collection (see Fig. 2). Firstly, the design quality domain evaluates the rigor of the mixed methods study design, before the data quality domain appraises the processes of data collection and analysis.

Design Quality

In terms of design quality, there are four criteria that need to be considered (O'Cathain 2010). Firstly, there is *design transparency* (Tashakkori and Teddlie 2008; O'Cathain 2010). While a mixed methods study must comprise quantitative and qualitative components within the single study and some form of integration of these components, the specific design used should be clearly defined (MacInnes 2009).

Fig. 2 Quality criteria for undertaking research phase

Undertaking	Design quality	Design transparency
		Design suitability
		Design strength
		Design rigor
	Data quality	Data transparency
		Data rigor/design fidelity
		Analytic adequacy
		Analytic integration rigor

Researchers can draw upon the range of "conventional" designs proposed for mixed methods research (see ▶ Chaps. 4, "The Nature of Mixed Methods Research," and ▶ 40, "The Use of Mixed Methods in Research"). In some cases, these may need to be adapted to the specific study. It is important, however, that aspects such as the sequencing of methods, priority of various components, and the stage of integration are clearly defined and explained (O'Cathain 2010; Collins et al. 2012).

Secondly, *design suitability* refers to the appropriateness of the chosen design to be congruent with the stated purpose of the study and to answer the specific research questions (Collins et al. 2012). The design must also be aligned appropriately to the stated paradigm. For example, in a study where the purpose of the interviews is to build upon the survey findings, it is not appropriate to have a concurrent design as it is necessary to have at least undertaken preliminary survey analysis in order for the interviews to ask questions to deepen the understanding of survey findings. In this case a sequential design would clearly be more appropriate.

Design strength focuses upon the extent to which the strengths of each of the qualitative or quantitative components of the study compensates for the weaknesses in the other component (Tashakkori and Teddlie 2008; O'Cathain 2010). For example, combining interviews or focus groups with survey data allows the richness of the qualitative interviews to provide depth to the large-scale survey findings.

The final consideration is *design rigor*, whereby consideration is given to whether the researcher stays true to the elements of the design in the implementation of the research methods. For example, if insufficient time is available to analyze the first phase of a sequential design, the second phase may not have sufficient understandings of the initial data to adequately build on the data (Halcomb and Andrew 2009).

Data Quality

O'Cathain (2010) identifies five criteria which comprise the domain of data quality. Firstly, *data transparency* refers to the adequacy of the description of the data collection methods. Such description should include specific details about the participant recruitment, data collection methods, sample size, and analysis techniques (Creswell and Plano Clark 2007, 2011, 2018). The reader should be able to understand what is being done in enough detail for them to be able to plausibly replicate the work. Stemming from the data transparency is *data rigor*. This refers to the rigor of the implementation of the research methods. Consideration should be given to whether any compromises in the rigorous conduct of each of the study components have been made to accommodate the mixed methods nature of the study (Dellinger and Leech 2007). For example, have any phases in a sequential study been cut shorter then would be considered reasonable in a single method study to fit the project within a constrained time frame?

The subsequent three criteria focus on specific aspects of the research process. *Sampling adequacy* refers to the appropriateness of the sampling method and sample size for each component of the research (Tashakkori and Teddlie 2008; O'Cathain 2010). As in single method studies, a sample size that is too small may impact on the inferences that can be made from the data. A further consideration in mixed methods studies is that the relationship between the sampling in each component should be clear to the reader.

Also important to evaluate is the *analytic adequacy* to ensure that the data analysis techniques are appropriate to both answer the research question and be coherent with the design and methods used (Collins et al. 2012). A key, but often largely forgotten, characteristic of mixed methods research is the integration of components (Halcomb and Hickman 2015). *Analytic integration rigor* refers to the quality of integration undertaken within the analysis phase of the study. While not all mixed methods studies integrate data at this stage, consideration of any integration should be included in any quality evaluation. Integration undertaken in the analysis stage may involve transformation of qualitative to quantitative data or using the findings of one component to inform the subsequent component (Andrew et al. 2008, 2011). In recent years, there has been greater attention paid to integration in the literature, and several authors offer strategies to guide integration methods (Andrew et al. 2008, 2011; O'Cathain et al. 2010; Halcomb and Hickman 2015).

2.2.3 Interpretation Phase

The quality of the inferences made from any study and the transferability of results are vital for readers in order to determine if the findings are sufficiently credible and trustworthy to inform practice and policy (Tashakkori and Teddlie 2008; O'Cathain 2010; Collins et al. 2012). Therefore, the quality of the interpretation phase can be evaluated by combining evaluation of interpretive rigor, the degree to which the conclusions are based on the findings, and inference transferability, or the extent to which findings can be applied in other contexts (Tashakkori and Teddlie 2008; O'Cathain 2010) (see Fig. 3).

Interpretive Rigor

Interpretation in mixed methods research involves the creation of what Tashakkori and Teddlie (2008) describe as "meta-inferences." That is, the conclusions drawn extend beyond the findings of each component of the study to provide an

Fig. 3 Quality criteria for interpretative phase

Interpreting	Interpretive rigor	Interpretive transparency
		Interpretive consistency
		Theoretical consistency
		Interpretive agreement
		Interpretive distinctiveness
		Interpretive efficacy
		Interpretive bias reduction
		Interpretive correspondence
	Inference transferability	Ecological transferability
		Population transferability
		Temporal transferability
		Theoretical transferability

overarching "meta-inference" or broad conclusions that overlay both the qualitative and qualitative findings (Collins et al. 2012). O'Cathain (2010) identifies eight criteria for evaluating interpretive rigor in mixed methods research. *Interpretive transparency* refers simply to the ability for the reader to follow which findings have emerged from the various components of the study (Teddlie and Tashakkori 2009). For example, in a study that combines interviews and survey data including open-ended questions, it should be clear to the reader which data collection method each data element is drawn from.

In addition to being able to follow the trail from data collection to specific study findings, *interpretive consistency* considers whether the conclusions drawn are consistent with the reported findings (Creswell and Plano Clark 2007; Greene 2007). Similarly, *theoretical consistency* refers to whether the conclusions drawn are consistent with contemporary understandings of the topic area (Dellinger and Leech 2007). This concept is similar to face validity where others can read the conclusions and deem them credible and reasonable when compared to what is known. Additionally, *interpretive agreement* refers to whether readers are likely to reach the same conclusions as the researcher (Creswell and Plano Clark 2007; Greene 2007; Collins et al. 2012). This can be demonstrated through strategies such as external peer review or member checking.

Teddlie and Tashakkori (2009) also highlight the need to consider *interpretive bias reduction*. This refers to the need to explore consistency and contradiction between the findings of different study components. If contradictory findings emerge, clear evidence should be available to demonstrate consideration of the potential impact of bias related to the particular method as a plausible explanation of the contradiction (O'Cathain 2010; Creswell and Plano Clark 2007, 2011, 2018).

In contrast, *interpretive distinctiveness* refers to the degree to which the conclusions drawn by the researchers are more credible than other plausible inferences (Collins et al. 2012). To this end, researchers should demonstrate that other possible interpretations are less plausible that their interpretation.

Interpretive efficacy relates to the adequacy with which the overall study conclusions incorporate the inferences of the qualitative and quantitative components of the study (Greene 2007; Teddlie and Tashakkori 2009; O'Cathain 2010; Collins et al. 2012). Achieving the right balance of the conclusions from each dataset is complex and requires skill on the part of the research team.

Consideration of how well the inferences flow from the initial study purpose and research question is termed *interpretive correspondence* (O'Cathain 2010). As in single method studies, it is important that the conclusions actually flow logically from the study purpose and address the initial research questions (MacInnes 2009).

Inference Transferability

In the same way that external validity and transferability demonstrate quality for quantitative and qualitative research respectively, *inference transferability* refers to the ability to apply the conclusions of the mixed methods research into another setting or context (Teddlie and Tashakkori 2009; O'Cathain 2010). Teddlie and Tashakkori (2009) describe transferability being evaluated in four ways, namely:

1. Ecological – to other contexts and practice settings
2. Population – to different participant groups
3. Temporal – to a point in the future
4. Theoretical – to other methods of measuring outcomes

2.2.4 Dissemination Phase

Dissemination is an important indicator of study quality as it reflects the conclusions drawn from the study findings and communication of these to the target audience. *Report availability* provides evidence of the completion of the project as proposed within the available resources and time frame (Tashakkori and Teddlie 2008; O'Cathain 2010). Importantly, in addition to report availability, *yield* refers to the additional insight and knowledge gained from a mixed methods study above that which would have been gained from either component in isolation (O'Cathain et al. 2007a). A low yield will be seen in studies that undertake limited integration or where the design fails to optimally overcome the weakness in one method with the strength of the other. The explicit reporting of key aspects of the study, including the integration, is evaluated under the *reporting transparency* criteria (O'Cathain 2010). Perhaps the most well-recognized tool for reporting quality in mixed methods research is the "Good Reporting of A Mixed Methods Study" (GRAMMS) tool (see Box 1). This was developed by O'Cathain et al. (2008) from a review of 118 mixed methods studies funded between 1994 and 2004 by the English Department of Health, which explored the quality of mixed methods research and identified current standards of reporting. Rather than being a checklist, this tool provides broad guidance around transparency in reporting mixed methods research.

> **Box 1 Good Reporting of a Mixed Methods Study (O'Cathain et al. 2008)**
> 1. Describe the justification for using a mixed methods approach to the research question
> 2. Describe the design in terms of the purpose, priority, and sequence of methods
> 3. Describe each method in terms of sampling, data collection, and analysis
> 4. Describe where integration has occurred, how it has occurred, and who has participated in it
> 5. Describe any limitation of one method associated with the present of the other method
> 6. Describe any insights gained from mixing or integrating methods

2.3 Central Criteria for Appraisal of Mixed Methods Research

In contrast to the complex framework for critical appraisal proposed by O'Cathain (2010), Bryman (2014) proposes a much broader set of six central criteria for the evaluation of mixed methods research. Unlike the comprehensive framework described by O'Cathain (2010), these criteria very broadly identify the key considerations in critically appraising mixed methods research. The broader nature of these considerations makes them more accessible to research consumers seeking to evaluate reports of mixed methods studies. However, understanding the range of issues explicated by O'Cathain (2010) puts these broad criteria into context.

> **Box 2 Criteria for Appraising Mixed Methods Research (Bryman 2014)**
> - Need for the quantitative and qualitative components to be implemented in a technically competent manner
> - Need for transparency
> - Need for mixed methods to be linked to research questions
> - Need to be explicit about the nature of the mixed methods design employed
> - Need for a rationale for the use of mixed methods research
> - Need for integration

2.3.1 Need for the Quantitative and Qualitative Components to Be Implemented in a Technically Competent Manner

There is little argument that the various elements of quantitative and qualitative research need to be executed rigorously across the research process from sampling, instrument design, and implementation to data collection and analysis and interpretation (Tashakkori and Teddlie 2010; Creswell and Plano Clark 2011). This requires mixed methods researchers or at the very least research teams, to have broad skills across both quantitative and qualitative techniques (Halcomb and Andrew 2009; Bowers et al. 2013). Tools to evaluate these two components are well established in the literature (see ▶ Chaps. 58, "Appraisal of Qualitative Studies," and ▶ 59, "Critical Appraisal of Quantitative Research").

2.3.2 Need for Transparency

In addition to having sufficient clarity about how the steps of the research process were executed in terms of both the quantitative and the qualitative components, mixed methods researchers need to also have transparency about mixed methods considerations. Such considerations include the timing of the phases, relative priority given to the various datasets, and the nature of the integration. Clear descriptions of how these steps have been executed allow the reader to evaluate the appropriateness and implications of choices made. The key to demonstrating rigor in mixed methods research is in providing the reader with a clear audit trail and well-considered and justified rationales for the decisions made throughout the research process (Lavelle et al. 2013).

2.3.3 Need for Mixed Methods to Be Linked to Research Questions

While the link between research questions and methodology/methods is important in all research, in mixed methods studies, it is essential to demonstrate the need to employ a mixed methods approach rather than a pure quantitative or qualitative design. There should be a clear and logical flow from the research questions to the data collection and interpretation of findings.

2.3.4 Need to Be Explicit About the Nature of the Mixed Methods Design Employed

Although mixed methods research can certainly incorporate elements of creativity in research design, a number of typologies exist that provide an overview of the most common designs (Creswell and Plano Clark 2011, 2018). These typologies extend beyond providing a framework for data collection and also explicate key mixed methods characteristics such as study purpose, integration timing, allocation of priority, and level of interaction (Halcomb and Hickman 2015).

2.3.5 Need for a Rationale for the Use of Mixed Methods Research

The use of mixed methods should be a deliberate choice to add value to the research above that which could be achieved by a single method study (Creswell and Plano Clark 2011; Scammon et al. 2013). Therefore, in order to allow the reader to understand what the researcher was trying to achieve by mixing methods, the rationale for the design choice should be clearly articulated. Within the mixed methods literature, there has been a significant discussion about the range of rationales for using mixed methods (Greene et al. 1989; Bryman 2006a; Wisdom et al. 2012; Halcomb and Hickman 2015). Where possible this consistent terminology should be used to reduce confusion and promote mutual understanding.

2.3.6 Need for Integration

A frequent criticism of mixed methods research is the absence of integration between the quantitative and qualitative strands of a mixed methods study (O'Cathain et al. 2008; Andrew and Halcomb 2009). This is often seen when the quantitative and qualitative components are reported separately and there is little attempt to explore the richness that could result from the linking of the two datasets (Bryman 2014). Others argue that without integration, the study does not meet the criteria of mixed methods research (Andrew and Halcomb 2009). While, in some cases, some data may be reported separately to meet journal requirements or to report a specific aspect of the dataset, there should be some integration and recognition of papers across a series (see, e.g., Ashley et al. 2018a, b, in press-a, in press-b). There is growing interest in the literature around specific aspects of integration (Bryman 2006a; Andrew et al. 2008, 2011; Zhang and Creswell 2013; Halcomb and Hickman 2015). This growing discourse can help researchers to appreciate the range of strategies available to combine the quantitative and qualitative strands of their mixed methods research to better answer the research question.

3 Conclusion and Future Directions

From the complexity of the quality considerations outlined in this chapter, it is apparent that rigorous quality appraisal of mixed methods research requires specific tools focused on the specific considerations of mixed methods research. While appraisal must incorporate both qualitative and quantitative elements, it must also encompass the characteristics of transparency of decision-making, integration, and yield that characterize mixed methods research. Given the heterogeneity of mixed methods research and the diversity of researchers employing this methodology, it is unlikely that a single tool will be developed to critically appraise all mixed methods studies. Additionally, some frameworks that have been proposed to date are very lengthy and complex, making appraisal unnecessarily time-consuming for the reviewer. Care must be taken to avoid the assumption that the use of mixed methods makes a study more rigorous. Understanding the considerations identified in this chapter will help the reader to make critical decisions about the quality of mixed methods research and its utility to inform and guide practice.

As mixed methods continue to grow in popularity in the health sciences, it is important that health services and health professional researchers seek agreement about ways of defining and measuring quality in the health sciences. Additionally, researchers embarking on mixed methods research in the health sciences should consider aspects of quality when designing and implementing their research. Embedding quality considerations is vital to ensure that mixed methods research is robust and adds to the body of knowledge informing health policy, education, and practice.

References

Andrew S, Halcomb EJ, editors. Mixed methods research for nursing and the health sciences. London, UK: Wiley-Blackwell; 2009.

Andrew S, Salamonson Y, Halcomb EJ. Integrating mixed methods data analysis using NVivo©: an example examining attrition and persistence of nursing students. Int J Multiple Res Approaches. 2008;2(1):36–43.

Andrew S, Salamonson Y, Everrett B, Halcomb EJ, Davidson PM. Beyond the ceiling effect: using a mixed methods approach to measure patient satisfaction. Int J Multiple Res Approaches. 2011;5(1):52–63.

Ashley C, Brown A, Halcomb E, Peters K. Registered nurses transitioning from acute care to primary healthcare employment: a qualitative insight into nurses' experiences. J Clin Nurs. 2018a;27(3–4):661–8.

Ashley C, Halcomb E, Brown A, Peters K. Experiences of registered nurses transitioning from employment in acute care to primary health care – quantitative findings from a mixed methods study. J Clin Nurs. 2018b;27(1–2):355–62.

Ashley C, Halcomb E, Brown A, Peters K. Exploring the reasons why nurses transition from acute care to primary health care: a mixed methods study. Appl Nurs Res. in press-a. Accepted 4 Sept 2017.

Ashley C, Peters K, Brown A, Halcomb E. Reflections on transitioning and future career intentions of experienced nurses new to primary health care nursing. J Nurs Manag. in press-b. Accepted 14 Nov 2017.

Barnat M, Bosse E, Trautwein C. The guiding role of theory in mixed-methods research: combining individual and institutional perspectives on the transition to higher education. Theory Method Higher Educ Res. 2017. https://doi.org/10.1108/S2056-375220170000003001.

Bowers B, Cohen LW, Elliot AE, Grabowski DC, Fishman NW, Sharkey SS, ... Kemper P. Creating and supporting a mixed methods health services research team. Health Serv Res. 2013;48(6 Pt 2):2157–80.

Bryman A. Integrating quantitative and qualitative research: how is it done? Qual Res. 2006a;6(1):97–113.

Bryman A. Paradigm peace and the implications for quality. Int J Soc Res Methodol. 2006b;9(2):111–26.

Bryman A. June 1989 and beyond: Julia Brannen's contribution to mixed methods research. Int J Soc Res Methodol. 2014;17(2):121–31.

Collins KMT, Onwuegbuzie AJ, Johnson RB. Securing a place at the table: a review and extension of legitimation criteria for the conduct of mixed research. Am Behav Sci. 2012;56(6):849–65.

Creswell JW, Plano Clark VL. Designing and conducting mixed methods research. 2nd ed. Thousand Oaks: Sage; 2011.

Creswell JW, Plano Clark VL. Research design: qualitative, quantitative, and mixed methods approaches. 5th ed. Los Angeles: Sage; 2018.

Creswell JW, Plano Clark VL. Designing and conducting mixed methods research. Thousand Oaks: Sage; 2007.

Critical Appraisal Skills Programme. CASP checklists. 2017. Retrieved from http://www.casp-uk.net/casp-tools-checklists.

Dellinger AB, Leech NL. Toward a unified validation framework in mixed methods research. J Mixed Methods Res. 2007;1(4):309–32.

Fàbregues S, Paré M-H, Meneses J. Operationalizing and conceptualizing quality in mixed methods research: a multiple case study of the disciplines of education, nursing, psychology, and sociology. J Mixed Methods Res, online early. 2018. https://doi.org/10.1177/1558689817751774

Greene JC. Mixed methods in social inquiry, vol. 9. San Francisco: Wiley; 2007.

Greene JC, Caracelli VJ, Graham WF. Toward a conceptual framework for mixed-method evaluation designs. Educ Eval Policy Anal. 1989;11:255–74.

Halcomb EJ, Andrew S. Practical considerations for higher degree research students undertaking mixed methods projects. Int J Multiple Res Approaches. 2009;3(2):153–62.

Halcomb EJ, Hickman L. Mixed methods research. Nurs Stand. 2015;29(32):42–8.

Heyvaert M, Hannes K, Maes B, Onghena P. Critical appraisal of mixed methods studies. J Mixed Methods Res. 2013;7(4):302–27.

Lavelle E, Vuk J, Barber C. Twelve tips for getting started using mixed methods in medical education research. Med Teach. 2013;35(4):272–6.

MacInnes J. Mixed methods studies: a guide to critical appraisal. Br J Cardiac Nurs. 2009;4(12): 588–91.

Moher D, Liberati A, Tetzlaff J, Altman DG, PRISMA Group. Preferred reporting items for systematic reviews and meta-analyses: the PRISMA statement. PLoS Med. 2009;6(7): e1000097.

O'Cathain A. Assessing the quality of mixed methods research: toward a comprehensive framework. In: Tashakkori A, Teddlie C, editors. Handbook of mixed methods in social and behavioral research. Thousand Oaks: Sage; 2010. p. 531–55.

O'Cathain A, Murphy E, Nicholl J. Integration and publications as indicators of "yield" from mixed methods studies. J Mixed Methods Res. 2007a;1(2):147–63.

O'Cathain A, Murphy E, Nicholl J. Why, and how, mixed methods research is undertaken in health services research in England: a mixed methods study. BMC Health Serv Res. 2007b;7(1):85.

O'Cathain A, Murphy E, Nicholl J. The quality of mixed methods studies in health services research. J Health Serv Res Policy. 2008;13(2):92–8.

O'Cathain A, Murphy E, Nicholl J. Three techniques for integrating data in mixed methods studies. BMJ. 2010;341:c4587. https://doi.org/10.1136/bmj.c4587

Onwuegbuzie AJ, Johnson RB. The validity issue in mixed research. Res Schools. 2006;13(1): 48–63.

Pace R, Pluye P, Bartlett G, Macaulay AC, Salsberg J, Jagosh J, Seller R. Testing the reliability and efficiency of the pilot mixed methods appraisal tool (MMAT) for systematic mixed studies review. Int J Nurs Stud. 2012;49(1):47–53.

Pluye P, Robert E, Cargo M, Bartlett G, O'Cathain A, Griffiths F, Boardman F, Gagnon MP, Rousseau MC. Proposal: a mixed methods appraisal tool for systematic mixed studies reviews. 2011. Retrieved from http://mixedmethodsappraisaltoolpublic.pbworks.com.

Sale JE, Brazil K. A strategy to identify critical appraisal criteria for primary mixed-method studies. Qual Quant. 2004;38(4):351–65.

Scammon DL, Tomoaia-Cotisel A, Day RL, Day J, Kim J, Waitzman NJ, … Magill MK. Connecting the dots and merging meaning: using mixed methods to study primary care delivery transformation. Health Serv Res. 2013;48(6pt2):2181–207.

Souto RQ, Khanassov V, Hong QN, Bush PL, Vedel I, Pluye P. Systematic mixed studies reviews: updating results on the reliability and efficiency of the mixed methods appraisal tool. Int J Nurs Stud. 2015;52(1):500–1.

Tashakkori A, Teddlie C. Quality of inferences in mixed methods research: calling for an integrative framework. In M. M. Bergman (Ed.), Advances in mixed methods research: Theory and applications. Thousand Oaks: Sage; 2008. p. 101–19.

Tashakkori A, Teddlie C. Handbook of mixed methods in social & behavioral research. Thousand Oaks: Sage; 2010.

Teddlie C, Tashakkori A. Foundations of mixed methods research: integrating quantitative and qualitative approaches in the social and behavioral sciences. Thousand Oaks: Sage; 2009.

Tong A, Sainsbury P, Craig J. Consolidated criteria for reporting qualitative research (COREQ): a 32-item checklist for interviews and focus groups. Int J Qual Health Care. 2007;19(6):349–57.

Wisdom JP, Cavaleri MA, Onwuegbuzie AJ, Green CA. Methodological reporting in qualitative, quantitative, and mixed methods health services research articles. Health Serv Res. 2012;47(2):721–45.

Zhang W, Creswell J. The use of "mixing" procedure of mixed methods in health services research. Med Care. 2013;51(8):e51–7.

Section II

Innovative Research Methods in Health Social Sciences

Innovative Research Methods in Health Social Sciences: An Introduction

61

Pranee Liamputtong

Contents

1 Introduction	1072
2 Innovative/Creative Research Methods	1073
3 The Innovative Researcher	1074
4 About the Innovative Research Methods in Health Social Sciences Section	1075
4.1 Theoretical Lens	1075
4.2 Arts-Based and Visual Methods	1076
4.3 The Body and Embodiment Research	1080
4.4 Digital Methods	1082
4.5 Textual (Plus Visual) Methods of Inquiry	1085
5 Conclusion and Future Directions	1087
References	1088

Abstract

Innovative, or creative research, methods have become increasingly popular in the last few decades. In this chapter, I will include several salient issues on which chapters in the section on "Innovative Research Methods in Health Social Sciences" can be situated. First, I discuss some ideas about innovative and creative methods. This is followed with the notion of those who practice innovative methods: the innovative researcher. I will then bring readers through a number of innovative and creative methods that researchers have adopted in their research. These include the theoretical lens, arts-based and visual research methods, the body and embodiment research, digital methods, and textual (plus visual) methods of inquiry. As an innovative researcher, our choice of innovative methods primarily depends on the questions we pose; the people who are involved; our moral, ethical, and methodological competence as researchers; and the sociocultural environment of

P. Liamputtong (✉)
School of Science and Health, Western Sydney University, Penrith, NSW, Australia
e-mail: p.liamputtong@westernsydney.edu.au

© Springer Nature Singapore Pte Ltd. 2019
P. Liamputtong (ed.), *Handbook of Research Methods in Health Social Sciences*,
https://doi.org/10.1007/978-981-10-5251-4_1

the research. As we are living in the world that continue to change, it is likely that health and social science researchers will continue to experiment with their creative methods in order to ensure the success of their research. I anticipate that in the future, we will see even more creative methods that researchers will bring forth.

> **Keywords**
> Innovative methods · Creative research methods · The innovative researcher · Arts-based research · Visual research · The body · Embodiment research · Digital methods · Textual method of inquiry

1 Introduction

> Research differs in the ways in which it is conducted and in the products that it yields. What one needs to research in a situation must be appropriate for the circumstances one addresses and the aims one attempts to achieve. (Eisner 2008, p. 4)

This section presents "new era" in research in the health and social sciences (Sinner et al. 2006). It encourages creative ways that researchers can adopt to access the experiential knowledge of research participants (as well as of their own). Innovative, or creative research, methods have become increasingly popular in the last few decades. They were created and promoted within the USA and the UK, but have now been widely adopted by researchers in other parts of the globe (Gwyther and Possamai-Inesedy 2009).

In 2005, Yvonna Lincoln and Norman Denzin (2005, p. 1116) point out that the current moment in research was "the methodological contested present." Even the next moment, they argue, will still be "methodological contested" moment as "struggle and contestation" will continue. Indeed, "we are in the new age where ... new experimental works will become more common" (Denzin and Lincoln 2005, p. 26). In this new age, we have witnessed more creative forms of research bubbled up in health and social sciences. Many of these are included in this section.

As a researcher, Elliot Eisner (2008, p. 5) suggests knowledge, or our understanding of the world, can come in many ways. He contends that knowledge is "not always reducible to language." Knowledge also "comes in different forms," and "the forms of its creation differ." This is also what Norman Denzin (2010, p. 425) has advocated. He encourages health and social science researchers to embrace "methodological diversity." This is because it will bring "new ways of knowing" which will lead to "new knowledge" (Simons and McCormack 2007).

Within the fractured world in which we are now living, I contend that there are many situations where no conventional method (constructivist, positivist, or mixed methods) will work and can be alienating for some individuals. It is essential that health and social science researchers adopt unconventional alternative approaches (Liamputtong 2007). Indeed, many researchers have increasingly realized the value of more creative inquiries in working with marginalized groups (Gillies and Robinson 2012; Baker et al. 2015). Research that involves children, for example, traditional research methods such as questionnaires, in-depth interviews, or focus groups may be problematic. Creative methods that treat children as research participants instead of research objects will allow children to play an important role in the research. These creative methods also allow researchers to gain a deeper insight into the understanding and experiences of children (Angell and Angell 2013).

Increasingly too, there have been many health and social science researchers who believe in the social justice value and attempt to change the social conditions of people and communities (Mertens et al. 2009; Denzin 2015; Bryant 2016; Denzin 2017). This is precisely what Denzin (2010) has encouraged researchers who are situated within the "moral and methodological community" to do. This has resulted in the development of innovative and creative approaches in many parts of the globe. These creative approaches are also in expansion. In this section, a number of innovative and creative ways that researchers have utilized in their research is presented.

In this chapter, I will include several salient issues on which chapters in the Innovative research methods in health social sciences section can be situated. First, I discuss some ideas about innovative and creative methods. This is followed with the notion of those who practice innovative methods: the innovative researcher. I will then bring readers through a number of innovative and creative methods that researchers have adopted in their research.

2 Innovative/Creative Research Methods

Innovative or creative research methods here refer to methods which are not situated neatly within the traditional research methods. Taylor and Coffey (2008, p. 8) suggest that innovative methods include "the creation of new designs, concepts and ways doing things." However, Rose Wiles et al. (2011, p. 588) contend that innovative methods are not necessarily restricted to "the creation of new methods." They can also refer to "advances or developments of 'tried and tested' research methods." Often, innovative methods are created in an attempt to ameliorate some facet of the research practice which may not work properly (Wiles et al. 2011).

Innovation has been classified into three main areas: new designs or methods, new concepts, and new ways of doing research (Taylor and Coffey 2008; Xenitidou and Gilbert 2009). New designs or methods include data collection and analysis methods as well as representation of research. New concepts embrace frameworks and methodological concepts. New ways of doing research refer to "new applications" as well as "crossing disciplines" (Xenitidou and Gilbert 2009, p. 6). In their report, Maria Xenitidou and Nigel Gilbert (2009, p. 7) define innovative research practices as "those which involve technological innovation, cross disciplinary boundaries and/or extend existing methodologies and methods."

Some researchers refer to innovative methods as "creative" research methods (Kara 2015; Bryant 2016). Creative methods entail "imagination" (Wilson 2010; Bryant 2016). Imagination is "transformative ways" that researchers perceive and practice their research (Bryant 2016). For Les Back (2012), imaginative methods can widen the sociological imagination of the researchers. They can also help in the pursuit to democratize the research process. Indeed, Caroline Ellis and Art Bochner

(2008) argue that in planning for research, imagination is as crucial as the rigor if researchers wish to produce ethical research. Nick Wilson (2010, p. 368) tells us that imagination "thrives at the edge of things, between the gaps." For health and social science researchers, this imagination helps to open up creative ways of conducting research that is departed from the traditional ways of doing research. As readers will see, contributors in this section have used their imagination to create innovative and creative methods that they adopted in their research.

In their research with children in a Scottish radiology department, Sandra Mathers et al. (2010) used "graffiti walls" as their creative research method as a means for the children to tell their own stories. Large sheets of paper were painted with "a breeze block pattern" to make it similar to a wall. A cartoon "worm" that invited the children with a phrase "Please use our graffiti wall" was used as starter graffiti because it would help the children to feel more relaxed about adding their own graffiti. Marker pens and crayons were supplied. The graffiti walls were on display for 7 days. The graffiti wall was not supervised and hence the children felt free to post anything on the wall. However, the content was checked daily to ensure that no offensive material was posted. All sheets were photographed at the end of this stage before storage. Mathers and colleagues contend that graffiti walls allowed the children to be able to express their thoughts and experiences. Although the method did not yield a large amount of information (reflecting the nature of graffiti), it was acceptable to the children, and they did participate in the activity.

Innovative research methods have also been coined by Sharlene Hesse-Biber and Patricia Leavy (2006) as "emerging research methods." These emergent research methods, according to Hesse-Biber and Leavy (2006, p. xi), are "the logical conclusion to paradigm shifts, major developments in theory, and new conceptions of knowledge and the knowledge-building process." Often, emergent methods are invented to examine research questions that orthodox research methods may not be able to sufficiently address. Theoretical paradigm shifts in the health and social sciences have allowed innovative methods to be developed and this has also resulted in the achievement of new theoretical perspectives within the disciplines. In order to cultivate rich new meanings, emergent methods alter traditional ways of knowing. Emergent methods demand researchers engage "at the border" of conventional methods. Often, we need to operate from a multidisciplinary or interdisciplinary ground (Hesse-Biber and Leavy 2006). According to Hesse-Biber and Leavy (2006), emerging methods can be seen as "hybrid" because often they are borrowed and adapted from different disciplines to generate new tools, or reshape existing tools in order to answer new and often complex questions. Emergent methods can be both qualitative and quantitative methods, or a combination of both approaches, as illustrated in chapters in this section.

3 The Innovative Researcher

I contend that the innovative researcher is an individual who Norman Denzin and Yvonna Lincoln (2005, p. 4) have referred to as the "bricoleur" or "quilt maker." Denzin and Lincoln suggest that a bricoleur "makes do by 'adapting the bricoles of the world.'" Bricoles are "the odds and ends, the bits left over" (Harper 1987, p. 74). Thus, the bricoleur is "a Jack of all trades, a kind of professional do-it-yourself" (Lévi-Srauss 1966, p. 17). Denzin and Lincoln contend that there are various types of bricoleurs: interpretive, narrative, theoretical, methodological, and political (p. 4). The result of bricoleur's method, what they coin as "the solution (bricolage)" is "an [emergent] construction that changes and takes new forms as the bricoleur adds different tools, methods, and techniques of representation and interpretation to the puzzle" (p. 4). Innovative researchers are energized by their curiosity and creativity (Jones and Leavy 2014).

As a quilt maker, the innovative researchers deploy "the aesthetic and material tools" of their discipline, using whatever methods, strategies, and research materials which are accessible to them (Becker 1998, p. 2). As Cary Nelson et al. (1992) contend, what researchers choose as their research practices are depended on the question that is asked and the research context. It also depends on "what the researcher can do in that setting" (Denzin and Lincoln 2005, p. 4). Thus, if the researchers have to initiate or construct new tools, as a quilt maker must do, they will do so (Denzin and Lincoln 2005). In essence, the innovative researchers are what Alasuutari (2007, p. 513) refers to as the "up-to-date, well-informed" researchers who make use of creative methods in their research. For example, nowadays, the availability of mobile phones means that most participants would already have the tools that they need (Phillips 2014). Hence, innovative researchers have made used of the mobile phone as a tool for research, including mobile interviews (see also ▶ Chaps. 80, "Cell Phone Survey," and ▶ 81, "Phone Surveys: Introductions and Response Rates").

4 About the Innovative Research Methods in Health Social Sciences Section

The section is loosely organized into two main parts. Part one includes some theoretical aspects which are relevant to creative methods that are discussed in the section. Part two embraces a number of innovative and creative research methods that the innovative researchers (the bricoleurs, the quilt makers) have used to answers their research questions (solve their puzzles). I will discuss them in turn in the following sections.

4.1 Theoretical Lens

Two chapters deserve to be treated as the theoretical lens on which chapters that follow are based. Viv Burr, Angela McGrane, and Nigel King introduce the personal construct psychology (PCP) theory developed by George Kelly that innovative researchers can adopt in their creative methods in ▶ Chap. 62, "Personal Construct Qualitative Methods." PCP was developed in clinical psychological practice and has now cultivated various techniques that allow individual to gain insight into their own perceptions and experiences. Within the research context, PCP methods allow research participants to articulate their own experiences and behavior, as well as to gain insight into their own constructions of the world. PCP methods can be viewed as "participant-led" methods. In this chapter, the authors contend that PCP methods offer many values to the qualitative researcher: "They have the advantage of being highly flexible and can be adapted for use in a wide variety of research topics and settings, providing opportunities for qualitative researchers to create innovative ways of researching."

In ▶ Chap. 63, "Mind Maps in Qualitative Research," basing on the visual approach, Johannes Wheeldon and Mauri Ahlberg write about mind maps that qualitative researchers can adopt in their research. Mind maps, the authors tell us, are valuable tools for researchers because they provide "a mean to address researcher bias and ensure data is collected in ways that privilege participant experience." Minds map refers to "diagrams used to represent words, ideas, and other concepts arranged around a central word or idea." They are more flexible than other types of maps and present ideas in a number of ways. Mind mapping is built on the idea that individuals learn in different ways and think using a combination of words, images, and graphics. Wheeldon and Ahlberg contend that visually oriented approaches such as mind maps can assist researchers to plan their research, collect and analyze data, as well as to present their research findings.

4.2 Arts-Based and Visual Methods

Arts-based research is an emergent, appealing, and expanding terrain (Chilton and Leavy 2014). According to Shaun McNiff (2008, p. 29), arts-based research refers to "the systematic use of the artistic process, the actual making of artistic expressions in all of the different forms of the arts, as a primary way of understanding and examining experience by both researchers and the people that they involve in their studies." It became to be known between the 1970s and the 1990s, and has now been adopted by many innovative researchers. Arts-based research embraces the assumptions of "the creative arts" in health and social science research. Chilton and Leavy (2014, p. 403) contend that "the partnership between artistic forms of expression and the scientific process integrates science and art to create new synergies and launch fresh perspectives."

Arts-based research, as Suzanne Thomas (2001, p. 274) writes, possesses "the power to provoke, to inspire, to spark the emotions, to awaken visions and imagining, and to transport others to new worlds." The arts can assist researchers as they attempt to "portray lives" and light up "untold stories" (Coles and Knowles 2001, p. 211; Chilton and Leavy 2014, p. 403). Through the arts, we can reach people's "inner life" through their "stories, metaphors, and symbols, which are recognised as both real and valuable" (Chilton and Leavy 2014, p. 403).

Norman Denzin (2000, p. 261) argues that arts-based research is essentially "a radical ethical aesthetic." Tom Barone (2001, p. 26) suggests that arts-based research methods are valuable for "recasting the contents of experience into a form with the potential for challenging (sometimes deeply held) beliefs and values." Thus, arts-based research methods have also become "socially responsible, politically activist, and locally useful research methodologies" (Finley 2005, p. 681).

The arts-based inquiry is situated within a tradition of participatory action research (PAR) in the health and social sciences (Higginbottom and Liamputtong 2015). Researchers adopting this line of inquiry call for a "reinterpretation of the methods" as well as its ethics concerning human social research (Finley 2005, p. 682). They attempt to develop inquiry involving action-oriented processes that provide benefits to the local community where the research is undertaken. Arts-based research, Susan Finley (2005, p. 686) maintains, is carried out to "advance human understanding." Primarily, arts-based researchers attempt to "make the best use of their hybrid, boundary-crossing approaches to the inquiry to bring about culturally situated, political aesthetics that are responsive to social dilemmas."

In ▶ Chap. 64, "Creative Insight Method Through Arts-Based Research," Jane Edwards suggests that arts-based research (ABR) "represents a way of using the arts to facilitate and enhance processes within research." ABR embraces "creating works" that health and social science researchers have adopted and these include poetry, narrative fiction, plays, painting, drawing, or song writing. This is in line with what Patricia Leavy (2015) has recently identified as arts-based research: poetry, narrative inquiry, music, dance, performance, and visual arts. Arts-based research offers researchers "boundless possibilities for inventiveness, discovery, and creativity" (Viega 2016, p. 3).

In the last few decades, researchers in the health social sciences have started to embrace the use of visual research so that their understanding of the human condition can be enhanced (Harper 2012; Rose 2012; Andriansen 2012; Gubrium et al. 2015; Kolar et al. 2015; Leavy 2015). There are a wide variety of visual forms that are available to researchers. Each of these visual forms can result in different ways of knowing. Thus far, we have witnessed visual forms such as photographs, drawings, cartoons, graffiti, maps, diagrams, films, video, signs, and symbols have been adopted in research in the health and social sciences (Weber 2008). Most often, however, health and social science researchers use visual methods together with some form of interviewing.

Images speak louder than words (Harper 2002, 2012) and there is a saying that "a picture is worth a thousand words" (see ▶ Chap. 65, "Understanding Health Through a Different Lens: Photovoice Method"). A picture can be captured instantly at a glance, but those thousand words would need time to read or to listen to (Weber 2008). Eisner (1995, p. 1) too argues that images provide an "all-at-once-ness" which help to explain things that would be difficult to capture through words and numbers. Images can also invoke social justice actions. Weber (2008, p. 47) invites us to visualize this:

> Take, for example, the powerful photograph taken by Nick Ut during the Vietnam War of an obviously terrified young Vietnamese girl running naked down a street to flee a napalm fire bomb. It may have done more to galvanise the antiwar movement in the West than all the scholarly papers on the horrors of war.

The use of visual images as a data collection tools in research can assist health social science researchers in many ways. Weber (2008) suggests a few. Images can assist us to capture knowledge that is hidden, elusive, or hard-to-put-into-words which would be ignored or remain hidden without the use of visual forms. Photographs, for example, can greatly invoke "information, affect and reflection" (Rose 2007, p. 238) that written texts may not be able to do. Similarly, drawing and painting can grasp "capacity, range, and emotions" which are not easily produced in words alone (Russell and Diaz 2013, p. 2). Images can assist researchers to pay attention to things in different ways. Ordinary things can become extraordinary with the use of images. This can make us embrace new ways of doing things. Often, images can invoke new research questions and inspire the research design (Weber 2008).

Authors of chapters in Part 2 in the Innovative research methods section make use of visual images in their research. In Understanding health through a different lens: Photovoice method, Michelle Teti, Wilson Majee, Nancy Cheak-Zamora, and Anna Maurer-Batjer tell us that in health research and practice, visual methods are common tools. Public health researchers have appreciated the values of photography as a means "to understand health issues from the perspectives of those living with health challenges, inform health interventions, and engage community members in identifying and solving health problems." In their chapter, however, Teti and colleagues write about the application of the Photovoice method to HIV/AIDS and Autism Spectrum Disorder research and practice. Within community-based participatory action research (PAR), the method of Photovoice has emerged as an innovative means of working with marginalized people and in cross-cultural research. Photovoice method allows individuals to record and reflect the concerns and needs of their community via taking photographs. It also promotes critical discussion about important issues through the dialogue about photographs they have taken. Their concerns may reach policy-makers through public forums and the display of their photographs. By using a camera to record their concerns and needs, it permits individuals who rarely have contact with those who make decisions over their lives, to make their voices heard (Wang 1999; Wang and Burris 1994, 1997; Wang et al. 2004; Lopez et al. 2005a, b; Castleden et al. 2008; Rhodes et al. 2008; Hergenrather et al. 2009; Teti et al. 2012; Sanon et al. 2014; Switzer et al. 2015; Maratos et al. 2016; Rose et al. 2016).

In ▶ Chap. 66, "IMAGINE: A Card-Based Discussion Method," Ulrike Felt, Simone Schumann, and Claudia Schwarz-Plaschg write about a card-based discussion method, what they also coin as the IMAGINE methodology, that they employed in their qualitative research and engagement in the Austrian context. This method extends the focus group method commonly adopted in qualitative inquiry in that several prepared cards are used to create some imaginations among the participants. They tell us that by providing different sets of cards for participants to work with, the method can "stimulate and support the process of developing imaginations with regard to emerging technoscientific and other complex social issues." In an interaction with other participants in the group, the method can "enhance the capacity to gradually assemble the 'building blocks' that are used to assess an issue and construct a position." The core interest of this card-based discussion method is "to investigate how people analyze and relate to specific matters of concern in the present and to their potential futures, i.e., how they think a specific part of 'the world' works and how they imagine it might or should work in the future."

E. Anne Marshall, in ▶ Chap. 67, "Timeline Drawing Methods," discusses the value of timeline drawing method in qualitative research. Timelines refer to "visual representations of particular and selected events or 'times' in a person's life" (see also Guenette and Marshall 2009; Adriansen 2012; Jackson 2012; Kolar et al. 2015; Rimkeviciene et al. 2016). A timeline is coined by Lynda Berends (2011, p. 2) as "visual depiction of a life history, where events are displayed in chronological order." Timelines, depending on the research focus, can cover a participant's lifetime, a particular number of months or years. Timelines can be created by a participant or a researcher, or collaboratively. They can also be constructed at different times in the research process. The authors suggest that timelining "adds a visual representation related to the experience that can anchor the interview and help focus the participant on key elements." The method is especially suitable for sensitive and complex research or when the oral language expression of the participants is limited due to a number of situations: "Timelining can provide participants with a way to engage their stories deeply and even help to create new meanings and understandings."

▶ Chapter 68, "Semistructured Life History Calendar Method" is written by Ingrid A. Nelson on semistructured life history calendar as a method in health social science research. The Life History Calendar (LHC) has been used as one key method for gathering quantitative life course studies (see Belli et al. 2009; Kendig et al. 2014; Morelli et al. 2016; Vanhoutte and Nazroo 2016). In LHC method, researchers utilize "a preprinted matrix, with time cues running horizontally across the page and topic cues running vertically down the page, to help respondents piece together their past." However, traditional LHC is highly structured. It does not allow researchers to obtain in-depth understanding about why or how the life story of the participants is unfolded as it is. Due to the need for comprehensive contextual data in qualitative life course research, Nelson created a semistructured adaptation of the LHC. Nelson contends that "this methodological innovation incorporates research participants' attitudes and aspirations, interpretations and explanations of life transitions, and major events that would not be captured in standardized event histories." The semistructured LHC method attains "nuanced longitudinal narratives" by marrying "the characteristic detail across multiple domains of the traditional LHC with in-depth narrative."

The last chapter in this section, ▶ Chap. 69, "Calendar and Time Diary Methods," is written by Ana Lucía Córdova Cazar and Robert F. Belli on time use research, focusing on calendar and time diary methods which have been shown to be particularly effective for assessing human well-being. Time use research methods are composed of both time diaries and life histories or event history calendars. It has been widely accepted that time use research allows researchers to understand human behavior and its intrinsic relationship with individual and social well-being. It has attracted the interest of researchers from many disciplines including the health and social sciences. The authors argue that "although these methods forego the standardization of question wording (the most prevalent approach in traditional survey interviewing), they are nevertheless able to produce reliable and valid responses, while also encouraging conversational flexibility that assists respondents to remember and correctly report the interrelationships among past events."

4.3 The Body and Embodiment Research

Corporeal realities, or embodiment, has become a site of attention among feminist and postmodern researchers (Gonzalez-Arnal et al. 2012; Lennon 2014). This has resulted in the advancement of creative methods that can be used to elicit the knowledge of the corporeality (the body) within the social sciences (Hesse-Biber and Leavy 2006; Gray and Kontos 2015). According to Hesse-Biber and Leavy (2006, p. xix), "all social actors are embodied actors – our experience, vision, and standpoint are embodied. We know and experience social reality from our embodied standpoints within the society." The embodiment is embraced by Elizabeth Grosz (1994) as the "lived" or "inscribed" body. The lived body symbolizes experiential knowledge that is connected with the physicality of an individual (Hesse-Biber and Leavy 2006; Tarr and Thomas 2011). It is through the lived body that meanings are brought about (Grosz 1994; Liamputtong and Rumbold 2008; Tarr and Thomas 2011; Gonzalez-Arnal et al. 2012; Lennon 2014). As a researcher, we can attain crucial knowledge by the lived body of the research participants. At the same time, we can also access this important knowledge through our own body (Hesse-Biber and Leavy 2006; Todres 2007; Gonzalez-Arnal et al. 2012; Lennon 2014).

In their conversation about how to obtain knowledge about the identities of individuals, David Gaunlett and Peter Holzwarth (2006, p. 8) contend that "we need research which is able to get a full sense of how people think about their own lives and identities, and what influences them and what tools they use in that thinking, because those things are the building blocks of social change." This has prompted many embodiment researchers to invent creative methods that can allow them to do so. Some of these are included in this section.

The body mapping method, as Bronwyne Coetzee, Rizwana Roomaney, Nicola, Willis, and Ashraf Kagee write in ▶ Chap. 70, "Body Mapping in Research," is a "research tool that prioritizes the body as a way of exploring knowledge and understanding experience." Body mapping is a creative method that really grabs the imagination of research participants (Brett-MacLean 2009; Orchard et al. 2014; de Jager et al. 2016; Ebersöhn et al. 2016). In the body mapping method, life-size body drawings are drawn (or painted) to visually portray "aspects of people's lives, their bodies and the world they inhabit" (see also Gastaldo et al. 2012). The method has also been coined by researchers in the human and social sciences as "body map storytelling" because the meaning of a body map can only be fully understood by the story and experience as told by the individual who creates the body map.

Anna Bagnoni, in ▶ Chap. 71, "Self-portraits and Maps as a Window on Participants' Worlds," presents two creative methods in her research regarding identities. One is the body mapping and the other the self-portrait method, which she created herself in her doctoral study. Within this method, the body is located at the center as in the body mapping method. She provides the participants with a blank sheet of paper, colored felt-tips, and pens. She then invites the participants to show who they are at the present moment in their lives. They are also asked to add anything that is important to them at that moment. In her study, the self-portrait method was invented as "a creative input to support in particular those who might not feel too comfortable with words when asked about personal and subjective issues, and might feel reassured by writing, drawing or doodling as alternative forms of self-expression."

Creative research methods that incorporate a bodily experience also include walking and talking together with the research participants (Brown and Durrheim 2009; Carpiano 2009; Evans and Jones 2011; Garcia et al. 2012; Begeron et al. 2014; Holton and Riley 2014). As Alexandra King and Jessica Woodroffe write in ▶ Chap. 72, "Walking Interviews", walking interviews involve researchers and participants talking while walking together. According to King and Woodroffe, "as a shared corporeal or bodily experience, the physical act of walking alongside someone shapes the research encounter, aiding the development of an intersubjective understanding of the physiological particularities of a respondent's lifeworld." Walking interviews are "a valuable means of deepening understandings of lived experiences in particular places." Walking interviews generate "rich, detailed and multi-sensory data." For health research, the authors contend, walking interviews permit researchers to "engage with the nature and meaning of bodily experiences including physical well-being, illness, ageing or disability, and to explore the ways in which these experiences are interwoven with the places in which people live and the meanings they have for their lives."

Similarly, in ▶ Chap. 73, "Participant-Guided Mobile Methods," Karen Block, Lisa Gibbs, and Colin MacDougall discuss what they coin as the participant-guided mobile method. Participant-guided mobile methods blend a participant-led guided tour with an in-depth interviewing method (see also Finlay and Bowman 2016). The authors suggest that the tour can occur "on foot or using a vehicle and can even be virtual, investigating participants' online worlds or using technologies such as Google Earth or Google Maps to explore otherwise less accessible places." With the participant-guided mobile method, researchers are able to access "multiple types of data simultaneously; adding contextual, observational, and potentially also visual data to interviews conducted in a naturalistic setting."

Voice, according to Hesse-Biber and Leavy (2006, p. xxv), is also a part of the corporeal realities because voice occurs "in a cultural context, in relation to self, and in relation to others." Voice is hinged on a mutual form of expectation. Voice, when it is expressed in certain ways, such as the digital storytelling method, allows individual's stories to be heard. In ▶ Chap. 74, "Digital Storytelling Method," Brenda Gladstone and Elaine Stasiulis present the digital storytelling method that they employed in their research with young people in Canada. Digital stories refer to "short (2–3 min) videos using first-person voice-over narration synthesized with visual images created in situ or sourced from the storyteller's personal archive." The method permits the first person narrative; the participants have an opportunity to write and use their own voice to tell their own story. This is indeed where the power of the method lies. The method is also situated within the "emergence of arts-based health research" and is adopted widely in community-based participatory research, public health and health promotion research and practice (see Alexandra 2015; Otañez and Guerrero 2015).

4.4 Digital Methods

Digital methods refer to the application of "online and digital technologies" that researchers utilize to gather and analyze research data (Snee et al. 2016, p. 1). Globally, the digital has become a significant part of our daily life and researchers within the health and social sciences have embraced it as part of their creative research methods (Turney and Pocknee 2005; Liamputtong 2006; Dillman 2007; Hewson 2014; Rogers 2013; Synnot et al. 2014; Halfpenny and Proctor 2015; Iacono et al. 2016). We have witnessed a number of research projects that make use of digital methods in recent time. Annette Markham (2004, p. 95) writes:

> [T]he internet provides new tools for conducting research, new venues for social research and new means for understanding the way social realities get constructed and reproduced through discursive behaviours.

Digital methods have many advantages over more conventional research methods. Digital communication can reach a large number of people across different geographical and sociocultural boundaries (Hessler 2006; Cater 2011; Sue and Ritter 2012; Iacono et al. 2016). Chris Mann and Fiona Stewart (2000, p. 80) argue that "the global range of the Internet opens up the possibilities of studying projects which might have seemed impracticable before." Gillian Dunne (1999), for example, was able to conduct in-depth interviews with gay fathers from different international locations including the UK, New Zealand, Canada, and the USA.

Digital methods provide possibilities to reach a terrain of vulnerable participants, such as people with disabilities, mothers at home with small children, older people, and people from socially marginalized groups such as gays and lesbians, who may not be easily accessed in face-to-face research methods (Mann and Stewart 2002; Elford et al. 2004; Seymour and Lupton 2004; Egan et al. 2006; Liamputtong 2006, 2013; Synnot et al. 2014; see ▶ Chap. 78, "Synchronous Text-Based Instant Messaging: Online Interviewing Tool" and ▶ Chap. 79, "Asynchronous Email Interviewing Method"). These vulnerable individuals can make contact with others from their familiar and physically safe locations. People with disabilities who have access to email and necessary online information can take part in research without having to leave home or be mobile (Seymore and Lupton 2004). Digital methods also permit health and social science researchers a possible vehicle for connecting with people situated within restricted access like schools, hospitals, cult and religious groups, bikers, surfers, punks, and so on.

In social science areas, digital methods provide the possibilities of carrying out research within politically sensitive or dangerous areas (Mann and Stewart 2002).

Due to the anonymity and physical distance, both the researchers and the participants are protected (see Coomber 1997, for example). Some highly sensitive and vulnerable participants, such as political and religious dissidents or human rights activists, will be more likely to participate in online research without excessive risk. Researchers can access censored and politically sensitive information without being physically in the field. People living or working in war zones, or sites of criminal activity, or places where diseases abound can be accessed without needing to combat the danger involved in actually visiting the area. Digital methods also permit researchers to distance themselves physically from research sites. This helps to eliminate the likelihood of suspicion that might alienate some participants.

In this section, several innovative methods that situated within the digital methods are included. In ▶ Chap. 75, "Netnography: Researching Online Populations," Stephanie Jong writes about netography in health social sciences. Netnography, according to Jong, can be seen "as a means of researching online communities in the same manner that anthropologists seek to understand the cultures, norms and practices of face-to-face communities, by observing, and/or participating in communications on publically available online forums" (see also Nelson and Otnes 2005; Bowler 2010; Kozinets 2010, 2015). In this chapter, Jong explores the transition of netnography, a consumer marketing method, to the field of health social science research using the example of her study related to fitness communities on social networking sites (SNSs).

In the past two decades, we have witnessed a growth in the area of online survey methodology (see Evans and Mathur 2005; Wright 2005; Lieberman 2008; Murray et al. 2009; Greenlaw and Brown-Welty 2009; Sue and Ritter 2012; Kramer et al. 2014). As the general population becomes increasingly made up of "digital natives" (Kramer et al. 2014), web surveys will become more prominent, particularly in the health domain (Riper et al. 2011; Kramer et al. 2014). In ▶ Chap. 76, "Web-Based Survey Methodology," Wright contends that an online survey method is a valuable tool for health researchers. Many health researchers have employed online surveys to access various population groups, including consumers, patients, caregivers, health care professionals, online support community participants, and policy-makers in the health care system. In his chapter, Wright provides salient issues relevant to the use of online surveys to reach a number of stakeholders in the health care system, including patients, caregivers, and health care providers.

Recently, we have witnessed an increase use of blogs in research (see Wakeford and Cohen 2008; Chenail 2011; McCosker and Darcy 2013; Harricharan and Bhopal 2014; Saiki and Cloyes 2014; Wilson et al. 2015; Genoe et al. 2016). In ▶ Chap. 77, "Blogs in Social Research", Nicholas Hookway and Helen Snee write about blogs in social research. Blogs, accordingly to the authors, are "the quintessential early twenty-first century text" that made "the boundary between private and public" obscured. Blogs refer to "interactive and multimedia, converging text, image, video, GIFS and other types of media into one space." For health social science researchers who are interested in the everyday life of individuals, blogs can offer rich and first-person textual accounts. Hookway and Snee contend that "embracing new confessional technologies like blogs can provide a powerful addition to the qualitative researcher's toolkit and enable innovative research into the nature of contemporary selves, identities and relationships."

Another digital method that has become popular in the health science is an online interviewing method, either synchronous of asynchronous. As we have witnessed, electronic mail (email) has been used widely as an effective tool of communication (Meho 2006; Burns 2010; Brondani et al. 2011; Cook 2012; Ratislavová and Ratislav 2014; Bowden and Galindo-Gonzalez 2015). In Synchronous Text-Based Instant Messaging: Online Interviewing Tool, Gemma Pearce, Cecilie Thøgersen-Ntoumani, and Joan Duda discuss online interviewing using synchronous text-based instant messaging. The synchronous text-based online interviewing method is "a method of interviewing participants online using an instant messaging service to type to each other (text-based) at the same time in a conversational style (synchronous)." Using an Instant Messaging service to conduct interviews is unique in its "ability to carry out a synchronous discussion with the participant." This method is essentially useful for situations where face-to-face or telephone interviews are problematic.

Similar to the previous chapter, in Asynchronous Email Interviewing Method, Mario Brondani and Rodrigo Mariño contend that "in the era of multimedia and *at-finger-tips* convenient information, electronic communication can provide answers to research inquiries in a timely manner." This is particularly so when the researcher does not need to meet face-to-face with the research participants or have difficulties in meeting them personally. In this chapter, the authors discuss the use of email interviews and offer readers nine steps for conducting an email interview which would help readers to get the most out of the method.

Digital methods also include the use of mobile or cell phone via wireless web devices in data collection (Casey and Turnbull 2011; Hesse-Biber 2011). Mobile devices including cell phones and smart phones have become an intrinsic element of peoples' daily lives (Casey and Turnbull 2011. As such, they have become a means for health and social science researchers to conduct their research. In ▶ Chap. 80, "Cell Phone Survey," Lilian A. Ghandour, Ghinwa El Hayek, and Abla Mehio Sibai write about cell phone survey research. In this chapter, the authors tell us that the increase in global cell phones usage has eroded traditional data collection means, particularly landline surveys. This has led to the development of novel survey methods and designs. They contend that using a single landline frame survey is problematic. As a result, researchers have invented methods that can integrate the cell phone and landline frames, and conducted "dual frame" surveys using either overlapping or nonoverlapping modes of integration. The authors also point out that cell phones are likely to become "an inevitable mode" for researchers to collect health survey data particularly when existing barriers are reduced.

The last chapter in this part, ▶ Chap. 81, "Phone Surveys: Introductions and Response Rates," is also on phone survey but its focus is on how to increase response rates. Jessica Broome contends that although telephone surveys have been declined due to the increase of web surveys, phone surveys "are far from becoming extinct." Due to the limited Internet access in some groups such as older and lower income people, a telephone is still a preferred method when researchers need to reach broad cross-sections of a population (Lepkowski et al. 2008; Tomlinson et al. 2009;

DeRenzi et al. 2011; Dillon 2012; Bradley et al. 2012; van Heerden et al. 2014). This is particularly so in the health research arena. Several large-scale surveys such as the Behavior Risk Factor Surveillance System (BRFS), California Health Interview Survey (CHIS), and the Canadian Community Health Survey (CCHS) still rely heavily on telephone surveys. But a phone survey has its own challenges (see O'Toole et al. 2008; Haberer et al. 2010). A critical aspect of phone surveys is the introduction as this can impact greatly on the response rates of the surveys. Broome contends that "introductions that are effective at convincing sample members to participate can help to improve shrinking response rates in this mode." In this chapter, Broome provides some creative suggestions for effective interviewer training that can increase the response rates in phone surveys.

4.5 Textual (Plus Visual) Methods of Inquiry

Researchers in the health and social sciences have used writing as a means for collecting research data (Warkentin 2002; Hesse-Biber and Leavy 2006). Writing can act "as a process of discovery" for both the researchers and the researched (Hesse-Biber and Leavy 2006, p. xxvii). As researchers have to create a new means to provide answers to their particular research questions, the new method that focuses on writing and the textual data has emerged (Hesse-Biber and Leavy 2006). Some of these new ways of collecting research data are included in this section.

Marsha Quinlan, in Freelisting Method, writes about the freelisting method which has been adopted widely in the social sciences and has recently become more popular in health research (see Ryan et al. 2000; Fiks et al. 2011; Huang 2014; Auriemma et al. 2015; Jonas et al. 2015) (▶ Chap. 82, "The Freelisting Method"). A freelist is "a mental inventory of items an individual thinks of within a given category." In a research using the freelisting method, research participants are asked to lists things (or persons) that they see to be part of a realm (for example, "ways to avoid HIV," "breakfast foods," "reasons to fear hospitals," or "treatments for a cough") in whatever order they can think off. The lists that the participants come up with can "tap into local knowledge and its variation" of the community under investigation. In health social science, Quinlan tells us that the freelisting method is "ideal if one wants to find the most culturally salient knowledge (e.g., cut treatment, mosquito control), attitudes towards, or associations with, an issue or topic (e.g., obesity, vaccinations, violence), or different ways locals do something (e.g., prepare a medicine or a food, decide on healthcare)." Data generated from the freelist method "allows the researcher to discover the relative salience of items across all respondents within a given domain." The freelisting method is mostly done through written data but it can also be oral and via the Internet as well.

In ▶ Chap. 83, "Solicited Diary Methods," Christine Milligan and Ruth Bartlett write about the diary method they used in their research. The diary has been adopted as a research method in health social science research (see Galvin 2005; Jacelon and Imperio 2005; Alaszewski 2006; Hyers et al. 2006; Gills and Liamputtong 2009; Nezlek 2012). There are two types of diary method: solicited and unsolicited (personal diaries). The solicited diary method is written for the purpose of research in minds (Elliott 1997; Jacelon and Imperio 2005; Nezlek 2012). The participants explicitly write their diaries as data for the researcher with a full knowledge that their writing will be used in research and will be read and interpreted by another person (Jacelon and Imperio 2005). Through diaries, the researcher can collect data about the day-to-day events of participants, and then further investigate those events in subsequent interviews. Diaries offer researchers hints about the events which are important for the participants as well as their attitudes toward those events (Jacelon and Imperio 2005). People may record their feelings, experiences, observations, and thoughts about a particular aspect of their lives in a diary (Hesse-Biber and Leavy 2006; Nezlek 2012). The method can provide researchers with in-depth understandings of, for example, the experience of living with HIV/AIDS, dealing with daily discrimination, and caring for a child with disabilities and so on. Hence, diary method can be an invaluable vehicle to gather information from some sensitive issues and with hidden and hard-to-reach populations. However, it is noted too that researchers may also combine textual data with visual and/or digital data. As Milligan and Bartlett suggest, a diary can be recorded digitally and visually.

In ▶ Chap. 84, "Teddy Diaries: Exploring Social Topics Through Socially Saturated Data," Marit Halder and Randi Wærdahl tell us about a creative method they used in their research on family lives; what they coin as teddy diaries. After the school reform in Norway in 1997, Teddy bears and teddy diaries were introduced as "a pedagogical device to ease the transition between a student's family and the first year of school." Each new school class receives a teddy bear who will visit every child's home in turn. The teddy bear carries a diary where the bear's experiences in the child's home are recorded. During the first school year, the teddy will visit each student a number of times. The children, or the children together with their parents, write a diary that describes activities that the bear is involved with the child and the family on a given day. The child and the family can decide on topics that they believe worth mentioning to teachers, classmates, and other families as these people would share the diary entries that the child and the parents have written. Teddy diaries, according to Halder and Wærdahl, "can be read as an exchange of normative everyday standards between different homes, and between home and the school public." Arguing from a research perspective, they contend that "researcher effect is relatively low, but the impact of the social, cultural and contextual on the data is very high. What we actually learn from these diaries are topics that a researcher would not necessarily ask about, yet that convey highly saturated information about norms and values and those that are socially accepted. What is exchanged and reinforced by the evaluation of others becomes the most interesting feature of the material." They conclude that "teddy diaries, as naturally occurring data, are a good source of knowledge about the norms, values and ideals in the social context we wish to examine."

Another textual method that is rather creative is the story completion method (SC). Virginia Bruan, Victoria Clarke, Nikki Hayfield, Naomi Moller, and Irmgard Tischner, in Qualitative Story Completion Method, introduce "a novel technique" which provides "exciting potential" to researchers, particularly qualitative researchers (▶ Chap. 85, "Qualitative Story Completion: A Method with Exciting

Promise"). In this method, a researcher writes "the start of a story", referred to as "a story 'stem' or 'cue'." Often, it will be an opening sentence (or two). The participants are then invited to continue or complete the story. The authors tell us that SC was "originally developed as a form of projective test, for use by psychiatrists and clinical psychologists, to assess the personality and psychopathology of clients." The authors also contend that this method holds much potential for researchers, as they have demonstrated vividly in this chapter.

5   Conclusion and Future Directions

In this chapter, I have presented readers with a number of innovative and creative research methods that researchers in the health and social sciences have used in their research. Sue Wilkinson (2004, pp. 271–272) contends that "a method is an interpretation." When researchers choose a method for their research, their decision is not only based on an epistemological and theoretical reason, but also moral and ethical considerations. The contributors to this section have shown this when they adopt innovative or unusual methods in their research. They are attuned to "the epistemological commitments and value assumptions they make" when proposing or using a particular creative research method. The contributors also use their ways of knowing and creativity to forge new ways or revamp old methods in order to meet the needs of their research environment and the people that involved. Research methods, as Hesse-Biber and Leavy (2006, p. xxx) tell us, are "not fixed entities." Often, research methods are "fluid." They "can bend and be combined to create tools for newly emerging issues and to unearth previously subjugated knowledge." This can also be attested in chapters in this section.

As for any research method, I am in no way suggesting that all of these innovative methods will suit all research projects and contexts, nor that these creative methods are better than the orthodox methods. I am not suggesting either that these innovative methods are without challenges. Indeed, each method presents some epistemological, practical and ethical challenges to the researchers and the research participants and these can vary according to different socio-political-cultural situations. We as innovative researchers must bear this in mind in adopting these methods. And of course, there are other creative and innovative methods that health and social science researchers may wish to experiment with that I have not been able to include in this section. There is always a space limit in a book. I would suggest that we do our own experiment with our creative methods and then document it so that other researchers may be able to see and follow our steps. This is the only way we can make our innovative research known by others and hope that someone will adopt it in their future research.

As an innovative researcher, our choice of innovative methods primarily depends on the questions we pose; the people who are involved; our moral, ethical, and methodological competence as researchers; and the sociocultural environment of the research. As we are living in the world that continue to change, it is likely that health and social science researchers will continue to experiment with their creative methods in order to ensure the success of their research. I anticipate that in the future, we will see even more creative methods that researchers will bring forth. It is really exciting indeed.

References

Adriansen HK. Timeline interviews: a tool for conducting life history research. Qual Stud. 2012;3(1):40–55.

Alasuutari P. The globalization of qualitative research. In: Seale C, Gobo G, Gubrium J, Silverman D, editors. Qualitative research practice. London: Sage; 2007. p. 507–20.

Alaszewski A. Using diaries for social research. London: Sage; 2006.

Alexandra D. Are we listening yet? Participatory knowledge production through media practice: encounters of political listening. In: Gubrium A, Harper KG, Otañez M, editors. Participatory visual and digital research in action. Walnut Creek: Left Coast Press; 2015. p. 41–56.

Angell RJ, Angell C. More than just "snap, crackle, and pop" "draw, write, and tell": an innovative research method with young children. J Advert Res. 2013:377–90.

Auriemma CL, Lyon SM, Strelec LE, Kent S, Barg FK, Halpern SD. Defining the medical intensive care unit in the words of patients and their family members: a freelisting analysis. Am J Crit Care. 2015;24(4):e47–55.

Back L. Live sociology: social research and its futures. Sociol Rev. 2012;60:18–39.

Baker NA, Willinsky C, Boydell KM. Just say know: engaging young people to explore the link between cannabis and psychosis using creative methods. World Cult Psychiatry Res Rev. 2015;10(3/4):201–20.

Barone T. Science, art, and the pre-disposition of educational researchers. Educ Res. 2001;30(7):24–9.

Becker HS. Tricks of the trade: how to think about your research while you're doing it. Chicago: University of Chicago Press; 1998.

Belli RF, Stafford FP, Alwin D, editors. Calendar and time diary methods in life course research. Thousand Oaks: Sage; 2009.

Berends L. Embracing the visual: using timelines with in-depth interviews on substance use and treatment. Qual Rep. 2011;16(1):1–9. Retrieved from http://nsuworks.nova.edu/tqr/vol16/iss1/1

Bergeron J, Paquette S, Poullaouec-Gonidec P. Uncovering landscape values and microgeographies of meanings with the go-along method. Landsc Urban Plan. 2014;122:108–21.

Bowden C, Galindo-Gonzalez S. Interviewing when you're not face-to-face: the use of email interviews in a phenomenological study. Int J Doctoral Stud. 2015;10:79–92.

Bowler GM. Netnography: a method specifically designed to study cultures and communities online. Qual Rep. 2010;15(5):1270–5.

Bradley J, Ramesh BM, Rajaram S, Lobo A, Gurave K, Isac S, Gowda GCS, Pushpalath R, Moses S, Sunil KDR, Alary M. The feasibility of using mobile phone technology for sexual behaviour research in a population vulnerable to HIV: a prospective survey with female sex workers in South India. AIDS Care. 2012;24(6):695–703.

Brett-MacLean P. Body mapping: embodying the self living with HIV/AIDS. Can Med Assoc J. 2009;180(7):740–1.

Brondani M, MacEntee M, O'Connor D. Email as a data collection tool when interviewing older adults. Int J Qual Methods. 2011;10(3):221–30.

Brown L, Durrheim K. Different kinds of knowing: generating qualitative data through mobile interviewing. Qual Inq. 2009;15(5):911–30.

Bryant L. Introduction: taking up the call for critical and creative methods in social work research. In: Bryant L, editor. Critical and creative research methodologies in social work (Chapter 1). Hoboken: Taylor and Francis; 2016.

Burns E. Developing email interview practices in qualitative research. Sociol Res Online. 2010;15(4). Retrieved 1 Feb 2014, from http://www.socresonline.org.uk/15/4/8.html

Carpiano RM. Come take a walk with me: the "go-along" interview as a novel method for studying the implications of place for health and well-being. Health Place. 2009;15(1):263–72.

Casey E, Turnbull B. Digital evidence on mobile devices. In: Casey E, editor. Digital evidence and computer crime. 3rd ed. Waltham: Academic; 2011. p. 2–45.

Castleden H, Garvin T, Huu-ay-aht First Nation. Modifying photovoice for community-based participatory indigenous research. Soc Sci Med. 2008;66(6):1393–405.

Cater JK. Skype a cost-effective method for qualitative research. Rehabil Couns Educ J. 2011;4:10–7.

Chenail RJ. Qualitative researchers in the blogosphere: using blogs as diaries and data. Qual Rep. 2011;16:249–54.

Chilton G, Leavy P. Arts-based research practice: merging social research and the creative arts. In: Leavy P, editor. The Oxford handbook of qualitative research. Oxford: Oxford University Press; 2014. p. 403–22.

Cole AL, Knowles JG. Qualities of inquiry: process, form, and "goodness". In: Nielsen I, Cole AI, Knowles JG, editors. The art of writing inquiry. Halifax: Backalong Press; 2001. p. 211–9.

Cook C. Email interviewing: generating data with a vulnerable population. J Adv Nurs. 2012;68(6):1330–9.

Coomber R. Using the Internet for survey research. Sociol Res Online. 1997;2(2). Internet acces http://www.socresonline.org.uk/socresonline/2/2/2.html

de Jager A, Tewson A, Ludlow B, Boydell KM. Embodied ways of storying the self: a systematic review of body-mapping. Forum Qual Soc Res. 2016;17(2). http://www.qualitative-research.net/index.php/fqs/article/view/2526/3986

Denzin NK. Aesthetics and the practices of qualitative inquiry. Qual Inq. 2000;6:256–65.

Denzin NK. Moments, mixed methods, and paradigm dialogs. Qual Inq. 2010;16(6):419–27.

Denzin NK. What is critical qualitative inquiry? In: Cannella G, Pérez M, Pasque P, editors. Critical qualitative inquiry: foundations and futures. Walnut Creek: Left Coast Press; 2015. p. 31–50.

Denzin NK. Critical qualitative inquiry. Qual Inq. 2017;23(1):8–16.

Denzin NK, Lincoln YS. Introduction: the discipline and practice of qualitative research. In: Denzin NK, Lincoln YS, editors. The Sage handbook of qualitative research. 3rd ed. Thousand Oaks: Sage; 2005. p. 1–32.

DeRenzi B, Borriello G, Jackson J. Mobile phone tools for field-based health care workers in low-income countries. Mt Sinai J Med. 2011;78(3):406–11.

Dillman DA. Mail and internet surveys – the tailored design method. 2nd ed. New York: Wiley; 2007.

Dillon B. Using mobile phones to collect panel data in developing countries. J Int Dev. 2012;24:518–27.

Dunne GA. The different dimensions of gay fatherhood: exploding the myths. Report to the Economic and Social Research Council. London: London School of Economics; 1999.

Ebersöhn L, Ferreira R, van der Walt A, Moen M. Bodymapping to step into your future: life design in a context of high risk and high diversity. In: Ronél F, editor. Thinking innovatively about psychological assessment in a context of diversity. Cape Town: Juta; 2016. p. 228–41.

Egan J, Chenoweth L, Mcauliffe D. Email-facilitated qualitative interviews with traumatic brain injury survivors: a new and accessible method. Brain Inj. 2006;20(12):1283–894.

Eisner E. What artistically crafted research can help us to understand about schools. Educ Theory. 1995;45(1):1–13.

Eisner E. Knowing. In: Knowles JG, Cole AI, editors. Handbook of the arts in qualitative research: perspectives, methodologies, examples, and issues. Thousand Oaks: Sage; 2008. p. 3–12.

Elford J, Bolding G, Davis M, Sherr L, Hart G. The internet and HIV study: design and methods. BMC Public Health. 2004;4:39. https://doi.org/10.1186/1471-2458-4-39.

Elliot H. The use of diaries in sociological research on health experience. Sociol Res Online. 1997;2(2). http://www.socresonline.org.ul/socresonline/2/2/7.html. Accessed 2 Mar 2005.

Ellis C, Bochner A. Foreword: opening conversation. Creat Approach Res. 2008;1(2):1–3.

Evans J, Jones P. The walking interview: methodology, mobility and place. Appl Geogr. 2011;31(2):849–58.

Evans JR, Mathur A. The value of online surveys. Int Res. 2005;15(2):195–219.

Fiks AG, Gafen A, Hughes CC, Hunter KF, Barg FK. Using freelisting to understand shared decision making in ADHD: parents' and pediatricians' perspectives. Patient Educ Couns. 2011;84(2):236–44.

Finlay JM, Bowman JA. Geographies on the move: a practical and theoretical approach to the mobile interview. Prof Geogr. 2016:1–12. https://doi.org/10.1080/00330124.2016.1229623

Finley S. Arts-based inquiry: performing revolutionary pedagogy. In: Denzin NK, Lincoln YS, editors. The Sage handbook of qualitative research, 3rd ed. Thousand Oaks: Sage; 2005. p. 681–94.

Finley S. Critical arts-based inquiry. In: Denzin NK, Lincoln YS, editors. The Sage handbook of qualitative research. 4th ed. Thousand Oaks: Sage; 2011. p. 435–50.

Galvin RD. Researching the disabled identity: contextualising the identity transformations which accompany the onset of impairment. Sociol Health Illness. 2005;27(3):393–413.

Garcia CM, Eisenberg ME, Frerich EA, Lechner KE, Lust K. Conducting go-along interviews to understand context and promote health. Qual Health Res. 2012;22(10):1395–403.

Gastaldo D, Magalhães L, Carrasco C, Davy C. Body-map storytelling as research: Methodological considerations for telling the stories of undocumented workers through body mapping. Facilitator Guide. 2012. http://www.migrationhealth.ca/undocumented-workers-ontario/body%20mapping. Accessed 10 Oct 2015.

Gaunlett D, Holzwarth P. Creative and visual methods for exploring identities. Visual Stud. 2006;21(1):82–91.

Genoe MR, Liechty T, Marston HR, Sutherland V. Blogging into retirement: using qualitative online research methods to understand leisure among baby boomers. J Leis Res. 2016;48(1):15–34.

Gills J, Liamputtong P. Walk a mile in my shoes: researching lived experiences of mothers of children with autism. J Family Stud. 2009;15(3):309–19. Special issue on "Parenting around the world".

Gillies V, Robinson Y. Developing creative research methods with challenging pupils. Int J Soc Res Methodol. 2012;15(2):161–73.

Gonzalez-Arnal S, Jagger G, Lennon K. Embodied selves. London: Palgrave; 2012.

Gray J, Kontos P. Immersion, embodiment, and imagination: moving beyond an aesthetic of objectivity in research-informed performance in health. Forum Qual Soc Res. 2015;16(2). Art no. 29. http://nbn-resolving.de/urn:nbn:de:0114-fqs1502290. Accessed 7 Apr 2016.

Greenlaw C, Brown-Welty S. A comparison of web-based and paper-based survey methods: testing assumptions of survey mode and response cost. Eval Rev. 2009;33(5):464–80.

Grosz E. Volatile bodies: towards a corporeal feminism. London: Routledge; 1994.

Gubrium A, Harper KG, Otañez M, editors. Participatory visual and digital research in action. Walnut Creek: Left Coast Press; 2015.

Guenette F, Marshall A. Time line drawings: enhancing participant voice in narrative interviews on sensitive topics. Int J Qual Methodol. 2009;8(1):85–92.

Gwyther G, Possamai-Inesedy A. Methodologies à la carte: an examination of emerging qualitative methodologies in social research. Int J Soc Res Methodol. 2009;12(2):99–115.

Haberer JE, Kiwanuka J, Nansera D, Wilson IB, Bangsberg DR. Challenges in using mobile phones for collection of antiretroviral therapy adherence data in a resource-limited setting. AIDS Behav. 2010;4(6):1294–301.

Halfpenny P, Proctor R. Innovations in digital research methods. London: Sage; 2015.

Harper D. Working knowledge: skill and community in a small shop. Chicago: University of Chicago Press; 1987.

Harper D. Talking about pictures: a case for photo elicitation. Visual Stud. 2002;17(1):13–26.

Harper D. Visual sociology. London: Routledge; 2012.

Harricharan M, Bhopal K. Using blogs in qualitative educational research: an exploration of method. Int J Res Method Educ. 2014;37(3):324–43.

Hergenrather KC, Rhodes SD, Cowan CA, Bardhoshi G, Pula S. Photovoice as community-based participatory research: a qualitative review. Am J Health Behav. 2009;33(6):686–98.

Hesse-Biber SN. Handbook of emergent technologies in social research. New York: Oxford University Press; 2011.

Hesse-Biber SN, Leavy P. Emergent methods in social research. Thousand Oaks: Sage; 2006.
Hessler R. The methodology of internet research: some lessons learned. In: Liamputtong P, editor. Health research in cyberspace: methodological, practical, and personal issues. New York: Nova Science Publishers; 2006. p. 105–20.
Hewson C. Qualitative approaches in internet-mediated research: opportunities, issues, possibilities. In: Leavy P, editor. The Oxford handbook of qualitative research. Oxford: Oxford University Press; 2014. p. 423–51.
Higginbottom G, Liamputtong P, editors. Participatory qualitative research methodologies in health. London: Sage; 2015.
Holton M, Riley M. Talking on the move: place-based interviewing with undergraduate students. Area. 2014;46(1):59–65.
Huang S. Using freelisting to examine the destination image of China among Australian residents. Paper presented at the 4th advances in hospitality & tourism marketing & management conference, Mauritius. 2014. 25–27 June 2014.
Hyers LL, Swim JK, Mallett RM. The personal is political: using daily diaries to examine everyday gender-related experiences. In: Hesse-Biber SN, Leavy P, editors. Emergent methods in social research. Thousand Oaks: Sage; 2006.
Iacono VL, Symonds P, Brown DHK. Skype as a tool for qualitative research interviews. Sociol Res Online. 2016;21(2):12. http://www.socresonline.org.uk/21/2/12.html
Jacelon CS, Imperio K. Participant diaries as a source of data in research with older adults. Qual Health Res. 2005;15(7):991–7.
Jackson KF. Participatory diagramming in social work research: utilizing visual timelines to interpret the complexities of the lived multiracial experience. Qual Soc Work. 2012;12(4):414–32.
Jonas JA, Davies EL, Keddem S, Barg FK, Fieldston ES. Freelisting on costs and value in health care by pediatric attending physicians. Acad Pediatr. 2015;15(4):461–6.
Jones K, Leavy P. A conversation between kip Jones and Patricia Leavy: arts-based research, performative social science and working on the margins. Qual Rep. 2014;19(19):1–7.
Kara H. Creative research methods in the social sciences: a practical guide. Cambridge: Policy Press; 2015.
Kendig H, Byles JE, O'Loughlin K, Nazroo JY, Mishra G, Noone J, Loh V, Forder PM. Adapting data collection methods in the Australian life histories and health survey: a retrospective life course study. BMJ Open. 2014;4:e004476. https://doi.org/10.1136/bmjopen-2013-004476.
Kolar K, Ahmad F, Chan L, Erickson PG. Timeline mapping in qualitative interviews: a study of resilience with marginalized groups. Int J Qual Methods. 2015;14(3):13–32.
Kozinets R. Netnography: doing ethnographic research online. London: Sage; 2010.
Kozinets R. Netnography: redefined. 2nd ed. London: Sage; 2015.
Kramer J, Rubin A, Coster W, Helmuth E, Hermos J, Rosenbloom D, ... Brief D. Strategies to address participant misrepresentation for eligibility in Web-based research. Int J Methods Psychiatr Res. 2014;23(1):120–9.
Leavy P. Method meets art: arts-based research practice. London: Guilford Publications; 2015.
Lennon K. Feminist perspectives on the body. In Zalta EN, editor. The Stanford encyclopedia of philosophy. 2014. http://plato.stanford.edu/archives/fall2014/entries/feminist-body. Accessed 18 Nov 2015.
Lepkowski JM, Tucker C, Brick JM, de Leeuw ED, Japec L, Lavrakas PJ, Link MW, Sangster RL, editors. Advances in telephone survey methodology. Hoboken: Wiley; 2008.
Lévi-Strauss C. The savage mind. 2nd ed. Chicago: University of Chicago Press; 1966.
Liamputtong P, editor. Health research in cyberspace: methodological, practical and personal issues. New York: Nova Science Publishers; 2006.
Liamputtong P. Researching the vulnerable: a guide to sensitive research methods. London: Sage; 2007.
Liamputtong P. Qualitative research methods. 4th ed. Melbourne: Oxford University Press; 2013.

Liamputtong P, Rumbold J. Knowing differently: arts-based and collaborative research methods. New York: Nova Science Publishers; 2008.

Lieberman DZ. Evaluation of the stability and validity of participant samples recruited over the internet. Cyberpsychol Behav Soc Netw. 2008;11(6):743–5.

Lincoln YS, Denzin NK. Epilogue: the eighth and ninth moments-qualitative research in/and the fractured future. In: Denzin NK, Lincoln YS, editors. The Sage handbook of qualitative research. 3rd ed. Thousand Oaks: Sage; 2005. p. 1115–26.

Lopez EDS, Eng E, Randall-David E, Robinson N. Quality-of-life concerns of African American breast cancer survivors within rural North Carolina: blending the techniques of photovoice and grounded theory. Qual Health Res. 2005a;15(1):99–114.

Lopez EDS, Eng E, Robinson N, Wang CC. Photovoice as a community-based participatory research method: a case study with African American breast cancer survivors in rural Eastern North Carolina. In: Israel B, Eng E, Schulz AJ, Parker E, Satcher D, editors. Methods for conducting community-based participatory research for health. San Francisco: Jossey-Bass; 2005b.

Mann C, Stewart F. Internet communication and qualitative research: a handbook for researching. Online. London: Sage; 2000.

Mann BL, Stewart F. Internet interviewing. In: Gubrium JF, Holstein JA, editors. Handbook of interview research: context and method. Thousand Oaks: Sage; 2002. p. 603–627.

Maratos M, Huynh L, Tan J, Lui J, Jar T. Picture this: exploring the lived experience of high-functioning stroke survivors using photovoice. Qual Health Res. 2016;26(8):1055–66.

Markham A. Internet communication as a tool for qualitative research. In: Silverman D, editor. Qualitative research: theory, method and practice. London: Sage; 2004. p. 95–124.

Mathers SA, Anderson H, McDonald S, Chesson RA. Developing participatory research in radiology: the use of a graffiti wall, cameras and a video box in a Scottish radiology department. Pediatr Radiol. 2010;40:309–17.

McCosker A, Darcy R. Living with cancer: affective labour, self-expression and the utility of blogs. Inf Commun Soc. 2013;16:1266–85.

McNiff S. Art-based research. In: Knowles JG, Cole AI, editors. Handbook of the arts in qualitative research: perspectives, methodologies, examples, and issues. Thousand Oaks: Sage; 2008. p. 29–40.

Meho LI. E-mail interviewing in qualitative research: a methodological discussion. J Am Soc Inf Sci Technol. 2006;57(10):1284–95.

Mertens D, Holmes HM, Harris RL. Transformative research and ethics. In: Mertens DM, Ginsberg PE, editors. The handbook of social research ethics. Thousand Oaks: Sage; 2009. p. 85–191.

Morselli D, Berchtold A, Granell J-CS, Berchtold A. On-line life history calendar and sensitive topics. J Comput Hum Behav. 2016;58:141–9.

Murray E, Khadjesari Z, White IR, Kalaitzaki E, Godfrey C, McCambridge J, Thompson SG, Wallace P. Methodological challenges in online trials. J Med Int Res. 2009;11(2):e9. https://doi.org/10.2196/jmir.1052.

Nelson C, Treichler PA, Grossberg L. Cultural studies: an introduction. In: Grossberg L, Nelson C, Treichler PA, editors. Cultural studies. New York: Routledge; 1992. p. 1–16.

Nelson MR, Otnes CC. Exploring cross-cultural ambivalence: a netnography of intercultural wedding message boards. J Bus Res. 2005;58(1):89–95.

Nezlek JH. Diary methods. Thousand Oaks: Sage; 2012.

O'Toole J, Sinclair M, Leder K. Maximising response rates in household telephone surveys. BMC Med Res Methodol. 2008;8(71). https://doi.org/10.1186/1471-2288-8-71.

Orchard T, Smith T, Michelow W, Salters K, Hogg B. Imagining adherence: body mapping research with HIV-positive men and women in Canada. AIDS Res Hum Retrovir. 2014;30(4):337–8.

Otañez M, Guerrero A. Digital storytelling and the hepatitis C virus project. In: Gubrium A, Harper KG, Otañez M, editors. Participatory visual and digital research in action. Walnut Creek: Left Coast Press; 2015. p. 57–70.

Phillips BD. Qualitative disaster research. In: Leavy P, editor. The Oxford handbook of qualitative research. Oxford: Oxford University Press; 2014. p. 533–56.
Ratislavová K, Ratislav. Asynchronous email interview as a qualitative research method in the humanities. Hum Aff. 2014;24(4):452–60.
Rhodes SD, Hergenrather KC, Wilkin AM, Jolly C. Visions and voices: Indigent persons living with HIV in the southern United States use photovoice to create knowledge, develop partnerships, and take action. Health Promot Pract. 2008;9(2):159–69.
Rimkeviciene J, O'Gorman J, Hawgood J, Leo DD. Timelines for difficult times: use of visual timelines in interviewing suicide attempters. Qual Res Psychol. 2016;13(3):231–45.
Riper H, Spek V, Boon B, Conijn B, Kramer J, Martin-Abello K, Smit F. Effectiveness of e-self-help interventions for curbing adult problem drinking: a meta-analysis. J Med Internet Res. 2011;13(2):e24. https://doi.org/10.2196/jmir.1691.
Rogers R. Digital methods. Cambridge, MA: The MIT Press; 2013.
Rose G. Visual methodologies: an introduction to the interpretation of visual materials, 2nd ed. London: Sage; 2007.
Rose G. Visual methodologies: an introduction to the interpretation of visual materials, 3rd ed. London: Sage; 2012.
Rose T, Shdaimah C, de Tablan D, Sharpe TL. Exploring wellbeing and agency among urban youth through photovoice. Child Youth Serv Rev. 2016;67(2016):114–22.
Russell AC, Diaz ND. Photography in social work research: using visual image to humanize findings. Qual Soc Work. 2013;12(4):433–53.
Ryan G, Nolan J, Yoder S. Successive free listing: using multiple free lists to generate explanatory models. Field Methods. 2000;12(2):83–107.
Saiki LS, Cloyes KG. Blog text about female incontinence. Nurs Res. 2014;63:137–42.
Sanon MA, Evans-Agnew RA, Boutain DM. An exploration of social justice intent in photovoice research studies from 2008 to 2013. Nurs Inq. 2014;21(3):212–26.
Seymour W, Lupton D. Holding the line online: exploring wired relationships for people with disabilities. Disab Soc. 2004;19(4):291–305.
Simons H, McCormack B. Integrating arts-based inquiry in evaluation methodology: challenges and opportunities. Qual Inq. 2007;13(32):292–311.
Sinner A, Leggo C, Irwin RL, Gouzouasis P, Grauer K. Arts-based educational research dissertations: reviewing the practices of new scholars. Can J Educ. 2006;29(4): 1223–70.
Snee H, Roberts S, Watson H, Morey Y, Hine C. Digital methods as mainstream methodology: an introduction. In: Snee H, Roberts S, Watson H, Morey Y, Hine C, editors. Digital methods for social science. London: Palgrave Macmillan; 2016.
Sue VM, Ritter LA. Conducting online surveys. 2nd ed. Thousand Oaks: Sage; 2012.
Switzer S, Guta A, de Prinse K, Chan Carusone S, Strike C. Visualizing harm reduction: methodological and ethical considerations. Soc Sci Med. 2015;133:77–84.
Synnot A, Hill S, Summers M, Taylor M. Comparing face-to-face and online qualitative research with people with multiple sclerosis. Qual Health Res. 2014;24(3):431–8.
Tarr J, Thomas H. Mapping embodiment: methodologies for representing pain and injury. Qual Res. 2011;11(2):141–57.
Taylor C, Coffey A. Innovation in qualitative research methods: possibilities and challenges. Cardiff: Cardiff University; 2008.
Teti M, Murray C, Johnson L, Binson D. Photovoice as a community based participatory research method among women living with HIV/AIDS: ethical opportunities and challenges. J Empir Res Hum Res Ethics. 2012;7(4):34–43.
Thomas S. Reimagining inquiry, envisioning form. In: Nielsen L, Cole AL, Knowles JG, editors. The art of writing inquiry. Halifax: Backalong Books; 2001. p. 273–82.
Todres L. Embodied enquiry: phenomenological touchstones for research, psychotherapy and spirituality. New York: Palgrave Macmillan; 2007.

Tomlinson M, Solomon W, Singh Y, Doherty T, Petrida Ijumba C, Tsai AC, Jackson D. The use of mobile phones as a data collection tool: a report from a household survey in South Africa. BMC Med Inform Decis Mak. 2009;9:51. https://doi.org/10.1186/1472-6947-9-51.

Turney L, Pocknee C. Virtual focus groups: New frontiers in research. Int J Qual Methods. 2005;4(2). Available at http://www.ualberta.ca/~ijqm/backissues/4_2/pdf/turney.pdf

van Heerden AC, Norris SA, Tollman SM, Richter LM. Collecting health research data comparing mobile phone-assisted personal interviewing to paper-and-pen data collection. Field Methods. 2014;26(4):307–21.

Vanhoutte B, Nazroo J. Life-history data. Publ Health Res Pract. 2016;26(3):e2631630. https://doi.org/10.17061/phrp2631630.

Viega M. Aesthetic sense and sensibility: arts-based research and music therapy. Music Ther Perspect. 2016;34:1–3.

Wakeford N, Cohen K. Field notes in public: using blogs for research. In: Fielding N, Lee R, Blank G, editors. The Sage handbook of online research methods. Thousand Oaks: Sage; 2008. p. 307–26.

Wang C. Photovoice: a participatory action research strategy applied to women's health. J Womens Health. 1999;8(2):185–92.

Wang C, Burris MA. Empowerment through photo novella: portraits of participation. Health Educ Q. 1994;2(2):171–86.

Wang C, Burris MA. Photovoice: concept, methodology, and use for participatory needs assessment. Health Educ Behav. 1997;24(3):369–87.

Wang C, Morrel-Samuels S, Hutchison PM, Bell L, Pestronk RM. Flint Photovoice: community building among youths, adults, and policymakers. Am J Public Health. 2004;94(6):911–3.

Warkentin E. Writing competitions as a new research method. Int J Qual Methods. 2002;1(4):10–25.

Weber S. Visual images in research. In: Knowles JG, Cole AI, editors. Handbook of the arts in qualitative research: perspectives, methodologies, examples, and issues. Thousand Oaks: Sage; 2008. p. 41–53.

Wiles R, Crow G, Pain H. Innovation in qualitative research methods: a narrative review. Qual Res. 2011;11(5):587–604.

Wilson N. Social creativity: re-qualifying the creative economy. Int J Cult Policy. 2010;16(3):367–81.

Wilson E, Kenny A, Dickson-Swift V. Using blogs as a qualitative health research tool: a scoping review. Int J Qual Methods. 2015:1–12. https://doi.org/10.1177/1609406915618049.

Wright KB. Researching internet-based populations: advantages and disadvantages of online survey research, online questionnaire authoring software packages, and web survey services. J Comput Mediated Commun. 2005;10, Article 11. Retrieved from http://jcmc.indiana.edu/vol10/issue3/wright.html

Xenitidou M, Gilbert N. Innovations in social science research methods. Surrey: ESRC National Centre for Research Methods, University of Surrey; 2009.

Personal Construct Qualitative Methods

62

Viv Burr, Angela McGrane, and Nigel King

Contents

1 Introduction 1096
2 A Constructivist Epistemology 1096
3 Constructs and Construing 1097
4 PCP Methods: Key Features 1098
5 Eliciting and Using Constructs in Research 1099
    5.1 Eliciting Constructs in an Interview: Cross-Cultural Perceptions 1099
    5.2 The Triadic Method of Construct Elicitation: The Self in Relationship 1100
    5.3 Interviewing Using Elicited Constructs: Footwear and Women's Identities 1101
    5.4 Laddering: Reflective Practice in Social Work 1102
    5.5 Analysis of Constructs 1105
6 The Pictor Technique: Experiences of Care 1105
    6.1 The Pictor Technique: Procedures 1106
    6.2 Analysis of Pictor Charts 1107
7 The Self-characterization Sketch: Work Placements and Students' Perceptions of Self . 1109
    7.1 Analysis of the Self-characterization Sketch 1109
8 Reflecting on the Use of PCP Methods 1110
9 Conclusion and Future Directions 1111
References 1111

V. Burr (✉)
Department of Psychology, School of Human and Health Sciences, University of Huddersfield, Huddersfield, UK
e-mail: v.burr@hud.ac.uk

A. McGrane
Newcastle Business School, Northumbria University, Newcastle-upon-Tyne, UK
e-mail: angela.mcgrane@northumbria.ac.uk

N. King
Department of Psychology, University of Huddersfield, Huddersfield, UK
e-mail: n.king@hud.ac.uk

© Springer Nature Singapore Pte Ltd. 2019
P. Liamputtong (ed.), *Handbook of Research Methods in Health Social Sciences*,
https://doi.org/10.1007/978-981-10-5251-4_23

> **Abstract**
>
> In this chapter, we examine a number of research methods arising from personal construct psychology (PCP). Although its techniques are often relatively unfamiliar to qualitative researchers, we show how PCP provides opportunities to extend and enrich the predominant methods currently used by them. PCP adopts a constructivist epistemology and offers a number of techniques for enabling people to gain insight into their own and others' constructions and perceptions. We outline several of these techniques in this chapter, illustrating them through examples of our own research and providing some guidelines for data analysis.

> **Keywords**
>
> Construct elicitation · Constructivism · Construing · Laddering · Personal construct psychology · Pictor · Self-characterization sketch

1 Introduction

Although social scientists often have some familiarity with personal construct psychology (PCP), many are not aware of the range of qualitative methods it offers. PCP originated in clinical psychological practice with the work of George Kelly (1955) and has since developed a number of techniques for enabling people to gain insight into their own and others' thinking which have been successfully adapted and applied in organizational, educational, and health and social care research. The Repertory Grid is probably the most familiar and widely used PCP method, but it has been principally used as a quantitative technique. Although the potential of this and other PCP methods for qualitative research is now being acknowledged (see Fransella 2005), PCP methods are still relatively unknown to qualitative researchers.

In this chapter, we will examine a number of research methods arising from PCP and argue that they provide opportunities to extend and enrich the predominant methods currently used by qualitative researchers. We illustrate these methods through examples of our own research and provide some guidelines for data analysis.

2 A Constructivist Epistemology

PCP can be thought of as one of a number of approaches referred to as "contextual constructionism" (Madill et al. 2000), where reality is seen as actively constructed through our interpretative processes. Since there are many different ways of interpreting the world characteristic of different societies, cultures, groups, and individuals, there are multiple constructed realities and alternative perspectives on the world. Kelly (1955) referred to this as "constructive alternativism." Contextual constructionism differs from, and can be thought of as lying somewhere between, realism (the idea of a single, objectively defined reality) and radical constructionism (the idea that our constructions of the world are constructed without reference to any presumed external "reality").

Like other constructivist theories, PCP argues that one account of reality cannot be regarded as any more "accurate" than another, locating it as a relativist approach. Relativism is the idea that points of view have no absolute truth or validity in themselves; they can only have relative, subjective value, so some accounts may be more useful or facilitative than others for the person. In this emphasis upon individual experience and knowledge as a useful construction, PCP is philosophically grounded in both pragmatism and phenomenology (Butt 2008). Pragmatism emphasizes the utility rather than the truthfulness or accuracy of people's knowledge, and PCP is phenomenological since it is concerned with the world as it is perceived by the person, the "phenomena" that present themselves to consciousness.

At the heart of PCP is the idea that people actively construct themselves and their psychosocial world in the course of daily life, using their subjective experience and its personal and social meanings. PCP sees people as conducting themselves and making choices according to these meanings and thus stands in contrast to theories that regard the person's qualities or behavior as determined by internal or external forces. It also rejects causality as an explanatory concept, since human beings are seen as having choice and agency. PCP theory and methods are, therefore, epistemologically compatible with approaches that take seriously subjective experience and/or that challenge deterministic or essentialist models of the person.

3 Constructs and Construing

Kelly was working as a clinical psychologist in the USA in the 1930s and developed PCP as an alternative to the mainstream psychologies of the day, behaviorism and psychoanalysis, with which he became dissatisfied. Rather than life events causing people to be traumatized, Kelly felt that it is our interpretation of events, the meaning they hold for us rather than the events themselves, that can be problematic. How people perceive or "construe" events is seen as key to understanding them.

Kelly argued that people construe the world in their own idiosyncratic way, using a system of meaning that each individual builds for themselves. This "construct system" is like a lens through which we perceive the world, giving it a particular appearance and significance. A person's construct system is thought of as a set of bipolar dimensions ("constructs") that each take the form of a contrast, such as "vulnerable vs. resilient" or "independent vs. sociable," and the person constantly uses these to interpret their experience. However, we are often not consciously aware of our construing; it is our taken-for-granted way of seeing things.

Construing, therefore, frames and gives meaning to our experience, powerfully influencing our interactions and relationships. For example, when one nurse meets a new patient for the first time, they may be (nonconsciously) asking "are they going to be compliant or refuse their medication?" whereas another may be asking "will I be able to help, or will I feel frustrated?" The nature of their relationship and interaction with the patient is inevitably shaped by this construing; the two nurses may be expected to have somewhat different perceptions of and interactions with the same patient. In PCP terms, if we want to understand a person, we must gain some insight into their construing.

But our many constructs do not exist independently of each other; they are linked together in a system. They are arranged hierarchically, with relatively concrete and mundane constructs, used in relation to quite narrow aspects of life, at the bottom of the hierarchy and more abstract constructs relating to our overarching values and beliefs at the top. For example, lower level constructs for a person might include "rhythmic vs. melodic" (in relation to music) or "talkative vs. reserved" (in relation to people). A construct higher up their system, to which the lower ones will be related, might be "stimulating vs. dull" as this could apply to both music and people. This might in turn be subsumed under an even higher-order (or "superordinate") construct such as "life-enhancing vs. life-limiting." This relationship between constructs means that they have meaning implications for each other. In the hypothetical example above, the person may be drawn to "talkative" people as "talkative" to some extent also implies "life-enhancing" for them. This hierarchical organization of constructs is explicitly used in the method of "laddering" which we explore later.

4 PCP Methods: Key Features

Kelly devised methods to enable him and his clients to gain insight into their construing, and other PCP clinicians and researchers have since developed a considerable number of further techniques for this purpose. Such methods may be used within a PCP theoretical framework but can also be adopted in a wider range of approaches that sit at the intersection between constructivism and phenomenology, where subjective experience and perceptions are the focus.

The aim in PCP psychotherapy is to enable the client to articulate and inspect their own construing in order to allow them to understand and overcome their psychological difficulties. Similarly, in the research context, PCP methods focus on enabling people to reflect upon their own experience and conduct and gain insight into their own construing. In PCP therapy, it is important to gain an in-depth understanding of the client's worldview, seeing and describing the world in their terms rather than as seen by the therapist. Similarly, PCP research privileges the "voice" of participants. The PCP researcher is careful to describe events in terms used by participants themselves; in giving verbal and written labels to their constructs, care is taken to adopt the words and terms used by the participant. The participant's perspective always remains the priority, ensuring that the interpretative process remains in their control rather than being taken over by the researcher. PCP methods, therefore, explicitly "democratize" the relationship between client and practitioner, researcher and researched. Whereas other methods of analysis, such as IPA (Smith and Osborn 2003), rely principally upon the researcher's interpretation of an interview transcript some time after the interview, a characteristic of PCP methods is the greater time spent during data gathering in agreeing construct labels and their meanings with the participant.

In this sense, PCP methods can be described as "participant-led." Participant-led techniques such as photo-elicitation and audio diaries are often adopted in difficult or sensitive areas of research so that participants from marginalized or vulnerable groups are given greater agency than they would in a conventional interview (Liamputtong 2007; Pink 2007; Johnson 2011; Sargeant and Gross 2011; see also "Understanding Health Through a Different Lens: Photovoice Method" and "Digital Storytelling Method"). But PCP methods have three potential additional benefits. Firstly, although they are intrinsically participant-led, they are used in collaboration with the researcher. This avoids participant worries about "doing it right" when asked to produce the required material unaided. Secondly, PCP methods are less reliant on the verbal fluency of participants than methods where people are asked to explain the meaning of what they have produced; PCP methods can be particularly effective in researching experiences that are hard for participants to access and articulate, as it focuses on concrete examples from their experience. This focus on the concrete can enable participants to overcome the difficulties of expressing abstract ideas. Thirdly, PCP methods tend to be very efficient; in our experience, participants are generally able to carry out the exercises in a relatively short time while still producing rich data. Furthermore, with some participant groups it can be difficult to avoid socially desirable but readily available responses, for example, where practitioners are keenly aware of what is regarded as "best practice" in their field. Using concrete examples provides a means of accessing accounts reaching beyond socially desirable or common-sense responses since they avoid asking participants direct questions about what they feel is important.

In the remainder of this chapter, we will illustrate several methods derived from PCP, chosen to indicate something of the range of issues that they may be used to address. Some of these methods (the various forms of construct elicitation) are explicitly designed to identify participants' bipolar constructs, while others (Pictor and the self-characterization sketch) aim to gain access to the participant's broader construal of their role in their social world.

5 Eliciting and Using Constructs in Research

There are numerous methods for helping participants to articulate their construing and for gaining access to the bipolar constructs that inform their meaning-making activities. We will first describe how construct elicitation may be used as part of the familiar in-depth interview, describe the commonly used "triadic" method of construct elicitation, and show how "laddering" may be used to help participants to articulate constructs akin to their core values and assumptions.

5.1 Eliciting Constructs in an Interview: Cross-Cultural Perceptions

A common method of data collection in qualitative research is the in-depth interview. However, interviewing requires skill in the effective use of probes and prompts (King and Horrocks 2010). Moreover, where the research topic is psychologically and socially complex, participants can struggle to articulate their experience.

Literature on interviewing about difficult and sensitive topics (e.g., Lee 1993; Liamputtong 2007; Mercer 2008) suggests that it can be time-consuming, requiring the researcher to explore more accessible aspects of experience first, and to use multiple probes to "get below the surface." But, using a PCP approach to interviewing can help to quickly bring significant issues into focus for discussion.

We will illustrate this through a research study in which we invited participants to compare the perceived characteristics of English and Italian people (Burr et al. 2014a). Small groups of participants in the UK and Italy were interviewed with questions such as "What comes to mind when you think of someone as 'typically Italian/English'?" and "What do you think Italians/the English) imagine when they think of someone as being 'typically English/Italian'?" We audio-recorded the interviews and also used a simple flip chart to write down the first pole of emerging constructs, such as "passionate and romantic" and "self-contained," and then asked for contrast poles for these. For example, the English participants suggested that Italians have a "musical, expressive language." They were then asked "as opposed to what? How are English people by contrast?" and they suggested "loud" and "raucous." The interviewers then discussed with the group all the responses on the flip chart, clarifying the constructs and, in particular, their contrast poles. Care was taken to use construct labels with which all participants in the group felt comfortable and that they felt represented the range of views expressed. The audio recordings were later used as a check on these constructs. This method enabled us to identify interesting similarities and differences in construing between the two cultures. There was a good degree of shared construing, referred to in PCP as "commonality"; for example, both English an Italian participants used constructs around hospitality, involving a contrast between "giving of oneself" to strangers as opposed to being more "detached," and around family life with the contrast of being family-oriented as opposed to being more independent of family. But subtle differences in the constructs used by the two cultures were often very informative. While the English participants saw English people as not very hospitable or "giving" of themselves, the Italian participants construed this English reserve as "valuing privacy"; the English participants envied the Italian culture in its family focus, but the Italians felt their own close family involvement may cast them as "Mummy's boys" in the eyes of others.

Constructs previously elicited, often using the "triadic" method, can also be used as the starting point for an in-depth interview. Before going on to provide an example of this, we will therefore describe the triadic method of construct elicitation using a research example.

5.2 The Triadic Method of Construct Elicitation: The Self in Relationship

This popular method typically entails the comparison of things, people, or events, referred to as "elements," in groups of three ("triads") and asking the participant to think of any way in which two are similar and different from the third (Kelly's definition of a construct). A pool of elements is first developed. Where the topic under investigation concerns our construing of people and relationships, these elements would typically consist of people known to the individual. The number of elements used is not important, but there should be enough to enable a range of comparisons to be made. They might be asked to compare, for example, their mother, their best friend, and their brother. They may answer "Two of these people are outgoing but the other is shy," so the bipolar construct elicited here is "outgoing – shy." Comparison of different triads of elements would produce further constructs, such as "tidy – chaotic" or "thorough – hurried."

Inspection of the list of constructs obtained in this way can itself provide insight into the individual's worldview and promote useful discussion with them. Butt et al. (1997) used this method to examine people's sense of self in their relationships with others. Participants were asked to think of themselves in relationship with a variety of others in their lives, and all of these different "selves" constituted the elements in the construct elicitation process. They were asked to compare these selves in triads, producing a range of constructs related to their sense of self. An example of the constructs produced by one participant, "Katherine," can be seen in Fig. 1, and we will comment on these constructs in the section on construct analysis.

5.3 Interviewing Using Elicited Constructs: Footwear and Women's Identities

Using elicited constructs as the starting point for an interview can enable the researcher to focus on and probe significant issues. The interview can be used for refining the elicited constructs with the participant, and the interview transcript can additionally be analyzed in the usual way, with the possibility of using some constructs as a-priori codes. Burr et al. (2014b) show how this method was used to explore four women's identities using shoes. Pictures of a wide range of footwear

```
           feel defensive......................not defensive
       feel uncomfortable......................feel comfortable
           I'm guarded....................I'm open
       feel disinterested......................feel concerned
              unsure......................relaxed
   wouldn't share my feelings......................would share my feelings
       want to protect myself......................feel comfortable
          feel resentful......................don't feel resentful
           unstimulated......................stimulated
          feel vulnerable......................don't feel vulnerable
           distrust them......................trust them
```

Fig. 1 Katherine's Constructs

formed the pool of elements. Triads of images were presented to the women and they were asked to compare these in terms of the "personality" the shoes might be said to have, or the kind of woman who might wear them. Elicited constructs included a number relating to sexuality and "sexiness," issues that were clearly of importance to the women; but these were subtly different for each participant. For one participant, the contrast to "sexy" was "old-fashioned" whereas another contrasted "overtly sexual" with "toned down sexual." The women were then interviewed, probing into the meaning of the elicited constructs and producing rich insights into their sense of self.

5.4 Laddering: Reflective Practice in Social Work

Burr et al. (2016) used a laddering exercise with undergraduate social work students to enable them to reflect upon the more value-laden constructs that might be said to lie towards the "top" of their construct hierarchy.

Social work practitioners are expected to reflect on their practice, to gain insight into their reactions to clients, their own decision-making, and the assumptions and values that drive these. Reflexive thinking is, therefore, explicitly addressed in the education of social work students. Students are often required to write reflective assignments to demonstrate the insights they have gained.

However, reflecting upon one's assumptions and beliefs is not necessarily easy to do, since these are often "ground" rather than "figure" for us; we may need assistance to render them "visible" to us and therefore available for inspection. With some participant groups, it can also be difficult to avoid socially desirable responses, for example, where practitioners are keenly aware of what is regarded as "best practice" in their field, and this is arguably the case with social work students. A structured exercise can help participants to articulate underlying values and assumptions, and can help avoid the tendency to simply report socially desirable ideas. First described by Hinkle (1965), laddering takes previously elicited constructs and uses these to explore the person's superordinate constructs, those that say something about the values and taken-for-granted assumptions that inform their conduct. It is especially helpful when the participant has difficulty in articulating their abstract values and beliefs, or explaining the reasons for their behavior.

One of the person's previously elicited constructs is selected and written down, preferably one that is relatively "subordinate" or "concrete," for example, "reserved vs. chatty." They are then asked to say which pole of the construct they would prefer to see themselves at and to say why. Their response is written above that pole of the construct. They are asked what the contrast would be and their response is written above the other pole. In this example, they might say they would prefer to be "chatty" because chatty people are "socially adept" as opposed to being "hard to relate to." These two responses now effectively constitute another, more abstract, construct, and the two constructs constitute the first two "rungs" of a ladder (see

Fig. 2 Two "rungs" of a ladder

Fig. 3 Jenny's constructs

senior level – entry level
caring – cold
highly strung – calm
harsh – warm
client centred – business centred
detached – mindful of clients
immature – mature
complacent – super keen

Fig. 2). The same process is applied to the new construct and so on until the person cannot meaningfully "ascend" any further.

However, it is important not to think of the identification of constructs as an end in itself. The process of reflecting in-depth upon one's construing is likely to be more important than the specific outcomes of the exercises. Therefore, in our research, we were relatively unconcerned with exploring students' construct systems, focusing instead on the effectiveness of their self-reflection.

The students were asked to think of a number of people whose social work practice they were familiar with and these, including themselves, formed the "elements" of a construct elicitation task using the triadic method as described earlier. In making their comparisons, they were asked to focus on the behavior and practice of the individuals. Jenny's constructs (see Fig. 3) are an example.

We then chose relatively "concrete" or subordinate constructs from each student's list for laddering as described above. For example, Jenny's construct "senior level – entry level" appears very concrete and would be a better candidate for laddering than "client centered – business centered," which is more abstract. The students each produced between two and four ladders, using different constructs as the starting point each time.

The ladders sometimes produced counter-intuitive results. For example, Fig. 4 shows Alice's ladder, which prompted a good deal of self-reflection.

The laddering began with Alice's construct "too nosey vs. not nosey enough." The concept of "nosiness" usually carries pejorative meanings, and it seemed that Alice was unsure of just how "nosey" a social worker ought to be. But in completing this ladder, it was clear that Alice was reflecting on her social work values. As she

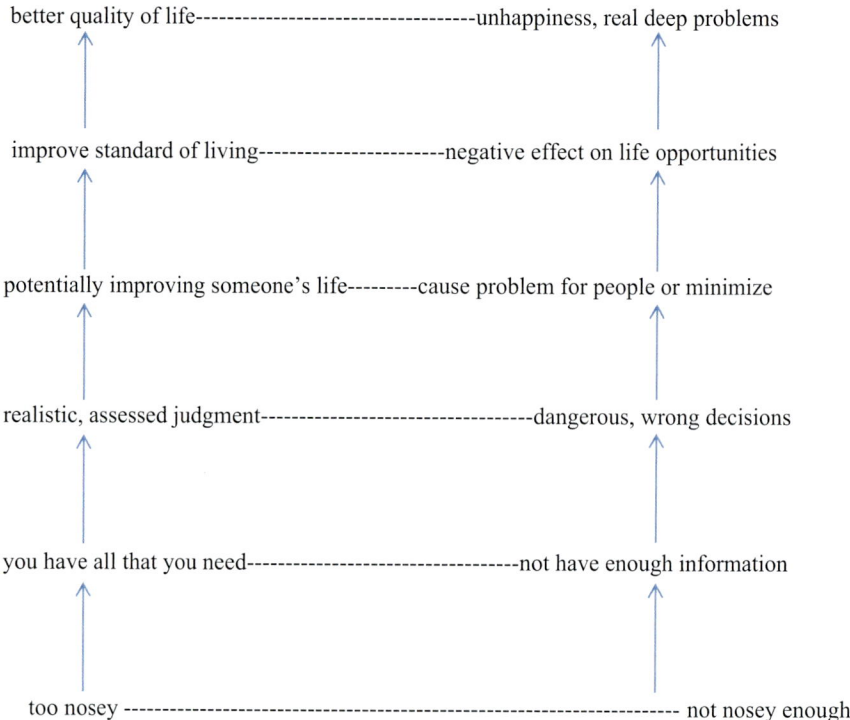

Fig. 4 Alice's ladder

proceeded with the exercise, she reflected on the implications of these two kinds of behavior and concluded that it was important for a social worker to be highly inquisitive about a service user's situation in order to realistically assess the problem and, ultimately, give the service user a better quality of life.

Laddering is generally accepted as a technique that allows researchers to see the organization of a person's construct system, helping us to access their superordinate constructs. However, Butt (2007) and Bell (2014) have challenged this. They suggest that construct systems are not so neatly organized, and that conceptualizing them in this hierarchical way leads to confusion about what laddering can achieve. They acknowledge that laddering sometimes does lead "upwards" to constructs that are more abstract and value-laden than those we began the exercise with, as was clearly the case for some of our participants. But, sometimes it does not, and we can end up "going round in circles," with the constructs elicited later looking similar to those we started with. This happened with some of our participants. Nevertheless, what laddering does is to help us explore the implications of some of our construing. It gives us insight into how our constructs are related to each other, and in so doing frequently offers us food for reflective thought. So "going round in circles" is not necessarily a problem- interesting issues may nevertheless be revealed in the process.

5.5 Analysis of Constructs

As with other qualitative methods, analysis of the data from construct elicitation is largely interpretative. One might look for "themes" in the constructs as one would with interview data, the difference being that the researcher need not examine large amounts of material to find issues of significance (Braun and Clark 2006; see also "Qualitative Story Completion Method"). A quick inspection of Katherine's constructs (Fig. 1) suggests a theme of wariness and self-protection. Her constructs around feeling defensive, guarded and distrustful, and feeling vulnerable and needing to protect herself suggest a relational world fraught with threat and potential danger. This is certainly not all that can be said about Katherine as a person; the construct elicitation exercise produces a vivid picture of one aspect of her sense of self. But we might assume that such construing would be important in understanding the nature of Katherine's interactions and relationships with others.

Although the focus of construct elicitation is usually idiographic, it is also possible to examine the construing of a wider sample of participants. In our research on women's identities, we went on to elicit constructs about shoes with over 20 further women using the same images. We then performed a content analysis on these constructs, identifying clusters of constructs that appeared to hold common significance for the women. They provided 215 constructs in total. Although many of these appeared to be idiosyncratic, 162 constructs were judged to show some commonality and were coded into categories. These categories included femininity, elegance, sexuality, boring, exhibitionism, conformity, and individuality. But, as in the interview stage of the research, it was an examination of the contrast poles of the constructs that was particularly informative. For example, the contrasts to "boring" supplied by the women suggested a preference for fun and frivolity, glamor and excitement, creativity and expressiveness, interest and vibrancy. The data also suggested a tension for women between expressing themselves as individuals and being seen as attention-seeking and exhibitionist. Likewise, it suggested a desire to be "sexy" but sexiness also often involved less welcome descriptions such as "tarty" and "slutty." The analysis, therefore, identified some interesting tensions and faultlines in the women's identities.

6 The Pictor Technique: Experiences of Care

Pictor is a visual method based on a technique used in PCP family therapy (Hargreaves 1979). The client writes the names of family members on arrow-shaped cards and lays them out in a way that illustrates the perceived nature of relationships between them. Ross et al. (2005) developed this for research purposes in a study of collaborative working between health and social care professionals. In their previous, traditional semistructured interviews, some participants had presented a rather sanitized and idealized version of collaboration that appeared to reflect the rhetoric of their professional training rather than lived experience. Borrowing from Hargreaves, Ross et al. (2005) tried asking participants to represent, with a name on a cardboard arrow, each of those involved in a case and to place them on a large sheet of paper, using features such as direction and proximity of arrows to indicate significant aspects of roles and relationships. They found that the "charts" thus produced served as very effective facilitators to detailed discussion of the case, encouraging a focus on the concrete and specific rather than the abstract and generalized. A further advantage was that it helped both interviewer and interviewee to keep in mind a wide range of individuals and services involved in a case; in conventional interviews this had proved hard to do and consequently there was a tendency to concentrate just on a few main players

With some refinements to the procedure, discussed below, and the discovery of arrow-shaped sticky "Post-it" notes that were more convenient to work with, this technique was developed into Pictor by King and colleagues at Huddersfield (King et al. 2013). We have used Pictor in several studies looking at the complex collaborative working that occurs for people with life-limiting illnesses (e.g., King et al. 2010, 2017; Noble et al. 2014). Hardy et al. (2012) extended the use of the technique to patients and their primary lay carers, to explore how they saw themselves in a network of care and support; and Elliott et al. (in preparation) have used it with parents of children with Autism-Spectrum Disorders. We have found Pictor successful in eliciting rich and detailed accounts, and in the great majority of cases, participants – whether professionals or lay people – have responded very positively to it. While our research has been in health and social care settings, the technique can be used in any circumstances where people need to collaborate to achieve a goal or to provide some kind of support to someone, for example in business, education, and the criminal justice system.

6.1 The Pictor Technique: Procedures

Turning now to the procedure for using the technique, the participant is asked to think of a particular "case" of collaborative working. If they are a patient or client, this would be their own case. They would then be asked to think of all those involved in the case and write a pseudonym or other identifier on a sticky arrow for each person/agency. The arrows are normally available in three different colors, and participants are told they can use the colors to indicate something if they wish. They then lay out the arrows on a large sheet of paper in a way that helps them tell their story of the case. Participants are told that features such as the direction of the arrows, the spaces between them, and the way they are grouped can be used to indicate particular aspects of roles and relationships, but there is no "right" way to do it – above all, they should carry out the task in the manner that best enables them to describe the case. The researcher may leave the room while they are completing the chart, but sometimes participants prefer them to stay – often because they are concerned about getting it "wrong." While there is a risk of the interviewer influencing chart construction, there are often benefits in hearing (and recording) participants vocalizing their reasons for placing arrows as they do. Once the chart is complete, it is used in an interview as the starting point for probing the participant about their

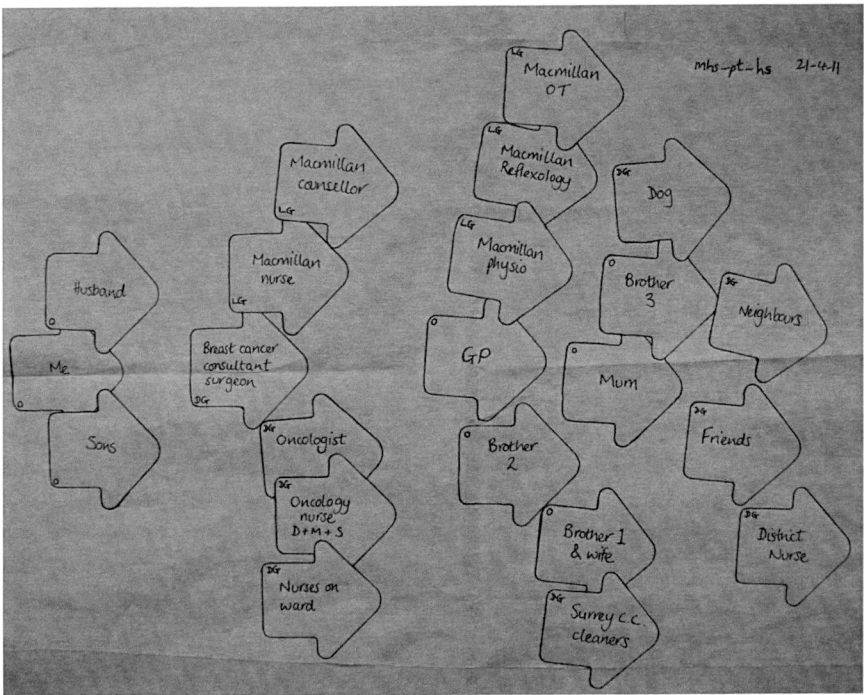

Fig. 5 Example of a Pictor chart

perspective on the story. At the end of the interview, it is best to draw around the arrows, as they can be prone to falling off in transit! If real names are used by participants, they should be anonymized at this point. A good-quality digital photograph of the chart is a simple and practical way to keep a record of it.

Fig. 5 shows a Pictor chart from a study funded by Macmillan Cancer Support, which evaluated an innovative community-based palliative care service in Midhurst, West Sussex. Pictor was used to examine how professionals, patients and carers viewed experiences of care. This chart was created by "Ella," a woman in her forties with terminal breast cancer.

6.2 Analysis of Pictor Charts

As in the case of other PCP methods, the Pictor chart is examined together with the participant, the arrow patterns noted by the researcher serving as interview prompts.

The way in which Ella laid out the arrows for her chart is very clearly structured. She presents herself on the left of the chart, overlapped by her husband and sons, with four distinct rows of arrows moving away from her. It is important when discussing the chart with a participant not to make assumptions about what particular positions might mean. In this case, Ella was quite explicit that proximity reflects how important particular individuals or agencies were in her care. For example, pointing to the "Macmillan Counsellor" arrow she says:

> ...and [name], that's [name]; I'd say [name] is very important as well. These come down like that [pointing along the line]. I'd say that's probably my main support team, they're the ones that I go to most.

Immediately after this, she points to the second line of arrows and says:

> And then I'd probably put my GP [places arrow] who's got more involved lately. See, I don't see these [Macmillan OT, Reflexology, Physio] a lot but when I have met them they've been important.

It is interesting to note that her family members are either presented as if part of a unit with her (husband and sons), or are in the second or third "lines" of support. From her comments, this clearly reflects their less central roles than the (literally) closest family members. For instance, she says "...my Mum bless her, Mum tries and my other brother [brother 3]." She explains her placing of the district nurses in the furthest position in line four rather apologetically, but with the clear intent of indicating a degree of disappointment:

> I'm not sure if I'm doing the district nursing justice, I don't know, it's just they're probably the ones I feel probably could have been around a bit more.

While we developed Pictor to help elicit rich and detailed interview responses, the Huddersfield team quickly recognized that the charts should be considered as part of the data along with the interview transcripts. They can, for example, help identify key themes in the way a participant construes their situation, such as Ella's perception of a "main team" in contrast to secondary supporters. It may also be valuable to compare features of charts across groups of participants to see whether there are differences that might be meaningful. For example, do carers tend to use the arrows in different ways from patients? Comparisons based on the charts should not be treated as if they were hypothesis tests, of course, but direct the analyst's attention to the relevant transcripts to interrogate these further.

The Pictor technique has, in our view, some clear strengths for constructivist, qualitative research. Most participants engage well with it, reporting that they enjoy doing it; many have described the process as "illuminating" or "surprising," suggesting it succeeds in bringing to awareness taken-for-granted ways of construing the world. Using the chart in the interview achieves a strong emphasis on concrete examples of lived experience. Also, the chart often appears to help the interview process seem less interrogative to participants; both interviewer and interviewee can focus together on examining the chart. A small proportion of participants struggle to understand the task, so it is a good idea to have an example to show – ideally not on the same topic as the project at hand, to avoid leading the participant. The technique requires space in the interview setting to lay out a large sheet of paper, so improvization may be necessary!

7 The Self-characterization Sketch: Work Placements and Students' Perceptions of Self

The self-characterization sketch as a method for construct elicitation arose in Kelly's clinical practice and is described in detail by him (Kelly 1955, p. 323). The participant is asked to produce a description of themselves, written in the third person, as if they were "the principal character in a play." They are also asked to write it from the perspective of a friend who knows them intimately and sympathetically. The idea is that this sketch provides an insight into the person's construing of themselves and others. The researcher then inspects the sketch to create a basis for dialogue which can clarify the world view of the writer. In addition to their use in therapeutic work self-characterization sketches (e.g., Fransella and Dalton 2000; Androutsopoulou 2001), they have also been used in education research (e.g., Pope and Denicolo 2001).

In the example presented here, they were used to explore undergraduate students' views of themselves in relation to a work role, looking at how this changes and develops through time and with experience. Joanne, a second year social work undergraduate, was invited to complete a self-characterization sketch imagining herself in a future job. In order to encourage her to explicitly think of her work self, the instructions were adapted from Kelly's original script:

> Write a character sketch of yourself as you think of or imagine yourself in a work role, ideally in the type of role you think you would like to take up after graduation. Write just as if you are the principal character in a play. Write it as it might be written by a friend who knows you very intimately and very sympathetically, perhaps better than anyone ever really could know you. Be sure to write it in the third person. For example, start out by saying "Joanne as a social worker is..."

7.1 Analysis of the Self-characterization Sketch

Kelly (1955, pp. 330–340) recommends a number of different "readings" of the sketch in order to draw out issues of significance. This is an interpretative process and involves, for example, analyzing how the sketch is organized, identifying themes and pulling out significant words and phrases.

Some initial analysis of Joanne's self-characterization sketch took place before discussing it in depth with her. The aim was to understand how she saw her anticipated role within the world of work she described. "Where did *she* place *herself*?" and "what did *she* see as important?" were key questions for this reading. The first and last sentences of the account can give particular insights, the first sentence setting out the safest and most secure place for the writer to start describing themselves and the final sentence providing a glimpse of the construed future. Joanne started her sketch by saying she had worked hard to achieve her degree and become a social worker, and finished by saying she would like to make a difference within her field of work and these were both useful areas to explore with her.

In imagining herself as a newly qualified social worker, Joanne also said:

...she has done an excellent job of adapting to the job requirements. She demonstrates an ability to manage workload and asks for help if required during supervision. She gets on well with colleagues and supports them during difficult times. Whilst the job hasn't been easy and there has definitely been stressful situations, Joanne has used her initiative to deal with the situation.

From just this short extract, constructs related to her adaptability, being able to manage workload, asking for help, using initiative, getting on well with colleagues, being supportive, finding the job difficult and using her initiative can be seen. These were used in the interview for discussion and further elaboration. These constructs, together with other impressions gained from analysis of the sketch, provided a rich source for discussion with Joanne. For example, would she say that, now, she is someone who asks for help or not? How has her construal of herself in relation to this changed since she started her degree, and how does she think it might change by graduation?

Most participants produce a relatively short self-description. In Joanne's case, the entire sketch was only about 270 words, and yet, used as a basis for the subsequent interview, it provided a rich insight into her anticipations and hopes for herself as a social worker.

8 Reflecting on the Use of PCP Methods

PCP methods may initially be experienced as rather different from conventional qualitative methods. They impose a degree of structure on the interaction between researcher and participant, requiring tasks to be carried out in a certain order, for example eliciting constructs and then laddering them. This may appear restrictive to qualitative researchers, but our experience is that participants actually find this structure helpful. Data gathering with PCP methods tends to have a lively and dynamic feel to it, with participants taking a very active role; they are 'doing something' more than just sitting and answering questions, and often report that they found the experience not only interesting or revealing but also fun.

While recommending PCP methods, we acknowledge that they have challenges and potential drawbacks. They can seem rather game-like to some participants, suggesting that their experiences are being trivialized. This can usually be addressed by taking time to explain why the technique is useful for the particular research project in which they are involved. Another difficulty can be deciding how much to intervene in the process of generating data. For example, when using Pictor to explore collaborative working, Bravington (2011) found that some ways of laying out the arrows tended to produce richer descriptions than others. The researcher might, therefore, decide to reflect this in their guidance to participants, but then too much intervention by the researcher might undermine the essential participant-led nature of PCP methods.

9   Conclusion and Future Directions

We have demonstrated some of the PCP methods that are likely to be of value to the qualitative researcher. They have the advantage of being highly flexible and can be adapted for use in a wide variety of research topics and settings, providing opportunities for qualitative researchers to create innovative ways of researching.

Our experience is that participants find PCP methods engaging and interesting, and that they have a number of advantages compared with more familiar qualitative methods. They are essentially participant-led, enabling participants to remain in control of the research process while benefitting from guidance from the researcher; they enable participants to quickly focus on issues of importance through the use of concrete examples, which is also helpful in topics where participants find it hard to abstract from and articulate their experience. In reflecting upon the data produced, participants produce accounts that are particularly rich and the process often results in the participant gaining a new awareness and insight. They help to avoid common-sense or party-line responses, which is especially useful when exploring sensitive issues. And where appropriate, they can enable the researcher to handle data from larger samples than is usual in qualitative research by searching, say, self-characterization sketches or interview transcripts for construct dimensions rather than performing a thematic analysis.

The principle use of PCP methods in qualitative research has been the exploration of personal experience and selfhood. But, in common with narrative psychology and social constructionism, PCP emphasizes the constructed nature of our psychological and social worlds and there is therefore no reason why PCP methods should not also be used by discourse analysts in researching social constructions. PCP's clinical origin also means that its methods are particularly effective for addressing issues of change, making them particularly appropriate for action research where social or community change is the desired outcome.

References

Androutsopoulou A. The self-characterization as a narrative tool: applications in therapy with individuals and families. Fam Process. 2001;40(1):79–94.

Bell R. Did Hinkle prove laddered constructs are superordinate? A re-examination of his data suggests not. Pers Constr Theory Pract. 2014;11:1–4.

Bravington A. Using the Pictor technique to reflect on collaborative working in undergraduate nursing and midwifery placements. Masters thesis, University of Huddersfield; 2011.

Burr V, Giliberto M, Butt T. Construing the cultural other and the self: A Personal Construct analysis of English and Italian perceptions of national character. Int J Intercult Relat. 2014a;39:53–65.

Burr V, King N, Butt T. Personal construct psychology methods for qualitative research. Int J Soc Res Methodol. 2014b;17(4):341–55.

Burr V, Blyth E, Sutcliffe J, King N. Encouraging self-reflection in social work students: using personal construct methods. Br J Soc Work. 2016;46:1997. https://doi.org/10.1093/bjsw/bcw014.

Butt TW. Personal construct theory and method: another look at laddering. Pers Constr Theory Pract. 2007;4:11–4.

Butt TW. George Kelly: the psychology of personal constructs. London: Palgrave Macmillan; 2008.

Butt T, Bell R, Burr V. Fragmentation and the self. Constructivism Hum Sci. 1997;2(1):12–29.

Elliott D, Brooks J, King N. Support networks of young people with autism spectrum disorder using Pictor: exploring multiple perspectives. (In preparation).

Fransella F. The essential practitioner's handbook of personal construct psychology. London: Wiley; 2005.

Fransella F, Dalton P. Personal construct counselling in action. 2nd ed. London: Sage; 2000.

Hardy B, King N, Firth J. Applying the Pictor technique to research interviews with people affected by advanced disease. Nurse Res. 2012;20(1):6–10.

Hargreaves CP. Social networks and interpersonal constructs. In: Stringer P, Bannister D, editors. Constructs of sociality and individuality. London: Academic Press; 1979. p. 153–75.

Hinkle D. The change of personal constructs from the viewpoint of a theory of implications. Unpublished PhD thesis, Columbus: Ohio State University; 1965.

Johnson K. Visualising mental health with an LGBT community group: method, process, theory. In: Reavey P, editor. Visual methods in psychology: using and interpreting images in visual research. Hove: Psychology Press; 2011. p. 173–89.

Kelly GA. The psychology of personal constructs. New York: Norton; 1955.

King N, Horrocks C. Interviews in qualitative research. London: Sage; 2010.

King N, Melvin J, Ashby J, Firth J. Community palliative care: Role perception. Br J Community Nurs. 2010;15(2):91–8.

King N, Bravington A, Brooks J, Hardy B, Melvin J, Wilde D. The Pictor technique: a method for exploring the experience of collaborative working. Qual Health Res. 2013;23(8):1138–52.

King N, Brooks J, Bravington A, Hardy B, Melvin J, Wilde D. The Pictor technique: exploring experiences of collaborative working from the perspectives of generalist and specialist nurses. In: Brooks J, King N, editors. Applied qualitative research in psychology. Basingstoke: Palgrave Macmillan; 2017.

Lee R. Doing research on sensitive topics. London: Sage; 1993.

Liamputtog P. Researching the vulnerable: A guide to sensitive research methods. London: Sage; 2007.

Madill A, Jordan A, Shirley C. Objectivity and reliability in qualitative analysis: realist, contextualist and radical constructionist epistemologies. Br J Psychol. 2000;91:1–20.

Mercer B. Interviewing people with chronic illness about sexuality: an adaptation of the PLISSIT model. J Clin Nurs. 2008;17(11c):341–51.

Noble B, King N, Woolmore A, Hughes P, Winslow M, Melvin J, et al. Can comprehensive specialised end of life care be provided at home? Lessons from a study of an innovative consultant-led community service in the UK. Eur J Cancer Care. 2014;24:253–66.

Pink S. Doing visual ethnography. London: Sage; 2007.

Pope M, Denicolo P. Transformative education: personal construct approaches to practice and research. London: Whurr; 2001.

Ross A, King N, Firth J. Interprofessional relationships and collaborative working: Encouraging reflective practice. Online J Issues Nurs. 2005;10(1):4. http://www.nursingworld.org/MainMenuCategories/ANAMarketplace/ANAPeriodicals/OJIN/TableofContents/Volume10 2005/No1Jan05/tpc26_316010.html

Sargeant S, Gross H. Young people learning to live with inflammatory bowel disease: working with an "unclosed" diary. Qual Health Res. 2011;21(10):1360–70.

Smith JA, Osborn M. Interpretative phenomenological analysis. In: Smith JA, editor. Qualitative psychology: a practical guide to research methods. London: Sage; 2003.

Braun V, Clarke V. Using thematic analysis in psychology. Qual Res Psychol. 2006;3(2):77–101.

Mind Maps in Qualitative Research

63

Johannes Wheeldon and Mauri Ahlberg

Contents

1 Introduction	1114
2 Background: Mind Maps and Qualitative Research	1115
3 Mind Maps: Theory and Methods	1116
4 Mind Maps and Qualitative Research: Applications and Examples	1117
4.1 Planning Research	1117
4.2 Collecting Data	1118
4.3 Analyzing Data	1118
4.4 Presenting Data	1119
5 Assessing Mind Maps: Value(s), Limitations, and Challenges	1120
6 Mind Maps and Mixed Methods	1123
7 Case Study: Mind Maps: Priming the Pump, Depth, and Detail	1123
8 Mind Maps and Health and Social Sciences	1125
9 Conclusion and Future Directions	1126
References	1127

Abstract

Traditionally, qualitative data collection has focused on observation, interviews, and document or artifact review. Building on past work on visual approaches in the social sciences, in this chapter we consider the value(s) of mind maps for qualitative research. Mind maps are useful tools for qualitative researchers because they offer a mean to address researcher bias and ensure data are collected in ways that privilege participant experience. Qualitative researchers can benefit from visually oriented approaches to research by using them to assist them to plan

---

J. Wheeldon (✉)
School of Sociology and Justice Studies, Norwich University, Northfield, VT, USA
e-mail: jwheeldo@norwich.edu

M. Ahlberg
Department of Teacher Education, University of Helsinki, Helsinki, Finland
e-mail: mauri.ahlberg@helsinki.fi

© Springer Nature Singapore Pte Ltd. 2019
P. Liamputtong (ed.), *Handbook of Research Methods in Health Social Sciences*,
https://doi.org/10.1007/978-981-10-5251-4_7

their research, collect qualitative data, analyze what they have collected, and present findings. Of particular interest in this chapter is how mind maps can offer a graphic and participant-centric means to ground data within theory, assist participants to better frame their experience, and can be used as part of the design and development of additional data collection strategies and mixed methodological approaches. While future applications of mind maps are likely to use technological tools and techniques, there is value in the original approach of putting pen to paper and engaging in a creative and tactile process of outlining ideas and recounting experiences.

Keywords

Mind mapping · Qualitative research · Theory · Constructivism · User generated data collection · Mixed methods

1 Introduction

Qualitative research provides an interpretation of the social world of research participants by concentrating on their "experiences, perspectives, and histories" (Ritchie and Lewis 2003, p. 3; Liamputtong 2017). Prioritizing individualistic accounts of knowledge, experience, and perception, meaning discovered through social interactions privileges the constructed realities of individuals when reporting social science research findings. In addition to more traditional approaches to qualitative research, new approaches have sought to explore how visual approaches can make more evident how an individual constructs, frames, and describe one's experience (Wheeldon and Faubert 2009).

Although widely used data gathering techniques for human subjects research such as participant observation, interviews, and focus groups are still important (Wolcott 1990), mapping provides a means for participants to personally construct a graphic representation of their experiences (Wheeldon 2010). Focused on precision (Winter 2000) and credibility (Hoepf 1997), qualitative researchers have also begun to acknowledge that the approach chosen by the researcher shapes subsequent research interactions (Feyerbend 1978). These choices mean that the role of the researcher in assigning value to one of what may be many possible meaningful interpretations of the same data must be confronted (Guba and Lincoln 1989).

The trend toward reflexivity (MacBeth 2001) has helped to explicitly outline the role of the researcher in qualitative analysis. To address this concern, researchers often explicitly consider personal and epistemological reflexivity to acknowledge their own possible biases (Willig 2001). Cognizant that the approach chosen by the researcher will shape any interaction between the phenomena studied and the data collected (Feyerbend 1978), qualitative researchers acknowledge the importance of the role they themselves play within their own research (Guba and Lincoln 1989).

New approaches of data collection include vignette responses, subject-operated cameras/videos/sound recordings, focus groups, journaling, and visual life histories (Wheeldon and Faubert 2009). Mind maps offer yet another means. This chapter explores mind mapping, outlines its key features, and provides practical guidance for employing it in fieldwork, outlining some epistemological, ethical, and practical considerations for researchers. Finally, it canvasses its varied applications for qualitative health research.

Key aims of this chapter are to:

- Provide some background and context to the emergence of the method
- Situate the method within methodological and theoretical approaches
- Provide consideration of the strengths and advantages, weaknesses, and limitation of the method
- Look at its compatibility and use with other methods
- Briefly explore the key issues and practicalities of using the method
- Provide case study examples of mind maps in practice
- Outline areas in which mind maps are used within the health and social sciences

2 Background: Mind Maps and Qualitative Research

Traditionally, qualitative data collection methods include observational methods, in-depth interviews, and group-based approaches, such as focus groups (Liamputtong 2013). Interviews are one of the most common data collection methods in social research (Denzin 2001; Serry and Liamputtong 2017). An interview, which is a verbal exchange between an interviewer and one or more interviewees (Varga-Atkins and O'Brien 2009), can be conducted in a variety of ways. While there are numerous variations of the application of an interview (open or closed questions, in person or over the phone), there is a common understanding about what the data collection approach of "interviewing" generally means across research disciplines. This is not the case when it comes to the data collection approach of using maps.

A major issue complicating knowledge translation regarding visual methods in the natural and social sciences is the different terminology used to describe diagrams (Umoquit et al. 2011). To establish an overarching "umbrella term" that groups the different uses of diagrams in the data collection process, the boundaries of the larger category of graphic communication and where the term diagram fits in must first be clarified. Mind maps are more flexible than other forms of visual approaches (Wheeldon and Ahlberg 2012). Defined as diagrams used to represent words, ideas, and other concepts arranged around a central word or idea (Buzan 1991), mind maps are structurally more flexible than other sorts of maps and present ideas in a variety of ways (Buzan 1974). Figure 1 provides an example.

The re-emergence of visual maps in social science research has occurred at a time when there appears to be a desire to develop data collection methods that are either more explicitly user-generated or which are less influenced by what may be sui generis research/participant interactions. For qualitative researchers, they offer a way for researchers to "... search for codes, concepts and categories within the data... based on how the participant(s) frame(d) their experience" (Wheeldon and Faubert 2009, pp. 72–73). Since 1997, maps have been used with increasing

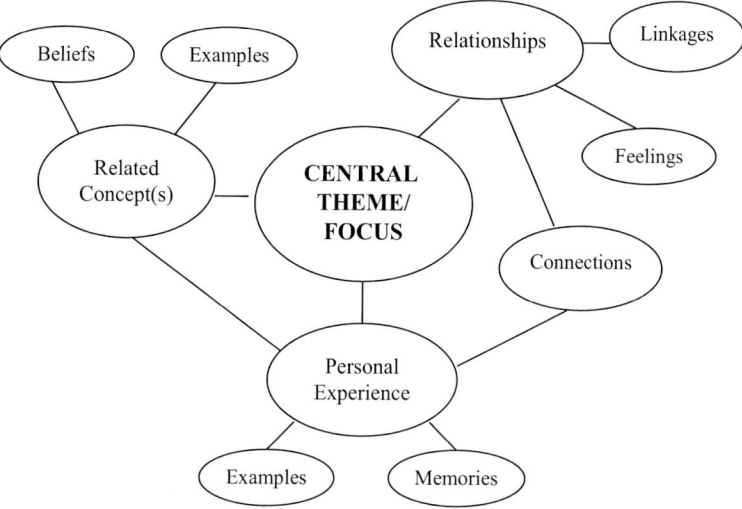

Fig. 1 Constructing a Mind Map (Wheeldon 2010, p. 91)

frequency in fields such as health, education, sociology, and engineering (Nesbit and Adescope 2006).

3 Mind Maps: Theory and Methods

Mind maps are best suited to qualitative research because they can be connected to the theoretical starting place generally associated with qualitative research (Wheeldon and Ahlberg 2012). For those who hold constructivist accounts of knowledge, meaning is assumed to be highly subjective and best understood through social interaction, and personal histories and experiences (Creswell and Plano Clark 2007; see also ▶ Chap. 7, "Social Constructionism"). As a result, knowledge is inherently localized. Linked concepts can uniquely demonstrate how participants connect their experiences (Daley 2004). Using maps in this way may provide a means to prompt research participants to consider past experience in more depth and detail (Legard et al. 2003).

The theoretical value of mapping is based on the acknowledgment that people learn in different ways and think using a combination of words, graphics, and images. For qualitative researchers, the use of interviews as the sole means of data collection may be relying on psycho-linguistic assumptions about the role of syntax, semantics, and context to guide their construction of meaning (Cassirer 1946). Because people live their lives both in their own head and as part of a social, cultural, and linguistic collective (Habermas 1976), consciousness is both something that people experience on their own and through their interactions with others (Husserl 1970). Because maps offer a unique way for research participants to represent their experiences, they may provide a means for individuals to think more clearly by avoiding the assumptions built into language (Korzybski 1933). They may provide one strategy to break out of conventional and linguistically limited representations of experience, rehearsed narratives, and canned responses (Hathaway and Atkinson 2003).

4 Mind Maps and Qualitative Research: Applications and Examples

Mind maps have a variety of uses in qualitative research from planning a project, collecting data, analyzing narrative and visual data, and presenting analysis strategies and findings. Figure 2 presents some uses.

4.1 Planning Research

Maps can be used to plan research. By outlining the various steps in a research project, researchers can more easily see the various tasks before them, and activities required. This might include literature review, participant identification, data collection design, obtaining institutional review board approval, conducting data collection activities, analyzing collected data, and writing up and/or presenting the data. Maps can help researchers brainstorm and plan research. They can be used before, in the place of, or even after an interview to provide additional data. Comparing interviews with mind maps from the same participant may offer a useful means to validate the data collected. Maps can also be used to demonstrate an analysis strategy

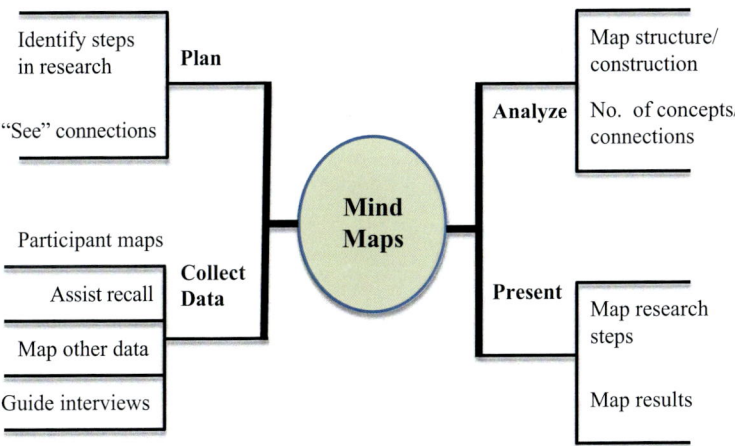

Fig. 2 Uses of mind maps in qualitative research (Wheeldon and Ahlberg 2012, p. 88)

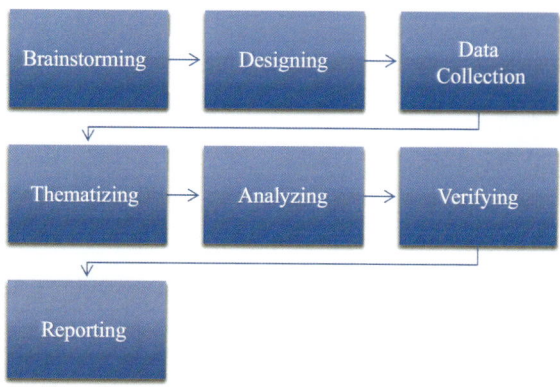

Fig. 3 Mapping stages of qualitative research

or to present findings of interviews in a visually appealing and accessible way. Figure 3 provides one way to conceive of the stages in qualitative research.

4.2 Collecting Data

Perhaps the most obvious use of mind maps may be as a means to collect personalized and individualistic data from research participants (Tattersall et al. 2007; Wheeldon and Faubert 2009). To make one, take a blank piece of paper and turn it on its side so that it is wider than it is high. Starting in the middle of the page, make a central image or write an important word. Draw a line out from the central image in each direction and write an important theme that is connected to your central image or word. Write other words or images that are connected to this theme and continue to build your map outward (Buzan and Buzan 2000). Buzan suggests focusing one and only one word for each association and to use color, the thickness of lines, boxes, or pictures and graphics to make your map unique and expressive (Buzan 1991). Figure 4 provides an example.

There are perhaps two key aspects to mind maps that make them valuable for data collection in qualitative research. The first is that they focus on a central theme or idea and "radiate" associated ideas outward, based on related ideas and examples. The second is that they be as much as possible free form, creative, and developed by participants with minimal instruction by the researcher. This can allow the participant's individual creative process to emerge.

4.3 Analyzing Data

Another way mind maps could be useful in qualitative research analysis of collected data. Instead of simply reading transcriptions, researchers might read, map, read, and add to their original map or decide to create another. This can allow for more flexibility to draw out different sorts of connections, relationships, or themes (Tattersall et al. 2007). It may also assist the researcher to break out of their own

Fig. 4 Example of data in a mind map (Wheeldon and Ahlberg 2012, p. 103)

cycle of assumption or expectation by forcing them to graphically represent what they are reading.

This approach to data analysis is in line with more traditional strategies derived from Pope et al. (2000). The suggestion that researchers immerse themselves in the data, identify key themes, connect these themes to concepts in the data set, and then rearrange the data based on the themes is often much easier to do graphically than textually. One approach used mind maps to show the research process from the researcher's point of view (Wheeldon 2012). This may also have the added benefit of serving as a reflexive demonstration of how themes emerged from the data (Kelle 2005). This process is represented in Fig. 5.

4.4 Presenting Data

Mind maps can also be used to present research data. They might be part of broader efforts to streamline research for policy-makers which would be an improvement over the provision of long complex data tables and the problematic assumption that readers can make sense of the complex analysis provided.

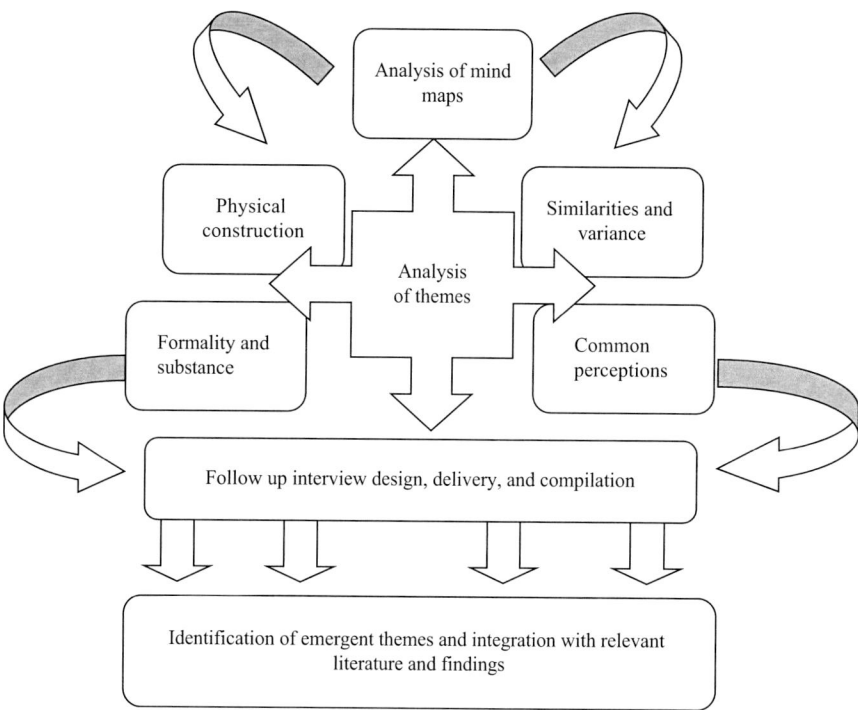

Fig. 5: Analyzing maps and interviews (Wheeldon 2012)

Beyond visual representations of research findings, maps can also be used to help students and researchers understand the nature of the questions at the heart of qualitative research. Figure 6 explores some key questions that are relevant for researchers.

5 Assessing Mind Maps: Value(s), Limitations, and Challenges

Part of the assessment of the value of mapping in qualitative research must involve to what extent they can be used to address outstanding conceptual issues. Of specific interest here is the value of maps offer a middle ground between the alternative analysis strategies offered by Glaser (1992) and Strauss (1987). Since grounded theory emerged (Glaser and Strauss 1967) as a theoretical frame in qualitative research, it has played an important role in specifying how a qualitative approach to data analysis can privilege localized understanding in theory creation. Silverman (2005) has suggested that grounded theory in general involves an initial attempt to develop categories from the data, locating the data within these categories to demonstrate relevance and developing these categories into a more useful framework for general understanding.

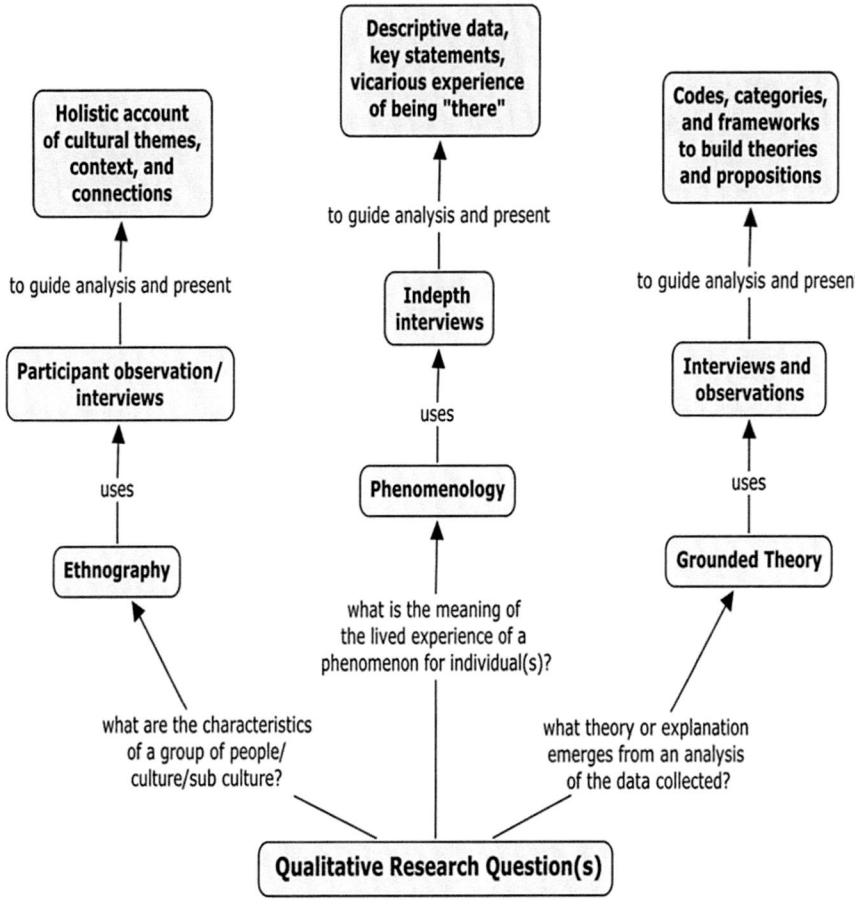

Fig. 6 Key qualitative questions (Wheeldon and Ahlberg 2012, p. 99)

However, the general agreement between Glaser and Strauss about how to analyze data from the ground up has been challenged by specific differences about how the analysis should unfold and the role and relevance of structured versus more ad hoc coding. On one view, the controversy between Glaser and Strauss boils down to the question of whether the researcher uses a well-defined coding paradigm and always looks systematically for "causal conditions," "phenomena/context, intervening conditions, action strategies," and "consequences" in the data, or whether theoretical codes are employed ad hoc as they emerge in the same way as substantive codes emerge, but drawing on a huge fund of "coding" families (Kelle 2005, p. 20).

Using maps in multistage data collection allows for middle ground in grounded theory. Instead of looking to the researcher to search for codes, concepts, and categories within the data, maps allow for the identification of concepts and connections based on how the participant frames their experience. As a result, subsequent data strategies remain based on codes and concepts as demonstrated through the participant-generated maps. Although this approach might initially appear closer to the structured Glaserian strategy, subsequent data collection strategies can allow for the participant-generated framework to be tested, explored, and further detailed and delineated through interviews, surveys, or focus groups.

Of interest is that the maps provided an explicit basis for additional data collection strategies, but the subsequent themes that emerged challenged the initial analysis of the maps. In these cases, the use of maps can provide a flexible and unsolicited participant-led approach to assist coding and analysis. Because any emergent codes can be tested through subsequent data collection, the use of participant maps might be seen as a unique and innovative data-gathering instrument of interest to a new generation of more visually oriented researchers.

As described in this chapter, a variety of options exist for researchers who seek to use mind maps in the planning, analysis, and presentation in qualitative research. For those seeking to use participant-completed maps as part of data collection in qualitative research, the limitations associated with their use should not be discounted. There is evidence that important differences between people, groups, personalities, and learning styles can limit the utility of concept maps in gathering data (Rohm 1994). For example, Czuchry and Dansereau (1996) found that women identified the mapping assignment to be easier than did men. However, in my own work, some participants did suggest that they found the initial request for a map "odd," and the process of creating a map "strange."

Another limitation is based on the definitional confusion that exists in studies and discussions of concept maps and mind maps. Often, it appears, these two distinct types of maps are combined and referred to as though they are the same (Nesbit and Adescope 2006). This is a problem given the theoretical basis for mind maps, and the need to balance the practicalities of collecting and assessing data, with the need in qualitative research to ensure the prioritization of the participant's experience. In some studies, the use of maps as part of qualitative data collection resulted in the return of maps that did not conform to traditional definitions (Wheeldon and Faubert 2009). This may be a training issue, but it also may be related to how people view experience. As Tomas (1997, p. 75) has suggested, "recall of experience is always selective and there will be many absences or gaps. People forget things or choose not to tell things or are not aware of things – for all sorts of reasons."

This highlights a central issue for visual approaches in data collection. On the one hand, understanding how data based on a common definition is important. On the other, definitional requirements may be at odds with a focus on precision, credibility, and transferability as they relate to capturing individual experiences and perceptions in qualitative research (Tattersall et al. 2007).

This definitional concern is also shared by those who draw a sharp distinction between data visualization based on rigorous statistical models and infographics, which may be exciting and/or visual appealing but not logically organized or consistently designed (Gelman 2011). The question is not which is better, but which is best suited to what sort of goal. This distinction is important for concerns related to visual data collection. Different assumptions must inform the consideration about which types of maps are best suited to address specific research designs and research questions (Wheeldon and Ahlberg 2012).

6 Mind Maps and Mixed Methods

Few doubt the value of mixed methods approaches to combine the clarity of counts with the nuance of narrative reflection (Wheeldon and Ahlberg 2012). New perspectives have also emerged that consider the challenges posed by partnerships between qualitative and quantitative approaches through the construction of mixed methodologies (Johnson and Onwuegbuzie 2004; Poole and Davis 2006; see also ▶ Chap. 4, "The Nature of Mixed Methods Research"). Using maps as part of mixed methods research can help fuse qualitative assumptions about value with quantitative approaches to capturing frequency. For example, using mind maps to collect data ensures the reflections that emerge are coming from the participant themselves rather than being prompted through series of survey or interview questions.

This could involve using mind maps as a pre−/posttool to compare number of concepts and/or complexity of the map's construction. It might involve the use of mind maps as a qualitative data collection tool that could be used to identify a list of key concepts for inclusion in a survey that would lend itself to numeric analysis. One approach involves the use of mind maps to create a mixed methods measure (Wheeldon 2010). Imagine 4 phases of data collection. First data are collected through open-ended mind maps from participants. These data are then sorted and organized to identify key themes that might be explored through more general open-ended and then more specific close-ended interview questions. As part of a final conclusionary or summative data collection phase, participants could be invited to revisit key ideas that they would want to make sure were captured as part of the research. These different phases are more or less participant centric. This allows a weighting process to privilege user-generated concepts consistent with qualitative researcher's focus on the "experiences, perspectives and histories" (Ritchie and Lewis 2003, p. 3) of participants.

For example, concepts that emerged through mind maps could be scored at 4 points. Concepts that emerged through open-ended questions could be worth 3 points, and those that emerged through more close-ended or specific questions worth 1. Concepts that re-emerged during the summative stage of data collection could be worth 2 points. A concept that emerged through all four phases would be scored at 10, while those that emerged from a close-ended question only would be scored at 1. In general, the more user generated and unprompted, the higher the score of salience.

7 Case Study: Mind Maps: Priming the Pump, Depth, and Detail

The Latvian Legal Reform Project (LLRP) was a 20-month initiative funded by the Canadian International Development Agency (CIDA). It ran from 2002 to 2004 and offered targeted legislative support, institutional capacity development, and human resources training to the Latvian Ministry of Justice as it established the National Probation Service (NPS). Ministry officials developed new laws, amended existing laws, trained staff, assisted in policy development, and provided support to leading officials from other related Latvian ministries. The Latvian Probation Service established in 2003 has since grown year by year in number of staff, programs, and services delivered (Jurevičius 2008).

In one study, a mind map methodology was used as part of a 4-stage data collection approach (Wheeldon 2009). Originally, participants were asked to complete mind maps between September and December 2007. This proved impractical. At the advice of senior colleagues, instead of gathering data for all participants using mind maps, researchers split participants into two groups. One group completed a mind map before data collections through a series of interviews, and one group did not. While not initially envisaged, this iterative approach allowed another sort of map-based analysis. The first contained 14 participants who completed both stages of the data collection. The second contained five other participants who completed only the interviews.

In past projects, the maps themselves served as a valuable source of data (Wheeldon and Faubert 2009). This study focused on the responses of participants through notes I took during the interviews. Of specific interest in this analysis was whether the completion of a map would impact the depth and detail of individual reflections. In this study, depth and detail were defined using the following criteria. Detail was related to the number of unique concepts provided within all interviews, including the reflective or summative responses. By contrast, depth was connected to the nature of the interview responses based on the length of responses, complexity of connections made, and specificity of the examples provided.

The results of the interviews among both groups were compared, based on the number of individual concepts identified through the interviews, and the length, detail, and number of specific examples provided through the participants' responses. By combining an analysis of the presence and frequency of unique individual concepts, along with the specificity of participant reflections, broader comparisons between these groups became possible. In this study, mind maps appeared to "prime the pump" of participant reflection. By requiring research participants to first frame their experience, maps may help researchers in the refinement of other data collection strategies. In this study, the group who had completed the maps before the interviews identified, on average, seven more individual concepts than the group who did not first complete a map, as presented in Table 1 (Wheeldon 2011).

Table 1 Average number of concepts identified by map and nonmap groups

Map status	Average number of concepts
Map	16.57
Nonmap	9

In addition, participants who completed maps provided more depth when recounting their experience during the interviews. This included a greater number of specific examples, memories, and suggestions (Wheeldon 2011). When asked to reflect on the experience of creating a concept map, virtually all participants in the map group identified the maps as a "useful way to see experience." Some suggested this was because making a map "helped them to remember events from years ago" and "organize their thoughts about the experience systematically." Others suggested that, as a visual aid, it helped put the experience in "context" and provided a "clearer view" by looking at events again, realizing how much had happened, and helped them to "focus on the key experiences, concepts and connections."

The notion that qualitative researchers ought to consider the value of numeric-based means of analysis in their research is controversial. As Sandelowski (2001) points out, a perception exists that real qualitative researchers do not count. While simplistic, given the utility of numbers to present what is known about a problem, and describe research samples, this aversion to numbers speaks to broader political and ideological differences between qualitative and quantitative research agendas (Jick 1979). While the hallmark of past academic debates, it seems quaint and somewhat outdated given the complexity of emergent research problems social science researchers face today. Instead of propagating antinumeric myths, qualitative researchers would do well to use numbers to showcase the labor and complexity of qualitative work and examine in more detail the meanings that may emerge from qualitative data. While the over-reliance on numbers and counts may be problematic, an under reliance may be just as dangerous.

8 Mind Maps and Health and Social Sciences

Visual maps have become a central methodological tool in the fields of nursing and health studies. Studies have employed maps as part of nursing education (Rooda 1994), care planning (Mueller et al. 2001; Kern et al. 2006) and for reflection and evaluation (Jenkins 2005). This trend can be connected to what is known as a genogram. Defined as a pictorial display of a person's family relationships and medical history, it may include visualized hereditary patterns and psychological factors that inform relationships (McGoldrick et al. 1999). Today, mind maps are essential tools in qualitative research in these fields (Tattersall et al. 2007). They can be differentiated by those who view maps as a tool to assist in teaching and learning in nursing and medicine, those who have used maps to understand professional values and views, and those who have outlined the application of mapping techniques as part of qualitative methodologies and approaches.

For teaching and learning in medicine, using maps improved recall the massive amounts of content and complex inter relationships between them (Farrand et al. 2002). Maps have been shown to be one example of a creative strategy to make learning both more interesting and more fun as part of nursing education (Rooda 1994). Mind mapping assists students to engage in a unique method of learning that can expand memory recall and help create a new environment for processing information (Kotcherlakota 2013). For example, research by Farrand et al. 2002) explored mapping as a memory aid among medical students. Students who used mind maps saw a 10% increase in recall over baseline versus students who used their preferred study methods who only had a 6% increase over baseline.

Others have explored how maps can assist students to understand nurses' attitudes, values, and behaviors. For example, Stephens (2015) found that nursing students felt that by sharing initial mind maps in a content area and receiving feedback from peers and faculty, they were able to better understand both key concepts and attitudes and values congruent with the profession. This might be connected to how mind maps have been shown how to establish trust between those collecting health data and participants and/or patients. Mind maps can serve as a means to obtain participant verification of an emerging theoretical framework and establishing trust between the research and participant (Whiting and Sines 2012).

Perhaps the most detailed use of mind maps as a tool is by Tattersall et al. (2007). They identify transcribing and analysis. Mind maps can serve assist traditional transcribing, as a complementary approach used in conjunction with traditional data gathering techniques in health sciences. For example, to augment data collected through written transcription or interviews, the researcher can use mind maps to document or capture nonverbal communication such as eye contact, posture, facial expression, personal space, and touch. While mind mapping and transcribing verbal interactions would be difficult, a mind-mapping research assistant could observe the interaction, and add important, relevant data (Tattersall et al. 2007).

The use of maps in analysis could involve using visual techniques to bracket preconceived ideas and thus, limits the potential they will interfere with subsequent analysis. If conducted correctly, mind mapping should lend itself well to this style of thought process, as it allows freethinking or, as Buzan and Buzan (2000) state when discussing the aims of creative mind mapping, "to clear the mind of previous assumptions about the subject" (Tattersall et al. 2007, p. 33). Beyond bracketing, maps can be used to identify themes including issues, concepts, and ideas derived from the raw data. Mind maps can also be used to index these themes by linking them throughout all respondents' data. Tattersall et al. (2007, p. 33) also identify charting, whereby the data are rearranged into "charts" containing the relevant data from various respondents.

9 Conclusion and Future Directions

Traditionally, qualitative data collection has focused on observation, interviews, and document or artifact review. Building on past work on visual approaches in the social sciences, in this chapter I considered the value(s) of mind maps for qualitative research. Qualitative researchers can benefit from visually oriented approaches to research, especially when they offer a graphic and participant-centric means to ground data within theory, focus on how user-generated maps can assist participants to better frame their experience, and justify the design and development of additional data collection strategies. This is based in part on an imperative to develop models of knowledge that capture data that are freely "given" by participants (Drucker 2009).

The visual turn in social science research is likely to continue and embrace more technological tools and techniques. At the same time, many may continue to use mind maps as a process of artistic and imaginative reflection to capture one's experience. While future applications of mind maps are likely to use technological tools and techniques, there is value in the original approach of putting pen to paper and engaging in a creative and tactile process of outlining ideas and recounting experiences (Buzan 1974).

Mind maps can be connected to broader trends away from the traditional positivistic practices of researcher-led data collection and toward more subjective explorations of phenomena including how creative explorations spark new ways to interrogate meaning (Lippens and Hardie-Bick 2013). Given the myriad of research decisions (Palys 2003) that are made in the construction and analysis of any study, the acknowledgment of the potential for researcher bias is an important contribution to social science research. New visual approaches to data collection might offer another means to explore reflexive analysis within qualitative research.

It is worth repeating that perhaps one of the most interesting features of mapping is that it can allow participants to break out of the rehearsed narratives of their daily lives (Hathaway and Atkinson 2003) and provide an entry point into the unadulterated views of participants. Through the graphic construction of experience, researchers can get another view of how participants see the world (Wheeldon and Faubert 2009). Such approaches are of interest in many fields and disciplines in the social sciences. As I have demonstrated in this chapter, they have particular relevant to the health sciences.

References

Buzan T. Use of your head. London: BBC Books; 1974.
Buzan T. Use both sides of your brain. New York: Plume; 1991.
Buzan T, Buzan B. The mind map book. London: BBC Books; 2000.
Cassirer E. Language and myth. New York: Harper & Brothers; 1946.
Creswell JW, Plano Clark VL. Designing and conducting mixed methods research. Thousand Oaks: Sage; 2007.
Czuchry M, Dansereau D. Node-link mapping as an alternative to traditional writing assignments in undergraduate psychology courses. Teaching of Psychology 1996;23(2):91–96.
Daley B. Concept maps: theory, methodology, and technology. Paper presented at the proceedings of the First International Conference on Concept Mapping, 14–17 Sept 2004, Pamplona; 2004.
Denzin NK. The reflexive interview and a performative social science. Qual Res. 2001;1:23–46.
Drucker J. SpecLab: digital aesthetics and speculative computing. Chicago: University of Chicago Press; 2009.
Farrand P, Hussain F, Hennessy E. The efficacy of the mind map study technique. Med Educ. 2002;36(5):426–31.
Feyerbend P. Science in a free society. London: New Left Press; 1978.
Gelman A. "Information visualization" vs. "Statistical graphics." 2011. Retrieved from http://andrewgelman.com/2011/07/22/information-visualization-vs-statistical-graphics/.

Glaser BG. Basics of grounded theory: emergence vs. forcing. Mill Valley, CA: Sociology Press; 1992.

Glaser BG, Strauss A. Discovery of grounded theory. strategies for qualitative research. Chicago: Aldine Press; 1967.

Guba E, Lincoln Y. Fourth generation evaluation. Beverly Hills: Sage; 1989.

Habermas J. Communication and the evolution of society. London: Polity Press; 1976.

Hathaway AD, Atkinson M. Active interview tactics in research on public deviants: exploring the two-cop personas. Field Methods. 2003;15:161–85.

Hoepf MC. Choosing qualitative research: A primer for technology education researchers. J Technol Educ. 1997;9(1):47–63.

Husserl E. The crisis of European sciences and transcendental phenomenology (trans: Carr D). Evanston: Northwestern University Press; 1970.

Jenkins A. Mind mapping. Nurs Stand. 2005;20:85.

Jick TD. Mixing qualitative and quantitative methods: triangulation in action. Adm Sci Q. 1979;24(4):602–11.

Johnson RB, Onwuegbuzie AJ. Mixed methods research: a research paradigm whose time has come. Educ Res. 2004;33(7):14–26.

Jurevičius I. Creation of the state probation service [Unpublished master's thesis]. Riga: University of Latvia; 2008.

Kelle U. Emergence vs. forcing of empirical data? A crucial problem of grounded theory reconsidered. Forum Qual Soc Res. 2005;6(2):27. [On-line Journal]

Kern CS, Bush KL, McCleish JM. Mind-mapped care plans: integrating an innovative educational tool as an alternative to traditional care plans. J Nurs Educ. 2006;45:112–9.

Korzybski A. Science and sanity: an introduction to non-Aristotelian systems and general semantics. Fort Worth: Institute of General Semantics; 1933.

Kotcherlakota S. Developing scholarly thinking using mind maps in graduate nursing education. Nurse Educ. 2013;38(6):252–5.

Legard R, Keegan J, Ward K. In-depth interviews. In: Ritchie J, Lewis J, editors. Qualitative research practice: a guide for social research students and researchers. Thousand Oaks: Sage; 2003. p. 138–69.

Liamputtong P. Qualitative research methods. 4th ed. Melbourne: Oxford University Press; 2013.

Liamputtong P. The science of words and the science of numbers. In: Liamputtong P, editor. Research methods in health: foundations for evidence-based practice. 3rd ed. Melbourne: Oxford University Press; 2017. p. 3–28.

Lippens R, Hardie-Bick J. Can one paint criminology? J Theor Philos Criminol. 2013;1(1):64–73.

MacBeth D. On reflexivity in qualitative research: Two readings, and a third. Qual Inq. 2001;7(1):35–68.

McGoldrick M, Gerson R, Schellenberger S. Genograms: assessment and Intervention. 2nd ed. New York: Norton; 1999.

Mueller A, Johnston M, Bligh D. Mind-mapped care plans: a remarkable alternative to traditional nursing care plans. Nurse Educ. 2001;26(2):75–80.

Nesbit JC, Adescope O. Learning with concept and knowledge maps: a meta-analysis. Rev Educ Res. 2006;76(3):413–48.

Palys T. Research decisions. Toronto: Thompson Canada; 2003.

Poole D, Davis T. Concept mapping to measure outcomes in a study abroad program. Soc Work Educ. 2006;25(1):61–77.

Pope C, Ziebland S, Mays N. Qualitative research in health care: analysing qualitative data. Br Med J. 2000;320:114–6.

Ritchie J, Lewis J, editors. Qualitative research practice: a guide for social science students and researchers. Thousand Oaks: Sage; 2003.

Rohm R. Positive personality profiles. Atlanta: Personality Insights; 1994.

Rooda L. Effects of mind mapping on student achievement in a nursing research course. Nurse Educ. 1994;19(6):25–7.

Sandelowski M. Real qualitative researchers do not count: the use of numbers in qualitative research. Res Nurs Health. 2001;24(3):230–40.

Serry T, Liamputtong P. The in-depth interviewing method in health. In: Liamputtong P, editor. Research methods in health: foundations for evidence-based practice. 3rd ed. Melbourne: Oxford University Press; 2017. p. 67–83.

Silverman D. Doing qualitative research. Thousand Oaks: Sage; 2005.

Stephens M. Changing student nurses values, attitudes, and behaviours: a meta ethnography of enrichment activities. J Nurs Care. 2015;5:320–30.

Strauss A. Qualitative analysis for social scientists. Cambridge: Cambridge University Press; 1987.

Tattersall C, Watts A, Vernon S. Mind mapping as a tool in qualitative research. Nurs Times. 2007;103(26):32–3.

Tomas A. The visual life history interview. 1997. Retrieved 12 July 2009, from http://www.colinwatsonleeds.co.uk/RMarticles/ReadingG.pdf.

Umoquit MJ, Tso P, Burchett HED, Dobrow MJ. A multidisciplinary systematic review of the use of diagrams as a means of collecting data from research subjects: application, benefits and recommendations. BMC Med Res Methodol. 2011;11(11):1–10.

Varga-Atkins T, O'Brien M. From drawings to diagrams: maintaining researcher control during graphic elicitation in qualitative interviews. Int J Res Methods Educ. 2009;32:53–67.

Wheeldon JP. Mapping knowledge transfer: Latvian-Canadian cooperation and justice reform [Unpublished doctoral dissertation]. Burnaby: Simon Fraser University; 2009.

Wheeldon JP. Mapping mixed methods research: methods, measures, and meaning. J Mixed Methods Res. 2010;4(2):87–102.

Wheeldon JP. Is a picture worth a thousand words? Using mind maps to facilitate participant recall in qualitative research. Qual Rep. 2011;16(2):509–22.

Wheeldon JP. After the spring: probation, justice reform, and democratization from the Baltics to Beirut. Den Haag: Eleven International Publishers; 2012.

Wheeldon JP, Ahlberg M. Visualizing social science research: maps, methods, and meaning. Thousand Oaks: Sage; 2012.

Wheeldon JP, Faubert J. Framing experience: concept maps, mind maps, and data collection in qualitative research. Int J Qual Methods. 2009;8(3):68–83.

Whiting M, Sines D. Mind maps: establishing 'trustworthiness' in qualitative research. Nurs Res. 2012;20(1):21–7.

Willig C. Introducing qualitative research in psychology: adventures in theory and method. Buckingham: Open University Press; 2001.

Winter, G. (2000). A comparative discussion of the notion of validity in qualitative and quantitative research. The Qualitative Report. 4(3–4). Retrieved from http://www.nova.edu/ssss/QR/QR4-3/winter.html.

Wolcott HF. Writing up qualitative research. Newbury Park: Sage; 1990.

Creative Insight Method Through Arts-Based Research

64

Jane Marie Edwards

Contents

1. Introduction: What is Arts-Based Research? 1132
2. ABR Methodology in Healthcare 1133
3. The Theoretical Location of ABR 1134
4. The Arts and Health 1136
5. Arts-Based Research in the Creative Arts Therapies 1136
6. Arts-Based Research: A Narrative Example 1138
7. ABR: Where Does the New Researcher Begin? 1139
8. The Foundations of Creative Insight Method 1140
    - 8.1 Narrative 1141
    - 8.2 Poetry 1142
    - 8.3 Visual 1142
9. Conclusion and Future Directions 1142
References 1143

Abstract

Arts-Based Research (ABR) has gradually developed momentum across many disciplines. It represents a way of using the arts to facilitate and enhance processes within research and even to conduct entire research studies. It can be used within healthcare research and in research studies with participants who are vulnerable. ABR approaches use any type of arts creation processes including poetry, narrative fiction, painting, drawing, or song writing. This chapter presents three levels of ABR researcher – novice, emergent, and expert. Examples from these three levels are presented and proposals for tasks and exercises to begin ABR as a novice to this method are described. A critical awareness of the ways in which ABR exists as a stand-alone method, and also aligns with other

J. M. Edwards (✉)
Deakin University, School of Health and Social Development, Geelong, VIC, Australia
e-mail: jane.edwards@deakin.edu.au

contemporary developments in qualitative healthcare research, is presented and discussed. The choice of studies reported here reflects the author's expertise in creative arts therapy. Providing this context and background helps new health researchers using ABR to position and expand their approach to the methods of their inquiry. Processes in choosing a method are described, with reference to multiple opportunities afforded by ABR. The chapter additionally elaborates a contemporary approach within ABR developed by the author: Creative Insight Method. This method will be presented in terms of ways of facilitating safe and collaborative arts-based processes with groups.

Keywords

Arts-based research · Music therapy · Arts therapy · Expressive therapy · Constructivist epistemology

1 Introduction: What is Arts-Based Research?

Arts-Based Research (ABR) represents a way of using the arts to facilitate and enhance processes within research, to advance knowledge (Malchiodi 2017). ABR has gradually developed momentum and impact across many disciplines. ABR approaches can involve creating works such as poetry, narrative fiction, plays, painting, drawing, or song writing. In her ground-breaking text, Leavy (2015) identifies six types of Arts-Based Research: narrative inquiry, poetry, music, performance, dance, and visual arts.

Boydell et al. (2012, np) describe ABR as being both "(1) a process to produce knowledge, and (2) a product to disseminate results." In the broadest conception of ABR, the method can be used as the primary approach to research within a project or can be used within a specific part of a research study. ABR has been used in healthcare research (for example, McCaffrey and Edwards 2016) and in research studies beyond healthcare with participants who are vulnerable (Osei-Kofi 2013). The researcher can be responsible for generating the arts-based components, perhaps in response to narrative data, or the participants can create the art materials. The researcher can also use the arts to reflect on and amplify data, deepening insights and reflections.

Jones (2015) has recommended that engaging the arts, and artistic processes, is a much needed *fresh inspiration* in the ongoing development of qualitative methods research. Chilton and Leavy (2014, p. 403) have proposed the appeal of ABR results from the capacity of the arts to

> …promote autonomy, raise awareness, activate the senses, express the complex feeling-based aspects of social life, illuminate the complexity and sometimes paradox of lived experience, jar us into seeing and thinking differently, and transform consciousness through evoking empathy and resonance.

At its core, ABR rejects any distinction between the sciences and art. It is proposed, instead, that all forms of inquiry and processes of knowing require, and result from, curiosity and creativity (Jones and Leavy 2014). Equally, ABR, perhaps due to its emergent character, currently remains free of disciplinary prejudice or any kind of theoretical expectations (McCaffrey and Edwards 2015).

In this chapter, three levels of ABR are presented with the encouragement that the researcher chooses the ABR approach most suitable to their journey as a researcher. Level one is conceptualized as referring to the novice ABR researcher. In level two, the emerging ABR researcher is represented. Level three refers to the experienced and/or expert ABR researcher. Distinguishing these levels is intended to provide support to researchers looking for a way *in* to ABR methods, and for those who are reviewing ABR proposals, or scholarly papers, who may not have encountered the approach before. The idea of boxing up ABR into levels of practice is somewhat at odds with the freedom of ABR methods. Therefore, the caveat that this is just meant as a useful device for people new to this field of research, and something I have found helpful in ABR method with university students (Table 1).

At Level 1, the processes in ABR are presented as arts and action based and may exist somewhat apart from theory. However, even while the researcher at the earliest stage may not be engaging theory development, it is important to have a senior adviser/mentor or research supervisor. This person must have theoretical competence to help fan the tiny sparks of creativity that may, at a later stage of the research journey, blaze into highly promising theoretical contributions. At Level 2, identified as *emergent*, the arts become more central to the ABR method chosen, and at Level 3, an *expert* level, the qualities of a fully integrated ABR project are available to the researcher.

2 ABR Methodology in Healthcare

Qualitative methods in healthcare developed as a reaction to the dominance of empirical methods. Many authors claimed that insights gained from service user engagement with treatments, and services, were marginalized and silenced (for example, Dyck and Kearns 1995). Finding ways to represent these hidden aspects of treatments, patient experiences, and services was needed (Pope and Mays 1995).

Some authors using ABR methods have noted the hesitancy of acceptance of these methods (Jones 2015). Reflecting on their experience of developing ABR projects, Einstein and Forinash (2013, p. 84) identified that they were aware of their fear, "that art is not enough and will not be understood." Within the rationality of healthcare decision-

Table 1 Descriptions of ABR levels

Level 1 – novice	Level 2 – emergent	Level 3 – expert
Arts informed	Arts based	Performative
The arts are used as a way to reflect on research data – to contribute to deeper understanding of the phenomenon under investigation	The arts form part or all of the data used to discover and develop the topic under investigation	The arts are both the process of the inquiry and the outcome. The result is a live or recorded performance

making, there is suspicion of the emotion or sentimentality that might be associated, or even evoked, through the research outcomes which involve the arts (Jones 2015).

In spite of the observations that ABR has been met with skepticism, and is not easily accepted in healthcare, multiple ABR studies based in healthcare sites or dealing with healthcare issues have been conducted. A scoping review by Boydell et al. (2012) found 71 healthcare studies that met their documented criteria for ABR. The most common rationale they found for employing arts-based process in research was with the goal to highlight and explore the illness experience of patients. A secondary intention they identified was to challenge existing practice in healthcare, with the goal of improving services to patients. A report of a play about the experiences of men with prostate cancer and their partners (Gray et al. 2003) is an example of achieving these goals through conducting ABR in a healthcare context. The play was performed for health professionals, who were then interviewed about their responses to the play. Participants reported that they experienced a new awareness or understanding about the issues facing men with prostate cancer; the play reinforced health workers' positive attitudes and behaviors in relation to patients; attending the play resulted in participants having an increased sense of connection with ill people; for some health worker participants, the play prompted them to alter their clinical practices to more effectively meet their patients' needs.

A similar review of ABR studies in healthcare found 30 studies (Fraser and Sayal 2011). Drawing out patients' experiences of illness and treatment were found to be the primary reasons for undertaking ABR studies. The authors noted that many of the studies they reviewed involved seriously ill patients with life-threatening conditions. They indicated that the arts were engaged because for such patients "their experience, associated feelings, and perceptions are not easily described in words" (Fraser and Sayal 2011, p. 138). They observed the growth in popularity of these methods, but they also documented their concern regarding the absence of critical debate about the applicability of these methods.

3 The Theoretical Location of ABR

The novice ABR researcher may wish to skip over the short theoeretical section presented here. This is understandable. You are new, enthusiastic, and you want to *get on with it*. You cannot understand why so much is written in this section and most of it in a scholarly language that takes time and effort to understand. Feel free to skim this section and move on. At some stage during your research reporting process, you may be challenged by a reviewer, or the question of a conference atendee, or even an examiner for a thesis, and you then might find this tiny section on theory helpful in further developing your understanding and preparing your response.

The process of research inquiry should draw you into deeper reflection, and it is common that this process may well bring confusion, and even uncertainty, before clarity emerges (Edwards 2012). Courageous exploration into the theoretical unknown is encouraged, whether you are an undergraduate student skimming this while trying to finish an essay or an experienced researcher seeking fresh territory to explore.

Bourriaud's *relational aesthetics* (Bourriaud 2002) is increasingly relevant to the philosophical parameters of all practical endeavors in ABR. In relational aesthetics, all participants in the arts are active in the creation of a work. These active participants include the viewer or audience and the primary creator. Bourriaud described this process of engagement in art between multiple parties including viewer and creator as a *transitive ethic,* that is, in order to complete its meaning, the work of art must be viewed or experienced beyond the sole perception of the artist. For Jones (2015, p. 87), this relationality has the quality of,

> ...inter-subjectivity, being-together, the encounter and the collective elaboration of meaning, based in models of sociability, meetings, events, collaborations, games, festivals and places of conviviality.

As research methods are increasingly constrained by needing to align with either qualitative (social science), quantitative (empirical, experimental), or what is termed *mixed methods*, it is refreshing to find that ABR is unable to be easily categorized within any of these. ABR accesses the potential fun and playful messiness of the arts and offers what Viega (2016, p. 3) has described as "boundless possibilities for inventiveness, discovery, and creativity." A constructivist sensibility can provide the researcher with new experiences leading to greater insights into the creative potential of our humanity.

Many expert ABR projects do not easily fit within a single *genre* or type*,* instead combining, or interweaving, multiple perspectives from a range of forms. For example, in McClaren's study (McLaren 2000) of menopause as experienced by 12 artists, she used multiple data forms including visual diaries, interviews, and art works. Her interpretation and processing of the materials was examined using a range of lenses including "feminist post-structuralism, phenomenology, narrative inquiry, art, and medicine" (Hopkins 2008 p. 259).

In their introduction to an inspiring book on ABR, Liamputtong and Rumbold (2008) describe how, at the time of writing, reports of ABR projects focused on *propositional* knowing rather than the more congruent experiential or *presentational* knowing that first-hand engagement with the arts involves. In the time since they wrote, *experiential knowing* could be described as a new force in analysis of data and representation of results. The signs for this emergence are clearly outlined in the text they developed and edited.

Woods (2011a), a medical humanities researcher, has suggested that the limitations of narrative need to be acknowledged when presenting the experiences of healthcare service users. Woods' concern lies with the assumption that all humans have a narrative self and can express their experiences in narrative. She proposed that her field is in *thrall* to narrativity (Woods 2011b). In relation to the arts, many art forms are predominantly nonnarrative such as visual arts, music, and dance. However, written or spoken narrative can become a strong force in explaining or presenting such works within ABR. There is more terrain to be explored in relation to the complexity of assumptions within ABR and their relevance for greater criticality within ABR that can include, but also exclude, research practices based on an extant philosophy of experience.

Many ABR projects at the expert level combine, or interweave, multiple perspectives from a range of forms, creating new forms which need new theoretical language in description of the processes engaged. For example, Allegranti (2013, p. 397), in presenting *Becoming Bodies,* describes how her research process "demonstrated a material-discursive engagement where I encompassed a recursive loop between felt-sense movement improvization and language."

ABR is, and should continue to be, theoretically diverse, engaging complexity and contributing to theory building in healthcare practice.

4 The Arts and Health

ABR has developed alongside the arts and health movement. The arts have a long history within the provision of healthcare in many countries. However, with the rise of the dominance of rationality in evidence-based healthcare services, the arts were increasingly relegated to entertainment and diversion rather than included as a way to offer psychological support to people coping with illness, hospitalization, and even pending death. In recent times, the arts have had greater impact and engagement within healthcare, notably through some art councils having programs devoted to arts and health with accompanying policy documents. For example, in Ireland, http://www.artscouncil.ie/Arts-in-Ireland/Arts-participation/Arts-and-health/. Some practitioners have considered the need for the arts to address *aesthetic deprivation* in healthcare, proposing that "arts and health programmes...provide the nexus for reflection and action in implementing a range of aesthetic enrichment in health-care setting" (Moss and O'Neill 2014, p. 1033).

In Western healthcare, multiple examples can be found where the arts have been promoted by dynamic and passionate individuals. For example, Cunningham Dax, a psychiatrist working first in London, and then in Australia, through the second half of the twentieth century, collected patient's art works on paper with a view to using the imagery to better understand the lifeworld of the patient (Harris 2014). Originally called *psychiatric art,* the images provided insights into diagnoses, experiences of living with debilitating conditions, and were categorized according to conditions, but also experiences. The work on display in the Cunningham Collection in Melbourne is only permitted to appear with the approval of the art maker.

Recently, the emerging field of arts and health has appeared in medical journals (for example, Moss and O'Neill 2014). Arts and Health has emerged as a field of its own with journals, conferences, and disciplinary groups existing in, and across, many countries.

5 Arts-Based Research in the Creative Arts Therapies

Creative arts therapist have demonstrated reluctance to describe their research method as ABR. I have previously explored this reluctance by undertaking a review of ABR in the creative arts therapies with a colleague (Ledger and Edwards 2011).

Our findings reported multiple examples of arts-based processes in music therapy research projects that had not mentioned, nor aligned with ABR. In considering why this occurred, we suggested that it is:

> ...possible that music therapy researchers are reluctant to explicitly adopt arts-based research practices. This could be due in part to a desire to ensure that music therapy research is accepted as scientific and scholarly within the dominant traditions of healthcare research (Ledger and Edwards 2011, p. 314).

In further commentary, I have suggested that music therapy can be observed as a profession which has demonstrable social status anxiety (Edwards 2015). This was in response to concerns raised in a paper about the way that disability studies could benefit considerations of music therapy practice (Cameron 2014). This status anxiety may have contributed to the absence of the creative option of ABR within research studies and the reluctance to name it as such when it appeared.

However, times change. Recently, I accepted an invitation to serve as guest editor for the *Journal of Music Therapy*, the first music therapy journal to focus an entire issue on ABR. I was able to call upon my colleagues and showcase the uniqueness of ABR; integrating and honoring the multiple ways human knowing and perceiving can be explored, represented, and enacted.

One of the papers in this special issue, Carolyn Kenny's *performative writing* piece (Kenny 2015) reflected her extensive research and practice experience as an Indigenous scholar and a creative arts therapist. She engaged imagined dialogues with the many scholars she has collaborated with, and undoubtedly influenced, while acknowledging her muses and intellectual roots. Performative writing can be provocative, and even *difficult*, while also having the quality of bringing the reader into close contact with the experience of another (Rath 2015). Readers who are not familiar with reading philosophical texts, or who have not encountered postmodern concepts in their research to date, might find the genre of performative writing challenging to engage. However, bringing curiosity and an open mind to reading, such works will bring rich dividends. Be prepared to understand that this way of writing is not always immediately self-explanatory; it wends and weaves through many complex concepts. The reader is encouraged to engage states of contemplation and stillness to gain the best insight from such work.

Also in the same special issue of JMT, Gilbertson's (2015) creative engagement with ABR was demonstrated through his research study of music therapy practitioner experiences. During an interview in which they were asked about an experience from their practice, clinicians' hands were submerged into silicon solution and cast in the position that they recalled from that prior moment. The reader is drawn deeply into consideration of the impacts of the findings.

Ledger (2010, 2016) used poetry writing to develop a response to narrative materials generated by participants' descriptions of their experiences founding new programs in healthcare. She described her process of developing the poems as follows:

I re-read a narrative, identifying words or phrases that seemed particularly significant or meaningful to me. I then wrote one or two poems in response to the words or phrases that I had previously identified. By the end of this procedure, I had written twelve poems in all (Ledger 2010, p. 121).

As Ledger's doctoral supervisor, I remember her first poem sent to me and the responsibility to encourage her efforts with delight. She often said that my excitement and enthusiasm for the poems as she created them inspired her to keep going with the work. The delicate work of encouraging creative flow while supporting scholarly depth and exploration is a key to collaboration and support for ABR projects.

McCaffrey (2014) demonstrated how song writing acted as a means to process her uncertainty about a self-portrait image drawn by a participant in an arts-based focus group about mental health services participation (McCaffrey and Edwards 2015). Similarly to Ledger (2010), she used her own reflections on the materials to generate an arts-based response.

Within ABR, there is acknowledgement of the important place for research methods that can facilitate and honor thinking, feeling, and reflecting for all research participants, including the person developing the research and the people who participate or co-research with them. This orientation within the ABR epistemology offers a way to bring multiple perspectives into research processes and to deepen reflection on data and findings. In the midst of the *playful messiness* (Edwards 2015) that ABR inspires, novel insights, and new dynamics of scholarship, can be engaged.

6 Arts-Based Research: A Narrative Example

Inspired by Bochner's considerations of narrativity (Bochner 2012), I have recently prepared research reports using narrative inquiry (for example, Edwards 2016; Edwards et al. 2016). I have also coauthored a chapter about this research approach (Hadley and Edwards 2016). Ideally, ABR should be determined as belonging to the research processes from the planning stage. However, similarly to Ledger's experiences (Ledger 2010), I recently found in analyzing interviews with child play therapists about their understanding of the United Nations Convention on the Rights of the Child (Edwards et al. 2016), that narrative materials might evoke a response that was not anticipated at the planning stage of the study. I prepared four stories or *vignettes* based on the stories told by the therapists during the interviews. I carefully checked the ethics of this procedure. To assure confidentiality of the children in the stories, I changed certain details and checked back with the interviewees that they agreed the stories they told about children's experiences were adequately disguised in order to be publicly available.

Below is one of these vignettes, *Baby sister*. In the story, I reanimated various aspects of interview data from play therapists voicing the story from the child's perspective.

Baby sister:
    When my new baby sister was born, I had great fun playing with her. When she looked at me, she always smiled or laughed. I could put my head down on hers and then breathe and make gurgly noises and she would giggle. One day, I saw Mum crying and then heard sirens and a policeman came and sat on the couch. I don't know what happened but Mum was always sad after that. The policeman told me my sister was dead. I asked him when she would be coming home again and even though he was a policeman, he couldn't answer the question. I had to go and live with my Dad because Mum was too sad. Dad took me to a place where I met Sally who had a big room full of toys I could play with in any way I liked. Sometimes, I bashed and crashed things in there which made both of us laugh a lot. Sometimes, I wanted to set up the toys so one doll was my sister who died and then one doll was Dad, until I had the dolls and teddies all lined up with their names. My Mum came to see me at Dad's and she always brought me a present. One day, I asked Mum why she was getting really fat. She laughed and said that she was having a baby. I felt scared. What if this baby died too and would it be my fault? I asked Sally whether this baby would die too. Sally said some babies die but not all babies. That was good to know. She said I had been a baby too when I was younger and I am alive. I did not realize that. I asked Mum if there were some photos of me as a baby. She searched for a long time, and then she found one. I looked at it and thought I looked a bit like my baby sister who died but maybe babies all look the same. When my new baby sister was born, I loved playing with her. She liked it when I put my head down onto hers and breathed and made sounds. She would laugh and laugh. One day I was told I had to say goodbye to her. She was going to live with another family. I really could not believe it. It was like being dead but in a different way. I said goodbye to her and played all our usual games. When I was back in the car with Dad, I didn't know whether I was going to cry or laugh so I ended up just sitting there saying nothing. My baby sister made me laugh so much so it was sad to think I would never see her again. Sally helped me to write a story about things like how I had played with my baby sister and how much I loved her. I drew the pictures, and at the end of the story, there is a photo that Mum took of me and my baby sister. I hope one day I will see her again. I hope it is not my fault she will not be part of our family any more.

Narrative researchers have described vignettes written in this way as *factional stories* (for example, Kallio 2015). A merger of *fact* and *fiction*, the story is marked as an imaginative invention based on factual details of a real situation. The term *creative nonfiction* has also been used indicating that writing in this way can have the purpose of "... engaging readers' emotions [to] enable vicarious experiencing of reported events – to feel them, and their outcomes – with a view to enhanced understanding" (Vickers 2014, p. 961).

7 ABR: Where Does the New Researcher Begin?

Ledger and McCaffrey (2015) propose four questions which are helpful for the new researcher approaching ABR with curiosity: (1) When should the arts be introduced? (2) Which artistic medium is appropriate? (3) How should the art be understood? and (4) What is the role of the audience? Each of these questions assists commencement of an ABR study. It may be that the researcher only wishes to employ arts-based methods after the research has commenced. The introduction of the arts might align with a wish to deepen an understanding of some aspect of the data, or it may only be realized post hoc, that the project has engaged an arts dimension. For example, if a reflexive journal is kept, it may be that this document includes narrative or poetic forms, and these might be reworked to further inform the analysis, and reflect on the findings.

8 The Foundations of Creative Insight Method

My approach to ABR, termed Creative Insight Method, was developed through work with groups of therapy trainees in my role as Head of Training in music therapy at the University of Queensland for 7 years and at the University of Limerick where I was Course Director of the Music Therapy training for 14 years. I currently teach at Deakin University where I have developed new curriculum in teaching therapy trainees about use of the self in practice. I have developed a series of exercises that can assist trainees to expand their concept and techniques of self-awareness. I focus on enhancing insight as the goal of the work within the context of self-development that is an essential part of training for therapy work (Edwards 2013). I have also had the opportunity to apply these skills with practitioner groups in Australia, Ireland, and the UK who have invited me to lead self-development sessions for staff teams.

I draw on participants' creativity by offering the opportunity to make work together or individually in the group. My intention is to promote insight and self-awareness through *experiential knowing*. Insight is described in the therapy literature as a way of perceiving what is happening in therapy that is distinct from an intellectual understanding or theoretical interpretation (Jørgensen 2004). As Jørgensen (2004, p. 529) has indicated:

> Part of what the good therapist has to offer is different perspectives on the self and reality – perspectives that open up new forms of behavior and new ways of relating to others...

I would also add to this that the therapist's capacity to listen closely, attentively, and with reference to bodily sensations, impressions, and even fleeting thoughts allows a person who is often quite distressed to feel psychologically *held* and supported. The ability to experience and discern one's reactions, and emotional states, and use these to reflect upon the interpersonal dynamics occurring within therapy sessions is a key to competent practice within relational therapy traditions (Bateman and Fonagy 2011). As Kumari (2011, p. 213) has identified:

> A heightened awareness of the self is seen as a fundamental aspect of the majority of approaches to therapy ... Adequate self-awareness is also essential to ensure that the therapist does not become completely overwhelmed by seeing clients who are often extremely distressed.

This self-awareness needs to be *cultivated* in an ongoing process of committed practice and habits (Geller and Greenberg 2002). In the section that follows, suggested tasks for research classes, students, or practitioners new to this approach are presented. These are all processes I have engaged in teaching ABR to groups, or in facilitating group processes through *Creative Insight Method*, the approach to ABR that I have developed. For the purposes of using this approach, it is presumed that group leaders, and academic teachers, will have at least basic knowledge of psychological safety when working with groups. That is, participating and sharing are optional, students are asked to pay attention when someone is sharing (no digital distraction permitted), and facilitators will remain connected and sensitive to what the group produces during the creation process, and when reporting. The tasks cycle through *action* then *reflection*. The participants are encouraged to share but their safety is more important than overt participation. If people are shy, or reluctant to present for some reason, this should be respected.

Usually, I start with a short presentation in which I introduce myself and talk briefly about Creative Insight Method. I work to set a contemplative mood by grounding myself, speaking calmly and creating a quiet atmosphere. After this introduction, I ask people to sit quietly and listen attentively to all the sounds they can hear inside and outside the room. I ask them to gradually bring their attention into the room and then their bodies, focusing on their breathing, slowing and deepening their breath. I spend a few minutes doing this and then ask them to bring their attention back into the room. After setting up this atmosphere of relaxation and thoughtful focus, I introduce the arts process. I have described some options below that readers may wish to try.

Challenges I have experienced include people arriving late and disrupting the flow of the group, the group not being the right level of maturity to engage the work, or individuals in the group being overwhelmed in some way by the task set for them. Sometimes, participants have cried when they start to write or paint, or for some reason they cannot tolerate the solemn feeling of quietness and need to disrupt it. When participants are not able to engage in a more contemplative way, I might switch to a more playful style with humor and lightness of focus. I am always careful when such a switch has occurred not to try to go back to the deeper work. However, when I have been invited to work with a group and I do not undertake the deeper work, the person who has invited me can feel let down in some way. It is, therefore, worth discussing with them in advance about how the group might progress is deeper and more reflective work is not able to be tolerated.

8.1 Narrative

8.1.1 Action

Ask members of the group to write without stopping for 5 min on a theme that you have chosen in advance. Hint: sometimes I read out a newspaper story about an event, or I read one of my own narrative pieces; see for example, http://bit.ly/2aNCNgk; I always consider whether there might be any vulnerability within the group. If I have not met the group before, I try to choose a *safe* topic remembering that the first time I meet a group, I will have no idea what creates safety, and what increases their sense of threat and fear.

As leader, be observant of the group as they write and encourage individual participants to keep writing if they stop – NB: I do this gently in mime so as not to disturb the other writers. At the end of the period, ask them to share with the person beside them, and if they feel they are able to read out the piece they have just written, then ask for volunteers to read out their work to the class.

8.1.2 Reflection

Ask the group to reflect on what, in their opinion, writing in this way offered that just talking about it would not have done. What did it feel like to share? Were there any parts that were surprising?

8.2 Poetry

8.2.1 Action

Depending on the experience of the group, either use existing poetry to encourage reflection and further poetry or narrative writing or present a topic, encourage discussion and ask the group members to write a poem in response. Sometimes, I have introduced the Japanese poetry form *haiku* and encouraged group members to use the form to write a poem on the topic.

8.2.2 Reflection

Encourage the group to reflect on one poem they experienced as highly charged in some way – emotionally, viscerally, or through memory. They can do this in small groups, in discuss as a larger group, or through further writing.

8.3 Visual

8.3.1 Action

Bring multiple copies of magazines of all types – sport, fashion, travel. Ask participants in groups to create a montage using cut outs from the magazines that reflects a feeling or idea, or a contemporary issue (for example, climate change or the refugee crisis).

8.3.2 Reflection

Ask the groups to present their montage to the larger group. Encourage discussion of the challenging aspects of presenting the topic visually. What did the group learn? How was the process of cooperating in the task together?

9 Conclusion and Future Directions

Elsewhere, I have pointed to the current freedoms enjoyed in ABR (McCaffrey and Edwards 2015). Although there are a relatively low number of reports of research studies conducted through ABR in health, the advantage of the developing and evolving nature of this space is that those of us championing the approach have the luxury of a wide remit; we can choose or develop methods and approaches that are unbounded by conventions and disciplinary or dogmatic regulations. However, this does not mean that laxity is part of the process. The main requirement of reporting qualitative studies is that the process by which the materials were generated and analyzed is clearly explained. In ABR, this can be a challenge as creative processes are not always linear or clear cut. Using a journal to keep track of developments in the research process and to document the steps in analysis can be useful.

One of the claims across many who research using ABR methods is the ability of arts outcomes to reach wider audiences than other types of research (Chilton and Leavy 2014). However, this claim has not been formally tested. Researchers reporting ABR studies are, therefore, encouraged to include the data regarding audience participation. For example, if the work has resulted in a play, the author/s should describe the number of shows and audience numbers for each performance. Additionally, it may be useful to describe the audience demographic details where these are sought during the research process or available by other means such as observation and description of the audience.

As ABR increases scope and impact, researchers, audiences, and participants will have valuable opportunities to reflect on insights gained and potential for further development. Within ABR, descriptions of unique approaches such as Creative Insight Method allow contemplation, testing, and elaboration to further the utility of techniques within the approach. Scholars at any level are encouraged to be courageous and excited at the opportunities that can be offered through Art-Based Research.

References

Allegranti B. The politics of becoming bodies: sex, gender and intersubjectivity in motion. Arts Psychother. 2013;40:394–403.
Bateman A, Fonagy P, editors. Handbook of mentalizing in mental health practice. Arlington: American Psychiatric Publications; 2011.
Bochner AP. On first-person narrative scholarship: Autoethnography as acts of meaning. Narrat Inq. 2012;22:155–64.
Bourriaud N. Relational aesthetics. Dijon: Les presses du réel; 2002.
Boydell KM, Gladstone BM, Volpe T, Allemang B, Stasiulis E. The production and dissemination of knowledge: a scoping review of arts-based health research. Forum qualitative sozialforschung/forum: qualitative social research, 13(1). 2012. Online http://www.qualitative-research.net/index.php/fqs/article/view/1711/3328
Cameron CA. Does disability studies have anything to say to music therapy? and would music therapy listen if it did?. In voices: A world forum for music therapy 2014;14:3.
Chilton G, Leavy P. Arts-based research practice: merging social research and the creative arts. In: Leavy P, editor. The Oxford handbook of qualitative research. New York: Oxford University Press; 2014. p. 403–22.
Dyck I, Kearns R. Transforming the relations of research: towards culturally safe geographies of health and healing. Health Place. 1995;1(3):137–47.
Edwards J. We need to talk about epistemology: orientations, meaning, and interpretation within music therapy research. J Music Ther. 2012;49:372–94.

Edwards J. Examining the role and functions of self-development in healthcare therapy trainings: a review of the literature with a modest proposal for the use of learning agreements. Eur J Psychother Couns. 2013;15(3):214–32.

Edwards J. Getting messy: Playing, and engaging the creative, within research inquiry. J Music Ther. 2015;52:437–440.

Edwards J. Narrating experiences of sexism in higher education: a critical feminist autoethnography to make meaning of the past, challenge the status quo, and consider the future. Int J Qual Meth Educ. 2016;30(7):621–34.

Edwards J, Parson J, O'Brien W. Child play therapists' understanding and application of the United Nations Convention on the Rights of the Child: a narrative analysis. Int J Play Ther. 2016;25:133–45.

Einstein T, Forinash M. Art as a Mother tongue: staying true to an innate language of knowing. J Appl Arts Health. 2013;4(1):77–85.

Fraser KD, al Sayah F. Arts-based methods in health research: a systematic review of the literature. Arts Health. 2011;3:110–45.

Geller S, Greenberg L. Therapeutic presence: therapist's experience of presence in the psychotherapy encounter. Per-Cent Exp Psychother. 2002;1:71–86.

Gilbertson S. In visible hands: the matter and making of music therapy. J Music Ther. 2015;52 (4):487–514.

Gray RE, Fitch MI, LaBrecque M, Greenberg M. Reactions of health professionals to a research-based theatre production. J Cancer Educ. 2003;18(4):223–9.

Hadley S, Edwards J. Narrative inquiry. In: Wheeler B, Murphy K, editors. Music therapy research. 3rd ed. Gilsum: Barcelona; 2016. p. 527–37.

Harris JC. Cunningham Dax collection. JAMA Psychiatry. 2014;71:1316–7.

Hopkins L. Women's studies and arts-informed research: some Australian examples. In: Knowles J, Cole A, editors. Handbook of the arts in qualitative research: perspectives, methodologies, examples, and issues. London: Sage; 2008. p. 558–69.

Jones K. A report on an arts-led, emotive experiment in interviewing and storytelling. Qual Rep. 2015;20(2):86–92.

Jones K, Leavy P. A conversation between Kip Jones and Patricia Leavy: arts-based research, performative social science and working on the margins. Qual Rep. 2014;19(19):1–7.

Jørgensen C. Active ingredients in individual psychotherapy: searching for common factors. Psychoanal Psychol. 2004;21:516–40.

Kallio AA. Factional stories: creating a methodological space for collaborative reflection and inquiry in music education research. Res Stud Music Educ. 2015;37:3–20.

Kenny C. Performing theory: Playing in the music therapy discourse. J Music Ther. 2015;52:457–486.

Kumari N. Personal therapy as a mandatory requirement for counselling psychologists in training: a qualitative study of the impact of therapy on trainees' personal and professional development. Couns Psychol Q. 2011;24:211–32.

Leavy P. Method meets art: arts-based research practice. London: Guilford Publications; 2015.

Ledger A. Am I a founder or am I a fraud? Music therapists' experiences of developing services in healthcare organizations [Unpublished doctoral thesis]. Limerick: University of Limerick; 2010. https://ulir.ul.ie/bitstream/handle/10344/1131/Ledger_PhD_2010.pdf?sequence=6

Ledger A. Developing new posts in music therapy. In: Edwards J, editor. The Oxford handbook of music therapy. Oxford: Oxford University Press; 2016. p. 875–93.

Ledger A, Edwards J. Arts-based research practices in music therapy research: existing and potential developments. Arts Psychother. 2011;38:312–7.

Ledger A, McCaffrey T. Performative, arts-based, or arts-informed? Reflections on the development of arts-based research in music therapy. J Music Ther. 2015;52:441–456

Liamputtong P, Rumbold J, editors. Knowing differently: arts-based and collaborative research methods. New York: Nova Science Publishers; 2008.

Malchiodi CA. Creative arts therapies and Arts-Based Research. In P. Leavy (Ed). Handbook of Arts-based Research, New York: Guilford Press; 2017. pp. 68–87.

McCaffrey TM. Experts' by experience perspectives of music therapy in mental health care: A multimodal evaluation through art, song and words [Unpublished doctoral thesis]. Limerick: University of Limerick; 2014.

McCaffrey T, Edwards J. Meeting art with art: arts based methods enhance researcher reflexivity in research with mental health service users. J Music Ther. 2015;52:515–32.

McCaffrey T, Edwards J. "Music therapy helped me get back doing": perspectives of music therapy participants in mental health services. J Music Ther. 2016;53:121–48.

McLaren R. Rethinking the body: spaces for change. Women artists discuss the menopause. Soc Altern. 2000;19(3):48–51.

Moss H, O'Neill D. Aesthetic deprivation in clinical settings. Lancet. 2014;383(9922):1032–3.

Osei-Kofi N. The emancipatory potential of arts-based research for social justice. Equity Excell Educ. 2013;46(1):135–49.

Pope C, Mays N. Reaching the parts other methods cannot reach: an introduction to qualitative methods in health and health services research. Br Med J. 1995;311(6996):42–5.

Rath C. Emergent writing methodologies in feminist studies. Gend Educ. 2015;27(4):465–6.

Vickers MH. Workplace bullying as workplace corruption: a higher education, creative nonfiction case study. Adm Soc. 2014;46(8):960–85.

Viega M. Aesthetic sense and sensibility: arts-based research and music therapy. Music Ther Perspect. 2016;34:1–3.

Woods A. The limits of narrative: provocations for the medical humanities. Med Humanit. 2011a;37:73–8.

Woods A. Post-narrative – an appeal. Narrat Inq. 2011b;21:399–406.

Understanding Health Through a Different Lens: Photovoice Method

65

Michelle Teti, Wilson Majee, Nancy Cheak-Zamora, and Anna Maurer-Batjer

Contents

1 Introduction	1148
2 Photovoice Procedures	1150
3 Photovoice Data Analysis	1151
4 Project 1: *Picturing New Possibilities*	1152
4.1 Project Summary	1152
4.2 Key Results: Visual Transformations of Self-Acceptance	1152
4.3 Implications of Findings for Practice	1154
5 Project 2: The *Snapshots Project*	1155
5.1 Project Summary	1155
5.2 Key Results: Impact of Posters on Viewers	1155
5.3 Implications of Findings for Practice	1158
6 Project 3: *A Picture Is Worth 1,000 Words*	1158
6.1 Project Summary	1158
6.2 Key Results: Growing up, Facing Loss, and Finding Strength	1159
6.3 Implications of Findings for Practice	1161
7 Discussion	1161
8 Methodological Challenges	1162
9 Conclusion and Future Directions	1164
References	1164

Abstract

This chapter describes the application of photovoice, a method that enables people to identify, share, and address their lived experiences with photographs and discussions, to HIV/AIDS and Autism Spectrum Disorder (ASD) research. The use of photovoice has expanded tremendously over the last decade. The case studies in this chapter highlight the varied and important methodological uses

M. Teti (✉) · W. Majee · N. Cheak-Zamora · A. Maurer-Batjer
Department of Health Sciences, University of Missouri, Columbia, MO, USA
e-mail: tetim@healthmissouri.edu; majeew@missouri.edu; cheakzamoran@health.missouri.edu; ammqpd@mail.missouri.edu

© Springer Nature Singapore Pte Ltd. 2019
P. Liamputtong (ed.), *Handbook of Research Methods in Health Social Sciences*,
https://doi.org/10.1007/978-981-10-5251-4_4

of photovoice in public health. The cases feature photovoice projects that facilitate women with HIV's expression of their illness narratives, youth with ASD's accounts of growing up, and HIV patients' stories of medication adherence. Although the photovoice methodology varied slightly in each case, each project included: (1) introduction and camera tutorial, (2) group photo sharing and discussion sessions, (3) individual photo reflection interviews, and (4) public photo sharing. Theme and story analysis of project session transcripts and photographs revealed projects' process and outcomes. Across each case example, photovoice proved to be versatile and adaptable to different participants' needs. Photos enhanced dialogue by giving participants a way to creatively and concretely express themselves and share their ideas. The method served as a way to collect data, but also as an enjoyable and empowering process for participants with HIV and ASD. Photovoice also supported individual and group development and action such as increased self-awareness and education for picture viewers. Photovoice can be a powerful method for vulnerable populations. Care must be taken to match the method to research questions and participants' needs, create effective collaborations between researchers and participants, and attend to project ethics.

Keywords

Autism · HIV/AIDS · Photovoice · Public health · Qualitative · Visual research methods

1 Introduction

Visual artistic methods are common tools of health research and practice (Fraser and Al Sayah 2011). Public health researchers have capitalized on the benefits of photography, in particular, as a way to understand health issues from the perspectives of those living with health challenges, inform health interventions, and engage community members in identifying and solving health problems (Catalani and Minkler 2010; Fraser and Al Sayah 2011; Switzer et al. 2015). Although photography has been used in various ways in health research, this chapter focuses on the application of photovoice to HIV/AIDS and Autism Spectrum Disorder research and practice.

Photovoice is a research method that enables people to identify, share, and address their lived experiences with photographs and discussions (Wang 1999). The use of photovoice in public health originated with Wang and her colleagues' efforts to understand the everyday health experiences of Chinese women through participant-generated images (Wang 1999; Wang and Burris 1994, 1997; Wang et al. 2004; Wang and Pies 2004). Wang and colleagues described three goals of using photovoice – to enable participants to record their community strengths, to promote critical dialogue and knowledge about personal and community issues through the discussion of images, and to reach policy makers via these images (Wang 1999). The photovoice method relies on the ability of an image to act as a concrete starting point for shared conversations and interpretations, draw attention, motivate dialogue, and present stories across diverse levels of participant language and literacy (Wang and Burris 1994; Wang 1999; Hergenrather et al. 2009).

The theoretical basis of photovoice includes three related concepts: empowerment education for critical consciousness, feminist theory, and documentary photography. These foundations share common assumptions about the importance of participants' voices (Wang and Burris 1994). Through focusing on and portraying participants' experiences, photovoice empowers individuals to understand and critically discuss the contextual issues that affect their health and well-being. This approach of collecting knowledge that is grounded in experience enables participants to collectively communicate their shared concerns/needs/vision to each other and to those with authority to allocate resources towards creating needed change (Wang and Burris 1994, 1997). From a feminist approach, photovoice promotes the value of women's and other unrepresented groups' experiences. Photovoice fosters conditions in which underrepresented groups are able to influence action (Wang and Burris 1994, 1997). Finally, in contrast to conventional documentary photography, photovoice puts cameras in the hands of participants (not professional photographers) and trains them to capture and use their own experiences to gain insights about their lives and inform decision-making processes in their communities. When project participants record their lived experiences and tell their stories, they are "often imaginative and observant" in ways that provide more richness to the data (Wang and Burris 1994, p. 177). Original photovoice methods (via Wang and colleagues) included a set of procedures to both gather data and translate research findings into action and practice. As a first step, projects included the identification of both a target audience – who could help put participants' ideas into action – and a group of interested photo-takers. Participants would then undergo a camera tutorial, a photo-ethics training, and an introduction to the photovoice method. A cycle of photo taking, discussion, and sharing followed. Discussions were guided by the "SHOWeD technique": What do you **S**ee here? What is really **H**appening here? How does this relate to **O**ur lives? **W**hy does the situation or concern exist? What can we **D**o about it? The facilitators and participants then worked together to share their photos and stories with policy makers via slide shows or exhibits (Wang 1999).

To highlight the varied and important methodological uses of photovoice in public health, we summarize the key aspects of three case studies in detail below. Although the study methodology varies slightly based on participant needs, several core components were consistent across the projects. These are outlined in Table 1. The use of photovoice in public health has expanded tremendously since its early applications. Numerous reviews (Riley and Manias 2004; Hergenrather et al. 2009; Catalani and Minkler 2010; Martin et al. 2010; Lal et al. 2012; Sanon et al. 2014; Han and Oliffe 2015; Switzer et al. 2015) document the expansive use of Photovoice and its application to women's health, chronic illness, mental health, and health disparities, among other public health topic areas (Riley and Manias 2004; Hergenrather et al. 2009; Catalani and Minkler 2010; Martin et al. 2010; Lal et al. 2012; Sanon et al. 2014; Han and Oliffe 2015; Switzer et al. 2015). Across diverse applications, the end goals of photovoice have centered largely on the understanding of health needs, challenges, and assets; health action and advocacy; individual and

Table 1 Project Descriptions

Project	Participants	Goal	Dissemination	Main results
Picturing new possibilities	30 women with HIV	Participants: tell story of HIV	Exhibits at local HIV community events	Participant experiences: empowerment, social support, way to communicate health transformations, help with HIV stigma
		Researchers: understand impact of photovoice on participants		
Snapshots	16 people with HIV	Participants: capture medication adherence challenges and motivators	Posters in HIV clinic	Patients who viewed the posters described them as comforting, relatable, inspiring, and as creating a culture of caring in the clinic
		Researchers: explore how disseminated photos affect other patients' adherence practices		
A picture's worth 1,000 words	11 young adults with autism	Participants: express transition to adulthood	Exhibit in autism center	Youth expressed meaning of adulthood, need for independence, employment desires
		Researchers: understand youth's priorities and service needs		

community empowerment; and health intervention (Catalani and Minkler 2010). The original steps of the process have been modified by researchers to adapt to specific participant needs and research goals. For instance, Switzer et al. (2015) adapted their photovoice project with drug users to accommodate participants' concerns about confidentiality, completing individual interviews with participants, and forgoing group discussions.

2 Photovoice Procedures

Each project included four major components: (1) project introduction and camera tutorial, (2) two group photo sharing and discussion sessions, (3) individual photo reflection interviews, and (4) public photo sharing. Introductory and photo discussion sessions were facilitated by one to two trained facilitators (including the first and third authors). The facilitators delivered the sessions to participants in small groups of four to eight. Participants attended the sessions weekly over the course of 1 month. During the first session, the facilitator explained the purpose of the project and the participants brainstormed different potential ideas for photographs and discussed the ethics of picture taking (Teti et al. 2012). For instance, participants identified appropriate and inappropriate photographs (e.g., illegal, unsafe), how to obtain consent to photograph another person, and how and where to access support for issues raised in the taking and discussing of photographs. At the end of the first session, each participant received a digital camera to keep and practiced basic camera functions (e.g., on, off, shoot, save, edit, zoom).

During the second and third meetings, the participants reconvened to review and discuss their photographs. The facilitator led these discussions with a laptop computer to download and display photos via a projector. Each participant presented several (range 2–10) photos of his/her choice to the group and discussed what each photo meant. We facilitated these discussions using a semi-structured guide that helped participants to express themselves. Sample questions included: "What does this photo capture about your life or story with [illness]?" and "What does the picture or issue mean to you?" and "What challenges or strengths does the image convey?" After each participant presented and discussed his or her photos, the group responded to the images.

During each session, participants discussed and brainstorms opportunities to share their photos with others in their families or communities. The facilitators finalized exhibit planning. As part of the third session, participants chose photos for review and helped to develop or edit captions for the photos. Each participant presented one to three photos of their choice at public exhibits or public displays (see Table 1).

Following the exhibit, photovoice project staff (including the first and third authors) conducted 1 h individual-level interviews with each participant to explore his or her experience in the project and give participants a chance to reflect on their pictures individually. Example questions were: "What did you learn about yourself in the process?" and "How did you decide what to photograph?" and "What was it like to share your pictures with others?" All participants were compensated monetarily for their time in the project sessions. We digitally recorded each group and individual intervention session to capture data for analysis.

3 Photovoice Data Analysis

Project data included photographs and group and individual interview transcripts. We transcribed the interviews verbatim and analyzed them using Atlas.ti (Scientific Software Development 2011), a qualitative software analysis package. To analyze the data, we employed a mix of theme (Boyatzis 1998) and narrative/story (Riessman 1993) analysis – a common strategy used in public health research with vulnerable and complex populations whose data transcripts reveal rich concepts that need to be understood in the context of their overall stories (Rich and Grey 2005; see also ▶ Chap. 48, "Thematic Analysis" and ▶ 49, "Narrative Analysis").

First, we analyzed the data for key themes (Boyatzis 1998). Two researchers reviewed the transcripts and photos multiple times to become familiar with the data and key themes. We created a codebook describing the most salient themes of interest and analyzed the data via coding and analytical memos (Boyatzis 1998; Charmaz 2006). Coding matched text to themes and progressed in two stages, open or more general coding – which included coding for broad codes, which was followed by selective or more specific coding which broke down the broad codes into more detailed concepts. Selective coding resulted in the final themes presented in the results (Charmaz 2006). We wrote analytical notes throughout all phases of coding to highlight key questions about relationships in the data and to refine codes. After coding the data, we generated coding reports that collated the evidence for each theme.

Comments about themes often occurred within participants' stories. To ensure that we understood the themes with the context of these larger narratives, we also constructed a narrative table for each participant (Riessman 1993). Each narrative table summarized each participants' overall story conveyed in both the group and individual sessions in a linear fashion (e.g., was diagnosed in 2005, had to quit job, struggles with stigma, is still fearful about illness but feels healthy, etc.) and included a list of pictures that elaborated on specific aspects of the overall story. Then we drafted and revised the results section using the theme coding reports and the narrative table to ensure proper context of the thematic examples.

4 Project 1: *Picturing New Possibilities*

4.1 Project Summary

Picturing New Possibilities was implemented to meet HIV-positive women's expressed need for more opportunities to share their stories of HIV and relate those stories to their health (Teti et al. 2010). Thus, the goal of the project was for women to tell their stories of HIV through photovoice. The research questions driving the project were "What stories do women tell through photovoice?" and "How does photovoice support or affect women's story-telling process?" Participants included 30 women, recruited from HIV clinics and AIDS Service Organizations. The majority of the participants were poor. Over half ($n = 18$) of the women reported earning less than $10,000 a year. Most of the participants identified their race/ethnicity as Black ($n = 25$), three identified as White, and two as "Other." On average, the participants were living with HIV for 11 years (6 months to 17 years).

Analysis of the data indicated that one of the key ways that photovoice supported women was to give them a way to visually demonstrate various positive transformations in their lives. There is no doubt that women with HIV face various challenges. Popular media and scientific research generally dwell on these challenges (e.g., women with HIV as sick, poor, down and out, struggling). Photovoice offered women a way to define themselves differently. With concrete images women shared positive changes in their lives, including their journey to self-acceptance with HIV.

4.2 Key Results: Visual Transformations of Self-Acceptance

Women presented visual analogies of reaching a point where they: accepted their diagnosis, believed that they could live with or survive HIV, created a life and an identity *beyond* HIV, and took control over managing HIV stigma. For instance, a participant who had HIV for 17 years said she initially had an "internal stigma" and could not accept that she had HIV. She said she "kept [having HIV] inside of me for so many years and then finally, I [decided], I'm going to talk out about it, and when I talked out about it, I felt liberated, I was free." Sharing a picture of her face, she said she became "comfortable with [HIV] in [her] own skin." She said she learned to "accept HIV for what it is" and describes herself as a different person:

> I've chosen not to be ashamed in my life...I'm free to be whoever I want to be. I tell people, "I'm grateful for [HIV]. I truly am." It's like living a whole new life...HIV really made me know who I was overall...I believe that HIV has freed me to be.

For other women, acceptance of their HIV meant recognizing that they wanted to share their life with HIV with others more openly. For instance, one participant said that through her pictures she showed that she "is strong enough to let everybody know what's going on with me in my life and that I have HIV. I'm not in the closet about it." She said when she showed others her pictures, "I was pulling people to look – 'Look at me. Look at me.'" Similarly, another woman shared a picture of her in the newspaper and said "That picture means the struggle's over and I don't have a problem putting my name to nothing. I came a long way."

Sometimes reaching a point of acceptance with HIV meant making peace with the struggle. Living with HIV for 8 years, another woman described how she wanted to tell other people about her HIV status, but "wasn't there yet." To explain this visually, she shared a photograph of wires, tangled and attempting to connect with each other (see Fig. 1). She said she was taking baby steps though. She shared several pictures, including this flower growing out of a rock (Fig. 2), to explain that with her art work and her life with HIV, she "created something new with something that had died," and that she was looking forward to "growth" and how "life will find a way" despite HIV and HIV stigma.

Still other women portrayed their transformation to HIV acceptance with visual representations of the realization that they wanted to focus less on HIV in their life.

Fig. 1 Wires demonstrate how the participant feels disconnected and confused about sharing her HIV status with others

Fig. 2 Growth exemplifies that the participant is looking forward to sharing her HIV status

One participant explained that she wanted to continue to take pictures to capture new beginnings in her life – like her school. She said, "I'm not HIV anymore. It's not 'my everything.' It is part of me and it's becoming kind of more in the background of my life." Another noted, as she shared a photo of the city that she lived in, that her transformation was becoming just "like everybody else."

> I don't think about my HIV when I'm walking down the street or when I'm talking to somebody, "Do they know I got it, or they scared to shake my hand?" I'm just a person like everyone else in the world, trying to survive – who's going to survive. HIV – it don't have me. I got it, but it's working with me.

HIV prompted all of the participants to consider their lives with HIV in new ways. Although acceptance of HIV was different for everyone – some women moved or made steps to move out of HIV-related denial and secrecy, other women learned strategies to confront stigma, and still others expressed a desire to move beyond HIV – reaching their version of acceptance was an important transformation in women's lives, particularly given the barriers that denial, secrecy, and stigma can pose to women living with HIV and their mental, physical, and social health. Through using new ways of expression and open dialogue, photovoice supported women's presentation of this experience.

4.3 Implications of Findings for Practice

All of the participants could recognize positive changes and transformations and all of them said that they were rewarded by this process. Their stories of transformation also highlighted the lack of spaces for women with HIV to talk about positive aspects of their lives, such as their efforts and journey to accept HIV. Health programs and interventions also tend to unintentionally discount women's positive experiences. HIV interventions, for example, typically help people at risk for and living with HIV address what they may be doing wrong – like having unsafe sex or failing to take medications, versus helping them to build what they do right (Herrick et al. 2011, 2014). This is important and understandable because these behavior changes are paramount to the health of women and their partners and because the challenges women face to living healthily – like poverty, discrimination, violence, and stigma – are so daunting. Our findings serve as a reminder to practitioners developing and implementing programs that women with HIV are not a monolithic group and experience successes as well as challenges (Goggin et al. 2001), and that women feel prideful about these successes.

Understanding women's transformations may also be a potentially helpful way to understand women's motivation to make healthy behavior changes, such as safe sex, medication adherence, and engagement in HIV care. Meeting these goals can prolong WLH's lives and also help curb the spread of HIV to others. Behavior change, particularly change that needs to last over the life course, is difficult, complex, and hard to understand. *Picturing New Possibilities* provided a glimpse into women's experiences of making notable changes in their lives.

5 Project 2: The *Snapshots Project*

5.1 Project Summary

Snapshots was implemented to improve medication adherence among people living with HIV. A small sample of people living with HIV used photovoice to share how they experienced life with HIV, managed consistent adherence to their HIV medications, and problem-solved medication-related challenges. They used these photos and stories to create ten different medication themed posters (versus a "traditional" photovoice exhibit). The posters were then displayed in the participants' HIV clinic to help improve adherence attitudes and communication with health care providers about adherence. The research questions driving the analysis of this project were "What stories do participants tell about medication adherence?" "How does using photovoice to create the posters affect participants?" "How does viewing the posters created through photovoice affect the viewers?" Photovoice participants, who made the posters, included 17 people with HIV, 10 of whom were women. The majority of participants were poor, over half reporting earning less than $10,000 a year. Ten participants were Black and seven were White. On average the participants were living with HIV for 13 years (range 3–33 years).

5.2 Key Results: Impact of Posters on Viewers

Three key themes in photovoice participants' images and discussions captured their motivations to become and remain adherence including important people or things,

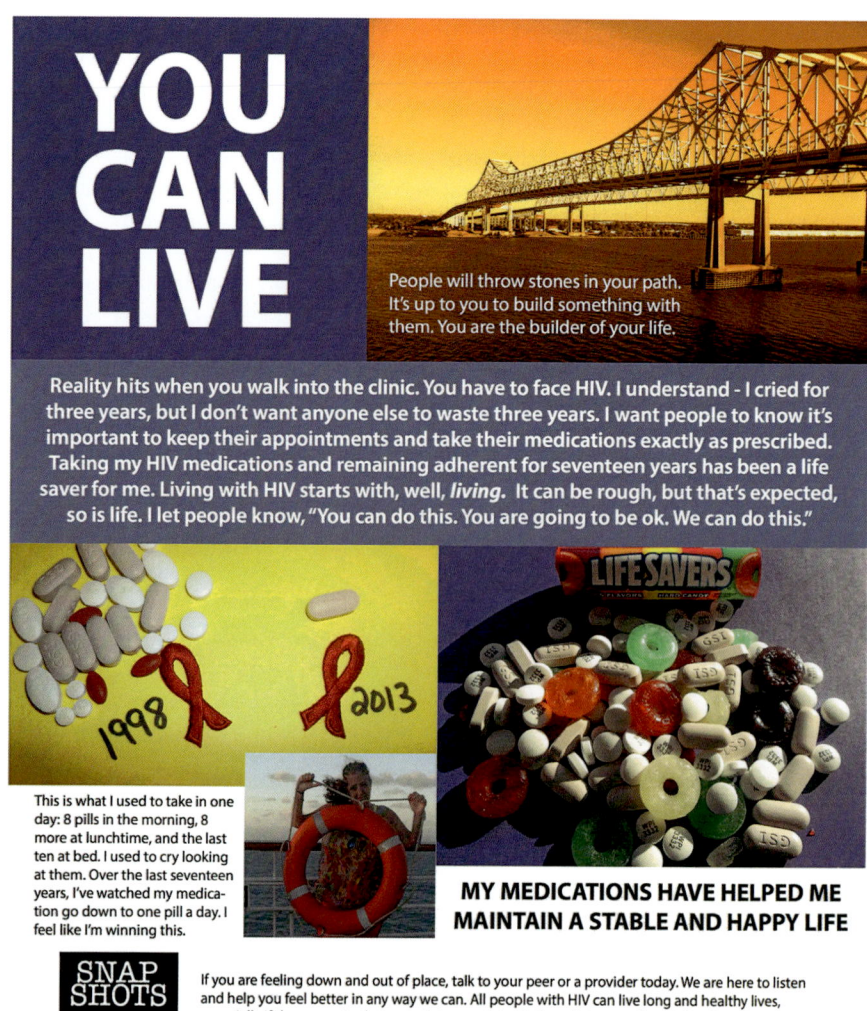

Fig. 3 Poster example – hung in clinic to inspire others with HIV

symbols of hopefulness about the future, and adherence as a way to take control over one's health. Example images corresponding to these themes included medication routines, support systems, pets, self-portraits, and nature. The research team and the participants worked together to create posters that showcased these themes (Figs. 3 and 4). Our primary analysis question focused on understanding how the posters affected the viewers who saw them. Participants told us that the posters were powerful because they were relatable and also because they helped to

Fig. 4 Poster example – hung in clinic to inspire others with HIV

create a culture of caring in the clinic. For instance, one viewer noted, "I was inspired, touched. I felt and saw me in these pictures [on posters] – I could relate [to them]." Another viewer called the posters "comforting." Others commented on the benefit of seeing other's faces, saying, "They'll get your attention, especially when you start seeing faces" and that the posters were helpful because "Each one of these posters is different, you can see a little bit of each person's life inside these pictures."

Other viewers appreciated the posters' value for the clinic overall. One viewer noted, "[The posters] let me know that the providers are willing to showcase what people are going through because sometimes when you come to the clinic you have people scared." Another called the posters "poems in the entrance area" of the clinic. Several viewers remarked that the posters would be important for others to see, especially newly diagnosed patients. For example, one viewer said:

> If they had these in the rooms while we're sitting back and waiting, I can just see that person who's very fearful, and scared...Being alone in the room while you're waiting, we do stare at the walls. This can be artwork for the doctor's offices where people can relate. They're powerful. This here is us. My doctor goes home every day. My doctor doesn't have to take a regimen. This here will talk to you in a different way than what our providers can teach us. This is real. This is someone who understands, and knows, and deals with it.

5.3 Implications of Findings for Practice

Snapshots findings suggested that patients appreciate education materials that are relatable and easy to understand. They like seeing other people like themselves, and hearing other stories like their own stories, on educational materials. These findings support a growing body of research that stories and narratives are more effective than didactic information in promoting both health education (Jibaja-Weiss et al. 2006) and health behavior change (Houston et al. 2011a, b; Murphy et al. 2013, 2015). In the *Snapshots* study, stories helped viewers with HIV to feel comforted and supported. The posters themselves reinforced, via the images of real people with HIV, that if they take their medicines regularly they will be okay. Thus, our findings suggest that these posters can improve confidence towards taking medicine. Seeing the posters in the clinic also sent a message that the clinic cares about people with HIV, the stories of people with HIV, and in creating an open dialogue about those stories. This indeed could improve communication between patients and providers. Harnessing photovoice to capture the stories and images of people with HIV can lead to the development of easy to understand, accessible, and powerful health education materials.

6 Project 3: *A Picture Is Worth 1,000 Words*

6.1 Project Summary

Youth with autism spectrum disorder (ASD) experience great difficulty in the transition to adulthood. Most research on youth's transition focus on poor achievement like low college graduate and employment rates (Shattuck et al. 2012). There is a paucity of research on how youth with ASD experience this time in their lives and navigate and address challenges. In this project, 11 youth with ASD used photovoice to express their experiences of becoming adults. Youth with ASD are particularly challenged by this transition due to their increased dependence on their families, need for repetition and consistent schedules, and difficulty in establishing and maintaining friendships (Cheak-Zamora and Teti 2014). Although the disorder affects each individual in different ways, the combination of comorbid physical conditions, mental health deficits, and developmental disability makes transition into adult life difficult for all youth with ASD (Beresford et al. 2013). Given that many youth with ASD also face challenges communicating their opinions, feelings, and needs to other people, the purpose of this project was to use photovoice to facilitate concrete communication and discussion of youth's priorities. The mean age of the sample for this project was 20 years (range 18–23 years), 64% of the participants were male, and 91% were Caucasian. Four of the 11 (36%) participants were in high school, 3 (27%) were in college, and 4 (36%) were not in school at the time of the study. Although we did not specifically measure functioning level, 27% of the participants required assistance to complete the demographic survey, which may suggest lower functioning.

6.2 Key Results: Growing up, Facing Loss, and Finding Strength

The youth mainly described adulthood in terms of "accepting responsibilities." One youth stated "[by assuming] certain responsibilities [one] would infinitely detach you from being described as a kid." Examples of images that youth used to help them explain adulthood and what responsibility meant to them included pictures of a money saving system in which the youth separated money into different jars, and a picture of a youth taking on the responsibility of exercising at the gym. The latter picture stressing specifically that youth with ASD may interpret responsibility in nontraditional ways, which could be hard to explain verbally.

Pictures and discussions of adulthood also focused on jobs – since youth were also concerned with living on their own and "moving out of the nest." Although the majority of youth were employed at the time of the study ($n = 7$), they expressed dissatisfaction with these experiences. Several youth remarked or hinted that they believed they had more skills and potential than working in janitorial services, a common place of youth employment. For example, one youth who disliked his job as a janitor stated that he wished he could "stop cleaning up crap" and "[does not] want to be there [at work] sometimes." When describing one of his pictures, a youth said that his facial expression conveys, "Get me out of this building. I don't want to work here anymore. ... I don't want to work there anymore." As opposed to their current jobs, many of the youth described their ideal job as "being an author" or "thinking about becoming a veterinarian." One youth took a picture of a bird he was taking care of in order to explain his love of exotic animals and drive to become a veterinarian. Jobs represented a route to independent living, socialization, and a new burgeoning identity.

Unfortunately, youth also associated growing up with experiences of loss. Multiple youth noted that they felt sad due to losing a loved one. Roger lost a grandfather shortly before the study, when asked about how this affected him, he stated, "It's

Fig. 5 Table representing Grandfather's life and death

weird because it just feels like a piece missing." He also took a picture that represents his feelings surrounding losing his grandfather (Fig. 5):

> This [picture] is the table with the last memory of seeing my grandfather alive, and he passed away this October (date). .... And I'm sitting where he was sitting and just remembered that I took him that day. ... [I talked to] my mom, my grandma [about my grandfather's death], and it's weird because my aunt's husband's father passed away the week before my grandfather died. That Thursday after the (date), my grandpa's sister died from cancer. So it was a very [depressed] couple of weeks and drained a lot of energy out of us, but we made it."

Laura used pictures to explain that she experienced both sadness and anxiety around her grandmother's death. She experienced stress over her grandmother's house, saying that, "I'm just worried about that house, her house. I really love that house. I don't want any stranger living in that or changing it. I just want maybe a relative might stay..., just to keep it safe." Several youth talked about sadness related to losing a pet, sharing pictures of the animals to express their grief.

Youth also explained that showing their lives through pictures allowed them to view ASD as a special experience, and not just a challenge. One youth noted people with autism are "gifted." Through photographs the youth highlighted some of the positive characteristics, abilities that individuals with ASD possess, and how they were "defying the label." For example, one youth took photographs of sporting events to show that she engages in social activities, another youth photographed his teacher to represent "the progress I'm making to better myself," and yet another showed a photo-shopped picture of himself shooting an apple off of his own head. Highlighting that he is a "very artistic person," one youth stated that people with ASD are "very detailed" and thereby could capture detailed pictures that others could learn from or enjoying viewing.

6.3 Implications of Findings for Practice

Health care and service providers who work with youth with ASD want to provide good care, but it can be difficult to understand and meet youth's needs. Photovoice may offer youth and those who care for them an alternate way to communicate. Further it helped youth get passed the surface (rote answers) that they generally provide to alternatively describe what they really needed and wanted in their live. Our findings suggest that youth were able to explain themselves and even their emotions through images. They were confident when explaining their pictures and took pride in sharing their views and thoughts. Integrating photovoice into medical services, for example, could allow youth to create images to form a bridge to communicate with providers and help to form their care plans. Social and employment support services could also gain from photovoice endeavors that give youth a chance to express how they feel about their social situations, what they want from work, and their personal goals. The camera and the pictures also allowed youth to express some of the positive aspects of their lives, which could otherwise go unnoticed. Thus, photo-stories could also serve as a way for youth to demonstrate and celebrate their strengths more confidently than they may be able to do than with words alone.

7 Discussion

In this chapter, we have presented three different applications of photovoice in public health. Although the projects included different populations and had different goals, each project used the same basic photovoice process. Similarities across the projects shed light on some of the most important aspects of the method.

Photovoice is versatile. Original applications of the method included a structured set of procedures (Wang and Burris 1994) which have been adapted or used to varying degrees over time. Given that one of the key tenets of photovoice is its focus on participants' experiences, it is logical that the method should adapt to meet participants needs. However, our work confirms that the method does include several core concepts. Fundamentally, the method includes photos as catalysts for identifying and describing experiences, reflecting on experiences, and sharing those with others. Photos contribute more than words alone. For example, in Project 2 (*Snapshots* – men and women with HIV), the photos (e.g., "familiar faces") were what captured viewers' attention and helped them to relate to HIV medication adherence messages. In Project 3, participants with ASD may not have been able to express themselves as clearly without the assistance of photos. Photos help participants to be concrete, to see their environment in new ways – or to see new things in their everyday surroundings – and to reflect on their experiences in different ways than through words alone.

Secondly, photo-sharing is essential for group dialogue, which can enhance both individual's and groups' ability to identify and solve problems or difficult scenarios that evolve through the photos. The photo provides a shared vision or a starting point for conversation. In each project, the participants shared their photos with a group and one-on-one with a facilitator. By sharing, participants were able to discuss different meanings in the images and further create their stories. In project 1, for example, each HIV-positive woman's journey was complex. By sharing it they were able to make sense of it, identify similarities with others, and problem-solve challenges that arose, such as how to manage HIV stigma. Within project 3, youth with ASD were able gain understanding of others that they hadn't prior, "I think just hearing everybody else's stories...It was nice to hear, you know, people like me, what they're dealing with and how they operated." Photovoice is thus particularly relevant for vulnerable groups of people who often face stigma or misunderstandings about their illness. This is true in the case of both HIV and Autism, despite the fact that they are very different health experiences.

Third, as a participatory method, photovoice is a data collection tool that can enhance participants' lives. In project 1, for example, the researchers gathered important data about the implications of women's transformation experiences for HIV interventions. Women themselves benefitted from telling these stories and capturing their positive transformations. In project 2, the research team was able to design and test educational posters using the results of photovoice. Participants themselves were able to share their challenges and successes, which they reported helped them to make important reflections and help others. In project 3, the researchers learned about the transition needs of youth with ASD that they previously were unable to communicate with their families. Youth were able to express themselves, and like women with HIV, they were able to access positive parts of their experiences in particular through photos.

Lastly, photovoice is about action. Such action can take different forms, however. Women with HIV in project 1, for example, took action by redefining themselves as capable of making positive life changes with HIV. Men and women with HIV in project 2 embodied action by turning their stories into ways to educate other people with HIV. For youth with Autism in project 3, action was being able to more clearly and confidently express themselves.

8   Methodological Challenges

Photovoice does pose methodological challenges. Photovoice projects are only appropriate for participants who are interested in the method and able to express themselves with pictures. This means that participants need to be willing to take pictures, disclose their identity to a small group, self-identify with their illness (i.e., not be in denial about having HIV, for example), and be willing to share their illness experiences publically via exhibits. They also need to be able to operate a camera and use it to talk about their health. All of these tasks can be daunting, especially when the method is used with populations who may face stigma or misunderstanding or with underserved populations who may be uncomfortable with learning new things and with opening up to "strangers."

Photovoice results in complex data in multiple forms, like text and images, or individual and group-based discussion data. Thus, the data are timely to analyze and data transcription can be costly. These types of data are obviously suited to qualitative analysis. While the identification of qualitative themes is helpful, it is usually hard to make any comparisons or associations qualitatively and with a small sample. When examining the impact of the process on participants, it is hard to distinguish the effects of different aspects of the process on participants, like taking pictures versus discussing pictures versus exhibiting pictures. Experiences that participants do report as resulting from the projects, such as "empowerment" are also generally broad and multifaceted.

Ethical concerns arise when using photographs as data as well. Participants may take or want to take pictures of themselves or others in compromising positions. Similarly, participants are challenged to explain the project and seek consent if they want to include others in their pictures. If participants engage in illegal or unsafe activities as part of their illness experiences (e.g., drugs and alcohol), they must find ways to safely capture and portray these experiences to others. Participants may fear disclosure of themselves with their illness as their images are shared via exhibits. Additionally, both participants and researchers need to decide on fair rules for photo dissemination. Academic standards for publications such as the use of pseudonyms may not make sense to participants who want to take ownership over their work, for example.

The continued and growing use of technology may also change the future of photo projects and complicate ethical and other project procedures. People of all income levels now routinely have phones. These phones, however, may take pictures of varied quality. To ensure that all photos can be used a process may be necessary to standardize the photographs. Social media platforms such as Facebook or Instagram that rely on photos, or phone apps, may be logical places to integrate photovoice. The more immediate and more widespread sharing of photos may indeed require enhanced ethical protections for participants.

Despite these limitations, we believe that photovoice is an excellent way to collect nuanced information about health experiences. Based upon the results of the projects described in this chapter, we recommend the following for using photovoice successfully:

- Before beginning a photovoice project, decide if photovoice is the right method for the research question at hand. Photovoice is best suited for exploratory research questions and for use with marginalized populations who may have lacked access to expressing their needs historically.
- Determine how photos will add value to the data collection and participant experience. How will the research question be more deeply answered by the photos? How will participants express themselves differently via photos than with words alone?
- Photovoice is meant to be a partnership. Participants will drive the data collection with their photos and discussions and the data, in turn, will benefit research. The goals of the project and the pros and cons for each group should be obvious and

apparent before the project starts. Determine ways for participants to view and member check the data before it is exhibited.
- Ensure that the project is properly budgeted for and staffed. Will participants use their own cameras? How will participants be compensated for their time? Where will the exhibit take place? What is the purpose of the exhibit?
- Ethical guidelines and expectations need to be established at the project start. At a minimum, these need to include plans for taking picture of others, private and illegal activity, disclosure, and photo ownership. Support systems for participants should be in place if the process brings up issues that require further assistance or steps for action.
- The design process should allow for flexibility. The purpose of photovoice is to adapt to participant needs and give them a way to identify, express, and prioritize their needs. Although original photovoice projects had very specific protocols, participant needs, such as to participate in one on one versus group discussions, should be honored above following a strict protocol.
- The ways that technology can assist the process should be considered and tried at participant request.

9 Conclusion and Future Directions

Photovoice is a flexible method that public health researchers can use to uncover rich data about participants' lived experiences of their health. These narratives can help us better align public health research and practice with people's lives and priorities. The method is also beneficial in that it can help participants gain skills and embrace confidence about their health stories. The core components of photovoice – images, group discussion, participant growth, and action – can be adapted to meet individual project and population needs. Although challenges to using Photovoice do exist, careful planning and project development can lead to insightful visions of participant's lives and health needs.

References

Beresford B, Moran N, Sloper P, Cusworth L, Mitchell W, Spiers G, Weston K, Beecham J. Transition to adult services and adulthood for young people with autistic spectrum conditions. Working Paper, no: DH 2525. York: Social Policy Research Unit, University of York; 2013.

Boyatzis RE. Transforming qualitative information: thematic analysis and code development. Thousand Oaks: SAGE; 1998.

Catalani C, Minkler M. Photovoice: a review of the literature in health and public health. Health Educ Behav. 2010;37(3):424–51.

Charmaz K. Constructing grounded theory: a practical guide through qualitative analysis. London: SAGE; 2006.

Cheak-Zamora NC, Teti M. "You think it's hard now … It gets much harder for our children": youth with autism and their caregiver's perspectives of health care transition services. Autism. 2014;19(8):992–1001.

Fraser KD, Al Sayah F. Arts-based methods in health research: a systematic review of the literature. Art Health Int J Res Policy Pract. 2011;3(2):110–45.

Goggin K, Catley D, Brisco ST, Engelson ES, Rabkin JG, Kotler DP. A female perspective on living with HIV disease. Health Soc Work. 2001;26(2):80–9.

Han CS, Oliffe JL. Photovoice in mental illness research: a review and recommendations. Health (London). 2015. https://doi.org/10.1177/1363459314567790.

Hergenrather KC, Rhodes SD, Cowan CA, Bardhoshi G, Pula S. Photovoice as community-based participatory research: a qualitative review. Am J Health Behav. 2009;33(6):686–98.

Herrick AL, Lim SH, Wei C, Smith H, Guadamuz T, Friedman MS, Stall R. Resilience as an untapped resource in behavioral intervention design for gay men. AIDS Behav. 2011;15 (Suppl 1):S25–9.

Herrick AL, Stall R, Goldhammer H, Egan JE, Mayer KH. Resilience as a research framework and as a cornerstone of prevention research for gay and bisexual men: theory and evidence. AIDS Behav. 2014;18(1):1–9.

Houston TK, Allison JJ, Sussman M, Horn W, Holt CL, Trobaugh J, . . . Hullett S. Culturally appropriate storytelling to improve blood pressure: a randomized trial. Ann Intern Med. 2011a;154(2):77–84.

Houston TK, Cherrington A, Coley HL, Robinson KM, Trobaugh JA, Williams JH, . . . Allison JJ. The art and science of patient storytelling-harnessing narrative communication for behavioral interventions: the ACCE project. J Health Commun. 2011b;1–12. https://doi.org/10.1080/10810 730.2011.551997. 937140203 [pii].

Jibaja-Weiss ML, Volk RJ, Granch TS, Nefe NE, Spann SJ, Aoki N, . . . Beck JR. Entertainment education for informed breast cancer treatment decisions in low-literate women: development and initial evaluation of a patient decision aid. J Cancer Educ. 2006;21(3):133–9.

Lal S, Jarus T, Suto MJ. A scoping review of the photovoice method: implications for occupational therapy research. Can J Occup Ther. 2012;79(3):181–90.

Martin N, Garcia AC, Leipert B. Photovoice and its potential use in nutrition and dietetic research. Can J Diet Pract Res. 2010;71(2):93–7.

Murphy ST, Frank LB, Chatterjee JS, Baezconde-Garbanati L. Narrative versus non-narrative: the role of identification, transportation and emotion in reducing health disparities. J Commun. 2013;63(1):116. https://doi.org/10.1111/jcom.12007.

Murphy ST, Frank LB, Chatterjee JS, Moran MB, Zhao N, Amezola de Herrera P, Baezconde-Garbanati LA. Comparing the relative efficacy of narrative vs nonnarrative health messages in reducing health disparities using a randomized trial. Am J Public Health. 2015;105(10):2117–23.

Rich JA, Grey CM. Pathways to recurrent trauma among young black men: traumatic stress, substance use, and the "code of the street". Am J Public Health. 2005;95(5):816–24.

Riessman CK. Narrative analysis. London: SAGE; 1993.

Riley RG, Manias E. The uses of photography in clinical nursing practice and research: a literature review. J Adv Nurs. 2004;48(4):397–405.

Sanon MA, Evans-Agnew RA, Boutain DM. An exploration of social justice intent in photovoice research studies from 2008 to 2013. Nurs Inq. 2014;21(3):212–26.

Scientific Software Development. Atlas.ti (version 6.2). Berlin; 2011. http://atlasti.com/

Shattuck PT, Narendorf SC, Cooper B, Sterzing PR, Wagner M, Taylor JL. Postsecondary education and employment among youth with an autism spectrum disorder. Pediatrics. 2012;129(6):1042–9.

Switzer S, Guta A, de Prinse K, Chan Carusone S, Strike C. Visualizing harm reduction: methodological and ethical considerations. Soc Sci Med. 2015;133:77–84.

Teti M, Bowleg L, Cole R, Lloyd L, Rubinstein S, Spencer S, . . . Gold M. A mixed methods evaluation of the effect of the protect and respect intervention on the condom use and disclosure practices of women living with HIV/AIDS. AIDS Behav. 2010;14(3):567–79.

Teti M, Murray C, Johnson L, Binson D. Photovoice as a community based participatory research method among women living with HIV/AIDS: ethical opportunities and challenges. J Empir Res Hum Res Ethics. 2012;7(4):34–43.

Wang C. Photovoice: a participatory action research strategy applied to women's health. J Womens Health. 1999;8(2):185–92.

Wang C, Burris MA. Empowerment through photo novella: portraits of participation. Health Educ Q. 1994;2(2):171–86.

Wang C, Burris MA. Photovoice: concept, methodology, and use for participatory needs assessment. Health Educ Behav. 1997;24(3):369–87.

Wang C, Pies CA. Family, maternal, and child health through photovoice. Matern Child Health J. 2004;8(2):95–102.

Wang C, Morrel-Samuels S, Hutchison PM, Bell L, Pestronk RM. Flint photovoice: community building among youths, adults, and policymakers. Am J Public Health. 2004;94(6):911–3.

IMAGINE: A Card-Based Discussion Method

Ulrike Felt, Simone Schumann, and Claudia G. Schwarz-Plaschg

Contents

1 Introduction	1168
2 A Method as a Response to Three Major Challenges	1169
2.1 Rendering Nondebated Issues Accessible	1170
2.2 Participatory Justice	1170
2.3 Countering the Lay-Expert Divide	1170
3 IMAGINE: A Card-Based Discussion Method	1171
4 Creating the Cards and Deciding on Stages	1172
4.1 Exploring and Analyzing the Issue	1172
4.2 Card Types and Choreography	1174
4.3 Validating the Card Sets and Making Choices in the Choreography	1176
5 Conducting IMAGINE Discussion Groups	1177
5.1 Assignment of Turn-Taking	1178
5.2 Keep the Discussion Going	1178
5.3 Encourage Discussion Among Participants: Not with Moderator	1178
6 Analyzing IMAGINE	1179
7 Conclusion and Future Directions	1180
References	1181

U. Felt (✉)
Department of Science and Technology Studies, Research Platform Responsible Research and Innovation in Academic Practice, University of Vienna, Vienna, Austria
e-mail: ulrike.felt@univie.ac.at

S. Schumann
University of Vienna, Vienna, Austria
e-mail: simone.schumann@univie.ac.at

C. G. Schwarz-Plaschg
Research Platform Nano-Norms-Nature, University of Vienna, Vienna, Austria
e-mail: claudia.g.schwarz@univie.ac.at

© Springer Nature Singapore Pte Ltd. 2019
P. Liamputtong (ed.), *Handbook of Research Methods in Health Social Sciences*,
https://doi.org/10.1007/978-981-10-5251-4_9

Abstract

This chapter introduces IMAGINE – a card-based group discussion method for qualitative research and engagement processes. IMAGINE was developed as a response to three major challenges that tend to emerge in discussion groups and participatory exercises. First, it renders new or complex issues accessible by offering participants a broad repertoire of structured resources without pre-configuring the issue too much. Second, it seeks to contribute to participatory justice by assuring that all participants get time and space for expressing their visions. Third, the cards allow the introduction of expert opinions without expert presence, thus avoiding the emergence of strong lay-expert divides. The method consists of a number of different card sets and a specific choreography. We explain the rationale behind different card types and how researchers can go about creating their own card sets. The contribution also includes suggestions for how to conduct and analyze IMAGINE discussion groups so as to harness their full potential. It concludes by pointing towards potential future directions in which the method could be developed.

Keywords

Card-based discussion method · Group discussions · Deliberation · Engagement · Participation · Participatory justice

1 Introduction

Over the past decades, the use of discussion groups as a research method has resurged in the social sciences. Simultaneously, there have been numerous efforts to create new formats that hold the promise of being better adapted to explore and engage with new problem formations that emerge in relation to technological and scientific developments. Through rethinking engagement, they try to address the changing social fabric of contemporary societies, the diversified information infrastructures, as well as the shifting perceptions of how knowledge, values, and choices are or should be interrelated. These methodological innovations are of particular interest as they often cross the boundaries between academia and nonacademic environments, being used in multiple sites and for a broad range of purposes. In policy-related fields that deal with questions of science and technology governance and decision-making about public goods and (healthcare) services, for example, diverse forms of discussion and engagement methods, have become central to stimulate citizen, patient, or stakeholder engagement (Chilvers and Kearnes 2016).

The reasons for the development of such participatory processes and a broader engagement with the spectrum of positions held within diverse societal collectives are manifold (Stirling 2008). In an instrumental way, they are often perceived as means to build public trust in institutions and systems, which are fairly distant for most members of society, as well as to gain acceptance for technoscientific or biomedical innovations. From a normative point of view, they are meant to constitute an essential and intrinsic part of a democratic society with openness, transparency, and plurality as its core values. Finally, including visions and values of various societal groups (e.g., patients, user groups, consumers, concerned citizens) may open up new perspectives on innovation as well as address relevant ethical and social issues that foster the design of more socially robust processes and products.

In order to realize these different goals, a range of discussion oriented participatory methods have been developed and applied, among them most prominently citizen juries, citizen panels, consensus conferences, or citizen advisory committees (Bowman and Hodge 2007; Delgado et al. 2011). These methods vary in structure, group composition, and output-orientation, but they all tend to share the ideal of making the discussion of technoscientific issues and/or decisions more inclusive, empowering, and deliberative. What such events of participation and engagement aim for, ranges from a clear statement towards the issue at stake (sometimes based on the analysis of present social scientists, sometimes written by participants themselves) to the wish for a deeper understanding of the interactive process itself and how positions in the group (can) emerge.

In this chapter, we present a *card-based group discussion method* (IMAGINE) whose development was driven by an STS (science and technology studies) understanding of social scientific methods as embodying a normative and performative power (Law 2004). By using different sets of cards for participants to work with, the method seeks to stimulate and support the process of developing imaginations with regard to emerging technoscientific and other complex social issues. IMAGINE aims to enhance the capacity to gradually assemble, in interaction with other participants, the elements or "building blocks" that are used to assess an issue and construct a position. Its core analytic interest is to investigate how people analyze and relate to specific matters of concern in the present and to their potential futures, i.e., how they think a specific part of "the world" works and how they imagine it might or should work in the future. The method can thus be used in two ways: (1) to inform policy or other decision-making authorities about the processes, resources, and value systems people, both individually and collectively, relate to and draw upon when developing a position towards a problem at stake; (2) to support social scientists in developing a more fine-grained understanding of processes of public imagination and assessment with regard to scientific, technological, or other complex and nascent issues in society.

2 A Method as a Response to Three Major Challenges

IMAGINE was developed as a response to three major challenges that we identified in observing discussion processes in participatory exercises (Felt and Fochler 2010) but also in performing public engagement exercises in the framework of research projects on nanotechnology (Felt et al. 2014, 2015) and biomedicine (Felt et al. 2008, 2009).

2.1 Rendering Nondebated Issues Accessible

The first challenge is related to the matter of concern at the core of the engagement exercise. Sometimes, we have to analyze and assess a development at a point in time when it is not yet widely present in the public arena. Then most participants lack broader frames of reference. They have no examples of how different relevant societal actors conceptualize and narrate what is at stake, and thus, have difficulties to develop and unfold their thinking beyond rather vague reflections. To counter this, we deliberately did not want to present ready-made scenarios (Bennett 2008; Türk 2008) to participants, as we think this might not sufficiently foster their own creative work. We, thus, needed to find a way to bring dispersed and often not easily accessible information (scientific, technological, social, political, and so on) and societal positions to the table. We decided to prepare different sets of cards, which participants could use as starting points for their thinking and deal with in a flexible manner.

2.2 Participatory Justice

If one works with heterogeneous groups – as it is often the case in public engagement initiatives – people with different backgrounds, agendas, relations to the topic, and knowledge forms (e.g., professional expertise, experiential knowledge of patients) will meet. Due to this diversity, the degree of participation may vary largely among participants and create a situation of participatory injustice. For instance, people with more experience in discussion settings or with higher education often find it easier to express their position than others who are not well acquainted with such situations. Thus, a central question in discussion groups is how to assure the inclusion of all participants and to create a space where a broad spectrum of positions could be expressed. This means both assuring moments in the process where everybody gets voice and offering support to express their views openly, even if it was a clear minority position.

2.3 Countering the Lay-Expert Divide

Research into participation processes with lay and expert interaction shows that the inclusion of experts may lead to lay participants being co-opted by and following expert framings (Kerr et al. 2007; Felt and Fochler 2010). This "stakeholder capture" effect (MacLean and Burgess 2010) reproduces the power of representatives from influential groups, who often have a privileged access to technological or scientific knowledge. In order to minimize this effect, we refrained from inviting experts and instead thought about alternative ways to bring expert opinions to the table.

These three concerns led to the development of a card-based discussion method we call IMAGINE (Felt et al. 2014), and which we describe and reflect on in the following sections.

3 IMAGINE: A Card-Based Discussion Method

The general aim of this card-facilitated discussion approach is to create a flexible space that takes shape through a series of individual choices (of a number of cards) and subsequent phases in which participants are encouraged to develop and negotiate individual and collective positions in a discussion. Several shorter and diverse inputs in the form of different sets of cards stimulate and guide this process.

Card-based methods have been used in both participatory settings and qualitative research. For instance, the PlayDecide card game (PlayDecide is based on the Democs cards, a card-based public engagement tool (see http://www.playdecide.eu, accessed 01 July 2016).) has been developed as a public engagement tool. It shares with IMAGINE the idea of providing distinct sets of cards, but its goal is neither research-oriented nor do the stages and cards follow the same logic: PlayDecide emphasizes the need to "inform" participants with the cards, and it often ends with a voting on several, predefined solutions/options. Cards have also been applied in focus group research to discuss sensitive issues (Chang et al. 2005; Sutton 2011) or to observe how people rank topics in a collective manner (Kitzinger 1994; Bloor et al. 2001). IMAGINE differs from these approaches in the following way: it is strongly process-oriented; card-choice is more individual as each participant has her/his own sets of cards and a board for "documenting" choices; it does not aim at information or decision-making. Furthermore, IMAGINE does not solely encourage participants to assess the issue at stake at present (and reflect on past developments), but to also engage in imagining where a specific development might lead to and how this future potential might impact present choices. The method should, thus, trigger the imagination of participants of where we stand in a development and where it might lead us, i.e., embed a temporal dimension in their reflection. We consider imagination here as a productive cognitive but also social process that includes both retrospective and prospective dimensions (for a summary of research on imaginaries see McNeil et al. 2017).

We now want to briefly describe the central elements of the method. IMAGINE follows a multistage choreography, which consists mostly of three to four stages. Each stage sheds light on the matter of concern from a different perspective, lasts approximately the same length (in our case 45 min), and comes with a specific set of cards (see next section for more on card types and stages). This choreography provides a clear structure without predefining or restricting what can be discussed, invites to shift perspective, and allows some moments of individual reflection when choosing the cards for the respective next stages. A discussion group would in our case last for approximately four hours (including a break), which is about twice as long as traditional focus groups. This length is feasible because the method alternates between more active discussion phases and moments in which the participants read and choose new cards in a silent manner. This process allows refocusing their attention, to rethink the discussion process so far, and decide where they would like to take it in the next stage.

IMAGINE was originally designed for around *six to eight participants*. We recommend keeping the number of participants rather low in order to allow a maximum of time for interactions and the development of positions. A facilitator should guide the participants through the process and should ideally not give an introduction to the issue at stake in order to avoid presenting him/herself as defining the content of what is to be discussed. If necessary, a short video clip that lays out the topic and its dimensions could be shown.

Each participant has a personal board that materializes the four different stages with boxes on which the chosen cards in each stage should be put (see Fig. 1). Next to the board, four piles of cards are placed flipside up. The participants are asked to pick up one pile at the beginning of each stage, to go through the cards and decide on a predetermined number of cards by applying their own rationale for choice (e.g., dislike or agree with a statement, be astonished, find it puzzling, and so on). When every participant has placed the chosen cards on the board in front of him/her (this may take 5–10 minutes, depending on the respective number of cards), the moderator can start to inquire about their choices.

4 Creating the Cards and Deciding on Stages

In this section we address some questions that arise for researchers who would like to adapt IMAGINE to a specific topic: What should I consider when creating the cards, deciding on the exact number of stages, and the thematic focus of these stages?

4.1 Exploring and Analyzing the Issue

In order to come to these decisions, a detailed analysis of the topic to be debated is a necessary prerequisite. This is a crucial element and essential first step in the IMAGINE methodology. In contrast to focus groups, where usually just a discussion guide consisting of themes and questions is developed, more detailed prior research is needed to develop the comprehensive and quite work intensive card material. This research may also include the use of other qualitative research methods such as interviews, media, and document analysis; ethnographic methods; and participant observation or online research. It is not sufficient to roughly survey the topic, but it is necessary to identify the spectrum of concrete arguments, explanations, actors, regulations, images, and examples that together constitute the topic.

In what follows, we will present how we concretely developed the IMAGINE card types and their contents. In the specific case of our research project, the aim of the preceding analysis was to explore the existing technoscientific, political, and cultural discourse and meaning production on nanotechnology. ('Making Futures Present. On the Co-Production of Nano and Society in the Austrian Context', funded by the Austrian Science Fund, project number P20819, PI: Ulrike Felt, see http://sts.univie.ac.at/en/research/completed-projects/making-futures-present-nano-and-society/). Since the research we wanted to undertake with IMAGINE was to investigate the public perceptions on nanotechnology in the Austrian context, our analysis had a specific focus on the Austrian context – but we also drew on international resources to give a

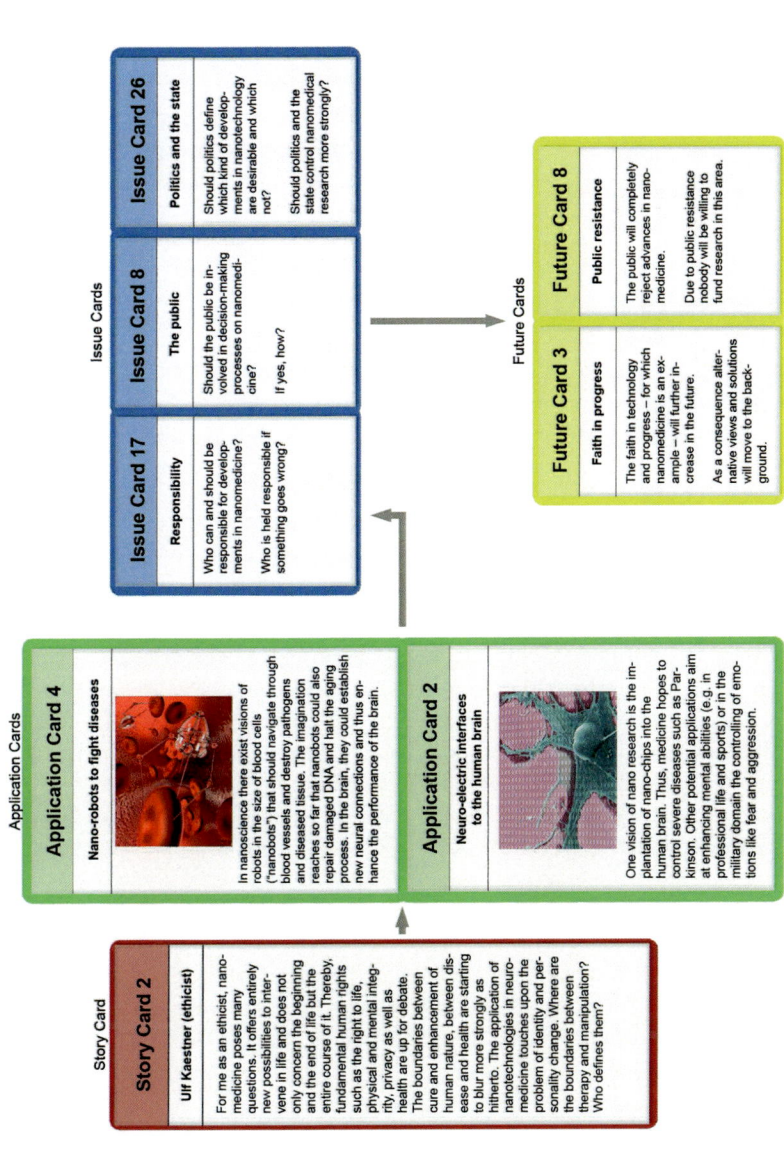

Fig. 1 Schematic representation of a board with a selection of cards. The topic discussed here was nanomedicine (Felt et al. 2014)

broader picture. Our main methods were literature (especially on social and ethical issues) and media research, analysis of policy documents, and qualitative interviews, which we conducted with Austrian stakeholders (among them primarily scientists, policymakers, and representatives from nongovernmental organizations). Such interviews are essential to interpret some of the material found and to add further perspectives. But, we also looked into other projects on nanotechnology and their methodological tools for enabling participation, such as the *concept boards* from the DEEPEN (Deepening Ethical Engagement and Participation in Emerging Nanotechnologies) project (https://www.dur.ac.uk/geography/research/research_projects/?mode=project&id=241, accessed 25 June 2016.) and a PlayDecide card set on nanotechnology.

4.2 Card Types and Choreography

In order to structure the discussion process, we decided to develop four different card types. Each of it addressed a specific angle of the issue to be explored, and their subsequent introduction allowed to gradually build a discussion through different stages. Below, we present the card types in the chronological order of their appearance in the discussion. We generalize their aim beyond the concrete case of the original IMAGINE cards. For specific examples of the cards we used in the project on nanotechnology see Fig. 1 (see also Felt et al. 2015).

4.2.1 Story/Statement Cards
These cards are the first that participants are asked to read and choose from. They contain short statements or narratives as expressed by different actors and are meant to stand for the spectrum of existing positions towards the issue. Calling them "story cards" points to the fact that they are not an enumeration of facts but are meant to show the multiple ways how people create meaning and coherence of a complex phenomenon by developing a position. These cards look like short newspaper columns in which people convey their view on a specific topic in a concise manner. The material for these cards primarily stemmed from qualitative interviews, quotes of relevant actors in media articles, and policy documents. However, the text has not been taken word for word from these sources. In most cases, it was necessary to rephrase and condense statements to get them into an easily graspable format. Moreover, we decided to present all these statements in an anonymous way – that is, we made the actor status visible (e.g., politician, scientist, legislator, doctor, affected person) but did change the name.

4.2.2 Application/Situation Cards
The second stage of the debate is introduced by cards presenting a set of specific applications or situations in which the issue at stake materializes. In the case of nanotechnology, we presented applications already existing for purchase and visions of applications from researchers or nano-visionaries. In other contexts, for example medical ones, these cards could for instance represent key-moments in a healthcare

situation, when choices have to be made by doctors and patients. More recently, in discussion groups on responsibility in research, a range of concrete research situations were presented using three sets of cards through which a debate on these issues in research practice was made possible (Felt et al. 2017; for this version of a card-based discussion method developed to discuss with scientists see also http://rri.univie.ac.at/aktivitaeten/series-of-group-discussions/ (accessed 10 October 2016). The sources used for the creation of these cards should also derive from prior analysis of the topic, public debates, policy documents, webpages, and many more. This type of cards is meant to render participants' positioning work more concrete and detailed. It should allow to elicit a nuanced view on the situatedness and multiplicity of assessments, i.e., participants might come to rather divergent assessments of the issue when they are confronted with different application contexts or situations. In the project on nanotechnology, these cards included each an illustrative image and a short description of the application. The cards had a promissory tone but also included potential risks and negative implications. In the case of responsibility in research situations, cartoons were used to express the diverse situations in which responsible choice can come to matter.

4.2.3 Issue/Context Cards

After having already discussed societal attitudes towards the topic as well as more concrete applications/situations, the third card type, the issue or context cards, takes the debate to a level at which broader themes and contextual aspects are addressed. Therefore, in the nano case, these cards would address in one or two short sentences/questions a variety of ethical, social, political, environmental, health, economic, and legal issues. These issues were primarily extracted from literature research and were the outcome of our analysis of the collected material. In contrast to the previous card types, much more analytical work was going into their development.

The issue/context cards should include as much potentially relevant issues with the aim to foster a deeper discussion in the IMAGINE groups. The cards mostly included a more abstract question or raised a specific argument. Issues covered in our research were, for instance, questions with regard to the role of the state in governing emerging technologies, who should take responsibility if developments go wrong, or to what degree citizens should be involved in decision-making. This allowed the participants to either try to develop a position to a raised question, to pose it to the group or to agree with or critique specific statements. Generally, phrasing the issues as questions seems to bear the most potential for the discussion. Our experience was that many of these issues already appeared in a more latent form in the first two stages, but that this stage was relevant to make them more explicit and to also raise awareness for issues that were not so evident at first. In this stage particularly, shared concerns were often evident by participants' similar card choice.

4.2.4 Future Cards

These cards mark the final stage of discussion and should stimulate the collective imagination of how the issue at stake would continue to develop, what actor constellations might matter, and what this would mean in terms of present choices.

While past and future developments have been discussed more or less explicitly in the previous stages, this stage focuses explicitly on how the discussants imagine and assess the possibility of potential developments. Whereas the story/statement and application/situation cards are more descriptive, and the issue/context cards are more analytical, this last set of cards invites participants to make more general projective statements. This comprises questions such as where developments could lead, how the future may be shaped by various actors and forces, what this might mean for the individual and society, and how one could intervene in this making of a potential future. Strong utopian/dystopian or highly speculative scenarios, which lacked any rooting in present developments, have not been included.

Additionally, we also included blank cards, on which the participants could write their own issues/situations and futures, for the last two card types.

4.3 Validating the Card Sets and Making Choices in the Choreography

A more practical but nevertheless essential task in creating the cards is to fill them with content that is not only comprehensible for participants coming from diverse backgrounds but also adequately represents the multiplicity of narratives, situations/applications, issues, as well as possible future developments. At the same time, the number of cards should be limited as participants have to read them all before making a choice. In the nano-project we used 6–8 story cards, about 7–9 application cards, 25–30 issue cards and 15–20 future cards. We created much more cards at first and then went through a selection and validation process. First, all members of the research team as well as noninvolved colleagues should read the cards and give their feedback. Second, professionals or experts in the researched field can be consulted to check statements and if the field and its scope has been represented in a balanced and comprehensive way. The feedback process can be carried out in diverse ways, ranging from mere textual editing to more elaborate ranking exercises. Cards can be ranked with regard to different attributes like relevance, readability, value of information, and so on with the aim to condense the card set by sorting out those at the end of the scale. In a next step, the cards should be tested with a user group to ensure that they are understandable and without unwanted ambiguities. Generally, at the end of an IMAGINE discussion it is also recommended to allow a feedback round on the method and card content. Since the actual participants have experienced how the cards work for them, their feedback can help to re-work and refine the cards.

The content of the IMAGINE cards and how they should be written is also affected by group-related design choices. Designers of IMAGINE need, from the beginning, a vision of the participants they want to engage. A decision for homogeneous or heterogeneous groups, for example, often derives from and is aligned to a specific research question and methodology but has also major implications for the creation of the cards. Social groups such as adolescents, elderly people, people with lower formal education, or participants with diverse cultural backgrounds will demand careful reflection of both cards and contents. In general, this means considering the following points, which all follow the principle of "low entry level participation":

- *Texts* should be easily understandable; if technical terms are to be used at all, they should always be accompanied with an explanation.
- *Sentences* should be short and straight forward.
- *A larger font* should be preferred to make the reading quick and easy for everybody.
- The *number of cards* should be adapted to the group composition.

The choice of groups also impacts the duration of the discussion and the overall choreography. If IMAGINE discussions are conducted in institutions such as schools, the method might have to be adapted to institutional timeframes. In general, if there exists a shorter timeframe for discussion, the types and number of cards should be reduced. Moreover, specific groups might have a different attention span than others, which then impacts the duration of the discussion.

Another issue that influences the choreography or course of discussion is the question of the aim of the whole exercise. There are two ways to approach this: one that is usually practiced in qualitative research and another that is typical for public engagement. A lot of public engagement initiatives seek for a group decision or shared statement at the end of a discussion process, which most of the time means that some kind of consensus has to be reached. Qualitative research, on the other hand, is more interested in the process as such and the themes that are discussed. Here, leaving things open, even if positions remain incompatible at the end, is not problematic. The former also implies that time has to be planned for an extensive final round, in which people are enabled to formulate collectively a recommendation paper, whereas IMAGINE groups in qualitative research contexts may just end with a short round of reflection among participants.

5 Conducting IMAGINE Discussion Groups

In this section, we reflect on three practical issues with regard to conducting IMAGINE discussion groups: place, moderation, and documentation. First, it is necessary to keep in mind how the place where discussion groups are held, shapes the way the debate may unfold. No setting is "neutral" and thus it is important to reflect on what a specific place or setting might enable or close down in terms of what and how opinions can be expressed. While this is relevant for every research method, IMAGINE calls for a more specific consideration due to its materiality; the fact that it includes a board and several piles of cards that need to be placed in front of every discussant and needs face-to-face interaction between participants. Hence, IMAGINE requires in general a setting in form of a round table.

The choreography and cards act as strong structuring and facilitating tools. This is why there is in principle less need for moderator involvement than in focus groups.

Nevertheless, a moderator is important for the smooth conduct of an IMAGINE discussion group. The main role of the moderator is to know and explain the stages and guide the group through each of them, which includes time management tasks such as keeping an eye on the duration of individual reading and decision-making phases in each stage. While the mode of moderation is always open for choice, the following three elements are important for a good functioning of such an approach:

5.1 Assignment of Turn-Taking

In the first phase of each stage, the participants are asked to choose cards and then present and explain their card choices. Here, the moderator should make sure that all participants get their turn and encourage quieter participants to contribute. Ideally, the assignment of turn-taking becomes a self-directed process among the participants without much intervention by the moderator, especially in the more discussion-oriented periods.

5.2 Keep the Discussion Going

Although the explanation of card choices is one central aspect of the choreography, the main interest of IMAGINE lies in the discussion among the participants that follows. The benefit of alternating more individual and collective phases, however, also poses the risk that the discussion process does not get started. If this is the case, the moderator is required to come up with open-ended questions and comments that stimulate the interaction in the group. It has proven helpful to ask about the participants' opinion about specific cards. For instance, the moderator may suggest the discussion about cards that have been chosen by several participants or inquire why specific cards were not chosen at all. Part of the moderator's task is also to allow room for flexible timekeeping whenever it is in the research interest to keep the discussion going.

5.3 Encourage Discussion Among Participants: Not with Moderator

The moderator should encourage participants to discuss the topic with each other and not enter in a dialogue with individual participants. A general guideline to achieve this is to address the group and not individual people. Moreover, the moderator should not present himself/herself as an expert, because this would prompt participants to ask the moderator questions instead of debating with each other.

Depending on the research question and analytical approach, one may opt for different means to document the discussion process. In general, we recommend to audiotape the discussions and transcribe them (or relevant passages) in a way that includes some details on turn-taking, overlapping speech, pauses, and annotation of nonverbal activity. Such a transcription system allows making the most of the discussion data without going into too much detail. An alternative or complementary option is to make a video recording. The advantage of visual documentation is that it allows analyzing body language, particularly if there is an interest in how the card choice process unfolds. However, the presence of a video recorder might also affect the discussion in a different way than an audio recorder. Another "cost" is that the analysis of videotapes generally represents a more complex analytical undertaking than an analysis of discourse. A compromise is to complement the audiotape with ethnographic observations, which are also able to capture body language and the appropriation of the card materiality (for such an analysis see Felt et al. 2014, p. 241ff.).

6   Analyzing IMAGINE

In terms of analysis, we suggest a close examination of interactive processes and their role in the construction of meaning. This approach relies on the premise that knowledge, experience, and memory cannot be regarded as fixed entities existing independently in peoples' minds but are situated enactments emerging within specific circumstances and contexts. It follows that the analysis needs to move from a focus on what is said – the positions and understandings people articulate – to how content emerges through interaction. Here, discourse analytic methodologies can help to explore dynamic meaning-making processes as well as the more tacit moral and social assumptions of utterances (for examples from the analysis of IMAGINE discussion groups see Felt et al. 2015; Schwarz-Plaschg 2016).

In contrast to public engagement, where the focus tends to lie on fast output (e.g., a recommendation paper written by the participants or a consensus on a controversial issue), a discourse analytical approach requires time and social scientific qualifications to produce results. However, IMAGINE can also be carried out as part of a more inclusive and participatory research process that invites participants to co-create the cards and/or involves them in the analysis. We utilized IMAGINE in this way in a participatory school project ("Nanomaterials: Possibilities and risks of a new dimension," 2010–2012, funded by the Austrian Federal Ministry of Science, Research and Economy under the program "Sparkling Science") that sought to engage students around the age of 17 in reflecting on the risks and benefits of nanotechnology. In four workshops, we first developed an IMAGINE card set on nanotechnology together with the students, which was then used for the discussion with other groups of students. Then, we jointly analyzed the discussion groups and the students presented the findings at a "Young Researcher Symposium" to other students and stakeholder representatives. One aim of the project was to familiarize the students with the way qualitative research is conducted. Beyond this learning and empowering potential, such participatory research can also ensure the validity of findings and provide rich new analytic material.

As mentioned above, IMAGINE allows to integrate the materiality of and interaction with the cards into the analysis when video recordings or ethnographic observations are made. Here, our analysis has already shown that the cards stimulate specific practices and serve a variety of purposes for the participants (Felt et al. 2014, p. 241ff). One option is to analyze how the individual participants order and choose cards in the "silent" card choice phase – and to relate these practices to the discursive explanations they give afterwards. Another perspective would be to trace how the cards are used as mnemonic tools to remember the course of the discussion and the positions of other participants. We furthermore found that the frequency in which cards are chosen might display tendencies of those issues that are of central concern, which would need further qualitative exploration.

7 Conclusion and Future Directions

In this chapter, we introduced the card-based discussion method IMAGINE, which we developed as a response to three major challenges in participatory research settings: (1) Rendering issues that are not publicly debated accessible to a wide range of participants; (2) enabling participatory justice, i.e., to partly compensate for differences among the participants in discussion groups that may affect their ability to participate; (3) countering the classic lay-expert divide by bringing expertise without experts to the table.

The preparation of sets of cards allows to present and structure an unfamiliar or complex topic and to discuss it from a variety of perspectives. The potential of IMAGINE lies in the flexible card material and the way in which specific card types are utilized in the various stages of discussion. This means that it manages to structure debate, while leaving it open at the same time. As the cards bring a diverse spectrum of opinions, possibilities, and problems into the discussion, it has proven easier for participants to embrace a position which is marginalized, stigmatized, or a taboo in the wider public debate via a card. The variation of angles from which to address a matter of concern further adds a developmental perspective to the discussion as people can gradually construct and work on their assessments, revise positions, or reinforce them.

This approach, therefore, also seems fruitful when an issue is already strongly polarized in the public space. In this case, the cards, as material-semiotic resources, could be used to deconstruct fixed positions into smaller entities and, allow to open up the possibility of reconfiguration. The cards, thus, render visible the complexity of any scientific and technological issue in society, point at the large network of actors involved, and thus, make an issue more broadly assessable for the discussants.

The method is particularly apt for understanding how people order a specific topic, which becomes visible through the card selection processes. It allows to better understand the frames of references people work with but also what they need for developing, defending, or adapting a position towards an issue. This becomes palpable through the individual arguments justifying their choice of cards but also through the follow-up interaction. The method further invites to explicitly relate past experiences with future expectations. This is done as the final phase of a longer argumentative process is less open for speculation but constitutes more an argumentative projection exercise. Yet, it also enables the moderator as well as later the analyst to observe the nonchosen cards/issues and thus to address and reflect on absent presences (Law 2004).

Overall, the method is particularly apt for projects in which the core interest is to get a better understanding of how participants articulate their positions, where zones of conflict arise and where agreements are emerging, and what things can be said in such a semi-public setting and which are more contested. It is less fitting if the expectation is to have one opinion at the end of the process.

Most importantly, IMAGINE is *not a ready-made tool*. It is an open, process-oriented method that needs careful adaptation to the situation or issue to be discussed. This means considering the multiplicity of ways in which to assess any matter of concern, the diversity of participants, and the specific cultural context in which it is performed, to mention a few obvious factors. The card-based method, therefore, demands considerable input from the social scientists who want to use IMAGINE in their qualitative research and/or engagement exercise, as they have to perform a careful qualitative analysis of diverse sets of available background materials as well as produce case-specific data such as interviews. IMAGINE, thus, has to be seen as an integral part of a broader process of understanding an issue and of triggering engagement with it, rather than an isolated method.

References

Bennett I. Developing plausible nano-enabled products. In: Fisher E, Selin C, Wetmore JM, editors. The yearbook of nanotechnology in society. Volume I: presenting futures. Dordrecht: Springer; 2008. p. 149–55.

Bloor M, Frankland J, Thomas M, Robson K. Focus groups in social research. London: SAGE; 2001.

Bowman DM, Hodge GA. Nanotechnology and public interest dialogue: some international observations. Bull Sci Technol Soc. 2007;27(2):118–32.

Chang JC, Cluss PA, Ranieri L, Hawker L, Buranosky R, Dado D, McNeil M, Scholle SH. Health care interventions for intimate partner violence: what women want. Womens Health Issues. 2005;15(1):21–30.

Chilvers J, Kearnes M, editors. Remaking participation: science, environment and emergent publics. London: Routledge; 2016.

Delgado A, Kjølberg K, Wickson F. Public engagement coming of age: from theory to practice in STS encounters with nanotechnology. Public Underst Sci. 2011;20(6):826–45.

Felt U, Fochler M. Machineries for making publics: inscribing and de-scribing publics in public engagement. Minerva. 2010;48(3):219–38.

Felt U, Fochler M, Mager A, Winkler P. Visions and versions of governing biomedicine: narratives on power structures, decision-making and public participation in the field of biomedical technology in the Austrian context. Soc Stud Sci. 2008;38(2):233–55.

Felt U, Fochler M, Müller A, Strassnig M. Unruly ethics: on the difficulties of a bottom-up approach to ethics in the field of genomics. Public Underst Sci. 2009;18(3):354–71.

Felt U, Schumann S, Schwarz CG, Strassnig M. Technology of imagination: a card-based public engagement method for debating emerging technologies. Qual Res. 2014;14(2):233–51.

Felt U, Schumann S, Schwarz CG. (Re)assembling natures, cultures, and (nano)technologies in public engagement. Sci Cult. 2015;24(4):458–83.

Felt U, Fochler M, Sigl L. IMAGINE RRI. A card-based method for reflecting responsibility in life science research. Under Review. 2017.

Kerr A, Cunningham-Burley S, Tutton R. Shifting subject positions: experts and lay people in public dialogue. Soc Stud Sci. 2007;37(3):385–411.

Kitzinger J. The methodology of focus groups: the importance of interaction between research participants. Sociol Health Illn. 1994;16(19):103–21.

Law J. After method. Mess in social science research. London/New York: Routledge; 2004.

MacLean S, Burgess MM. In the public interest: assessing expert and stakeholder influence in public deliberation about biobanks. Public Underst Sci. 2010;19(4):486–96.

McNeil M, Arribas-Ayllon M, Haran J, Mackenzie A, Tutton R. Conceptualizing imaginaries of science, technology. In: Felt U, Fouché R, Miller C, Smith-Doerr L, editors. Handbook of science and technology studies. Cambridge, MA: MIT Press; 2017. p. 435–63.

Schwarz-Plaschg C. Nanotechnology is like…The rhetorical roles of analogies in public engagement. Public Underst Sci. 2016. Online first. https://doi.org/10.1177/0963662516655686.

Stirling A. "Opening up" and "closing down": power, participation, and pluralism in the social appraisal of technology. Sci Technol Hum Values. 2008;33(2):262–94.

Sutton B. Playful cards, serious talk: a qualitative research technique to elicit women's embodied experiences. Qual Res. 2011;11(2):177–96.

Türk V. Nanologue. In: Fisher E, Selin C, Wetmore JM, editors. The yearbook of nanotechnology in society. Volume I: presenting futures. Dordrecht: Springer; 2008. p. 117–22.

Timeline Drawing Methods

67

E. Anne Marshall

Contents

1 Introduction 1184
2 Definitions and Formats 1185
3 Advantages of Using Timelining Methods 1187
   3.1 Richer and More Complete Data 1187
   3.2 Establishing Rapport 1188
   3.3 Participant Engagement 1188
   3.4 Flexibility 1188
   3.5 Language 1189
   3.6 Power Differences 1189
   3.7 Professional Settings 1189
4 Challenges of Using Timelines 1189
   4.1 Personal Content 1190
   4.2 Confidentiality 1190
   4.3 Assumptions of Linearity 1190
   4.4 Analysis 1190
   4.5 Reporting 1191
5 Review of Timeline Research 1191
6 Four-Phase Model for Using Timelines 1197
7 Conclusion and Future Directions 1198
References 1198

Abstract

Qualitative researchers are continually seeking approaches that will yield in-depth and high-quality interview data. Incorporating a timelining method adds a visual representation related to the experience that can anchor the interview and helps focus the participant on key elements. Timelining can provide participants

E. Anne Marshall (✉)
Educational Psychology and Leadership Studies, Centre for Youth and Society, University of Victoria, Victoria, BC, Canada,
e-mail: amarshal@uvic.ca

© Springer Nature Singapore Pte Ltd. 2019
P. Liamputtong (ed.), *Handbook of Research Methods in Health Social Sciences*,
https://doi.org/10.1007/978-981-10-5251-4_10

with a way to engage their stories deeply and even help to create new meanings and understandings. Timelining is particularly appropriate with sensitive and complex topics or when interviewees' oral language expression is limited due to a variety of circumstances. This chapter begins with definitions and descriptions of timeline methods used in qualitative inquiry. Potential advantages and limitations are discussed and several specific applications illustrate how different methods have been utilized. A five-phase model of the timelining process is proposed: Introduction, Methodological Decisions, Application, Immediate Reflection, and Analysis/Reporting. Finally, future directions and ideas related to timelining methods in qualitative research are suggested.

> **Keywords**
> Complementary research methods · Creative research methods · Qualitative research methods · Timeline drawings · Timelines · Timelining methods

1 Introduction

Interviews are the main means of data collection in qualitative research. Whether structured, semistructured, or open format, interview data form the basis. However, several authors have called for more varied and creative approaches to qualitative inquiry in order to better reflect the rich and multidimensional aspects of participants' experiences (Deacon 2000; Ponterrotto 2006; Adriansen 2012; Kolar et al. 2015). Increasingly, researchers are utilizing visual data collection methods such as timelines, life grids, mapping, drawing, poetry, video, photographs, graphs, and writing (Marshall and Guenette 2008; Blanchet-Cohen et al. 2003; Groenwald and Bhana 2015). Most often, researchers use visual methods in a complementary fashion, together with some form of interviewing. Researchers who have used visual methods have identified a number of advantages, including greater data depth and detail, enhanced rapport with participants, and the ability to conduct interviews with those whose expressive language is limited. See also chapters in Section 2 in this volume.

Timelining methods show particular promise for qualitative researchers and are the focus of this chapter. Timelines are visual representations of particular and selected events or "times" in a person's life. Depending on the focus of the research, timelines can span a participant's lifetime, a certain number of years, or even a few months. Timelines can be constructed by a single person (participant or researcher) or collaboratively and at different times in the research process. Analysis can be part of the interview or done later, and done in different ways. Timeline methods can be adapted to suit a broad diversity of research questions, interview formats, participant groups, and settings (see ▶ Chaps. 68, "Semistructured Life History Calendar Method," and ▶ 69, "Calendar and Time Diary Methods" in this volume).

Despite their utility, flexibility, and ease of application, there is surprisingly littl in the scholarly literature about how to use timelining methods (Berends 2011).

In this chapter, I begin with definitions and descriptions of timeline methods used in qualitative inquiry. Potential advantages and limitations are discussed next. Several specific applications of timelining are described in order to illustrate how different methods have been utilized. Based on a review of these and other studies in the literature, a four-phase model of the timelining process is proposed: Introduction and Methodological Decisions, Timelining Application, Immediate Reflection, and Analysis/Reporting. The chapter concludes with some suggestions for future directions related to timelining methods in qualitative research.

2 Definitions and Formats

Although researchers in diverse disciplines have used timelines for many years, there appears to be no universally accepted definition (Berends 2011; Kolar et al. 2015). Scholars and researchers who have employed timeline methods tend to define them in terms of their particular use. Adriansen (2012) describes timelines as collaboratively constructed visual representations of the main events of a person's life. Berends (2011, p. 2) defines a timeline as "visual depiction of a life history, where events are displayed in chronological order." A timeline can be as simple as a straight line with points or dates marked on it or it can be an elaborate depiction using colors and images. Timelines can also include text such as a description of the event, an explanation of the meaning of the event, or even feelings and thoughts associated with the event. With respect to format, timelines can be products and images that have already been prepared (by the researcher or others) as well as products created by participants in the interview, or what Jackson (2012, p.415) terms "respondent generated imagery." The possibilities are virtually endless, depending on researcher stance, research purpose, time available, resources, participant interest, and potential audience.

The most widely used timelining method is the classic *straight line*, with key events marked along it (see Fig. 1). This approach is easily understood and can be accomplished in as little as 10 or 15 min, yet can yield a surprising amount of data for exploration and reflection. Moreover, participants can construct this type of timeline on their own before or after an interview.

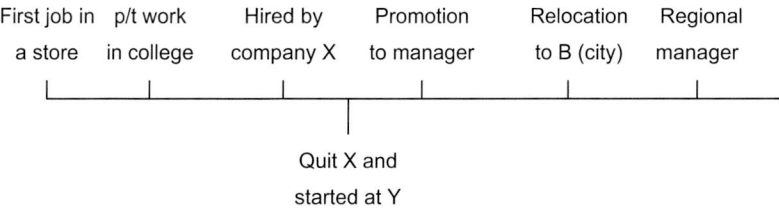

Fig. 1 Example of a simple timeline

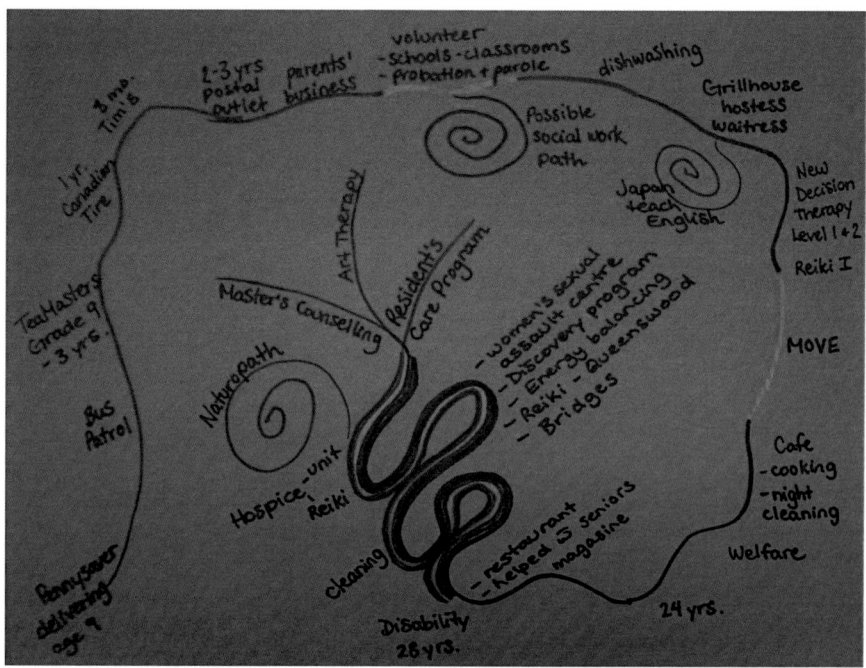

Fig. 2 Meandering timeline

A variation of the horizontal timeline is a *meandering line* that includes "peaks" and "valleys" or changes in direction to indicate "choice points", pivotal experiences, or events of particular significance. Text is often added and different colors may be used. Figure 2 illustrates an example of this type of meandering timeline, showing a participant's experiences related to history of surviving abuse. The colors had particular meanings that were described in detail during the interview.

Life grids are a form of timeline that are still linear but have two dimensions. Life grids have been used frequently in life course research (Nelson 2010; see also ▶ Chap. 68, "Semistructured Life History Calendar Method" in this volume). In her study of Indigenous helpers, Stewart (2007) used a life grid variation she called a *story map* to complement her in-depth interviews. The story maps had time on the vertical axis and literature-derived data themes on the horizontal axis. A summarized example is shown in Fig. 3.

There are also nonlinear or *drawing* type timelining methods. Jackson (2012, p. 415) describes this type of *participatory diagramming* as "drawings, sketches or outlines demonstrating or explaining a social phenomenon." Saarelainen (2015) created a life tree drawing process (described below) that was particularly suited to an exploration of life stories among young adults during cancer remission

	Participant's world				
	Self	Indigenous culture	Community	Mental health & healing	Counseling practice
Past experiences	Background self-identity roles	Roots Personal history Events	Setting the context Past connections	Incidents Sites	Past work experiences
Present experiences	Current status Level of awareness	Current support	Current connections	Community experiences	Current work experiences
Future intentions	Outcomes Personal development Self-identity	Future support	Future connections	Plans for future schooling	Future work expectations

Fig. 3 Story map template for data analysis

3 Advantages of Using Timelining Methods

Researchers who use timelining have identified a number of benefits associated with their use. These include richer and more complete data, enhanced rapport, greater participant engagement, a more collaborative interview process, appropriateness for sensitive topics, and mitigation of language limitations. Several of these factors are related; for example, better quality data is more likely when participants feel at ease and are actively engaged in the interview process.

3.1 Richer and More Complete Data

Qualitative research has been criticized for being too verbally focused; thus, including a visual element such as a timeline can make findings more accurate and complete (Berends 2011; Jackson 2012). Having more than one source of data assists in obtaining a more complete sense of participants' experiences. Multiple data sources (writing, drawing, and speaking) represent different aspects of the same stories that can be analyzed side by side, thereby creating triangulation for interpretive confirmation (Saarelainen 2015). Berends (2011) suggests that including a timeline procedure may uncover novel or expected aspects of the phenomenon being studied and prompt researchers to reconsider theme categories, while also providing a method of triangulating across data sources. Researchers who use timelines consistently attest to the richer and more in-depth data that result (Nelson 2010; Sheridan et al. 2011; Adriansen 2012; Groenwald and Bhana 2015). Laying the story out visually can draw attention to patterns in the narrative, as well as to contradictions and inconsistencies (Patterson et al. 2012). This can then lead to further explication and participant reflection on meaningful events and impacts. Moreover, shifting the focus in an interview to something novel or different can stimulate a new participant perspective or enhanced description of experiences (Guenette and Marshall 2009; Kolar et al. 2015).

3.2 Establishing Rapport

Many researchers report enhanced rapport with participants when using timelines (Jackson 2012; Kolar et al. 2015). Qualitative interviews often involve a chronological orientation or story-like aspects. Participants are familiar with describing events over time and usually feel comfortable with having a visual depiction of events included as part of the research process (Patterson et al. 2012). Rapport is particularly enhanced when the participant creates the timeline or works together with the researcher on it (Kolar et al. 2015). Timelines and other engaging "side-by-side" methods have been found to be particularly helpful with marginalized participant populations where there may be some mistrust or hesitancy (Jackson 2012) and with sensitive topics (Nelson 2010; Berends 2011; Saarelainen 2015; Rimkeviciene et al. 2016).

3.3 Participant Engagement

Related to rapport, timelining methods have been observed to stimulate a greater level of participant engagement that yields richer and more complete data (Guenette and Marshall 2009; Berends 2011). Co-constructive and collaborative narrative type designs are engaging because the participant and researcher interact as the timeline and the interview unfold (Marshall 2009; Patterson et al. 2012). Creative drawing methods such as Saarelainen's (2015) *life tree drawings* are particularly engaging because participants bring their own preferences and meanings to the timelining process.

3.4 Flexibility

Timelining methods are flexible and can be adapted to a multitude of diverse participants, purposes, settings, and research designs (Nelson 2010; Berends 2011). They are particularly useful for hard-to-reach and marginalized populations that often have been excluded from research (Jackson 2012). Timelines have been successfully applied in focus groups, child and youth populations, and cross-cultural settings. They are also amenable to structured and unstructured data analysis approaches (Marshall 2009).

3.5 Language

With visual data collection methods, there is less emphasis on verbal language skills, which is of benefit when working with younger participants or with those whose language or cultural backgrounds are different to those of the researchers (Jackson 2012; Nelson 2010). Even if participants are verbally skilled, nonverbal techniques can capture emotions and experiences in greater depth and detail than verbal descriptions alone. Visual methods such as timelining can be particularly helpful when qualitative interviews are not complete or fully coherent due to factors such as language difficulties, traumatic experiences, anxiety, or impaired memory (Patterson et al. 2012; Rimkeviciene et al. 2016). With timelines or drawings, "stories can be evaluated as a whole, even if a coherent meaning...has not presented itself" (Saarelainen 2015, p.73).

3.6 Power Differences

Using nonverbal data collection techniques such as timelines can help minimize the expert role of the interviewer and decrease perceived power differences between interviewer and interviewee (Patterson et al. 2012). Visual methods often involve a high degree of participant input and thus can facilitate a more collaborative or coconstructed experience. Adriansen (2012, p. 49) maintains that timeline interviews bring interviewees "closer in the landscape of power"; participants have more power because there are no set interview questions and they are able to select important events themselves.

3.7 Professional Settings

In educational settings, timelines have been used for personal and career-related exploration with individuals and groups of students (Guenette and Marshall 2009; Marshall 2009). For clinical investigators such as nurses, counselors, or psychologists who are affiliated with treatment settings and training programs, timelines can complement in-depth interviews with patients and clients. They can also be used in clinical supervisions and to assess program effectiveness. In alcohol treatment, Bryant-Jeffries (2001) observed that the use of timelines (or lifelines) could help bring client realities into the therapeutic setting and highlight the significance of events.

4 Challenges of Using Timelines

Some of the advantages of timelines also pose challenges – the very depth and detail valued by researchers can pose a threat to anonymity, for example. Other challenges include the very personal nature of timelining, the implication of linearity, data analysis, and reporting or publishing decisions.

4.1 Personal Content

Adriansen (2012) cautions that ethical issues can arise due to the personal nature of timelining. Participants may be sharing experiences and information that they have not divulged previously or be describing sensitive issues. Researchers have to be alert and willing to stop or divert the interview if the participant appears to be distressed. Timelining is not always successful with or appropriate for everyone – not all participants are comfortable with sharing personal information or actively collaborating with the researcher in the interview process. It is important that the timelining process and the potential impacts be clearly described in recruitment materials and during the informed consent phase and that potential participants are able to ask questions about the procedures.

4.2 Confidentiality

Timelines contain specific and sometimes detailed information about people and events. While this is desirable for researcher understanding, it does raise issues about confidentiality. For reporting and publication purposes, the names of people and locations can be disguised or changed to a more descriptive term; for example, "relative" or "boss" could be used in place of a person's name and "city" or "workplace" instead of location or company names. Even with these precautions, however, a participant might be identifiable, depending on the events depicted and the amount of detail. Adriansen (2012) contends that participants' timelines should not be published in papers; the information from them should be integrated into the data analysis. She also suggests that timelines are not appropriate for interviewing insiders.

4.3 Assumptions of Linearity

The visual timeline may appear to oversimplify or assume a linear sequence to participant experiences (Berends 2011). This can be particularly problematic if the timeline is completed before an interview or by the participant on his or her own – the tendency is to follow instructions (that are also linear step-by-step) and proceed in a chronological fashion that may emphasize time accuracy more than experiences or the significance of events. Overemphasis on linearity is easier to avoid when the researcher is present and able to probe for meaning or when the timelining procedure is embedded in the interview.

4.4 Analysis

Although there has not been much written in the literature about how to do timelines, there is even less written about how to analyze them. Kolar et al. (2015) point out that timelines can complicate data analysis because they are more visual than verbal; it can be difficult to convey the meaning or significance of timeline events in text. The type of analysis needed partially determines the way the interview is conducted. Berends (2011) cautions against over-reliance on timelines because the process can simplify a life into a linear sequence of events that can cause loss of depth; she recommends combining timelines with interviews.

4.5 Reporting

Challenges in reporting and publication are often related to confidentiality concerns. Program evaluation reports, professional practice investigations, and scholarly journals all have particular audiences whose members may be familiar with the program, treatment intervention, or research site in which timelines were gathered. The depiction of timeline examples or the description of specific events may provide enough information for participants to be identified. On the other hand, anonymyzing these types of data can rob them of their meaning and obscure thematic relationships. Berends (2011) suggests that depicting a composite representative timeline instead of individual timelines has the advantage of highlighting typical underlying factors and relationships between events without threatening confidentiality.

The specific studies described in the next section provide examples of how researchers and scholars have used timelining procedures in particular qualitative research contexts.

5 Review of Timeline Research

Adriansen (2012) writes about using timeline interviews as part of life history research. She maintains that life story research helps to more fully understand a person's perspective of their own life story as well as the patterns of their story in context. In her doctoral research in Africa, Adriansen found that that using timelines yielded the most interesting and in-depth life story data. However, she was surprised to discover that little had been written about this method of data collection, particularly how to actually do it.

In her paper, Adriansen describes conducting timeline interviews with research participants that typically take about 2 h. She provides a detailed description of the steps she uses in the timeline process. In her approach, the researcher often does the timelining in response to the participant's story; however, interviewees can participate in the drawing and/or the event writing. Usually, a timeline is drawn in the center of a large piece of paper and important events are marked on it with different-colored pens. Different lines can represent different perspectives. For example, there can be a line in the middle depicting the core or main story, with other lines branching off or drawn separately near the edges of the paper. Sometimes the paper is divided into two or more sections, representing different contexts such as family, work, neighborhood or local, national, political arenas. The researcher or interviewee can also add contextual information and explanations. As the interview unfolds, the timeline can change. Adriansen claims that using a timeline actually makes a story less linear because having the visual anchor allows the participant to jump around to different time points and construct multiple stories without losing key elements.

Kolar et al. (2015) maintain that little research has been done on using visual methods and that much that is available focuses on content rather than on form. Few studies include supporting examples or discussion of how the participatory space in created. Kolar and colleagues describe two of their studies using timelines to investigate resilience among marginalized groups. One study included male and female street-involved youth aged 18–26 and the other involved young South Asian women; all participants had experienced violence in their lives. The approach to timelining was flexible, depending on participant comfort and preference. Sometimes the timeline was created first, followed by questions referring back to the timeline to elicit contextual information; other interviews were more interactive, involving the simultaneous creation of the timeline along with the interview questions and probes about important events as they are shared. Kolar et al. observed that timelining helped to bring participants' life stories to the forefront and added contextual richness. Moreover, they found that the researchers were able to shift the topic away from sensitive or emotional material if needed and also to emphasize a focus on strengths and resilience. The timelines were coded for both content and form. Kolar et al. distinguish between *List-like timelines* (chronological and text-heavy lines with notes) and *Continuous-line timelines* (horizontal lines that go up and down). The authors suggested that timelining is a particularly useful method when there is only one chance to engage with participants due to confidentiality issues, the transient nature of a population, or financial constraints.

Drawing primarily from a study involving 112 mostly male participants in their 20s and 30s who were substance users, Berends (2011) describes the benefits and challenges of using timelines together with in-depth interviews. She maintains that timelines are useful for aiding memory, making comparisons to other data sources, allowing participants to construct their own reality, identifying common themes across participants, and situating a phenomenon in context. Focusing on exploring pathways to alcohol treatment, Berends used a timeline tool, a sheet of paper with a horizontal axis where drug use was plotted above the line and treatment experiences below the line. Timeline data were compared to interview data for consistency and recurrent themes (such as age when drug use started, types of drugs used, and treatment settings) were identified. With regard to analysis and interpretation, Berends points out that timelines and interviews are not necessarily records of fact but a way of constructing meaning to bridge memory with reality. To preserve participant anonymity, she constructed composite timelines that reflected more than one participant's experience. In that process, the voice of individual participants is then lost; however, it can be reintroduced by using carefully chosen illustrative quotes in the report or publication. Berends recommends, "generalities drawn about participant experiences should be reflected in individual journeys" (p. 8); thus, common themes across participants should be complemented by quotes and examples from individuals in reports and papers for publication.

Jackson's 2012 study focused on identity development with multiracial participants in a clinical social work context. Jackson contends that the aim of timelining in clinical practice is to help client reinterpret the past and their self-perceptions in context, while remaining sensitive to cultural factors. In this study, participants spent 20–25 min completing timelines with minimal instructions and only colored markers and blank paper as materials. Then they participated in an audiotaped interview to discuss the experience. The timelines and interviews were analyzed for themes based on what the literature suggested about the research topic. Jackson identified three levels or *sites* related to data collection and analysis. The *production site* of the timeline, or the research context, involved reassuring participants about the process, and giving them control by having them select a key event and starting point. The *image site* was analyzed based on the organization, content, and colors of the timeline. The *audience site* referred to the researcher's position in the research. The researcher created some distance or safety by leaving while timelines were created and by providing minimal instruction; however, the researcher's position as a knowledgeable insider resulted in personal connection to the participants' stories. An insider position, however, poses some challenge for analysis and interpretation. Jackson observed, "as an *insider*, I must be conscious of and reliant on the meaning participants attach to their visual timelines, instead of my own interpretations" (p. 426). Visual methods are empowering and participant-centered; they elicit a more complete illustration of a phenomenon because they are able to evoke participants' emotional experiences that do not always emerge in interviews. There are also some cautions: the linearity of timelining can be constraining, some participants may be uncomfortable drawing, and the method has not been well described or researched.

In their study with substance using adolescents and their parents, Groenwald and Bhana (2015) used the *lifegrid* timelining method. These authors describe the *lifegrid* (or LG) as a matrix of rows and columns that represent years and events, respectively. Relevant events are selected for inclusion in the lifegrid based on the interview topic, thus, that the finished product shows how and when the participant's life changed over a particular period of time. The LG process also included a "formative phase" intended to put participants at ease and to identify key areas of focus. Groenwad and Bhana conducted two interviews with each participant; the first involved filling in the lifegrid and the second was a review of the lifegrid, using it to help the participant recall events and chronology in more detail. Parent and children were interviewed concurrently to cross-reference for accuracy. The researchers found that most participants wanted the researcher to fill out the LG, which had the unintended benefits of facilitating the interview flow and minimizing concerns about whether they were doing it correctly. Written chronologically, the LG permits flexibility and allows participants to jump back and forth when talking about their experiences. Researchers make notes of the event dates for coding accuracy.

Groenwald and Bhana claim that the advantages of using lifegrids in qualitative research include participant engagement, support for sharing experiences, and detailed descriptions of retrospective experiences. Accuracy and reliability of memory is not as important as the meaning participants ascribe to events.

Life story mapping has some similarity to the life grid method. Stewart (2007) explored Indigenous helpers' stories of mental health and healing using story maps (see Fig. 3 above for a summarized map) to complement in-depth narrative interviews. The story maps served to organize participants' recounting of past and present experiences and future intentions as they related to the focus of the inquiry – how professional counselors developed their approaches to Indigenous mental health counseling and healing. Along the horizontal axis were five structural categories chosen a priori, based on literature reviews and the research question: self, Indigenous culture, community, mental health & healing, and counseling practice. On the vertical axis were three time dimensions: past, present, and future, yielding 15 cells in total. Working with transcribed first interviews in which these elements had been explored in depth, Stewart coded the data using the categories and dimensions above, then constructed a map by inserting the specific participant data content codes into the different cells of the map. The map thus created a visual depiction of each participant's specific and particular story as it related to the content categories and time dimensions. In addition, the map resulted in a more penetrating analysis because it allowed the researcher to use a format that revealed patterns within the story. In a second interview, the map and emerging patterns enabled both the participant and the researcher to extend, amend, reconstruct and make sense of the story in a deeper way than simply rereading the story and identifying a list of content themes. All participants observed that the story maps helped them to *see* their story in a more holistic manner over time and facilitated their identification of missing or incomplete elements.

Nelson (2010) adapted the typically quantitative and structured Life History Calendar (LHC) method into a semistructured qualitative interview. The LHC is a printed matrix with temporal cues (days, months, years) presented horizontally and domain cues (categories related to marriage, living arrangements, educational histories, work histories) presented vertically. To counter this limiting structure, Nelson's 2010 adaptation started with a blank page that participants filled in with colored markers and stickers, beginning wherever they liked. This open format helped with rapport, depth of sharing, and identification of context. Participants were Mexican or Mexican-American college students aged 19–22 years who had taken part in an extracurricular program at the beginning of high school. Most were young women from working class families. In a 2-h interview, Nelson explored the impact of their earlier extracurricular activities and how these influenced their educational paths. She observed that giving the participants choices encouraged feelings of ownership and allowed for the delay of emotionally painful stories until rapport was established. See also ▶ Chap. 68, "Semistructured Life History Calendar Method" in this volume.

Saarelainen's (2015) creative and nonlinear approach to timelining utilized *life tree drawings* with young women and men aged 18–34 years old who were interviewed during the time of cancer remission. The interviews focused on "how cancer impacts the patient from the perspective of one's life story" (p.68). Using a method called *visual narrative inquiry*, the participant and researcher coconstructed meaning from the participant's experiences, which were described verbally and also depicted visually with a drawing of a tree. The idea of a tree as a metaphor for life is familiar to many people; the task was thus accessible and easily understood by participants. They were given a blank sheet of paper and a black marker, and then simply asked to think about their life as a tree and to draw what that looked like. Other colored markers were placed on the table. Participants were asked to explain the meaning of the signs and symbols they used and how the components of the tree were related. Saarelainen noted that color was used symbolically: green and pink denoted hope and healing; black represented illness and negative events. There were three life tree types drawn by participants: *neutral trees* used normal tree colors, *multiple-element trees* used colors that symbolized emotional meaning, and *thematic trees* included words that referred to periods of life or specific events. Themes were either written by participants on trees themselves or assigned later by the researcher. Data analyzed included the life tree drawings, interview recordings and transcripts, and blog posts written by the participants. Saarelainen used a critical approach for visual analysis of the drawings, focusing on the image production, the image itself, and the audience. With this particular population of cancer survivors, the life tree drawing was considered to have an advantage over liner-type timelines because it did not have an ending point that could be a reminder of death for participants. For future research, Saarelainen recommended making the life-tree drawing process more co-constructive, allowing participants a more active role and chance to give feedback, and getting participants together for group discussion of their experiences.

In their study of weight loss, Sheridan et al. (2011) created a *timelining over time* method. Their nine participants had once been obese but had lost between 23 and 62 kg (27–44% of weight) and maintained the weight loss for 5 years or more. The study was termed a *graphic elicitation,* defined as a visual method based in drawing or art that create diagrams or drawing for the specific purpose of the research. Participants were interviewed four times over a period of 2–4 weeks; they chose the time frame for the timeline graph, focusing on times when their weight was of particular concern. A graph was constructed, with time depicted horizontally and weight vertically. In this study, the researchers drew the graph, but participants decided what to include and in what order to plot events they had chosen. Describing and drawing the graph provided a preliminary framework that facilitated the sharing of the story. Although the original intention was to do the graph first and then discuss it, researchers found that the two parts melded together and that completion of the graph involved a great deal of storytelling. Participants kept the graph between sessions and were instructed to add to it or change it based on their own reflection or input from significant others. The interviews and timelines were supplemented with objects brought in by participants, such as photos, clothing, journals, and medical documentation; these helped to add details.

The graphing process in Sheridan et al.'s (2011) study was described as largely nonlinear, but a central storyline appeared to hold the different threads of the narrative together. The method provided insight into the different ways participants experienced time (historical, circular, cyclical, spiral, personal, and future). The authors observed that "time is not simply a 'series of "nows," instant that exist along a timeline' but an abstract (re)presentation of time...*doing timelining* plays with and manipulates this linear (pro)portioning of time." (p.560). Sheridan and colleagues thought that the visual elements in timelining helped participants to focus on the topic and comprehend the scope of the project, as well as elicit layers of experience that might not otherwise have emerged.

Patterson et al. (2012) interviewed 31 male and female participants aged between 26 and 66 years of age about their experiences of homelessness and mental illness. The authors maintain that most homelessness studies emphasize individual factors and daily experience rather than structural inequality and social context. Their study involved narrative interviews and timeline construction. Content analysis of the 1–2-h interviews yielded three main themes: Longstanding Social Devaluation, Feeling Trapped, and Lac of Autonomy. Next, timelines were constructed by the researchers to illustrate the personal stories shared by participants regarding access to social determinants of health and resulting impacts. The timelines were analyzed for timing and order of events as well as context and the researchers constructed an aggregated timeline mapping out key themes and common patterns that provided a generalized view of participants' experiences. Patterson and her colleagues found that constructing timelines in conjunction with narrative interviews facilitated rich data and allowed them to examine trajectories of events and experiences.

These researchers asserted that the timeline method helped them move toward an insider perspective, helped organize a wealth of data, and revealed aspects that could not be accurately described in words. The limitations noted included some underreporting of stigmatized behaviors, power imbalances, the retrospective creation of timelines without participant involvement, a false implication of linearity, and oversimplification in the aggregated timeline. Despite these limitations, Patterson and colleagues claim that timelines anchor life experiences in the context of developmental periods and can effectively map stories of cumulative adversity and resilience; the research issue or question is placed in the context of a broader life story.

Research by Rimkeviciene et al. (2016) involved participants that could be termed at-risk or marginalized – people who had attempted suicide. Two studies were conducted: the first was with eight male and female participants between the ages of 20 and 50 who had attempted suicide at least twice and the second study involved 49 participants aged 18–35 who were in hospital after recent suicide attempts. The research focus was on understanding suicide as a process and clarifying the role of impulsivity in the decision to attempt. The aim was to obtain rich qualitative description of thoughts, feelings, interpretations, and reactions to significant events contributing to the choice to attempt suicide. Timelines were drawn on blank paper, starting and ending wherever was relevant for the participant. The researchers found that usually the most important events were described first; more detail emerged as the interview proceeded.

Similar to other studies involving sensitive topics, Rimkeviciene and colleagues observed that giving the participants greater control over the interview direction allowed them to shift away from difficult topics when needed. Some participants struggled to recall accurately and this brought up emotions such as annoyance or fear of disappointing the researcher – it was suggested that the interviewer should normalize these feelings, so participants do not feel pressured to make things up. Final narratives generally capture that participant's *narrative truth*; facts can sometimes be triangulated with other sources. The timelining process captures timing of significant events, thought processes, decision-making, and emotional fluctuations. The participants benefitted by gaining a greater understanding about the interrelatedness of events, actions, and emotions. The authors also note that when talking about sensitive topics in research interviews, it is important to prioritize the participants' well-being while avoiding straying into a therapeutic role. Strategies used to protect participant well-being included ending with a nonemotional task, leaving space for processing the experience, and follow-up support from referring agencies.

6 Four-Phase Model for Using Timelines

Based on a review of the above and other studies in the literature, a four-phase model of the timelining process is proposed: Introduction and Process Elements, Timeline Application, Immediate Reflection, and Analysis/Reporting.

Introduction and Process Elements, similar to what Groenwald and Bhana (2015) call the "formative stage", is a key phase that sets out expectations for the timelining and interview process as well as the roles for participant and researcher. The process will be new for most participants, so it is particularly important that there be enough time for detailed explanations and time to answer questions. It is usually preferable that participants have as much information as possible in advance so that the informed consent at the beginning of the interview is not inordinately long (Marshall 2009). Rapport begins in the Introduction phase; this is a key element for participant commitment and engagement in the research. Depending on the research design, some *Process Elements* are also part of the Introduction: whether timeline and interview are sequential or occur together, who creates the timeline, how much description or narrative takes place, and (importantly) whether these elements are decided beforehand or whether the process can evolve more organically. Experienced researchers tend to offer more flexibility, guided by participant preferences and judgment about how well the interview is proceeding. In contrast, some research designs and timelining methods work best with a set structure that is clearly laid out and followed.

The *Timelining Application* is the second phase, where the chosen interview and timeline method takes place. Because this process can be long and taking place over two or more sessions, reminders and explanations about what is coming next can be helpful – Munhall (1988) terms this "process consent." If there are changes in sequencing or format of the process, they should be clearly acknowledged and consented to by the participant. Variations should also be mentioned in reports and publications describing the research.

The *Immediate Reflection* phase is a familiar one for qualitative researchers; field notes and reflection sessions are widely used practices. Timelining is more likely novel for the participants and may be for the researcher as well; observations, thoughts, and constructive suggestions can assist in improving its effectiveness and lead to further methodological innovations and implementations. Participants' experiences will probably be diverse and may be relevant to data analysis and interpretation. As noted by many researchers, there is very little written about analysis and interpretation of timelining data; systematic reflections on form and content will further refine the process (Patterson et al. 2012; Kolar et al. 2015).

Analysis/Reporting is the final phase, informed by the preceding ones. Decisions regarding the timelining process and any changes that occurred will have an effect on analysis of the data. For example, if the participant created and drew the timeline by him or herself and wrote in explanations, this type of participant-generated data would be analyzed differently than a timeline or life grid that was filled in by the researcher. A timeline completed beforehand and debriefed in an in-depth interview would be analyzed differently than one co-created with a researcher during an interview session. Participant and researcher reflections from Phase 3 would be important information for some of the analysis and interpretation decisions in Phase 4.

The final report or paper/chapter for publication should outline the steps and procedures in the timelining process. In the qualitative research tradition, writing would ideally include *in vivo* observations as well as participant quotes to illustrate data themes and categories in order that the reader can understand the process followed. Representations of an individual or composite timeline are particularly helpful in depicting the process, although considerations of anonymity as described above by Berends (2011) are important to address, perhaps together with participants.

7 Conclusion and Future Directions

As suggested by several authors, more research and more specific descriptions of timeline processes, adaptions, and applications are needed. Graduate students can be encouraged to include visual methods such as timelines in their thesis research. The publication and dissemination of research studies and practical "how-to" tips will contribute to the further development of timelining and other visual methods and contribute to new ones. It is hoped that the application of the four-phase process model will generate more thorough information about timelining methods and stimulate further refinement of the model.

Timelining methods are adaptable to a diverse array of theoretical frameworks, epistemological approaches, and participant contexts. They show great promise for extending and enriching qualitative interviewing.

References

Adriansen HK. Timeline interviews: a tool for conducting life history research. Qual Stud. 2012;3(1):40–55.

Berends L. Embracing the visual: using timelines with in-depth interviews on substance use and treatment. Qual Rep. 2011;16(1):1–9.

Blanchet-Cohen N, Ragan D, & Amsden J. (2003). Children becoming social actors: using visual maps to understand children's views of environmental change. Child Youth Environ. 2003;13(2). Retrieved from http://colorado.edu/journals/cye.

Bryant-Jeffries R. Counseling the person behind the alcohol problem. Philadelphia: Jessica Kingsley Publishers; 2001.

Deacon, S. (2000). Creativity within qualitative research on families: new ideas for old methods. Qual Rep. 2000;4(3&4). Retrieved July 2, 2007 from http://www.nova.edu.ssss/QR/QR4-3/deacon.html.

Groenwald C, Bhana A. Using the lifegrid in qualitative interviews with parents and substance abusing adolescents. Forum: Qual Soc Res. 2015;16(3):24.

Guenette F, Marshall A. Time line drawings: enhancing participant voice in narrative interviews on sensitive topics. Int J Qual Methodol. 2009;8(1):85–92.

Jackson KF. Participatory diagramming in social work research: utilizing visual timelines to interpret the complexities of the lived multiracial experience. Qual Soc Work. 2012;12(4):414–32.

Kolar K, Ahmad F, Chan L, Erickson PG. Timeline mapping in qualitative interviews: a study of resilience with marginalized groups. Int J Qual Methods. 2015;14(3):13–32.

Marshall EA, Guenette F. Possible selves mapping process: a culturally relevant process for life-career counselling and decision-making. DVD and workbook available from the authors. Victoria: University of Victoria; 2008.

Marshall A. Mapping approaches to phenomenological and narrative data analysis. Encyclopaideia. J Phenomenol Edu. 2009;25(XIII):9–24.

Munhall PL. Ethical considerations in qualitative research. West J Nurs Res. 1988;10(2):150–62.

Nelson IA. From quantitative to qualitative: adapting the life history calendar method. Field Methods. 2010;22(4):413–28.

Patterson ML, Markey MA, Somers JM. Multiple paths to just ends: using narrative interviews and timelines to explore health equity and homelessness. Int J Qual Methods. 2012;11(2):133–51.

Ponterrotto JG. Qualitative research in counseling psychology: a primer on research paradigms and philosophy of science. J Couns Psychol. 2006;54(2):126–36.

Rimkeviciene J, O'Gorman J, Hagwood J, De Leo D. Timelines for difficult times: use of visual timelines in interviewing suicide attempters. Qual Res Psychol. 2016;13(3):231–45.

Saarelainen S-MK. Life tree drawings as a methodological approach in young adult' life stories during cancer remission. Narrat Works: Issues Investig Interv. 2015;5(1):68–91.

Sheridan J, Chamberlain K, Dupuis A. Timelining: visualizing experience. Qual Res. 2011;11(5):552–69.

Stewart, SL. (2007). Indigenous mental health: Canadian native counsellors' narratives. Unpublished doctoral dissertation, University of Victoria, Victoria, BC.

Semistructured Life History Calendar Method

68

Ingrid A. Nelson

Contents

1 Introduction	1202
2 Life Course Theory	1203
3 The Life History Calendar	1204
4 The Evolution of the Life History Calendar	1205
5 Adapting the Life History Calendar for Qualitative Studies	1207
6 Implementation	1210
6.1 Time Cues	1210
6.2 Domain Cues	1211
6.3 Starting Point	1211
6.4 Materials	1212
6.5 Data Entry and Analysis	1215
7 Conclusion and Future Directions	1215
References	1217

Abstract

The Life History Calendar (LHC) methodology was pioneered for large-scale quantitative life course studies but has since been adapted for various research goals across the health-related fields and social science disciplines. This chapter explores the potential of a semistructured Life History Calendar administered in tandem with open-ended interviewing to facilitate qualitative life course research. By merging the depth across multiple social contexts characteristic of the Life History Calendar method with explanatory data gleaned from interviews, this semistructured protocol succeeds at producing nuanced longitudinal data. In this chapter, I highlight the benefits and limitations of the semistructured Life History

---

I. A. Nelson (✉)
Sociology and Anthropology Department, Bowdoin College, Brunswick, ME, USA
e-mail: inelson@bowdoin.edu

© Springer Nature Singapore Pte Ltd. 2019
P. Liamputtong (ed.), *Handbook of Research Methods in Health Social Sciences*,
https://doi.org/10.1007/978-981-10-5251-4_22

Calendar method, address common implementation issues, and offer examples of how the semistructured Life History Calendar method has been used to study trajectories of education, employment, health, and other topics, among diverse populations.

Keywords

Ecological systems theory · Event history analysis · Life course theory · Life History Calendar method · Qualitative social research · Social pathways

1 Introduction

In response to many of the challenges that arise from studying individuals as they age and progress through various stages of development – from childhood into adolescence, adulthood, and beyond – the interdisciplinary life course framework emerged in the late 1990s and has spread across many social science and health-related fields (Elder et al. 2015). The concept of the life course refers to a series of age-graded events and roles that shape the individual's biography and are, in turn, shaped by both micro- and macrolevel social contexts. Although longitudinal panel studies comprise the gold standard in life course research, logistical matters often limit the viability of large-scale data collection over time. Thus, researchers have worked to develop valid and reliable methods of collecting retrospective life course data that capture the ways that biography, history, and social contexts interact to influence development over time.

The Life History Calendar (LHC) has emerged as one key method for gathering reliable retrospective event timing and sequence data for quantitative life course studies. In this method, researchers use a preprinted matrix, with time cues running horizontally across the page and topic cues running vertically down the page, to help respondents piece together their past. The grid aids event recall because it matches the structure of autobiographical memory and improves data quality because researchers can easily pinpoint omissions and contradictions (see also ▶ "Calendar and Time Diary Methods"). Its highly structured format, however, does not permit researchers to capture explanatory data on why or how the respondent's life story has unfolded as it has. Because of the need for detailed contextual data in qualitative life course research, I developed a semistructured adaptation of the LHC.

The semistructured LHC discussed in this chapter builds on Martyn and Belli's (2002, pp. 271–272) assertion that life history calendars "could be used in qualitative research to stimulate discussion about past experiences and underlying processes that help explain behavior, attitudes, and emotions." As such, this methodological innovation incorporates research participants' attitudes and aspirations, interpretations, and explanations of life transitions and major events that would not be captured in standardized event histories. By merging the characteristic detail across multiple domains of the traditional LHC with in-depth narrative, the semistructured LHC method achieves nuanced longitudinal narratives. In this chapter, I review the history and evolution of the LHC, as well as practical advice for implementation of a semistructured LHC protocol for qualitative research in the social sciences.

2 Life Course Theory

The tradition of life course research highlights the interplay between cultural background, social ties, human agency, and timing (Giele and Elder 1998). These factors intertwine such that individuals' actions and attitudes depend not only on microlevel situations (e.g., parent-child interactions) and macrolevel structures (e.g., federal policy) but also on historical context and individuals' constructive activity (Bronfenbrenner 1979; Mortimer and Shanahan 2003). Thus, life course theory serves to unite ecological systems theory (Bronfenbrenner 1979; Bronfenbrenner and Morris 2006) with a temporal approach (Elder 1974). In contrast to psychological approaches that tend to parse out distinct facets of the developing individual, such as emotion, cognition, or motivation, life course theory views each person as a dynamic whole playing an active role in his or her own development (Elder et al. 2015). Although we experience personal agency, our choices are always constrained by the pathways made available to us by historical and social circumstances. For example, the career choices available to women in New York City during the 1950s were markedly different than the career options available to women in rural Oregon during the same time period, and both sets of choices are different from the options available to American women in the 1990s. Social pathways, a key concept in life course theory, are defined as age-graded sequences of social positions that reflect the macro- and microlevel contexts of development.

Developmental psychologists have often used Bronfenbrenner's ecological systems theory (1979) and expanded bioecological theory (Bronfenbrenner and Morris 2006) as a framework for examining human development. Ecological approaches to development argue that context matters. Early iterations of the theory positioned the individual at the center of four contextual layers – much like the smallest piece in a set of Russian nesting dolls – through which each plays a key role in development. The first layer – the microsystem – refers to face-to-face settings where a person spends time, such that each setting (e.g., home, classroom, and workplace) comprises a distinct microsystem. The next layers contain connections between microsystems – with mesosystems comprised of linkages between settings that include the individual in question, and exosystems comprised of linkages between microsystems where one does not include the individual. For example, the interaction between a parent's work schedule and their child's ability to participate in an after school club is part of an exosystem for the child as one of these settings – the parent's work – does not include the child. Meanwhile, the interaction between the child's school and the after school club are of part of the mesosystem, as both settings include the child. These three layers of context – microsystems, mesosystems, and exosystems – are all nested within a macrosystem, which is comprised of societal beliefs and values.

Recent iterations of ecological theory recognize more than just context and argue that development is the product of multiple layers of context interacting with person, process, and time (Bronfenbrenner and Morris 2006). Person refers to individual-level characteristics, including social identities such as age and race, and developmental traits like temperament and personality. Process refers to individuals' experiences within a microsystem or the ways that person and context transact. Finally, time refers to minute-by-minute exposure to processes within microsystems, duration of participation in a given microsystem, and historical changes in society across generations.

Life course theory brings together Bronfenbrenner's conception of multilevel social context with sociological understandings of the life course as a pattern of age-graded events and roles. Life course theory positions human development as a lifelong process situated in historical time and place and compelled by human agency. This framework highlights the ways that the timing of particular events and transitions drives cumulative effects over time, by underlining the importance of transitions (i.e., exit from one role and entry into another) and turning points (i.e., marked shifts into new behavioral patterns). While focusing on the ways macro- and microlevel factors shape social pathways, and individuals' behaviors and achievements over time shape their own trajectories, life course theory also recognizes the ways that relationships with significant others play a role in development. Specifically, the principle of "linked lives" highlights how social networks exert informal social control and how intergenerational transmissions contribute to the social reproduction of education, occupation, and beliefs. Research on the life course is becoming more prominent across various fields of study (Mortimer and Shanahan 2003), including the health social sciences.

3 The Life History Calendar

The Life History Calendar (LHC) is ideally suited to life course research as it is designed to situate the study of individual agency within multiple social contexts and historical time. The traditional LHC is a preprinted matrix with time cues organized horizontally and domain cues relevant to the study running vertically. Researchers may partition temporal cues into months, years, or any time unit appropriate for the study at hand. Domain cues refer to indicators, such as births, marriages, living arrangements, and educational transitions, and vary according to each study's intention. The LHC may be completed by the interviewer (in person or via phone), or filled out by the respondent (Furstenberg et al. 1987; Freedman et al. 1988; Mortimer and Johnson 1999). The LHC not only spotlights processes of engagement in and disengagement from activities, groups, and behaviors – thus exposing human agency – it also uncovers patterns of behavior over time, thus revealing continuity and change (Laub and Sampson 2003). In addition, the LHC exposes the heterogeneity of behaviors across a population that may lead to a common outcome, helping researchers to unpack complex phenomena (Laub and Sampson 2003). Finally, this approach brings to light the ways that individuals' behaviors are the product of social and historical contexts and the ways that those contexts and individuals' reactions to them change over time (Laub and Sampson 2003).

The LHC has emerged as a reliable method for collecting retrospective data on multiple simultaneous event histories and offers various benefits during the data collection process (Freedman et al. 1988; Belli et al. 2009). Use of the matrix has been shown to ease respondents' event timing recall (Freedman et al. 1988; Caspi et al. 1996), as more memorable events, such as the birth of a child, act as reference points that help subjects accurately place the timing of less memorable events, such as job changes. The visual nature of the matrix also supports event recall (Meltzer 2001), while providing respondents an opportunity to reflect on their own experiences (Feldman and Howie 2009). Furthermore, the LHC provides a simple format for recording detailed sequences of events (Freedman et al. 1988), allowing interviewers and participants to identify gaps and contradictions within each event history. For example, by mapping marriages chronologically, the interviewer can confirm that marriage and divorce dates for a first marriage do not overlap with dates given for subsequent marriages.

In addition, studies have shown that the LHC method elicits more accurate retrospective data than traditional questionnaires (Belli et al. 2001; Van der Vaart 2004; Van der Vaart and Glasner 2007; Yoshihama et al. 2005). Two studies find at least 90% agreement between retrospective reports of activities for a given month on LHCs and concurrent reports obtained 3–5 years earlier (Caspi et al. 1996; Freedman et al. 1988). Belli (1998) posits that LHCs promote accuracy because they map to the structures of autobiographical memory. Both the calendar and the human mind activate recall through multiple pathways: in time order within a life theme and across parallel themes involving simultaneous or sequential events. While survey questions tend to segmentrelated aspects of autobiographical events, LHCs actually draw upon the interrelatedness of events to improve recall and accuracy (Belli 1998). LHCs improve data quality by providing visual and conversational cues that cater to the way the brain retrieves information and improving cognitive and conversational engagement (Belli et al. 2004).

Since its inception, the LHC has been used primarily for large-scale quantitative studies (Axinn et al. 1999). One of the earliest uses of the LHC was in a large-scale study examining race and the timing of education, employment, and family events (Blum et al. 1969). Since then, the LHC methodology has been used to study myriad topics, including many relevant to the health social sciences: adolescent transitions to adulthood (Freedman et al. 1988), psychopathology (Kessler and Wethington 1991), marital transitions (Thornton et al. 1992), organizational dynamics (McPherson et al. 1992), drug dependence (Hser et al. 1997), couples' social relations (Munch et al. 1997), agricultural chemical exposure (Hoppin et al. 1998), domestic violence (Yoshihama et al. 2002), sexual behaviors (Martyn et al. 2006), and welfare transitions (Harris and Parisi 2007).

4 The Evolution of the Life History Calendar

Like any methodology, the Life History Calendar (LHC) demands careful implementation and, often, updates and alterations. One of the first methodological articles on the quantitative LHC examined a sample of 900 young people between ages 15 and 23, born in the same city during the same month (Freedman et al. 1988). The article highlighted implementation considerations, such as: specifying time units, identifying domains, streamlining use of symbols, and designing an easy to use matrix. It also cited logistical complications including the ways the size of paper, tables, and clipboards used for each calendar impacted data collection procedures (Freedman et al. 1988). The common birth month and location, as well as the respondents' young age, however, spared these researchers from some of the logistical difficulties that other scholars would subsequently encounter.

In their study of farmworkers, Hoppin et al. (1998) introduced a number of modifications to the LHC to accommodate older respondents and increased age diversity within their sample. These researchers increased the time unit on the preprinted matrix from months to years, which allowed more room on the page for larger numbers of years of lived experience. In casual interviews prior to beginning the formal study, the authors noticed that respondents used world and personal events to recall farm events. Thus, through focus groups, Hoppin et al. identified world events that were relevant to their sample – such as elections and wars – and preprinted those events on the LHCs used during formal interviews. Having landmark events spaced evenly across time aided recall for older participants, as life events – such as graduations, marriages, and births – tended to cluster together. The inclusion of landmark events has since become common practice in LHC research. Notably, Hoppin et al.'s use of the LHC to study employment histories has been adopted by the field of occupational epidemiology, where this methodology has become known as the Occupational History Calendar. Several studies in this field have corroborated the utility of the Occupational History Calendar in capturing complex work histories among the general public (Lilley et al. 2011) and among special populations such as migrant farmworkers (Zahm et al. 2001).

In their study of rural Nepalese communities, Axinn et al. (1999) revised the standard LHC to accommodate culturally situated notions of time. Their study included multiple temporal cues: the Western year, the year on the local calendar, the animal year, and a blank line for the respondent's age. In addition to including world events identified by focus groups, the authors included local "landmarks" as identified by neighborhoods (Axinn et al. 1997). Further, the matrix and recording symbols were modified to accommodate culturally acceptable behaviors, such as the practice of having multiple spouses simultaneously (Axinn et al. 1999). These innovations opened the door to large-scale event history analyses with populations that do not adhere to Western time-keeping conventions.

Such adaptations of the LHC have improved the methodology and diversified its uses, yet these modification have done little to illuminate the processes of individual decision-making in the context of life transitions (Harris and Parisi 2007). To understand how personal circumstances might play a role individuals' interactions within the welfare system, Harris and Parisi (2007) paired the traditional LHC with a series of open-ended questions and analyzed responses using qualitative data analysis techniques. The authors found that opened-ended questioning exposed nuanced circumstances leading to welfare entry and helped interviewers engage more openly with respondents. In one example, the authors identified two women with similar LHCs, yet open-ended questioning revealed that the women had led decidedly different lives. Thus, Harris and Parisi were able to overcome one of the key limitations of the traditional LHC by adding open-ended interviewing to the standardized protocol.

Respondents' explanations, however, were still bounded by the categories established on the preprinted matrix. Although limiting the categories of response eases data coding, entry, and analysis, and thus may be desirable in quantitative or large-scale studies, qualitative research depends upon the breadth and depth of respondents' narratives being unhindered by researchers' preconceived categories. Therefore, the very nature of the preprinted matrix contradicts one of the primary goals of qualitative research. By developing a life history protocol that does not rely on a preprinted matrix, I bring many of the advantages of the LHC method to qualitative research.

5   Adapting the Life History Calendar for Qualitative Studies

Initial development of the semistructured Life History Calendar (SSLHC) method for qualitative research occurred within a longitudinal study exploring how afterschool programs and extracurricular activities might influence working class and poor Mexican American students' pathways to college (Nelson 2017). Interviews had been conducted while respondents were eighth grade students (approximately age 12), and the Life History Calendar (LHC) method was adapted for use during follow-up interviews with the same respondents during their young adult years (ages 19 to 22). All respondents attended the same urban California middle school and continued to reside within a 1-hour drive of that school at the time of the follow-up interviews. Ten of the twelve respondents were Mexican or Mexican American and nine were female. Approximately 40% of the respondents attended college full-time, while the remainder attended community college or trade school intermittently or part-time. Each young adult interview lasted an average of 2 hours.

I began the project by pilot testing a structured LHC protocol coupled with open-ended questioning, as described by Harris and Parisi (2007), among a population of 18- to 22-year-old full-time college students from working class or poor family backgrounds. Pilot interviews, however, failed to elicit in-depth responses; instead, participants interpreted the structure of the matrix as a signal to limit their responses to event timing information and only shared events that fell within the domains preprinted on the matrix. During semistructured interviews conducted immediately after the LHC was completed, respondents revealed critical life events that were not explicitly part of the matrix. Such events were influential in their educational trajectories, yet the preprinted matrix had not left space for events outside the domains predetermined by the researcher at the onset of the study. For example, one participant spoke about her transition from living with her birth parents to a foster family; although residential moves were accounted for on the LHC, the motivation for the moves was not. In this case, the change of guardianship would have been a critical life event omitted from the respondent's biography because of the structure of the LHC matrix. Another participant spoke about the deportation of a parent by immigration authorities during his elementary school years. Since this change was not accompanied by a residential move, it too was omitted from the structured LHC. Since the goal of my research was to generate in-depth qualitative data that accounted for all of the influences on students' educational pathways over time, I began pilot testing less structured adaptations of the LHC.

After multiple rounds of pilot testing, each with less structure to the matrix than the previous, I determined that the most detailed data emerged from the interviews I had begun with a blank page. The respondent and I subsequently coconstructed time cues horizontally – from birth to present day – and domain cues vertically. Due to the age of my respondents, I determined that school experiences and home life should each constitute one domain and drew a line representing school events and relationships across the top of the page, and a line representing home events and relationships across the bottom of the page. I indicated to the respondent that the middle band of the page could be used for extracurricular activity participation (the topic of interest in this particular study), as well as any other relationships, experiences, or events. This provided each respondent with a designated space to include topics that fell outside the predetermined domains. How respondents filled this space varied dramatically and underscored the importance of moving away from a preprinted matrix. Instead of pencils, interviewers and respondents used colored markers and children's stickers to map out the respondent's life course. See Fig. 1.

This semistructured LHC (SSLHC) method has many of the same benefits as the traditional LHC, including mapping out processes of engagement in and disengagement from activities, groups, and behaviors; exposing patterns of behavior over time; and bringing to light the ways that individuals' behaviors are the product of ever-changing social and historical contexts. The SSLHC also relies on the matrix format

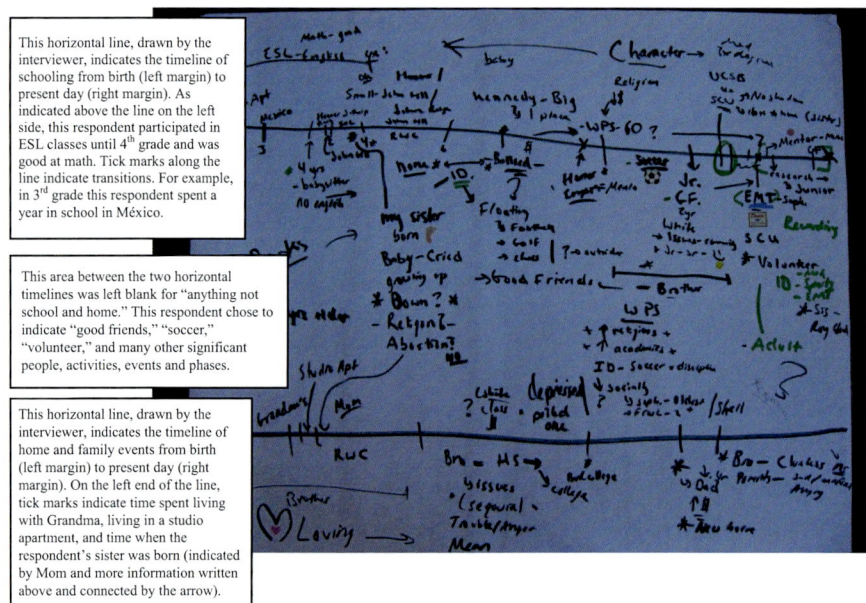

Fig. 1 Sample LHC adapted for qualitative research

to ease respondents' event timing recall and provide a simple format for recording detailed sequences of events. In addition, the SSLHC elicited in-depth longitudinal narratives.

One benefit specific to the SSLHC method stems from the fact that the respondent is empowered to take ownership of the interview by dictating the order in which topics and time periods are discussed and the way in which the matrix is constructed. Rather than taking charge of which questions were asked and in what order, the interviewer's role consisted of outlining the parameters of the interview – including time periods and topics of interest – giving the respondent permission to begin with any topic and time period, then following up with probing questions as the respondent narrated their own life course. The SSLHC scaffolds the interview by acting as a visible reminder of which fields and time periods have been covered and which have yet to be explored, and the domain cues prompt respondents to nest their educational trajectories in context. Because respondents were not confined by narrow lines of questioning, they were empowered to talk about their lives by highlighting the events, people, and places which, in their view, were most significant.

This adaptation of the LHC provides a specific benefit to researchers studying emotion-laden topics. Because the respondent directs the sequence of the interview, the interviewer and the respondent tended to develop a strong rapport that encouraged elaborated narratives. Furthermore, since respondents begin with any time and domain, respondents may delay speaking about emotionally challenging experiences until they have settled into the interview and feel more comfortable with the interviewer. During my pilot study, respondents consistently opted to begin recounting their lives with narratives of times during which they felt successful and delayed recounting difficult periods. For example, one respondent who immigrated from México as a young child and was in eighth grade when her mother died, chose to begin the interview with her elementary school years, followed by her time in high school, and returning to her early childhood and middle school years – her most trying times – near the end of the interview. Since interviewers cannot know which periods of a respondent's life were most difficult prior to embarking on the interview, letting the respondent to dictate his or her own sequence of events, within the structure of the SSLHC matrix, provides researchers with an opportunity to establish rapport early in the interview, thus positioning themselves to capture vivid and valid narratives of emotionally sensitive events. For this and other reasons, Carpenter (2015, p. 75) posits that scholars researching sexualities over the life course may find the semistructured LHC method "particularly helpful."

Another benefit of the SSLHC method is the ability to offer open-ended domain fields, thus capturing events and relationships outside the researcher's preconceived notions of what might be important. In the case of my study, I designated the center section of the page for "anything not school or home," opening the matrix up for unlimited potential topics. For example, one respondent's recalled a time during his elementary school years when his mother was pregnant and feared the baby would have Down syndrome. The pregnancy was a turning point in the respondent's trajectory because it brought his family together and strengthened their religious practice. The subsequent birth of a healthy child prompted the family to become even more enmeshed in their church community and steered the respondent toward private Catholic schools. Standard questionnaires recounting students' pathways to college likely would have recorded that this respondent attended private Catholic schools and that those schools influenced his decision to apply to and attend a competitive religiously affiliated university. Standard questionnaires, however, likely would not have captured the influence of his mother's pregnancy.

Since the development of the SSLHC for qualitative research (Nelson 2010), this methodological innovation has been used in various studies across social science and health-related fields. For example, Thompson (2015) used the technique to develop case studies of individual students as she investigated the costs and benefits of being labeled a long-term English learner (LTEL). In her study, each respondent completed the SSLHC during their first interview. The respondent's parent subsequently added to and modified the SSLHC during their interview. Finally, the respondent reflected on and updated the SSLHC in a follow-up interview. In another example, Smith (2015) referred to her SSLHCs as "chronologies of loss," as she used the technique to study the frequency and timing of homicide deaths across the life course among low-income Baltimore adolescents and young adults. Vermeer et al. (2016) adopted the SSLHC method to study sexual dysfunction and psychosexual support among cervical cancer survivors and their partners, adding disease diagnosis as a timing cue in addition to calendar time. Reisner et al. (2016) used the SSLHC method to explore sexual and gender development in a pilot study examining STI and HIV prevention interventions among transgender men. Rimkeviciene et al. (2016) utilized the SSLHC to study the timeline of events leading up to suicide attempts. Each of these studies supports the utility and flexibility of the SSLHC method for qualitative research.

6 Implementation

In the sections that follow, I provide logistical advice on implementing the semi-structured Life History Calendar (SSLHC), including the selection of time cues and landmarks, domain cues, starting points, materials, and data entry and analysis.

6.1 Time Cues

Time cues vary across studies and should be adapted to fit the goals of each research project. Cues can include Western or non-Western calendar time, age, educational, or employment landmarks, health or disease-related markers, or any other unit of time or combination of time cues the researcher and respondent deem relevant. Due to the age of the respondents in my study of educational trajectories (Nelson 2017), I employed school transitions as the most salient "landmarks" across the temporal line of the SSLHC (Axinn et al. 1997; Hoppin et al. 1998). Thus, the matrix, like the interview questions, was divided into five sections: before elementary school, elementary school, middle school, high school, and after high school. Since not all participants started elementary school on time or completed high school, it made sense to begin the interview with a blank page and add the school landmarks in the presence of the interviewee. For example, one respondent moved to USA at the age of 7 and began school in the second grade. Prior to that time, she had not attended school. Another respondent dropped out of high school after tenth grade when his girlfriend gave birth to their son. In his case, the "after high school" section began at age 16. In Vermeer et al.'s (2016) study of cancer survivors' sexual experiences, time cues were constructed relative to disease diagnosis rather than across the entire life course, including "1 year before diagnosis; diagnosis; treatment; 3, 6, and 12 months after diagnosis; until 5 years after diagnosis" (p. 1680). In Rimkeviciene et al.'s (2016) study of suicide attempts, the researchers used the attempted suicide as the end point of the timeline and allowed the respondent to fill in the important events in their life that led up to the attempt.

6.2 Domain Cues

Domain cues, like time cues, were introduced by the interviewer at the onset of the interview and were not preprinted on the page. The domains for my study were: school (across the upper third of the page), home and family (across the bottom third of the page), and "everything that's not school and not home" (across the middle third of the page). This final category was intentionally open and helped to record many idiosyncratic events across interviews. My intended purpose for this domain was to capture extracurricular activities, including sports, paid work, and community programs, yet this section provided space for other topics as well, such as religious involvement, gang affiliations, and chronic illnesses. By leaving this domain cue open to interpretation, the category was not limited to my conceptualization of extracurricular activities, and the SSLHC served to capture a variety of meaningful events and experiences. O'Connor et al. (2015) posit that this open-ended approach allows the researcher to shift from a variable-centered approach to a person-centered approach and provides a better understanding of why individuals follow different trajectories.

Other studies, based on the topic of interest and the age of the respondents, have used other domain cues. In their study of cervical cancer survivors, Vermeer et al. (2016) used work, relational status, holidays, disease and treatment, sexual functioning and received information and care as their domain cues. They also included an open-ended domain cue – important life events – to leave room for respondents to tailor the calendar to their own experiences. Smith's (2015) study of homicide among young adults relied on one primary domain cue – death-related loss. Similarly, Porcellato et al.'s (2016) study of employment used work experience as the only domain. Like time cues, domain cues can vary in content and quantity, as applicable to the study at hand.

6.3 Starting Point

Respondents were invited to dictate where they wanted to begin their life history. Some opted for a chronological approach starting with birth, others began at the present day and worked backwards, and some skipped around in time, as described earlier in this chapter. This flexibility gifted respondents with ownership of the interview and license to begin where they felt comfortable. Generally, respondents chose to begin at the time furthest from their most difficult years. By integrating the SSLHC and interview questions, the scope of the interview was apparent to respondents, thus both the interviewer and respondent could monitor which periods of life had yet to be covered regardless of where the interview began. Others who have implemented this methodology have adopted a similar approach. This flexibility, enabled by the coconstructed matrix, stands out as one of the unique features of this methodology.

6.4 Materials

As in previous incarnations of the LHC (Freedman et al. 1988), I found the size of the paper to be important. In pilot interviews, respondents began to truncate their responses when the paper appeared to be full; the larger the paper (and the more blank space), the more detailed the respondents' answers became. In my first pilot interviews, I used a standard letter size paper (8.5 × 11 in.), then grew to a legal size (8.5 × 14 in.), and then to subsequently larger and larger sizes of drawing paper. Ultimately, I settled on a standard easel pad (25 × 30 in.) consisting of large, high-quality paper with a cardboard backing. The sturdiness of the cardboard backing made the paper easier to transport to interviews and suitable for use even on smaller-sized tables. The size of the paper, however, had implications for the location of the interview. Ideally, the SSLHC matrix would sit on a table facing the interviewer and respondent so that both can add to the calendar easily and write right-side-up. Libraries and conference rooms proved to be suitable locations as they often contained larger sized tables.

In my study, the respondent and interviewer coconstructed entries on the SSLHC using colored markers and children's stickers. To indicate length of involvement in an activity, organization, or relationship, I drew horizontal lines with a vertical tick to indicate starting and stopping points. Although others have standardized color codes and symbols (e.g., Freedman et al. 1988), I allowed the respondent to use colors, symbols, and stickers freely. Rather, the interviewer and respondent employed symbols that resonated with the respondent. For example, if a respondent lived with her mother and two sisters she might write "M + 2S" at that point on the calendar, or she might draw stick figures instead. One respondent indicated a difficult period during high school by drawing a cloud around that entire period on the SSLHC. Granting each respondent permission to use colors, symbols, and words in their own fashion allowed respondents the freedom to express not only events in their lives but also their reactions and interpretations of those events. This kind of expression and interpretation cannot be achieved on the traditional LHC. See Figs. 2 and 3.

Stickers were another tool which helped some respondents engage with the LHC protocol. I offered respondents a variety of stickers representing different activities, events, affiliations, places, people, and animals. As with the colored markers, I did not prescribe each sticker a meaning; I offered the stickers to respondents at the beginning of the interview and encouraged them to place stickers along the calendar

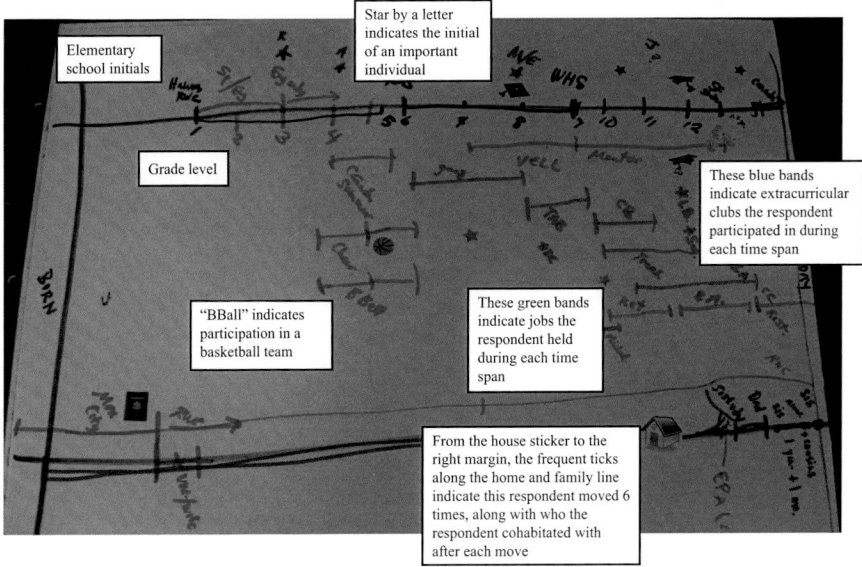

Fig. 2 Sample of adapted LHC symbols

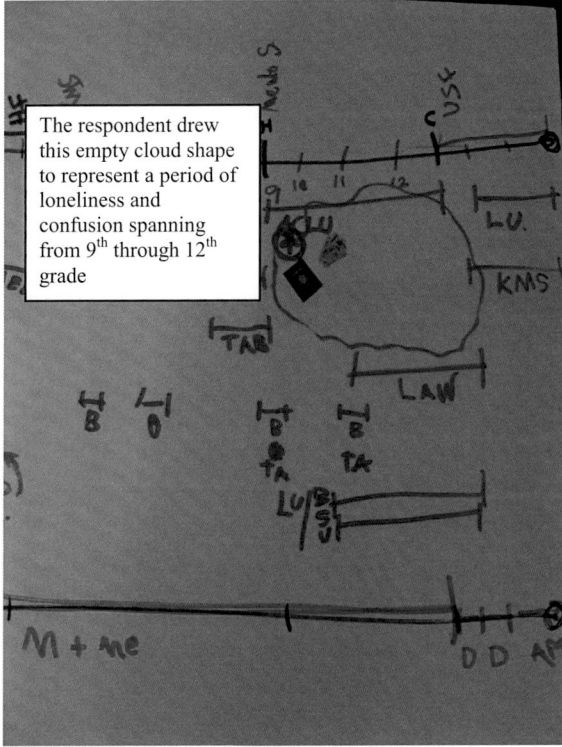

Fig. 3 Sample of interpretation of life events

as they saw fit. Many respondents used the stickers in a traditional way – if they played soccer in middle school, they used a soccer sticker, or if they studied to be a doctor, they used medical stickers. Other respondents used stickers to creatively signify life events. For example, while answering a question about elementary school, one respondent began sticking medical stickers in the high school section of the SSLHC next to a house sticker with black bars across it. I did not change the course of the interview, but when she began talking about high school I inquired about the stickers. The respondent revealed that when she was a high school student she had been addicted to drugs. She then locked herself in her house for many months in order to kick the habit (see Fig. 4). Another respondent piled on suitcase and passport stickers near the end of the high school section of the SSLHC. She reported she had been on a trip abroad at that time which had changed her life. To this respondent, a larger quantity of stickers came to indicate a more influential event.

Some respondents, particularly young men, did not choose to use stickers, and it is likely that stickers may be more useful for some respondents than others. Even when respondents chose not to use stickers during the interview, however, the

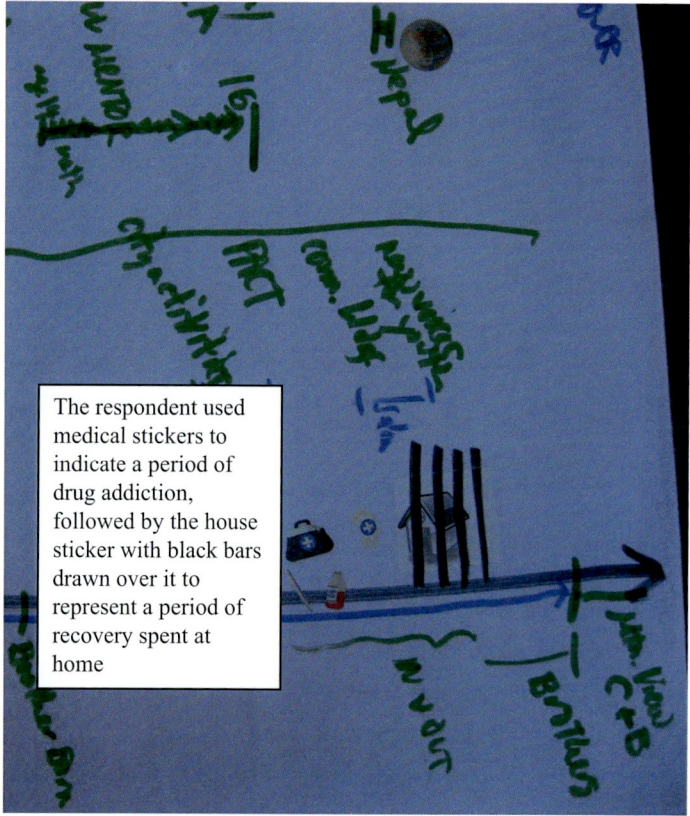

Fig. 4 Sample of use of stickers

interviewer's introduction of the stickers helped break the ice and establish rapport. Many respondents noted that having stickers (even if they did not use them) brought back images of elementary school and helped set the mood for conversations about their childhood. Overall, the stickers enabled greater participation for some and improved rapport for all respondents. While other researchers have used stickers to indicate specific events on a traditional preprinted LHC matrix (e.g., Hoppin et al. 1998) or occupational history calendar (e.g., Porcellato et al. 2016), the open-ended use of stickers lends itself particularly well to the SSLHC.

6.5 Data Entry and Analysis

The primary source of data used for analysis was the transcribed audio recording of each interview. The purpose of the SSLHC was to collect both event timing information and individual explanations of behaviors and attitudes, nested within overlapping social contexts. As such, the SSLHC was an effective tool for easing retrospective recall, building rapport between interviewer and respondent, and easily checking to make sure all desired time periods and domains had been covered during the interview. Beginning with a blank page and allowing respondents to coconstruct their matrix using markers and stickers supports these goals. But because each respondent created a unique system of symbols and colors, the SSLHCs were of limited analytical use. Although this incarnation of the SSLHC method succeeds at providing in-depth interview narratives, it does not offer the same ease of data entry as preprinted LHC matrices.

7 Conclusion and Future Directions

In recent years, the semistructured adaptation of the Life History Calendar (SSLHC) for qualitative research has been used to study a range of topics across health- and social science-related fields. This adaptation allows for many of the same benefits of the traditional LHC. Yet, unlike the traditional LHC, the SSLHC captures not only the starting and stopping of activities but also the how's and the why's behind each transition. This coupling of event timing with explanatory data provided more nuanced insights into marginalized students' pathways to college than either the structured LHC or qualitative interviewing could produce independently. For example, when a respondent reported changing schools at the end of third grade without a housing move, the SSLHC captured both the change and the explanation:

> We had this…Oregon Trail project and…the teacher assigned each one of us to be a pioneer. I told [the teacher] that I didn't think that this game or simulation was a fair representation. What about the Indians? And…that was not part of her happy pioneer project. I was totally defiant. I said I didn't want to participate and I didn't want to be a pioneer because they killed the Indians and they exploited them and I didn't think it was doing justice and we weren't representing all parties involved and representing it historically accurately. I didn't want to participate. So in the end, we had 19 pioneers and one little Indian. It was just a big fight. From then on…well it was time I went to a different school.

In this case, as in many others, the cause of the school transition is more meaningful in her overarching educational trajectory than the transition itself. Although frequent school changes are one predictor of academic failure among Latina/o students, it is assumed these changes are the product of household disruptions or residential moves. This example suggests why understanding the cause of the move is critical to understanding whether the change of schools might have positive or negative consequences within an individual's educational trajectory.

My study highlighted many examples of how the SSLHC produced invaluable explanatory data in tandem with in-depth interviews. For instance, one respondent stopped participating in extracurricular activities at the time of her mother's death. In the traditional LHC method, researchers might assume the respondent discontinued participation due to grief. However, the SSLHC brought to light that her change of routine was only indirectly attributable to her mother's passing. In her traditional Mexican family, this respondent was now the oldest female and – at age 12 – was expected to take on the family's cooking, cleaning, child care, and shopping. The increased level of responsibility at home accounted for her subsequent lack of participation. Given the tendency in the current literature on Latina/o students to model pathways to college as straight lines, predicted by easily operationalized variables, the SSLHC has the potential to further nuance both research and theory.

While the SSLHC method is likely not well suited to large samples due to the potential complexity of data entry and coding, this method shows promise for life course research with smaller samples across a variety of fields. For example, O'Connor et al. (2015) posit that this qualitative life history method offers an important complement to statistical modeling in the study of individual trajectories in the field of clinical psychology. The authors argue that even the most sophisticated trajectory models adhere to inappropriate assumptions about human behavior and suggest a paradigm shift toward thinking about transitions and turning points within a person-centered approach using the SSLHC to begin identifying key life events. O'Connor et al. (2015, p. 20) write:

> The life calendar method of Nelson (2010) offers a key to anchoring recall in key landmark events, rather like squares on a game board. The person is likely to recall their detailed experiences around these key points. One could then calculate the most likely events that were concurrent with the turning point, including cognitive and emotional reactions to outside events, and identify the a posteriori likelihood of the early patterns reported as early transition points reflecting current patterns.

In clinical psychology, the SSLHC stands out as a particularly useful methodology because it situates data on turning points within overlapping social contexts. Carpenter (2015) posits that the qualitative LHC may prove particularly useful for sexualities researchers as well, because the nature of the matrix facilitates recall of both highly salient events and less salient events that may play a role in an individual's developing sexuality across the life course. These are just two examples in a world of possibilities of how the SSLHC may prove useful for life course research across a variety of fields in the years ahead.

References

Axinn WG, Barber J, Ghimire D. The neighborhood history calendar: a data collection method designed for dynamic multilevel modeling. Sociol Methodol. 1997;27:355–92.

Axinn WG, Pearce LD, Ghimire D. Innovations in LHC applications. Soc Sci Res. 1999;28:243–64.

Belli RF. The structure of autobiographical memory and the event history calendar: potential improvements in the quality of retrospective reports in surveys. Memory. 1998;6:383–406.

Belli RF, Shay WL, Stafford FP. Event history calendars and question list surveys: a direct comparison of interviewing methods. Public Opin Q. 2001;65:45–74.

Belli RF, Lee EH, Stafford FP, Chow CH. Calendar and question-list survey methods: association between interviewer behavior and data quality. J Off Stat. 2004;20(2):185–218.

Belli RF, Stafford FP, Alwin D, editors. Calendar and time diary methods in life course research. Thousand Oaks: Sage; 2009.

Blum Z, Karweit N, Sorensen A. A method for the collection and analysis of retrospective life histories. Report no. 48. Baltimore: Johns Hopkins University, Center for the Social Organization of schools; 1969.

Bronfenbrenner U. The ecology of human development. Cambridge: Harvard University Press; 1979.

Bronfenbrenner U, Morris P. The bioecological model of human development. In: Damon W, Lerner RM, editors. Handbook of child psychology. 6th ed. Hoboken: Wiley; 2006. p. 793–828.

Carpenter LM. Studying sexualities from a life course perspective. In: DeLamater J, Plante RF, editors. Handbook of the sociology of sexualities. Cham: Springer; 2015. p. 65–89.

Caspi A, Moffitt TE, Thornton A, Freedman D, Amell JW, Harrington H, et al. The LHC: a research and clinical assessment method for collecting retrospective event-history data. Int J Methods Psychiatr Res. 1996;6:101–14.

Elder G Jr. Children of the great depression: social change in life experience. Chicago: University of Chicago Press; 1974.

Elder G Jr, Shanahan M, Jennings J. Human development in time and place. In: Lerner RM, editor. Handbook of child psychology and developmental science, vol. 4. Hoboken: Wiley; 2015. p. 6–54.

Feldman S, Howie L. Looking back, looking forward: reflections on using a life history review with older people. J Appl Gerontol. 2009;28:621–37.

Freedman D, Thorton A, Camburn D, Alwin D, Young-DeMarco L. The LHC: a technique for collecting retrospective data. Sociol Methodol. 1988;18:37–68.

Furstenberg FF, Brooks-Gunn J, Morgan SP. Adolescent mothers in later life. Cambridge: Cambridge University Press; 1987.

Giele JZ, Elder GH, editors. Methods of life course research: qualitative and quantitative approaches. Thousand Oaks: Sage; 1998.

Harris DA, Parisi D. Adapting LHCs for qualitative research on welfare transitions. Field Methods. 2007;19(1):40–58.

Hoppin JA, Tolbert PE, Flagg EW, Blair A, Zahm SH. Use of a life events calendar approach to elicit occupational history from farmers. Am J Ind Med. 1998;34:470–6.

Hser YI, Anglin MD, Grella C, Longshore D, Prendergast ML. Drug treatment careers: a conceptual framework and existing research findings. J Subst Abus Treat. 1997;14(6):543–58.

Kessler RC, Wethington E. The reliability of life event reports in a community survey. Psychol Med. 1991;21:723–38.

Laub JH, Sampson RJ. Shared beginnings, divergent lives: delinquent boys to age 70. Cambridge: Harvard University Press; 2003.

Lilley RC, Cryer PC, Firth HM, Herbison GP, Feyer AM. Ascertainment of occupational health histories in the working population: the occupational history calendar approach. Am J Ind Med. 2011;54:21–31.

Martyn KK, Belli RF. Retrospective data collection using event history calendars. Nurs Res. 2002;51:270–4.

Martyn KK, Reifsnider E, Barry MG, Trevino MB, Murray A. Protective processes of Latina adolescents. Hispanic Health Care Int. 2006;4(2):111–24.

McPherson JM, Popielarz PA, Drobnic S. Social networks and organizational dynamics. Am Sociol Rev. 1992;57(2):153–79.

Meltzer P. Using the self-discovery tapestry to explore occupational careers. J Occup Sci. 2001;8:16–24.

Mortimer J, Johnson M. Adolescents' part-time work and post-secondary transition pathways in the United States. In: Heinz WR, editor. From education to work: Cross-national perspectives. Cambridge: Cambridge University Press; 1999. p. 111–48.

Mortimer J, Shanahan M, editors. Handbook of the life course. New York: Kluwer Academic/Plenum Publishers; 2003.

Munch A, McPherson JM, Smith-Lovin L. Gender, children, social contact: the effects of childrearing for men and women. Am Sociol Rev. 1997;62(4):153–79.

Nelson IA. From quantitative to qualitative: adapting the life history calendar method. Field Methods. 2010;22(4):413–28.

Nelson IA. Why afterschool matters. Camden, NJ: Rutgers University Press; 2017.

O'Connor K, Robert M, Pérodeau G, Séguin M. Trajectory-based methods in clinical psychology: a person centred narrative approach. New Ideas Psychol. 2015;39:12–22.

Porcellato L, Carmichael F, Hulme C. Using occupational history calendars to capture lengthy and complex working lives: a mixed method approach with older people. Int J Soc Res Methodol. 2016;19(3):269–86.

Reisner SL, Hughto JMW, Pardee DJ, Kuhns L, Garofalo R, Mimiaga MJ. LifeSkills for men (LS4M): pilot evaluation of a gender-affirmative HIV and STI prevention intervention for young adult transgender men who have sex with men. J Urban Health. 2016;93(1):189–205.

Rimkeviciene J, O'Gorman J, Hawgood J, Leo DD. Timelines for difficult times: use of visual timelines in interviewing suicide attempters. Qual Res Psychol. 2016;13(3):231–45.

Smith JR. Unequal burdens of loss: examining the frequency and timing of homicide deaths experienced by young black men across the life course. Am J Public Health. 2015;105(S3):S483–90.

Thompson KD. Questioning the long-term English learner label: how categorization can blind us to students' abilities. Teach Coll Rec. 2015;117(12):1–50.

Thornton A, Axinn WG, Hill DH. Reciprocal effects of religiosity, cohabitation, and marriage. Am J Sociol. 1992;98(3):628–51.

Van der Vaart W. The time-line as a device to enhance recall in standardized research interviews: a split ballot study. J Off Stat. 2004;20(2):301–17.

Van der Vaart W, Glasner TJ. Applying a timeline as a recall aid in a telephone survey: a record check study. Appl Cogn Psychol. 2007;21(2):217–38.

Vermeer WM, Bakker RM, Kenter GG, Stiggelbout AM, ter Kuile MM. Cervical cancer survivors' and partners' experiences with sexual dysfunction and psychosexual support. Support Care Cancer. 2016;24:1679–87.

Yoshihama M, Clum K, Crampton A, Gillespie B. Measuring the lifetime experience of domestic violence: application of the LHC method. Violence Vict. 2002;17(3):297–317.

Yoshihama M, Gillespie B, Hammock A, Belli RF, Tolman R. Does the life-history calendar method facilitate the recall of domestic violence victimization? Comparison of two methods of data collection. Soc Work Res. 2005;29(3):151–63.

Zahm S, Colt J, Engel L, Kiefer M, Alvarado A, Burau K, Blair A. Development of a life events/icon calendar questionnaire to ascertain occupational histories and other characteristics of migrant farmworkers. Am J Ind Med. 2001;40:490–501.

Calendar and Time Diary Methods

69

Ana Lucía Córdova-Cazar and Robert F. Belli

Contents

1 Introduction 1220
2 Time-Use Research 1221
   2.1 A Brief Historical Account of Time-Use Research 1222
   2.2 A "New" Paradigm to Measure Well-Being 1223
3 Well-Being and Surveys: Calendar and Time Diaries as the Tools to
   Measure Well-Being 1224
   3.1 The Role of Flexible Conversation Techniques in Reducing Error in Timeline
      Surveys 1225
   3.2 The Role of (Autobiographical) Memory in Reducing Error in
      Timeline Surveys 1226
4 Applying Time Diary and Calendar Methods to the Analysis of Well-Being 1228
   4.1 The Currency of Life: Measuring Well-Being Through the National Time
      Accounting Framework 1229
   4.2 Event History Calendar Methods to Assess Health-Related Issues 1231
5 Conclusion and Future Directions 1233
References 1234

Abstract

For decades, individual and social well-being have been reduced to material well-being. Given that economic or financial success cannot fully capture an integral concept of human well-being, a new paradigm has emerged, one which focuses on the nonmarket contribution and behavior of households. This new paradigm goes

---

A. L. Córdova-Cazar (✉)
Colegio de Ciencias Sociales y Humanidades, Universidad San Francisco de Quito, Diego de Robles y Vía Interoceánica, Quito, Cumbayá, Ecuador
e-mail: alcordova@usfq.edu.ec; al.cordovacazar@huskers.unl.edu

R. F. Belli
Department of Psychology, University of Nebraska-Lincoln, Lincoln, NE, USA
e-mail: bbelli2@unl.edu

© Springer Nature Singapore Pte Ltd. 2019
P. Liamputtong (ed.), *Handbook of Research Methods in Health Social Sciences*,
https://doi.org/10.1007/978-981-10-5251-4_2

beyond the typical scope of economic science in which only the activities that can be measured in monetary terms are included. A critical aspect of this paradigm is that the way people spend their time needs to be taken into account in a rigorous and scientific way, as it is only by understanding previous life events that present and future life course development can be predicted. Thus, valid retrospective and behavioral reports are needed from individuals. When the goal, however, is not only to conduct qualitative research (e.g., through in-depth interviews), but to draw quantitative inferences from samples to the general population, survey interviewing is inevitably involved. In this chapter, we will examine calendar and time diary methods which have been shown to be especially effective when it comes to assessing well-being. Although these methods forego the standardization of question wording (the most prevalent approach in traditional survey interviewing), they are nevertheless able to produce reliable and valid responses, while also encouraging conversational flexibility that assists respondents to remember and correctly report the interrelationships among past events. Further, in this chapter we also discuss the application of the calendar and time diary methods in several health and social sciences fields of research, through which well-being is evaluated.

Keywords

Calendar method · Time diary method · Time-use research · Well-being

1 Introduction

One key concern of behavioral, social, and health scientists is to assess people's well-being. Within the field of modern economics, national income statistics such as the Gross National Product (GNP) and the Gross Domestic Product (GDP) have not only been looked upon as reliable measures of economic success or failure, but also have increasingly been thought of as measures of societal well-being (Stiglitz et al. 2010). During the 1980s, a growing number of voices (economists and social scientists alike) have expressed discontent with the notion that an increase in GDP could be equated with an increase in well-being (e.g., see Juster and Stafford 1985). The creation of a Commission to "measure economic performance and social progress" in 2008, spearheaded by the economist Nobel laureates Joseph Stiglitz and Amartya Sen, is a clear sign of such a dissatisfaction. Its main recommendations included emphasizing the household perspective and broadening income measures to nonmarket activities. Although the Commission's creation was a highly visible initiative within the international political arena, the topic was not new. Indeed, the idea of devising an augmented system of economic and social accounts that would recognize the nonmarket activities of households had already been proposed decades before (Juster 1985). Understanding how people spend their time is at the heart of such an endeavor (Stiglitz et al. 2010).

In this chapter, we argue that time-use research can provide hard and replicable data that reflect people's decisions, preferences, attitudes, and environmental factors (Pentland et al. 1999), with which the many aspects of well-being can be assessed. In Sect. 2, we first discuss the developments of time-use research and we introduce a new paradigm to measure and analyze well-being. In Sect. 3, we discuss the need for well-being researchers to be able to draw quantitative inferences from samples to the general population through the use of surveys. We draw particular attention to time diary and calendar methods, as these constitute feasible alternatives to traditional ways of survey interviewing on well-being issues. The fourth section looks at ways in which time diaries and calendar methods have been applied to the measurement of well-being. We first look at the "National Time Accounting" framework to measure well-being through the use of the evaluated time-use approach. We then review some examples of how calendar methods have been used to assess health-related issues, arguably one of the most important predictors of well-being. In the final section, we present our conclusions and ideas for future directions.

2 Time-Use Research

The ability of time-use research to understand human behavior and its intrinsic relationship with individual and social well-being has been widely accepted and has garnered the interest of researchers from a broad range of disciplines including economics, gerontology, political science, nursing and medicine, psychiatry, health education and research, sociology, psychology, education, social epidemiology, criminology, demography, social work, and survey methodology (Pentland et al. 1999; Belli et al. 2009b).

Time-use research methods consist of both time diaries and "life histories" or event history calendars (calendars from this point forward). Both are similar in that they collect timeline data – for diaries the timeline is a 24-h day; for calendars it can range from months to years or even longer sections of the life course (Belli et al. 2009b; see also ▶ Chaps. 67, "Timeline Drawing Methods," and ▶ 68, "Semistructured Life History Calendar Method"). Both diaries and calendars examine the allocation of time into different activities, including paid work, personal care, leisure, childcare, and, increasingly, a wide range of health-related behaviors by the different population groups (e.g., women, the elderly, persons with disabilities), at the daily, weekly, monthly, or yearly levels. During the first half of the twentieth century, the majority of time studies were conducted in the Soviet Union, Great Britain, and the United States, with some studies conducted in France, Germany, and Japan (For a detailed historical account of time use studies during the first half of the twentieth century see Pentland et al. 1999). Currently, almost every nation in the world conducts time-use studies of some sort, suggesting that from early on, researchers interested in examining those key events that govern people's behavior, health, and social interactions have drawn on the study of how people use their time (Belli et al. 2009b). A brief historical account of time-use research follows.

2.1 A Brief Historical Account of Time-Use Research

Time-use research, in the form of diaries, emerged during the second decade of the past century in the context of early studies of the living conditions of the working class. That is, the original time-use research studies emerged as a response to industrialization and its ensuing pressures on people's daily lives. Interestingly, calendars also appeared in connection to social issues, specifically with the objective of investigating migration processes to the United States. The two first published works that gave an account of the daily lives of working-class families were published separately in the same year of 1913 in the United States and the United Kingdom (Bevan 1913 and Pember-Reeves 1913 cited in Pentland et al. 1999). These works, however, did not include systematically collected and representative data from diaries. The earliest sophisticated diary study belongs to the Soviet S.G. Strumlin, which was intended for use in governmental planning (Pentland et al. 1999). Several other smaller and isolated efforts were launched in Japan, the United Kingdom, and the United States, until the mid-1960s when Alensander Szlai launched a very ambitious program to systematically obtain time diary data from 13 countries around the world.

The first study where quantitative information coming from calendars was collected, processed, and analyzed dates from 1969, when the Argentinean Jorge Balán collected 1640 life histories of men aged 21–60 in Monterrey, Mexico. Before that, Thomas and Znaniecki (1918) had strongly advocated for an intensive use of life histories, but did not pursue systematic data collection because of insurmountable technical difficulties at that time. For Balán et al. (1969, p. 107), their ability to take advantage of "the possibilities opened up by large-capacity computers" (which, at that moment, involved the use of punch cards) made it possible to systematically analyze a large number of life histories. In the same year, researchers from Johns Hopkins University conducted the first calendar survey in the United States, which looked at socio-economic well-being (Belli and Callegaro 2009); specifically, its purpose was to empirically examine how social groups and individual households attained social mobility in order to identify alternative intervention directions (Blum et al. 1969).

Almost two decades elapsed before the next systematic study using calendars was conducted by the demographers Freedman et al. (1988). They sought to accurately measure the trajectories and event transitions that shape life course (Belli et al. 2009b) and especially to understand the processes that govern the transition from adolescence into adulthood. Freedman et al. (1988, p. 38) were able to estimate, through the use of sophisticated statistical methods, dynamic causal interrelations among several aspects of the life course, through which they concluded that the life course is not a unidimensional series of events unfolding, but a "simultaneous unfolding of many dimensions, all interwoven temporally and causally in complex ways."

Many more studies came afterwards, which have shown that time-use research not only contributes to explain people's current condition, but also the long-term consequences of their daily decisions on their later well-being (Belli and Callegaro 2009).

2.2 A "New" Paradigm to Measure Well-Being

The time-use and well-being paradigm presented in 1981 by economists Thomas F. Juster, Paul N. Courant, and Greg K. Dow was the first formal conceptual system to measure and analyze well-being through the use of time. The key idea of their paradigm is that "the ultimate constraints determining the level of individual well-being are the availability of human time and the set of factors that determine the effectiveness with which time is used" (Juster and Stafford 1985, p. 1). Although they included material or intellectual resources in their set of factors, they also included the *capability* individuals have of enjoying and utilizing those resources throughout their life course, which is shaped by levels of physical and mental health.

This "new" social accounting system linking time-use and well-being develops from a long-established conceptual framework in economics that is based on the measurement of tangible resources. Monetary values are assigned to each element of the system and concepts such as wages, prices, profits, interest rates emerge. The new social accounting system appears as a general critique to that system and consists of three main critiques.

The first critique centers on the sole focus on flows of material goods and services, whereas well-being of individuals and societies is determined by the combination of available goods and leisure. In particular, the traditional economic welfare function, where leisure and goods are the only elements to be considered, lacks any appreciation for a positive connotation for time that is spent working: time at work is always a "bad" and only leisure time constitutes a "good" (Juster and Stafford 1985).

A second critique is that the traditional model accounts for the so-called "value added" to products at the different market stages (i.e., extraction, manufacture, and distribution), but expressly excludes any type of value added within the household. For instance, costs of manufacturing and distributing food are accounted for, but those costs related to the time spent in preparing (nutritious) food are not considered. Therefore, one of the main differences between both systems is that the social accounting system focuses not only on resource inputs, but on the changes in output. The result of incorporating a measure of the real output (such as a well-nourished family) allows for distinguishing between intermediate and final output (Dow and Juster 1985; Juster and Stafford 1991; Gershuny and Halpin 1996; Krueger 2009b). By avoiding an exclusive focus on resource inputs, the social accounting framework captures other phenomena that also impact well-being, but which traditional economic systems of welfare conceal. In sum, in the social accounting framework, material or "objective" conditions are "intermediate" outputs, while subjective measures are the "ultimate outputs" of interest.

Finally, the movement towards a more comprehensive system of accounts emerged from a deeper understanding of the concept of utility (Juster (1990, p. 156) reminds us of the original Benthamite concept of utility as the "cardinally measurable psychological flow of satisfactions attached to goods and services purchased in the market."), which has provided the intellectual basis to incorporate the role of time in the measurement of well-being. From the social accounting perspective, utilities do not just depend on the final product that results from a certain personal activity, but also on the *enjoyment* of the time spent in that activity. The theoretical implication is that the way the time is used – its level of enjoyment – needs to be taken into account (Juster et al. 1981; Gershuny and Halpin 1996; Krueger et al. 2009a, b). Hence, what becomes of interest are the so-called "process benefits," that is, the extent to which a person actually enjoys the activity regardless of its price.

This new framework has been crucial for the measurement of well-being. It provides support for the creation of quality of life and social indicators and for the notion that goods and services are instruments for the (subjective) enjoyment of activities rather than ends in themselves. The "final output" can thus be registered by looking at indicators of subjective satisfaction with the various domains of life (Juster et al. 1985).

3 Well-Being and Surveys: Calendar and Time Diaries as the Tools to Measure Well-Being

Constructing a system with which to measure well-being entails understanding not only how people use their time but also how previous life events may predict future developments. For instance, evidence of the future detrimental consequences of tobacco use can be better supported with studies that look at the health status of smokers and nonsmokers who have been followed throughout a long period of time. Thus, time-use researchers need to have accurate information about the past and the present; that is, valid retrospective and behavioral reports from respondents. This ambitious goal cannot be based on qualitative research analysis (e.g., through in-depth interviews) that cannot be generalized. On the contrary, statistical-based analysis is needed for researchers to be able to draw quantitative inferences from samples to the general population. For that reason, interviewing a representative sample of the population is inevitably involved in such an endeavor.

Panels that use standardized methods and which follow a cohort across several year have been regarded as a reliable way of obtaining life course information. However, panels are costly, and thus other methods have been devised. Time diary and calendar methods have been proposed as feasible alternatives to traditional ways of survey interviewing, especially regarding different aspects of people's well-being, including their emotional and physical health (Juster and Stafford 1985; Agrawal et al. 2009; Belli et al. 2009b; Belli and Callegaro 2009; Martyn 2009). Although these methods forego the standardization of question wording and encourage a more flexible interviewing approach, they are able to produce reliable and valid responses. Furthermore, calendar and time diary methods can be applied to a wide range of disciplines in health and social sciences research, allowing for a more systematic study of the many components of well-being.

The quality and validity of reports also hinge upon the way the information is collected. When studying the specific ways in which people use their time, conventional standardized questionnaires have been used. For instance, respondents may be asked to estimate how much time they allocate to their different activities using a stylized list of activities. Yet it has been found that it can be very difficult for respondents to produce accurate responses using this approach and measurement error is likely to arise. From the time-use research perspective, two different mechanisms can be associated with error when using stylized questions: (a) the lack of flexibility that prevents the conversation to flow in a natural way (Jabine et al. 1984; Suchman and Jordan 1990), and (b) the fact that activity frequency and duration surveys using stylized sets of possible activities provide reports that are episodic and may be taken out of context (Pentland et al. 1999). Time diaries and calendar-based interviews, which ask about time-use using a conversational approach, have been proposed as an alternative to overcome such complications, mainly because of their ability to encourage respondents to incorporate in their cognitive processing temporal changes that serve as cues that assist providing a more accurate reporting of events (Belli et al. 2009b).

3.1 The Role of Flexible Conversation Techniques in Reducing Error in Timeline Surveys

Conventional standardized interviewing is the most widely practiced technique as it purportedly reduces variance in responses due to interviewers and maximizes variance attributable to the actual differences in respondents' circumstances (Fowler and Mangione 1990; Fowler and Cannell 1996; see also ▶ Chaps. 80, "Cell Phone Survey," and ▶ 81, "Phone Surveys: Introductions and Response Rates"). The first mechanism that may produce error in retrospective survey reports, namely the lack of conversational flexibility, can be attributed precisely to the standardization of the questionnaire. Time diary and calendar methods address this source of error by allowing interviewers to lead the conversation in a natural manner (Houtkoop-Steenstra 2000; Maynard et al. 2002). Although in conventional interviewing the *wording* of questions is standardized, there is no guarantee of a nonambiguous and consistent *understanding* of questions by respondents (Suchman and Jordan 1990; Houtkoop-Steenstra 2000). By assuming an interview is nothing more than a neutral measurement instrument, conventional standardized interviewing suppresses the elements of ordinary conversation, compromising both the understanding of the intended meaning and the validity of the answers.

Numerous studies have demonstrated how interviewers frequently cannot maintain the rules of standardized interviewing (Houtkoop-Steenstra 2000; Belli et al. 2001b, 2013). The conversational interviewing technique accepts that an interview involves an interaction between the participants, in which the rules of conversation must be respected (Schwarz 1996; Schwarz et al. 2009). Following Clark and Schober (1992), this technique recognizes that language is not about the literal meaning of words but about people and what they mean. The coveted goal of providing greater consistency to question meaning may be better reached by allowing interviewers to clarify the concepts and assist respondents when doubts of any sort arise (Schober and Conrad 1997; Conrad and Schober 2000).

Calendar and time diary methods take advantage of the conversational survey technique by disregarding the need to use fixed words and phrases and permitting flexibility to interviewers as long as they complete the diaries or the calendars in the way they are intended. The benefits to data quality due to the use of the conversational technique in diaries and calendars are further enhanced by the memory cues that are encouraged.

3.2 The Role of (Autobiographical) Memory in Reducing Error in Timeline Surveys

Answering survey questions necessarily involves cognitive and memory processes and their limitations are associated with error in survey reports (Belli 2013; Belli and Al Baghal 2016). For that reason, survey methodologists have incorporated cognitive science perspectives into their field of study (Jabine et al. 1984; Tanur 1992) (The initiative to incorporate cognitive science knowledge into survey research started in the 1980's was called the Cognitive Aspects of Survey Movement (CASM) and continues to this day. For a brief history on the topic, see the Preface of Tanur (1992).). An example is the classic question response model proposed by Tourangeau (1984), which involves question comprehension, memory retrieval, the judgment of the relevance of retrieved information, and the selection and editing of the final response (Tourangeau et al. 2000). Although the role of memory is important in the answering of any question, in the case of time studies as time diaries or calendars, the role of memory is crucial as respondents are queried about past events that they have experienced. For instance, the American Time Use Survey (ATUS), which is a time diary survey conducted by the US Census Bureau to understand the time-use patterns of the population of the United States, asks the following question: "Now I'd like to find out how you spent your time yesterday, from 4:00 in the morning until 4:00 AM this morning. I'll need to know where you were and who else was with you" (Phipps and Vernon 2009). Likewise, a great reliance on memory processes will occur with questions asked about longer periods of time, such as with the Panel Study of Income Dynamics, a prospective national study of life course socioeconomics and health, which asks the following question: "I'd like to know about all of the work for money that you have done since January 1, [Past year]. Please include self-employment and any other kind of work that you have done for pay." In both circumstances, respondents will need to retrieve information from their autobiographical memory. Importantly, if the past 24 h or the past year consisted of complex experiences by including several different activities or jobs, respondents will have to make a considerable memory effort to derive a complete and accurate answer.

Belli and colleagues have noted that calendar questionnaires encourage the use of cues that exist in the structure of autobiographical knowledge which, together with flexible interviewing, enhance the quality of retrospective reports in comparison to conventional standardized questionnaires (Belli 1998; Belli et al. 2001b). Importantly, they have shown that improvements in retrospective reporting also occur when collecting subjective assessment information, such as health status over the life course (Belli et al. 2012). In particular, they have noted that events that are more easily remembered can become memory cues that will help respondents to remember events that are more difficult to remember (see Belli 2013). There are three possible types of cueing: top-down, sequential, and parallel cueing. Top-down cuing occurs when more general events serve as cues to remember more specific ones: remembering the name of an employer helps remembering more specific details such as weekly pay. Sequential cueing occurs when a remembered event is used as an anchor that aids in the remembering of a temporally adjacent event within the same life domain: remembering that one worked for one employer during a period of time helps remembering the name of the employer one worked with afterwards. Finally, with parallel cueing, a remembered event in one life domain assists in the remembering of an event from a different life domain that occurred contemporaneously or nearly so: remembering a change of residence helps remembering that one changed employment.

Time diaries also take advantage of these cueing properties, although differently. Whereas calendars encourage respondents to report about periods of stability and points of transitions in different life domains (work, relationships, health), time diaries ask about transitions between the different activities one engaged in during the day. In calendar surveys, respondents provide information about a number of timelines covering events from the different life domains of interest, and reference periods can range from months, to years, and up to the entire lifetime. In time diaries, respondents provide information about activity sequences, and the context in which these occurred; their reference period is generally 24 h. In terms of cuing, given that time diaries are driven by location, transitions between activities generally involve a change in location (within the house or traveling to a different place) (Stafford 2009). Hence, top-down cuing may occur in which the more general event (the activity) will trigger one to remember the more specific detail of "where." A bottom-up cueing can also occur as one may first remember "where" before remembering the activity. Likewise, sequential cueing may occur regarding the details and context of the activity in which remembering one activity may assist in the remembering the next. Finally, parallel cueing may occur when a person reports a secondary contemporaneous activity.

Research has found that timeline methods (especially calendars) do enable more complete reconstructions of one's past and an enhancement in retrospective reporting data quality (Belli et al. 2004, 2013; Bilgen and Belli 2010). Given that social, behavioral, and health scientists will continue to administer surveys that ask respondents about their pasts, such interviewing methods are encouraged in order to produce more valid scientific inferences about individual life course trajectories and social interrelationships (This section draws heavily on Belli 2013).

4 Applying Time Diary and Calendar Methods to the Analysis of Well-Being

An important feature of data collected through a time diary or a calendar is that they can be used not only to create simple individually aggregated summaries, but summaries that can give account of "complex constructions" (Stafford 2009). For instance, in the area of employment histories, calendars can be used to construct summary employment measures for the entire 52 weeks of the year (such as employment periods, sick periods, being on vacation, and being out of the labor force). In addition, complex constructions can be created for calendars by analyzing descriptors. For instance, by asking about hours worked in more than one job or about a second job at a lower wage rate, calendars can permit the descriptive characterizations of multiple job holding status over the course of a determined amount of time. Likewise, descriptors can be used to characterize a given spell or episode, as well as to track their sequence. For example, complex constructions can be captured regarding the type of employment spell, such as full-time or part-time, or whether one was unemployed or out of the labor force due to an injury. Descriptors can also tell us whether the weeks of unemployment during a calendar year correspond to one spell or a number of shorter spells, as well as their timing. These distinctions, which are critical for the correct analysis of labor economics, are not available from summary measures using noncalendar methods.

Similarly, the data obtained from individual 24-h time diaries can be processed into overall time-use measures with which to account for how a society as a whole allocates time into the different activities and how these are distributed across different subgroups (e.g., housework for males and females). Complex constructions can also be created through microdiary activity data, in which descriptors of the activity (where, who with, secondary activities) can be included. Such descriptors are critical to understanding the patterns of social interactions, such as child or elderly care. Similarly to calendars, the resulting complex timelines overcome the limitation of time-use aggregate measures which are not able to track the sequencing of activities. For instance, in a diary using stylized lists, one may know the total amount of hours of television watching or sleep, but their sequencing remains unknown. Finally, diary activity records are also amenable to the inclusion of subjective descriptors such as affect, whereby elapsed time can be characterized as productive, enjoyable, unpleasant, or meaningful. Including affect descriptors can be extremely useful in the assessment of well-being, even in spite of the fact that diaries only provide a small snapshot of the life of respondent (Stafford 2009).

In summary, time diaries and calendars are a valuable tool for the measurement and analysis of many aspects that impact well-being. Their capability of including descriptors reduces the potential for biases that can arise from respondents' direct reports of activities or spells over a "typical month," a "typical week," or a "typical day," where averages and timing of events can be considerably imprecise due to socially desirable answers or the reliance on stereotypic responses (e.g., 40 h of work per week). The literature has shown several examples of how time diary and calendar methods can provide a theoretic framework for the analysis of well-being through the collection of valid and reliable information.

4.1 The Currency of Life: Measuring Well-Being Through the National Time Accounting Framework

Based on the ideas proposed by Juster et al. (1981, 1985), the pursuit for a new system of accounts that will allow the measurement of well-being in a more comprehensive fashion – one which would include how time is experienced – gained momentum. During the first decade of the twenty-first century, the idea of establishing a system of "National Time Accounting" (NTA) was presented by Krueger and colleagues (Krueger 2009a, b; Krueger et al. 2009a, b). This new system of accounts is dependent on self-reported data of subjective outcomes ("subjective well-being") that is to be measured and reported *in tandem* with traditional national estimates of a country's economic activity in order to measure overall well-being. The NTA is a "framework for measuring, comparing, and analyzing the way people spend their time across countries, over historical time, or between groups of people within a country at a given time" (Krueger 2009a, p. 2). It is based on time-use and its affective (emotional) experience. The novelty of this method is that time-use evaluation does not rely on researchers' or coders' judgments about whether an activity constituted enjoyable leisure or hard or tedious work or home production. The NTA approach, referred to as the "evaluated time-use" approach, relies on individuals' own evaluations of their emotional experiences during their various uses of times. In it, respondents can express emotions in a multidimensional fashion: they can be happy, tired, and stressed, all at the same time, during a certain activity or situation.

Just as their predecessors from the mid-1980s, the proponents of the NTA base their theory on a critique to the National Income Accounts (NIA) system, as such measures (e.g., GDP per capita or consumption per capita) only represent a piece of total welfare. For them, well-being is more than economic output and material consumption. Notably, certain factors that may contribute to economic output (pollution due to increased production) may actually be detrimental to people's well-being. In contrast, factors that contribute to well-being (socializing with friends) are not measured by national income. Similarly to Juster and colleagues, Krueger et al. (2009b) do not intend to substitute the NIA, but to complement it, recognizing that the NTA is still incomplete and only provides a partial measure of society's well-being. Nonetheless, their evaluated time-use approach does provide an additional valuable indicator of society's well-being. Moreover, the NTA can offer analytical and political advantages that may not be available from other type of measures of subjective well-being including those of overall life satisfaction.

The NTA framework is built on Juster's concept of the "process benefits" of activities (Juster 1985), defined as the set of satisfactions that emerge from activities themselves. Juster, however, did not link the satisfaction evaluation to the specific activity reported, but to how individuals enjoyed the activity or situation *in general*. Such an approach failed to capture what people actually experienced and resulted in profound discrepancies between concurrent and retrospective reports of affective experience (Schwarz et al. 2009). Therefore, in the NTA framework, time diary methods are utilized in a way that respondents connect specific events that actually occurred to the way they affectively experienced them. Four potential biases are prevented: (1) respondents do not need to develop a theory of how much they should be enjoying that activity in general in order to construct an answer to the question; (2) respondents will be less sensitive to the interviewers' opinion about them, as one is only talking about an specific instance (i.e., respondents will feel less self-conscious by reporting that one particular time they were not happy while taking care of their children); (3) respondents are not put in the position of needing to accurately aggregate their experiences over many times they engaged in a particular activity in order to provide a "general activity judgment"; and (4) the potential for selection bias will be less likely, as respondents will not need to choose from their past memories the best or worst moments of a particular type of activity in which they were engaged (Krueger et al. 2009b).

Finally, within the NTA approach, the uncertainty of whether individuals interpret enjoyment scales in an interpersonally comparable way is potentially better handled. The NTA framework proposes the U-index, where the U stands for "unpleasant" or "undesirable," with which the authors address the problem of comparability by focusing on measuring the proportion of time an individual spends in an unpleasant state. This approach allows for the computation of an average U-index for a group of individuals. According to its proponents, the virtues of this statistic are that it can be immediately understandable and serves as an ordinal measure *at the level of feelings*.

Two serious critiques have emerged toward the NTA's approach. First, the fact that reducing emotional experiences to a dichotomous characterization (pleasant or unpleasant) reduces the amount of information about the intensity of positive or negative emotions. The second critique is that the NTA only provides information about episodic feelings and cannot account for people's general sense of satisfaction with their lives as a whole. The best example for this is given by Loewenstein (2009) who argues that traveling to a new a different country is fraught with uncomfortable and unpleasant situations (e.g., airport long lines). However, the experiences gained from traveling may be extremely valuable and can make one a happier and wiser person. In spite of these criticisms, Krueger and his colleagues (2009a, b), along with numerous other prominent authors, have been able to provide empirical evidence that self-reported affect, even reduced to a binary scale, can predict important (objective) life outcomes, especially with regard to the quality of individuals' social life, work stability, longevity, and the quality of health. Significantly, in the areas of health psychology and behavioral medicine, it has been shown that positive and negative affect play a central role in health outcomes, particularly in connection with the translation of the psychosocial environment into physiological states. In sum, collecting emotional experiences directly connected to actual occurrences has proven to be useful in terms of predicting future crucial life outcomes, and thus, there is "signal" in people's self-reports of their affective experiences which is possible to be analyzed and interpreted.

4.2 Event History Calendar Methods to Assess Health-Related Issues

Calendars have the capacity to more fully capture concurrent activities or events, as they not only capture incidence of events, but also their timing and patterns (Stafford 2009; Barber et al. 2016). Indeed, calendars allow for the examination of timing and sequencing of events in different domains and thus provide a rich picture of potential causal mechanisms in the development of a person's well-being (Barber et al. 2016). Likewise, calendar interviewing methodology has been shown to produce high-quality retrospective reports. Such positive outcomes result from the ability of calendars to encourage respondents to reconstruct periods of social (e.g., residence, marriages), economic (e.g., employer names) or health-related (e.g., tobacco use) episodes of activity or statuses, by using chronological time and their previous own experiences memory cues (Belli et al. 2009b). Studies using calendars can include a variety of modalities, including face-to-face and telephone modes, paper-and-pencil and computer-assisted interviewing methods, and life course and shorter reference periods (Belli et al. 2001b, 2007; Yoshihama 2009).

Because of these advantageous properties, calendars have been applied to several areas of research related to the diverse components of well-being, such as education, employment, and, more recently, to the consequences of exposure to political violence (McNeely et al. 2015; Barber et al. 2016). Among the various components of well-being, using calendars to measure health conditions (e.g., women's sexual health, alcohol abuse, mental health, adolescent health, health status of the elderly) has been dominant.

One area of study using the calendar method is that related to adolescent health research (Martyn 2009; Luke et al. 2011; Martyn et al. 2013). This topic has garnered interest because of the high rates at which adolescents get involved in risky sexual behaviors. The 2013 CDC Youth Risk Behavior Surveillance Survey of US high school students showed that 47% engaged in intercourse, 15% had intercourse with four or more people, and 22.4% drank alcohol or used drugs before their last sexual intercourse (Centers for Disease Control and Prevention 2014). Ensuing problems include the increasing rates of STDs among the 15- to 19-year-old population as well as unwanted pregnancies (Martyn et al. 2013). In an effort to improve adolescent report and awareness of their sexual risk histories, and to improve the communication between adolescents and their health care providers, Martyn and colleagues (Martyn et al. 2006; Martyn 2009) developed a self-administered calendar to study this issue. In addition to the privacy benefit of the self-administered questionnaire, by which adolescents felt freer to report potentially sensitive information without concerns about being embarrassed, the studies were able to show that one of the main advantages of using a calendar was that it provided adolescents with the opportunity to not only record and view their behavior and life events over a 3-year period, but also reflect on the interrelationships between behaviors and events. According to Martyn and colleagues, the calendar not only allowed the collection of risk behaviors, but also fulfilled the objective of achieving an increased risk perception and communication among the adolescents involved in the study. Importantly, such increased awareness is associated with the ability of calendars to visually show information that is contextually and temporally linked, encouraging adolescents to put their lives in context, and allowing them to see more clearly the consequences of their decisions across their different, though interrelated, life domains. Additionally, the self-administered calendar enabled a rich qualitative discussion about the reasons they engaged in risky behaviors, as well as their future behavioral intentions. What is more, Martyn et al. (2006) showed that adolescents reported decreased risk intentions and behaviors. Indeed, in a 3-month follow-up survey, adolescents reported increased abstinence from sexual activity and of the use of tobacco, alcohol, or marihuana (Martyn 2009). Calendars were shown to be a powerful tool to not only analyze current adolescent health risk behavior and predict future developments, but to also devise possible intervention research studies that can have an impact on adolescent's future health and well-being.

The second example of the use of calendars related to health and well-being is the Pitt County Study in North Carolina, an investigation of social, economic, and behavioral precursors of cardiovascular disease in Southern African Americans. The objective of the study was to link information on childhood socioeconomic condition, obtained retrospectively, to health outcomes in African American adults (Belli et al. 2009a). Importantly, this was the first time in which a calendar was used to measure economic and psychosocial factors relevant to life course social epidemiology research. The calendar instrument that was utilized included eight conceptual life domains, designed to collect periods of stability and transitions between these periods for the entire life course. Smoking was the domain corresponding to the health behavior of interest. The other domains tried to capture key dimensions of social and economic well-being and included the following: household utilities, the family's exposure to economic hardship, social and material life conditions, organizational memberships, relationships with parents/guardians during the respondent's childhood, relationships with other family members, and the individual's exposure to unfair treatment (Belli et al. 2009a).

In the Pitt County study, the calendar was able to produce retrospective data of reasonably high quality, as was expected from an instrument designed to optimize the accuracy of autobiographical recall. Additionally, it was found that the calendar also contributed valuable data to other studies with original research questions within a population with many preventable risk factors for chronic diseases (See, for instance, James et al. 2006). Finally, the average administration time was of 30 min, which was confirmation of the practicality of this instrument for the use in large-scale, community-based health research. The Pitt County study was innovative by implementing a computerized calendar. Design improvements are still possible, especially to overcome the disadvantage that computerized instruments cannot be readily shared with respondents for them to view the calendar elements together with the interviewer. However, in spite of such challenges, it was demonstrated that it is feasible and useful to utilize computerized calendar elements to conduct life course epidemiological research, as the retrospective data produced by the calendar method were able to expose important associations among early life socioeconomic conditions and later life risks damaging to well-being such as obesity, hypertension, and cigarette smoking.

5 Conclusion and Future Directions

The predominant economic tradition has tended to equate an increase in national income statistics with an increase in well-being. This perspective can be misleading as it does not acknowledge that well-being is much more than material welfare and is shaped by a multidimensional set of factors. Notably, what ultimately matters when it comes to "quality of life" are not objective wealth conditions, but rather the degree of enjoyment of engaging in activities associated with production and consumption (Juster 1985; Gershuny and Halpin 1996). This realization paved the way to the introduction of a new paradigm to assess well-being, one which involves the analysis of how time is experienced in people's lives.

Such analyses necessarily involve methods that can reliably capture complex interrelations and dynamics, as well as subjective perceptions of personal and societal well-being. Empirical research has demonstrated advantages of calendar and time diary methods in the collection of well-being data. In contrast to conventional standardized questionnaires, time diaries and calendars place activities in context. In consequence, a broad range of subjective and contextual data can be collected through these methods (e.g., where, who with, perceived satisfaction, level of stress, and so on), with the advantage that respondents and interviewers can use their own words and repair misunderstandings as afforded by a flexible approach to interviewing. Moreover, such additional contextual details are not only important because of the extra information they provide, but because they contribute to enhance the quality of the data collected. In effect, Belli and colleagues (2009b) have concluded that compared to standardized interviewing methods, the temporal cues that emanate from calendar and time diary interviews assist respondents in providing retrospective reports more completely and accurately, even when these are of a subjective nature (Belli et al. 2012).

The use of time diaries and calendars to assess well-being is an emerging method that can be further developed. Comparable information on well-being across time and countries collected using these methods is still lacking. Efforts from international bodies such as the Better Life Index from Organization for Economic Co-operation and Development (OECD) are noteworthy in that they wish to incorporate more dimensions to well-being besides material welfare. However, they are still based on averages and not on reliably and validly measured indicators of subjective satisfaction with the various domains of life, indicators that may capture the "final output" of what makes life worthwhile.

References

Agrawal S, Sobell MB, Carter Sobell L. The timeline followback: a scientifically and clinically useful tool for assessing substance use. In: Belli RF, Stafford FP, Alwin DF, editors. Calendar and time diary methods in life course research. Thousand Oaks: Sage; 2009. p. 57–68.

Balán J, Browning HL, Jelin E, Litzler L. A computerized approach to the processing and analysis of life histories obtained in sample surveys. Behav Sci. 1969;14(2):105–20.

Barber BK, McNeely C, Olsen JA, Belli RF, Doty SB. Long-term exposure to political violence: the particular injury of persistent humiliation. Soc Sci Med. 2016;156:154–66.

Belli RF. The structure of autobiographical memory and the event history calendar: potential improvements in the quality of retrospective reports in surveys. Memory. 1998;6(4):383–406.

Belli RF. Autobiographical memory dynamics in survey research. In: Perfect TJ, Lindsay DS, editors. The sage handbook of applied memory. Thousand Oaks: Sage; 2013. p. 366–84.

Belli RF, Al Baghal T. Parallel associations and the structure of autobiographical knowledge. J Appl Res Mem Cogn. 2016;5(2):150–7.

Belli RF, Callegaro M. The emergence of calendar interviewing: a theoretical and empirical rationale. In: Belli RF, Stafford FP, Alwin DF, editors. Calendar and time diary: methods in life course research. Thousand Oaks: Sage; 2009. p. 31–52.

Belli RF, Lepkowski JM, Kabeto MU. The respective roles of cognitive processing difficulty and conversational rapport on the accuracy of retrospective reports of doctor's office visits. Proceedings of the Seventh Conference on Health Survey Research Methods, Hyattsville; 2001a. p. 197–203.

Belli RF, Shay WL, Stafford FP. Event history calendars and question list surveys: a direct comparison of interviewing methods. Public Opin Q. 2001b;65(1):45–74.

Belli RF, Lee EH, Stafford FP, Chou C. Calendar and question-list survey methods: association between interviewer behaviors and data quality. J Off Stat. 2004;20(2):185–218.

Belli RF, Smith LM, Andreski PM, Agrawal S. Methodological comparisons between CATI event history calendar and standardized conventional questionnaire instruments. Public Opin Q. 2007;71(4):603–22.

Belli RF, James SA, Van Hoewyk J, Alcser KH. The implementation of a computerized event history calendar questionnaire for research in life course epidemiology. In: Belli RF, Stafford FP, Alwin DF, editors. Calendar and time diary methods in life course research. Thousand Oaks: Sage; 2009a. p. 225–38.

Belli RF, Stafford FP, Alwin DF. The application of calendar and time diary methods in the collection of life course data. In: Belli RF, Stafford FP, Alwin DF, editors. Calendar and time diary: methods in life course research. Thousand Oaks: Sage; 2009b. p. 1–9.

Belli RF, Agrawal S, Bilgen I. Health status and disability comparisons between CATI calendar and conventional questionnaire instruments. Quality & Quantity. 2012;46(3):813–28.

Belli RF, Bilgen I, Al Baghal T. Memory, communication, and data quality in calendar interviews. Public Opin Q. 2013;77(S1):194–219.

Bevans GE. How working men spend their spare time. New York: Columbia University Press; 1913.

Bilgen I, Belli RF. Comparison of verbal behaviors between calendar and standardized conventional questionnaires. J Off Stat. 2010;26(3):481–505.

Blum ZD, Karweit NL, Sørensen AB. A method for the collection and analysis of retrospective life histories. Baltimore: John Hopkins University; 1969.

Centers for Disease Control and Prevention. Youth risk behavior surveillance – United States, 2013. MMWR. 2014;63(SS-4):1–172.

Clark HH, Schober MF. Asking questions and influencing answers. In: Tanur JM, editor. Questions about questions: inquiries into the cognitive bases of surveys. New York: Russell Sage Foundation; 1992. p. 15–48.

Conrad FG, Schober MF. Clarifying question meaning in a household telephone survey. Public Opin Q. 2000;64(1):1–28.

Dow GK, Juster FT. Goods, time and well-being: the joint dependence problem. In: Juster FT, Stafford FP, editors. Time, goods, and well-being. Ann Arbor: Institute for Social Research, University of Michigan; 1985. p. 397–413.

Fowler FJ Jr, Cannell CF. Using behavioral coding to identify cognitive problems with survey questions. In: Schwarz N, Sudman S, editors. Answering questions: methodology for determining cognitive and communicative processes in survey research. San Francisco: Jossey-Bass; 1996. p. 15–36.

Fowler FJ Jr, Mangione TW. Standardized survey interviewing: minimizing interviewer-related error. Newbury Park: Sage; 1990.

Freedman D, Thornton A, Camburn D, Alwin DF, Young-DeMarco L. The life history calendar: a technique for collecting retrospective data. Sociol Methodol. 1988;18:37–68.

Gershuny J, Halpin B. Time use, quality of life and process benefits. In: Offer A, editor. In pursuit of the quality of life. Oxford: Oxford University Press; 1996. p. 188–210.

Houtkoop-Steenstra H. Interaction and the standardized survey interview. Cambridge: Cambridge University Press; 2000.

Jabine TB, Straf ML, Tanur JM, Tourangeau R, editors. Cognitive aspects of survey methodology: building a bridge between disciplines. Report of the advanced research seminar on cognitive aspects of survey methodology. Washington, DC: National Academies Press; 1984.

James SA, Van Hoewyk J, Belli RF, Strogatz DS, Williams DR, Raghunathan TE. Life-course socioeconomic position and hypertension in African American men: the Pitt county study. Am J Public Health. 2006;96(5):812–7.

Juster FT. Conceptual and methodological issues involved in the measurement of time use. In: Juster FT, Stafford FP, editors. Time, goods, and well-being. Ann Arbor: Institute for Social Research, University of Michigan; 1985. p. 19–32.

Juster FT. Rethinking utility theory. J Behav Econ. 1990;19(2):155–79.

Juster FT, Stafford FP, editors. Time, goods, and well-being. Ann Arbor: Institute for Social Research, University of Michigan; 1985.

Juster FT, Courant PN, Dow GK. A theoretical framework for the measurement of well-being. Rev Income Wealth. 1981;27(1):1–31.

Juster FT, Courant PN, Dow GK. A conceptual framework for the analysis of time allocation data. In: Juster FT, Stafford FP, editors. Time, goods, and well-being. Ann Arbor: Institute for Social Research, University of Michigan; 1985. p. 113–31.

Juster, F. T., Stafford, F. P. The Allocation of Time: Empirical Findings, Behavioral Models, and Problems of Measurement. Journal of Economic Literature. Vol. 29, No. 2 (1991), pp. 471–522.

Krueger AB. Introduction and overview. In: Krueger AB, editor. Measuring the subjective well-being of nations: national accounts of time use and well-being. Chicago: University of Chicago Press; 2009a. p. 9–86.

Krueger AB, editor. Measuring the subjective well-being of nations: national accounts of time use and well-being. Chicago: University of Chicago Press; 2009b.

Krueger AB, Kahneman D, Fischler C, Schkade D, Schwarz N, Stone AA. Time use and subjective well-being in France and the US. Soc Indic Res. 2009a;93(1):7–18.

Krueger AB, Kahneman D, Schkade D, Schwarz N, Stone AA. National time accounting: the currency of life. In: Krueger AB, editor. Measuring the subjective well-being of nations: national accounts of time use and well-being. Chicago: University of Chicago Press; 2009b. p. 9–86.

Loewenstein G. That which makes life worthwhile. In: Krueger AB, editor. Measuring the subjective well-being of nations: national accounts of time use and well-being. Chicago: University of Chicago Press; 2009. p. 87–106.

Luke N, Clark S, Zulu EM. The relationship history calendar: improving the scope and quality of data on youth sexual behavior. Demography. 2011;48(3):1151–76.

Martyn KK. Adolescent health research and clinical assessment using self-administered event history calendars. In: Belli RF, Stafford FP, Alwin DF, editors. Calendar and time diary methods in life course research. Thousand Oaks: Sage; 2009. p. 69–86.

Martyn KK, Reifsnider E, Murray A. Improving adolescent sexual risk assessment with event history calendars: a feasibility study. J Pediatr Health Care. 2006;20(1):19–26.

Martyn KK, Saftner MA, Darling-Fisher CS, Schell MC. Sexual risk assessment using event history calendars with male and female adolescents. J Pediatr Health Care. 2013;27(6):460–9.

Maynard DW, Houtkoop-Steenstra H, Schaeffer NC, Van der Zouwen J, editors. Standardization and tacit knowledge: interaction and practice in the survey interview. New York: Wiley; 2002.

McNeely C, Barber BK, Spellings C, Belli RF, Giacaman R, Arafat C, Mallouh MA. Political imprisonment and adult functioning: a life event history analysis of Palestinians. J Trauma Stress. 2015;28(3):223–31.

Pember-Reeves M. Round about a pound a week. London: Bell; 1913.

Pentland WE, Harvey AS, Lawton MP, McColl MA. Time use research in the social sciences. New York: Springer; 1999.

Phipps PA, Vernon MK. Twenty-four hours: hours: an overview of the recall diary method and data quality in the American time use survey. In: Belli RF, Stafford FP, Alwin DF, editors. Calendar and time diary: methods in life course research. Thousand Oaks: Sage; 2009. p. 109–20.

Robinson JP. The validity and reliability of diaries versus alternative time use measures. In: Juster FT, Stafford FP, editors. Time, goods, and well-being. Ann Arbor: Institute for Social Research, University of Michigan; 1985. p. 33–62.

Schober MF, Conrad FG. Does conversational interviewing reduce survey measurement error? Public Opin Q. 1997;61(4):576–602.

Schwarz, N. (1996). Cognition and communication. Judgmental biases, research methods, and the logic of conversation. New York: Psychology Press.

Schwarz N, Kahneman D, Xu J. Global and episodic reports of hedonic experience. In: Belli RF, Stafford FP, Alwin DF, editors. Calendar and time diary methods in life course research. Thousand Oaks: Sage; 2009. p. 157–74.

Stafford FP. Timeline data collection and analysis: time diary and event history calendar methods. In: Belli RF, Stafford FP, Alwin DF, editors. Calendar and time diary methods in life course research. Thousand Oaks: Sage; 2009. p. 13–30.

Stiglitz JE, Sen A, Fitoussi J. Mismeasuring our lives: why GDP doesn't add up. New York: The New Press; 2010.

Suchman L, Jordan B. Interactional troubles in face-to-face survey interviews. J Am Stat Assoc. 1990;85(409):232–41.

Tanur JM, editor. Questions about questions. Inquiries into the cognitive bases of surveys. New York: Russell Sage Foundation; 1992.

Thomas WI, Znaniecki F. The polish peasant in Europe and America. Monograph of an immigrant group. Boston: The Gorham Press; 1918.

Tourangeau R. Cognitive sciences and survey methods (appendix A). In: Jabine TB, Straf ML, Tanur JM, Tourangeau R, editors. Cognitive aspects of survey methodology: building a bridge between disciplines. Report of the advanced research seminar on cognitive aspects of survey methodology. Washington, DC: National Academy Press; 1984. p. 73–100.

Tourangeau R, Rips LJ, Rasinski K. The psychology of survey response. New York: Cambridge University Press; 2000.

Yoshihama M. Application of the life history calendar approach: understanding women's experiences of intimate partner violence over the life course. In: Belli RF, Stafford FP, Alwin DF, editors. Calendar and time diary methods in life course research. Thousand Oaks: Sage; 2009. p. 135–55.

Body Mapping in Research

70

Bronwyne Coetzee, Rizwana Roomaney, Nicola Willis, and Ashraf Kagee

Contents

1	Definition of Body Mapping	1238
2	The Process of Body Mapping	1239
3	History of Body Mapping	1240
4	When Is Body Mapping Used?	1240
5	The Varied Uses of Body Mapping	1241
6	Examples of Body Mapping Studies	1241
7	Things to Consider When Planning a Study Using Body Mapping	1243
	7.1 Will Body Mapping Facilitate Answering the Research Question?	1243
	7.2 Has the Purpose of the Body Map Been Made Clear in the Study?	1244
	7.3 Is Body Mapping an Appropriate Technique to Use with the Participants?	1244
	7.4 What Other Qualitative Methods will Be Used in Conjunction with Body Mapping?	1244
	7.5 How Many Contact Sessions will Be Required with Participants?	1245
	7.6 How Structured will the Body Mapping Sessions Be?	1245
	7.7 How and What Data will Be Analyzed?	1245
	7.8 How to Conduct a Body Mapping Study	1246
8	An Example of a Study Using Body Mapping	1246
	8.1 Subjective Experiences of Depression	1249
	8.2 Idioms of Distress Used by Adolescents Living with HIV	1249
9	Methodological Rigor	1251
10	Ethical Aspects of Body Mapping	1251
11	Conclusion and Future Directions	1253
References		1253

B. Coetzee (✉) · R. Roomaney · A. Kagee
Department of Psychology, Stellenbosch University, Matieland, South Africa
e-mail: bronwyne@sun.ac.za; rizwanaroomaney@sun.ac.za; skagee@sun.ac.za

N. Willis
Zvandiri House, Harare, Zimbabwe
e-mail: nicola@zvandiri.org

© Springer Nature Singapore Pte Ltd. 2019
P. Liamputtong (ed.), *Handbook of Research Methods in Health Social Sciences*,
https://doi.org/10.1007/978-981-10-5251-4_3

Abstract

This chapter describes the methodology of body mapping, a visual technique that is used to collect qualitative data from participants about their subjective experiences pertaining mainly to bodily experiences. We begin with a definition of body mapping and provide an account of its history. We describe the process of conducting a body mapping study and offer some examples of when this approach is used most appropriately in its various forms. In preparing to use a body mapping approach, researchers should be mindful of whether body mapping is the best approach to answer the research question; whether the purpose of the body map been made clear in the study; whether it is an appropriate technique to use with participants; what other qualitative methods will be used in conjunction with body mapping; how many contact sessions will be required with participants; how structured the body mapping sessions will be; and how the data will be analyzed. We provide a detailed example of how to conduct a body mapping study and call attention to important considerations such as ensuring methodological rigor and the ethical aspects of using this approach. Body mapping is an innovative methodological technique that is often able to capture the imagination of research participants. Our aim in this chapter is to convince readers that body mapping has its place as a methodological approach alongside a range of other approaches in social and behavioral research.

Keywords

Body mapping · Research · Methods · Qualitative · Visual methodology · Social science

1 Definition of Body Mapping

Body maps or body-mapping, as it is commonly referred to in the literature, is both a therapeutic technique and research tool that prioritizes the body as a way of exploring knowledge and understanding experience. Life-size body drawings are either drawn or painted to visually depict aspects of people's lives, their bodies, and the world they inhabit (Gastaldo et al. 2012). These life-size artworks have been likened to totems (sacred objects or symbols) as they often contain symbolic value. The meaning of a body map may be fully understood only by the accompanying story and experience of its creator.

Researchers in the human and social sciences often refer to body mapping in research as body-map storytelling. Body mapping is considered as "a data generating research method used to tell a story that visually reflects social, political and economic processes, as well as individuals' embodied experiences and meanings attributed to their life circumstances that shape who they have become" (Gastaldo et al. 2012, p.10). Body mapping is a visceral approach to data collection and elicits data pertaining to the emotions of the body (Sweet and Escalante 2015).

Sweet and Escalante (2015) have identified three broad elements that constitute a body mapping approach, namely, (1) the life sized body map, (2) the testimonio, and (3) the key. The body map is a pictorial outline of the participant's body and is therefore a visual representation of his or her physical form. The *testimonio* is a short story narrated by the participant that is activated by the body map and that pertains to the research question at hand. The narrative is recounted in the first person and is a detailed description of each visual element that is documented on the body map.

Body mapping falls under the umbrella of participatory qualitative research (see also ▶ Chap. 17, "Community-Based Participatory Action Research"). It is often used to complement traditional qualitative data collection techniques such as interviews and focus groups. The approach is rooted in a therapeutic process known as narrative therapy (Santen 2014). Narrative therapy seeks to conceptualize psychological problems as distinct from the individual person and assumes that patients have many skills, areas of competency, assumptions, values, beliefs, and abilities that may be harnessed to help them ameliorate the impact their problems have on their lives (Morgan 2000). The combination of body maps and narrative therapy gives rise to a creative therapeutic technique that allows for the expression of individual experience through visual art.

2   The Process of Body Mapping

Participants who are recruited into a study involving the technique of body mapping assume a horizontal lying position on a large sheet of paper which is placed on the floor. A second person, the researcher, an assistant, or co-participant, draws an outline of the person's body on the sheet of paper (see Fig. 1). The participant is then asked to examine the outline, which serves as a springboard for discussion in

Fig. 1 Body mapping example showing the activity of tracing outlines of the participant's body on large sheets of paper

the research interview. The researcher poses questions based on the outline, to which the participant responds by using colors, symbols, or words to represent their answers and thus also, their experiences. The evolving artwork provides a means for the individual to externalize somatic and emotional experiences in a way that is appropriate and understandable to them (Solomon 2002). The body map thus facilitates an interactional interviewing style where visual cues become the basis for probing (Cornwall 2002).

3 History of Body Mapping

In the early 1990s, Andrea Cornwall, an anthropologist from London, used body mapping with Zimbabwean women as a way to access local knowledge of reproductive health and contraception. In the context of ambiguity in language and the importance of indigenous knowledge when attempting to understand illness, Cornwall and her colleagues showed that body maps were useful ways to understand the terminology used and perceptions held by the participants regarding their bodily processes.

In South Africa in the early 2000s, body mapping was used among women living with HIV (Braque 2008; MacGregor 2009). A clinical psychologist, Jonathan Morgan, in collaboration with the AIDS and Society Research Unit at the University of Cape Town, used the technique as part of the Memory Box Project, which was a community outreach program organized through the AIDS and Society Research Unit (Morgan 2003). Rather than artistically drawn bodies, memory boxes (i.e., boxes made out of cardboard and other recycled materials) were used by the research participants to visually depict their stories. The first series of memory boxes were put on display at the 2000 Durban AIDS conference. Soon after and following the success of using visual methodologies to depict experiences, Jane Solomon, a South African artist, developed a guide for body mapping and used it as an advocacy tool to call attention to the magnitude of the HIV/AIDS pandemic in Africa.

4 When Is Body Mapping Used?

Body mapping is known for its use as a therapeutic tool, especially among individuals living with HIV (Solomon 2002). Increasingly, body mapping has been used as a research method and has been applied creatively and constructively across various academic disciplines (Brett-MacLean 2009; MacGregor 2009; Gastaldo et al. 2012; Griffin 2014). For example, in Canada, Crawford (2010) used body mapping as a therapeutic tool for people with alexithymia or somatic issues following a trauma. Also in Canada, Denise Gastaldo et al. (2012) used body mapping as a research tool to document health problems and migration experiences of undocumented laborers in Toronto. In both studies, body mapping was an appropriate method as it provided a platform for participants to engage with researchers, which may otherwise have been challenging if only interviews had been used.

Body mapping as a methodological approach to facilitate the collection of qualitative data is often appropriate among children and adolescents who are not yet able to coherently express their experiences with language alone. With the use of this method, children are able to respond creatively to interview questions on a full size paper outline of themselves.

5 The Varied Uses of Body Mapping

Body mapping, as described by Solomon (2002), may be used as a treatment information and support tool, as an advocacy tool, an intergenerational dialogue tool, a team building tool, an art making tool, and a biographical tool.

As a treatment information and support tool, body maps can be used in workshops to educate individuals about a particular treatment regimen. For example, healthcare workers or adherence counselors can use body maps with individuals enrolling on antiretroviral therapy (ART). The body maps may be used to illustrate how ART strengthens immune systems and to create awareness of its potential side effects.

As an advocacy tool, body maps can be displayed at exhibitions or published in books or online with the purpose of creating awareness of various personal, social, economic, health, or political issues, for example, the effect of pollution on health. These displays become a platform on which activists communicate their feelings, thoughts, and ideas about these issues.

As an inter-generational dialogue tool, body maps bridge intergenerational gaps. They can therefore be used as a means for young children to communicate with their parents or caregivers and to establish trust and deepen emotional connections.

As a team building tool, body maps can be used to establish trust and form relationships between people. It can be used to highlight the importance of embracing differences and working together towards a mutual goal.

As an art making tool, body maps can be used to facilitate art classes. As a form of art, the technique is also likely to foster creativity among research participants, thus potentially yielding rich data.

As a biographical tool, body maps can be used to construct biographies (personal narratives about life histories).

6 Examples of Body Mapping Studies

Most research utilizing body mapping has been conducted in the fields of HIV, trauma, and violence, but researchers have also used body mapping to investigate issues such as wellbeing, pain, and risky behavior. As mentioned previously, body mapping is typically used in conjunction with other qualitative methods such as interviews and focus groups in research. Table 1 contains a summary of nine studies that have used body mapping as a methodological tool. The table is not an exhaustive list, but is presented as an example to illustrate the broad range of topics, samples, and methodology that have used body mapping.

Table 1 Summary of studies that have employed body mapping as a method

Study	Year	Aim	Methods	Notes
Brett-Maclean	2009	To explore identity in PLWA	BM workshops conducted with 3 HIV-positive men in South Africa	Results published as an exhibition
Crivello, Camfield and Woodhead	2009	To determine suitable methods to explore well-being among children	Study reports on three qualitative methods used in project (BM, life-course timeline, group discussion)	BM provided visual stimulus
Joarder, Cooper, and zaman	2014	To explore meaning and perceptions of death of elderly people in a Bangladeshi village	BM used in conjunction with individual interviews, informal discussions and recording of daily routines BM was used to generate narratives Only narratives were analyzed	
MacGregor	2009	To trace personal and political dimensions of HIV/AIDS in South Africa	BM workshops conducted by peer counselors Participants reflected on BMs	Data analysis was not clearly outlined
Maina, Sutankayo, Chorney and Caine	2014	To understand how	BM workshops conducted with nursing students Data collected using individual interviews in which three participants reflected on BM Interviews were transcribed and thematically analyzed	BM was used as an intervention tool and no data related to the BM exercises were analyzed
Apiyo	2012	To explore personal stories of violence	BM workshops conducted with women who experienced violence Narratives elicited from BMs were reported	
Senior, Helmer, Chenhall and Burbank	2014	To explore young people's perceptions of risk from STIs in rural Australia	BM workshops conducted with participants Data collected via interviews, group discussions, and BM	No guidelines on analysis

(*continued*)

Table 1 (continued)

Study	Year	Aim	Methods	Notes
Silva-Segovia	2016	To explore self-perception of an imprisoned woman in Chile	Case study design with one participant BM was one of a number of methods used Study also relied on intertextual analysis, autobiographical fragments, self-interpretation, and reinterpretation of pictures	No guidelines to analysis and integration of analysis
Tarr and Thomas	2011	To explore body mapping as a method for representing pain and injury	Mixed methods: Questionnaires, interviews, BM 205 dancers	No details on analysis

7 Things to Consider When Planning a Study Using Body Mapping

While the above studies provide good examples of body mapping research, there are a number of questions to consider in planning such a study. Some questions are provided below:

7.1 Will Body Mapping Facilitate Answering the Research Question?

The most important consideration when selecting any methodological approach is the research question. Thus, the way the research question is asked will inform the method that is selected. Body mapping is, therefore, only useful if it is appropriately matched to the research question. The approach typically lends itself to qualitative research in which the task at hand is to generate a thick description of research participants' experiences. In our experience, typical research questions with which body mapping may be used often pertain to somatic concerns, stress and trauma, and health and illness. Thus, body mapping is well suited to the fields of health psychology, behavioral medicine, and medical anthropology. Using the body map as a springboard for the interview conversation, the interviewer can create a rich narrative that reflects the participant's subjective experience of the research topic of interest.

7.2 Has the Purpose of the Body Map Been Made Clear in the Study?

Body maps can be used for a number of purposes such as providing participants with a means of expressing their thoughts and feelings, as an intervention or educational tool, or as a stimulus. The reason for using body mapping in research needs to be made explicit to the participant. For example, it can be explained that the body map is intended to elicit a rich narrative from participants about their experiences and the meaning they create from these experiences. It can also be explained to the participant that the body map is the main tool of data collection and provides the data that will then be analyzed.

7.3 Is Body Mapping an Appropriate Technique to Use with the Participants?

Body mapping is especially appropriate when participants have difficulty expressing themselves verbally. The visual experience of appraising one's self-constructed body map provides a useful platform for verbal expression and elaboration. For example, Crivello et al. (2009) used body mapping to explore subjective well-being among children. In their study, body maps provided a useful starting point for discussions with children. The researchers stated that body mapping was a more appropriate technique than individual interviews as children found the novelty and creativity of body mapping to be more appealing than interviews on their own. The similarity and overlap of body mapping with play made the approach especially suited to the research question and the population of interest, i.e., children.

7.4 What Other Qualitative Methods will Be Used in Conjunction with Body Mapping?

One of the key components of qualitative research is to establish credibility and trustworthiness of the research findings. Using more than one qualitative method creates opportunities for such rigor to be applied, for example, by supplementing body maps with in-depth or semi-structured interviews or focus groups. Joarder et al. (2014) used body mapping in combination with in-depth individual interviews, informal discussions, and recording of daily routines in exploring a sample of elderly participants' perceptions of the meaning of death. Similarly, Silva-Segovia (2016) used body mapping, intertextual analysis, autobiographical fragments, self-interpretation, and reinterpretation of pictures in analyzing the case study of an imprisoned woman in Chile. These examples illustrate that a range of qualitative techniques can be useful in combination with body mapping. Tarr and Thomas (2011) combined qualitative and quantitative methods in their explorations of pain and injury among dancers in the United Kingdom. In their study, body mapping and interviews were used in conjunction with questionnaires to answer the research questions pertaining to pain and injury as a consequence of dance.

7.5 How Many Contact Sessions will Be Required with Participants?

It is necessary to decide on the number and length of the body mapping sessions and other forms of data collection. This decision is usually based on a number of factors, including the purpose of using body mapping, the characteristics of the participants, and the nature of the research question. If body mapping is used as part of an intervention or as part of an in-depth case study, the researchers may consider a series of body mapping workshops. For example, researchers using body mapping as a means of stimulating discussion may use one workshop, followed up by an interview or series of interviews. Main et al. (2014) used body mapping as an educational tool and conducted one 4-h long workshop with participants. Participants completed the body maps in groups of three to five participants, after which the researchers conducted in-depth interviews with each participant. There is no specific formula for making these decisions. The approach is fairly flexible and responsive to the research question at hand.

7.6 How Structured will the Body Mapping Sessions Be?

The structure of body mapping sessions can be designed to answer the research question. We suggest a semi-structured approach in which the content of the interview is guided by an interview schedule that uses open questions, probes, facilitative statements, paraphrasing, and reflection. The body map is therefore a springboard to the interview conversation and is intended to generate speech from the participant that forms the qualitative data that will then be analyzed.

7.7 How and What Data will Be Analyzed?

It is important for the researcher to consider whether to only analyze the body map, the accompanying narrative, or to include both forms of data in the analysis. One shortcoming of many existing body mapping studies is that researchers often failed to report how they analyzed the data. This omission was evident in a number of studies listed in Table 1 (MacGregor 2009; Tarr and Thomas 2011; Senior et al. 2014). In several studies, body mapping was used as a technique to elicit a narrative from participants, which was recorded, transcribed, and coded for analysis. For example, Apiyo (2012) conducted body mapping workshops with women who reported experiencing violence. The body maps themselves were not analyzed but instead used as a tool to elicit narratives from participants. Maina et al. (2014) used body mapping as an intervention tool with nursing students who were infected with HIV and taught others about the disease. In their study, the nursing students participated in a body mapping workshop that aimed to educate them about the social and personal aspects of HIV. The nurses were interviewed after the workshop and the interviews were transcribed and analyzed. The interviews were, therefore, the data collection tool, not the body mapping.

7.8 How to Conduct a Body Mapping Study

In Fig. 2, we describe seven steps to be taken when using body-mapping as a research method. These steps are: deciding on the research question, identifying the participants, developing the research protocol, collecting the data, practicing interview techniques, conducting the interviews, and recording and analyzing the data.

8 An Example of a Study Using Body Mapping

In this section, we provide an in-depth example of a study in which body mapping was used in an investigation of depression among adolescents in Zimbabwe. The study we describe was conducted among adolescents living with HIV in Zimbabwe, a group among whom depression has been found to be a serious mental health concern (Patel et al. 2007; Mellins and Malee 2013). In the context of limited research on the experience, impact, and manifestations of depression among Zimbabwean adolescents with HIV, one of the authors of this chapter, Nicola Willis, conducted a study using body mapping as a methodological technique. She sought to document the subjective experiences of depression among the adolescent participants, to identify the idioms of distress used by the participants, and to understand how the participants perceived their experiences as being addressed by families, communities, and service providers.

Nicola recruited 21 adolescents living with HIV who had a diagnosis of major depression from a community health program for children and adolescents living with HIV in Zimbabwe. The program provides community-based psychosocial support and health services to over 5,000 children and adolescents living with HIV in Zimbabwe (Jackson et al. 2015).

Body mapping was well suited to the adolescent population as this life stage is associated with significant physical and neurodevelopmental changes. In early adolescence, abstract and metaphorical cognitive abilities are still developing (Sawyer et al. 2012). Adolescents may be able to express themselves through body maps but may not be as able to reflect on or describe the meaning of what they have expressed. To this extent, body mapping methodology was well suited to investigate their subjective experiences and perceptions of the care they received at the community program.

Nicola sought to engage young HIV-positive research participants with depression in an exploratory process that would enable them to describe their experiences of depression and the ways in which they perceived care. She therefore selected a participatory, qualitative approach that drew on narrative therapy with body mapping as the methodological tool for data collection.

Narrative therapy is a therapeutic process that encourages the storyteller to regain a sense of authorship and re-authorship of their own experiences in the telling and retelling of their own story (White and Epston 1990) and to see their experiences as subjective and personal phenomena rather than as others perceive them. In narrative therapy, storytelling provides the opportunity for an exploration of the participant's experiences and provides in-depth insight into the ways in which participants interpret and understand their world. Narrative enquiry has been described as an effective

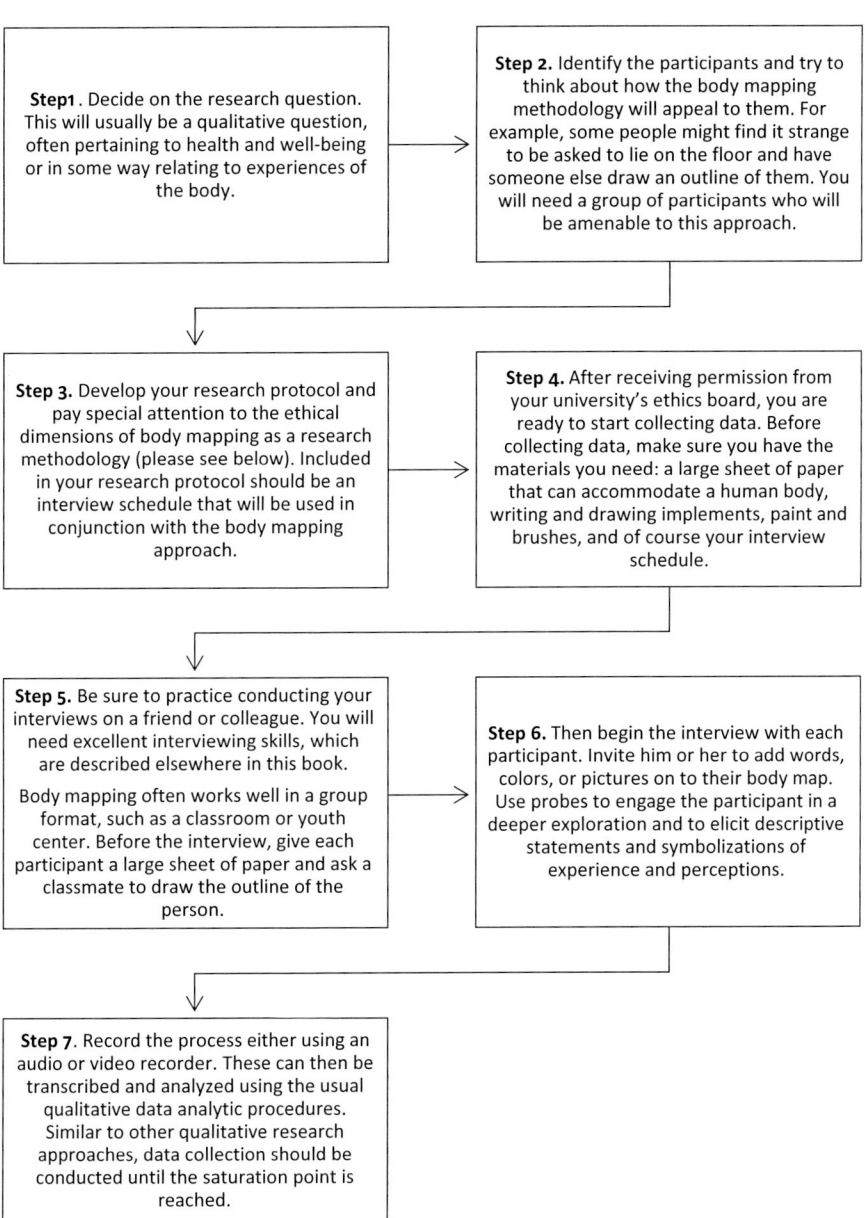

Fig. 2 Flow chart representing possible steps for undertaking research using body-mapping as a method

means to understand the complexities of the individual experience of health and illness. It has also been suggested that storytelling offers considerable value where the human experience of illness is unknown or unexplored (Greenhalgh 2001).

Nicola developed an in-depth interview schedule that was centered on the body mapping process. She used a semi-structured interview and body mapping guide (see below) to facilitate the interview so that she could engage participants in a creative dialogue around the research questions. The interview contained open-ended questions aimed at exploring participants' subjective experiences and perceptions of care. At the beginning of the process, she asked each participant to lie down on a large sheet of paper so that an outline of him or her could be drawn. She invited the participant to add words, colors, or pictures on to their body map and to assist them in responding to the questions. Nicola used probes to engage participants in an in-depth exploration of their experiences and to elicit descriptive statements and symbols of these experiences and perceptions. She developed the interview guide in consultation with one of her supervisors, a psychiatrist based in Zimbabwe, and counselors from the community program who had experience in body mapping.

As the interviews were based on the body mapping session, two types of data were generated, namely, audio data from the interview and visual data from the body map. Each interview was audio-recorded with the participant's consent. At the end of the interview, a photograph was taken of the body map. Two research assistants assisted with data collection throughout the study period. They were present during each interview to assist with field notes and practical issues associated with the body mapping itself, such as preparation of paints and water.

Each in-depth interview was transcribed and translated from Shona into English and the body maps were photographed. The transcripts were separately coded by two coders and the two sets of codes were compared and discrepancies resolved by discussion. Once the coding was consistent for both the transcripts, they were single-coded. Codes were then grouped into categories and emerging themes were identified following the general principles of thematic analysis (Attride-Stirling 2001; Braun and Clark 2006; see also ▶ Chap. 48, "Thematic Analysis").

The photographs of each body map were also analyzed along with the transcriptions. Common themes from the different interviews and body maps were then identified and were illustrated with quotes and images. Computer-assisted qualitative data analysis software (CAQDAS) programs, such as ATLAS.ti, are capable of managing several media including word and pdf documents, photo images, and audio files in nearly every available format. This software allows its user access to multiple media to formulate a meaningful interpretation of the data (Scientific Software Development 2003). In body mapping studies such as the one we describe, images from the body maps can be hyperlinked with quotations from the transcribed interviews. Codes emerging from the analysis and hyperlinked data can then be grouped together to reflect salient themes. The results of the study were organized in terms of the research questions, namely, to document the subjective experiences of depression among participants, to identify cultural idioms of distress, and to understand how depressed adolescents perceived their experiences as being addressed by families, communities, and service providers.

8.1 Subjective Experiences of Depression

Participants in the study described their lives as characterized by negative and traumatic experiences. They conveyed these descriptions through both their verbal narratives and the evocative imagery which they chose to paint on their body maps to illustrate their experiences and emotions. Seven main themes emerged from the data: (1) being different from others, (2) learning of their HIV status, (3) isolation and rejection, (4) loss and grief, (5) low self-worth, (6) lack of protection, and (7) the future. In addition, idioms of distress also emerged from the data.

8.2 Idioms of Distress Used by Adolescents Living with HIV

Thinking too much. Participants' idioms of distress were conveyed through both their verbal narratives and the words and images which they chose to paint on their body maps. The most commonly used term to describe depression was "thinking too much" or being "lost in thought" as a result of the events in their lives. Participants often chose to use a color they identified with depression such as black or red (see in Fig. 3 the painted parts of the body that they associated depression). The brain and heart were the parts of the body which they most commonly identified. One female participant used the color purple to link the different parts of her body affected by depression. She began by painting her head and her heart purple, explaining that these were the areas of the body that she associated with depression. However, she then painted her legs the same color, explaining that when she felt depression in her heart and head, it led her to wanting to walk to the dam to drown herself.

Stress. The next most commonly used term was "stress," which was represented by a specific color which participants chose and located on certain parts of their bodies. Participants tended to use black or red and painted this on the head of the drawn figure. Stress was often used in connection with "thinking a lot." One female participant explained that stress arises from deep thinking and in turn resulted in suicidal ideation.

Pain. Many participants referred to depression as the "pain" they experienced in their lives in relation to many of the themes reported above, including being different from others, their HIV status, their experiences of isolation, rejection, loss and grief, their feeling of being unprotected, and uncertainty for the future (see Fig. 3).

Darkness. Darkness was referred to widely when participants were describing their depression. It was illustrated by the use of color, with black being the most commonly used. They chose to paint specific places on the body with which they identified this darkness.

Suicidal ideation. Suicidal ideation was also described by several participants. Four participants named or painted the methods they considered to end their lives in the past, including poisoning, hanging, and drowning. In general, idioms of distress were identified and expressed through a combination of color, pictures, and words.

Perceptions of care. Participants identified four sources of care, namely, families, peers, and support groups, clinic staff. It was implicit in their narratives that psychosocial care could also help to prevent these difficult experiences and

Fig. 3 An 18-year-old female adolescent's body map depicting the words, "pain" and "suffer"

emotions. One female participant expressed her need for care and reliance on others for support by painting a stone in her heart which she colored brown.

Many participants referred to the care and support they had received from other young people living with HIV in the program. They illustrated these experiences by painting pictures of their friends and sharing the way peers had helped them to accept their HIV status and reduce their experiences of stress, isolation, and lack of hope for the future.

Participants frequently referred to the program's training and support center as a key source of care in their lives where they have found love and support. The program was often painted on to their maps and described as a place where young people were valued, loved, and supported.

At the end of each interview, participants were asked how they felt about having participated in the body mapping process and interview. All participants stated that they felt positively about the process, that they had found it helpful, and that it had helped them to share experiences and emotions which they had not previously done. They stated that they were able to express themselves and open up by drawing and writing what they had previously kept to themselves. Participants indicated that they recognized that this approach would help others to understand them better as it had enabled them to articulate thoughts and feelings which they had been unable to express to counsellors before.

The example above illustrates how body mapping may be used as a research tool to access thoughts and feelings related to their physical health.

9 Methodological Rigor

As body mapping is a method used to collect qualitative data, it is necessary to ensure trustworthiness of body mapping studies. Trustworthiness is central to qualitative research and researchers using body mapping are expected build trustworthiness into different aspects of their research design. According to Lincoln and Guba (1985), trustworthiness comprises four components namely credibility, transferability, dependability, and confirmability. Creswell (2014) has outlined eight procedures that can be used within qualitative research to improve trustworthiness. These procedures include prolonged engagement, triangulation, peer review and debriefing, negative case analysis, clarification of researcher bias, member checking, development of a rich, thick description, and external audit. (For a more detailed description, please see Liamputtong, 2013; Creswell 2014; see also ▶ Chap. 63, "Mind Maps in Qualitative Research").

In our assessment of studies that used body mapping as a research method, we noted that studies reported their methodological approaches inconsistently. For example, many studies did not explicitly state the purpose of using body mapping or did not provide sufficient detail of the body mapping process. More importantly, studies failed to name the methods used in analyzing the data and did not describe the process of analysis. The lack of detail regarding the rationale for using body mapping, the implementation of body mapping exercises, and the analysis of body mapping data may be viewed as threats to the trustworthiness of studies. As with all qualitative research, trustworthiness should be built into every phase of the research process, i.e., from conceptualization of a study to the reporting of results.

While no techniques for enhancing trustworthiness have been outlined specifically for body mapping, we make the following suggestions for increasing trustworthiness when conducting body mapping research. These techniques are not unique to body mapping but can also be found in other forms of visual methodologies and qualitative research in general. We have adapted Creswell's (2014) procedures to recommend the following (Table 2):

10 Ethical Aspects of Body Mapping

It is very important when explaining the body mapping technique to participants that they understand that they will be asked to lie down on the floor and that someone will outline their body using a pen marker. For some people, this will sound quite strange so the researcher might want to show them a picture of someone lying the ground and participating in such an exercise.

Similar to other research, informed consent is very important (see also▶ Chap. 106, "Ethics and Research with Indigenous Peoples"). This includes informed consent to participate in the research, to lie on the ground and for one's outline to

Table 2 Establishing rigor and trustworthiness in body mapping data

Recommendation	Example
Conduct member checks (credibility)	Discuss the results of your analysis with participants to see if they approve
Peer debriefing (credibility)	Discuss your study and specifically your findings with peers to see if they agree with your analysis
Triangulation (credibility)	Use other forms of data collection such as focus groups or interviews
Persistent observation (credibility)	Aim for prolonged interaction with participants
Admit lack of generalizability	Acknowledge that findings from the research may not be generalized, but can be transferable
Describe the process of data collection	Describe the process of body mapping
	1. Who conducted the exercise
	2. How many people attended
	3. Over how many sessions were the exercises conducted
	4. Describe instructions given to participants
	5. Describe interactions between data collector and participants
Describe the process of data analysis – Are the maps being analyzed or the narrative or both?	Describe which data are being analyzed, i.e., the body map, the narrative, or both
	If the body map was analyzed, describe the analysis process
	If the narrative was analyzed, describe the data analysis method used
Maintain an audit trail	Note all processes and attempt to verify as many aspects of the methodology as possible
Reflexivity	Practice reflexivity throughout the process

be drawn, to be interviewed, to be audio or video-recorded, and to have their body map photographed.

It is also necessary to obtain consent to use the data for specific purposes and to be specific about how data will be displayed. For example, participants need to consent to researchers displaying photos of the body maps either in journals or in exhibitions, even if their identity is disguised. Body maps cannot be exhibited unless researchers obtain explicit permission to exhibit the map.

Similar to any research, participating in body mapping research can cause participants to become distressed if their emotions are evoked during the process of data collection. It is important that the researcher puts in place a referral trajectory so that distressed persons can access psychological help if needed.

When referring to participants in the research report, the researcher should disguise their identities, for example, by using pseudonyms and removing any personal identifiers from the transcripts before these are analyzed.

Data labeled with an ID unique to that individual should be kept securely and separately from the interview transcriptions, audio data, and body maps. Names, addresses, and locator information should be kept securely for follow-up purposes in a lockable filing cabinet, accessed only by the research team.

11 Conclusion and Future Directions

Body mapping is an innovative methodological technique that is often able to capture the imagination of research participants. For many participants, it can seem playful and fun, thus making the process of participating in a research study a pleasant one. The process of having the participant lie down on a sheet of paper and having their outline drawn can also facilitate a warm and engaging professional relationship between the researcher and the interviewer. Such a relationship can be extremely helpful in eliciting useful data that may inform the findings of the research. We believe that body mapping has its place as a methodological approach alongside a range of others in social and behavioral research.

Body mapping can be of relevance when conducting cross-cultural studies and can bridge language barriers between researchers and participants. This chapter has demonstrated the usefulness of body mapping, especially when engaging children and adolescents in research. Body mapping is a relatively recent methodological innovation that remains underutilized in both therapeutic and research settings. Visual-based methods provide attractive ways to communicate human experiences to the public. We encourage researchers and therapists to consider body mapping as a creative, interactive technique when working with participants and patients.

References

Apiyo, N. Ododo wa: our stories, Voices magazine, 2, October 24, 2012, 2012. http://justiceandreconciliation.com/2012/10/ododo-wa-our-stories/. Accessed 18 Dec 2012.

Attride-Stirling J. Thematic networks: an analytic tool for qualitative research. Qual Res. 2001; 1(3):385–405.

Braque G. The moon, the stars and a scar: body mapping stories of women living with HIV/AIDS. Border Cross Mag Arts. 2008;27(1):58–65.

Braun V, Clarke V. Using thematic analysis in psychology. Qual Res Psychol. 2006;3(2):77–101.

Brett-Maclean P. Body mapping: embodying the self living with HIV/AIDS. CMAJ. 2009; 180(7):740–1.

Cornwall A. Body mapping: bridging the gap between biomedical messages, popular knowledge and lived experience. In: Cornwall A, Welbourne A, editors. Realizing rights: transforming approaches to sexual and reproductive well-being. London: Zed Books; 2002. p. 219–34.

Crawford A. If 'the body keeps the score': mapping the dissociated body in trauma narrative, intervention, and theory. UTQ Univ Tor Q. 2010;79(2):702–19.

Cresswell J. Qualitative inquiry and research design: choosing among five traditions. 3rd ed. Thousand Oaks: Sage; 2014.

Crivello G, Camfield L, Woodhead M. How can children tell us about their wellbeing? Exploring the potential of participatory research approaches within young lives. Soc Indic Res. 2009; 90(1):51–72.

Gastaldo D, Magalhaes L, Carrasco C, Davy C. Body-map story telling as research: Methodological considerations for telling the stories of undocumented workers through body mapping. Toronto.

2012. Available at: http://www.migrationhealth.ca/sites/default/files/Body-map_storytelling_as_reseach_HQ.pdf. Accessed 12 July 2015.

Greenhalgh T. Storytelling should be targeted where it is known to have greatest added value. Med Educ. 2001;35(9):818–9.

Griffin SM. Meeting musical experience in the eye: resonant work by teacher candidates through body mapping. Visions of Research in Music Education, 24. 2014. Retrieved from http://www.rider.edu/~vrme

Jackson H, Willis N, Dziwa C, Mawodzeke M, Pascoe M, Sherman J. Zvandiri: Supporting children and adolescents with HIV through the HIV care continuum. Harare, Zimbawe: Africaid; 2015.

Joarder T, Cooper A, Zaman S. Meaning of death: an exploration of perception of elderly in a bangladeshi village. J Cross Cult Gerontol. 2014;29(3):299–314.

Liamputtong P. Qualitative research methods. 4th ed. Melbourne: Oxford University Press; 2013.

Lincoln YS, Guba EG. Naturalistic inquiry. Beverly Hills: Sage; 1985.

MacGregor HN. Mapping the body: tracing the personal and the political dimensions of HIV/AIDS in khayelitsha, south africa. Anthropol Med. 2009;16(1):85–95.

Maina G, Sutankayo L, Chorney R, Caine V. Living with and teaching about HIV: engaging nursing students through body mapping. YNEDT Nurse Educ Today. 2014;34(4):643–7.

Mellins C, Malee K. Understanding the mental health of youth living with perinatal HIV infection: lessons learned and current challenges. J Int AIDS Soc. 2013;16(00):18593. https://doi.org/10.7448/IAS.16.1.18593.

Morgan A. What is narrative therapy. An easy-to-read introduction. Adelaide: Dulwich Centre Publications; 2000. p. 11–32.

Morgan J, the Bambanani women's group. Long life: positive HIV stories. Cape Twon: Double Storey Books; 2003.

Patel V, Flisher AJ, Hetrick S, McGorry P. Mental health of young people: a global public-health challenge. Lancet. 2007;369(9569):1302–13.

Santen B. Into the fear-factory: connecting with the traumatic core. Person-Centered Exp Psychother. 2014;13(2):75–93.

Sawyer SM, Afifi RM, Bearinger LH, Blakemore S-J, Dick B, Ezeh AC, Patton GC. Adolescence: a foundation for future health. Lancet. 2012;379(9826):1630–40.

Scientific Software Development. ATLAS.ti. Berlin. 2003. Retrieved from www.atlasti.com

Senior K, Helmer J, Chenhall R, Burbank V. 'Young clean and safe?' young people's perceptions of risk from sexually transmitted infections in regional, rural and remote Australia. Cult Health Sex. 2014;16(4):453–66.

Silva-Segovia J. The face of a mother deprived of liberty: imprisonment, guilt, and stigma in the norte grande, chile. Affilia J Women Soc Work. 2016;31(1):98–111.

Solomon J. A body mapping journey in the time of HIV and AIDS. A facilitation guide – psychosocial wellbeing series. Johannesburg: REPSSI; 2002.

Sweet EL, Ortiz Escalante S. Bringing bodies into planning: visceral methods, fear and gender violence. Urban Stud. 2015;52(10):1826–45.

Tarr J, Thomas H. Mapping embodiment: methodologies for representing pain and injury. Qual Res. 2011;11(2):141–57.

White M, Epston D. Narrative means to therapeutic ends. New York: WW Norton & Company; 1990.

Self-portraits and Maps as a Window on Participants' Worlds

71

Anna Bagnoli

Contents

1 Introduction 1256
2 Self-Portraits 1258
3 Maps 1262
4 Conclusions and Future Directions 1265
References 1266

Abstract

Visual and arts-based methods can be extremely beneficial to research investigating people's lives, subjectivities, and identities. Well suited to a participatory style of research, these methods work as an excellent support to an open style of interviewing and can help seeing the world from participants' own perspective, thus providing an insight into their own interpretation of their worlds. This chapter will review the use of two visual methods that I applied in the context of interviews in different research projects: a self-portrait with which I asked for a self-presentation narrative, and a map with which I encouraged participants to reflect on significant relationships in their lives. The use of visual methods as a support to interviewing can facilitate participants to think laterally and be more creative in their answers, and also enable them to take the lead in the interview and establish their own priorities. Simple drawing tasks and other creative arts-based methods can encourage reflection and help covering emotional and sensitive issues that might otherwise remain silent or underexplored. These methods also work well to make participants feel more at ease during an interview. The chapter will provide suggestions on how these methods could best be employed in a research study.

A. Bagnoli (✉)
Department of Sociology, Wolfson College, University of Cambridge, Cambridge, UK
e-mail: ab247@cam.ac.uk

© Springer Nature Singapore Pte Ltd. 2019
P. Liamputtong (ed.), *Handbook of Research Methods in Health Social Sciences*,
https://doi.org/10.1007/978-981-10-5251-4_5

Keywords

Self-portrait · Relational map · Visual methods · Arts-based methods · Participatory research

1 Introduction

Visual, arts-based methods and other creative approaches to research have much to offer to qualitatively oriented researchers. It is fair to say that the qualitative research community has over-relied on interviewing as their main source of data. While nobody would wish to deny the wealth of information and insight one can gain from a good interview, it is also clear that the choice of the interview as an instrument of data collection comes with many methodological limitations. A research interview assumes that the participating people will be able and willing to verbalize their thoughts and views, and that they will be comfortable with face-to-face interaction with a researcher (Mason 2002a). These are rather big assumptions to make, which actually work to the exclusion of a significant part of the population. Many people will feel uneasy sitting in front of a researcher to talk and this may concern, among others, children and young people, people who are not confident with their language skills, such as migrants, or people with intellectual disabilities.

Criticism to the efficacy of standard social science methods to adequately appreciate the changing realities of contemporary societies has been rife in recent years (Law 2004; Savage and Burrows 2007). As Heath and Walker point out (2012), some of these critical voices seem to have missed the emergence since the turn of the century of creative approaches to research, particularly in the field of youth studies, where the widely recognized need for more apt methods to study young people's everyday worlds has pushed researchers to be inventive about their craft (Bagnoli and Clark 2010).

Methodological innovation has especially moved in the direction of a pronounced engagement with visual methodologies (Harper 2012; Rose 2012) and more recently in a surge of interest with sensory approaches to research (Pink 2009), which aim to rely on the senses as a whole, rather than assigning vision the status of the main and privileged channel in the construction of knowledge. As part of this search for creative methods, many researchers have looked at the arts as a source of inspiration to develop novel and unconventional ways to look at the world (Leavy 2009; Knowles and Cole 2008). An interdisciplinary approach to research methods design, which blends the social sciences with the arts, can enliven social science methodologies and attune them to register everyday experience with a richness of detail and insight that eludes more conventional approaches. As Back (2012) argued, more imaginative and "artful" methods can help widen the sociological imagination and can very importantly be of crucial assistance in the attempt to democratize the research process (see also ▶ Chap. 62, "Personal Construct Qualitative Methods").

I see the arts as central to create participatory ways to do research, and my own interest in visual and arts-based methods has directly emerged from my commitment to seek novel and better ways to promote research inclusion. My research work has focused on the investigation of identities, which I have carried out holistically and with a participatory approach. Self-portraits and maps, two of the visual methods that I have applied, and the two methods that are the focus of this chapter, were specifically designed for visual elicitation (Rose 2012). They were, therefore, thought for use within the context of an interview, in order to promote better participation and to improve the quality of data that could be collected within a face-to-face interview session.

Both these methods rely on the use of respondent generated images (Prosser and Loxley 2008). The possibility of using images that have been created by participants, either contextually or before the data collection event, is an important feature of these methods. The grounding of data collection on the basis of participants' own images goes to emphasize the contribution and involvement of the people who take part in research. From this perspective, even a very simple drawing task can take on a huge significance in terms of making participants feel that they can contribute to shape the project. This is a crucial point on several levels. First, it enables participants to make the project their own and feel that they can decide and direct its contents. It is one small but significant way in which the research process can be opened up and made more democratic, and participants can be assigned a leading role, rather than merely being viewed as passive objects of study.

Second, widening the research scope to nonverbal dimensions, in this case specifically a visual dimension, is crucial to the production of a different kind of knowledge, a knowledge that is holistic and does not privilege words as the main format for its construction, sharing, and communication. The design of multiple ways to take part in research, which do not necessarily rely on the assumption that verbal interaction with an interviewer will always be the most appropriate channel for the communication of experience, is finalized to reflect people's own preferred expressive styles and thus to encourage participation (Bagnoli 2009). The visual angle of the two methods discussed in this chapter was specifically aiming to appreciate even those dimensions of experience that are not easily put into words.

Reliance on a participatory framework means that researchers will attempt to follow participants' guidance and will consider them as the experts on their own lives. This implies for a research team engaging in active reflexivity throughout the research project (Mason 2002b), critically looking at their own assumptions and examining how their own social positioning impacts on the research process and locates them in relation to the participants. As some scholars have noted (Luttrell 2010), the application of participatory methods opens important issues with regards to how participants' views are then considered in the final reading and interpretation of results. On an analytical level, the reading of data should distinguish between participants' and researchers' meanings, and ideally, take both into account. In a participatory perspective, the effort to provide a transparent picture of any differences or clashes in interpretation should then be at the forefront, if we properly aim to engage in a co-construction of meanings and in a dialogical and reflexive process of knowledge building (Hesse-Biber and Piatelli 2010).

2   Self-Portraits

While working on the methodology for my PhD project on the process of identity construction in young people, I became critical of the efficacy of an interview to collect good quality, reflective data. In order to get a good understanding of identities, it was important to make sure that participants could feel at ease and encouraged to reflect about themselves and their lives. The design of a method which would provide the means, time, and space for participants to carry out this reflection was, therefore, a priority. This methodological research resulted in the design of a multi-method approach, which crucially involved the use of visual methods, including a technique of my own creation which I called self-portrait (Bagnoli 2004).

The self-portrait presents participants with a blank sheet of paper, colored felt-tips, and pens, with the request to show who they are at the present moment in their lives. Participants are then also asked to add anything that is important to them at that moment (Bagnoli 2004, 2009). The instructions are left deliberately open and general, and the method is designed to be unstructured, inasmuch as people are comfortable to take part.

The idea behind this task, which I usually introduced around mid-interview, after participants had already been prompted for narratives about their lives through both general and more specific questions, was to offer a chance to take some time away from the potentially stressful face-to-face interview context. Engagement with the self-portrait I thought could provide a creative opportunity to expand and think about other areas of life that might perhaps have been left aside in the interview talk. The self-portrait was, therefore, designed to be an ice-breaker, which might help participants feel more at ease, and a creative input to support in particular those who might not feel too comfortable with words when asked about personal and subjective issues, and might feel reassured by writing, drawing, or doodling as alternative forms of self-expression.

When designing this method, my own expectations had been that participants would produce some kind of plan or schema illustrating different aspects of their identities. The data collected in my PhD fieldwork, however, turned out to be rather different than what I had envisaged. While the self-portraits followed a range of styles on a continuum from writing to drawing (Bagnoli 2004), which also included some plans, it was clear that most participants, perhaps also because of the felt tips they had been given, had interpreted my instructions as a request for a drawing. The way in which participants will interpret instructions and relate to the method is important to consider. Any means provided will go to affect the kind of data that is collected and even the degree to which participants may be happy to engage with the task or not. These issues obviously need to be taken into account on the basis of the specific sample to be involved.

Within the context of my PhD investigation, one participant declined to make a self-portrait, the only one to decline on a total sample of 41. A middle-class, well-educated young woman and high-achieving university student, she manifested uneasiness at a method that she perceived as invasive and leading to a potential loss of control. Most participants, on the contrary, were happy to do the task, even though they often felt they had to excuse themselves for their poor drawing abilities. Indeed, a potential problem with any drawing methods is that they may not be experienced as accessible enough. Providing reassurance that the artistic qualities of the portrait are not what one is after is important, as is clarifying that the "portrait" can in fact be made in any way one thinks best: a few key-words, a plan, or even an origami could have done just as well in my research. Such flexibility at the data collection stage then obviously needs to be reflected by an open mind also at the stage of analysis, with all that it concerns in terms of making sense of and integrating, or contrasting, mixed-source data within an overall interpretive framework.

The application of the self-portrait in the context of different research projects indicated that the type of data collected may vary significantly in relation to the sample involved. If the young people aged 16–26 of my PhD study had in most cases produced a drawing and had often even reported enjoying this task, with the younger sample that I involved in a more recent study (Bagnoli 2009), which saw the participation of 13–14 year-olds, drawing was even more predominant. The teenage sample did not need to justify themselves and were clearly more comfortable drawing, which appeared to be a task they were used to in their everyday life. On an analytical level though, this sample's drawings often seemed to include more clichés and standardized images than the portraits produced by the older sample and were far more rigid in their color palette according to gender stereotypical norms (Bagnoli 2009, 2012): a red heart like that drawn by 26-year-old Johnny (Fig. 1) would with great difficulty have appeared in the portrait of a teenage boy. In this sense, the portraits could be telling about the visual culture of a population; in this

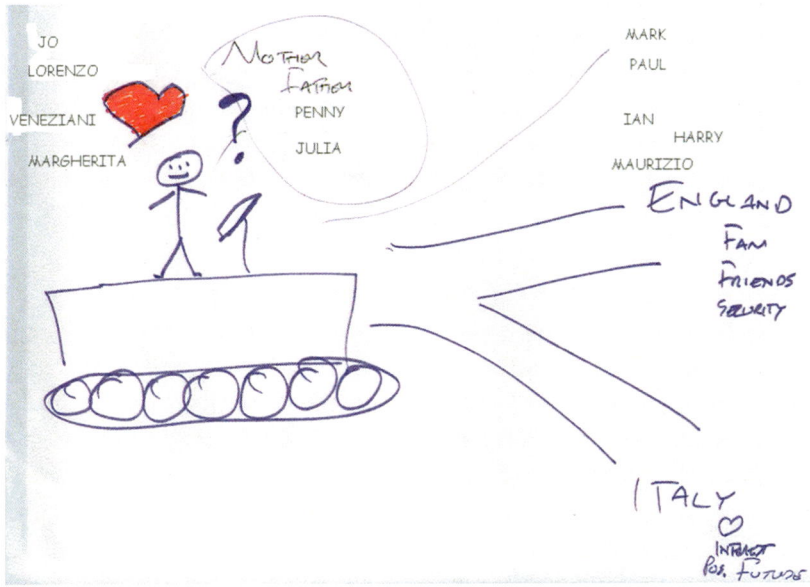

Fig. 1 Johnny's self-portrait

case a youth subculture. I noted, however, that it was with the older sample I researched that portraits seemed to have a deeper, more resounding meaning when read in context with the other biographical data collected.

The analytical value to attribute self-portraits has to be considered within the specific framework and aims of a research project. While obviously an example of visual data, a portrait is, however, more than just an image: it is a visual artifact produced in response to a researcher's request within a research interaction. It therefore carries with it also a link to verbal data, a narrative that is told contextually and which explains its meaning and the circumstances of its production. Analysis of self-portraits must note this connection and should pay attention to the researcher's role in their production. Portraits should, therefore, be looked at as data that are produced within the context of an interview situation and interaction.

The narrative collected at the moment of creation of self-portraits makes their meaning according to participants explicit. From a participatory point of view, this opens for people one channel to express their views from their own perspective. Qualitative research, however, is the art of interpretation, and there is usually more to meaning-making than a mere reliance on what is explicitly stated (Liamputtong 2013; Creswell 2014). While it is essential to make sure that participants can clearly communicate their message, interpretation does not just involve taking their words at face-value. The researcher's reading may point to other meanings in the portraits beyond what is said by participants and make connections to wider cultural repertoires. The analytical sensitivity one develops through the collection of self-portraits stimulates interpretation through comparison, with a focus that may be set on what is recurrent and what is not, what is contradictory, as well as what may be considered missing or unexpected.

As I have argued elsewhere (Bagnoli 2012), interpretation should best be considered as the result of a dialogue between researchers' and participants' views. A participatory perspective should, in my view, attempt to provide the instruments and possibilities for participants to effectively contribute to meaning-making, not only by providing their take on the world but also by getting the chance to comment on and respond to researcher's own interpretations. In this sense then, the construction of meanings may be seen as a collaborative effort, with the emergence and identification of different and even possibly divergent narratives from participants' and from researchers' perspectives.

Dedicated software such as Atlas.ti CAQDAS may be of great support in the analysis of self-portrait data. Its visual sophistication enables an easy coding of any images and facilitates the development of a coding structure that integrates visual as well as nonvisual data. In addition, it helps establishing hyperlink connections between visual and nonvisual data, such as between self-portraits and the narratives collected at the time of their production. Hyperlinks allow the simultaneous visualization of linked data and offer an alternative to explore the relationships among data to those researchers who do not wish to base their analysis on coding (Lewins and Silver 2007).

In my research experience, the self-portrait successfully aided the collection of reflective data and was an effective ice-breaker during the interview (Bagnoli 2004).

71 Self-portraits and Maps as a Window on Participants' Worlds

Some of the data provided by portraits were extremely useful in the construction of interpretations, and went on to acquire a driving role in the analysis, with the suggestion of insightful metaphors and powerful visualizations of what were participants' understandings of their own lives at that particular moment in time. The viewing of life as a journey, for instance, emerged as a widespread metaphor in my PhD data and was variously depicted in self-portraits through the images of a tree growing with fruits, a river running to the sea, one's self walking on a path or driving at a crossroads (Bagnoli 2012). See as an example Johnny's self-portrait in Fig. 1. Such images often corresponded to a prospective and fundamentally optimistic narrative about life, with an underlying idea of growth and development implicit to the narration of identity (Bagnoli 2012).

My PhD investigation included a longitudinal element, which required every participant to take part in two interviews. While for most people the time interval between the interviews was relatively short, of about 3 weeks' time, for some it amounted to up to 3 months. It was in those cases when the longitudinal interval was longer that the self-portrait demonstrated its interviewing potential. A few of these participants were by then living a very different reality from what they had shown in their portraits, which made these somewhat surprising, if not something they actively wanted to delete from their memory. Such was the case of Beatrice, who by the time of the second interview was in the process of divorcing her husband, who had instead been at the center of her self-portrait 3 months before (Fig. 2). The possibilities of applying a self-portrait in a longitudinal study are multiple and include, in addition to showing the portrait again at different times, asking for new portraits at different data collection stages.

This simple, low-tech method could lend itself to several potential applications, even with the use of different types of tools in place of paper, felt-tips, and pens. In

Fig. 2 Beatrice's self-portrait

my teaching of visual and arts-based research methods, I made a collective use of self-portraits and used this method as an ice/breaking task at the start of a training course, when asking participants to get to know each other and introduce themselves to the rest of the group. Group application in data collection could potentially be useful at the start of a workshop or focus group. While not everyone may be entirely comfortable with it, this is a creative method that can successfully collect novel and enlightening data.

3 Maps

The use of some form of maps during an interview can prove useful to help participants shape their thoughts and provide a narrative response to a question. Maps can be seen as visual scaffolding that support the construction of an answer and may also act as a memory aid, for both participants and researchers. Participatory mapping involves working with participants to collaboratively draw this map, relying as far as possible on participants' own ideas and links, as they emerge through the particular interview dialogue, rather than from a set of general and predefined research instructions valid for all (see also ▶ Chap. 65, "Understanding Health Through a Different Lens: Photovoice Method").

My research experience with mapping has involved asking for maps to show the important relationships in someone's life. Several researchers have used some form of maps to study relationships, among them Josselson (1996) and Roseneil (2006). The simple model that I used in a study with English teenagers asked the young people to place themselves in the center of the map and then show the important others in their lives indicating the different degrees of their importance (Bagnoli 2009). No more specific instructions were given. I was interested in keeping the method as unstructured as possible, in order to provide participants the freedom to interpret my own broad guidelines as it suited them best.

One common type of map that was produced in response to these guidelines looked like the spider diagram drawn by Grace in Fig. 3. A type of map that is often requested in school work, the spider diagram was part of this cohort of young people's visual culture. As the contents of Grace's map indicate, with the inclusion of the Beatles and Bob Dylan, my own research also inquired for the importance of relationships with imagined others, people whom the young participants admired and who might not directly be part of their everyday lives.

As I have noted elsewhere (Bagnoli 2009), the request to place the self at the center of the map, and to differentiate the level of importance associated with each of the relationships considered, does however provide some assumptions with regards to how the map should be drawn. In order to minimize the structure that I was implicitly requesting from participants, in another application of this visual method, as part of a study investigating the identities of young Italian migrants to the UK, I simply asked participants to show who their important others were, without providing any further guidance. My interest here was studying these migrants' social networks and the extent to which they spread transnationally.

Fig. 3 Grace's relational map

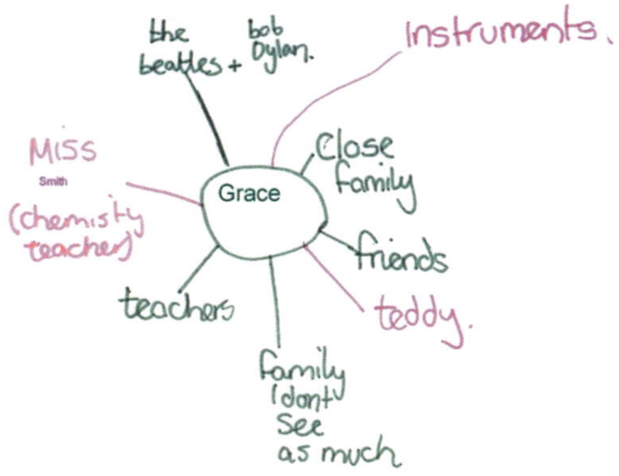

The provision of a looser structure, however, can have quite the opposite effect of making the task less accessible to some people, as Riccardo, one of the Italian migrants, commented in his feedback:

> Probably you should give more detail, saying 'imagine you are in the middle, and then just list the things that are important to you, basically starting from yourself, from the first one to the last one, or vice versa' or stuff like that. It depends on your study. Because when you do a psychological study, this is important to see, my personality, everything, the way, how do I think. But if I have a different target, probably you should give more details about the map. Because I found it hard, because it was too open.

It is often the case that people will have a need for structure and certainties and will happily follow whatever format a researcher has predefined for the study. An open structure may make them feel disoriented and possibly exposed and vulnerable. It is interesting to see that Riccardo himself came up with the guideline to put the self in the center, as an indication that this specific format may be an intuitive and often used type of map in a variety of different contexts. When designing a map for a research study, it is, therefore, crucial to consider the map accessibility, and the degree to which the way the map is introduced may be inclusive and appropriate to the sample of people to involve.

My own experience with both relational maps and self-portraits indicated that these methods could be significantly more inclusive than other approaches with young people with intellectual disabilities. The open structure of the methods and the flexibility of the guidance I gave ensured that these young people could well relate to the task at hand and properly take part in the research project.

While a useful task to structure one's thoughts, a map can also perform well as an aide-memoire. This may be true for researchers, as well as for participants, who by paying attention to the drawing that is being produced during an interview may be able to better follow the thinking of the interviewee and notice whether some areas are adequately being covered or not, or may require further prompting or clarification. The map can, thus, provide an ongoing check on one's interview agenda, which may be especially useful in the context of a very long interview.

Fig. 4 Rebekah's relational map

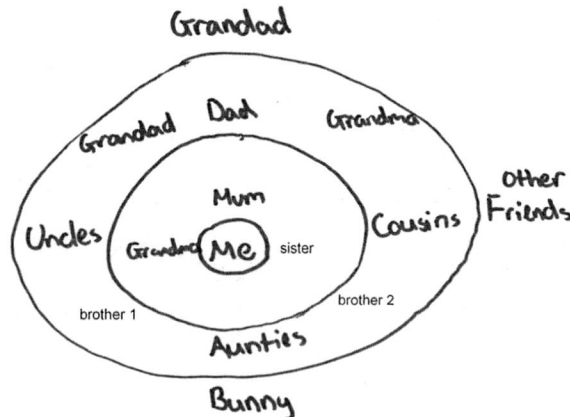

My research with young people showed that a relational map was on some occasion crucial to collect sensitive data that had not previously sifted through by interviewing alone. This was the case of Rebekah, whose map follows the standard concentric circles structure which is widely used in research on relationships (see Fig. 4). Rebekah's map showed evidence of a side of her family, her father and her two step-brothers, who were not as close to her as other people in her everyday life. She remembered to include them and felt comfortable talking about her relationship to them only through engagement with this visual method (Bagnoli 2009).

While my own research has employed maps to inquire about relationships and social networks, maps could have a much wider use in connection to multiple areas of interest. Similarly to what I argued earlier with regards to self-portraits, longitudinal use of maps could be useful to check on participants' changing views and circumstances, and this could include both revisiting previously produced maps and asking for new ones to be drawn at different points in time.

Analysis of maps will consider what areas the participant has chosen to include and will have to be put in relation to any other data provided. Absences, as well as inclusion, will be very revealing, as well as the particular emphasis attributed to any specific item in the map. In addition to a case-by-case consideration, a cross-sectional look at maps could be very interesting. Comparison to other participants' maps could be analytically productive, and this comparison could even be carried out at the data collection stage. Showing participants someone else's map, or even an ideal-type map, which may for instance correspond to some extreme case, could give rise to pertinent probing questions and suggest further elaboration of answers.

Potential re-use of any maps will obviously have to take into account ethical issues, and it should be pointed out that anonymization of maps and drawings is far more straightforward then other visual data (Wiles et al. 2008). This will practically facilitate their further use in data collection, even with other participants or with a different sample, as well as their dissemination.

Group-based data collection could be thought of for maps too, particularly as a task to accompany a focus group. The map would in this case provide a basis for sharing of participants' experiences and views, as well as an ice-breaking activity to promote group interaction. Just like with any other data, the interpretation of a map produced within a group session will have to take into account the particular data collection context. If the interaction with the researcher should always be kept in mind, for a map created in a group context the participation of a whole group in the activity and in the interaction will become salient.

4 Conclusions and Future Directions

This chapter has reviewed my research experience with two visual methods, the self-portrait and the relational map, which I have applied in the context of different research projects to facilitate elicitation during an interview and promote participation. Design of these methods was intended to provide a pause for reflection, a break from the interview interaction, and an opportunity for participants to give shape to and explore the unsaid, what they would not have yet had the chance to verbalize. Application in different projects proved both methods to be successful ice-breakers, which in most cases could enable participants to gather their thoughts and take time to develop their answers and thus engage in a wider and more reflective consideration of issues.

The visual scaffolding provided by these methods may be useful for participants to structure their thinking and may at the same time allow the researcher clearer access to their views. It can also usefully work as an aide-memoire for both parties in the interaction, particularly in the context of a long interview session. My fieldwork experience indicated that engagement with a visual task can even be a valid aid in getting participants to discuss potentially sensitive issues or other data that may not easily be collected through other, more standard methods.

In my design and application of self-portraits and relational maps I made a point to provide rather general, open instructions and keep the structure of any visual task as flexible as possible. The intention behind this was to enable these visual methods to adapt to participants' own expressive styles and to ensure inclusiveness. This flexibility was particularly successful to ensure the participation of children and of young people with intellectual disabilities. While it worked most times, it is important to consider that a loose structure can, not infrequently, be experienced as an obstacle by some people, who would prefer the reassurance of clear-cut instructions. It is obviously essential to adapt any methods to the specific sample to be involved in research. Flexibility and open instructions can facilitate the collection of participants' own preferred patterns and ultimately even provide an insight into their visual cultures. But it is important that a balance is struck with the need to design a method that is experienced as accessible by the target population.

Accessibility of methods should obviously be a fundamental point within a participatory approach to research. Several possible variations of self-portraits and maps may be thought in relation to different project needs, and longitudinal and even group applications may have considerable research potential. While the methods I proposed in my projects were low-tech and of straightforward application, more sophisticated developments could be appropriate in other contexts.

Since these methods were thought for use within an interview setting, the interactional context in which they were applied should be considered when making sense of the data that they collected. Such data are obviously visual but also include the narrative data with which participants describe and explain what the images show. Analysis will have to look at the links between images and words, and at the contextual dialogue and interaction from which portraits and maps emerged. These data channel through participants' own understandings and their participatory design can make them an open window on their worlds and meanings. Interpretation does, however, involve something more than merely reporting participants' views and requires researchers to provide their own understandings as well, which may even contradict what participants say. The researcher's point of view is quite obviously privileged by the possibility of drawing comparisons between data and contrasting different visualizations and narratives and will be informed by references to the wider social and cultural context in which data are produced. It will, thus, be important in the analysis to recognize whose input is being followed, what views are coming from participants, and what from the researcher's own understanding, instead of assuming an unproblematic main voice that obscures different interpretations.

Recent methodological developments in the social sciences have increasingly shown attention for the potentialities of creative, sensory, and arts-based approaches to research. These innovative methods are often proposed within a participatory research framework. As it has been noted, the way project results are presented often lacks a true appreciation of the meaning of participation at different stages of the research process. This ultimately goes to the expense of participants' voices, which end up being deleted from view, in favor of the researcher's perspective which becomes the only version available. It is clear that, for the analysis stage to properly have a participatory approach, a huge amount of effort must go into making the process more democratic. Consideration of the impact of different research actors' perspectives on the interpretations we construct is but a start in this direction.

Finally, it should be said that the use of visual methods such as self-portraits and maps brings numerous exciting possibilities as far as dissemination is concerned. These novel methods also suggest creative and multimedia forms of outreach that can stretch far beyond the traditional academic outputs we are used to, and which can potentially have a much higher impact on the wider community, as well as offering additional forms of reward for participating people.

References

Back L. Live sociology: social research and its futures. Sociol Rev. 2012;60:18–39.
Bagnoli A. Researching identities with multi-method autobiographies. Sociol Res Online. 2004;9(2). http://www.socresonline.org.uk/9/2/bagnoli.html.

Bagnoli A. Beyond the standard interview: the use of graphic elicitation and arts-based methods. Qual Res. 2009;9(5):547–70.
Bagnoli A. Making sense of mixed method narratives: young people's identities, life-plans and time orientations. In: Heath S, Walker C, editors. Innovations in youth research. Basingstoke: Palgrave Macmillan; 2012. p. 77–100.
Bagnoli A, Clark A. Focus groups with young people: a participatory approach to research planning. J Youth Stud. 2010;13(1):101–19.
Creswell JW. Research design: qualitative, quantitative and mixed methods approaches. 4th ed. Thousand Oaks: SAGE; 2014.
Harper D. Visual sociology. London: Routledge; 2012.
Heath S, Walker C, editors. Innovations in youth research. Houndmills: Palgrave Macmillan; 2012.
Hesse-Biber SN, Piatelli D. The synergistic practice of theory and method. In: Hesse-Biber SN, editor. Hanbook of feminist research: theory and praxis. London: SAGE; 2010. p. 176–86.
Josselson R. The space between us. Exploring the dimensions of human relationships. London: SAGE; 1996.
Knowles JG, Cole AL, editors. Handbook of the arts in qualitative research: perspectives, methodologies, examples and issues. London: SAGE; 2008.
Law J. After method: mess in social science research. London: Routledge; 2004.
Leavy P. Method meets art: arts-based research practice. London: the Guildford Press; 2009.
Lewins A, Silver C. Using software in qualitative research. A step by step guide. London: SAGE; 2007.
Liamputtong P. Qualitative research methods. 4th ed. Melbourne: Oxford University Press; 2013.
Luttrell W. 'A camera is a big responsibility': a lens for analysing children's visual voices. Vis Stud. 2010;25(3):224–37.
Mason J. Qualitative interviewing: asking, listening, and interpreting. In: May T, editor. Qualitative research in action. London: Sage; 2002a. p. 225–41.
Mason J. Qualitative researching. 2nd ed. London: Sage; 2002b.
Pink S. Doing sensory ethnography. London: Sage; 2009.
Prosser J, Loxley A. Introducing visual methods. ESRC National Centre for Research Methods Review Paper. 2008. http://eprints.ncrm.ac.uk/420/1/MethodsReviewPaperNCRM%2D010.pdf.
Rose G. Visual methodologies. An introduction to researching with visual materials. 3rd ed. London: Sage; 2012.
Roseneil S. The ambivalences of angel's "arrangement": a psychosocial lens on the contemporary condition of personal life. Sociol Rev. 2006;54(4):847–69.
Savage M, Burrows R. The coming crisis of empirical sociology. Sociology. 2007;41(5):885–99.
Wiles R, Prosser J, Bagnoli A, Clark A, Davies K, Holland S, Renold E. Visual ethics: ethical issues in visual research. NCRM Review Paper, NCRM/011. 2008. http://eprints.ncrm.ac.uk/414/1/Report_on_Reflections_onTCB_Activities.pdf.

Walking Interviews

72

Alexandra C. King and Jessica Woodroffe

Contents

1 Introduction	1270
2 Background to Walking Interviews	1271
3 Key Strengths and Characteristics of Walking Interviews	1273
3.1 Walking Interviews are Flexible, Adaptive, and Dynamic	1273
3.2 Walking Interviews Engage with Place and Encourage Collaboration	1275
3.3 Walking Interviews are Sociable and "Everyday" in Nature	1277
3.4 Walking Interviews Are Collaborative and Embodied	1277
3.5 Walking Interviews Are Compatible with Other Research Approaches and Methods	1278
4 Critical Considerations and Limitations of Walking Interviews	1279
4.1 Are Walking Interview Really "Natural" and Collaborative?	1279
4.2 Are Walking Interview Really Shared Experiences of Embodiment and Lifeworlds?	1280
5 The Practice of Walking Interviews: Advanced Techniques, Data Collection, and Other Considerations	1281
6 Walking Interviews in Practice: Introducing Mrs. Brown and Mr. Mitchell	1283
6.1 Walking with Mrs. Brown	1283
6.2 Walking with Mr. Mitchell	1284
6.3 Reflections on the Case Studies	1286
7 Conclusion and Future Directions	1287
References	1288

A. C. King (✉)
Rural Clinical School, Faculty of Health, University of Tasmania, Burnie, Tasmania, Australia

School of Pharmacy, Faculty of Health, University of Tasmania, Hobart, Tasmania, Australia
e-mail: alexandra.king@utas.edu.au

J. Woodroffe
Access, Participation, and Partnerships, Academic Division, University of Tasmania, Launceston, Australia
e-mail: jessica.whelan@utas.edu.au

© Springer Nature Singapore Pte Ltd. 2019
P. Liamputtong (ed.), *Handbook of Research Methods in Health Social Sciences*,
https://doi.org/10.1007/978-981-10-5251-4_28

Abstract

Walking interviews or "go-alongs" are an innovative qualitative research method which has recently gained popularity among cross-disciplinary researchers. Walking interviews entail researchers and participants talking while walking together. Informed by "the new mobilities paradigm" and "the spatial turn" within the social sciences, this method has been employed in various settings and with participants of all ages. Walking interviews are a valuable means of deepening understandings of lived experiences in particular places. The rich, detailed, and multisensory data generated by walking interviews demonstrates that they are a valuable, valid, feasible, and empowering means of conducting qualitative inquiry. They can also be employed concurrently with other qualitative methods such as in-depth interviews and ethnographic observation. The use of walking interviews in health research, with its potential to capture the lived experiences of health and illness, has so far been relatively limited. This chapter introduces the reader to the walking interview method, provides practical guidance for its use, outlines epistemological, ethical, and practical considerations for researchers, and canvasses its varied applications for qualitative health research.

Keywords

Walking interviews · Go alongs · Mobile research methods · Mobilities · Emplacement

1   Introduction

Walks. The body advances, while the mind flutters around it like a bird.
~Jules Renard, 1907 (Renard 2008, p. 266)

Human mobility, specifically achieved through the act of walking, is a somewhat habitual and taken-for-granted part of our individual experience and social existence. For those who are able, walking enables and presents many constant and unique opportunities and meanings on a daily basis. On one hand, it allows for physical movement through and among space and place and is seen to promote physical health and well-being. On another, it can be used for autonomy, reflection, and solace or for interaction, companionship, and social activity. Often the subject of poetry in decades past, in recent years, three major movies have been released which center on a transformative, long walk taken by its protagonists – *The Way* (2010) with Martin Sheen, *Wild* (2014) with Reese Witherspoon, and *A Walk in the Woods* (2015) with Robert Redford. This recent preoccupation with the symbolism of a long walk as a personal inner journey or a pilgrimage among filmmakers has arisen in parallel with a burgeoning interest among researchers in utilizing walking as a research method, aiming to enter new landscapes for generating understanding within qualitative enquiry.

Walking interviews are a relatively new and innovative qualitative method, with roots in ethnography, anthropology, and geography, which can effectively minimize some of the perceived limitations of more traditional, stationary sit-down interviews. Referred to by a variety of terms including "walk-alongs" and "go-alongs," walking interviews largely involve a researcher walking with one or more participants while conducting an interview. As a natural fusion of interviewing and participant observation, walking interviews are a powerful and unique method for engaging with space and place, and the important and nuanced meanings, experiences, values, and understanding of individuals in these domains. While used effectively within the disciplines of geography and social sciences for some time, the opportunities and strengths of this relatively new method to the health sciences are only now starting to emerge. This chapter calls for a greater appreciation that experiences and meanings of health, well-being, and illness at both community and individual levels can be explored and illuminated through this method.

The chapter introduces the reader to the method, outlining its key features and canvassing its varied applications for qualitative health research. It provides practical guidance for employing it in fieldwork, outlining some epistemological, ethical, and practical considerations for researchers. The key aims of the chapter are to:

- Provide some background and context to the emergence of the method
- Situate the method within methodological and theoretical approaches
- Provide consideration of the strengths and advantages, weaknesses, and limitation of the method
- Look at its compatibility and use with other research methods
- Briefly explore the key issues and practicalities of using the method
- Provide case study examples of walking interviews in practice
- Canvas areas in which this method could be used productively with the health and social sciences

2 Background to Walking Interviews

> Life itself is as much a long walk as it is a long conversation, and the ways along which we walk are those along which we live ~ Tom Ingold & Jo Lee Vergunst (Ingold and Vergunst 2008, p. 1)

A walking interview is a qualitative research method with roots in ethnography, anthropology, and geography, which involves a researcher walking with one or more participants while conducting an interview (see also ▶ Chap. 26, "Ethnographic Method"). They are naturally person-centered and interactive and are designed to understand human experiences and social action (Kusenbach 2012). Writers and academics have sought the generative effects of walking in their work for centuries, including the walking philosophers of the Aristotelian school in Ancient Greece, and the Romantic poets Wordsworth and Coleridge, who took long walks in the countryside to nurture their writing (Solnit 2001). In the early twentieth century, walking was perhaps first used as a research method by the Chicago School of sociologists, who conducted mobility studies of homeless men, street gangs, and sex workers (Büscher and Urry 2009), and for seminal ethnographies by Clifford Geertz, who used walking during his phenomenologically inspired fieldwork in Indonesia (Jackson 1996).

The burgeoning interest in employing walking interviews for empirical research was borne from the emergent field of mobility studies. Informed by "the spatial turn" in social sciences and the increasing mobility of human societies (Ricketts Hein et al. 2008), this enhanced attention on mobility as a key part of contemporary life has been labeled "the new mobilities paradigm" (Sheller 2014, p. 789). It is a diverse field, traversing the disciplines of geography, ethnography, sociology, transport, and tourism, and even sports studies and arts practice. Utilizing different forms of movement, mobility studies explore a variety of mobilities, ranging from those which are mechanized and global such as air travel, to those which are embodied and local such as walking (Cresswell 2012).

In recent years, researchers have employed a diverse set of mobility-focused methods and approaches, which can be loosely categorized according to whether the researcher remains stationary and observes participants' movements, or instead goes "on the move" alongside participants, whether in trains, cars, or on foot (Ricketts Hein et al. 2008; Büscher and Urry 2009). Within walking studies, a distinction can be made between those studies which employ walking methods to explore a range of phenomena other than walking, and research which focuses on walking itself, such as those conducted into the walking practices of children (Horton et al. 2013). And finally, a further distinction can be made between walking-based methods in which a participant primarily narrates a walk while accompanied by a researcher such as a guided tour, or those studies in which a participant primarily answers a researcher's questions, such as a walking interview (Stroud and Jegels 2014).

Researchers apply a variety of labels to walking interviews but perhaps the most common alternative label is "go-alongs" (Kusenbach 2003; Carpiano 2009; Bergeron et al. 2014). Other more idiosyncratic terms include "walk-alongs" (Rose et al. 2010), "mobile interviewing" (Brown and Durrheim 2009), "walking probes" (De Leon and Cohen 2005), "walking fieldwork" (Irving 2010), "dwelling-in-motion" or "stretched out belonging" (Edensor 2010), "'walking with' ethnography" (Peyrefitte 2012), "shadowing" (Jirón 2010), and finally, more playful terms such as "pedestrian enquiry" (Hall et al. 2009) and "pace in place" (Hitchings and Jones 2004) (see also ▶ Chap. 73, "Participant-Guided Mobile Methods").

Walking interviews are compatible with a number of traditions within interpretative and constructionist research approaches which aim to understand human behavior and experiences from the viewpoint and interpretations of those being studied, including phenomenology, social constructionism, ethnography, participatory and action research, feminism, and critical theory (see also ▶ Chaps. 26, "Ethnographic Method," and ▶ 17, "Community-Based Participatory Action Research").

3 Key Strengths and Characteristics of Walking Interviews

> ...the moment my legs begin to move, my thoughts begin to flow.
> ~Henry David Thoreau, 1851 (Thoreau 1960, p. 64)

There are many characteristics of walking interviews which make them innovative, unique, and worthy of consideration in the design of new qualitative enquiry. This section presents an overview of some of the key merits and strengths of walking interviews, including the flexibility of the method and its dynamic and adaptive nature, its ability to capture meaning while "in-place" and in the everyday context, its collaborative, sociable, and embodied nature and its compatibility with other methods.

3.1 Walking Interviews are Flexible, Adaptive, and Dynamic

Walking interviews are a flexible, adaptive, and dynamic research method. Unlike traditional sit-down or sedentary interviews, walking interviews literally transcend the boundaries of interview rooms and move the art of conversation and qualitative interviewing to the "outside." During walking interviews, participants can provide not only recollections of place but also experience and describe immediate "rich and varied perspectives of environment" (Garcia et al. 2012, p. 1) stimulated by the simultaneous act of walking and talking.

Walking interviews are readily adaptable to different local contexts, research topics, and participants' needs. Variations of the method might involve participants narrating their thoughts about the unfolding environment while walking (Irving 2010) or participants taking the researcher on a guided tour of a particular locality or setting (Carpiano 2009; Garcia et al. 2012). Walking interviews can be used to explore a variety of phenomena in a range of settings including people's feelings about large shopping centers in the United Kingdom (Rose et al. 2010), residents' perceptions of their neighborhoods (Kusenbach 2003), people's perceptions of local parks and streets (Moles 2008; Vergunst 2010), and intersections of ethnicity and place in South Africa (Brown and Durrheim 2009). In qualitative health research, walking interviews have been employed to enquire into health inequalities in neighborhoods (Carpiano 2009), an experience of being diagnosed as HIV positive (Irving 2010), the sexual health resources available on a university campus (Garcia et al. 2012), and older adults' experiences of eating and aging in rural Australia (King 2014; King et al. 2015).

In the burgeoning area of qualitative health research, uncovering the meanings which are attached to people's lived understandings of health and illness, including their experiences of health and social care, is now acknowledged as contributing to evidence-based practice (Liamputtong 2016, 2017). While health research has traditionally been dominated by biomedical and reductionist discourses of health which are focused on medicalization through diagnoses and cure (Nettleton 2013), there is increasing interest in engaging with the multifaceted experiences, feelings, emotions, and meanings of individuals for understanding health and illness (Grbich 1999; Liamputtong 2013; Green and Thorogood 2014). In this regard, the flexibility and dynamic nature of walking interviews within qualitative enquiry offers opportunities for health researchers to incorporate explorations of space and place and the meanings attached to particular settings and contexts (e.g., hospitals, outpatient clinics, waiting rooms, parks or recreational spaces, or community infrastructure) into studies which aim to develop understandings of individuals' lived experiences of their health and environment.

Walking interviews provide researchers with unique opportunities to describe the setting and contexts of research, as well providing insight into environmental and locational influences that can impact significantly on how individuals perceive, experience, and exercise agency over their wellbeing, health, and care. As Garcia et al. (2012, p. 2) argue, walking interviews assist in capturing "the natural relationship between health and place in a participatory manner." Walking interviews have also proven to be feasible and effective in research conducted with people of different ages, including children (Hitchings and Jones 2004), young people (Anderson and Jones 2009; Hall et al. 2009; Ross et al. 2009; Garcia et al. 2012; Holton and Riley 2014), and older adults (Riley 2010; King 2014).

Researchers have also noted the usefulness of walking interviews for conducting research with hard-to-reach and mobile populations, including disengaged youth and homeless people (Ross et al. 2009) who may be reluctant, intimidated, or simply unable to participate in "traditional" research conducted outside of their immediate environments which offers them security, safety, or meaning. Similarly, the simple practice of "going to" people and speaking with them while they interact with others and move about within their own environments, so as to explore experiences, ideas, and constructions of place, has been used by the authors during participatory research into a number of areas, including social inclusion, health promotion, aging, youth engagement, and local government planning, in which place-based participatory research is valuable and effective.

By moving out of the confines of the interview room, walking interviews "allow the environment and the act of walking itself to move the collection of interview data in productive and sometimes entirely unexpected directions" (Jones et al. 2008, p. 8). Walking interviews are often dynamic in nature and do not strictly adhere to methodological rules, such that the exact route, walking pace, duration, mood, and content of the interview often unfolds "on the go" in response to the changing landscape, weather, interactions with other walkers, or other unplanned events (Jirón 2010). This gives walking interviews a style which is more improvisational than highly structured or predetermined.

Walking interviews also inform the nature of the research encounter by "loosening up" the interview experience, allowing data to be more readily elicited. Both spontaneous talking and questions tend to arise more often, stimulated by the changing environment, such that interviews often flow more easily, with fewer awkward gaps in dialogue than might occur in more traditional, sit-down interviews (De Leon and Cohen 2005; Ross et al. 2009; Riley 2010; Evans and Jones 2011). This method tends to generate a more natural or conversational style of interview.

The "plethora of encounters, diversions and disruptions" that comprise a walking interview allows the conversation to naturally move between topics and moods, "incorporating the intimate and the mundane, the near and present, remembered and imagined in the free flowing movement of the walk and talk" (Ross et al. 2009, p. 619).

3.2 Walking Interviews Engage with Place and Encourage Collaboration

Walking interviews are sensitive to local context; they enable researchers to simultaneously engage with people and the places that have meaning within their lives. Conducted in situ, they deeply engage with concepts of place and thus provide qualitative researchers with an invaluable opportunity for generating rich and unique data. Walking and talking in different settings opens up opportunities to understand the complex, multilayered, and textured social world of individuals sought by qualitative researchers (Mason 2002).

In doing so, walking interviews assist in overcoming one of the key disadvantages of traditional sit-down qualitative interviews which take participants away from their day-to-day activities and locate them within a highly constructed dialogue setting which can inhibit natural conversation, stultify interaction, and thus constrain the building of rapport and trust advocated for and sought by qualitative researchers. In contrast, locating research encounters in outdoor or recreational spaces can help to generate relaxed interactions and dialogue between researchers and participants, which can more readily produce unstilted and spontaneous verbal data (Ricketts Hein et al. 2008). These places do not need to be publically significant or monumental in any way, as the "mundane locations and the events that occur at them can elicit rich responses" (De Leon and Cohen 2005, p. 201). In our own research, the use of walking interviews allowed older adults to share their valued places, for example, their homes, gardens, and towns. This method helped to create an atmosphere which allowed the researcher to appear and feel like a friendly visitor in participants' lives and less like an interrogator or indifferent observer.

As well, separating people from their natural environments runs the risk that "important aspects of lived experience may either remain invisible, or, if they are noticed, unintelligible. This is especially true for the spatial footings of experience and practices in everyday life" (Kusenbach 2003, p. 459). By situating people in particular places, walking interviews produce a "decidedly spatial and locational discourse of place", in which participants can readily talk about the places in which they live or act (Evans and Jones 2011, p. 856). Participants' oral histories or personal biographies tend to unfold more easily during walking interviews than sit-down interviews. Their recollections may be stimulated by the surrounding environment, such as old buildings they have worked in or houses of former friends, such that "the spaces prompt recollection of place attachment, demonstrating an active connection between body, landscape and memory" (Jones and Evans 2011, p. 2323; Kusenbach 2003; Vergunst 2010). Walking interviews thereby produce accounts that constitute an "interweaving of personal biography and individual experiences with collective (social) memories and spatial histories" (Clark and Emmel 2010, p. 5). This aspect of the walking interview method makes it particularly valuable for research with people who have longstanding work or family ties to a particular area, such as farmers (Riley 2010).

In our above-mentioned research with older adults, participants were readily able to point out changes in place and space, including changes to streetscapes, landscapes, weather, or gardens, most of which were intimately connected with significant people or events in their everyday lives, both past and present (King 2014). In this sense, walking interviews can be said to encourage participants to express "place-bound meanings and values of places", thereby revealing personally held and highly localized "micro-geographies of meanings" (Bergeron et al. 2014, p. 108). As such, they are a "unique tool for examining how physical, social, and mental dimensions of place and space interact within and across time for individuals" (Carpiano 2009, p. 264), which allow researchers to examine, first-hand, the informant's experiences, interpretations, and practices within their environment.

In one of the earliest articles on walking interviews as a research method, the sociologist Margarethe Kusenbach explicitly linked walking interviews with phenomenological research enquiry. Kusenbach argued that by "exposing the complex and subtle meanings of place in everyday experience and practices, the go-along method brings greater phenomenological sensibility to ethnography," thus labeling the method "street phenomenology" (2003, p. 455). Certainly, walking interviews support phenomenological concerns by generating deeper understandings of unique and meaningful relationships between people and particular places (Moles 2008). As such, walking interviews help researchers to explore phenomenological notions of emplacement and nonrepresentational theory, which are concerned with embodied and sensory experiences of place (Casey 1996; Jackson 1996; Thrift 2008; Vergunst 2010). A walking interview is a multisensory research encounter for both participants and researchers, in which landscapes and streetscapes are seen; pavement or soil is felt underfoot; odors and aromas of gardens, streets, and farms are smelled; and background noises of birds, children playing, or cars are heard (Anderson and Jones 2009; Ross et al. 2009).

Perhaps arising from these kinds of sensory experiences, walking interviews readily elicit participants' verbal expressions of emotions and affect about particular places. In research into visitors' experiences of a botanical garden, participants found it easier to verbalize their feelings during their walking interviews than in their earlier sit-down interviews (Hitchings and Jones 2004). In another study, walking interviews helped uncover the emotional ties that young people had to particular places, with the researchers noting that these kinds of "individualized meanings and identifications often remain out of view through other techniques" (Anderson and Jones 2009, p. 299). The enhanced capacity of participants to express their emotions which is evident during walking interviews grants researchers valuable opportunities to engage, albeit temporarily, in the experiential lifeworld of their research participants, by sharing in their sensations and witnessing their emotions (Rose et al. 2010). In research conducted with a young man recently diagnosed as HIV-positive, walking fieldwork effectively uncovered "realms of inner expression" that might have otherwise remained hidden (Irving 2010, p. 34).

3.3 Walking Interviews are Sociable and "Everyday" in Nature

Walking with others is a profoundly social activity which requires close awareness of another's movements while also engaging in conversation (Ingold and Vergunst 2008), and it is such a common practice among humans that it is arguably "fundamental to the everyday practice of social life" (Lee and Ingold 2006, p. 67). The deeply social nature of walking aids the mood of a research interview. Walking interviews are, in this sense, a socially familiar method; when visiting a friend, one might take a conversational walk around their property or neighborhood, which is not altogether dissimilar to a walking interview.

Importantly, a walking interview also has a particularly prosaic, everyday or pedestrian content or character which is generated by "talking on the go, moving through local spaces amidst all the mundane distractions and reassurances of place" (Hall et al. 2009, p. 552).

The method thus illuminates the "incidental, trivial, mundane and often dull dimensions" of phenomena which are enacted in place but are not necessarily readily verbalized during traditional sit-down interviews (Anderson and Jones 2009, p. 299).

3.4 Walking Interviews Are Collaborative and Embodied

Walking interviews have greater potential for collaborative construction of meaning and enquiry between researchers and their participants, than other more sedentary, sit-down interviews provide. It is important here to make a clear distinction between walking-based research methods (Evans and Jones 2011) and conducting research while mobile, such as "ride-alongs" in cars (Büscher and Urry 2009). Arguably, walking interviews are qualitatively different because walking is an active, embodied practice which contrasts with undertaking mechanized travel, during which one is essentially sedentary while in motion (Edensor 2010; Evans and Jones 2011). Various experiential benefits have been attributed to the physicality of walking, and it is thought to impact directly upon the experience of the interview. While Solnit (2001) argues that "...the rhythm of walking generates a rhythm of thinking" in walkers, others see walking's effect as more metaphorical than physical: "through walking, the researcher and the participant bimble into new narratives, and discover and construct new spaces together as a result" (Moles 2008, p. 13).

Certainly, as a shared corporeal or bodily experience, the physical act of walking alongside someone shapes the research encounter, aiding the development of an intersubjective understanding of the physiological particularities of a respondent's lifeworld (Pink 2007). Unlike a more traditional interview, the researcher and participant are side-by-side rather than face-to-face, such that direct eye contact is reduced but their view is now shared (Lee and Ingold 2006; Lorimer 2011). Through side-by-side engagement and a shared experience of sensory data, this method allows for a deeper understanding of the "here and now." For example, a participant being able to physically show the researcher where and why they feel unsafe in a public space enables the researcher to share in the emotions of the participant, as well as being able to physically experience the setting for themselves. As such, it moves beyond a recollection and explanation of space and place, to one which is real, "in-time" and observable, and serves as a point of further conversation. This arguably facilitates a greater empathic and collaborative encounter, in which each other's experiences and emotions are understood, acknowledged, and shared.

As well, walking together further aides collaboration by generating a sort of intimacy or empathy, as the shared practice "gives an immediacy as well as a kinaesthetic rhythm which makes for a different experience" between researcher and participant (Evans and Jones 2011, p. 850). This effect is reciprocal, such that "the walk shapes the rhythm of the talk, and the talk shapes the rhythm of the walk" (Lorimer 2011, p. 29). A degree of sociability is engendered by walking with another person, perhaps arising from the physical and cognitive process of developing a shared walking rhythm (Lee and Ingold 2006), thus enabling a form of collaborative interaction.

Walking interviews can thus serve as a "rapport builder" (Carpiano 2009, p. 67), working to minimize any discomforting differences in education, age, ethnicity, or gender between researchers and participants. Walking interviews can also partially "disrupt" traditional power relations within research or at least serve to encourage a more collaborative approach. The nature of the method allows for participants to become at least partial collaborators in the research, effectively co-generating meaningful data while engaging in a mutual dialogue with researchers, rather than merely responding to questions in an interrogation (Jones et al. 2008; Moles 2008; Brown and Durrheim 2009; Carpiano 2009; Clark and Emmel 2010). The flexible nature of walking interviews also allows researchers to cede some power or influence to participants in fieldwork, by inviting them to choose the walking route, determine the pace of a walk, avoid particular areas, or interrupt an interview to talk to other pedestrians encountered during the interview (Hall et al. 2009; Ross et al. 2009).

3.5 Walking Interviews Are Compatible with Other Research Approaches and Methods

As an amalgam of observation and traditional sit-down interviews, walking interviews offer new opportunities for qualitative researchers who want to explore, through movement and conversation, more multilayered and complex meanings and experiences within an individual's social world (see also ▶ Chap. 73, Participant-Guided Mobile Methods"). As such, they are suitable for both qualitative and mixed method designs and for use with other methods of data collection including surveys, focus groups, observations, and also repeat or serial interviews (see also ▶ Chaps. 63, "Mind Maps in Qualitative Research," and ▶ 4, "The Nature of Mixed Methods Research").

The last of these methods have been used in the social sciences for some time but remain relatively underused in the health and medical sciences (Murray et al. 2009). As a method, repeat interviews involve speaking with participants repeatedly over a period of time, usually months or years. Doing this provides more than a single snapshot of time, feeling or meaning, instead allowing researchers to explore changes in needs and experiences of people over time (Murray et al. 2009; King 2014) and to develop a closer relationship with participants, opening up greater rapport and trust. Within qualitative longitudinal health research, serial interviews have been fruitfully used to understand the shifting experiences and needs of people with progressive diseases such as lung and brain cancers, diabetes, and cardiac failure (see Murray et al. 2009 for examples).

The use of serial walking interviews alone or in combination with serial sit-down interviews is almost non-existent in the literature but offers great potential for the field of qualitative longitudinal enquiry into health. We have used walking interviews in combination with serial sit-down interviews to explore the meanings of food and aging in older rural Australians (King 2014). The study, using a phenomenological ethnographic design, showed that the passage of time within particular places significantly underpinned older adults' basic sense of confidence in themselves, which informed their food security and health and well-being while aging (King 2014).

Other researchers have combined walking interviews with various technologies, including photography, a variation of Photovoice (Irving 2010; Peyrefitte 2012; Bergeron et al. 2014; see also ▶ Chap. 65, "Understanding Health Through a Different Lens: Photovoice Method"), geographic information systems (GIS) mapping of walking routes (Jones et al. 2008; Bergeron et al. 2014), video ethnography (Pink 2007), and even bio-sensing, with heart-rate monitors worn by participants (Spinney 2014).

4 Critical Considerations and Limitations of Walking Interviews

I have walked myself into my best thoughts...
~Soren Kierkegaard, 1847 (Poole 1993, p. 172)

Although walking interviews have been written about in very favorable terms in recent years, additional rigorous examination of the benefits that this method brings to research encounters, over and above those of more sedentary methods, is warranted (Ricketts Hein et al. 2008; Merriman 2014). While this chapter highlights the innovative nature of walking interviews as a method and their potential for use in health and social sciences, there are also some key issues when considering the use of this method, from both a theoretical and practical perspective. Engaging in this critical reflection prompts several key issues for consideration by researchers when incorporating this method into health research frameworks and designs, as well as in regards to data collection.

4.1 Are Walking Interview Really "Natural" and Collaborative?

While there are many ways in which walking interviews can create greater ease and dialogue between researcher and participant, walking interviews are not entirely natural encounters because, like sit-down interviews, they are almost always planned, contrived, or "managed" events because the researcher has initiated the activity with the participant (Kusenbach 2003) and hence, "it can never be completely spontaneous" (Bergeron et al. 2014, p. 120). However, there is a kind of sliding scale of contrivance for walking interviews. At one end, there are guided tours of a city which are requested of participants by researchers, and at the other, researchers walking alongside participants while they walk a usual route as part of their usual routine (Kusenbach 2012).

Additionally, there is a need to acknowledge that all interactive research encounters, regardless of intent, inherently involve a power imbalance and that active reflexivity is required on the part of the researcher to minimize this as much as possible. It is well recognized among qualitative researchers that "the researcher is in a relatively powerful position vis-à-vis the interviewee" (Green and Thorogood 2014, p. 110). While this may not always be the case (e.g., when interviewing powerful or elite people), most interviews for health research do involve a power imbalance between researcher and interviewee which is generated in part by social differences, including socio-economic status, race and ethnicity, gender, and age (Grbich 1999). Even in cases where such characteristics are shared between researcher and participant, and thus assist in developing rapport and trust, the actual act of "being" the interviewer and initiating the interview through engaging in recruitment, obtaining consent and the formalities of asking questions, arguably still involves power and a level of perceived authority, which may inhibit the discussion and relationship.

While qualitative researchers generally strive to empower participants, and adhere to ethical standards of consent and the right of participants to withdraw from research, it is important to understand that with a walking interview, it might be more difficult or uncomfortable for a participant to stop an interview because of discomfort or discontent with the process, especially if a significant distance has already been traversed. As with all interview processes, power issues may be obvious within the walking interview; however, power may also be nuanced, subtle, and submerged – so much so that they may not become apparent to the researcher until after the walking interview is conducted. Significant power imbalances may be minimized through reflexivity on the part of the researcher and the use of serial interviews in order to develop trust between participant and researcher. To aid reflexive awareness, it may be useful for researchers to not only take field memos after walking, but to transcribe the walking interviews immediately after they are conducted (and before any others are conducted) in order to reflect on issues of power.

4.2 Are Walking Interview Really Shared Experiences of Embodiment and Lifeworlds?

The capacity of walking interviews to provide access to another's lifeworld is also worthy of critical reflection, particularly when applied in research informed by phenomenology. Some researchers have argued that walking interviews allow them to temporarily "live" participants' lives or even inhabit their bodies (Pink

2007; Anderson and Jones 2009; Rose et al. 2010). This is a significant epistemological claim which provokes questions about whether it is ever possible, even temporarily, to inhabit the body and life of research participants.

Phenomenological researchers Lee and Ingold (2006, p. 67) assert that "walking does not, in and of itself, yield an experience of embodiment", because this presupposes a level of shared bodily circumstance between researcher and participant which does not often exist. For example, in research with older adults described in the case studies below, the walking interviews revealed disjunctive rather than mutual corporal experiences. These provided insights into the participants' embodied life experiences, but did not extend to any sense of inhabiting their bodies or living their lives. Similarly, in "shadowing research" undertaken in urban Mexico, the researcher distinguished between achieving a closeness with participants' mobility and fully sharing in their existence, asserting that "...this 'being' is always someone else's" (Jirón 2010, p. 36). While it is not clear that walking interviews can produce fully shared, embodied understandings of others, it is nevertheless a valuable research method because it "opens up the possibility of other types of knowledge and appreciation that do not presuppose commonality" such as empathic understanding (Irving 2010, p. 35).

5 The Practice of Walking Interviews: Advanced Techniques, Data Collection, and Other Considerations

This section moves to practical considerations for conducting walking interviews. It notes that the method is an advanced technique which requires careful research practice in relation to the collection of data and the ethical involvement of participants.

Walking interviews are challenging to conduct and thus can be considered an advanced or complex qualitative research technique. Several tasks need to be successfully performed simultaneously which call on a researcher's physical, social, and cognitive skills. These include walking along a perhaps unfamiliar walking route, establishing and maintaining rapport with a participant, guiding the content and direction of the interview, noting the unfolding external environment, being aware of any physical, emotional, or mental discomfort of participants, negotiating a balance between a conversational and interview style of encounter, and responding flexibly to unexpected events, such as encountering neighbors or acquaintances. As such, walking interviews are perhaps better suited to researchers with some prior experience of conducting more traditional sit-down interviews. Even then, when undertaking this method for the first time or with an unfamiliar participant group, researchers might benefit from conducting several practice or trial walking interviews in order to hone their skills.

Walking interviews also pose challenges associated with data collection. As they are usually conducted outside and always conducted on the move, it is important to consider a number of issues. The first of these relates to capturing spoken words. Almost always, walking interviews are digitally recorded, rather than relying on field notes or memory, as writing while walking is rather awkward and memory is fallible. Recording technology has improved considerably in recent years, but it is important to obtain compact, lightweight digital recording equipment which has a sensitive microphone. Researchers may provide participants with a small lapel microphone or merely hold the recording device near to them. Wind and other sources of background noise such as loud traffic or music, which might obscure recorded speech, need to be taken into account when deciding when and where to conduct walking interviews.

Beyond recording the spoken words of participants, additional decisions need to be made in relation to collecting the spatial or place-based data generated by walking interviews. For example, researchers might wish to note particular locations during the walking interviews, so that during analysis, spoken words can be linked with particular places that are meaningful, for example, a cemetery or a tree. This might be done verbally during the walk, or by writing a brief note in a small notebook during the walk, or soon afterwards by making notes at time stamps in the transcript while listening to the recording. Other methods include using GIS devices to generate maps or taking photographs to generate visual images of the walking route taken. However, it is worth noting that while gathering this additional data can generate rich fieldwork results, it can also significantly complicate and magnify the work of data analysis and the presentation of findings.

As with all valid qualitative research, careful consideration should be given to both the practicalities of the method and the ethical principles behind conducting research, including respect for human beings, research merit and integrity, justice, and beneficence, all which help to shape the relationship between researcher and participant as one of trust, mutual responsibility, and ethical equality. In any qualitative research, the particular characteristics and needs of participants should be closely considered, especially when researching sensitive topics or engaging with vulnerable populations (Mason 2002; Liamputtong 2007; Green and Thorogood 2014).

Most ethical considerations for conducting walking interviews are shared by sit-down interviews. These include obtaining informed consent, being sensitive to the burden placed upon respondents' time and energy by the interview, being aware of the risk of emotional harm from respondents recalling and recounting negative experiences or emotions, and acknowledging any issues of confidentiality and anonymity (see also ▶ Chap. 106, "Ethics and Research with Indigenous Peoples"). In addition, walking interviews require researchers to consider a couple of ethical issues which are specific to walking-based methods.

Firstly, there is a small but not insignificant physical risk to participants (and possibly the researcher) from engaging in walks, particularly when participants are very young or old, unwell, or mobility-constrained; or when walks are conducted in insecure neighborhoods, on unsafe paths, or during inclement weather. In our recent research with older adults, we were cognizant of these risks and responded by incorporating significant flexibility in the research method. Depending on their personal circumstances, we either agreed with participants on a walking interview with a route and duration that was appropriate to their health, mobility, and level of interest in the activity, or for other participants, we invited them to consider declining a walking interview altogether and to opt instead for a sit-down interview (King 2014). There is no optimum length of time or distance for a walking interview; as illustrated in the case studies below, the distance, pace, and duration of walking interviews can be readily modified to meet participants' needs without impinging on the quality of the data generated during the research encounter. Indeed, short walks with many stops along the way may yield just as much high quality data as longer walks without breaks do, because the method's generative power lies in the shared mobility, experiences of place, and dialogue between researchers and participants, rather than the walk's duration or distance.

Secondly, both researchers and participants need to be aware that the confidentiality of participants' engagement in the research might be compromised during walking interviews in public places, by encounters with friends or neighbors who naturally enquire as to the nature of the walk. Such potential breaches of confidentiality may be of particular significance when the research engages with sensitive topics such as stigmatized health conditions. The somewhat novel nature of walking interviews may require researchers to be explicit in spelling out the advantages and key considerations of this method when complying with ethical approval processes that may be required of them.

6 Walking Interviews in Practice: Introducing Mrs. Brown and Mr. Mitchell

Walking is almost an ambulation of mind.
~ Gretel Ehrlich (Ehrlich 1992, p. 28)

This section explores some of the issues raised above through two illustrative case studies of walking interviews conducted with older adults. These walking interviews were conducted during fieldwork undertaken for qualitative health research exploring experiences of food security and insecurity among 21 older adults living in rural Tasmania, Australia (King 2014). The two older adults have been assigned pseudonyms to protect their anonymity.

6.1 Walking with Mrs. Brown

Mrs. Brown is a warm and talkative 77-year-old woman who was a dairy farmer for 45 years. Widowed 24 years ago, Mrs. Brown lives alone on her farm. She has several serious medical conditions including heart disease, spondylosis, and type 2 diabetes, and she now walks no further than her letterbox at the end of her short driveway. During the walking interview in her garden, Alexandra and Mrs. Brown walked a total distance of perhaps 20–30 meters over a period of 20 minutes, creating an open loop around three sides of her small home. This walk was characterized by a gentle pace and a stop-start rhythm, with Mrs. Brown stopping occasionally to rest or admire various flowers and plants in her garden.

Being outside with Mrs. Brown generated several useful visual prompts which were not available inside her home, and these provided a powerful means for

Mrs. Brown to talk further about her emotional attachment to her farm. She talked freely about both the immediate physical surroundings and our shared views of the landscape in the distance. At the beginning of the interview, she started by talking about the weeds in her garden but quite suddenly, and without prompting, she gestured at, and then described her enjoyment of, the soil and plants close by in her garden, as well as the surrounding paddocks and mountains. This walking interview also provided other visual prompts for talking about her past experiences and achievements. As a farmer and long-time resident, the work she and her husband did together on this farm was evident; for example, in the form of additional farming land they purchased and a dam they built together. The walking interview also provided visual evidence of the changes Mrs. Brown has experienced in her environment during her later life: from her garden, we could see the farm's homestead where she used to live before she retired from farming.

The following excerpts from the walking interview transcript reveal that Mrs. Brown spoke fluidly and volubly, and this talk was infused with expressions of her bodily sensations and her positive emotions about her rural home environment.

> **Mrs. Brown** – It is funny, it has got a feeling that you can't actually explain. It's just a feeling of... I don't know if it is contentment or what it is. But you just look at them [the cows] and you marvel. Oh, gee, they are beautiful, aren't they? You know?
>
> ....
>
> **Mrs. Brown** – Everything is being created all the time, and you see that. And you are growing with it, all the time. And I think you are blessed to be able to do it. Because... it doesn't matter what it is, it's all alive. You know what I mean? So, I think, when you walk on the farm, you can feel the life in it. Yeah, I take my shoes off and I walk around on this grass, and it is just like walking on a beautiful, soft carpet. You know? It is so lovely to walk on. ... Well, it is living.
>
> ....
>
> **Mrs. Brown** – It's peaceful. It's very peaceful looking like that. As I say, if you are uptight or anything, and you walk out and you look across that, you feel at peace. You know what I mean, don't you? See, even here, a tractor in the distance. It has got that sort of feeling of – you *belong*. So, you know... and look at the mountains. It is just as if they are asleep. It is beautiful.

Mrs. Brown's talk reveals how her emotional, sensory, and aesthetic enjoyment of her rural environment adds meaning to her daily life in older age, even after she has retired from active farming. Primarily, her emotions about where she lives are characterized by feelings of peace, belonging, contentment and "blessedness." Thus, this walking interview revealed that the rural environment Mrs. Brown has lived in for almost 60 years continues to nourish and support her in old age.

6.2 Walking with Mr. Mitchell

Mr. Mitchell is an inquisitive and well-educated man who tells detailed, lengthy, and lively stories. He has farmed his family's land for over 50 years. Now aged 77, Mr. Mitchell has been widowed twice. He has three adult children and two teenage children. He has had a heart attack and his hearing is low, but otherwise he is in good health and he is very physically active.

The walking interview with Mr. Mitchell was a quite different in character and content to the one conducted with Mrs. Brown. It took place over a distance of two kilometers, on a route which first went steeply uphill and then steeply downhill, over a period of about 45 minutes. This was primarily a goal-oriented walk for Mr. Mitchell. For the interview, he had chosen a walk on his farmland that he often takes; during it, he times how long it takes him to reach the top of a hill, which he then records in a notebook. Mr. Mitchell has been doing this timed walk for many years, and he sees it as a strategy for keeping his heart healthy and for loosening up his body, as well as providing a time-based benchmark for his overall well-being.

Mr. Mitchell and Alexandra set off at a rapid pace up a steep, rocky path which took us up a hill on his property. In contrast to our earlier in-depth interviews, it became clear that Mr. Mitchell was a little reluctant to talk much during this interview, as he was concentrating on keeping up his rapid walking pace. Answering questions very succinctly and without engaging in any extraneous speech, he walked up the hill without pausing, only stopping when we reached the predetermined halfway point of the walk: a tree at the top of the hill. The following excerpts from the interview transcript convey the succinct and unexpansive nature of the data generated.

> Alexandra – "So, does this feel like home to you, this land?"
> Mr. Mitchell – "Yeah. Yeah."
> Alexandra – "So, you like being outside?"
> Mr. Mitchell – "Yeah." [Silence while walking. Sound of interviewer unzipping her jacket. Heavy breathing of interviewer. Footsteps.]
> ....
> Mr. Mitchell – [He stops at the top of the hill and looks at his watch.] "24 minutes. That is not very good. On Saturday I did it in 22 and a half."

From the top of this hill, Mr. Mitchell pointed out the extent of his farmland, and once the timed aspect of the walk was complete, he became noticeably more talkative on the return journey. However, asking Mr. Mitchell questions was somewhat constrained by his fast pace; he was often several meters in front and could not always hear the questions being asked. Fortunately however, the digital recorder was capable of picking up his voice from this distance so most of the verbal data was recorded. On our downhill journey, Mr. Mitchell pointed out various physical landmarks and features in a primarily factual manner, apparently aiming to impart information rather than communicate feelings.

> Mr. Mitchell – "There are quite interesting things to look at. In summer, there is the butterflies. That is kangaroo poo. That is wallaby. Echidna holes."
> ....
> Mr. Mitchell – "Yeah, I can tell you all the things as we go back down."
> ....
> Mr. Mitchell – "I actually like to do [this walk] earlier in the morning. Because if I am working for the rest of the day, I have loosened myself up so that I can get a lot more done."

6.3 Reflections on the Case Studies

These two case studies illustrate the extent to which walking interviews can differ, both with regards to the experience itself and the mood of the data generated by the research encounter. However, these differences do not pose a challenge to the validity of walking interviews for explorations of how people experience particular places. Instead, these differences help to underline the specificity of each walking interview, which is heavily informed by the idiosyncratic characteristics and preferences of each individual who participates in the method. For example, these case studies illustrate that a walking interview was not capable of transmogrifying an action-oriented and analytically minded man, Mr. Mitchell, into a rhapsodic and emotional communicator about his relationship to his farm land. And during Mrs. Brown's sit-down interviews prior to the walking interview, she had already demonstrated a good capacity to articulate her emotions about her surrounds. Hence, the mood of the interview and the nature of the data generated reflect the unique personalities and life approaches of the respondents.

The value of these walking interviews lies in their ability, when used in addition to more sedentary methods such as in-depth interviews and ethnographic observation, to provide valuable insights into older adults' distinctly different engagements with rural places and their distinctly different strategies for healthy aging. Mr. Mitchell's walking interview was highly task-oriented in terms of his explicit aim to reach the top of a hill and achieve a specific time-related performance goal. This took primacy over the researcher's desire to make this a communicative encounter. When Mr. Mitchell did talk, it was often quite brief – in contrast to his more expansive conversation during his earlier in-depth interviews – and his talking was usually directed towards imparting knowledge about his land. Hence, one might reasonably conclude that Mr. Mitchell's engagement with his farmland is primarily motivated by intellectual curiosity and a desire to know what occurs there, as well as a desire to achieve specific health-related goals in that environment.

In contrast, Mrs. Brown's walking interview was primarily directed towards communicating with the researcher, as well as towards reveling in and expressing her positive emotions about her farmland. Mrs. Brown did not traverse much distance during the walking interview, as the slow pace was determined both by her low mobility and her desire to stop and quite literally "smell the roses" along the way. This walking interview, when considered in conjunction with her other in-depth interviews and ethnographic observations of her home, provided insights into Mrs. Brown's emotional responses to the land and her prioritizing sensory pleasure and comfort, over and above achieving perhaps more tangible outcomes, such as walking a long distance or living a long life.

While quite different in mood and content, these two walking interviews reveal a common capacity of the method to generate in-depth and in vivo data about people's engagement with place, and in this particular case, older adults' engagement with their rural environments. These case studies also allow further engagement with earlier questions about whether researchers can experience embodied empathy during walking interviews. In the first case study, the very low mobility of Mrs.

Brown was evident, which generated feelings of sympathy rather than empathy in the researcher; if anything, she felt more agile and more capable than she usually does, by comparison with Mrs. Brown. In the second case, the researcher felt quite distinctly physically different to the older adult, who was over 30 years' her senior: less sure of her footing, less confident she could reach the top of the hill, and less capable than he. Albeit for different reasons, neither of these walking interviews generated a sense of a shared bodily experience of aging or physical agility, but they did provide an invaluable opportunity to engage in in-depth qualitative research with participants in an empathic and flexible manner.

7 Conclusion and Future Directions

Walking is an activity which is simultaneously mundane and rich with meaning. Part of most people's everyday experience of life, walking can be either necessary or pleasurable, task-oriented, or experiential; but walking on two legs is always recognizable as a particularly human activity. Harnessing the dichotomous power of walking as both everyday and metaphorically rich, walking interviews are ideally suited for qualitative health research which seeks to engage with the nature and meaning of bodily experiences including physical well-being, illness, aging, or disability and to explore the ways in which these experiences are interwoven with the places in which people live and the meanings they have for their lives. Through physical movement and empathic social interaction, walking interviews help to generate "a deeper, experientially rich understanding" (Irving 2010, p. 26) of people's lives, their bodies, and their relationships to the places they inhabit, much of which lies in substrates of people's inner lives and hence is often inaccessible by quantitative health research methods.

Importantly, walking interviews also engage with the metaphorical meaning that walking holds for human lives (Kusenbach 2012). It is not merely a means of physically transporting ourselves; it carries a symbolic weight too, such that it can be seen to represent our life's journey. Hence, when "putting one foot in front of the other," one is not merely walking, but is also moving forward with life, sometimes in the face of hardship or pain. If life is indeed a long walk, then researchers can gain significant insights into people's lives by taking a short walk with them.

However, a balance needs to be struck between enthusiasm and caution in relation to walking interviews. While they hold considerable promise for researchers aiming to achieve physical closeness and gain insights into bodily experiences of place, walking methods are not necessarily superior to other, more traditional qualitative health research methods like sit-down interviews and ethnographic observation. Merriman (2014, p. 183) usefully advises researchers to avoid spurious claims that mobile methods are superior to traditional methods and to focus instead on "adopting modest, 'weak', open, non-representational epistemologies and ontologies – not as a means to grasp and represent elusive practices, but as a means to experiment and *move with*."

While they are open and modest in their intent, walking interviews do provide valuable opportunities for researchers to move with participants, engaging in collaborative dialogue while shoulder-to-shoulder, thereby facilitating rapport, empathy, and in situ insights into human experiences. Although social and cultural geographers have increasingly employed walking interviews in recent years, considerable potential remains for health researchers to do likewise. That is, for them to employ walking interviews either alone or in combination with other methods, whether in rural environments, urban streetscapes, or healthcare settings such as hospitals, in order to uncover and explore human beings' intangible, nuanced and meaningful experiences of illness, wellness, or healthcare.

A possible future direction for mobilities research lies in one of the criticisms it receives. In drawing attention to the rich potential of walking interviews for qualitative health research, we might run the risk of generating "an over-animated mobile subject", in which movement and action are, perhaps inadvertently, elevated above other human experiences of stillness, waiting, or immobility (Merriman 2014, p. 167). Particularly with respect to human experiences of illness, disability, and aging, such stationary states of "being" also carry significant experiential and metaphorical weight and, as such, are also worthy of close and empathic enquiry. So as to deepen understandings of experiences of health and illness, researchers might fruitfully illuminate the nature of human experiences of being immobile in a highly mobile world, by developing innovative research methods which engage with waiting, sitting, and even lying down.

References

Anderson J, Jones K. The difference that place makes to methodology: uncovering the 'lived space' of young people's spatial practices. Child Geogr. 2009;7(3):291–303.

Bergeron J, Paquette S, Poullaouec-Gonidec P. Uncovering landscape values and micro-geographies of meanings with the go-along method. Landsc Urban Plan. 2014;122:108–21.

Brown L, Durrheim K. Different kinds of knowing: generating qualitative data through mobile interviewing. Qual Inq. 2009;15(5):911–30.

Büscher M, Urry J. Mobile methods and the empirical. Eur J Soc Theory. 2009;12(1):99–116.

Carpiano RM. Come take a walk with me: the "go-along" interview as a novel method for studying the implications of place for health and well-being. Health Place. 2009;15(1):263–72.

Casey ES. How to get from space to place in a fairly short stretch of time: phenomenological prolegomena. In: Basso KH, Feld S, editors. Senses of place. Sante Fe: School of American Research Press; 1996. p. 13–52.

Clark A, Emmel N. Realities toolkit #13: using walking interviews. Manchester: ESRC National Centre for Research Methods, University of Manchester; 2010.

Cresswell T. Mobilities II: still. Prog Hum Geogr. 2012;36(5):645–53.

De Leon JP, Cohen JH. Object and walking probes in ethnographic interviewing. Field Methods. 2005;17(2):200–4.

Edensor T. Walking in rhythms: place, regulation, style and the flow of experience. Vis Stud. 2010;25(1):69–79.

Ehrlich G. Islands, the universe, home. New York: Penguin Books; 1992.

Evans J, Jones P. The walking interview: methodology, mobility and place. Appl Geogr. 2011;31(2):849–58.

Garcia CM, Eisenberg ME, Frerich EA, Lechner KE, Lust K. Conducting go-along interviews to understand context and promote health. Qual Health Res. 2012;22(10):1395–403.

Grbich C. Qualitative research in health: an introduction. St Leonards: Allen & Unwin; 1999.

Green J, Thorogood N. Qualitative methods for health research. 3rd ed. London: SAGE; 2014.

Hall T, Coffey A, Lashua B. Steps and stages: rethinking transitions in youth and place. J Youth Stud. 2009;12(5):547–61.

Hitchings R, Jones V. Living with plants and the exploration of botanical encounter within human geographic research practice. Ethics Place Environ. 2004;7(1–2):3–18.

Holton M, Riley M. Talking on the move: place-based interviewing with undergraduate students. Area. 2014;46(1):59–65.

Horton J, Christensen P, Kraftl P, Hadfield-Hill S. 'Walking ... just walking': how children and young people's everyday pedestrian practices matter. Soc Cult Geogr. 2013;15(1):94–115.

Ingold T, Vergunst JL. Introduction. In: Ingold T, Vergunst JL, editors. Ways of walking: ethnography and practice on foot. Aldershot: Ashgate; 2008. p. 1–19.

Irving A. Dangerous substances and visible evidence: tears, blood, alcohol, pills. Vis Stud. 2010;25(1):24–35.

Jackson M, editor. Things as they are: new directions in phenomenological anthropology. Bloomington: Indiana University Press; 1996.

Jirón P. On becoming 'la sombra/the shadow'. In: Buscher M, Urry J, Witchger K, editors. Mobile methods. Hoboken: Taylor & Francis; 2010. p. 36–53.

Jones P, Evans J. Rescue geography: place making, affect and regeneration. Urban Stud. 2011;49(11):2315–30.

Jones P, Bunce G, Evans J, Gibbs H, Ricketts Hein J. Research design: exploring space and place with walking interviews. J Res Pract. 2008;4(2):1–9.

King AC. Food security and insecurity in older adults: a phenomenological ethnographic study. Unpublished Doctor of Philosophy thesis (Rural Health), University of Tasmania, Hobart. 2014.

King AC, Orpin P, Woodroffe J, Boyer K. Eating and ageing in rural Australia: applying temporal perspectives from phenomenology to uncover meanings in older adults' experiences. Ageing Soc FirstView. 2015;1–24. https://doi.org/10.1017/S0144686X15001440.

Kusenbach M. Street phenomenology: the go-along as ethnographic research tool. Ethnography. 2003;4(3):455–85.

Kusenbach M. Mobile methods. In: Delamont S, editor. Handbook of qualitative research in education. Cheltenham: Edward Elgar Publishing; 2012. p. 252–64.

Lee J, Ingold T. Fieldwork on foot: perceiving, routing, socialising. In: Coleman SM, Collins P, editors. Locating the field: space, place and context in anthropology. Oxford: Berg; 2006. p. 67–85.

Liamputtong P. Researching the vulnerable: a guide to sensitive research methods. London: SAGE; 2007.

Liamputtong P. Qualitative research methods. 4th ed. Melbourne: Oxford University Press; 2013.

Liamputtong P. Qualitative research and evidence-based practice in public health. In: Liamputtong P, editor. Public health: local and global perspectives. South Melbourne: Cambridge University Press; 2016. p. 171–87.

Liamputtong P, editor. Research methods in health: foundations for evidence based practice. 3rd ed. Melbourne: Oxford University Press; 2017.

Lorimer H. Walking: new forms and space for studies of pedestrianism. In: Cresswell T, Merriman P, editors. Geographies of mobilities: practices, space, subjects. Farnham: Ashgate Publishing Ltd.; 2011. p. 19–33.

Mason J. Qualitative researching. 2nd ed. London: SAGE; 2002.

Merriman P. Rethinking mobile methods. Mobilities. 2014;9(2):167–87.

Moles K. A walk in thirdspace: place, methods and walking. Sociol Res Online. 2008;13(4):2. Retrieved from http://www.socresonline.org.uk/13/4/2.html.

Murray SA, Kendall M, Carduff E, Worth A, Harris FM, Lloyd A, ... Sheikh A. Use of serial qualitative interviews to understand patients' evolving experiences and needs. Br Med J. 2009;339:b3702. https://doi.org/10.1136/bmj.b3702.

Nettleton S. The sociology of health and illness. 3rd ed. Cambridge, UK: Polity Press; 2013.
Peyrefitte M. Ways of seeing, ways of being and ways of knowing in the inner-city: exploring sense of place through visual tours. Sociol Res Online. 2012;17(4):11. Retrieved from http://www.socresonline.org.uk/17/4/11.html.
Pink S. Walking with video. Vis Stud. 2007;22(3):240–52.
Poole R. Kierkegaard: the indirect communication. Charlottesville: University of Virginia Press; 1993.
Renard J. The journal of Jules Renard. Edited and translated by Louise Bogan and Elizabeth Roget. Portland: Tin House Books; 2008.
Ricketts Hein J, Evans J, Jones P. Mobile methodologies: theory, technology and practice. Geogr Compass. 2008;2(5):1266–85.
Riley M. Emplacing the research encounter: exploring farm life histories. Qual Inq. 2010;16(8): 651–62.
Rose G, Degen M, Basdas B. More on 'big things': building events and feelings. Trans Inst Br Geogr. 2010;35(3):334–49.
Ross NJ, Renold E, Holland S, Hillman A. Moving stories: using mobile methods to explore the everyday lives of young people in public care. Qual Res. 2009;9(5):605–23.
Sheller M. The new mobilities paradigm for a live sociology. Curr Sociol. 2014;62(6):789–811.
Solnit R. Wanderlust: a history of walking. New York: Penguin; 2001.
Spinney J. Close encounters? Mobile methods, (post) phenomenology and affect. Cult Geogr. 2014;22(2):231–46.
Stroud C, Jegels D. Semiotic landscapes and mobile narrations of place: performing the local. Int J Sociol Lang. 2014;2014(228):179–99.
Thoreau HD. Thoreau: a writer's journal. Selected and edited by Laurence Stapleton. Mineola: Dover Publications; 1960.
Thrift N. Non-representational theory: space, politics, affect. Abingdon: Routledge; 2008.
Vergunst J. Rhythms of walking: history and presence in a city street. Space Cult. 2010;13(4): 376–88.

Participant-Guided Mobile Methods

73

Karen Block, Lisa Gibbs, and Colin MacDougall

Contents

1	Introduction	1292
2	Power and Control	1293
	2.1 Informal Play Areas: The Creek	1295
	2.2 Informal Play Areas: Vacant Land	1295
3	Spatially Situated Research	1295
4	Eliciting Different Data While Mobile	1297
5	Some Additional Risks to Consider with Mobility!	1298
6	Generating Rich, Contextually Informed Data from Diverse Participants	1298
7	Mobile Methods Enabling Complex Insights	1299
8	Participant-Guided Mobile Methods: Limitations and Challenges	1300
9	Conclusion and Future Directions	1301
References		1301

Abstract

Health research is increasingly concerned with tackling health inequalities and inequities. Given that poorer health outcomes are often experienced by those who are suffering a degree of socially, economically, or environmentally determined disadvantage, it is incumbent on us as researchers to include the views and voices of diverse and sometimes marginalized or vulnerable population groups. Challenges which may accompany this imperative include engaging so-called hard-to-reach populations, and addressing an imbalance of power that often occurs between researcher and participant. Participant-guided mobile methods

K. Block (✉) · L. Gibbs
Melbourne School of Population and Global Health, The University of Melbourne, Melbourne, VIC, Australia
e-mail: keblock@unimelb.edu.au; lgibbs@unimelb.edu.au

C. MacDougall
Health Sciences Building, Flinders University, Bedford Park, SA, Australia
e-mail: colin.macdougall@flinders.edu.au

© Springer Nature Singapore Pte Ltd. 2019
P. Liamputtong (ed.), *Handbook of Research Methods in Health Social Sciences*,
https://doi.org/10.1007/978-981-10-5251-4_25

are one strategy for rebalancing this power differential when undertaking qualitative research. In this chapter, we describe the method and several case study examples where the authors have used it. We also discuss the types of research questions for which it is particularly well-suited along with its benefits and its challenges. When compared with a more traditional face-to-face interview, participant-guided mobile methods allow participants more power and control over the interview process. In addition, the method can yield observational and visual data as well as interview data, and is useful for including children and other participants who may be less articulate or lack proficiency in the language of the interviewer as it provides opportunities to "show" as well as "tell."

Keywords

Mobile interviews · Qualitative · Place · Space · Neighborhood · Walking interviews · Power

1 Introduction

Participant-guided mobile methods combine a participant-led guided tour with in-depth interviewing. The tour can take place on foot or using a vehicle and can even be virtual, investigating participants' online worlds or using technologies such as Google Earth or Google Maps to explore otherwise less accessible places. Combinations of walking and talking, sometimes described as "go along interviews," while still relatively uncommon in health research, have been used within a number of social science disciplines to investigate interactions between people and their social and physical environments (see, for example, Carpiano 2009; Kusenbach 2003; see also ▶ Chap. 72, "Walking Interviews"). The route for such walking interviews can potentially be determined to a greater or lesser extent by either the interviewer or the interviewee (Evans and Jones 2011). Our preferred term – participant-guided mobile methods – emphasizes a number of important attributes: the value of participants choosing the route, that interview data are only one of the multiple types of data that can be collected with this approach, and the fact that this is not a single method but rather a flexible set of methods that share particular characteristics. We discuss these attributes in more detail below.

The approach is best characterized as a type of *rapid ethnography,* which typically draws on a variety of data sources "to understand the social meanings and activities of people in a given 'field'" (Brewer 2000, p. 11). As with participant observation (but without the extended time commitment that method often entails), mobile methods provide an opportunity for researchers to gather multiple types of data simultaneously, adding contextual, observational, and potentially also visual data to interviews conducted in a naturalistic setting. There is, however, a critical difference between participant observation and participant-guided methods in the degree to which it is the participant or participants who decide what the researcher is shown and to what they should attend.

This characteristic of the method – the comparatively greater degree of control apportioned to participants – provides several advantages (and some challenges). In this chapter, we will discuss some of these methods' key features: a rebalancing of power and control between participants and researchers, recognition that people's lives are conducted in space, and the capacity of the method to yield rich, contextually informed data from a diverse range of participants. The discussion will draw on several case study examples where we have used this approach.

2 Power and Control

Health research is increasingly concerned with tackling health inequalities and inequities. Given that poorer health outcomes are experienced predictably along a gradient of socially, economically, or environmentally determined disadvantage, it is incumbent on researchers to include the views and voices of diverse and sometimes marginalized or vulnerable population groups. Challenges which may accompany this imperative include engaging so-called hard-to-reach populations, and addressing an imbalance of power that often occurs between researcher and participant. While research methods *per se* cannot reverse material, political, and social distances associated with different class and educational backgrounds, participant-guided mobile methods are one strategy for rebalancing some of the manifestations of these power differentials when undertaking qualitative data collection.

Bourdieu (1996, p. 19) has argued that researchers need to practice reflexivity along with "active and methodical" listening to avoid perpetrating "symbolic violence" on research participants who are frequently "disempowered" in the research process. He expands on this idea as follows:

> It is the investigator who starts the game and who sets up its rules: it is most often she who, unilaterally and without any preliminary negotiations, assigns to the interview its objectives and uses, and on occasion these may be poorly specified – at least for the respondent. This asymmetry is underlined by a social asymmetry which occurs every time the investigator occupies a higher place in the social hierarchy. . ..

Thus, if we wish to avoid inflicting such real – albeit intangible – harms on our participants, it is incumbent on the researcher to seek to redress this asymmetry in power as far as possible. Qualitative research paradigms have developed, in part, from researchers' discomfort with power dynamics in research and have been positioned by Denzin and Lincoln (2008) as democratically revolutionary. They take as a starting point the notion that research should be conducted *with participants* rather than *on research subjects*. Participatory methodologies take this idea further, seeking to engage participants throughout the research process including in conceptualizing the research problem or question, collecting and analyzing the data, and disseminating the findings (Cacari-Stone et al. 2014; International Collaboration for Participatory Health Research (ICPHR) 2013). It has been argued, however, that such practices do not automatically disrupt the power gradient between academic institutions and the communities they study (Janes 2016).

Qualitative and participatory approaches are influenced by feminist methodologies that "emphasize non-hierarchical interactions, understanding and mutual learning" with particular attention paid to the way in which data collection methods may reflect unequal power relations (Sultana 2007, p. 375). Participant-guided mobile methods sit within this broad family of qualitative participatory practices and, we would suggest, can reduce the power differences between researchers and researched. Inviting participants to take one on a guided tour of their community or environment, generally with a broad stated purpose associated with the research topic, explicitly casts them as the experts, with the researcher in the position of guest in their territory. Participants plan and control the route and point out features of importance. The interview and tour take place simultaneously, with prompts varying according to what the participant is showing the researchers. Along with observational and recorded interview data, visual data such as photographs or maps (produced manually or with digital technology) can also be generated. With participants "shaping the direction (both literally and metaphorically) of the interview" a "more democratic (co)construction of knowledge" is thus facilitated (Holton and Riley 2014, p. 60).

The case study below, *Mobile Me*, presents two themes from a study using mobile methods with Australian metropolitan children, focusing on their accounts of what assists or impedes their mobility when they are not in the direct care and supervision of their parents. The control that participating children were able to exercise during the data collection process was fitting, given the concerns driving the research, which were about understanding how and why adults ration children's independence in everyday life.

Mobile Me

Children's independent mobility – considered important for promoting health and well-being – refers to the freedom to travel or move about neighborhoods without adult supervision (Nansen et al. 2015). It has been argued that in recent times, we have been facing a moral panic, driving an unnecessary preoccupation with taking the risk out of childhood (Gill 2007), and research has found that parents determine boundaries for their children partly in response to fears about potentially dangerous people (MacDougall et al. 2009).

This study was conducted in the northern suburbs of Adelaide, the capital city of South Australia. Four boys and six girls aged 8–14 were recruited from a government primary school that had participated in a broader study (MacDougall et al. 2009). We chose to travel by minibus, rather than use walking tours, because of the distance between the places that children wanted us to see, and the fact that we wanted to explore places that were potentially unsafe. We conducted separate tours for boys and girls because, in earlier research using focus groups, we had observed boys talking over the girls. We recognized children as active social agents by asking them to plan the route for the bus driver, based on places that they considered important. They drew the route on an enlarged extract from a local street map.

During the bus tours, the children were interviewed at places that they had nominated. Children also took photographs or suggested photo opportunities to the researchers. Two of the themes are described below.

2.1 Informal Play Areas: The Creek

Boys and girls of all ages described to us a local creek that they saw as important. In Australia, a creek usually connotes a river or stream in rural areas. When the children directed the minibus tour to this creek, however, it was apparent they had a different meaning. To the eyes of the adult researchers on the minibus, it was a concrete drain on the side of a busy road with houses backing onto one side and open farmland on the other. The children, however, directed us to trees in which they had constructed cubby houses. They showed us the bushes where they played "chasey" and ran around. A 10-year-old boy responded to a question of why he likes coming here by saying that "it's like a big area and my backyard is not big enough to play chasey in." They showed us ladders which had been placed over the fences of gardens of houses which backed on to the creek, allowing children easy access to the creek. Another 10-year-old boy said "it's pretty much like my backyard" and added "last time I came here I spent probably about four hours here." Children of all ages reported that sometimes their parents allowed them to go there by themselves to meet other children. Motorbikes were driven by young adults off the road on a track next to the creek, and some children said this stopped their parents from allowing them to go there without their supervision. The 10-year-old who described the motorbikes was asked "who decides whether you can play here by yourself or not?" He answered that it was mum, and the interviewer asked "do you think that's fair that you can sometimes but not other times?" He replied "yes it is fair."

2.2 Informal Play Areas: Vacant Land

The younger boys directed the minibus tour to small patches of vacant land which were odd shaped, smaller than a house block, and with bushes and trees. There was enough open space here for them to come with their friends and play improvised games such as football and cricket. These were not formal parks, but smaller areas maintained by the local government that were very important for these children because they were local. They did not have to cross busy roads to get there and their parents felt that it was safe enough to allow them to go there with friends.

3 Spatially Situated Research

As is evident from the case study described above, participant-guided mobile methods also acknowledge that people's lives take place in spaces and involve travel. The health impacts of place, both direct and indirect – through its influence on social contexts and health-related behaviors – are well recognized. Scholars have argued that in order to gain a deeper understanding of place-based determinants of health, researchers need to engage explicitly with the "spatiality of social life" and peoples' experiences of their social and physical environments (Rainham et al. 2010, p. 668; Finlay and Bowman 2016).

To this end, quantitative researchers have begun utilizing technologies such as GIS (Geographical Information System) and GPS (Global Positioning System) to track people's movements and interactions with the spaces in which they live (Rainham et al. 2010). Qualitative inquiries are also complementing these studies and providing a deeper understanding of the meaning of these interactions. Participant-guided mobile methods are particularly suited to studies concerned with people's relationships with their communities and their environments and to studies concerned with how people interact and are influenced by their space and place. These qualitative methods also obviate some of the ethical and privacy dilemmas associated with tracking technologies, which some participants might regard as constituting excessive and unwanted surveillance.

Mobile methods are also aligned with the "new mobilities paradigm" in social research, the proponents of which argue that social science has historically ignored people's systematic movements and corresponding continuous reconfigurations of networks of people, objects, and environments as foci of study (Sheller and Urry 2006). Underlying this paradigm are understandings of the embodiment of human experience in space and time, and of a fundamental difference between being in motion and being static, "both in terms of the kinds of engagement with the world that it prompts, and the kinds of knowledge and identities that it therefore engenders" (Hein et al. 2008, p. 1268). The mobilities paradigm emphasizes that all places are tied into at least thin networks of connections that stretch beyond each such place and mean that nowhere can be an island. It begins from the complex patterning of people's varied and changing social activities, which mean that travel is necessary for social life and connections. It examines the proliferation of places, technologies, and gates that enhance the mobilities of some while reinforcing the immobilities of others. A corresponding mobilized ethnography involves walking or traveling with people as a form of deep engagement in their worldview (Sheller and Urry 2006; see also ▶ Chap. 72, "Walking Interviews").

Studies using mobile methods have also demonstrated that there is a relationship between what people talk about and the places in which they are talking. Geographers Evans and Jones (2011) compared discourses produced in sedentary and walking interviews in the Rescue Geography project exploring peoples' personal connections to their area. They found that participants in the walking interviews focused more on environmental features while the sedentary interviews produced more narratives about people. Thus, mobile methods are most appropriate when place is a significant part of the research question and less so if researchers are primarily interested in people or narratives located remotely.

Participant-guided mobile methods are clearly suited to engaging with people's spatially situated and mobile experiences. In the case study below, *Beyond Bushfires: Community, Resilience, and Recovery*, participant-guided mobile methods were used as one part of a larger mixed-methods investigation into the medium- to longer-term impacts of the "Black Saturday" bushfires, which devastated many communities across the State of Victoria, Australia, in February 2009. The research was focused on the interaction between individual and community trajectories in the years following the fires.

Beyond Bushfires: Community, Resilience, and Recovery
The Black Saturday bushfires were responsible for the loss of 173 lives, the destruction of 3500 buildings including 2133 homes and profound and prolonged social disruption in many affected rural communities. For this large mixed-methods study (Gibbs et al. 2013), approximately 1000 people completed surveys 3–4 years after the fires and again 2 years later, and 35 participants from severely affected communities also took part in a qualitative study using participant-guided mobile methods. Participants ranged in age from 4 to 66 years of age and included two grandparents, 17 parents, five children, and two young adults who were teenagers at the time of the fires. Some chose to take part with family members, as a couple or parent with a child for example.

These participants were invited to take us on a guided tour of places in their community that were important to them, with questions focusing on why these places were significant. At the participant's discretion, some tours were conducted entirely on foot while others included a driving tour to explore more distant parts of the community. We took photographs of the significant places and objects, as indicated by the participants, during the tours, and the photographs served as additional data for analysis as well a useful way to illustrate the project. Given the potential to evoke past traumas, we did not ask participants directly about their fire experiences. Most, however, chose to describe those experiences in some detail and these stories were certainly rendered more vivid for the researchers by being told in situ.

4 Eliciting Different Data While Mobile

Perhaps because the participants felt like hosts – and because of the relevance of their houses, either surviving or rebuilt – in most cases, we were invited into people's homes as well as being taken on a tour. This meant that many of the interviews took place in two phases, beginning or ending with a (sometimes quite lengthy) additional sedentary interview over a cup of tea. When the interview data were analyzed, it was apparent that the topic of conversation almost invariably changed when the interview switched from sedentary to mobile (or vice-versa). In many cases, the sitting interview elicited more personal information and stories while the mobile phase tended to focus more on the broader community and the environment (Block et al. 2014).

5 Some Additional Risks to Consider with Mobility!

As already noted, our participants for this study included children and, in one case, a very articulate and independent 7-year-old boy was permitted to take us on a tour of his small town without a parent. He confidently took charge of the tour, instructing us as to which things we should and should not photograph on the way. At one point, however, our attempts to address the natural power imbalance between adults and children appeared to have been too successful. We were impelled to wrest back some of the control we had relinquished when he suddenly dashed across the main road, apparently without looking for cars, and we had to exert considerable authority to convince him that he should not do so again (Block et al. 2014).

6 Generating Rich, Contextually Informed Data from Diverse Participants

As noted earlier, participant-guided mobile methods comprise a flexible range of techniques that are useful for conducting research with a broad range of participants. The multiple simultaneously produced forms of data that are generated – observational, visual, and interview data – can be integrated and synthesized to build a rich and contextually informed understanding of people's lives in the spaces in which they are lived. These methods allow the researcher both to experience participants' environments themselves and to interrogate people's perceptions, interpretations, and navigations of those environments – providing a range of insights not generally accessed through (sit-down) interviews alone (Carpiano 2009). Concerning this last point, Evans and Jones (2011, p. 850) differentiate between walking (or cycling) interviews, whereby researcher and participant are exposed to the "multi-sensory stimulation of the local environment" and interviews that take place in a car or a train, for example, which are "essentially sedentary from [a] bodily perspective" although undertaken while in motion.

Many of the more common research techniques, such as interviews, focus groups, and surveys, are ideally suited for gathering data from relatively well-educated participants with opinions on topics that can be readily articulated. By way of contrast, participant-guided mobile methods, similarly to other innovative methods using photography or drawings (for example), can be used effectively to include participants who may be less articulate and can consequently be marginalized or disregarded in much research. As well as using these methods successfully with children (as in the case studies described here), we have also used them with recently arrived migrants, not yet fluent in the language of the researchers. The approach provides opportunities to *show* as well as *tell* and more readily accommodates nonverbal cues, pauses, and silences than a traditional interview.

Mobile methods can also be used to understand how places are diversely experienced, by comparing, for example, the way spaces are experienced and negotiated by different genders and races or by able bodied and empowered individuals compared with those with disabilities, the aged or the young (Finlay and Bowman 2016).

The third case study, *Stepping Out*, is described below and investigated how children negotiate their independent travel. The use of participant-guided mobile methods elicited more complexity and interdependency of contributing elements to those negotiations than was revealed by the other methods used in the study.

Stepping Out: Children Negotiating Independent Travel
This study, conducted in 2011–2012 in the inner northern suburbs of Melbourne, Australia, aimed to explore the role that children played in negotiating their own active and independent travel. Participants were children aged 10–12 transitioning from primary to secondary school, a time when children often begin to undertake travel – especially to school – that is less supervised by adults. The multimethods study included observation, focus group discussions, and interviews as well as participant-guided mobile methods. A small number of parents and teachers also took part in interviews to provide additional contextual information.

In keeping with principles of participatory research and a child-rights informed practice, a small group of child participants were recruited as research partners in the early stages of the study to inform subsequent stages. Without prompting, the group suggested we should go outside and trace the actual routes taken by children. Following one of these mobile methods journeys, a recommendation was made that researchers should accompany children on travel journeys using the same mode of transportation as participants, such as a bike. Consequently, ten participating children took us on a routine travel journey – in most cases to school but also in some cases to shops and parks – to demonstrate the way in which they usually traveled to that destination (Gibbs et al. 2012).

7 Mobile Methods Enabling Complex Insights

A striking consequence of using participant-guided mobile methods in this study was that it enabled us to identify a considerably greater degree of complexity in children's independent mobility than was articulated through the focus group discussion. We found, for example, that when children told us they walked to school with their parents or with younger siblings, they did not necessarily walk side by side, but instead, they were often spread out along the route, perhaps regrouping at intersections to ensure safe crossing. We found that children's mobility was negotiated through multiple features of people, events, and environments. It was characterized by gradual transitions, use of technology such as mobile phones, significant influence of travel companions including friends and relatives, and the use of passive surveillance by parents on visible routes (Nansen et al. 2015).

8 Participant-Guided Mobile Methods: Limitations and Challenges

This chapter has demonstrated that participant-guided mobile methods provide a valuable addition to the qualitative researcher's array of methods. Here, we discuss some limitations. Firstly, as already noted, these methods are useful when place is salient to the research question. If researches are seeking to explore conceptual narratives unrelated to place, sedentary approaches may be more productive (Evans and Jones 2011). As always, when undertaking research, reflexivity on the part of the researcher is recommended. While a search for greater authenticity may underpin researchers' motivations for undertaking interviews in a naturalistic setting that is familiar to participants, we should still be aware that research encounters invariably comprise an element of performance where the participant and researcher are coconstructing a narrative and identity. A participant-guided tour is no exception to this rule, and the onus is on the researcher to reflect on their role as well as the role of their chosen method in coconstructing knowledge (Holton and Riley 2014).

There are also a number of additional challenges associated with this approach compared with a sedentary method. Weather – either very hot, very cold, or wet – may limit its feasibility and/or willingness of participants to undertake a guided tour. Both researchers and participants need to maintain a degree of flexibility in order to adapt to inclemency as well as other local conditions. As also noted previously, participant-guided mobile methods may be suitable for investigating the way in which participants with a disability or otherwise limited mobility navigate their environments. Any limitations to mobility, however, obviously need to be thought through when planning how to conduct the research. Limitations to mobility may occur because of personal characteristics of the participant or because the places under consideration are inaccessible. In one case, we were conducting research with recently arrived refugee background youth who were able to use Google Earth to show us places they had lived before coming to Australia.

In order to maximize the value of collecting data that is spatially situated, researchers will need to consider how they will include locational identifiers that can be linked to the generated narrative. These could include photographs, notes, maps, or conversational prompts that take note of local environmental features (Carpiano 2009; Evans and Jones 2011; Finlay and Bowman 2016). Researchers also need to plan how they are going to manage technical issues associated with recording the interview. While we have had success using a high-quality digital recorder held by the researcher, ambient noise and/or multiple participants may compromise the sound quality. Having more than one recording device and asking participants to hold one as well may help overcome this problem but clearly adds to the subsequent labor required when transcribing.

Finally, while empowering the participant has positive ethical implications, these methods also raise some ethical complexities. Safety is of course a primary consideration and various combinations of terrain and participant attributes may raise concerns as demonstrated in the *Beyond Bushfires* case study described above. Similarly, in the *Stepping Out* study, researchers had to take care that they did not divert the child participants from attending to traffic and other hazards. If the tour is undertaken while driving, then clearly driving skill of either the participant or researcher is relevant as are distractions associated with the research process. Confidentiality is also potentially more difficult to manage when using these methods – particularly if the tour is taking place in an environment where the participant is likely to be recognized and the researchers are conspicuously unfamiliar.

9 Conclusion and Future Directions

In this chapter, we have discussed some key features of participant-guided mobile methods along with some of their limitations and challenges. When undertaking research where space and/or movement are relevant to the research questions they provide a valuable addition to the researcher's cache of methods. They can be combined with other qualitative or quantitative methods and generate rich and contextually informed data that can help us to understand the ways in which place shapes and interacts with people's experiences. In such cases, these methods pass the methodological test, that increasing the number of data collection methods used is justified if they provide additional and complementary information (Darbyshire et al. 2005). They are particularly useful for including the contributions of participants who may be less articulate and can help to shift the balance of power between researcher and participant towards the participant in those cases. This feature makes them of particular use for health researchers seeking to investigate health inequalities and inequities affecting marginalized or vulnerable population groups.

In future research, we expect that using participant-guided mobile methods will assist researchers to extend their investigations of social determinants of health to incorporate more explicitly environmental, place-based, and mobility-associated determinants as well. Given the flexibility of these methods, the potential for researchers to incorporate new technologies and adapt them to new research questions in creative ways we have not yet imagined is considerable.

References

Block K, Gibbs L, Snowdon E, MacDougall C. Participant guided mobile methods: investigating personal experiences of communities. Sage research methods cases. London: Sage; 2014.
Bourdieu P. Understanding. Theory Cult Soc. 1996;13(2):17–37.
Brewer JD. Ethnography. Maidenhead: Open University Press; 2000.
Cacari-Stone L, Wallerstein N, Garcia AP, Minkler M. The promise of community-based participatory research for health equity: a conceptual model for bridging evidence with policy. Am J Public Health. 2014;104(9):1615–23.
Carpiano RM. Come take a walk with me: the "go-along" interview as a novel method for studying the implications of place for health and well-being. Health Place. 2009;15(1):263–72.
Darbyshire P, MacDougall C, Schiller W. Multiple methods in qualitative research with children: more insight or just more? Qual Res. 2005;5(4):417–36.

Denzin NK, Lincoln Y. Introduction: the discipline and practice of qualitative research. In: Denzin NK, Lincoln Y, editors. Collecting and interpreting qualitative materials. 3rd ed. Thousand Oaks: Sage; 2008. p. 1–43.

Evans J, Jones P. The walking interview: methodology, mobility and place. Appl Geogr. 2011; 31(2):849–58.

Finlay JM, Bowman JA. Geographies on the move: a practical and theoretical approach to the mobile interview. Prof Geogr. 2017;69(2):263–74. https://doi.org/10.1080/00330124.2016.1229623.

Gibbs L, MacDougall C, Nansen B, Vetere F, Ross N, Danic I, McKendrick J. Stepping out: children negotiating independent travel. Jack Brockhoff Child Health and Wellbeing Program. Melbourne: The University of Melbourne; 2012.

Gibbs L, Waters E, Bryant R, Pattison P, Lusher D, Harms L, Richardson J, MacDougall C, Block K, Snowdon E, Gallagher HC, Sinnott V, Ireton G, Forbes D. Beyond bushfires: community, resilience and recovery - a longitudinal mixed method study of the medium to long term impacts of bushfires on mental health and social connectedness. BMC Public Health. 2013;13:1036. https://doi.org/10.1186/1471-2458-13-1036.

Gill T. No fear: growing up in a risk-averse society. London: Calouste Gulbenkian Foundation; 2007.

Hein JR, Evans J, Jones P. Mobile methodologies: theory, technology and practice. Geogr Compass. 2008;2(5):1266–85.

Holton M, Riley M. Talking on the move: place-based interviewing with undergraduate students. Area. 2014;46(1):59–65.

International Collaboration for Participatory Health Research (ICPHR). Position paper 1: what is participatory health research? Berlin: International Collaboration for Participatory Health Research; 2013.

Janes JE. Democratic encounters? Epistemic privilege, power, and community-based participatory action research. Action Res. 2016;14(1):72–87.

Kusenbach M. Street phenomenology: the go-along as ethnographic research tool. Ethnography. 2003;4(3):455–85.

MacDougall C, Schiller W, Darbyshire P. What are our boundaries and where can we play? Perspectives from eight to ten year old Australian metropolitan and rural children. Early Child Dev Care. 2009;179(2):189–204.

Nansen B, Gibbs L, MacDougall C, Vetere F, Ross NJ, McKendrick J. Children's interdependent mobility: compositions, collaborations and compromises. Children's Geograph. 2015;13(4): 467–81.

Rainham D, McDowell I, Krewski D, Sawada M. Conceptualizing the healthscape: contributions of time geography, location technologies and spatial ecology to place and health research. Soc Sci Med. 2010;70(5):668–76.

Sheller M, Urry J. The new mobilities paradigm. Environ Plan A. 2006;38(2):207–26.

Sultana F. Reflexivity, positionality and participatory ethics: negotiating fieldwork dilemmas in international research. ACME Int J Crit Geograph. 2007;6(3):374–85.

Digital Storytelling Method

74

Brenda M. Gladstone and Elaine Stasiulis

Contents

1. Introduction: What is Digital Storytelling? 1304
2. Defining Digital Storytelling: A "Codified" Process 1305
3. Producing Digital Stories: A Workshop-Based Practice 1306
4. Initiating Critical Dialogue About DST as a Research Method in Health and Social Science Research 1307
   - 4.1 Producing "Voice" 1309
   - 4.2 Sharing Digital Stories 1312
5. Sharing the Stories: The Digital Afterlife 1314
6. Conclusion and Future Directions 1316
References 1318

Abstract

Digital stories are short (2–3 min) videos using first-person voice-over narration synthesized with visual images created in situ or sourced from the storyteller's personal archive. Digital storytelling (DST) is a codified process, originating in the 1990s as part of a community development arts initiative to mobilize voices marginalized by dominant, institutionalized media. Rooted in the rapid emergence of arts-based health research, DST is used in health promotion research and practice, public health and community-based participatory research, and multi-disciplinary fields such as psychiatry, disability studies, and social work, covering

---

B. M. Gladstone (✉)
Dalla Lana School of Public Health, Centre for Critical Qualitative Health Research, University of Toronto, Toronto, ON, Canada
e-mail: Brenda.gladstone@utoronto.ca

E. Stasiulis
Child and Youth Mental Health Research Unit, SickKids, Toronto, ON, Canada

Institute of Medical Science, University of Toronto, Toronto, ON, Canada
e-mail: elaine.stasiulis@sickkids.ca; elaine.stasiulis@mail.utoronto.ca

a range of physical, mental, emotional, and spiritual health topics and perspectives. In this chapter, we describe a case study in children's mental health, which combined DST and traditional ethnographic and participatory analysis methods. Exemplars from the case study are used to initiate critical dialogue about DST, inspired by key methodological questions raised by education, cultural, and media scholars, but generally lacking in health and social science scholarship. We consider how multiple human and institutional actors negotiate and shape the story told and the extent to which narrative constraints imposed by DST problematize claims around "voice" and representation. We conclude with thinking about the "after life" of digital stories, including how singular narratives can connect the personal and the structural to effect real social change.

Keywords
Digital storytelling · Arts-based health research · Participatory analysis · Qualitative research · Youth · Voice

1 Introduction: What is Digital Storytelling?

Digital storytelling (DST) is an intensive, workshop-based practice to guide the production of short audio-visual vignettes, or digital stories, using a structured process (Vivienne 2011). Digital stories are 2–3 min in length. They combine first-person voice-over narration, based on an individual's story script, with music, video clips, digital photographs, text, and drawings, sourced from the storyteller's personal archive or produced in situ. Digital tools (personal computers, digital cameras, editing software) are used to produce the digital story, and digital media (DVD's and/or the Internet) are used to mobilize and distribute the story (Vivienne and Burgess 2013). The "story" is positioned as a highly personal and authentic autobiographic account of the lives of ordinary, often marginalized individuals, and characterized as "tales told from the heart" (Meadows cited in Rossiter and Garcia 2010, p. 37).

Rooted in the rapid emergence of arts-based health research (see for example, Boydell et al. 2012), DST is emerging as a research method alongside innovations in more traditional qualitative and participatory research approaches and in different health research contexts, such as public health (see for example, Gubrium et al. 2013). The growing interest in digital storytelling likely coincides with claims that arts-based health research enhances researchers' understanding of participants' "lived experiences," empowers and engages participants in new ways, addressing or rebalancing power relations, accounts for emerging skills and abilities, including the growth of visual literacy, particularly among vulnerable populations, and helps to initiate social change by creating a platform for dialogue about troubling health topics (Boydell et al. 2012; see also ▶ Chap. 64, "Creative Insight Method Through Arts-Based Research").

DST is taken up in a number of ways, as a practice in professional clinical education, and as a research method (Wyatt and Hauenstein 2008; Gubrium 2009; Fenton 2013). It is used in nursing research (Stacey and Hardy 2011) and disability studies (Rice et al. 2015), studies on mental health and psychiatric practices (Baker et al. 2015), and as a knowledge translation tool (Ferrari et al. 2015). As a research method, DST has focused on gender and health from a feminist perspective (see for example, Gubrium et al. 2011) and also on the purported health effects of participation in the digital storytelling process (Gubrium et al. 2016).

A variety of health topics have been studied using DST as a research method. See for example, research about the experiences of individuals across the life course living with chronic and acute illnesses such as HIV/AIDS (Willis et al. 2014), dementia (Stenhouse et al. 2013), and psychosis (Boydell et al. 2016) or with difficult transitions from nursing trainee to professional practice (Stacey and Hardy 2011). The therapeutic or health-promoting attributes of participating in DST are reported, including individuals' recognition of their strengths and resilience in spite of difficult circumstances (Willis et al. 2014), enhanced understanding of and ownership over life experiences and how they are told (Ferrari et al. 2015; Gubrium et al. 2016), and improved self-confidence and socially supportive connections (Stenhouse et al. 2013; Gubrium et al. 2016). Despite the growing popularity of digital storytelling as a research method, it is relatively new to health studies (Lal et al. 2015). The DST process and the digital story as a product, are mostly unexplored but inter-related topics (Rossiter and Garcia 2010), and key methodological questions raised by education, cultural, and media scholars (Alrutz 2013; Worcester 2012), are generally lacking in health and social science scholarship.

2 Defining Digital Storytelling: A "Codified" Process

It is difficult to reach consensus on a definition of DST because as an emerging method of communication, it is being used in the midst of an ever-changing landscape of digital production (Wales 2012). For this reason, we define digital storytelling in relation to a codified process used to produce stories, originating in the 1990s, as part of a community development arts initiative to mobilize "ordinary" people's voices considered marginalized by dominant institutionalized media (Worcester 2012). The Centre for Digital Storytelling (CDS) in Berkeley California developed the codified process underpinning the DST model in its "classic" form (Burgess et al. 2010) and this is generally considered to be a preeminent model in amateur digital video composition (Fulwiller and Middleton 2012). The DST codified process is designed to help individuals produce, rather than simply consume media stories, democratizing the process of media production as people learn to tell their own stories (Alrutz 2013). Expert power is subverted in this process as individual, "ordinary" stories are purported to represent a more amplified and "truthful" voice, which can then be directed toward initiating social change (Worcester 2012).

The production of a "good" story is essential to the objective of DST in order to link the individual, personal narrative to the socio-political sphere. An esthetically appealing story is likely to maximize its relevance by having a greater impact on audiences. Poletti (2011) refers to this goal as "coaxing" for a good story. Gubrium (2009) raises a similar point, describing digital stories as "shaped" by instructional standards on how to tell a compelling story, which are known as the seven elements of

Table 1 The seven elements of digital storytelling

1. Point of view	What do you want to say?
2. Dramatic question (story structure)	How can you say it best?
3. Emotional content	Helping your audience care.
4. Voice	Only you can tell your story.
5. Soundtrack	Add music and other auditory elements for impact.
6. Economy	Keep it simple when it comes to words & images.
7. Pacing	Give viewers time to take in the story.

DST, a hallmark of the California model (see Table 1). The purpose of the seven story elements is to ensure that novice storytellers will be able to create a successful story because the narrative will "satisfy, surprise and engage the viewer" (Poletti, p.78).

The power of the digital story lies in its first person narrative, the story as written and "voiced" by the participant who tells their own story. Initially, priority is given to story content and how it is expressed rather than other audiovisual material (Worcester 2012). Script writing is foundational for the process of DST and the seven elements teach participants how to write a focused story, containing a point of view or dramatic question and emotional content, alongside participant-generated voice-over narration. The digital story may include an additional soundtrack. All of this must be accomplished economically, using a brief storytelling format that is paced to engage the audience.

The first step in the codified process involves choosing a particular moment from a larger story that could be told, and this choice is the basis for story content. This decision creates space for a resolution of some kind in response to a position or point of view the storyteller has taken. This choice must be articulated through a strong story structure that engages an audience by learning "how to say it best" (see Table 1). The concept of "voice" in DST is literal (the voice-over narration) but also metaphoric; the story is invested with emotion that can speak to and engage the listener, by personalizing the story. "Voicing a story" is the means by which audiences identify with the storyteller and the story is supposed to foster empathy and resonate with the listener. Economy and pacing refer to how the multimedia story is managed through editing to create a particular effect. Economy references the narrative's potential to produce closure in a short amount of time by generating implicit meanings through symbolism and metaphor that link back to the storyteller's point of view. Pacing governs how stories are structured to allow viewers to contemplate and potentially project themselves into the story.

3 Producing Digital Stories: A Workshop-Based Practice

DST is a workshop-based practice, which sets it apart from other methods for generating visual images (Vivienne 2011). Stories are produced during interactive group sessions with a small number of participants (typically 5–10), facilitated by professional digital storytellers over a three- to four-day period. This concentrated time is important for participants to experience full immersion in the workshop experience

(Gubrium 2009). The process follows a sequence of steps beginning with an overview of digital storytelling and a screening of exemplars from other DST projects.

The story circle is an important first step; participants share stories and receive feedback in a space that is meant to be comfortable and safe. The story circle builds individual and group confidence in learning how to tell and listen to stories at the same time that it provides mutual mentoring and fosters group cohesion and interconnection (Vivienne 2011). DST workshops, and the story circle, have been compared to Paulo Freire's educational process of building critical consciousness (Freire 1970, cited in Gubrium et al. 2013) because it opens up reflexive space for participants to think analytically about the place their stories have in the world. Engaging in a collaborative group process of shared media production actively constructs individual and collective agency, as participants consider how their own stories represent individual experiences brought together to speak back to social structures of power (Vivienne 2011; Worcester 2012; Gubrium et al. 2013)

An individual script writing exercise (using the seven story elements) and the creation of a visual storyboard to accompany the text follows the story circle. Several drafts of the script might be produced, which is highly dependent on the individual participant's skills and desire to revise the story. Participants take photos, shoot video, or record audio segments as part of the workshop, and they may produce drawings in situ or bring material from their personal archive to edit into their digital stories. The technical aspects of DST come into play as participants record voice-over narration and learn how to use computer software to design, edit, and assemble their stories. The workshop typically concludes with a celebratory screening of the digital stories to all in attendance (Willox et al. 2012).

4 Initiating Critical Dialogue About DST as a Research Method in Health and Social Science Research

This section introduces a case study in youth mental health research, which combined DST and traditional ethnographic and participatory analysis methods. Exemplars from the case study are used to initiate critical dialogue about DST, inspired by methodological questions raised in educational, cultural, and media scholarship (Poletti 2011; Worcester 2012) but generally lacking in the health and social sciences. A central tenet of DST is that it democratizes media production by facilitating the creation of mediatized stories by "ordinary" people about their own lives (Alrutz 2013). With the proliferation of digital storytelling projects in a range of institutional contexts, how DST is understood and used may be appropriated by the attending institution, and constrained by the seven story elements and the workshop-based practice. All of which may be combined to shape the DST process such that the final product is considered co-created, and sometimes predictable in form (Burgess 2006; Worcester 2012).

Our project examined help-seeking narratives created by young people using DST as a participatory research approach to foster critical reflection and facilitate expression and communication about (often sensitive) aspects of their experience. Young people were asked to focus the substantive content of their stories on how they manage everyday life when a parent has a mental health problem or diagnosed illness. This is a significant public health issue worldwide (Reupert et al. 2015), and while (under) estimates suggest that 12.1% of all Canadian children *under 12* have a mentally ill parent (Bassani et al. 2009), in this complex and under-serviced area little is known about how young people cope with the challenges they encounter and respond to services they receive, or how they would like to be supported in this context. A second objective of the study was to examine how young people acquire further competencies and skills through guided participation and active engagement in the research process, by documenting the process of arts-based knowledge production and dissemination.

Participants: Ten young people (13–19 years.; six females and four males) currently receiving community-based mental health services and self-identifying as having a parent with a mental illness participated in the project. Our sampling strategy was purposive (Patton 2015), to ensure variation among participant ages, gender, parental psychiatric diagnoses, and the type and length of young people's involvement with mental health services, including a range of referral pathways that could reflect different help-seeking experiences. Young people in the study were referred for help from school and hospital staff, a homeless shelter for youth, a children's crisis phone line and by family members, including in one case, the parent with mental health difficulties. Most ($n = 7$) lived at home with their parent(s), or other family members, and the remaining ($n = 2$) resided in residential group homes and one young person lived in supported housing on their own. Participants reported their parents experiencing a variety of mental health challenges including depression, bi-polar disorder, schizophrenia, psychosis, anxiety, personality disorder, and posttraumatic stress disorder.

Study Design: Participants worked together as a group, longitudinally over the course of the study, guided by adult facilitators to clarify study objectives and research questions (Project Introduction); produce 3–5 min digital stories (DST Workshop); analyze study data; provide feedback on project findings; and recommend knowledge translation strategies to reach and engage particular audiences (Analytic and End-of-Project Discussions) (see Fig. 1). Explicitly engaging young people in participatory analysis is a novel approach to DST and a significant but missing dimension of participatory research with young people generally (Nind 2011). Participant observation and informal interviewing methods enabled documentation of project activities and participant engagement in the research process.

Ethical Considerations: Prior to obtaining informed consent, the project manager (ES) met separately and at length with each potential participant to ensure they understood the study objectives, and the implications for participation, according to the research design. Mindful that participants could feel differently about sharing stories upon completion (Dush 2012), a two-stage consent process was incorporated. During the second stage, participants were asked to consider the extent to which they were willing to share their stories (the broadest possible choice being the Internet). This second consent process occurred during the end-of-project discussion (Fig. 1) and four participants chose to share their stories broadly. Recognizing that DST participants sometimes choose to tell painful stories, it was important to consider whether the process would be "triggering" for young people (Gubrium et al. 2013). We worked closely with the clinicians involved

Fig. 1 The study design

in recruitment to minimize the risk of harm and in our face-to-face meetings prior to obtaining consent to screen young people and carefully explain the potential risks of participating in the study, although we acknowledge that it is not always possible to gauge how participants will be affected. All young people were connected to a clinician they received support from and were encouraged to contact this person if they experienced distress during the course of the study, or afterwards.

The case example is presented here to raise questions about the co-created aspects of DST, to theorize how "voice" is produced in the digital story and as it circulates and is shared and interpreted by others (Yates 2010). A key purpose is to consider how multiple human and institutional actors negotiate and shape the story, and the extent to which narrative constraints imposed by DST problematize claims around "ordinary voices" as they are re/presented, re-contextualized, and re-mediated beyond the workshop setting (Matthews and Sunderland 2013). This is significant in thinking too about what it means to work the limits of voice in conventional, interpretive, and critical conceptions of voice in qualitative inquiry more broadly (Mazzei and Jackson 2009)

4.1 Producing "Voice"

4.1.1 Project Introduction: Bringing Voices to the Table

Participants were introduced to the project the day before the 3-day DST workshop (see Fig. 1). Members of the research team took care to introduce themselves and the research participants to one another, hoping to build familiarity and reduce any unspoken anxieties at the outset of the project. Participants were introduced to an emerging literature on children's "lived experiences" of parental mental health problems (Gladstone et al. 2011). The group worked toward consensus around project goals, addressed outstanding questions, and identified potential barriers to participation, in order to establish a protocol for working together. Discussion revolved around the following reflexive questions:

> How do my experiences shape my participation in the project? What experience have I had with (arts-based) research? What do I hope to get out of the research project? What concerns do I have or anticipate having about the challenges involved in being part of this project?

Young people wanted opportunities to educate others through participation in the project, particularly to dispel negative perceptions of mental illness. They thought adding a question about "what happens after" (living with a parent with a mental illness) was important because it addressed their own experiences more directly, focusing on their insights and expertise about the family situation and the "intergenerational" experiences of mental health and illness. Many ideas raised in this early discussion re-emerged in the digital stories created by individual participants. At this point, the stories were being shaped in multiple ways, through the introductory discussion that included the researcher's objectives for the project and findings from the empirical literature on families living in this context.

4.1.2 DST Workshop: Negotiating Voice

Two DST facilitators from a professional organization that offers media and digital storytelling services for community development purposes worked with the participants to design and produce digital stories during an intensive 3-day workshop (Fig. 1). The process included: script writing and voice-over recordings of first-person narratives; creating visual story content through images brought from home or photos and drawings created in situ; choosing (copyright-free) music and sound-effects to illustrate the stories; and editing visual and audio material to produce the final story using specialized software.

Participants first learned how to structure their story according to the seven story elements (Table 1), inspired by digital stories screened from previous projects. One participant, Justin, asked whether their stories had to "have a happy ending," sensing the expectation in DST to provide closure. The facilitator pointed to one story about addiction that was not resolved because the storyteller was "still drinking and waiting for recovery." She explained,

> She [the storyteller] is still in the middle of what she is going through. You don't have to tie everything up into a pretty bow. I won't ask you that, but I am asking you to draw on your own strength and power, your own insight. It doesn't have to have an ending.

Participants were also asked to consider research questions to create their stories:

> What would you like others to know about your experiences coping with difficulties because your parent has a mental illness? What led to the challenges you and your family experience? What could be done to prevent these difficulties? What are your feelings about the types of help that you have received? What type of support would be most useful to you and your family?

The extent to which participants had choice in responding to the research questions was difficult to sort out and the research team and the DST facilitators from the beginning of the project openly discussed it. It remained a challenging issue that was addressed with some ambivalence by the DST facilitators, but also with flexibility and a willingness to negotiate around this on everyone's part as in the following directive to the participants:

> These are the questions that the researchers want to explore, but you don't need to pick one and answer it. You may want to answer all of them, and it might be what happens next – some might want to be positive, others might say screw this I'm in the middle of something. The researchers want to know this, but don't feel constricted. (Facilitator, Story Circle, Day 1)

The shaping of the story continued during the story telling circle. Participants were encouraged to tell a story and to receive feedback or questions from group members. Young people were quick to respond, finding they had many similar experiences compared to their peers, responding with understanding or giving advice about how they might think about the digital story they wanted to tell. Facilitators also played an active role in shaping participants' stories during this exercise. For example, using traffic lights as a metaphor for types of stories that could be told, participants were challenged to tell a (yellow light) story that may be difficult, but not so demanding that they would "have a nervous breakdown" (red light), or one told so often that it was too easy (green light) and lacked impact. Alex shifted her story from one she told frequently about her mother's experience of schizophrenia and how it affected the family, to a story about her own sexual abuse that she had not wanted to burden her mother with when she was not well.

The tensions that are part of negotiating "voice" were present throughout the early part of the DST process. The facilitators struggled between wanting to ensure participants had choice over the stories they would tell and giving enough direction that would result in a "good" story, which followed the structure outlined by the seven story elements for writing a DST script (see Table 1). This is illustrated in the facilitators" struggle to provide guidance around storytelling:

> Process of storytelling is part of adult education process. We will give you feedback but it is a skill that needs to be learned. It can be hard. We are giving feedback out of respect because we want to help you have a story that is as good as it can be. You can take the feedback and consider it, or you can choose to not pay attention to it, because it is your story. If you feel we are pushing you, tell us to back off. I might push you hard because I think it may help your story. (Facilitator, Day 1)

The facilitators' success in meeting this challenge could be measured by the participants' adherence to the expectations of the seven story elements, evident in the brevity of the stories created (2–3 min), the use of first person narratives and a closed structure, with many story endings tending toward a more positive resolution.

Despite some of the constraints imposed on DST participants, young people seemed pleased with the stories they were able to tell. Some produced stories that were not clearly aligned with the research objectives, at least at first glance. The medium of digital storytelling seemed to enable participant "voice" because young people were able to express themselves in ways that aligned with how they wanted to be represented to others. The following examples from the analytic discussions illustrate this point, albeit slightly differently in each extract:

> Yeah, I think we did choose how we wanted to represent our story. That's like the difference between telling it to a therapist or telling it in a group. Because the story has a creative component to it, so that's like how we chose to represent our story, how we chose to represent our self... A common thing from all of our videos is I think, a loss of choice. Like none of us really had choices about where, what happened to us, like what happened with our parents. Like we don't have a choice. And this time we actually do have a choice about what we're going to say about it. So, I think that's important. (Fia)

> Being able to say exactly what you want to say and then having the images to back it up, make it go all that much deeper. I think it's a good medium to use. (Alex)
>
> It was rather cathartic to sort of um, maybe just get things out that wouldn't otherwise have been said in normal conversation. Right? Some of the things, most of the things, a lot of the things, in the videos either deal with heavy subject matter or very personal subject matter which doesn't necessarily come out all the time within conversation, not everyone's going to understand or going to deal well with that. So I think that for at least me, you know, me personally, it was a nice little release from my otherwise normal everyday scheduled programming. (Kaye)
>
> Art is in the eye of the beholder whereas a multi-media project, we're telling you straight up what we're saying. (William)

Overall, participants felt the digital storytelling method allowed for a deeper, more complex way of telling (and showing) a story. Fia who explained, "There is a whole world in just one story," best sums up the potential richness of the digital story. This is significant for young people who may be used to having to tell their story so often (to health and social care professionals in particular) that many routinely tell only easy, "green light" stories, even though others may become inured to hearing them. Participants saw their everyday experiences as valuable knowledge and felt a sense of mastery or agency in the choice and control they felt they had over the story they produced, if not over the events that had inspired them.

4.2 Sharing Digital Stories

4.2.1 Structured Group Analysis Discussions

From the outset of the project, beginning with recruitment, participants understood that the purpose of digital stories was to share them with others, although the extent of this sharing was to be determined later, once the stories were complete. They were eager to help other people by sharing their experiences and this contributed in large part to the reason they chose to take part in the project. Young people anticipated sharing their stories with particular audiences and some had these groups in mind when they produced the digital story. The study was designed to provide a space where participants could hold their stories at "arm's length," to analyze them individually and as a group and discuss who they wanted to share their stories with and why. Following the workshop, young people participated in four, three-hour group analysis discussions over a two-week period, in addition to a final end-of-project session, 2 months later (Fig. 1). The sessions were co-facilitated by the DST storytelling facilitators in collaboration with the researchers (BG and ES). Analytic activities involved small and whole-group exercises, including "fun" activities such as a rhythm circle and collective poem exercises that did not rely heavily on scholarly skills in order to mitigate power differentials between young people and adults and those with higher degree training during this process (Thomas and O'Kane 1998).

During a "head, heart, and feet" exercise, participants were asked to reflect on their thoughts and emotions during the DST workshop process and why their cognitive and emotional responses mattered to the stories they had created and were now thinking about sharing with others. Other exercises were used to help them think about the risks and benefits of telling stories, the challenges and strengths of each story, how visual and audio content deepen the interpretation of the story, as scripted, and spoken, and what it was like to share their stories with each other for the first time. Participants added more information about themselves to each of their stories. They were asked to provide a story synopsis and assign key words and discussion questions for each story. The end-of-project discussion involved reviewing a preliminary analysis and interpretation of the study data, which the researchers (BG and ES) added to and brought back to the group for reflection after a two-month hiatus. The discussion included initial responses from the first viewing of their digital stories by the project advisory group in a separate meeting (clinicians, educators, a policy maker, and a young person with a parent with mental health challenges). During the final meeting with the participants, they were asked to recommend knowledge dissemination strategies for their stories and which audiences they wanted the researchers to address.

The participatory analysis added participants' voices to the project in a different way, by creating a space for critical self-reflection and collaborative sense making, and this enabled each of them to further shape how their stories might be heard. The participants' *and* the researchers' understanding of the "lived experience" of young people managing in the context of having a parent with mental health challenges was deepened through the interactive and dynamic analytic discussions (Nind 2011). As researchers who participated in and observed and documented the entire DST process, we were also aware of an analytic responsibility we had undertaken to ensure that participants' voices were reflected in the study results and dissemination and included in our scholarly (and interpretive) engagement with one another (Drew and Guillemin 2014). Joint analytic activities were meant to facilitate whole-group interaction and to enable a mutual sharing of ideas, although we recognized the impossibility of a "level" playing field with respect to power differentials between young people and adults who have multiple forms of privilege and power. And yet the participants also recognized a distinct role for us as researchers, explicitly directing us to share their stories, what to emphasize, as illustrated in the following excerpt:

> Alex: Just make sure our strengths our emphasized. The fact that we do have strengths, it's not just like, you know, we're just these weak vulnerable kids who have had rough lives, like that's not all that we are at all.
> Researcher: Yeah, that's not the story you want to tell.
> Alex: Okay, um... I think with me it's more just like, that's not really what I'm trying to make them understand, why I share my story. It's not like, oh, look at how bad my life has been, feel sorry for me. That's not what I'm trying to get to. I'm trying to get to, "Yeah, this happened but I got through it and this is what I learned from it and this is the person I am because of what happened." Like it's not all bad, necessarily (End-of-Project Discussion)

It was ethically important to have a space to talk about their stories which emerged later in discussion with the participants who wanted to discuss them and debrief about the process that led to their creation.

Because after we talked about it, it kind of like.... cause when you made the videos you remembered everything and talking about it made you feel better. (Suzanne, End-of-Project Discussion)

Because it's kind of like opening up a wound and irritating it, like you need some closure. It kind of added meaning to this whole project. Cause, like if you had just like, wanted our stories and said "bye bye", here's your money, it would be kind of weird (Justin, End-of-Project Discussion)

Yeah, it would feel like you're bleeding was dry and then hang us up to dry. (William, End-of-Project Discussion)

Despite their willingness to disclose difficult aspects of their "lived experience(s)," young people talked about the emotional investment involved in telling stories that re-opened old wounds and memories of past events. There were tensions, between experiences of storytelling that were cathartic (e.g., Justin felt "less bitter" about his own experience after working with the DST group), but also risky due to the challenging and potentially re-traumatizing or stigmatizing narratives they produced. They acknowledged the risk as something that could undermine their intentions, and efforts to educate others and create change, or to evoke emotions that exoticize young people and their families as "other." This was significant for young people accustomed to being labeled "at risk," which as Alex said, "doesn't tell anyone anything about you." The process of examining and analyzing deeply significant and similar experiences, and learning from and observing each other's strengths, growth, and change over the course of the project, served to strengthen their connection and to mitigate some of the discomfort and worry they experienced during the DST process and now as they anticipated sharing stories.

5   Sharing the Stories: The Digital Afterlife

The "digital afterlife" refers to the expectation that digital stories are produced so that they will be shared, which raises questions about how stories are listened to and understood as they circulate beyond the workshop setting. However, the "digital afterlife" has received less attention compared to the emphasis placed on producing digital stories (Matthews and Sunderland 2013). The inclusion of analytic discussions as part of the study design opened up a space to think more concretely about the digital afterlife with participants who had clear and specific ideas about which audiences and messages they wanted to address and their desire to influence others to make change. Several entered the study with specific messages for particular audiences, which had its own shaping effect on the story that was created. For example, Jamie had strong ideas about wanting to communicate to other transgendered youth and the service providers who work with them. Kaye wanted to talk to other young people who had lost parents to suicide and to contribute to suicide prevention efforts by sharing her story about the need for mental health resources with health professionals and policy makers. Several participants, like Lee, wanted to deliver a hopeful message to young people in similar circumstances.

During the end-of-project meeting, we discussed the "digital afterlife" to think about how stories might be received outside the study group. The project's advisory group had viewed stories during the hiatus and provided feedback that re-ignited a heated discussion about the risk of telling stories that could lead others to feel sorry for them. This led to a robust discussion over more than one session about the concept of pity and a manuscript currently underway on this topic. Digital stories may draw on dominant discourses and clichés, which do more to conceal than to confront socially structured problems and power relations (Gladstone et al. 2012). Most of the participants did not show their digital stories to anyone outside the group, describing their reluctance and ambivalence to do so (often to protect family members, including parents). Jamie was an exception because he chose to show his story to his residential care workers. He was pleased by their response and felt his story had a significant influence on their understanding of his situation, despite the number of times he had tried to tell his story previously. His digital story seemed to produce a "voice" that could be heard by others when his own seemed to fail him. Jamie hoped this would change how residential care workers would treat others "like him" in the future (identifying himself as a transgendered male who had experienced difficulties related to his own mental distress as well as that of his mother).

> This is something we didn't actually know about you. This is something that we wish we could have known to help you more." And it's funny that I had spent two years trying to tell people this stuff or they had been there through most of this and didn't actually get it until they actually saw... like visuals to... like hearing it to understand the actual impact of a person's life. I'm definitely glad I did [show the story], especially because now, this situation that I've moved out. About a month before I left, not even a month, like maybe three weeks... um another transman moved into the house and now that they've seen my story and seen the impact that coming out can have on a family and a person themselves, I feel a lot more comfortable because I've paved the way for fifteen months in that house for what you should do, what you shouldn't do. And they've seen the actual impact on me, that I feel a lot more comfortable when it comes to him [the new resident] because he'll be a lot more respected, treated properly and actually have access to the resources that I've laid out for them. (Jamie, End-of-Project Discussion)

While researchers set the objectives and methods for the study, young people had their own motives for participating that can add to our understanding of what they hoped their stories would achieve. They talked about generating relevant substantive content based on personal information and knowledge situated in their own (intergenerational) experiences of mental health and illness. They wanted to tease out complex causes for the problems they faced and felt they had much to contribute to new knowledge in the field because they have something to say that only they can say as young people. As Alex argued, "I'm still in the middle of [the story] and I have a very different experience of it [than someone who is older]"). Young people believed in the power of knowledge for change and their stories as a way to counter stigmatizing and discriminatory narratives, and as an avenue for developing a public critique of experiences that are not "just a personal story, but one that would educate, prevent others from going through the same difficulties."

6   Conclusion and Future Directions

The case example presented in this chapter raises questions about DST in health social science research, particularly questions about how "voice" is produced by the multiple human and institutional actors that shape the story. Recall that digital stories are often positioned as highly personal, emotive, and authentic autobiographic accounts of the lives of ordinary, often marginalized individuals. In the same way that the seven story elements facilitate the production of a "good" story, this narrative structure may be formally constraining in how "experience" is or can be articulated (Worcester 2012; Alrutz 2013). From this perspective, the narrative schema may produce a predictable story because it stresses clarity, resolve, and closure, leaving less room for ambiguity and unsettled endings. This may pose challenges for participants in how they confront or redefine meaning attributed to their life experience (Poletti 2011). These are significant questions for scholars who want to take participatory research methods with young people or other "vulnerable" populations (Komulainen 2007; Nind 2011), as well as arts-based health research methods (Boydell et al. 2012), in a more critical, theoretical direction, to think about what it means to "have" or to "give" voice and to analyze and interpret visual images in addition to text-based data. Some general direction on how to think further about participatory and DST methods and the concept of "voice" is forthcoming from other scholars like Komulainen (2007, p. 22) who cautions against a "too simplistic and/or sensationalized usage of the term voice" because young peoples' voices should be subject to a thorough-going analysis that considers human communication dynamic, context-bound, and interactive and recognizes that there are always ambiguities present in adult/child encounters.

Similarly, Mazzei (2009, p. 7) and others (see for example, Yates 2010) argue for transformative research "practices that [can] elicit and account for the shifting and uncertain voices, spoken by participants with words, with images and with silences toward more subtle, more nuanced, more startling meanings...[an] undisciplining of voice that does not make easy sense and that transgresses the domesticated voice that we are accustomed to hearing, knowing and naming. Such an undisciplining results in the claiming of an excessive silent voice that we cannot 'hear' but that speaks to us nonetheless." As a site of cultural production rather than reproduction (Guillemin 2004), DST has much to gain and also to contribute to current scholarly debates about "voice" and other (e.g., data analysis and interpretation) methodological challenges in qualitative inquiry more broadly.

A key consideration is the extent to which DST liberates and/or constrains "voice" with respect to representation, which has been problematized more widely in critical and poststructural approaches to qualitative inquiry and "voice research" (see for example, Jackson and Mazzei 2009). Inherent to the participatory design of our study example was the shifting power dimensions among participants, researchers, and facilitators and how this influenced decisions made throughout the DST workshop and during the group analysis discussions (which moments of the larger story were selected for the digital story; how visual elements were chosen to represent the narrative; how the story should be shared and who it should be shared with and why). Not typically reported in descriptions of DST workshops is the role that facilitators play in working with participants to shape the stories (Gubrium et al. 2013) and also researchers, who also have a vested interest in the study.

A second question concerns what happens after once the digital stories begin to circulate outside the setting in which they were produced. Our case example goes beyond the production of digital stories (and how voice is produced) to consider the "digital afterlife," asking how singular narratives can connect the personal and the structural (or political) to effect social change (Matthews and Sunderland 2013). Despite the significance that sharing individual stories can have in local settings, as Jamie's experience illustrated, it is important not to assume that stories shared in broader (online) contexts will find their way to appropriate audiences to instigate the type of change DST participants (and researchers) are hoping for (Matthews and Sunderland 2013).

In thinking about how stories circulate, we can question not only how "voice" is produced, but consider also the ways in which voice itself may be productive (Mazzei and Jackson 2009), influencing how others understand and create meaning out of digital stories. Matthews and Sunderland (2013) alert us to the possibility that digital stories require translation, into other forms of representation that can speak to different audiences and to audiences differently. The researcher may be charged with particular responsibility regarding the role of "translation" in DST, including expectations by storytellers like those in our study, who directed us to use their stories, but did not always want this obligation themselves. Matthews and Sunderland (2013, p. 102) describe "re-contextualization" and "remediation" as way of thinking about consciously re-shaping stories for other modes of representation. They describe the "inherently political and interpretive shifting of meanings, between genres and contexts of social activity, such as personal storytelling, research, and policy making."

In our case, the digital stories have been re-contextualized and remediated in several ways. At the time of writing, a digital storytelling website is under construction. It is an evolving context in which stories with an online presence are remediated in a way that leads to a potential reshaping of the participant "voice" from the one produced during the interactive, face-to-face setting of the DST workshop. We have encountered challenging questions about "how much" context is required, and appropriate, for viewers who migrate to the website so that they understand how and why and in what context these digital stories were made. This is implicated also in thinking about "voice" and how others, including various (and also unseen) audiences, may have their own interpretations of the stories and what they mean (Gladstone et al. 2012). Audience "readings" will also contribute to how the young peoples' narratives are re-shaped, as the digital stories are re/viewed and re/told in other contexts. The risk in any re-telling or remediated context is the possibility that the voice of the storyteller will be obscured, misinterpreted, compromised, or even lost altogether.

References

Alrutz M. Sites of possibility: applied theatre and digital storytelling with youth. RiDE J Appl Theatre Perform. 2013;18(1):44–57.

Baker NA, Willinsky C, Boydell KM. Just say know: engaging young people to explore the link between cannabis and psychosis using creative methods. World Cul Psychiat Res Rev. 2015;10(3/4):201–20.

Bassani DG, Padoin CV, Philipp D, Veldhuizen S. Estimating the number of children exposed to parental psychiatric disorders through a national survey. Child Adol Psych Ment Health. 2009;3(6).

Boydell KM, Gladstone BM, Volpe T, Allemang B, Stasiulis E. The production and dissemination of knowledge: a scoping review of arts-based health research. Forum Qualitative Sozialforchung/Forum: Qualitative Social Research. 2012;13(1), Art. 32. http://nbn-resolving.de/urn:nbn:de:0114-fqs1201327

Boydell KM, Gladsotone BM, Stasiulis E, Cheng C, Nadin S. Co-producing narratives on access to care in rural communities: using digital storytelling to foster social inclusion of young people experiencing psychosis. Stud Soc Justice. 2016;10(2).

Burgess J. Hearing ordinary voices: cultural studies, vernacular creativity and digital storytelling. J Media Cult Stud. 2006;20(2):201–14.

Burgess J, Klaebe H, McWilliam K. Mediatisation and institutions of public memory: digital storytelling and the apology. Aust Hist Stud. 2010;41:149–65.

Drew S, Guillemin M. From photographs to findings: visual meaning-making and interpretive engagement in the analysis of participant-generated images. Vis Stud. 2014;29(1):54–67.

Dush L. The ethical complexities of sponsored digital storytelling. Int J Cult Stud. 2012;16(6):627–40.

Fenton G. Involving a young person in the development of a digital resource in nurse education. Nurse Educ Pract. 2013;14:49–54.

Ferrari M, Rice C, McKenzie K. ACE pathways project: therapeutic catharsis in digital storytelling. Psychiatr Serv. 2015;66(5):556.

Freire P. Pedagogy of the oppressed. 30th Anniversary ed. New York: Continuum International Publishing Group; 2000/1970.

Fulwiller M, Middleton K. After digital storytelling: video composing in the new media age. Comput Compos. 2012;29:39–50.

Gladstone BM, Boydell K, Seeman M, McKeever P. Children's experiences of parental mental illness: a literature review. Early Interv Psychiatry. 2011;5:271–89.

Gladstone BM, Volpe T, Stasiulis E, Boydell KM. Judging quality in arts-based health research: the case of the ugly baby. International journal of the creative arts in interdisciplinary practice. 2012; Spring Supplementary Issue,11, http://www.ijcaip.com/archives/IJCAIP-11-paper2.html

Gubrium A. Digital storytelling: an emergent method for health promotion research and practice. Health Promot Pract. 2009;10(2):186–91.

Gubrium A, DiFulvio GT. Girls in the world: digital storytelling as a feminist public health approach. Girlhood Stud. 2011;4(2):28–46.

Gubrium AC, Hill AL, Flicker S. A situated practice of ethics for participatory visual and digital methods in public health research and practice: a focus on digital storytelling. Am J Public Health. 2013;104(9):1606–14.

Gubrium A, Fiddian-Green A, Lowe S, DiFulvio G, DelToro-Jejias L. Measuring down: evaluating digital storytelling as a process for narrative health promotion. Qual Health Res. 2016;26:1787. https://doi.org/10.1177/1049732316649353.

Guillemin M. Understanding illness: drawings as a research method. Qual Health Res. 2004;14:272–89.

Komulainen S. The ambiguity of the child's "voice" in social research. Childhood. 2007;14(1):11–28.

Lal S, Donnelly C, Shin J. Digital storytelling: an innovative tool for practice, education, and research. Occupat Therapy Health Care. 2015;29(1):54–62.

Matthews N, Sunderland N. Digital life-story narratives as data for policy makers and practitioners: thinking through methodologies for large-scale multimedia qualitative datasets. J Broadcast Elect Media. 2013;57(1):97–114.

Mazzei LA. An impossibly full voice. In: Mazzei LA, Jackson AY, editors. Voice in qualitative inquiry: challenging conventional, interpretive, and critical conceptions in qualitative research. London: Routledge; 2009. p. 45–62.

Mazzei LA, Jackson AY. Voice in qualitative inquiry: challenging conventional, interpretive, and critical conceptions in qualitative research. London: Routledge; 2009.

Nind M. Participatory data analysis: a step too far? Qual Res. 2011;11(4):349–63.

Patton MQ. Qualitative methods & evaluation methods: integrating theory and practice. Sage Publishing; 2015.

Poletti A. Coaxing an intimate public: life narrative in digital storytelling. Continuum J Media Cult Studies. 2011;25(1):73–83.

Reupert A, Maybery D, Nicholson J, Seeman M, Göpfert M, editors. Parental psychiatric disorder: distressed parents and their families. 3rd ed. Cambridge: Cambridge University Press; 2015.

Rice C, Chandler E, Harrison E, Liddiard K, Ferrari M. Project re-vision: disability at the edges of representation. Disab Society. 2015;30(4):513–27.

Rossiter M, Garcia PA. Digital storytelling: a new player on the narrative field. New Direct Adult Contin Edu. 2010;126:37–48.

Stacey G, Hardy P. Challenging the shock of reality through digital storytelling. Nurse Educ Pract. 2011;11:159–64.

Stenhouse R, Tait J, Hardy P, Sumner T. Dangling conversations: reflection on the process of creating digital stories during a workshop with people with early-stage dementia. J Psychiatr Ment Health Nurs. 2013;20:134–41.

Thomas N, O'Kane C. The ethics of participatory research with children. Child Soc. 1998;12(5):336–48.

Vivienne S. Trans digital storytelling: everyday activism, mutable identity and the problem of visibility. Gay Lesbian Issues Psychol Rev. 2011;7(1):43–53.

Vivienne S, Burgess J. The remediation of the personal photograph and the politics of self-representation in digital storytelling. J Mater Cult. 2013;18(3):279–98.

Wales P. Telling tales in and out of school: youth performativities with digital storytelling. RiDE J Appl Theatre Perform. 2012;17(4):535–52.

Willis N, Frewin L, Miller A, Dziwa C, Mavhu W, Cowan F. My story – HIV positive adolescents tell their story through film. Child Youth Serv Rev. 2014;45:129–36.

Willox AC, Harper SL, Edge VL. Storytelling in a digital age: digital storytelling as an emerging narrative method for preserving and promoting indigenous oral wisdom. Qual Res. 2012;13(2):127–47.

Worcester L. Reframing digital storytelling as co-creative. IDS Bull. 2012;43(5):91–7.

Wyatt TH, Hauenstein EJ. Pilot testing okay with asthma: an online asthma intervention for school-age children. J Sch Nurs. 2008;24(3):145–50.

Yates L. The story they want to tell, and the visual story as evidence: young people, research authority and research purposes in the education and health domains. Vis Stud. 2010;25(3):280–91.

Netnography: Researching Online Populations

75

Stephanie T. Jong

Contents

1	Introduction	1322
2	Netnography Defined	1323
3	Netnography in Health Social Science Research	1324
4	The Process of Netnography	1325
5	The Study	1325
	5.1 Phases 1: Planning	1325
	5.2 Phase 2: Entrée	1326
	5.3 Phase 3: Data Collection	1327
	5.4 Phase 4: Data Analysis	1328
	5.5 Phase 5: Research Representation and Evaluation	1329
6	Conducting Ethical Netnography	1329
	6.1 Online informed consent	1330
	6.2 Private Versus Public Medium	1331
7	Practical Implications	1332
	7.1 Lack of Face-to-Face Interaction	1332
	7.2 Researcher as Participant	1333
	7.3 Size of Data Set	1334
8	Conclusion and Future Directions	1334
	References	1335

Abstract

This chapter explores the transition of netnography, a consumer marketing research method, to the field of health social science research. In contemporary society, the Internet has become an essential communication and information medium. Researchers are increasingly using the Internet as a research medium for participant recruitment and data collection. Netnography, an adaptation of

---

S. T. Jong (✉)
School of Education, Flinders University, Adelaide, SA, Australia
e-mail: styj2@medschl.cam.ac.uk

© Springer Nature Singapore Pte Ltd. 2019
P. Liamputtong (ed.), *Handbook of Research Methods in Health Social Sciences*,
https://doi.org/10.1007/978-981-10-5251-4_17

ethnography, is primarily concerned with online communication as a source of data to form an understanding of a cultural phenomenon. It is through the use of this qualitative research method that holistic research about online cultures and communities can be conducted. In the provision of a common set of methodological procedures and protocols, netnography contributes to the debate of researching online populations, and innovation in appropriate settings. Using the example of a study related to fitness communities on social networking sites (SNSs), this chapter will identify key strengths, practical implications, and ethical considerations of netnography. Discussion focuses on netnography as a dynamic adaptation of a research method emerging in the field of health social sciences research.

Keywords

Netnography · Qualitative research · Innovative methods · Health research · Social networking sites · Online research

1 Introduction

The large number of active users of SNSs and the widely recognized ability of the Internet to influence society has resulted in researchers increasingly utilizing the Internet as a medium for research and the collection of data (Stewart 2005). The variety of online data sources and the emergence of online communities on SNSs has necessitated the innovative adaptation of traditional research methods, which have previously been used for studying the "real world" offline (Hine 2000). For example, traditional methods of participant and nonparticipant observation, interviewing, and survey research have been used with varying degrees of success to collect data from this "virtual reality" (Stewart 2005). The "survivability" of these methods within an online environment is dependent upon their ability to "adapt to the technology that facilitates human interaction online" (Stewart 2005, p. 395).

Researchers in marketing and consumer research have focused on the development of an innovative research method based on the 'traditional' research method of ethnography, 'netnography'. Netnography captures archival and emergent social and individual online interactions. This emerging method is being used to understand the online world, interaction styles, and lived experiences of online users (Kozinets 2015). Netnography also has the potential to blend with other research methods which involve individuals posting interactions, an evolvolution of the practice of ethnography. Researchers note the prime advantages of netnography to include the ability to conduct 'fieldwork' from researcher offices (Hine 2000), the ease and low cost of data collection, the ability to connect with geographically dispersed online community groups, and the ease of collecting different types of data (Kozinets 2010). In utilizing these strengths, there has been a growing use of netnography by researchers from diverse fields (1,300 results in a systematic search of Google Scholar conducted by Bengry-Howel et al. 2011). A common set of procedures and protocols helps to aggregate common knowledge derrived from netnography

(Kozinets 2015); it contributes to the debate of researching online communities and cultures, and innovation in appropriate settings.

Online research methods, such as netnography, provide novel opportunities for understanding new forms of interaction and how people create and maintain personal relationships online (Beneito-Montagut 2011) and the development of online communities and cultures. However, netnography poses several methodological challenges and a robust discussion about designing meaningful, useful and ethical online research processes is needed. This chapter will focus on: practical strengths, practical implications, and online ethical considerations of netnography, using the example of a netnographic study related to fitness communities on SNSs.

2 Netnography Defined

Robert Kozinets, the founder and leader of netnography in marketing and consumer research, presents netnography as a method which draws upon computer-mediated communications or network-based data, textual and visual, to arrive at an ethnographic understanding of an online social experience or cultural phenomenon (1997, 2002, 2010). Kozinets (2015, p. 3) describes netnography as "rooted to core ethnographic principles of participant-observation while also seeking to selectively and systematically incorporate digital approaches such as social network analysis, data science and analytics, visualization methods, social media research presence and videography." Netnography can be explained as a means of researching online communities in the same manner that anthropologists seek to understand the cultures, norms, and practices of face-to-face communities, by observing, and/or participating in communications on publically available online forums (Nelson and Otnes 2005; Sandlin 2007). It involves ethical online conduct, online interaction, downloading, and reflection with the aim to express and help others express and share thoughts, opinions, and experiences (Kozinets 2015). This method has the potential to gather first-hand naturalistic data from computer-mediated communication.

Netnography offers researchers the opportunity to focus on new areas of social life (Nind et al. 2012). The concepts of community and culture are at the center of netnography. It is vital to understand that these concepts are unstable, transformative, and fluid "worlds of meaning" (Kozinets 2015). To explore how community and culture are adopted, these worlds of meaning, created by interacting individuals, need to be examined in order to understand their meaningfulness, and continuance. Justifiably, the emphasis of netnography is on understanding what is shared between people, the "momentary construction of common ground" (Amit and Rapport 2002, p. 11).

The data created by online communities is vast, abundant and takes many forms. Online data can be created by an individual or by a group (e.g., interactions), or even with the use of software. Digital archives are also commonly used by researchers as data. Both textual and visual data (including drawings and pictures) are common.

Sound files, audiovisual productions, and websites are also considered data and their online prevalence is rapidly increasing.

It is vital to note that netnography is not purely the downloading of data. Netnography differs from data mining or big data analysis due to the human-to-human interaction, interpretation, and integration (Kozinets 2015). Reflecting on the continuum of participation (Kozinets 2010), netnographers have the ability to negotiate their level of participation in online research. For example, netnographers have the ability to be "lurkers" within an online community without evident researcher copresence. Importantly, attention must be given to the ethnographic style of netnography. This means that the researcher should inspect, index, interpret, and expand on the data, linking to specific research positions and theoretical constructs to develop a representation of understanding (Kozinets 2015). Irrespective of how data is gathered or sourced, the netnographer's aim is to reflectively and respectfully tell of people's experiences, illuminated with personal stories, created artifacts and images.

3 Netnography in Health Social Science Research

Significant efforts are underway to utilize netnography in health social science research using research data from blogosphere (blogging), video casting, podcasting, forums, SNSs, and so on. Blending netnography with quantitative measures, Berger et al. (2008) explored adolescent sport participation. After using national statistics, netnography was used to identify the meaning of trends in sport participation and sport behaviors of adolescents as revealed online. In research conducted on the life experiences of infertile women going through infertility treatment and their need for social and psychological support, Isupova (2011) deemed netnography as the best method to explore online support offered to patients. Isupsova found that online support had both positive and negative aspects, but the most beneficial support is that received from people who are in the same life situation. Here, netnography allowed for the investigation of physically dispersed groups who were in the same life situation. In an exploration of an online gluten-free community, Bean (2014) used a qualitative inquiry to investigate the key characteristics of this community and gain a deeper understanding of member purposes for participation. In this study, netnography offered an unobtrusive exploration of the community, yet rendered a thick description of the unique culture created online.

As the aforementioned studies demonstrate, netnography can bring many benefits to health social science research. Developing the use of netnography within these areas will allow researchers to examine human society and social relationships online, activity of communities, and the development of cultures, providing insight into people's online behavior and an understanding of how people negotiate their Internet activity. Further strengths include the ability to investigate online communities by observing and/or participating in communications on a publically available online forum in a cost-effective manner, the ability to explore sensitive topics, and the first-hand naturalistic investigation into computer-mediated communication.

4 The Process of Netnography

In 2010, Kozinets offered a rigorous set of guidelines for the conduct of netnography, developed from six steps of ethnography. These netnographic phases include: research planning, entrée, data collection, interpretation/analysis, and research representation and evaluation. Underpinning these steps is ensuring ethical standards. These phases allow an applied and systematic approach intended to address many of the procedural, ethical, and methodological issues specific to online research. Furthermore, having guidelines allows reviewers of academic papers to have clear standards in evaluating such research (Kozinets 2015).

Adhering to the 2010 guidelines, a study about fitness on SNSs, underpinned by a social construction theoretical framework, will be used as an example (outlined below). Since completing the study, recent developments on the netnography method have been published. In 2015, Kozinets wrote on the evolution of the original phases from 5 to 12: introspection, investigation, information, interview, inspection, interaction, immersion, indexing, interpretation, iteration, instantiation, and integration. These further phases include some additions, more explicit descriptions and subdivisions of the originals, for example, greater emphasis on narrowing the community group of interest, and the use of an interactive researcher website.

5 The Study

The aim of this research was to explore the way in which online communities contribute to young women's perspectives of health and how they may be barriers or facilitators to health-oriented behaviors among this online community. Netnography was identified as the ideal approach to capture meaningful data from participants within this community. This study followed the five phases of the Netnography method. This method assisted in establishing a boundary of thinking about online fitness communities and culture, through the observation of photographs, videos, and comments posted on the SNSs, Facebook and Instagram. This study also blended netnography with individual interviews of 22 participants from the online fitness community.

5.1 Phases 1: Planning

The first step of netnography involved the selection of online fitness communities. There are various sources for exploring online communities such as bulletin boards or forums, chat-rooms, SNSs, and so on. After commencing a fundamental literature review of the research area, I became an observational participant within the online fitness community on Instagram and Facebook for 1 year, prior to data collection. This observational role allowed me to start from the outside and move on the inside to become familiar with the cultural meanings and understandings, language, interests, practices, and the rituals of being involved within the online fitness community.

Throughout this year, I was able to familiarize myself with several online fitness communities and online fitness accounts that were considered for inclusion in the study. This process ensured that the sites for netnographic fieldwork were relevant, active, substantial, heterogeneous, and data-rich. Once involved, my experiences and reflections were recorded. This is vital and will be further discussed in field noting in Phase 3.

Following the identification of the online communities, I devised appropriate research questions for inquiry. Open-ended questions were developed to allow flexibility and further discussion and inquiry. These questions provided the foundation for the research; however, they were altered over time, molded by my interaction and interpretation. An ethics application followed the decision about which online forum would be used, and the research questions for inquiry.

5.2 Phase 2: Entrée

During the observational year prior to commencing the study, I was able to firmly conceptualize how I would approach the community. As commonly used for an exploration of online culture, the researcher created an alias account, deciding on minimal one-on-one interaction with online fitness participants. It is important to consider this "spectrum of participation and observation" (Kozinets 2010, p. 75) as Kozinets stresses that the participant role provides opportunity to experience entrenched cultural understanding within the specific community of interest, an ethnographic insight. Conversely, other scholars have reasoned the value of "covert studies" of online communities (Langer and Beckman 2005; Brotsky and Giles 2007) or "observational" netnographies (Beaven and Laws 2007; Brownlie and Hewer 2007; Füller et al. 2007).

After ethical approval was granted, the Facebook and Instagram alias researcher accounts were created in order to have full access to the SNSs to conduct the netnography without the use of a personal profile, a method suited to netnography (Kozinets 2010) or online ethnographies (Grbich 2013). For this study, I followed Lamb's (2011) recommendations from an online research project based in the United Kingdom, as well as Kozinets' advocacy on a transparent social networking site profile. The alias researcher account provided open and truthful information about myself as a PhD student, and provided extensive information about the netnographic inquiry and university affiliation. This was also provided through comments posted to the communities on walls and pictures. Furthermore, existing academic research on fitness culture, and online health and fitness communities was offered to those interested in the account. There was no encountered resistance or negativity about the fact that I was "studying the community." Instead, I was greeted with an interest in the research reflected by a number of online community members following my page, as well as people commenting and tagging friends.

Throughout this first phase, I found it useful to reflect on what I was studying and how I was going to study it. This also involved considering what would constitute data and how I would collect this data. It is also important at this stage to consider the future phases of the research regarding data analysis, ethical concerns, and the overall benefits of the research (Kozinets 2015).

5.3 Phase 3: Data Collection

Kozinets (2010) describes three different types of data collection that can be involved in netnographic research: archival data, elicited data, and field note data. Archival netnographic data are saved social networking site interactions stretching back throughout time. Elicited data is new data created through researcher and participant interaction, and field note data are observations from the researcher. The online study of fitness culture included both archival data and field note data. Although data collection and data analysis are separate steps, it is important to note early that they do not occur in isolation of one another.

Once connected, I searched Facebook and Instagram for fitness accounts and pages by using a hash-tag word search. A prevalent fitness "thread" on SNSs about fitness, "#fitfam" or the "fitness family" was purposely selected for analysis due to its potentially rich field of data. Other hash-tag searches included: "#fitspo" (fitness inspiration), "#fitness", "#girlswithmuscle", "#inspiration", "#femalefitness", "#fitmotivation", "#fitgirls", and "#strongisthenewskinny." These hash-tags were made popular by the posts observed in the year prior to data collection. These searches led me to a pool of pictures and comments ranging from motivational statements to pictures of "fit" bodies, revealing vast numbers of publically available posts, updating in real-time. This process linked me to fitness accounts and pages where these images and comments were abundant. Although it was not explicit what would be found by the hash-tag search, I looked for general fitness pages showing a number of female fitness photographs. Further investigation of different fitness pages on the SNSs provided me with a number of online fitness female users who were active within online communities. Although considered limited interaction, I "followed" a number of online fitness pages, involving me as a "user" in the fitness community by observing posts with other members. I followed females who used Instagram to communicate in fitness communities.

Interaction with the culture and downloading archival data occurred for four months. In line with the data "saturation" principle (Liamputtong 2013; Creswell 2014), downloading archival data continued until the investigation did not provide new insights on theoretically important topical areas or additional themes. Although recorded, the date of data collection is superfluous as the data is logged on SNSs, with photographs and comments accessible from the past as well as present.

Forum posts were separated into those that were relevant to the research questions, and those that were off topic. Textual and visual data were also captured through NVivo (a computer-assisted software analysis tool) for Facebook and Evernote (a screen capture tool) for Instagram, including field notes. All relevant data were uploaded to Microsoft Excel. The data collected included dates of observation and retrieval, comments from community members on a post, the number of likes attributed to a post, how many times the post had been shared, and how many posts were linked to a hash-tag. This allowed the highly pertinent voices of participants to be heard, offering important perspectives to discussions of online fitness. If the post came from a specific online fitness account, their followers were documented, as well as how many people were "talking" about them at the time, which is documented on Facebook. The names of pages where I collected data were posted to my alias accounts for social networking site users to view.

As mentioned above, field notes are an important way to collect data. These field notes are introspective reflections based on my personal observations, interactions, experiences of being involved within the community, and learning about particularities of the culture (Emerson et al. 2011). I found reflective prompting questions for descriptive netnographic field noting effective. Examples include: What is new? What is meaningful? What is absent from the findings that you expected to find? What do I not understand? What is it like to connect to the community members? This process enabled me to detect what was going on, and to document the journey learning of the practices, languages, rituals, members, and so on. I found this valuable in conducting the research as once I was more involved with the community and culture, I found it difficult to decipher what was "new" in the online social experiences. Field notes were documented in Word, but any software program can be used, even notes alongside the downloaded communication.

5.4 Phase 4: Data Analysis

Archival and field notes were used to organize distinct purposes and themes that emerged from the analysis of the online community. The data were analyzed using qualitative and ethnographic thematic content analysis techniques. Qualitative content analysis extends the scope of inquiry to examine meanings, themes, and patterns that may be apparent or concealed in a text (Grbich 2013). As netnography involves an inductive approach to the analysis of qualitative data, thematic content analysis was deemed an acceptable method for a textual inquiry (Silverman 2006). Thematic content analysis is the detailed examination of coding and categorizing textual information into constituent trends of consensus (Kozinets 2010; Grbich 2013; see also ► Chap. 47, "Content Analysis: Using Critical Realism to Extend Its Utility" and ► Chap. 48, "Thematic Analysis"). This research utilised a thematic content analysis following Braun and Clarke's (2006) six-step analysis model.

The textual content of the posts was analyzed to determine the kind of "fitness-" or "health"-related discussion that emerged from the community relevant to the theoretical framework of social constructionism. This analytical approach aimed to allow the communicative meaning latent within the Instagram and Facebook data to emerge, and to witness experiences and the formation of "fitness culture" via this media. These themes and patterns informed interview questions for the second part of the study.

In qualitative research, coding can be performed either manually or electronically using software (Liamputtong and Serry 2017; Serry and Liamputtong 2017). Numerous computer-assisted qualitative data analysis software (CAQDAS) programs (e.g., NVivo, QualPro, NUD.IST) are available and widely used (see also ▶ Chap. 52, "Using Qualitative Data Analysis Software (QDAS) to Assist Data Analyses"). There is debate about the use of electronic coding/software; arguments against the use of software include distancing the researcher from the data through the use of "auto-coding" (Fossey et al. 2002), the lack of meaning making, and reflective engagement with the textual data (Kidd and Parshall 2000) and concerns about validity and reliability (Welsh 2002). However, when working with large datasets, as with a netnography, electronic coding software such as CAQDAS provides a means of analyzing a greater quantity of data (Serry and Liamputtong 2017). Other advantages of using CAQDAS are efficiency in the coding process, and the ability to efficiently manage, code, and analyze large and complex datasets (Kozinets 2010; Serry and Liamputtong 2017). Auto-coding electronic techniques were not used within the study. NVivo was used to manage and catalogue the large dataset and, for ease, in organizing the data in to codes and categories.

5.5 Phase 5: Research Representation and Evaluation

Standards of excellence need to be adhered to in order to write, present and report research findings from netnography. Previously, evaluative standards for qualitative research, and particularly those pertaining to developing online research, have been criticized for being unclear (Kozinets 2010). Kozinets (2010) builds on Denzin and Lincoln's (2005) evaluative positions for judging qualitative research to develop a set of netnographic quality standards aimed at inspiring netnographic quality. Within the current study, the ten recommended evaluation standards were reflected upon throughout the course of the research. They are named: coherence, rigour, literacy, groundedness, innovation, resonance, verisimilitude, reflexivity, praxis and intermix (Kozinets 2010). These underpinned my actions as a netnographic researcher, prompted discussions, helped to build ideas, and acted as a toolkit to aid with evaluation prior to representing the research through writing.

6 Conducting Ethical Netnography

The Internet and SNSs are proving to be 'ethically problematic' fields for researchers. Although ethics and policies developed in the context of offline research apply to online investigations, questions about the need for and means of obtaining informed consent, anonymity, and the conceptualization of public versus private information pose certain problems for conducting research online. Some emerging literature is beginning to provide useful insights into social media and ethical guidelines (Fielding et al. 2008; Zimmer 2010; James and Busher 2015).

In this study, Kozinets' (2002) four recommendations of ethical research procedures when using netnography were followed. These were: (1) The researchers should fully disclose their presence, affiliations, and intentions to online community members during any research; (2) the researchers should ensure confidentiality and anonymity of informants; (3) the researchers should seek and incorporate feedback from members of the online community being researched; and (4) the researcher should take a cautious position on the private-versus-public medium issue. Ethical clearance for the project was gained by following the procedures outlined by the associated institution.

6.1 Online informed consent

The inability of participants to provide autonomous consent as part of online research has been widely and duly noted (Zimmer 2010; Lamb 2011). The accessibility of updated and instant data relating to every social phenomenon on the Internet creates concerns that the researchers could be tempted to use it as a "research playground" (Hine 2008, p. 316) without due regard to the respect of the people involved.

Kozinets' (2002) four recommendations of ethical research procedures provides an understanding of how to ethically contact community members in order to obtain their permission, or informed consent, and to directly quote any specific postings in the research. Debates arise over the idea of archived messages viewed as human subjects' social research (Kozinets 2015). Furthermore, it is not always possible to gain consent from the many users of an online community. Some may change or delete their account, or may no longer be involved with the community. Nevertheless, Bassett and O'Riordan (2002) suggest that this should not prevent the researching of this material.

Langer and Beckman (2005) justify their use of covert netnography where participants are not informed of the researchers' presence; they argue that by revealing themselves as researchers in a study on a 'sensitive' research topic, the research project could be potentially threatened. Other researchers have also used a netnographic approach where participant statements have been used without a discussion on ethics (see Smith and Stewart 2012). Conversely, Kozinets (2015) views netnographers as cultural participants, interacting with the community, undertaking human subject's research. Hine (2008) and Kozinets (2015) share similar views of online consent. They affirm that, where possible, researchers should ask for consent in order to lead to interesting insights that help interpret data.

Interestingly, Hine (2008) claims that there are circumstances where informed consent may not be appropriate. For example, where information is publically available such as a website or discussion forum, where "the topic is not intrusive or troubling" were the person to recognize themselves in the research, and where there is "no foreseeable harm" to possible participants, "then it might be justifiable to go ahead and collect data without seeking informed consent" (p. 317).

Within my study, my alias social networking accounts stated the nature and purpose of the research and type of data collection. Pages where data collection occurred were documented on my alias accounts. Posts were also made on each of the places of data collection informing people that posts were used from the thread. To respect the users and their data, discussion was read with reference to the context of the posts, in order to not misconstrue the meaning of their online discussions.

This transparency allowed people within the online fitness community to connect with me, but also provided documentation to see what sites had been visited and where data had been gathered. In line with Facebook's terms and conditions (Facebook 2014), I made it clear through a statement on my alias profiles, that I was an independent researcher, and that Facebook was not collecting the data. The information on these accounts also noted that data attained for the research would be made anonymous with pseudonyms. I made it a priority to maintain a high level of ethical sensitivity and adhered to the values and principles of ethical conduct to ensure expectations of consent and privacy caused no, or limited potentially damaging effects for participants (Flicker et al. 2004; Hine 2008). Whilst taking action to be respectful, and to consider the welfare of participants, the risks associated with the project, including the potential risk associated with consent, anonymity and private vs. public data were minimised as much as possible, and the potential benefits for participants and broader society were clarified to participants (National Health and Medical Research Council et al. 2007 updated May 2015). In developing a greater understanding of online fitness culture and the impact that SNS influence has on health perceptions and behaviours, the study has the potential to add to knowledge on the effect of online fitness culture on constructions of health and fitness, body image and ideals, and consequently, has the potential to positively impact future health promotion projects.

6.2   Private Versus Public Medium

On the Internet or SNSs, the lines between what is private and what is public are blurred, and ownership of data is contentious (Henderson et al. 2013). Some participants who use publicly available communication systems on the Internet have an expectation of privacy, which, as stated by Walther (2002, p. 207), is "extremely misplaced." In their research, Kozinets (2015) and Bassett and O'Riordan (2002) recognize the Internet as text-based and space-like medium, where participants create cultural artifacts. With this view, the Internet can thus be "perceived as a form of cultural production, in a similar framework to that of the print media, broadcast television, and radio" (Bassett and O'Riordan 2002, p. 235), a medium for publication, or a "public document" (Kozinets 2015, p. 136).

According to Rosenoer (1997, cited in Jacobsen 1999), posts to publicly accessible forums or sites are not private and are not protected by privacy laws. Although people know their postings are public in their accessibility through a simple Internet search, users may not expect researchers to be gathering their exchanges with others, nor does it automatically lead to the conclusion that users grant automatic consent to researchers using this data. However, as Paccagnella (1997, p. 83) notes, "that doesn't mean that they can be used without restrictions, but simply that it shouldn't be necessary to take any more precautions than those usually adopted in the study of everyday life."

The debate over who owns the data posted to public forums is important when referring to an online setting. If information is posted on a public forum or open social networking account, the question of ownership is valid. The person who posted to a social networking site is not the only one to take ownership of a photograph. It must be recognized that the SNS also maintains part ownership (see Facebook 2014). This information is integral to netnographers who wish to undertake research requiring the use of photographs from SNSs. Ownership issues also vary due to Fair Use Doctrine in different locations around the world. Ethics for accessing comments on SNSs has been granted for other research on SNS data in a public space (Attard and Coulson 2012; Barnes et al. 2015).

To surmount these ethical dilemmas, researchers are advised to implement strategies aimed to decrease the accessibility of data by random searches. Posts made by participants that pertain to personal accounts, including names, pseudonyms, and faces, should be de-identified; i.e., removed or blurred. This will enhance the privacy and anonymity of the users. However, as a researcher, one must note that given the public nature of these pages in the first place, "there is no reasonable expectation of privacy in these conceptual spaces" (Jacobson 1999, p. 135), especially since posts may be located by Internet searches.

7 Practical Implications

The innovation and adaptation of method techniques for the study of online communities and cultures may also bring methodological challenges. These are often focused on the lack of nonverbal data. Throughout the netnographic research, I also found issues pertaining to my role as a researcher and as a participant in the online world, and with the potential size of the data set.

7.1 Lack of Face-to-Face Interaction

The Internet has created a vast increase in new online social spaces, "devoid of physicality" (Stewart 2005, p. 413). Lack of face-to-face interaction is a common objection to online research (Beaulieu 2004; Liamputtong 2013). Some researchers suggest that the online environment, where the data is represented as "text-only," reduces social cues such as expression, emphasis, and movement (Mann and Stewart 2000). The limitations created by the lack of face-to-face interaction are deemed insignificant by Kozinets (2010) who argues that where the research focus and questions are specific to online content, a netnographic approach is sufficient. Furthermore, these limitations can be somewhat ameliorated by a blended netnography where the data collection methods connect online and offline research in a systematic manner (Kozinets 2002). Kozinets (2015) strongly advocates the inclusion of an interview stage within the new 12 phases of research. My study used this blended approach, including interviews with online fitness participants; I found this integration to offer a deeper understanding of the culture.

7.2 Researcher as Participant

Through the use of the Internet as a new medium for research, researchers now have instantaneous raw data on demand and the technological advancements to collect and analyze data. Given the open nature of many SNSs, online participant observation theoretically allows a covert position (Mann and Stewart 2000). A netnography provides the option of allowing the researchers to identify themselves among the online community, or remain a "lurker." This can be considered problematic within netnography as the method has evolved from the assimilated nature of ethnography. It is up to the researcher to explore, interpret, and negotiate the meaning underpinning these various forms of "researcher as a participant" and to recognize the limitations associated with each process.

Within an online community, there are a diverse range of users who consider participation very differently. For example, there may be participants who play an active role in the community, or there may be those who play a marginal role in their online community, often labeled a "lurker." This individual will usually follow the online interaction of the community without contributing to it. Other members may register some level of presence from the lurker (such as being on a "friend's list" or "follower"), but at times (for example, when in a large online community), a lurker will go unnoticed and their role will purely involve reading posts. Noteworthy, many lurkers still think of themselves as community members (Nonnecke et al. 2006), despite their nonpublic presence.

Some researchers raise ethical issues when acting as lurkers within the online community. Specifically, this relates to the potentially unnoticed invasion of users' privacy (Stewart 2005) (discussed later). Additionally, Beauleiu (2004) argues that by avoiding interaction, the researcher may miss information or interpretations, specifically relating to a nonvisible part of a phenomena (Hine 2000). Furthermore, Hine (2005) and Kozinets (2015) advocate active participation on the researcher's behalf in order to access and understand the lived experiences of online community members. Contrastingly, other researchers argue that observations via lurking can reduce distorting naturally occurring behavior, and therefore data (Paccagnella 1997; Schaap 2002), and minimize the presence of the researcher (Paccagnella 1997; Grbich 2013). Within the current study, the profile pages on the SNSs selected for data collection have up to 1.5 million "followers" from around the world. As I "followed" a page, the likelihood of the page administrator or participants on the site seeing me was low. In other words, as there are high volumes of people within the community, my presence would likely be unknown. However, in order to be as transparant as possible, I constructed alias profile accounts documenting information about the current research. I also posted into community groups about my research.

As with ethnography, the netnographer becomes a vital part of the research. Through reflective field notes, the netnographer documents his/her experiences within the culture. The netnographer's decisions of what/where to examine become data and should be documented. Through participation within the online community, the netnographer also becomes vital in the creation and analysis of data

(Kozinets 2015). Although some may consider this presence of the researcher in the field as "contamination" or that the researcher may involve oneself to the extent that objectivity is lost (Paccagnella 1997), Kozinets describes this researcher involvement as "the true nature of observation" (2015, p. 164). Paccagnella (1997) reinforces this view by advocating the role of the researcher in understanding the social reality behind the phenomena and by documenting real emerging experiences, stating, "...it is not safe to think of these data as some sort of objective reality frozen by the computer; Archived messages and logs are representations of the online phenomena as perceived by participants" (p. 87). Paccagnella further reflects, "...it is not intended for people uninvolved directly in interaction, and it loses part of its sense and meaning when re-read afterward by neutral observers" (p. 87).

7.3 Size of Data Set

In her article on the methodological issues arising from researching sensitive issues online, Paechter (2012) notes the potential size of the data set as a methodological challenge faced by researchers concerned with studies of online communities. In my research, I found that because the category selected was too broad, the potential size of data collection was difficult. Searching hash-tags and key words also proved difficult with the abundance of data available. Each page and profile that was visited would create new circles within the fitness community to look at, creating a larger data set.

To handle the sometimes overwhelming amounts of data that netnography can generate, Dholakia and Zhang (2004) suggest researchers be specific about what community will be under investigation. For future netnographic researchers, it is suggested to select one community group to engage the netnography with, and to specify the research questions. These decisions will limit the potential data set size and allow researchers to explore the community in greater detail in a designated time frame. Once the community is selected and specified, the researcher may classify messages by relevance to the topic, leading to a refocus on the research questions.

8 Conclusion and Future Directions

While imitative of traditional qualitative research, online qualitative research exhibits fundamental adaptations of aspects of traditional qualitative research. For example, watching what people do, listening to what they "say" through their typing and clicking, and collecting and interpreting this as data. Utilizing the strengths of technological developments, online research allows for connection to dispersed networks around the world, larger participant populations, asynchronous (not an immediate reply online, such as emails) or synchronous (immediate) communication, and cost-effective research. While researching online, social spaces inspires new methods and promotes innovative means of adapting traditional methods, it also poses significant challenges to researchers (Stewart 2005). Although these aforementioned challenges are not necessarily new, the application of netnography has raised important questions about researching in the online environment.

The aim of this chapter has been to provide a brief overview of netnography, and to raise a number of practical implications and considerations for researchers to reflect on when considering the use of this method. In particular, a discussion has been raised about the implications of the steps of the research process. In following the steps of netnography in the study of online fitness communities and culture, I found that netnography guided the research successfully, providing an adaptable framework suitable for the disciplines of health social science research. Furthermore, by following the ethical netnography recommendations from Kozinets (2010), and other developing online ethics texts (see Mann and Stewart 2000; Fielding et al. 2008), this type of online research is becoming more achievable for researchers. An expansion of dialogue on the analysis and interpretation of digital data could benefit researchers using this method.

It is important to continue to invest in netnography as a research method in health social science research. Netnography holds promise in exploring the perspectives of online interaction, the participants in online community groups, the development of social norms and their circulation within online communities, and potentially, the voice of consumers of health interventions. Collectively, the application and integration of netnography will improve research on a new area of social life, but must be responsive and adaptive to be effective in a rapidly changing environment. The method itself is developing as evidenced by the changes from 5 to 12 phases, proving it to be responsive to the evolving research environment.

Acknowledgements I acknowledge the receipt of the Australian Government Research Training Program Scholarship.

References

Amit V, Rapport N. The trouble with community: anthropological reflections on movement, identity and collectively. London: Pluto; 2002.

Association of Internet Researchers Ethics Working Group. Ethical decision-making and internet research: recommendations from the AoIR Ethics Working Committee (Version 2.0). 2012. http://www.aoir.org/reports/ethics2.pdf. Accessed 21 Jan 2016.

Attard A, Coulson N. A thematic analysis of patient communication in Parkinson's disease online support group discussion forums. Comput Hum Behav. 2012;28(2):500–6.

Barnes N, Penn-Edwards S, Sim C. A dialogic about using Facebook status updates for education research: a PhD student's journey. Educ Res Eval. 2015;21(2):109–21.

Bassett EH, O'Riordan K. Ethics of internet research: contesting the human subjects research model. Ethics Inf Technol. 2002;4:233–47.

Bean EA. Man shall not live by bread, at all: a netnography of the key characteristics and purposes of an online gluten-free community. Brigham Young University, all theses and dissertations. Paper 4082. 2014.

Beaulieu A. Mediating ethnography: objectivity and the making of ethnographies of the internet. Soc Epistemol. 2004;18(2–3):139–63.

Beaven Z, Laws C. 'Never let me down again'1: loyal customer attitudes towards ticket distribution channels for live music events: a netnographic exploration of the US Leg of the Depeche Mode 2005–2006 World Tour. Manag Leis. 2007;12(2–3):120–42.

Beneito-Montagut R. Ethnography goes online: towards a user-centred methodology to research interpersonal communication on the internet. Qual Res. 2011;11(6):716–35.

Bengry-Howell A, Wiles R, Nind M, & Crow G. A review of the academic impact of three methodological innovations: netnoraphy, child-led research and creative research methods. ESRC National Centre for Research Methods. University of Southampton. NCRM Hub. 2011.

Berger IE, O'Reilly N, Parent MM, Seguin B, Hernandez. Determinants of sport participation among canadian adolescents. Sport Manage Rev. 2008;11:277–307.

Braun V, Clarke V. Using thematic analysis in psychology. Qualitative Research in Psychology 2006;3(2):77–101.

Brotsky SR, Giles D. Inside the "pro-ana" community: a covert online participant observation. Eat Disord. 2007;15(2):93–109.

Brownlie D, Hewer P. Cultures of consumption of caraficionados. Int J Sociol Social Policy. 2007;27(3/4):106–19.

Creswell JW. Qualitative inquiry and research design: choosing among five aproaches. 3rd ed. Thousand Oaks: Sage; 2014.

Denzin NK, Lincoln YS. (Eds.). The Sage handbook of qualitative research (3rd ed.). Thousand Oaks, CA: Sage Publications Inc; 2005.

Dholakia N, Zhang D. Online qualitative research in the age of e-commerce: data sources and approaches. Forum Qual Soc Res Sozialforschung. 2004;5(2), Art. 29. http://www.qualitative-research.net/index.php/fqs/article/view/594/1289. Accessed 25 Apr 2014.

Emerson RM, Fretz RI, Shaw LL. Writing ethnographic fieldnotes. 2nd ed. Chicago: University of Chicago Press; 2011.

Facebook. Statement of rights and responsibilities. 2014. http://www.facebook.com/legal/terms. Accessed 10 June 2015.

Fielding N, Lee RM, Blank G, editors. The Sage handbook of online research methods. London: Sage; 2008.

Fossey E, Harvey C, McDermott F, Davidson L. Understanding and evaluating qualitative research. Aust N Z J Psychiatry. 2002;36(6):717–32.

Füller J, Jawecki G, Mühlbacher H. Innovation creation by online basketball communities. J Bus Res. 2007;60(1):60–71.

Grbich C. Qualitative data analysis: an introduction. 2nd ed. London: Sage; 2013.

Henderson M, Johnson NF, Auld G. Silences of ethical practice: dilemmas for researchers using social media. Educ Res Eval. 2013;19(6):546–60.

Hesse-Biber S, Griffin A. Internet-mediated technologies and mixed methods research: problems and prospects. J Mixed Methods Res. 2013;7(1):43–61.

Hine C. Virtual ethnography. London: Sage; 2000.

Hine C. Internet research and the sociology of cyber-social-scientific knowledge. Inf Soc. 2005;21(4):239–48.

Hine C. The internet and research methods. In: Gilbert N, editor. Researching social life. 3rd ed. London: Sage; 2008. p. 304–20.

Isupova OG. Support through patient internet-communities: lived experience of Russian in vitro fertilization patients. Int J Qual Stud Health Well-being. 2011;6(3). https://doi.org/10.3402/qhw.v6i3.5907.

Jacobson D. Doing research in cyberspace. Field Methods. 1999;11(2):127–45.

James N, Busher H. Ethical issues in online research. Educ Res Eval. 2015;21(2):89–94.

Kidd PS, Parshall MB. Getting the focus and the group: enhancing analytical rigor in focus group research. Qual Health Res. 2000;10(3):293–308.

Kozinets R. 'I want to believe': a netnography of the X-philes' subculture of consumption. Adv Consum Res. 1997;24:470–5.

Kozinets R. The field behind the screen: using netnography for marketing research in online communities. J Mark Res. 2002;39:61–72.
Kozinets R. Netnography: doing ethnographic research online. London: Sage; 2010.
Kozinets R. Netnography: redefined. 2nd ed. London: Sage; 2015.
Lamb R. Facebook recruitment. Res Ethics. 2011;7(2):72–3.
Langer R, Beckman SC. Sensitive research topics: netnography revisited. Qual Mark Res Int J. 2005;8(2):189–203.
Liamputtong P. Qualitative research methods. 4th ed. Melbourne: Oxford University Press; 2013.
Liamputtong P, Serry T. Making sense of qualitative data. In: Liamputtong P, editor. Research methods in health: foundations for evidence-based practice. 3rd ed. Melbourne: Oxford University Press; 2017. p. 421–36.
Mann C, Stewart F. Internet communication in qualitative research: a handbook for researching online. London: Sage; 2000.
Nelson MR, Otnes CC. Exploring cross-cultural ambivalence: a netnography of intercultural wedding message boards. J Bus Res. 2005;58(1):89–95.
Nind M, Wiles R, Bengry-Howell A, Crow G. Methodological innovation and research ethics: forces in tension or forces in harmony? Qual Res. 2012;13(6):650–67.
Nonnecke B, Andrews D, Preece J. Non-public and public online community participation: needs, attitudes and behavior. Electron Commer Res. 2006;6(1):7–20.
Paccagnella L. Getting the seats of your pants dirty: strategies for ethnographic research on virtual communities. J Comput Mediat Commun. 1997;3(1). https://doi.org/10.1111/j.1083-6101.1997.tb00065.x.
Paechter C. Researching sensitive issues online: implications of a hybrid insider/outsider position in a retrospective ethnographic study. Qual Res. 2012;13(1):71–86.
Sandlin JA. Netnography as a consumer education research tool. Int J Consum Stud. 2007;31(3):288–94.
Schaap R. The words that took us there: ethnography in virtual ethnography. Amsterdam: Aksant Academic Publishers; 2002.
Serry T, Liamputtong P. Computer-assisted qualitative data analysis (CAQDAS). In: Liamputtong P, editor. Research methods in health: foundations for evidence-based practice. 3rd ed. Melbourne: Oxford University Press; 2017. p. 437–50.
Silverman D. Interpreting qualitative data: methods for analysing talk, text and interaction. London: Sage; 2006.
Smith ACT, Stewart B. Body perceptions and health behaviors in an online bodybuilding community. Qual Health Res. 2012;22(7):971–85.
Stewart K. Researching online populations: the use of online focus groups for social research. Qual Res. 2005;5(4):395–416.
Walther J. Research ethics in Internet-enabled research: human subjects issues and methodological myopia, vol. 4 (pp. 205–216). 2002. http://www.nyu.edu/projects/nissenbaum/ethics_wal_full.html: Ethics Information Technology.
Welsh E. Dealing with data: using NVivo in the qualitative data analysis process. Forum Qual Social Res Sozialforschung. 2002;3(2), Art. 26. http://www.qualitative-research.net/index.php/fqs/article/view/865/1880. Accessed 18 Apr 2015.
Zimmer M. "But the data is already public": on the ethics of research in Facebook. Ethics Inf Technol. 2010;12:313–25.

Web-Based Survey Methodology

76

Kevin B. Wright

Contents

1 Introduction	1340
2 Advantages and Problems Associated with Online Survey Methods	1341
3 Critiques of Online Surveys	1342
4 Sampling, Measurement, and Enhancing Response Rates in Online Survey Research: Promises and Pitfalls	1344
4.1 Sampling Issues and Online Surveys	1344
4.2 Measurement Issues with Online Surveys	1345
4.3 Enhancing Response Rates Using Online Surveys	1346
5 Resources for Creating/Managing Online Surveys	1347
6 Overseeing Web-Based Survey Data Collection and Analysis Issues	1347
7 Pros and Cons of Various Web-Based Survey Platforms and Services	1348
8 Conclusion and Future Directions	1349
References	1351

Abstract

This chapter examines a number of issues related to online survey research designed to access populations of various stakeholders in the health care system, including patients, caregivers, and providers. Specifically, the chapter focuses on such issues as finding an adequate sampling frame for obtaining samples of online populations, measurement issues, enhancing response rates, overseeing web-based survey data collection, and data analysis issues. Moreover, it examines issues such as measurement validity and reliability in web-based surveys as well as problems with selection biases and generalizability of study findings. Finally, the chapter assesses the pros and cons of using *SurveyMonkey* and *Qualtrics* as

K. B. Wright (✉)
Department of Communication, George Mason University, Fairfax, VA, USA
e-mail: kwrigh16@gmu.edu

© Springer Nature Singapore Pte Ltd. 2019
P. Liamputtong (ed.), *Handbook of Research Methods in Health Social Sciences*,
https://doi.org/10.1007/978-981-10-5251-4_18

web-survey platforms/services and their utility for studying various online contexts that may be of interest to social science and health scholars.

Keywords

Online data collection · External validity · Response rates · Sampling · Survey research · Websurveys

1 Introduction

Over the past two decades, we have seen considerable growth in the area of online survey methodology, particularly in the areas of online survey development and implementation (Dillman 2000; Wright 2005; Lieberman 2008; Murray et al. 2009; Greenlaw and Brown-Welty 2009; Kramer et al. 2014). Paralleling the growth of online survey methodology, scholars have engaged in research to evaluate the relative merits and problems associated with online survey methods within a broad array of academic disciplines and areas (Wright 2005), including health social science researchers (Shaw and Gant 2002; Wright 2000, 2011; Konstan et al. 2005; Wright et al. 2010a). Recent trends suggest that web surveys will become increasingly prominent in the future and result in higher response rates as the general population becomes increasingly made up of "digital natives" (Kramer et al. 2014).

Understanding how to effectively design and implement online surveys as well as some of the many advantages and disadvantages of this research method can help health social science researchers study a variety of health contexts and issues. In short, understanding online survey methods is a helpful set of skills to add to one's research method toolbox. Health social science researchers have used online surveys to access a number of segments of the population, including access to physicians, other health care workers, online support community participants, hospital staff members, patients, lay caregivers, and a variety of other web-accessible stakeholders in the health care system (Owen and Fang 2003; Konstan et al. 2005; Wright et al. 2010a; Siegel et al. 2011; Wright and Rains 2013).

Given the ubiquity of popular web survey platforms and services, such as *SurveyMonkey* and *Qualtrics*, it is easy to forget that online survey research is relatively young and constantly evolving. Until recent years, creating and conducting an online survey was a time-consuming task requiring familiarity with web authoring programs, HTML code, and scripting programs. Today, various survey authoring software packages and online survey services like *SurveyMonkey* have made online survey research much easier and faster. Yet many researchers who have been slow to move to online survey research from traditional paper and pencil survey research may be unaware of the advantages and disadvantages associated with online survey research. For example, previous research has identified numerous advantages to using online surveys over traditional survey methods, including access to individuals in distant locations, the ability to reach difficult to contact participants, and the convenience of automated data collection (Greenlaw and Brown-Welty 2009; Kramer et al. 2014). However, disadvantages of online survey research include uncertainty over the validity of the data and sampling issues and concerns surrounding the design, implementation, and evaluation of an online survey (Kramer et al. 2014).

While a number of health social scientists currently use online surveys, many others may be unfamiliar with the process as well as the major promises and pitfalls of this research method. Other researchers may have been trained to use other researcher methods (e.g., experiments, content analysis, and so on), but they may be interested in expanding their research skills to include the use of online surveys. This chapter should be of interest to both types of scholars.

Toward that end, this chapter examines a number of issues related to using online surveys to access populations of various stakeholders in the health care system, including patients, caregivers, and providers. Specifically, the chapter focuses on such issues as finding an adequate sampling frame for obtaining samples of online populations, measurement issues, enhancing response rates, overseeing web-based survey data collection, and data analysis issues. Moreover, it examines issues such as measurement validity and reliability in web-based surveys as well as problems with selection biases and the generalizability of study findings. Finally, the chapter assesses the pros and cons of using web-based survey platforms and services like *SurveyMonkey* and *Qualtrics* and their utility for studying various online health-related communities and web portals.

2 Advantages and Problems Associated with Online Survey Methods

Research conducted on the Internet provides expanded opportunities for reaching populations of interest to health social scientists, including various stakeholders in the health care system (Eysenbach and Wyatt 2002; Wright 2005; Lieberman 2008). Emerging methodological research suggests that the Internet is an appropriate venue for survey data collection, including within health contexts (Riper et al. 2011; Kramer et al. 2014). Online surveys offer some key advantages over traditional surveys (see ▶ Chaps. 80, "Cell Phone Survey," and ▶ 81, "Phone Surveys: Introductions and Response Rates"). For example, studies have found that recruitment advertisements on Facebook have been used successfully to recruit "hard-to-reach" populations, such as sexual minorities, people with rare diseases, veterans with post-traumatic stress disorder (PTSD), and a variety of other participants who may be of interest to health communication researchers that are not easily accessed through traditional recruitment strategies (Pedersen et al. 2015). In terms of longitudinal surveys, recruiting participants via social networking sites, like Facebook, may also benefit longitudinal retention in research, which is often affected by inability to locate participants who have moved or changed contact information (Pedersen et al. 2015). Another major advantage of web-based survey research is that it allows researchers to conveniently access populations in ways that bypass spatial, chronological, and material constraints (Evans and Mathur 2005). Moreover, web-servers are capable of collecting large numbers of data from participants who are accessing the online survey at the same time (Evans and Mathur 2005). As a result, relatively large sample sizes can be attained within very short periods of time. For example, in one of my recent graduate research methods seminars, I had a group of Air Force Officers use their military social networks to obtain a sample of 870 people in less than 24 h. Social networking sites and other online platforms allow researchers to draw upon existing online social networks to reach large numbers of people quickly. Participants can respond to online surveys at a convenient time for themselves, and they may take as much time as they need to answer individual questions. Broadband access to the Internet also facilitates the transmission of multimedia content, which can enhance the sophistication of online surveys. For qualitative researchers, improved broadband access also allows for online focus groups and chat rooms where participants interact with each other and the interviewer/facilitator in a multimedia setting (Wright 2005). Multimedia capabilities also allow survey researchers to embed video or add more interactive measurement features to the survey (e.g., sliding bars and fuel gauge images to help them visualize different perceptions, etc., they are being asked about on the online survey).

Another important advantage of online surveys is lower cost to researchers and their institution (Evans and Mathur 2005). Compared to traditional survey methods, online surveys are much cheaper to construct and implement. In addition, more sophisticated experimental designs can take advantage of the online sphere in terms of randomly linking participants to different stimuli (e.g., YouTube videos with a specific persuasive appeal in a health message design experiment) and then having members of the control and treatment groups complete a postexperimental survey. Such features can reduce the need for physical laboratory space, the cost of incentives to get participants to come to the lab, and the cost of transferring responses to data analysis programs. Moreover, online surveys help reduce the environmental burden since participants do not need to travel to take a survey, and there is no need to print pencil-and-paper surveys (Wright 2005). Features such as automated data collection, skip patterns, automated reminders to participate survey instruments, and easy-to-download data files into SPSS or other statistical analysis software make online survey methods an attractive and less expensive way to engage in survey research compared to traditional survey methods (Couper 2008; Greenlaw and Brown-Welty 2009; Kramer et al. 2014).

3 Critiques of Online Surveys

Despite the many advantages of using online surveys, there have also been numerous critiques of online survey methods, including data security issues, sampling issues, and ethical concerns (Manfreda et al. 2008; Payne and Barnfather 2012; Curtis 2014). This section focuses on common concerns about using online survey research that stem from these criticisms.

During the first several years when online surveys started to become more popular (during the mid to late 1990s), many early studies that compared online surveys to paper-and-pencil surveys were concerned with the issue of measurement equivalency. At the time, researchers worried that online surveys may invite measurement validity and reliability problems. However, a number of studies have been published in the past decade support an emerging consensus that both modes of data collection are generally comparable in terms of reliability and validity (Johnson 2005; Wright 2005; See Fan and Yan 2010).

Another major concern that has often been raised regarding online surveys is that online samples are not representative of the general population (Dillman 2000; Wright 2005). While this was certainly true in the early days of the Internet, in recent years more and more diverse groups of individuals (including populations of all types of patients, health-related support groups, health information communities, and so on) have found their way to cyberspace due to the reduction in costs of smart phones and other devices as well as the increased affordability of high-speed online Internet access. Studies have found that individuals do not seem to differ on many psychological and communication measures when comparing online surveys to traditional paper-and-pencil surveys (Bosnjak and Tuten 2001; Fan and Yan 2010). Another problem that can occur among researchers who use longitudinal online surveys (i.e., repeated measures) is participant attribution (Wright 2005; Murray et al. 2009). However, studies suggest that the attrition in online longitudinal surveys does not differ much from traditional surveys (Murray et al. 2009; Fan and Yan 2010). Moreover, automated email reminders appear to be a cheap and convenient way to reduce attrition in online surveys.

For researchers who use online surveys as a component of an online experimental design, the lack of experimental control can become a serious issue depending upon the nature of the experiment and the inability to have control over manipulating the environment beyond random assignment to different experimental conditions online (Wright 2005). Participants may need guidance in filling out the questionnaire, and there may be little or no opportunity for real time questions from participants. People who participate in online experiments or online surveys that ask sensitive questions may be hesitant to participate if they believe that a researcher will use a person's IP address or other information that could be used to identify their particular responses. A careful online informed consent form that contains reassurances about confidentiality and security can help to increase online survey response rates. However, a researcher needs to make sure that safeguards are built into surveys and the way in which they are disseminated that will help keep the identity of participants anonymous and secure (Couper 2008). Institutional Review Boards (IRB) at most major universities typically have some guidelines regarding the conduct of web-based research, particular in terms of participant confidentiality and privacy issues. However, depending upon the sophistication of the survey design, the IRB may have additional concerns or questions for a researcher to address. It is important for researchers to clarify the IRB guidelines for online survey research at their particular institution early in the conceptualization process to avoid added delays in terms of launching the survey.

4 Sampling, Measurement, and Enhancing Response Rates in Online Survey Research: Promises and Pitfalls

4.1 Sampling Issues and Online Surveys

Global Internet use increased around 400% between 2000 and 2009 and, today, it is a common reality in affluent Western societies (Kramer et al. 2014). In terms of sampling frames, the Internet provides many possibilities in terms of reaching potential international participants for an online survey. Such surveys are important to cross-cultural research in the health social sciences as well as building research teams comprised of scholars from different nationalities. Some scholars have argued that online surveys allow for more efficient implementation of psychological assessments when compared with traditional assessment procedures (Lieberman 2008; Shih and Fan 2008; Kramer et al. 2014). In some cases, participant recruitment on the Internet may be the most appropriate way to reach the target population. For example, if the long-term goal of a study is to establish effectiveness of an online health intervention in online cancer support communities, then a sample recruited directly from online communities would more accurately represent that population.

However, there are a number of researchers who have warned about threats to external validity (due to sampling problems) in online survey research, and they recommend certain practices to avoid (Murray et al. 2009). For example, researchers should avoid posting an open invitation link on a forum or sending out invitations to the entire target population (a census). Moreover, many respondents have a tough time distinguishing between a legitimate survey and a spam message (especially if a person is fearful that clicking on the link to the survey might infect their computer with a virus). In addition, some individuals are more drawn to a survey topic than others or may have more time than others in terms of being able to complete an online survey questionnaire. Such issues often lead to selection biases in the sampling process. Researchers need to consider how recruitment and enrollment on the Internet may present unique challenges to sample validity and representativeness (Shih and Fan 2008; Murray et al. 2009). Whenever feasible, online survey researchers should attempt to use available sampling frames to generate a probability sample of potential participants who will be invited to participate in the online survey.

Due to the global reach of the Internet, additional sampling frames of participants can be accessed conveniently and cheaply within relatively short periods of time if a questionnaire is translated and/or adapted for use in other cultures. For example Wright and colleagues (2015) conducted an online survey in China and Korea about media use, willingness to communicate about health, and weight-related stigma associated with US fast-food restaurant chain food consumption behaviors. Using contacts at universities in China and Korea, the research team was able to recruit a relatively large convenience sample of participants from both countries. However, it is important to rely on native speakers (who are members of the culture being investigated) in the construction of translated online surveys so that more nuanced regional differences in language and language use can be included in the survey

(Payne and Barnfather 2012). Researchers should not rely on more simplistic language translation tools that can be found online.

However, it is also important to point out that true global research is difficult to obtain since Internet access and use are not equally distributed worldwide, and a substantial digital divide exists between privileged and underprivileged socioeconomic groups and countries (Pullmann et al. 2009). In general, the countries with the greatest Internet access are typically more affluent, better educated, and have a higher gross domestic product (GDP) rate. Moreover, online platforms that exist in one country (e.g., Facebook in the USA) may not be available (for political or legal reasons) in other countries (such as China).

4.2 Measurement Issues with Online Surveys

Online surveys do not appear to compromise the psychometric properties of common quantitative measures (e.g., Likert-type scales, and so on), and studies have found that participants are typically not less representative of the general population than those of traditional studies (Denissen et al. 2010). The anonymity of the Internet-based platforms has been found to have a positive influence on communication behaviors that have important implications for the ubiquitous use of self-report questions on online surveys. Online anonymity has been linked in a variety of studies to feelings of reduced risk when communicating with others (Wright and Miller 2010a), increased self-disclosure of thoughts and feelings (Valkenburg and Peter 2009), reduced stigmatization of visible disabilities and health conditions (Simon Rosser et al. 2009; Wright and Rains 2013), and greater willingness to communicate with others (van Ingen and Wright 2016). Shy or anxious individuals are faced with fewer inhibitions to participate in online surveys compared to face-to-face surveys, and sensitive topics can be addressed confidentially. Because many health issues are highly sensitive and a number of diseases and health conditions are negatively stigmatized, online surveys can help health communication scholars gain access to individuals living with stigmatized health problems, people who have limited mobility due to their health issues, and people who are apprehensive about discussing sensitive health issues (who may be less willing to participate in traditional surveys).

However, it is important to recognize that creating an online survey questionnaire is not simply a case of reproducing an e-version of the paper-and-pencil survey. Formatting may need to be changed to simplify data entry, to clarify possible responses, or to avoid the possibility of submitting data before completing the survey. Due to the diversity of participants' access to the Internet, computer or smartphone differences, software differences, researchers need to make decisions about the complexity of visual design, the potential speed differences when downloading the survey on different devices, and the ability to view the whole questionnaire on a range of screen settings. Each of these design decisions can potentially influence the measurement reliability and validity of key variables on the online survey.

4.3 Enhancing Response Rates Using Online Surveys

Researchers have identified several factors that appear to increase response rates in online surveys, including personalized email invitations, follow-up reminders, pre-notification of the intent to survey, and simpler/shorter web questionnaire formats (Cook et al. 2000; Porter and Whitcomb 2003; Galesic and Bosnjak 2009). Other factors that increase response rates include: incentives, credible sponsorship of the survey, and multimodal approaches (Johnson 2005; Fan and Yan 2010). When online surveys initially appeared in widespread form in the 1990s, many researchers were concerned about inferior responses rates of online surveys (compared to mailed surveys). However, a number of studies have since found online surveys to be similar to traditional mailed surveys in terms of response rates (see Dillman 2000; Kaplowitz et al. 2004; Manfreda et al. 2008).

For example, Kaplowitz et al. (2004) found that a web survey application achieved a comparable response rate to a mail hard copy questionnaire when both were preceded by an advance mail notification. In addition, reminder mail notification had a positive effect on response rate for the web survey application compared to a treatment group in which respondents only received an e-mail containing a link to the Web survey. In terms of health social science research, van Ingen and Wright (2016) examined online coping and social support following a major life crisis using a large, representative web-based panel study in the Netherlands that yielded a response rate of 83% (2,544 respondents). Reminder emails and easy-to-use web questionnaire formatting were used in this survey, and the researchers were able to obtain a diverse sample of participants.

Another factor that appears to influence response rates in online survey is the convenience for participants. Participants can take an online survey in the comfort of their home environment. In addition, web survey questionnaire programs (e.g., *SurveyMonkey*) provide easy to navigate Likert-type, semantic differential, scales that allow participants to quickly click on a choice using a computer mouse (compared to a cumbersome phone survey or a lengthy paper-and-pencil survey). Such convenience features of online surveys appear to increase readiness to participate and may lower the compensation necessary to convince members of the target population to participate (Wilson et al. 2010).

Online surveys offer several other advantages in terms of the recruitment of participants. Researchers can rely on easy to create Internet advertisements and use online community and mailing lists, which are less time-consuming to produce and less costly to distribute, than posters, flyers, newspaper, TV, and radio advertisements. This can help extend the reach of an online survey. In addition, online surveys appear to be well suited in terms of their ability to attract greater diversity in sample by encouraging recipients (who share characteristics of interest to the researcher) to forward the message to potentially suited and interested participants. For example, including a request to forward a message about an online survey to other senior citizens if a research advertises the survey within an online community for older adults may lead to additional older adults becoming aware of (and potentially participating in) the survey. Other web communities can be used to access large numbers of individuals based on sex, race, nationality, and other demographic variables of interest. Some online survey services (such as *SurveyMonkey*) will help researchers reach certain demographic groups via databases of people who have completed *SurveyMonkey* surveys in the past. Automatic emails can be sent to remind participants to participate in a cross-sectional survey or they can be set to remind participants in a longitudinal study to complete an online survey during designated time frames.

5 Resources for Creating/Managing Online Surveys

The particulars involved in designing, implementing, and managing an online survey are beyond the scope of this chapter. However, for scholars who are new to online survey methods, there are a number of web resources containing helpful information for scholars who want to conduct online studies, such as online pdf guides (Couper 2008), and the online course hosted by the University of Leicester (http://www.geog.le.ac.uk/ORM/site/home.htm) or the Web Survey Methodology project (http://www.websm.org/). In addition, popular online survey services like *SurveyMonkey* and *Qualtrics* offer online tutorials as well as customer support via email or phone. While such online survey services can make things easier for a researcher who is new to online survey research, they can come at a steep price, such as high subscription rates and "add on" fees for requesting particular features. Additional issues regarding online survey services will be discussed in more detail later in this chapter.

6 Overseeing Web-Based Survey Data Collection and Analysis Issues

Once the online survey has been launched, it is important for researchers to be diligent in terms of monitoring recruitment emails and postings to assure a sufficient number of recruitment messages have been sent/posted to the target population members. When posting recruitment advertisements to online communities, it is common community moderators to remove messages that have not been approved by the moderator or the community members. For researchers who are interested in surveying members of such groups, it is important to secure permission to post recruitment messages in advance of launching the survey. Many IRBs require evidence of permission from an online community moderator or leader in the form of an email or an attached letter to post recruitment messages for the study. In my own research (see Wright 2000, 2011; Wright et al. 2010b), I have found that providing community members with a link to a webpage that discusses the results of the survey once it is completed (in layperson's terms) is a helpful way to gain access to online health-related communities.

Online survey research allows for communication between the research and participants via email if questions about particular items surface during the data collection phase. In addition, features such as the amount of time it takes a participant to complete the survey and the time of day when the survey was taken are typically included when downloading survey data files from services like *SurveyMonkey* or *Qualtrics*. This can help a researcher decide whether or not to include or exclude data from a participant who took an extremely short (e.g., 30 s) or long (e.g., 3 weeks) time to complete the survey. Moreover, such services also include the IP addresses of the respondents' computers. *SurveyMonkey* and *Qualtrics* survey templates will recognize duplicate IP addresses, and it will not allow someone from the same IP address to submit more than one response to the online survey. As mentioned earlier, these services also include the ability to send automated reminder emails to potential participants, although it is important for the researcher to verify whether these emails have actually been sent (especially since systems can go offline due to unexpected power outages and maintenance issues).

Although it may take less time to reach a sufficient sample size using online surveys, it is important to realize that many responses from online participants may be left blank (unless the research requires participants to complete every question). As a result, what looks to be an initial sample of 300 people based on a *SurveyMonkey* data overview report, it is possible that there are large numbers of unusable responses from participants. I typically over-sample so that I receive 20–30% additional responses from participants over my target sample size goal. Most data from online surveys need to be cleaned, recoded, or transformed in some way. Most common statistical software programs (e.g., SPSS, SAS, and so on) make it relatively easy to perform these tasks. Data from surveys that use a large number of filter questions and that incorporate skip logic on the online survey may be more cumbersome to clean and organize once it has been collected.

7 Pros and Cons of Various Web-Based Survey Platforms and Services

New web-based survey platforms and services appear online each month, and so, it is difficult to provide a comprehensive list of all of the choices researchers have in terms of finding a web-based survey platform or service that will be most useful for the research projects. Certainly, *SurveyMonkey* and *Qualtrics* appear to be the two popular platforms/services for creating and distributing online surveys in the USA. However, there are many other platforms/service available online that range from relatively low cost to expensive, "full service" options (in which the company helps a researcher design the online survey, recruit participants, analyze data, and so on). In this section, I will discuss several pros and cons of various types of platforms and services in general as opposed to critiquing specific platforms/services.

Web-based survey platforms and services, such as *SurveyMonkey* and *Qualtrics*, provide an easy way for researchers to engage in online survey research. Standard subscription plans for these companies offer a variety of templates for different types of online surveys, tutorials, customer support, a wide range of online survey measures (e.g., short answer, Likert-type scales, semantic differential scales) and the ability to conveniently track and download response to a data analysis program (like SPSS or

SAS). However, the standard plans typically have a limit on the number of responses you can collect, and they do not include additional services, such as help with sampling, recruitment advertisement development, consultations, and data analysis. *SurveyMonkey* charges an extra fee (beyond the basic subscription plan) to download data from an online survey into SPSS or another data analysis program format. For additional fees, *SurveyMonkey* offers researchers access to a wide range of online populations (e.g., databases of people who are willing to complete online surveys that have been created by both companies) and help recruiting these individuals. *SurveyMonkey* offers services that will allow a researcher to narrow the range of online participants he or she would like to access for a particular online survey based on a multitude of demographic characteristics (i.e., age, sex, occupation, region of the country, and so on) and various other segmentation variables that are collected by *SurveyMonkey* for their participant databases. Of course, these types of potential participant databases suffer from problems such as selection bias and relevance issues (e.g., certain surveys may not be of concern or interest to people in the databases, but they may be willing to take the survey if they are being compensated by *SurveyMonkey*).

Individual subscriptions to *SurveyMonkey* and *Qualtrics* (as well as similar companies) can be expensive, but site licenses for universities and units within them tend to be reasonable in terms of cost. For researchers who wish to download qualitative data directly into programs like *Invivo* will not find this type of option when using *SurveyMonkey* or *Qualtrics*. However, there are ways to cut and paste qualitative data from SPSS or Excel into this type of qualitative data analysis program. Moreover, these companies continue to add new features for consumers on a regular basis, so they will likely become more flexible when it comes to the types of services and data management options that will be available in the future.

8 Conclusion and Future Directions

The purpose of this chapter was to introduce health social scientists to the pros and cons of conducting online survey research, including sampling, measurement, and response rate issues. Moreover, it briefly examined some resources for getting started with online survey research, best practices in terms of overseeing and managing online surveys, and some advantages of using *SurveyMonkey* and *Qualtrics* as web-based survey platforms/services. This section briefly discusses the implications of this research method for health social scientists.

For health social science researchers, there are clearly several benefits of conducting online surveys of various health care system stakeholders via the Internet which make it an attractive alternative to traditional survey methods. These include the relative ease of survey design and implementation (especially when using platforms/services like *SurveyMonkey* or *Qualtrics*) and the potential to conduct relatively large-scale surveys while eliminating the costs of stationery, postage, and administration. Most online survey creation tools and/or use of SurveyMonkey and Qualtrics do not require any programming skills, and the cost of sending multiple e-mail invitations and reminders is negligible. More sophisticated features of online surveys allow validation checks as data are collected or randomization of respondents to different versions of the questionnaire (for experimental designs). However, it is important to remember that the cost of online survey design and implementation may increase as the complexity/sophistication of the online survey increases (especially when using *SurveyMonkey* or Qualtrics).

Researchers should always be concerned with sample representativeness and other factors that may undermine the external validity of the data obtained. As with traditional survey methods, studies that can use probability samples will have better external validity than nonprobability samples. Online health organization websites, such as hospitals and physician groups, often have detailed lists of providers, staff, and other key members of the organization that can be used as a sampling frame when conducting a probability sample. Patients are harder to reach through health care organizations due to patient privacy laws (e.g., HIPPA) and organizational practices that are designed to protect patient confidentiality. However, online support communities and health information websites (such as WebMD) are helpful portals for gaining access to people who are living with a variety of health problems.

Online surveys have also been found to have issues with selection bias and the inability to reach individuals who may not have quality Internet access (e.g., high speed Internet) or the latest technology (e.g., I-Phone, tablets, and so on). Unfortunately, this may include a lot of older adults and people who face a number of health disparities due to socio-economic factors. However, one promising trend appears to be the gradual adoption of computers and devices in economically disadvantaged regions of the world. While people lack access to the most up-to-date technology, they may be able to access online surveys with the technology they have (especially if the online survey questionnaire uses a more simplistic design). Other problems, such as low response rates, can often be remedied by sending multiple email reminders, reducing the complexity of the online survey instrument, and by finding key opinion leaders who are members of the online population (or segment) of interest who may be willing to promote the survey on behalf of the researcher. Such collaborations can enhance a researcher's credibility with a specific segment of the population and ease the burden of participant recruitment.

Despite these problems, online surveys allow social scientists to access unique populations of individuals facing health concerns (e.g., people who seek online support for a rare disease), people who may be difficult to survey in other contexts, and a variety of health care professionals (since most physicians and other providers can be reached online). Online surveys also allow social scientists to research populations at a quasi-global level (i.e., the global south tends to have lower access to the Internet than other regions of the world), and this may open the door to international research collaboration on a variety of health issues.

In short, online surveys offer many advantages many advantages to health social science researchers. However, the technical and methodological implications of using this approach should not be underestimated. Additional research is needed to be better understand the pros and cons of online surveys and to find designs/approaches that improve their external validity, including approaches that increase the representativeness of invited samples and limit response bias.

References

Bosnjak M, Tuten TL. Classifying response behaviors in web-based surveys. J Comput Mediated Commun. 2001;6. Retrieved from http://www.ascusc.org/jcmc/vol6/issue3/boznjak.html.

Cook C, Heath F, Thompson R. A meta-analysis of response rates in web or internet based surveys. Educ Psychol Meas. 2000;60:821–36.

Couper MP. Designing effective web surveys. New York: Cambridge University Press; 2008.

Curtis BL. Social networking and online recruiting for HIV research: ethical challenges. J Empir Res Hum Res Ethics. 2014;9(1):58–70.

Denissen JJ, Neumann L, van Zalk M. How the internet is changing the implementation of traditional research methods, people's daily lives, and the way in which developmental scientists conduct research. Int J Behav Dev. 2010;34(6):564–75.

Dillman DA. Mail and internet surveys: the tailored design method. New York: Wiley; 2000.

Evans JR, Mathur A. The value of online surveys. Internet Res. 2005;15(2):195–219.

Eysenbach G, Wyatt J. Using the internet for surveys and health research. J Med Internet Res. 2002;4(2):e13.

Fan W, Yan Z. Factors affecting response rates of the websurvey: a systematic review. Comput Hum Behav. 2010;26:132–9.

Galesic M, Bosnjak M. Effects of questionnaire length on participation and indicators of response quality in a web survey. Public Opin Q. 2009;73(2):349–60.

Greenlaw C, Brown-Welty S. A comparison of web-based and paper-based survey methods: testing assumptions of survey mode and response cost. Eval Rev. 2009;33(5):464–80.

Johnson JA. Ascertaining the validity of individual protocols from web-based personality inventories. J Res Pers. 2005;39(1):103–29.

Kaplowitz MD, Hadlock TD, Levine R. A comparison of web and mail survey response rates. Public Opin Q. 2004;68(1):94–101.

Konstan JA, Simon Rosser BR, Ross MW, Stanton J, Edwards WM. The story of subject naught: a cautionary but optimistic tale of internet survey research. J Comput Mediated Commun. 2005;10(2):Article 11. http://jcmc.indiana/edu/vol10/issue2/konstan.html.

Kramer J, Rubin A, Coster W, Helmuth E, Hermos J, Rosenbloom D, . . . Brief D. Strategies to address participant misrepresentation for eligibility in web-based research. Int J Methods Psychiatr Res. 2014;23(1):120–29.

Lieberman DZ. Evaluation of the stability and validity of participant samples recruited over the internet. Cyberpsychol Behav Soc Netw. 2008;11(6):743–5.

Manfreda KL, Bosnjak M, Berzelak J, Haas I, Vehovar V, Berzelak N. Web surveys versus other survey modes: a meta-analysis comparing response rates. J Mark Res Soc. 2008;50(1):79.

Murray E, Khadjesari Z, White IR, Kalaitzaki E, Godfrey C, McCambridge J, Thompson SG, Wallace P. Methodological challenges in online trials. J Med Internet Res. 2009;11(2):e9. https://doi.org/10.2196/jmir.1052.

Owen DJ, Fang MLE. Information-seeking behavior in complementary and alternative medicine (CAM): an online survey of faculty at a health sciences campus. J Med Libr Assoc. 2003; 91(3):311.

Payne J, Barnfather N. Online data collection in developing nations: an investigation into sample bias in a sample of South African university students. Soc Sci Comput Rev. 2012;30(3):389–97.

Pedersen ER, Helmuth ED, Marshall GN, Schell TL, PunKay M, Kurz J. Using Facebook to recruit young adult veterans: online mental health research. JMIR Res Protocol. 2015;4(2):e63.

Porter SR, Whitcomb ME. The impact of contact type on web survey response rates. Public Opin Q. 2003;67(4):579–88.

Pullmann H, Allik J, Realo A. Global self-esteem across the life span: a cross-sectional comparison between representative and self-selected internet samples. Exp Aging Res. 2009;35:20–44.

Riper H, Spek V, Boon B, Conijn B, Kramer J, Martin-Abello K, Smit F. Effectiveness of e-self-help interventions for curbing adult problem drinking: a meta-analysis. J Med Internet Res. 2011;13(2):e24. https://doi.org/10.2196/jmir.1691.

Shaw LH, Gant LM. In defense of the internet: the relationship between internet communication and depression, loneliness, self-esteem, and perceived social support. Cyberpsychol Behav. 2002;5(2):157–71.

Shih TH, Fan X. Comparing response rates from web and mail surveys: a meta-analysis. Field Methods. 2008;20(3):249–71.

Siegel MB, Tanwar KL, Wood KS. Electronic cigarettes as a smoking-cessation tool: results from an online survey. Am J Prev Med. 2011;40(4):472–5.

Simon Rosser BR, Gurak L, Horvath KJ, Michael Oakes J, Konstan J, Danilenko GP. The challenges of ensuring participant consent in internet-based sex studies: a case study of the men's INTernet sex (MINTS-I and II) studies. J Comput-Mediat Commun. 2009;14(3):602–26.

Valkenburg PM, Peter J. Social consequences of the internet for adolescents. Curr Dir Psychol Sci. 2009;18:1–5.

van Ingen EJ, Wright KB. Predictors of mobilizing online coping versus offline coping resources after negative life events. Comput Hum Behav. 2016;59:431–9.

Wilson PM, Petticrew M, Calnan M, Nazareth I. Effects of a financial incentive on health researchers' response to an online survey: a randomized controlled trial. J Med Internet Res. 2010;12(2):e13.

Wright KB. Perceptions of on-line support providers: an examination of perceived homophily, source credibility, communication and social support within on-line support groups. Commun Q. 2000;48:44–59.

Wright KB. Researching internet-based populations: advantages and disadvantages of online survey research, online questionnaire authoring software packages, and web survey services. J Comput Mediat Commun. 2005;10:Article 11. Retrieved from http://jcmc.indiana.edu/vol10/issue3/wright.html.

Wright KB. A communication competence approach to healthcare worker conflict, job stress, job burnout, and job satisfaction. J Healthc Qual. 2011;33:7–14.

Wright KB, Miller CH. A measure of weak tie/strong tie support network preference. Commun Monogr. 2010;77:502–20.

Wright KB, Banas JA, Bessarabova E, Bernard DR. A communication competence approach to examining health care social support, stress, and job burnout. Health Commun. 2010a;25(4):375–82.

Wright KB, Rains S, Banas J. Weak tie support network preference and perceived life stress among participants in health-related, computer-mediated support groups. J Comput-Mediat Commun. 2010b;15:606–24.

Wright KB, Rains S. Weak tie support preference and preferred coping style as predictors of perceived credibility within health-related computer-mediated support groups. Health Commun. 2013;29:281–287.

Blogs in Social Research

77

Nicholas Hookway and Helene Snee

Contents

1	Introduction	1354
2	Blogging and the Confessional Society	1355
3	Blogs as Documents of Life: Two Research Case Studies	1356
4	Anonymity, Audiences, and Online Face-Work	1358
5	Data Collection	1360
6	Data Analysis	1361
7	Ethics	1362
8	Conclusion and Future Directions	1364
References		1365

Abstract

Blogs are the quintessential early twenty-first century text blurring the boundary between private and public. In this chapter, we approach blogs as contemporary "documents of life" and offer our reflections on what blogs can offer social researchers based on our own research experiences. Blogs offer rich first-person textual accounts of the everyday, but there are practical, methodological, and ethical issues involved in doing blog research. These include sampling, collecting, and analyzing blog data; issues of representation; and authenticity; whether blogs should be considered private or public, and if the people who create them are subjects or authors. The chapter also critically reflects on the methodological and ethical implications of the different decisions we made in our own research projects. We conclude that embracing new confessional

N. Hookway (✉)
University of Tasmania, Launceston, Tasmania, Australia
e-mail: nicholas.hookway@utas.edu.au

H. Snee
Manchester Metropolitan University, Manchester, UK
e-mail: H.Snee@mmu.ac.uk

© Springer Nature Singapore Pte Ltd. 2019
P. Liamputtong (ed.), *Handbook of Research Methods in Health Social Sciences*,
https://doi.org/10.1007/978-981-10-5251-4_19

technologies like blogs can provide a powerful addition to the qualitative researcher's toolkit and enable innovative research into the nature of contemporary selves, identities, and relationships.

> **Keywords**
> Blogs · Documents · Online ethics · Online methods · Qualitative research

1 Introduction

Recent decades have witnessed the emergence of the social web which enables users to produce and consume content at the same time as communicating and interacting with one another (Beer and Burrows 2007). The rise of Web 2.0 has reshaped social and cultural life but also significantly shifted the practice of doing social research (Liamputtong 2013; Fielding et al. 2016). This chapter addresses one aspect of Web 2.0 culture – the weblog or blog – and how it can be employed by social researchers as an innovative qualitative research method.

There is now over 15 years of work which considers the possibilities of online research methods; early influential texts include Jones (1999) and Mann and Stewart (2000). Internet research continues to offer an exciting frontier for innovative methods, as demonstrated by recent edited collections (Hine 2013; Fielding et al. 2016; Snee et al. 2016). This work continues to make practical, theoretical, and methodological contributions to within and beyond the field of online research: "these researchers study digital phenomena because they are social, and as such deserving of attention and significant within the overall concerns of their home disciplines" (Snee et al. 2016, p. 4; see also ▶ Chap. 75, "Netnography: Researching Online Populations," ▶ 76, "Web-Based Survey Methodology," ▶ 78, "Synchronous Text-Based Instant Messaging: Online Interviewing Tool," and ▶ 79, "Asynchronous Email Interviewing Method").

Quantitative survey research was an early adopter of data-gathering via the internet, and this has now become embedded within mainstream social science (Coomber 1997; Dillman 2007; see also ▶ Chap. 76, "Web-Based Survey Methodology"). In addition, qualitative researchers have explored how traditional techniques can be transferred to the online context, including interviews (O'Connor and Madge 2001; James and Busher 2009; Salmons 2015), focus groups (Gaiser 2008), and ethnography (Murthy 2008; Kozinets 2009; Hine 2015; see also ▶ Chaps. 75, "Netnography: Researching Online Populations," ▶ 78, "Synchronous Text-Based Instant Messaging: Online Interviewing Tool," and ▶ 79, "Asynchronous Email Interviewing Method"). Naturally occurring data – produced without the intervention of a researcher – has also been analyzed using large-scale quantitative approaches (Thelwall 2009; Bruns and Stieglitz 2012) and in-depth qualitative analysis (Herring 2004; Hookway 2008; Lomborg 2012). Some social scientists have put forward a case for "native" digital methods that repurpose already-existing methods and tools from online platforms and devices, rather than translating "offline" methods to internet spaces (Rogers 2013). However, this chapter puts forward the case for the benefits of blog analysis as an accessible online method for social scientists who wish to explore "naturalistic" data drawing on traditional qualitative approaches such as thematic analysis. Blogs offer rich, first-person textual accounts for health and social researchers interested in everyday life.

This chapter introduces the "blogosphere" and the nature of blogs and blogging, before discussing how we might see blogs as contemporary "documents of life." It suggests some benefits of blog analysis and draws on two case studies to reflect on why researchers might utilize blogs as data. The chapter then acknowledges some potential methodological issues associated with using blogs through a consideration of bloggers' presentation of self and the implications for trustworthiness and authenticity. Guidance on data collection and analysis is offered, along with an outline of ethical considerations for researchers interested in blog analysis.

2 Blogging and the Confessional Society

Blogs became a key player in online culture in the late 1990s and early 2000s. The growth of blogging was linked to the rise of free and user-friendly blog creation services such as *Blogger* and its rising cultural popularity, demonstrated by the emergence of "A-list" celebrity bloggers like Salam Pax (aka the "Baghdad Blogger") and media and gossip blog Gawker (Thompson 2006). The rise of blogging, like other Web 2.0 applications such as Facebook and Instagram, reflect a broader move to a "confessional society" where private sentiment has come to colonize the public sphere (Bauman 2007; Beer 2008). Personal blogs are the quintessential early twenty-first century new media, generating data that is simultaneously private and public.

Blogs are interactive and multimedia, converging text, image, video, GIFS, and other types of media into one space. The blogged about world can be almost anything. Blog styles differ within and between blogs, shifting from a confessional style to anguished vents, therapeutic self-writing, emotional outpourings, and advocating for social causes. There are different types of blogs, from "mommy," diet, and fitness blogs to food, travel, corporate, and educational blogs. Blogs vary in style and degree of reflexivity. Some bloggers recount experiences with little self-reflection while others are highly self-reflective and self-analytical. Although the blogged-about-universe can be on almost anything, the majority of the more diary-style bloggers are writing about ordinary relationships – work, family, friendship, neighbors – and the emotional and moral lives, these relationships are embedded within. Rather than the fragmented and distracted "whatever being" some blog theorists have proposed (Dean 2010), bloggers appear to be deeply engaged in tracing and evaluating their worlds and the ethical entanglements and relationships in which these evaluations occur. While new social media like *Facebook* and *Twitter* are characterized by the brevity of the "status update" or "the tweet," the blog format encourages a deeper engagement with self, personal expression, and community (Marwick 2008).

3 Blogs as Documents of Life: Two Research Case Studies

There are a number of practical benefits to conducting research using blogs. Blogs provide similar but far more extensive possibilities for social science researchers than offline diaries (see also ▶ Chap. 83, "Solicited Diary Methods"). They are publicly available, low cost, and allow researchers to gather substantial datasets. Blog data is available as immediate texts without the need for tape-recorders and transcription (Liamputtong 2013). As discussed in more detail below, online anonymity can mean bloggers can be less self-conscious about what they disclose, and blogs also enable access to harder-to-reach populations (Mann and Stewart 2000; Hessler et al. 2003; Liamputtong 2007, 2013). As with other online sources, blogs are a global phenomenon that can be utilized for small-scale comparative research as well as studies concerned with globalizing trends. As blogs are archived documents, they can be used to examine social processes over time. Blogs are, thus, a valid addition to the qualitative researcher's toolkit and address issues of access, practicality, and capacity. Most importantly, however, they enable access to first person and spontaneous narratives of experience and action.

Blogs are part of the generation of vast "archives of everyday life" via social media which offer unique access into biographical experience and subjective understandings. In this way, blogs can be understood as a contemporary "document of life" (Plummer 2001). Documents of life are expressions or artifacts of personal life produced in the course of everyday life such as diaries, letters, biographies, self-observation, personal notes, photographs, and films. The benefits of documents of life for qualitative researchers is that they offer insight into how the social world is experienced and creatively expressed from the perspective of the individual.

In this context, blogs share similarities with diaries, and blog analysis is an analogous method to diary research. Like diaries, blogs are personal documents produced in real time, with no precise addressee (Arioso 2010). As such, blog researchers may take inspiration from "offline" diary research. Plummer (2001, p. 49) suggests that "diaries may be one of the better tools for getting at the day-to-day experiences of a personal life." Through diary research, social actors can be understood as both observers and informants (Toms and Duff 2002).

There are a variety of diaries which can be used as raw material for research. They can be clustered into two main types: unsolicited "documents of life" (Allport 1943, p. xii), which are spontaneously maintained by respondents, and solicited "researcher-driven diaries" (Elliott 1997, p. 22), which are created and maintained at the request of a researcher. However, both forms of diaries present challenges (see also ▶ Chap. 83, "Solicited Diary Methods").

In the case of unsolicited diaries (those spontaneously maintained without the researcher's involvement), it can be difficult to identify suitable participants and ensure content meets the aims of the research. Solicited diaries, which are written for purposes of a research project, may overcome these issues but then pose additional problems in finding participants willing to create and maintain a diary over a period of time. Blogs offer a helpful solution. Blogs possess the spontaneity of naturally occurring diaries while being easier to find and access than unsolicited personal documents. The narratives found in personal blogs are spontaneous in the sense that they are documents produced by people "carrying out their activities ... without any link with research goals or aims" (Arioso, 2010, p. 25). It is these elements in particular that offered innovative ways of addressing the research concerns of the following case studies of qualitative blog analysis. For both projects, blogs promised a new type of "document of life" that enabled access to first person and spontaneous narratives of experience and action.

> **Box 1 Nicholas Hookway's Study of Australian's Expressions of Everyday Morality Using Blogs**
> Nicholas's research explored everyday Australian moralities: the sources, strategies, and experiences of modern moral decision-making. The study focused on everyday moral worlds, something that is difficult to explore using traditional qualitative methods such as interviews that ask people *directly* about their moral beliefs (Phillips and Harding 1985). Nicholas felt it was hard to contextualize such a topic so that it is meaningful for the participant, and he was concerned that it could also result in people attempting to present themselves in a specific moral light, abstracted from the way that morality is grounded in their day-to-day lives. Blogs offered Nicholas an alternative way to "get at" spontaneous accounts of everyday morality. The study was based upon 44 Australian blogs sampled from the hosting website LiveJournal, along with 25 online interviews. Nicholas found that morality was depicted by the bloggers as an actively created and autonomous do-it-yourself project and suggested that self, body, emotions, and authenticity may play an important role in contemporary moralities Hookway (2017).

> **Box 2 Helene Snee's Study of British Young People's Gap-Year Experiences Using Blogs**
> Helene's study into overseas "gap years" by British youth was driven by a similar concern with how experiences are understood and represented. Gap years – a period of "time out" overseas at transitional moments – are now a well-established activity, particular for young people before starting higher education. Helene's interest was in representations of cultural difference, the drawing of distinctions of taste, and the implications for identity work for this potentially cosmopolitan activity. The study drew on the concept of "frames" (Goffman 1974) to consider how bloggers understand their gap years and make them meaningful for audiences. Blog analysis allowed Helene to consider what young people themselves considered important to share about their gap years. Thirty-nine blogs written by "gappers" to document their journeys were sampled, which were supplemented with nine interviews. Her findings

(continued)

> **Box 2 Helene Snee's Study of British Young People's Gap-Year Experiences Using Blogs** (continued)
> suggest that gap years tend to follow fairly standard "scripts" and reproduced ideas about value and worth that question the status of the gap year as a progressive, cosmopolitan enterprise Snee (2013b).

4 Anonymity, Audiences, and Online Face-Work

While blogs are naturally occurring text, they are still typically written for an implicit, if not explicit, audience (Hookway 2008). It is this presence, or at least potentiality of an audience, that renders blogs distinct from traditional forms of personal diary keeping. In this way, blogs are similar to other types of public text, shaped by imagined audiences as bloggers choose, select, and even inflate what they believe to be important to record and communicate. Moreover, if these are public blogs, they are visible to anyone with internet access and are interactive (Arioso 2010). The role of potential discursive display or performance is something that needs to be thought about if researchers are considering using blogs as a data source.

One such issue is how to recognize the public nature of self-display and expression that blogging entails. Blogging might be conceptualized as a disembodied form of "face-work" (Goffman 1972), concerned with the art of self-representation, impression management, and potentially self-promotion. Bloggers may strategically select and write into existence convincing life-episodes that frame themselves as "good," "moral," and "virtuous" subjects. Blogging in this scenario is just another "stage" for what Goffman (1959, p. 244) refers to as the "the very obligation and profitability of appearing always in a steady moral light."

Online anonymity, however, can disrupt the "blogging as face-work" model. The anonymity of many blogs, as opposed to trends in other social media where the standard practice is to use real-names, affords the opportunity for bloggers to write more honestly and candidly. The anonymity of blogs troubles Goffman's analysis of face-work which is premised on the social practices of face-to-face interactions. The anonymity of the online context means that bloggers may be relatively unselfconscious about what they write since they remain hidden from view. This was evident in Nicholas's research on blogging and morality, where participants often mentioned that they revealed "more of themselves" in their blog than what they called "normal life." Conversely, in Helene's research, some of the young people on gap years consciously did not disclose all in their blogs, as they were written for friends and family to follow their trips from home. The relationship with the blog audience and the level of anonymity adopted thus contextualizes the presentation of self in blog accounts.

There is a paradox built into blogging: bloggers are writing for an audience and are, therefore, potentially engaged in a type of "face-work" but at the same time they can be anonymous or relatively unidentifiable. This tension between visibility and invisibility gives blogging a confessional quality, where a less polished and even uglier self can be verbalized. One can express one's faults, one's mishaps – whatever might be difficult to tell as we "enter the presence of others" (Goffman 1959, p. 1) in face-to-face relations. As one blogger wrote in Nicholas's study: "the whole point is to have some place where you can be anonymous in front of people you know...I don't know of any time that someone's mentioned in real life a thing that's been written on LJ (LiveJournal)."

Although the potential online anonymity of blogging may sidestep problems of "face-work," the flipside is that it raises issues about possible identity play and deception. This is of course not specific to blogs as an online genre in the context of a moral panic regarding the predatory potential of the internet. Here, people going online, typically imagined as pedophiles or old men, can disguise their identities in order to prey on vulnerable young people. The image of the predatory "stranger" lurking in the dark alleys of the internet has arguably been fuelled by a lot of early internet research, which focused on the social implications of online anonymity, particularly in terms of "simulated" identity production (Turkle 1995; Danet 1998). In these accounts, the internet provides a space where an illusory, playful, and deceptive self can dominate – for example, men can pretend to be women and vice-versa.

How trustworthy then are the expressions of self that bloggers provide? How do you know what bloggers are telling you is true? They could be an elaborate fiction. Scott (1990) asks researchers when evaluating the authenticity of documentary sources to ask if the data is credible, i.e., that it is truthful and genuine. Both of us were routinely questioned on the truthfulness of blog data when presenting our research. The online anonymity of blogging raised issues about potential identity play and deception. How do you know the bloggers are "telling the truth," was a typical question. These concerns are rooted in the mediated nature of online representations, where "[a]nonymity in text-based environments gives one more choice and control in the presentation of self, whether or not the presentation is perceived as intended"(Markham 2005, p. 809).

The question of the importance of "truthfulness" again depends on the aims and objectives of the research. While it seemed unlikely that our blog data was "faked," this was not of crucial methodological concern. We approached the blog data as providing insight into the stories told about gap years or moral life rather than transparent representations of actual experience. Like other forms of qualitative research, this approach to blog analysis recognizes that there may be more than one equally credible account (Heath et al. 2009). Even if bloggers do not tell the "truth," these "fabrications" would still tell us something about the manner in which specific social and cultural ideas about travel or morality are constructed. Consequently, qualitative blog analysis has much in common with wider quality concerns such as Lincoln and Guba's (1986) concept of "trustworthiness." Moreover, concerns regarding the "authenticity" of blog accounts in terms of genuine authorship could be replaced with attempts to ensure that the bloggers are fairly represented by the researcher – an alternative interpretation of authenticity suggested by Lincoln and Guba (1986).

However, the issue of truthfulness may be an important consideration for a researcher wanting to read off external "truths" from the textual data – for example, the researcher seeking trustworthy accounts of weight-loss or becoming a parent. Using the multimedia elements of blogs such as images and video, and the links a blogger may post to other online content or social media, can help to build up a "picture" of the events in question. Another strategy to alleviate concerns around the veracity of the data is to supplement blog data with interviews. As discussed, both the gap-year and everyday morality project combined blog data with blogger interviews. As the blogs were limited to whatever the author had chosen to record, interviews provided a means to seek clarification, to explore absences and implicit meanings, and to contextualize online representations in terms of articulations of offline experience. This form of triangulation can also provide a technique to reinforce the "trustworthiness" of the blog analysis (Lincoln and Guba 1986; Liamputtong 2013).

5 Data Collection

The first step is to establish where and how to locate blogs. The majority of weblogs are hosted by specific blog platforms. There is a range of blog platforms available. The blog landscape is dynamic, with new platforms and technologies constantly entering and evolving. Popular platforms are Blogger, WordPress, Tumblr, LiveJournal, Medium, and Weebly. Blogger and LiveJournal have existed since 1999 while Tumblr and Medium are examples of newer offerings. Different blog platforms are orientated toward different purposes. For example, Tumblr is geared toward short-form blogging, typically around reposting web-content, while Blogger and WordPress are orientated toward long-format blogging.

Most blog platforms include a search feature which allows readers to find bloggers according to demographic information such as age and location as well as interests and hobbies. This feature can also be appropriated by social researchers to sample participants. A social researcher may use a blog platform search engine to search for key terms that bear on a particular social process or phenomenon. For example, Helene searched for blogs that contained the phrase "gap year" across the blog platforms MySpace, LiveJournal, and Globenotes. General blog search engines can also be used to locate blogs. For example, researchers can use internet wide search engines such as BlogSearchEngine.org or Ice Rocket Blog Search. These too are subject to a high turnover. For example, two of the search engines – Google Blog Search (A work-around for using Google blog search is explained here: http://www.netforlawyers.com/content/google-kills-blog-search-engine-109) and Technorati – used on the Gap Year project are now defunct.

Platform search engines (e.g., *LiveJournal* search) and specific blog websites (e.g., weight-loss blog sites like "The 100 most inspirational weight-loss blogs") are useful for projects focused on analyzing a definite type of experience or process such as those provided in travel, weight-loss, or parenting blogs. However, blog researchers need to be cautioned that even when using search engines, a degree of blog "weeding" is needed as these searchers produce not only a range of irrelevant results but also spam blogs, fake blogs, discarded blogs, access-restricted blogs, and nontraditional blogs (e.g., blogs hosted on news websites or social networking sites) (Snee 2012).

Nicholas, in his research on everyday morality, used platform search engines. However, rather than searching using key phrases, Nicholas searched for a group of bloggers who met particular characteristics and then used developed selection criterion to identity suitable posts to include within the sample. Nicholas employed this technique in the everyday morality project to firstly search for bloggers within specific age ranges and locations, and then reading the blogs returned in the search results for posts that reflected on everyday moral decision-making. This approach proved to be relatively time-inefficient and produced limited results. Blogs offered rich insights into the nature of contemporary self-identity and experience, but it was frustrating matching relevant blogs to the research aims. Nicholas switched to soliciting bloggers through advertisements in *LiveJournal* communities. This solved the "digital needle in the haystack" issue where long periods of time were spent trawling blogs in the vain hope of finding relevant posts.

This approach involved inviting participants to identify their blog to the researcher using online advertisements (e.g., advertising on *LiveJournal* for bloggers who write on everyday moral concerns). This strategy is particularly appropriate for projects like Nicholas's where the content is not linked to explicit blog types (e.g., weight-loss blogs or travel blogs) or easily retrieved via web-based blog search engines like the former *Google Blog Search*. This blog solicitation resulted in more relevant data being collected as those who were interested in taking part in the research could then contact Nicholas and direct him to specific posts on moral issues. This approach has the benefits of identification and relevance associated with solicited "offline" diary research but avoids many of the problems, as these blogs are not created and maintained at the request of a researcher.

6 Data Analysis

Most blogs are text based and thus suit conventional qualitative methods of text analysis. Blog data can be easily converted into text files for analysis or imported into computer assisted qualitative analysis software (CAQDAS) tools like NVivo or ATLAS.ti (Liamputtong 2013; see also ▶ Chap. 52, "Using Qualitative Data Analysis Software (QDAS) to Assist Data Analyses"). Nicholas considered using NVivo7 for the everyday morality study but in the end adopted a "manual" approach to analysis. Nicholas's experience was that the fractured and unstructured nature of the blog data was unworkable with the linear and highly structured demands of the software. It seemed to do an injustice to the contextual richness of the blog narratives, "thinning" the data into fragmented codes (Liamputtong and Ezzy 2000, p. 118) and infringing the creative and playful dimensions of the research.

Helene, however, successfully used CAQDAS to analyze blog narratives. Projects dealing with more tightly defined research topics like Helene's work on gap-year travel may be more amendable to CAQDAS due to their more focused and structured nature.

Narrative analysis, discourse analysis, content analysis, and thematic analysis are all suitable for analyzing blog data (Huffaker and Calvert 2005; Tussyadiah and Fesenmaier 2008; Enoch and Grossman 2010; see also ▶ Chaps. 47, "Content Analysis: Using Critical Realism to Extend Its Utility," ▶ 48, "Thematic Analysis," and ▶ 50, "Critical Discourse/Discourse Analysis"). Nicholas and Helene's respective projects on gap years and everyday moralities are examples of thematic analysis. For the everyday morality study, narrative analysis was considered, but the segmented nature of the blogs did not seem to lend itself to a form of analysis premised on analyzing how the parts of a biographical past are "storied" into a meaningful and coherent whole (Chase 2003; Riessman 1993). For Nicholas, the blogs sampled exemplified narratives of self but they tended to develop as a "database narrative" (Lopez 2009, p. 738), where posted fragments of self are disconnected from each other.

Other researchers have analyzed the visual aspects of blogs. For example, Scheidt and Wright (2004) explored visual trends in blogs, and Badger (2004) investigated how images and illustration shape the construction and reception of blogs. Researchers need to consider whether nontextual elements such as image, video, and music are integral to the goals of the project and how these dimensions are to be best incorporated into the analysis.

7 Ethics

Drawing on naturally occurring, "found" online data presents a number of ethical issues. The key issue addressed in ethical guidelines rests on conceptualizations of the public/private divide in online spaces (Markham et al. 2012; BPS 2013; Liamputtong 2013; see also ▶ Chap. 75, "Netnography: Researching Online Populations"). This has implications for whether informed consent is required, and how the data is reported. There are three broad positions taken by researchers in response to this issue. The first suggests that, as documents such as blogs are publicly available and accessible, and consequently consent is not required from bloggers (Walther 2002; Liamputtong 2013). From this perspective, bloggers are treated as authors or producers – and should even be cited accordingly (Bassett and O'Riordan 2002). The second position argues that material posted online is written with expectations of privacy and should be treated accordingly (King 1996; Elgesem 2002). This means that just because blog data is available online, it should not be treated as "fair game." The differences between these first two stances thus engage with what Frankel and Siang (1999) note as the difference between technological and psychological privacy. Moreover, blogs can be viewed as part of a bloggers' identity, and the principles associated with human subjects research, such as informed consent and protection of identity need to be addressed (Markham et al. 2012). The two stances also require blog researchers to decide if the personal blogs they wish to analyze are representations of human subjects or texts produced by authors (Lomborg 2013).

The third position towards the public/private divide suggests that the very nature of such online spaces mean there is no clear demarcation. Early work considering these issues suggested that the internet is both "publicly-private and privately public" (Waskul and Douglas 1996, p. 131). With the advent of Web 2.0 onwards and the growth of the "confessional society" (Bauman 2007), these blurred boundaries have become more apparent. Consequently, ethical guidelines for online research highlight the importance of adopting contextual approaches to resolving these issues, depending on, for example, the vulnerability of the blogger and the expectations of the privacy of online spaces (Markham et al. 2012). Privacy may be of more concern for researchers interested in personal health narratives, given the sensitivity of the subject matter (Grinyer 2007). Conversely, the gap year blogs in Helene's study were often specifically written for a public audience, so the authors had fewer expectations of privacy.

Blog researchers must, therefore, grapple with some thorny issues regarding privacy and the protection of human subjects. The case studies from Nicholas and Helene provide one example of how to navigate these decisions but also offer alternative points to consider. In both studies, the blogs were technically private and in the public domain, hosted on platforms that offered privacy restrictions if required. The subjects discussed were public reflections on everyday experience with little potential for harm. Consent was, thus, not obtained in either study for the use of blog data.

Given the "traceability" of online material (Beaulieu and Estalella 2012, p. 24), quoting directly from the blogs could lead readers to the bloggers' accounts. In both case studies, decisions were made to adopt Bruckman's (2002) suggestion of "moderate disguise" when reporting the analysis. This meant that verbatim quotations were used but personable or identifiable details were disguised and no links to the blogs were provided. This addresses the Association of Internet Researchers' guidance regarding whether data can be linked to individuals (Markham et al. 2012). The moderate disguise approach taken by Nicholas and Helene offered an appropriate minimization of harm given the contexts they researched but still retained the meaning of the narratives. Bruckman (2002) offers a "continuum of possibilities" to help researchers decide appropriate levels or protection. Other researchers, for example, suggest paraphrasing online content in qualitative personal research (Wilkinson and Thelwall 2011). Such ethical decision-making requires ongoing ethical reflection throughout the research project, and these case studies can be open to reexamination to reflect on best practice in future research (Snee 2013). Draft digital research guidelines produced by the British Sociological Association reaffirm that prescriptive rules are not appropriate and advocates "situational ethics that can allow or discretion, flexibility and innovation" (BSA 2016, online). Useful resources therefore include not only guidelines but also contextual case studies to provide researchers with the tools to make decisions rather than definitive answers (Markham et al. 2012; BSA 2016; Townsend and Wallace 2016).

In addition to ethical considerations regarding quoting from blog source material, blog researchers must also recognize issues of copyright. If bloggers are to be regarded as authors, for example, then they should be appropriately cited. Indeed, bloggers have automatic rights regarding the reproduction of their work under US, Australian, and UK copyright law (US Copyright Office 2000; Australian Copyright Council 2005; UK Intellectual Property Office 2013). This could be a source of tension for those researchers who wish to offer greater protection to bloggers. There are allowances made under copyright legislation for "fair use" or "fair dealing" of material for the purposes of study or research, but researchers should take care regarding attribution from a legal as well as ethical standpoint.

8 Conclusion and Future Directions

This chapter has offered suggestions for social researchers who may be interested in expanding their methodological toolkit using qualitative blog analysis. It has advocated the benefits of blog data both in terms of practical advantages but also for the insights they provide into everyday life. Through gaining an understanding of the "blogosphere" and the sorts of narratives available in these contemporary "documents of life," social scientists can explore a range of experiences, processes, and practices. There are a number of methodological issues that can be presented by blog research, including potential concerns surrounding impression management and overall trustworthiness, along with practical issues of data collection and analysis. Moreover, there are some complex ethical decisions to be made. Nevertheless, unique perspectives are offered by these unmediated and naturally occurring first-person accounts. By providing guidance and suggesting strategies to mitigate any problems, along with case studies to illustrate the potential of blog analysis, this chapter has made the case for blog analysis as a powerful method to engage with these narratives of everyday life.

In looking to the future, technological trends tend to move so quickly that it is difficult for researchers to keep up. Even while the case studies discussed in this chapter were being conducted, a "blogging is dead" discourse was circulating, along with evidence that the numbers of people who wrote and read blogs was in decline (Halavais 2016). Instead, "blog-like" activity of recording everyday life becomes spread across social media platforms – on Twitter, Instagram, Tumblr, Facebook, and the like (Pinjamaa and Cheshire 2016; Rettberg 2014, forthcoming). At the same time, blogs that retain a more "traditional" format have become increasingly commercialized. Individual bloggers' roles shift from the personal to professional as they seek to monetize their blogs (Pinjamaa and Cheshire 2016), and companies/media organizations are increasingly producing content in a "blog" format (Rettberg 2014). There are political implications for the blogosphere; Halavias (2016) plots the decline of civic webspaces due to commercialization, centralization, and monopolization of blogging-like practices by large social media platforms. On a more personal level, blog audiences become more distrustful of popular bloggers in genres such as fashion and "mommy blogging" who make the transition to professionalization. Generating income through their blogging practices raises questions of the authenticity of the narratives presented (Hunter 2016; Williams and Hodges 2016).

Where does this leave qualitative blog research? Firstly, blog researchers need to acknowledge the potential shift from personal to professional blogging when analyzing narrative accounts. Secondly, those interested in contemporary documents of life may need to widen their perspective beyond the traditional blog format and find pragmatic new solutions to take account of various channels of expression across different formats. The human need to tell narratives endures however, as do the rich data available for researchers. As Rettberg (forthcoming, p. 16) suggests: "On another level, things haven't really changed that much. Online and offline, we record aspects of our lives." Many of the benefits of blog research – of engaging with first-person accounts of everyday life – and many of the challenges – dealing with trustworthiness and ethical negotiations of privacy – also remain the same.

References

Allport G. The use of personal documents in psychological society. New York: Social Science Research Council; 1943.

Arioso L. Personal documents on the internet: what's new and what's old. J Comp Res Anthropol Sociol. 2010;1(2):23–38.

Australian Copyright Council. Information sheet: an introduction to copyright in Australia. 2005. Retrieved from: http://www.copyright.org.au/pdf/acc/infosheets/G010.pdf

Badger M. Visual blogs. In: Gurak LJ, Antonijevic S, Johnson L, Ratliff C, Reyman J, editors. Into the blogosphere: rhetoric, community, and culture of weblogs. 2004. Retrieved from: http://blog.lib.umn.edu/blogosphere/visual_blogs.html

Bassett EH, O'Riordan K. Ethics of internet research: contesting the human subjects research model. Ethics Inf Technol. 2002;4(3):233–47.

Bauman Z. Consuming life. Cambridge: Polity Press; 2007.

Beaulieu A, Estalella A. Rethinking research ethics for mediated settings. Inf Commun Soc. 2012;15(1):23–42.

Beer D. Researching a confessional society. Int J Mark Res. 2008;50(5):619–29.

Beer D, Burrows R. Sociology and, of and in Web 2.0: Some initial considerations. Sociol Res Online. 2007;12(5). https://doi.org/10.5153/sro.1560.

British Psychological Society. Ethics guidelines for internet-mediated research. Leicester: British Psychological Society; 2013. Retrieved from: http://www.bps.org.uk/news/guidelines-internet-mediated-research

British Sociological Association. The BSA's draft guidelines and collated resources for digital research. 2016. Retrieved from: http://digitalsoc.wpengine.com/?p=51994

Bruckman A. Studying the amateur artist: a perspective on disguising data collected in human subjects research on the internet. Ethics Inf Technol. 2002;4(3):217–31.

Bruns A, Stieglitz S. Quantitative approaches to comparing communication patterns on twitter. J Technol Hum Serv. 2012;30(3–4):160–85.

Chase SE. Narrative inquiry: multiple lenses, approaches, voices. In: Denzin NK, Lincoln YS, editors. The Sage handbook of qualitative research. 3rd ed. Thousand Oaks: Sage; 2003. p. 651–81.

Coomber R. Using the internet for survey research. Sociol Res Online. 1997;2(2). Retrieved from: http://www.socresonline.org.uk/socresonline/2/2/2.html

Danet B. Text as a mask: gender, play and performance in the internet. In: Jones S, editor. Cybersociety 2.0: revisiting computer mediated communication and community. Thousand Oak: Sage; 1998. p. 129–58.

Dean J. Blog theory: feedback and capture in the circuits of drive. Cambridge: Polity Press; 2010.

Dillman DA. Mail and internet surveys: the tailored design method (2007 update with new internet, visual, and mixed-mode guide). Hoboken: Wiley; 2007.

Elgesem D. What is special about the ethical issues in online research? Ethics Inf Technol. 2002;4(3):195–203.

Elliott H. The use of diaries in sociological research on health experience. Sociol Res Online. 1997;2(2). Retrieved from: http://www.socresonline.org.uk/socresonline/2/2/7.html

Enoch Y, Grossman R. Blogs of Israeli and Danish backpackers to India. Ann Tour Res. 2010;37(2):520–36.

Fielding N, Blank G, Lee RM, editors. The Sage handbook of online research methods. 2nd ed. London: Sage; 2016.

Frankel M, Siang S. Ethical and legal aspects of human subjects research on the internet. American Association for the Advancement of Science workshop report. 1999. Retrieved from: http://www.aaas.org/spp/dspp/srfl/projects/intres.main.htm

Gaiser TJ. Online focus groups. In: Fielding NG, Lee RM, Blank G, editors. The Sage handbook of online research methods. London: Sage; 2008. p. 290–306.

Goffman E. The presentation of self in everyday life. Harmondsworth: Penguin; 1959.

Goffman E. Interaction ritual – essays on face-to-face behaviour. Harmondsworth: Penguin; 1972.

Goffman E. Frame Analysis: An Essay on the Organization of Experience. Boston, MA: Northeastern University Press; 1974.

Grinyer A. The ethics of internet usage in health and personal narratives research. Soc Res Update. 2007;49. Retrieved from: http://sru.soc.surrey.ac.uk/SRU49.html

Halavais A. The blogosphere and its problems: web 2.0 undermining civic webspaces. First Monday. 2016;21(6). https://doi.org/10.5210/fm.v21i6.6788.

Heath S, Brooks R, Cleaver E, Ireland E. Researching young people's lives. London: Sage; 2009.

Herring S. Computer-mediated discourse analysis: an approach to researching online behavior. In: Barab SA, Kling R, Gray JH, editors. Designing for virtual communities in the service of learning. New York: Cambridge University Press; 2004. p. 338–76.

Hessler R, Downing L, Beltz C, Pelliccio A, Powell M, Vale W. Qualitative research on adolescent risk using e-mail: a methodological assessment. Qual Sociol. 2003;26(1):111–24.

Hine C, editor. Virtual research methods (four volume set). London: Sage; 2013.

Hine C. Ethnography for the internet: embedded, embodied and everyday. London: Bloomsbury; 2015.

Hookway N. Entering the blogosphere: some strategies for using blogs in social research. Qual Res. 2008;8(1):91–103.

Hookway N. The Moral Self: Class, Narcissism and the Problem of Do-It-Yourself Moralities. 2017. The Sociological Review (online before print): http://journals.sagepub.com/doi/pdf/10.1177/0038026117699540.

Huffaker DA, Calvert SL. Gender, identity, and language use in teenage blogs. J Comput-Mediat Commun. 2005;10(2). https://doi.org/10.1111/j.1083-6101.2005.tb00238.

Hunter A. Monetizing the mommy: mommy blogs and the audience commodity. Inf Commun Soc. 2016;19(9):1306–20.

James N, Busher H. Online interviewing. London: Sage; 2009.

Jones S, editor. Doing internet research: critical issues and methods for examining the net. London: Sage; 1999.

King S. Researching internet communities: proposed ethical guidelines for the reporting of results. Inf Soc. 1996;12(2):119–27.

Kozinets RV. Netnography: doing ethnographic research online. London: Sage; 2009.

Liamputtong P. Researching the vulnerable: a guide to sensitive research methods. London: Sage; 2007.

Liamputtong P. Qualitative research methods. 4th ed. Melbourne: Oxford University Press; 2013.

Liamputtong P, Ezzy D. Qualitative research methods: a health focus. Melbourne: Oxford University Press; 2000.

Lincoln YS, Guba EG. But is it rigorous? Trustworthiness and authenticity in naturalistic evaluation. New Dir Prog Eval. 1986;30:73–84.
Lomborg S. Researching communicative practice: web archiving in qualitative social media research. J Technol Hum Serv. 2012;30(3–4):219–31.
Lomborg S. Personal internet archives and ethics. Res Ethics. 2013;9(1):20–31.
Lopez LK. The radical act of "mommy blogging": redefining motherhood through the blogosphere. New Media Soc. 2009;11(5):729–47.
Mann C, Stewart F. Internet communication and qualitative research: a handbook for researching online. London: Sage; 2000.
Markham AN. The methods, politics and ethics of representation in online ethnography. In: Denzin NK, Lincoln YS, editors. The Sage handbook of qualitative research. Thousand Oaks: Sage; 2005. p. 793–820.
Markham AN, Buchanan EA, The AoIR Ethics Working Committee. Ethical decision-making and internet research: recommendations from the AoIR Ethics Working Committee. 2012. Retrieved from: http://aoir.org/reports/ethics2.pdf
Marwick A. LiveJournal users: passionate, prolific, and private. LiveJournal research report. 2008. Retrieved from: www.livejournalinc.com/press_releases/20081219.php
Murthy D. Digital ethnography. Sociology. 2008;42(5):837–55.
O'Connor H, Madge C. Cyber-mothers: online synchronous interviewing using conferencing software. Sociol Res Online. 2001;5(4). Retrieved from: http://www.socresonline.org.uk/5/4/oconnor.html
Phillips D, Harding S. The structure of moral values. In: Abrams M, Gerard D, Timms N, editors. Values and Social Change in Britain. London: Macmillan; 1985. p. 93–108.
Pinjamaa N, Cheshire C. Blogs in a changing social media environment: perspectives on the future of blogging in Scandinavia. In Twenty-fourth European conference on information systems (ECIS) proceedings, Istanbul, June 2016. 2016. Retrieved from: http://aisel.aisnet.org/ecis2016_rp/17
Plummer K. Documents of life 2: an invitation to critical humanism. London: Sage; 2001.
Rettberg JW. Blogging. 2nd ed. Cambridge: Polity Press; 2014.
Rettberg JW (forthcoming). Online diaries and blogs. In: Ben-Amos B, Ben-Amos D, editors. The diary. Bloomington: Indiana University Press. Pre-print, September 2016. Retrieved from: http://dspace.uib.no/bitstream/handle/1956/12831/Online%20Diaries%20preprint.pdf
Riessman CK. Narrative analysis. Newbury Park: Sage; 1993.
Rogers R. Digital methods. Cambridge, MA: MIT Press; 2013.
Salmons J. Qualitative online interviews: strategies, design, and skills. 2nd ed. London: Sage; 2015.
Scheidt L, Wright E. Common visual design elements of weblogs. In: Gurak LJ, Antonijevic S, Johnson L, Ratliff C, Reyman J, editors. Into the blogosphere: rhetoric, community, and culture of weblogs. 2004. Retrieved from: http://blog.lib.umn.edu/blogosphere/common_visual.html
Scott J. A matter of record, documentary sources in social research. Cambridge: Polity Press; 1990.
Snee H. Youth research in web 2.0: a case study in blog analysis. In: Health S, Walker C, editors. Innovations in youth research. Baskingstoke: Palgrave; 2012. p. 178–94.
Snee H. Making ethical decisions in an online context: reflections on using blogs to explore narratives of experience. Methodol Innov Online. 2013;8(2):52–67.
Snee H. Framing the Other: Cosmopolitanism and the Representation of Difference in Overseas Gap Year Narratives, The British Journal of Sociology 2013b;64(1):142–162.
Snee H, Hine C, Morey Y, Roberts S, Watson H, editors. Digital methods for social science: an interdisciplinary guide to research innovation. Basingstoke: Palgrave Macmillan; 2016.
Thelwall M. Introduction to webometrics: quantitative web research for the social sciences. San Rafael: Morgan and Claypool; 2009.
Thompson C. A timeline of the history of blogging. New York Magazine. 2006. Retrieved from: http://nymag.com/news/media/15971/
Toms EG, Duff W. "I spent 1 ½ hours sifting through one large box...": diaries as information behaviour of the archives user: lessons learned. J Am Soc Inf Sci Technol. 2002;53(4):1232–8.

Townsend L, Wallace C. Social media research: a guide to ethics. 2016. Retrieved from: http://www.dotrural.ac.uk/socialmediaresearchethics.pdf

Turkle S. Life on the screen: identity in the age of the internet. New York: Simon & Schuster; 1995.

Tussyadiah IP, Fesenmaier DR. Marketing place through first-person stories – an analysis of Pennsylvania roadtripper blog. J Travel Tour Mark. 2008;25(3):299–311.

UK Intellectual Property Office. Copyright. 2013. Retrieved from: http://www.ipo.gov.uk/types/copy.htm

US Copyright Office. Copyright basics. 2000. Retrieved from: http://www.copyright.gov/circs/circ1.html#wci

Walther JB. Research ethics in internet-enabled research: human subjects issues and methodological myopia. Ethics Inf Technol. 2002;4(3):205–16.

Waskul D, Douglas M. Considering the electronic participant: some polemical observations on the ethics of on-line research. Inf Soc. 1996;12(2):129–39.

Wilkinson D, Thelwall M. Researching personal information on the public web: methods and ethics. Soc Sci Comput Rev. 2011;29(4):387–401.

Williams M, Hodges N. Are sponsored blog posts a good thing? Exploring the role of authenticity in the fashion blogosphere. In: Kim KK, editor. Celebrating America's pastimes: baseball, hot dogs, apple pie and marketing? Proceedings of the 2015 academy of marketing science (AMS) annual conference. New York: Springer International Publishing; 2016. p. 157–62.

Synchronous Text-Based Instant Messaging: Online Interviewing Tool

78

Gemma Pearce, Cecilie Thøgersen-Ntoumani, and Joan L. Duda

Contents

1 Introduction	1370
2 Why This Method Was Developed	1371
2.1 The Interviewer Effect	1371
2.2 Potential Interviewees	1372
2.3 Need for Online Interview to Be Synchronous	1373
3 How Do You Use It?	1373
4 Method Preference	1375
5 Conclusion and Future Directions	1378
References	1381

Abstract

This chapter presents an explanation of the use of a synchronous text-based online interviewing method, which is a method of interviewing participants online using an instant messaging service to type to each other at the same time in a conversational style. This method was originally designed to address the need to carry out interviews on a potentially sensitive subject while also achieving a continuity of private discussion. It is particularly useful for situations in which

---

G. Pearce (✉)
Centre for Advances in Behavioural Science, Coventry University, Coventry, West Midlands, UK
e-mail: gemma.pearce@coventry.ac.uk

C. Thøgersen-Ntoumani
Health Psychology and Behavioural Medicine Research Group, School of Psychology and Speech Pathology, Curtin University, Perth, WA, Australia
e-mail: c.thogersen@curtin.edu.au

J. L. Duda
School of Sport and Exercise Sciences, University of Birmingham, Birmingham, West Midlands, UK
e-mail: j.l.duda@bham.ac.uk

© Springer Nature Singapore Pte Ltd. 2019
P. Liamputtong (ed.), *Handbook of Research Methods in Health Social Sciences*,
https://doi.org/10.1007/978-981-10-5251-4_21

face-to-face or telephone interviews are inappropriate. Online text-based interviews can, therefore, act as a solution to a perennial concern in health-related research, and is an advance on the standard practices of open-ended questionnaires, email interviews, and online discussion boards. Synchronous text-based online interviewing can be offered as an additional choice to research participants, as well as used on its own if examining sensitive topics or if additional anonymity is suitable. The use of the method alongside its strengths and weaknesses will be outlined and discussed in this chapter.

Keywords

Online interview · Qualitative · Internet research · Innovative methods · Sensitive topic

1 Introduction

The synchronous text-based online interviewing method is a method of interviewing participants online using an instant messaging service to type to each other (text-based) at the same time in a conversational style (synchronous). Over the last decade, the use of technology as a research tool has begun to develop. There are three main methods of communication through a private instant messaging (IM) forum, such as Skype. These include microphones (aural-IM), web cameras (visual-IM), and an area on which to type on the screen in a conversational style (text-IM). Overall, IM is better suited for research when interview times can be organized in advance, as opposed to random cold calling. Aural-IM using a microphone is equivalent to an organized telephone interview (Feveile et al. 2007; Drew and Sainsbury 2010) and can be combined with visual-IM using a webcam in order to add communication through facial expression and body language (Hanna 2012). Although research has been investigating the use of text-only interviews through a range of methods, such as open-ended questionnaires, email interviews, and online discussion boards, these come with the limitation of asynchronicity (see ▶ Chaps. 76, "Web-Based Survey Methodology," and ▶ 79, "Asynchronous Email Interviewing Method"). The unique feature of using an Instant Messaging service to carry out interviews is the ability to carry out a synchronous discussion with the participant (Brewer 2000). This chapter, therefore, focuses on the text-based element that the IM services provide (text-IM).

In this chapter, we explain the use of instant messaging services to carry out text-based online interviewing that has the synchronous factor to allow for the conversation to flow during the interview. The aim of this chapter is to provide information to the reader about why and how to use this synchronous text-based online interviewing method. A case study is provided from the original development and implementation of this synchronous text-based online interviewing method. This is a piece of qualitative research exploring women's experiences of the menopausal transition, completed as part of Gemma Pearce's doctoral research (see Pearce et al. 2010, 2011, 2013, 2014).

2 Why This Method Was Developed

> **Box 1 Case Study**
> Imagine this situation...
> You want to carry out a qualitative research project with menopausal women. You are a 27-year-old female and the average age for menopause is 51. You decide not to recruit through the National Health Service (NHS) because that will mainly gather participants who have had a negative experience of the menopause and its related symptoms, and you want to talk to women who have had a range of experiences.
> You try to recruit by going to places where there will be lots of women, such as female-only charity events. Your mum offers to help you recruit. By the end of the day, women have been offended by you approaching them about the project but your mum has been very successful. You realize that you need to understand more about how menopausal women view you as a 27-year-old who has not been through the menopause and how you might recruit more women.
> You run a focus group with some menopausal women you have managed to find through your mum's friends. In this, they say that women might not want to open up to you because you are younger and haven't been through the experience yourself. They say that they may not want to talk to you face-to-face or on the phone about sensitive topics, like changes in sex drive, lubrication and pain during sex, embarrassing heavy menstrual bleeding, and changes to their bodies. They suggest that if you try and recruit them personally, they are less likely to say they are menopausal because of social stigma, and instead suggest recruitment posters in private spaces, like the back of toilet doors. They suggest that a form of online interview where they can feel more anonymous but still be engaged in a conversational form of dialogue would be better.
> This was the situation that led to the development of this text-based online interviewing method. The need for these methods with this case study was largely inspired for the need for anonymity and sensitivity. This can be applied to other research topics; however, it is not a necessary requirement to use text-IM. This method is also useful to interview some hard-to-reach populations, to save research costs, carry out more convenient interviews, or simply provide an additional choice for researchers and participants to use alongside other interview methods.

2.1 The Interviewer Effect

It is important to consider the impact that the interviewer can have when interacting with the research participant. Researchers need to consider how participants' self-presentation concerns might impact on the findings and, therefore, possibly change the accuracy of the results. Self-presentation is a complex topic based around how we judge ourselves and how we manage the image that we project to others emotionally, physically, financially, cognitively, and behaviorally (Leary 1996). This social desirability bias is especially relevant in qualitative studies, where the interviewer effect is more pertinent (Stacey and Vincent 2011). Participants may give answers to interviewers that make them appear like a "good" interviewee, seem an expert on the topic, or provide more socially acceptable responses (Smith 1995). Further, the researcher can be viewed as an authority figure, and the participant may wish to give a pleasing impression, or feel uncomfortable about opening up, when they perceive an unequal power relationship. Participants may be distracted by their interest in the researcher and start asking questions of them or be influenced by the researcher's personal factors, such as gender, age, and clothing, in addition to the actual questions asked and their phrasing (Lewis 1995). The researcher, therefore, needs to disentangle how their active participation influences the results of the study (Dockrill et al. 2000). This means that it is important to reflect upon the research design used in each study and the resulting influence on the participants and researchers involved in this process.

2.2 Potential Interviewees

Importantly, the use of this tool has the potential to dehumanize the interviewer and disperse any perceived unequal power relationship between the researcher and the participant, distorting any social desirability bias and reducing inhibition of the interviewee (Stacey and Vincent 2011). This, in turn, potentially results in a richer and more honest interview. Consequently, the synchronous text-IM method should be especially beneficial, when interviewing hard-to-reach populations, children, or when the interview topic is of a personal, illegal, or sensitive nature. Due to the complementarity between the participant, researcher, and the research measure being used, this extra guard of anonymity will reduce the possible biases involved between the researcher and the participant, therefore, potentially increasing the validity of the interview. Even in cases where additional levels of anonymity are not needed, using this technique could encourage participants to feel more at ease and comfortable to discuss the topic in more depth compared with a face-to-face interview, while keeping more verbal and affective intimacy than in an email interview (Hu et al. 2004).

Additionally, the synchronous online interview method has been found to be useful in educational settings examining informal mobile learning across an international context (Lambrecht 2015), and carrying out sensitive program evaluations, such as resource allocation of individual members of staff (Gruba et al. 2016). The convenience of the method is also a key consideration, as a range of participant groups may prefer communication online, for example, generations where technology is the top communication method, and people with hearing difficulties or that may prefer nonverbal communication (Ison 2009; Benford and Standen 2011).

2.3 Need for Online Interview to Be Synchronous

Over recent years, researchers have been exploring a range of data collection methods using computer technology and the internet. This has enabled ease of access to larger populations and increased response rates for survey-based studies. As online questionnaires became more accessible and reliable, researchers explored the use of open-ended questions more frequently within these surveys (Dillman et al. 2009; Reardon and Grogan 2011), and through other internet-based means, such as email interviews (Murray and Sixsmith 2002; Murray 2005; Meho 2006; James 2007), online discussion boards (Seymour 2001; Moloney et al. 2004; Im et al. 2008), and conferencing software (O'Connor and Madge 2001). None of which were synchronous because the interviewer and the participant were not online at the same time. Although these methods eliminate the need to set up a mutually convenient interview time, they reduced the continuity of interaction with the interviewer and the immediacy with the topic. These methods lack the continuity of discussion that is achieved through face-to-face, telephone, and online visual/audio interviews (James and Busher 2006).

Compared with email interviews (see also ▶ Chap. 79, "Asynchronous Email Interviewing Method"), this synchronous text-IM method takes on a style closer to that of a conversation, while also allowing the participants to see and potentially reflect upon the dialogue so far via visual written "record." Unlike using the visual/aural instant messaging during interviews, this method reduces the amount of personal contact, such as communication through body language, voice, and facial expression. Although this is possibly a disadvantage compared with a face-to-face interview, this technique might increase the validity through an increased level of anonymity between the participant and the researcher (Jowett et al. 2011), helping the participant feel more at ease, or providing a preferred method of communication that is more convenient to the participant (Pearce et al. 2014).

3 How Do You Use It?

Carrying out interviews using this synchronous text-based online interviewing method saves travel time and cost, and can also take place at the participant's convenient time and place, similar to telephone interviews (see also ▶ Chaps. 79, "Asynchronous Email Interviewing Method," and ▶ 81, "Phone Surveys: Introductions and Response Rates"). So they could be at home in their pyjamas with a cup of tea, or during a break in their working day if they wanted. We recommend that the interviewer familiarize themselves with instant messaging services prior to conducting the interviews.

Interviews can be conducted using a private instant messaging service, such as Skype. During the development of this method, participants preferred the instant messaging tools that were available through a web link, rather than as software that they needed to be downloaded on to their computer. The software option provided more effort from the participants and often required more assistance before the interview. Instructions need to be provided on their use in case this is not a tool the participant has used before. These instructions should also provide details on the use of emoticons (☺☹), so they can choose to use these to further express themselves during their typing.

To expand anonymity for the participant and so that interviewees do not need to use their own private instant messaging accounts, interviewers should set up an interviewee account. Interviewees can then be provided with the username and password to this independent research account. This can either be multiple accounts, one for each participant, or the same account can be used, but it is important to ensure the instant messaging service does need to keep a record of previous contact with that person; if so, this history needs to be deleted after each interview (ensuring you have saved the transcript). It is recommended that the interviewer is available online prior to the interview start time. Often, participants found this useful as they would sign in early to check that the login worked and ask any questions they had about the instant messaging service or the research.

Box 2 Example Instructions for Participants to Use Instant Messaging
Go to the website: www.skype.com
  Click on the option "use skype online."
  You should now reach a sign in page. Use these login details:
  Username: participant5000
  Password: Password5000
  Press the button to sign in and continue

  Click on "get started."
  This will open your profile as the participant, and you should be able to see "Researcher" as your contact.
  Note: It is recommended to provide print screen pictures so that the participant can double check if they are progressing correctly.

Participants should have been assigned an individual ID code during the consent stage, so that when they logon, the researcher can ask them to confirm this. The time the interview takes should be estimated as twice the length of a face-to-face interview. This allows for people to type their responses, not only because it takes longer to type than speak, but also because typing allows people to think more and generally tend to come up with longer answers. The data is text-based and, therefore, automatically transcribed, and can easily be copied, pasted, and saved into other electronic documents. As a result, if researchers wish to send the transcripts to participants, they can do this quickly. It is also optional following the interview if participants are given the opportunity to provide any further information they think of via email or another instant messaging conversation, which can then be added to the transcript.

> **Box 3 Example of the Beginning of a Synchronous Text-IM Interview**
> *Researcher says:*
>
> - Hello, it's Gemma Pearce, thank you for being part of this study ☺
> - Before we start, please can you confirm your ID code (the one that you put on the consent form)?
>
> *Participant says:*
>
> - Hi, no worries, I am signing in from home and I have my cup of tea, so I am ready to go!
> - Yeah, it is HJ011962Epsom
>
> *Researcher says:*
>
> - Perfect, thank you. Do you have any questions?
>
> *Participant says:*
>
> - No, I am happy with everything, thank you.
>
> *Researcher says:*
>
> - Great, if you need to leave the computer at any point (e.g., to pop to the loo or make yourself another cuppa), that's not a problem, just type 'brb' (stands for be right back).
> - Are you ready to begin?
>
> *Participant says:*
>
> - Oh thanks, that is useful to know....I might have to let the dog out at some point.
> - Yep, I am ready ☺

4 Method Preference

When this synchronous text-based online interviewing method was tested in the research with menopausal women, participants were asked which interviewing method they would prefer; 76% preferred the text-based instant messaging technique discussed in this book chapter, 12% would have preferred the face-to-face or webcam interview technique, and the remaining 12% had no preference and were

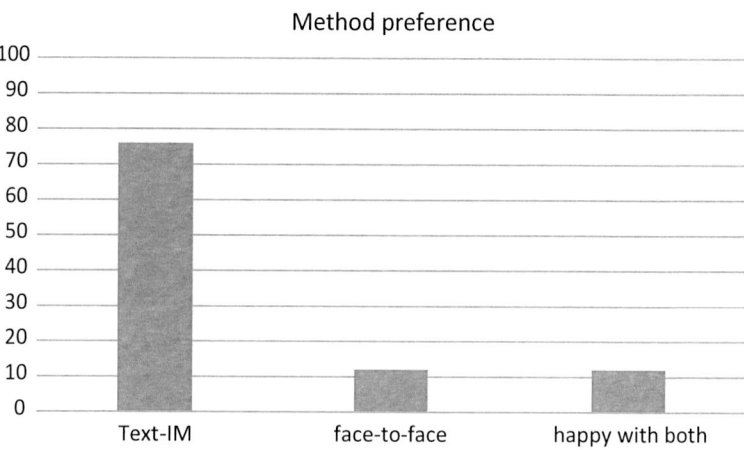

Fig. 1 Strengths and weaknesses of the method

happy using any (Fig. 1) (Pearce et al. 2014). This participant group were people who had already volunteered to participate in a synchronous online interview-based study. So, the potential for bias should be acknowledged, as those that did not want to take part were not included in the collection of this feedback. The participants put forward the suggestion that a range of methods could be offered to participants allowing them to choose their preference.

The lack of body language of voice tone meant that sentences may have been interpreted more at face value by both the interviewer and interviewee. However, as with face-to-face interviews, the interviewer could use reflexive prompts to double check they had interpreted the sentence correctly to how the interviewee intended it and encourage the interviewee to go in to more detail. Often, the participants used metaphors and similes and/or used the emoticons to help them express themselves, which often provided the researchers with rich, unique and interesting quotations about their individual experiences. Clearly, emoticons do not replace what is lost from interviews without body language, but the benefit from the additional cloak of invisibility (not seeing them or hearing their voice) provided by the tool may enable the participants to be themselves more and provide a rich interview in a different way.

During the actual process of the interview, interviewees needed more time to type their answers to the questions compared with a face-to-face interview. This allows a level of reflexivity not previously available during a synchronous interview. Participants can reflect critically on their narratives and interact in an interpretative interview developing a greater understanding of their experiences (James and Busher 2006). In addition, this enables the interviewer more time to ensure they were asking all intended questions in the interview schedule. It also allowed the researcher to double check that all relevant or unclear issues had been examined in depth. For example, sometimes an interviewee will answer a question with multiple interesting sentences, and the interviewer is spoilt for choice as to which sentence they explore first, which in turn can lead to more interesting topics. The online interview allowed the interviewer to double check that when each of these came to a natural end that other unexplored sentences could then be examined further. This was considered an advantage from the researcher's perspective, giving the feeling that the interviews were of rich quality (Pearce et al. 2014).

A limitation we found while using online methodologies is that some participants chose not to participate in the research due to unfamiliarity with technology. However, those more familiar with communication using technological tools may prefer this form of interview. Either way, the potential for bias in interview sampling and the methods used must be considered. A potential solution may be to offer a range of interview methods to the participants, when this is appropriate to the research topic. Logistically, the text-IM interview can be carried out from any computer with the internet, allowing the interview times to be much more flexible. The interviews could be carried out by the researcher from home or work and were offered to participants at their convenience any time of the day, any day of the week. Participants were told that if they wanted to go for a toilet break or to go and get a drink during the interview then they could, they just had to say they would "be right back" or "brb" for short. Many participants said that they liked the flexibility of this. As researchers, we viewed this as an advantage of the method, as participants were able to feel more in control of the interview (Pearce et al. 2014).

The interview transcript is produced as a result of the synchronous text-IM interview, and therefore, this saves manually typing up the transcription afterwards. Although this is positive as it can save time and money, it can also be considered a disadvantage as through the process of transcription a researcher can familiarize themselves with the data during a stage of their analysis. However, as some research teams pay assistants to transcribe their interviews for them, this replaces that need. Additionally, this enabled the transcript to be sent to the participant soon after the interview, while it is still fresh in their mind, providing them with an opportunity to add anything they felt they had forgotten to say in the actual interview.

Box 4 Strengths
- Conversational form of text-based online interview (synchronous).
- Out of the participants who have used this text-IM method, the majority preferred it to face-to-face interviews.
- Provides more time for participant and researcher to reflect during the interview – potential for more in-depth interview.
- Emoticons can be used to help interviewees and interviewers express themselves. Although this does not replace personal interaction, it does provide a useful tool for online text-based communication.
- Reduces the interviewer effect on the participant.
- Allows for researcher to scroll through prose during interview, and explore any outstanding topics or clarify any areas.
- Flexibility with where and when interview can be carried out – greater convenience. More comfort for the participant can make them feel more at ease to open up to interviewer.

(continued)

Box 4 Strengths (continued)
- Interviews can be carried out internationally, conveniently, and without cost (although language barriers may need to be addressed with international studies).
- Interviewees were more in control of their interview, such as they were able to leave the computer during the interview, and choose time and place – potentially more likely to open up to interviewer.
- Interview is transcribed as a result of text-IM interviews and so this saves transcription time and cost.
- Transcripts are available straight after interview and can be sent straight to participant or reviewed by the researcher.
- Opens up research to previous participant groups that may have been hard-to-reach or communicate with, such as people with hearing difficulties.
- Participants who prefer to use technology to communicate may find this an easier form of interview.
- Can be provided on its own or offered as an option of interview methods to the participants.

Box 5 Weaknesses
- Lack of body language and tone of voice.
- Text-based interview takes longer than verbal interview.
- If researcher prefers transcription to help them absorb and reflect on data, then they are unable to do this.
- Participants who do not like technology may not participate – this may introduce a potential source of bias.
- Adds the impact of technology to the interview dynamic.
- Researchers need to familiarize themselves with instant messaging services before conducting interviews.
- An instant messaging research account should be set up for participants to use (so they do not need to use their personal accounts).
- Instructions need to be made available for participants on how to use the instant messaging service, log in, and use emoticons.

5 Conclusion and Future Directions

This chapter has aimed to present the reader with an explanation of the use of the synchronous text-based online interviewing method and provide a discussion on its strengths and weaknesses to support future researchers in deciding this methodological tool is a suitable one to use for their research.

The veil of opaqueness from not being able to see or hear each other during the synchronous text-IM interview is an aspect that is usually impossible in synchronous qualitative methods. The synchronous text-IM tool ensures a degree of confidentiality and anonymity, while still gaining depth of enquiry, where other qualitative methods potentially risk invading a participant's privacy. While body language and other nonverbal cues are useful in developing rapport with the interviewee, it may also ruin it by undermining their sense of a nonjudgmental confidante and provides the risk of the interviewee being more conservative because of their biases towards the interviewer.

This methodological tool facilitates a level of invisibility for the participant, alongside the provision of more choice and convenience in terms of location and travel (Moloney et al. 2004). The synchronous text-IM interview is a convenient, flexible, and encouraging method, allowing participants to feel comfortable and relaxed so that they could open up and discuss personal matters. Where useful and relevant, the participant cannot only be anonymous in the research report, but to the researcher as well. However, it should also be acknowledged that there is the risk of embodied dislocation if participants seize the internet-mediated interaction as an opportunity to deliberately misrepresent themselves. The veil of opaqueness the text-IM method provides can enable the participants with the freedom to manage the image (Goffman 1959; Jones 1964; Leary 1996) they project to the interviewer without the accountability of identification. This may be less likely if participants are approached as the experts of their own experience and that their real-life experiences make a valuable contribution to the research topic. Participants understanding the importance of honesty and openness about their experiences in the research helps to reduce the risk of social desirability bias and increases validity (Dickson-Swift et al. 2007).

The inability to interpret each other's body language, tone of voice, and face-to-face facial expressions with text-IM interviewing was seen as both an advantage and a disadvantage by the researchers and participants (Pearce et al. 2014). Although the text-based interview does not replace the input of body language and tone of voice in face-to-face conversation, it does provide an additional layer of invisibility that other interviews cannot provide. The text-IM method is a great improvement on open-ended questions in surveys and email interviews due to the increase in conversational fluency and the ability to express oneself. Additionally, research that has assessed emoticon use, social interaction, and online message interpretation has concluded that emoticons can reduce misinterpretation of messages by emphasizing or clarifying the implied tone in a similar manner to facial expressions in face-to-face communication (Derks et al. 2007, 2008a, 2008b, 2008c).

The question of how researchers should interpret these emoticons still remains, and this will largely depend on the epistemology of the research, for example, conversation analysis is not possible. Emoticons do not compare to the complexity of facial expression, but they do provide the researcher with further clues as to how the participant feels about the comments made, allows the researcher to adjust their reply as a result, and aid rapport building. For example, if the participant tells a joke and puts a smiley face ☺, then the researcher can smile as well helping the participant to feel more at ease. Alternatively, if the participant shows a sad face ☹, then the researcher can take more caution and react empathetically.

Instant messaging is a promising methodological tool to use in a variety of communication styles for interviews conducted globally with a variety of populations and subject matters. There is much scope for further research to examine these media and allow researchers to offer more choice and comfort to their participants regarding the interview environment and level of anonymity. There is the potential for a reduction of biases between participant and researcher on some topics, such as genital mutilation or other sensitive cultural topics. It is important for researchers to consider the potential self-presentation and impact of the research tools and group of participants being researched before designing their studies.

Box 6 Summary of the Use of Synchronous Text-Based Online Interviewing

What is it?

– A method of interviewing participants online using an instant messaging service to type to each other at the same time in a conversational style.

When would you use it?

– To carry out a qualitative piece of research using interviews.
– It has also been suggested for areas of program evaluation that are sensitive, such as topics to do with resource allocation of individual members of staff (Gruba et al. 2016).

Why would you use it?

– To provide participants with an additional choice of interview method.
– To provide an additional level of invisibility (cannot see or hear each other) to help participants feel at ease and open up more, potentially increasing the validity.
– To provide a nonconfrontational interviewer–interviewee relationship and potentially encourage participants to express their thoughts more honestly.
– To provide an additional level of anonymity to participants if the research topic is sensitive, associated with embarrassment or shyness, or on legality or crime. Also if the interview may impact self-presentation or may be affected by interviewer.
– To carry out interviews further than you can travel to, while also reducing time and costs, while increasing the convenience for the participant.

With whom?

– This method could be offered to any participant groups, but may specifically be useful to offer to these specific populations.

(continued)

Box 6 Summary of the Use of Synchronous Text-Based Online Interviewing
(continued)
  Participant groups:

- Who may prefer communication online
- Who are hard-to-reach
- Who may be affected by the interviewer–interviewee relationship, such as children
- With hearing difficulties or that may prefer nonverbal communication (Benford and Standen 2011; Ison 2009)

Where?

- The participant can choose a location that is convenient to them independent to the researcher's location.

How?
Some key points include (while following ethical guidelines):

- The researcher should familiarize themselves with the instant messaging service.
- Set up an instant messaging account for participants and provide them with a login and password.
- Provide instructions to participants on using an instant messaging service.
- Provide guidelines on emoticons.
- Before the interview, the researcher can log on prior to the start time to provide any support needed from the participant.
- Ensure the interview prose is saved at the end of the interview before closing the instant messaging window.
- Make sure that the instant messaging service does not store the conversation history to ensure data is only saved in a secure location.

References

Benford P, Standen PJ. The use of email facilitated interviewing with higher functioning autistic people participating in a grounded theory study. Int J Soc Res Methodol. 2011;14(5):353–68.

Brewer M. Research design and issues of validity. In: Reis H, Judd C, editors. Handbook of research methods in social and personality psychology. Cambridge: Cambridge University Press; 2000. p. 11–26.

Derks D, Bos AER, von Grumbkow J. Emoticons and social interaction on the internet: the importance of social context. Comput Hum Behav. 2007;23:842–9.

Derks D, Bos AER, von Grumbkow J. Emoticons in computer-mediated communication: social motives and social context. Cyberpsychol Behav. 2008a;11:99–101.

Derks D, Bos AER, von Grumbkow J. Emoticons and online message interpretation. Soc Sci Comput Rev. 2008b;26:379–88.

Derks D, Fischer A, Bos AER. The role of emotion in computer-mediated communication: a review. Comput Hum Behav. 2008c;24:766–85.

Dickson-Swift V, James E, Kippen S, Liamputtong P. Doing sensitive research: what challenges do qualitative researchers face? Qual Res. 2007;7(3):327–53.

Dillman D, Smyth J, Christian L. Internet, mail, and mixed-mode surveys: the tailored design method. Hoboken: Wiley; 2009.

Dockrill J, Lewis A, Lindsay G. Researching children's perspectives: a psychological dimension. In: Lewis A, Lindsey G, editors. Researching children's perspectives. Buckingham: Open University Press; 2000. p. 46–58.

Drew I, Sainsbury R. Mode effects in qualitative interviews: a comparison of semi-structured face-to-face and telephone interviews using conversation analysis, Research Works. York/Heslington: University of York/Social Policy Research Unit; 2010.

Feveile H, Olsen O, Hogh A. A randomized trial of mailed questionnaires versus telephone interviews: response patterns in a survey. BMC Med Res Methodol. 2007;7:27. https://doi.org/10.1186/1471-2288-7-27.

Goffman E. The presentation of self in everyday life. Garden City: Doubleday; 1959.

Gruba P, Cárdenas-Claros MS, Suvorov R, Rick K. Blended language program evaluation. Hampshire: Palgrave Macmillan; 2016.

Hanna P. Using internet technologies (such as Skype) as a research medium: a research note. Qual Res. 2012;12:239–42.

Hu Y, Wood J, Smith V, Westbrook N. Friendships through IM: Examining the relationship between instant messaging and intimacy. Journal of Computer-Mediated Communication 2004;10, Article 6. https://doi.org/10.1111/j.1083-6101.2004.tb00231.x.

Im E-O, Liu Y, Dormire S, Chee W. Menopausal symptom experience: an online forum study. J Adv Nurs. 2008;62(5):541–50.

Ison NL. Having their say: email interviews for research data collection with people who have verbal communication impairment. Int J Soc Res Methodol. 2009;12(2):161–72.

James N. The use of email interviewing as a qualitative method of inquiry in educational research. Br Educ Res J. 2007;33:963–76.

James N, Busher H. Credibility, authenticity and voice: dilemmas in online interviewing. Qual Res. 2006;6:403–20.

Jones EE. Ingratiation: a social psychological analysis. New York: Appleton-Century-Croft; 1964.

Jowett A, Peel E, Shaw R. Online interviewing in psychology: reflections on the process. Qual Res Psychol. 2011;8:354–69.

Lambrecht H. Tackling low learning outcomes in South Africa: the contribution from informal mobile learning. Unpublished doctoral thesis: Lancaster University. 2015.

Leary M. Self-presentation: impression management and interpersonal behaviour. Oxford: Westview Press; 1996.

Lewis A. Children's understanding of disability. London: Routledge; 1995.

Meho L. E-mail interviewing in qualitative research: a methodological discussion. J Am Soc Inf Sci Technol. 2006;57:1284–95.

Moloney M, Strickland O, Dietrich A, Myerburg S. Online data collection in women's health research: a study of perimenopausal women with migraines. Natl Womens Stud Assoc J. 2004;16:70–92.

Murray C. The social meanings of prosthesis use. J Health Psychol. 2005;10:425–41.

Murray C, Sixsmith J. Qualitative health research via the internet: practical and methodological issues. Health Inf J. 2002;8:47–53.

O'Connor H, Madge C. Cyber-mothers: online synchronous interviewing using conferencing software. Sociol Res Online. 2001;5. Retrieved from: http://www.socresonline.org.uk/9/2/hine.html.

Pearce G, Thøgersen-Ntoumani C, Duda J, McKenna J. The menopausal transition in women who have had a hysterectomy: from embarrassment to liberation. In: Paper presented at the Midlands Health Psychology Network Conference: Coventry University; 2010.

Pearce G, Thøgersen-Ntoumani C, Duda J, McKenna J. Changing bodies: women's embodied experiences of the menopausal transition. In: Paper presented at the Division of Health Psychology Annual Conference: British Psychological Society, University of Southampton; 2011.

Pearce G, Thøgersen-Ntoumani C, Duda J. Changing bodies: symptoms, body image, health and wellbeing over the menopausal transition. Unpublished doctoral dissertation. University of Birmingham, Birmingham. 2013.

Pearce G, Thøgersen-Ntoumani C, Duda J. The development of synchronous text-based instant messaging as an online interviewing tool. Int J Soc Res Methods. 2014;17(6):677–92.

Reardon R, Grogan S. Women's reasons for seeking breast reduction: a qualitative investigation. J Health Psychol. 2011;16:31–41.

Seymour W. In the flesh or online? Exploring qualitative research methodologies. Qual Res. 2001; 1(2):147–68.

Smith J. The search for meanings: semi-structured interviewing and qualitative analysis. In: Smith J, Harre R, Van Langenhove L, editors. Rethinking methods in psychology. London: SAGE; 1995. p. 9–26.

Stacey K, Vincent J. Evaluation of an electronic interview with multimedia stimulus materials for gaining in-depth responses from professionals. Qual Res. 2011;11(5):605–24.

Asynchronous Email Interviewing Method

Mario Brondani and Rodrigo Mariño

Contents

1. Introduction 1386
2. Email as a Communication and a Research Tool 1387
3. Nine Steps to Administer Electronic Interview Via Email 1390
4. Our Experience in Using Email as an Interview Tool 1395
   - 4.1 Delayed Responses and Absence of Punctuation 1396
   - 4.2 Indirect Contact and Body Language 1396
   - 4.3 Validation 1397
   - 4.4 Informality 1397
   - 4.5 Sampling Possibilities and Cost-Effectiveness 1398
   - 4.6 Conclusion and Future Directions 1399
- References 1399

Abstract

This chapter explores the potential use of Internet-based communication applications (e.g., emails) as a method for gathering qualitative research data. In the era of globalized multimedia and *at-finger-tips* convenient information, electronic communication can provide answers to research inquiries in a timely manner, particularly in cases where the researcher is not required to meet face to face with the participants, or there is not need for audio-record the interview or conversation. We offer a nine-step process on how to administer an electronic interview, from selecting potential participants, interacting with them electronically, to closing the electronic encounter. We discuss the advantages and disadvantages

M. Brondani (✉)
University of British Columbia, Vancouver, BC, Canada
e-mail: brondani@dentistry.ubc.ca

R. Mariño
University of Melbourne, Melbourne, VIC, Australia
e-mail: r.marino@unimelb.edu.au

of such means while drawing from a brief experience in using email to interview older adults for a research study on aging partially published elsewhere. We use the existing literature to explore the benefits and limitations of email as a research tool. We close the chapter by inviting the reader to ponder about other data collection tools in today's evolving research arena as an alternative mean to conference calls or face-to-face interviews when time and resources are restricted.

Keywords
Email · Electronic mail · Communication · Research methods · Data collection · Qualitative research

1 Introduction

It is well agreed upon that a research question should be central to, and informative of, the best fitting research methodology. Traditional ways of collecting qualitative data has been composed mainly of audio-recorded face-to-face interviews and focus group discussions with or without video recording, or written field observations or responses to predetermined questions. Although the analysis of documents, memos and records is also used, in the era of electronic communication, however, conducting an interview via Internet-based communication applications (e.g., emails) can be convenient to both the interviewer and the interviewee, while obtaining the answers in a timely manner (Ahem 2005; see also ▶ Chap. 78, "Synchronous Text-Based Instant Messaging: Online Interviewing Tool"). In fact, electronic communication as the exchange of *meanings and information* between individuals via a common system of symbols has experienced a dramatic boom for the past two decades, not only in terms of developmental technology but also in terms of global presence (Charness et al. 2001; Adams and Neville 2012). Computer-assisted data collection as a research tool facilitated by the Internet has been used widely in product development and commercialization of goods, as well as in healthcare and sociology (de Leeuw and Nicholls 1996; Mann and Stewart 2003). Hence, the internet has experienced exponential growth as part of commerce and business (Simeon 1999; Whyte and Marlow 1999), politics (Plouffe 2009), dental and medical education (Mariño et al. 2012; Arnett et al. 2013; Cheston et al. 2013; Roy et al. 2016), and as a venue for both public and individual health promotion messages (Wong et al. 2008; Scanfield et al. 2010). In fact, more than three billion people currently use the Internet (Internet Live Statistics 2015). This boom has prompted some researchers to refer to the Internet as a cultural entity on its own, with its own environment and characteristics (Burkhalter 1999; Constantinides et al. 2010). The Internet may, for example, be used as "therapeutic" by and for older adults (Melenhorst and Bouwhuis 2004), provide support for caregivers of people with Alzheimer's disease (Alzheimer Society of Canada 2009), foster health promotion initiatives (Mariño et al. 2013), and may be a chosen environment to engage marginalized and vulnerable minority groups into research (Neville et al. 2015) (see also ▶ Chap. 78, "Synchronous Text-Based Instant Messaging: Online Interviewing Tool").

In turn, social media and the Internet are part of modern life where most individuals access electronic information or communicate using these means daily, almost anywhere at anytime. Although social media is most commonly used to facilitate communications, these technologies are having the same impact on research as they do in daily life with a number of existing mobile applications tailored to research (Merlien Institute 2016). A shift in the way one sees research being conducted is already occurring, which suggests that a move toward a growing number of alternative technologies to conventional pencil-and-paper or audio and video recorded qualitative data, for example. Technology will always evolve and has already changed the ways in which researchers propose and collect data where reaching participants is not bound to geographic access or temporality anymore. However, researchers need to be aware of its impact and understand the boundaries and limitations of such technologies.

Given the potential value of email as a research tool for collecting information, this chapter is organized in three parts. In the first part, we offer a more elaborated overview on the use of email as a research method for qualitative data collection with some focus on older adults. We proceed by offering a nine-step process on how to administer an electronic interview, from selecting potential participants to closing the electronic encounter. We then discuss the advantages and disadvantages of such means while drawing from a brief experience in using email to interview older adults for a research study on aging as presented in Brondani et al. 2010. We use the existing literature to explore the benefits and limitations of email as a qualitative research tool. We close the chapter by inviting the reader to ponder about other data collection tools in today's exploratory research arena as alternative means to conference calls or face-to-face interviews when time and resources are restricted.

2 Email as a Communication and a Research Tool

Associated with the Internet, the use of electronic mail (email) has increased dramatically as an effective tool of communication over the traditional *pen and paper* format (Meho 2006). Its popularity as a research tool has also increased and there is growing interest in assessing its effectiveness as such (Selwyn and Robson 1998; Benfield 2000; Cook 2012). As a research tool, email is a well-established means of distributing quantitative questionnaires (Mann and Stewart 2003; Meho 2006), interviewing people about their values and opinions (Selwyn and Robson 1998; Flowers and Moore 2003), and engaging marginalized communities (Cook 2012; Neville et al. 2015) and older adults (Brondani et al. 2010) in accessing health care and social services, for example. Nonetheless, email has been criticized as a "digital divide" by Lewis et al. (2005) since some groups may face potential barriers to internet use. It has also been thought to be abstract, impersonal, and insensitive to the nuances of nonverbal behaviors, rapport, and relationships (Melenhorst and Bouwhuis 2004); it is seen by many as a medium more appropriate to youth than to old age (McAuliffe 2003). Consequently, there are questions about the value of email as a sensitive and useful medium for interviewing people about their personal values and opinions (Freese et al. 2006) given that studies on the topic are relatively incipient. Yet (self) stereotypes still exist for some people, including older adults not having technological competence or savvy (Harwood 2007) or Internet being something more applicable to bussines or used as leisure by the youth. Such views, however, seem to be fading away given the widespread use of the Internet (Marx et al. 2002; Hage 2008; Brondani et al. 2010; Cook 2012).

Exchange of information through email requires a certain level of computer literacy that will enable participants to use such mediums efficiently and productively (Etter and Perneger 2001; Mann and Stewart 2003; Brondani et al. 2010). The precipitous increase in the use of computers and email has done much to develop this literacy in all segments of society, and electronic jargon is used widely throughout television, radio, newspapers, and text messaging via cellular (smart) phones. Moreover, the cost of computers has steadily decreased and software programs have become more accessible for use by the general public across the world. Furthermore, the cost of accessing the Internet is not a major financial barrier to most households and free world wide web access is available in many public spaces such as coffee shops and public libraries. As a result, the profile of Internet users has changed, and continues to do so. From a predominance of users being Caucasian men between 35 and 49 years with higher than average incomes, there are now more Asian Pacific females on the same age group using the Internet (National Center for Educational Statistics 2004; Internet Live Statistics 2015). More than 10% of all Internet users are people older than 60 years of age who own personal computers in North America and are online daily more than any other age groups (National Center for Educational Statistics 2004; EMarketer 2010). In fact, approximately 83% of the population in New Zealand use the internet including those who are older, who live in rural areas and those in lower socio-economic groups (Bell et al. 2010). As revealed by the Australian Bureau of Statistics report on Household Use of Information Technology, 79% of Australian households have home Internet access, and 83% of households have access to computers (ABS 2013). Interestingly, from 1998 to 2007, household access to the Internet in Australia has almost quintupled, from 16% to 79%. Moreover, during this period, access to computers increased to more than 80%. Nevertheless, as pointed out by Lewis et al. (2005), there are socioeconomic and regional disparities in household access to computer and the Internet across Australia that contributes to the potential *digital divide* nature of the Internet: households without children under 15 years, located in exmetropolitan or remote areas, and with low-household incomes are less likely to be connected to a computer and/or the Internet (ABS 2013). This situation may be the case in many countries although half of the 40% of the world's population that have access to the Internet are located in Asia. In fact, in emerging economies such as Brazil, about 52% of its population do not have access to the Internet at home despite Brazil being one of the largest internet markets in the world (Statista 2016). On the other hand, countries with a restrictive policy on the use of the world wide web like China may have Internet availability, but have differences compared to most Western countries including Wi-Fi connections not commonly available in hotels, and access to social media networks are somewhat limited (China Highlights 2016).

Either way, when discussing the involvement of older adults in electronic communication, there is a general assumption that they are neither computer literate nor familiar with email (Whyte and Marlow 1999; Harwood 2007), and that they are disinterested in the Internet. Consequently, electronic sampling for research used to be biased toward younger and relatively affluent segments of the population (Flowers and Moore 2003). However, these assumptions have been questioned given the advance of Internet use over the past decade (Harwood 2007; Brondani et al. 2010; Cook 2012). According to an email survey (Johnson and MacFadden 1997), 70% of the seniors who use the Internet claim to have intermediate computer skills, 60% use the Internet to keep their minds active, and more than 50% send and read emails regularly. Similar positive outcomes have been found by Freese et al. (2006) and Burns et al. (2012), whereas others have pointed out that some older adults might consider email to be somewhat impersonal (Melenhorst and Bouwhuis 2004). Nonetheless, among those senior citizens who have gone online, two-thirds believed that the Internet was a reliable source of health information. It has been shown that the Internet can be used as a tool to improve older adult's (and any other age group's) knowledge and awareness of their diagnosis, treatments, and options for care, and this could eventually become a key instrument in health promotion among the aging population worldwide (Kutz et al. 2013).

In Australia, for example, almost three-quarters of those aged 55–64 years (71%) and more than one-third of those older than 65 years of age (37%) reported having surfed the Internet at any location in a month period (home, workplace, house of a neighbor or a friend or a relative, and library) (Australian Bureau of Statistics 2013). Similar findings have been reported in Canada (Statistics Canada 2010), USA (National Center for Educational Statistics 2004), and New Zealand (Bell et al. 2010) possibly because of more leisure time and discretionary income to spend on computers (Pew Internet 2010). In fact, someone who is 65 years old today has witnessed the dramatic growth of the world wide web and the introduction of electronic social networks over the past 20 years (Harwood 2007; Ramsay 2010); social media has also grown exponentially over the same period to the extent that it is now a major form of academic research (Henry and Molnar 2013; Visser 2005). Hence, many of the today's Western older adults who have approached retirement age are likely to have worked with, or been exposed to, an online environment either at the workplace, at home, or in both locations. As Harwood (2007) has discussed, these individuals may feel comfortable with email for interview purposes.

As email seems to be a leading communication tool in both personal and work contexts, it also emerges as a potential means for interview especially if there is no requirement to meet with the participant face to face, or to audio-record the conversation. Email interviews may also be used effectively if there is a need to publish the interviews on a website or other form of digital media, or if there is simply the need to get expert advice on a particular topic. Moreover, the social media has become an increasingly popular means of data collection by advertisement companies and consumer services; it is rapidly becoming an irreplaceable part of research. Its use supplements traditional research methods and can provide comparable results. As researchers already use social media to conduct the various stages of research, email for data collection will only increase its importance while maximizing ways in which data is gathered and analyzed (see also ▶ Chaps. 78, "Synchronous Text-Based Instant Messaging: Online Interviewing Tool," ▶ 76, "Web-Based Survey Methodology," ▶ 77, "Blogs in Social Research," and ▶ 78, "Synchronous Text-Based Instant Messaging: Online Interviewing Tool").

Email is one of the three modes of gathering research data via the Internet. They include: (1) *online interviews*, which require real-time interaction and involve the researcher and the participant simultaneously engaging with a text-based method via "chat" or "instant messaging" software services (see ▶ Chap. 78, "Synchronous Text-Based Instant Messaging: Online Interviewing Tool"). WhatsApp, for example, may offer a valuable tool for research activities beyond personal messaging (https://www.whatsapp.com/). Online interviews generate mostly qualitative data (Jowett et al. 2011); (2) *online surveys*, which pose a series of standardized questions to the participants without real-time interaction with the researcher (see ▶ Chap. 76, "Web-Based Survey Methodology"). Online surveys have been administered traditionally using a template format as a link distributed to participants by email (Braun and Clarke 2013). SurveyMonkey®, for example, offers an online survey development cloud-based platform that is mostly free, and invited participants complete the assigned survey and return it usually by clicking "submit." It also allows users to design the surveys, collect responses, and conduct basic analysis on the data gathered, and (3) *email interview*, which is the focus of this chapter. As such, we will not discuss electronic surveys, texting message systems, or any other form of online and electronic data collection (see ▶ Chaps. 75, "Netnography: Researching Online Populations," ▶ 76, "Web-Based Survey Methodology," ▶ 77, "Blogs in Social Research," and ▶ 78, "Synchronous Text-Based Instant Messaging: Online Interviewing Tool"). An electronic interview via email is much like a conventional interview where the researcher collects information about a participant on a given topic, usually located remotely. Below, we offer a nine-step process to administer an email interview.

3 Nine Steps to Administer Electronic Interview Via Email

In order to properly and successfully administer an email interview and get the most out of it, we suggest the steps below to give readers the tools, although some of these steps are common to conventional face-to-face interviews. Some steps are straightforward while others may need to be adapted to an electronic interview platform including obtaining ethical clearance, obtaining informed consent, and familiarizing with guidelines and legal issues around appropriate research behaviors, and participants' confidentiality and anonymity. What follows are some steps drawn from our experience using electronic interviewing:

- **STEP 1: Prepare your interview questions**
  Before starting your research study, you must develop a list of interview questions for your participant based on the area or focus of your inquiry. Area of interest aside, if

you would like to use email as an exploratory qualitative interview tool, consider having opened-ended questions (e.g., *What are your thoughts on the use of email as a means to interview participants?*) instead of close-ended inquiries (e.g., *How often to you use emails on a given day? ___less than 5 times; ___between 5 and 10 times; ___more than 10 times*).

Have your list of email interview questions ready and be also prepared to adapt or change them depending on the type of answers you are getting (see STEP 4, ahead). As a rule, each question should contain only one inquiry or concept to keep the task clear and to the point. Avoid double-barreled questions. For example, if you are interested in the issue of homelessness, first ask your participant how s/he defines somebody as homeless to then follow-up with a second question about why s/he thinks somebody gets to be homeless. Do not ask both issues (e.g., defining homeless and getting to be homeless) in one question. Do not assume that your interviewee knows the concepts or the definitions you are inquiring about. You must also consider your participants' background (including writing and language skills). There are many published manuscripts, reports, and books as well as online information on how to properly develop and frame your interview questions, so we will not discuss this topic here (see Liamputtong 2013; Serry and Laimputtong 2017, for example).

- **STEP 2: Select your participant(s)**
  Based on your research area or topic, make sure your target participants can give valuable information. For example, if you would like to interview an animal activist for your class project, you would be more successfull in contacting the names of individuals under animal rights or vegan food websites than names listed as reviewers of restaurants. If possible, use the internet to research your potential participant(s)' background(s) prior to contacting. This will maximize the possibility of having a better fit between your topic of inquiry and their knowledge and potential interest to be part of your study . This background exercise will also provide you with the insight you need to tailor your questions accordingly. For example, when interviewing a paiter, research his/her background and accomplishments, and the types of paitings (it would help if you have seen them) to learn about their genre (modern, abstract, etc) or area (human, construction, nature, and so on), and content. Also, check on the reviews of the paitings, if available, to build your argument if that is the intention of your interview.

  Also consider issues of ethics and confidentiality. Within a research environment like an academic institution or university, you must obtain ethical approval from your Research Ethics Office/Board before commencing your research. That will make sure you adhere to ethical practices and principles that are acceptable and protective of the participants.

  Lastly, ponder about this question: *Is your potential participant available and willing to be interviewed via email?*

- **STEP 3: Contact your participants before administering the email interview**
  It is important to contact your potential participants(s) before administering the actual email interview from the get-go. This will not only allow you to introduce

yourself and/or your organization, but also to give you the opportunity to explain the reason(s) for the interview, gauge their interest and availability, and so on.

Set the tone of the first contact to be as *inviting* as possible while being polite yet straightforward. Make sure you write a friendly invitation on your subject heading: "*Your ideas are needed,*" or "*Potential participant for an interview,*" or "*We would like to hear what you think.*" Do NOT leave it blank. Remember that in times of electronic environments and busy lives, people may delete your email without even opening it. You must get their attention and interest "right-a-way."

Try to also do a good job in explaining how you came across their names and contact information so that your potential participants feel more at ease with you and with the request for an interview. Also, give them some ideas about you, your study/research, and the interview process itself. For example, consider something along the lines of:

'Greetings, Mr. Mario!

I'm Augusto from the University of British Columbia in Vancouver, Canada. I came across your name and contact information from your website/company/organization. I'm currently looking for people with your expertise, and I would like to know if you would be interested in being interviewed by email. I have 5 questions on the topic of _____ that would not take more than 20 minutes of your time to respond to. You can be brief or write as much as you want to. Your answers will be confidential to me.

If you are interested, please reply to this email in the next 5 business days, so I can send you the questions and provide you with more information about the study.

Looking forward to hearing from you,

Sincerely,

Augusto, Research Assistant for Gem Laboratories

The University of British Columbia

JBM 122/2199 Wesbrook Mall

Vancouver, BC, V6T 1Z3

P: _____ / F: _____

Also, there is a possibility that your email will end up in your interviewee's junk mail folder. Allow a week (5 business days) to send another first-contact email in case you do not hear back. There is also the possibility that the recipient of your email is not at all interested in your request.

Lastly, make sure you close the email with your contact information (and credentials) or have an email signature automatically attached to it with that information. Avoid ending the email just with your full name or worse, just with your nick name!

- **STEP 4: Pilot the interview questions**
  It is always a good idea to pilot your interview questions before you send them to your interviewee to check for clarity, comprehensiveness, and readability. You can approach a couple of coworkers or friends for their help and email them the interview guide. Ask them to not only answer the questions, but also to give their written feedback on the questions themselves in terms of wording, syntax, and meaning. If feasible, you can have a face-to-face chat with them to discuss the questions and take their feedback.

  This will save you time and hassle substantially since you do not want to start collecting your data realizing that your participants did not understand what you were actually asking about. Even worse is to realize this after all the textual data has been collected.

- **STEP 5: Interview your participant**
  Once you received the "green" light from your interviewee to go ahead (and you have proper ethical approval where needed), start your interview email by giving information regarding the nature of the interview and the study behind it. Here, you have the opportunity to offer further yet brief insights as to what and why you are conducting such interviews. Think about *"why would the topic of the interview be relevant to your participant?"* You may describe the benefits of exploring such a topic, for example, or the need to know more about it.

  Consider using the body of the email to pose your interview questions rather than as an attachment. People may not receive the attachment (in any format), may have trouble opening it, or may not want to open it.

  Although the number of questions may vary, avoid more than 10 open-ended questions per communication (think about 5 questions perhaps) and provide your participants with information about the length of the email interview. Start by writing 1 or 2 general questions to begin the interview, then continue into more specific questions or topics as the interview flows. For example, begin by asking a chocolatier why they chose that as a career, then ask them additional questions specifically about their line of chocolates, the process of making it, types of cocoa used, and the locations such chocolates are being sold.

  Send your interview questions all at once to avoid multiple emails (and annoying your interviewee). You can always ask for clarification on a follow-up email (see ahead). Ask your participants to take the time to answer your interview questions, but give them a timeline to send their answers back. Usually, a week or 5 business days will suffice, although you may consider an extension if they want to collaborate but need more time, or if the nature of the subject you are inquiring about requires further thinking and consideration. In all, as mentioned in STEP # 3 always provide your participant with a deadline; otherwise, you may be receiving the answers long after the research is done. Besides, a deadline often ensures that the participant completes the email interview on time, especially if there is strict deadline. Also advise your participant that if you do not hear back in the next 5 days (roughly 1 week), you will send a reminder. Ideally, you should aim to receive the answers prior to the deadline you gave them, but there is always the need to give your participants more time if that will allow them to give you more meaningful answers.

Lastly, let your participants know that, if necessary, you might send a couple of additional or clarification questions in order to fully explore the topic. This request should not come as a surprise to your participant: make sure you have this written when contacting them as per STEP # 3. We would caution you on an interview process that extents for more than two emails (the interview itself, and the follow-up clarification if needed) as people are busy and you do not want to annoy, or burn them out.

- **STEP 6: Be thankful to your participant**
  After the email interview has been completed, make sure you thank them for their time and effort. Your *thank you* can be in the form of an email or telephone call, and should convey a message along the lines of:

   Subject heading: *Thank you*, or *Gratitude*, or *We appreciate your help*

   Dear Mr. Mario,

   We would like to express our sincere gratitude and to thank you for your participation in your interview process. It was very helpful to get your insights on the questions we posed.

   Have a great day,

   Cheers,

   Augusto, Research Assistant for Frontier Laboratories

   The University of British Columbia

   JBM 4561/2199 Wesbrook Mall

   Vancouver, BC, V6T 1Z3

   P: _____ / F: _____

  Depending on the budget of your study, it is not uncommon to acknowledge participation by offering a token. Such acknowledgment can take many forms and be either a gift card, money order, a discount at a particular store, or a change to enter a draw for a given product. In fact, it is not uncommon to mention or advertise this token when you first approach your potential participants, usually within the head of your email. Using the heading examples from STEP # 3, you could reword them as: "*Your ideas are needed: enter to win a _____*," or "*Make $____ while being part of an email interview*," and so on.

- **Step 7: Save your interview material on a folder outside your mail box**
  As you receive your emails, do not leave them in your mail box. Electronic files are notoriously insecure, which poses a challenge to the confidentiality of your research responses, especially when computers are connected to a shared local network. To overcome this problem, you may consider using a password-protected file in your computer desktop to store communication from our

participants. This would store your data within your computer hard drive rather than on the Internet server itself. Alternatively, you may also save your email correspondences on a password-protected folder within your hard drive or external device. Either way, it is also wise to have backed up data in case your main computer crashes.

- **Step 8: Editing and analysis**
  Edit the email interview answers if needed. In most cases, especially if you are submitting the interview questions and answers to your boss or publishing the content to a website, you may need to make certain grammar and punctuation edits. Sometimes, you may need to reword the answers given in a manner that matches the style of your readership or publication, but without changing their content and meaning. Other times, you may leave "as is" if punctuation, unusual grammar and syntax, and colloquial expressions used are relevant to, or the focus of, your study. It also gives a more accurate idea of how the responses to your questions were written and interpreted.

  It is not unusual to review any major editing changes with your participant(s) prior to publishing the interview. For example, if you feel the need to edit a specific quote provided to you by the participant(s), contact the subject(s) before publication to clarify that you have their permission to edit their quote. That would avoid the "*he said, she said*" scenario and worse, any litigation about what was said that way it was said. But more often than not, participants' names are seldom used unless you got their approval to do so.

- **Step 9: Make yourself available**
  Provide your participants with information about yourself and your study. Also give them the time to research your background, if necessary. Some interviewees may want to verify your identity and credibility before they answer any of your interview questions, especially if the questions are more personal in any way. For example, provide the subject with links to other interviews you have administered by email and published on the Internet, or any of your work that you think is relevant.

  Make yourself available even after the study has been done, in case there is a need to follow up or give the participants feedback. Use different means for your participants to reach you if desired, not only via email but also by phone call or face-to-face interaction.

  Please note that this chapter does not present or discuss ways to analyze the information gathered or issues of anonymity and confidentiality of your participants. Nonetheless, we offer the following session to contextualize the use of email as a research tool within a brief study on aging involving older adults, and we may refer to analysis and confidentiality to illustrate a point or two.

4 Our Experience in Using Email as an Interview Tool

Given the nine-step process above, we now provide you with a critical evaluation of email as a research tool within a qualitative pilot study of feelings and experiences associated with aging as published elsewhere (and adapted in here from Brondani et al. 2010). The participants we encountered, an Italian-speaking man and a Portuguese-speaking woman, offered to participate only if we would communicate with them solely by email; we had not planned to use such a venue initially. After obtaining ethical approval from the University of British Columbia's Office of Research Services, we had a face-to-face meeting on obtaining signatures for informed consent (you do not necessarily need to do this). We then proceeded to communicate electronically with the two participants. Throughout that process, we observed several salient isues as presented below.

4.1 Delayed Responses and Absence of Punctuation

Most of the information was exchanged in a "received and answered format." Although there was no mandatory deadline for a reply or response (which we should have established), we received their responses within 72 h, as is suggested in the literature as sufficient time for a thoughtful reflection and response (McAuliffe 2003; Meho 2006).

Although punctuation is regarded as either a representation of a spoken word to convey intonation, duration, or stress or an integral part of the syntax, both participants emailed us back with responses using irregular or random absence of grammatical punctuation: either commas or periods. The absence of grammatical pauses is not uncommon within an online environment such as instant messenger and text services, particularly commas and full-stops when space is limited or cost per text is an issue. Even though such apparent freedom in punctuation rules can confuse the meaning of a sentence, the informality and ease of interactions make it relatively easy for us to seek clarification when needed (Flowers and Moore 2003; Meho 2006). More often than not, punctuation is avoided at best, or misplaced at worst, in transcribed texts from regular face-to-face interactions to the extent that free software (Ginger® Software) and statistical methods have been proposed to assert that the text is correctly punctuated (O'Kane et al. 1994). However, third-party initiatives like these can unintendedly change the meaning of what was originally conveyed, or change the significance of an entire sentence. Either way, it remains in the interpretation of the individual analyzing the textual information to make sense of sentences or excerpts given the context in which they occur.

4.2 Indirect Contact and Body Language

Forget about human interaction. There is a standardized online conversational behavior called "Netiquette" among most Internet users that substitutes paralinguistic cues and nonlinguistic body language with specific symbols to express feelings, emotions, and sentiments. For example: ":)" indicates a happy face; ";)" indicates a wink; and "lol" indicates a laugh-out-loud (King 1996; Selwyn and Robson 1998; McAuliffe 2003; Shea 2004). In fact, there has been a proliferation of emoticons (An *emoticon*, "etymologically a portmanteau of emotion and icon, is a metacommunicative pictorial representation of a facial expression that, in the absence of body language and prosody, serves to draw a receiver's attention to the tenor or temper of a sender's nominal nonverbal communication, changing and improving its interpretation." (Wikipedia)) available in *smartphones* and in almost any online communication (Wikipedia 2015); they even became a movie in 2017 - The Emji Movie by Sony Pictures Animation. Our two participants seemed comfortable with the focus of our enquiries, and, with no hint of distress or difficulty, occasionally used the symbols to tell their stories more emphatically. Apparently, we created an easy, trusting, and friendly relationship as evidenced by the informality and general tone of our exchanges. This easiness in online communication has also been documented by others (Melenhorst and Bouwhuis 2004; Reisenwitz et al. 2007).

Despite the availability of symbols to express emotions and compensate for the lack of visible body language, email as a one-dimensional textual data continues to pose limitations on the detection and interpretation of emotions (Haythornthwaite 2000), especially if that is a part integrant of, or informant to, the research being conducted. High-resolution web cameras with voice and video capabilities would probably help to overcome these limitations. In addition, it can be difficult for interviewers to acquire the skills needed to probe for responses when the emotional environment of direct human contact is missing (Flowers and Moore 2003).

4.3   Validation

The validation of qualitative data obtained by face-to-face interviews usually occurs while probing the participants during the interview, or later, through *member checks* (Liamputtong 2013; Creswell 2014). Despite the challenges in establishing or even in actually needing member checks, we argue that the same validation could happen for electronic interviewing. In turn, the informality of the interactions offered us the opportunity to validate and enhance the trustworthiness of data analysis as communications continued (Meho 2006). It should also offer the benefit of being able to easily prompt participants to express further feelings, thoughts and perceptions; it also increases rapport between the interviewer and participant (Patton 2015). Similar to either electronic or in-person interaction, it may be difficult for some participants to respond adequately if the question posed is too short, ambiguous, or unnecessarily succinct and they do not understand it as we have originally planned. Due to the informality of electronic communication, however, we are easily able to provide clarification as often as needed, and to gather the most information from the electronic interaction.

4.4   Informality

It has been suggested that email provides a context for a noncoercive and anti-hierarchical dialogue to promote equal opportunity and reciprocity (Brondani et al. 2010), which constitutes an ideal situation free of internal or external intimidation

(Creswell 2014). From the beginning, our participants adopted informal language when greeting the interviewer. When asked about what a normal day looks like to them, for example, they embellished their responses: "*... my day is really good (and I tell you... it is a way better than I thought it would be) ...,*" and "*first, I have my cappuccino around 8:00 in the morning ... well, actually the coffee here is not as good as in the village where I used to live, but ...*" (adapted from Brondani et al. 2010). These spontaneous expressions provided information similar to that which one would expect on a face-to-face context (Patton 2015), as off-topic comment, or as a way to embellish their remarks.

4.5 Sampling Possibilities and Cost-Effectiveness

E-communication opens up the possibility of sampling on a very large scale globally with relatively low administrative costs (Selwyn and Robson 1998; Harwood 2007; Hackworth and Kunz 2010). It potentially mitigates conventional constraints of spatial and temporal proximity between interviewer and subject, and offers the possibility of a relatively unobtrusive and communicative environment. There is less concern for social hierarchy, and it may decrease the uneasiness caused by a dominant interviewer confronting a shy respondent, or a young interviewer with an elderly subject (Selwyn and Robson 1998; Etter and Perneger 2001). This may be especially relevant when engaging with marginalized and vulnerable participants when a power relationship becomes an issue (Cook 2012; Neville et al. 2015).

There is clear cost-effectiveness from e-interviews because they provide written information directly without the costs of transcribing oral interviews and the textual data is ready to be analyzed (Flowers and Moore 2003). Furthermore, email interviews eliminate the need for tape-recorders and audiotapes, and for specific time, place and travel arrangement, which are required when conducting face-to-face interviews.

Although our participants themselves offered the use of emails, response rates to structured surveys in general have declined over the last few years, either paper format or online (Funkhouser et al. 2014). However, the response rates are also related to the size of the population under study – larger populations require smaller response rates (Nulty 2008) and in the case of a qualitative inquiry, sample size is not a major issue (Liamputtong 2013). In particular, response rates to online surveys about oral health are within the range of 2.5–26% (Goodchild and Donaldson 2011; Henry et al. 2012). Methods for boosting online survey response rates would include extending the period of data collection, repeat reminder emails to nonrespondents, repeat reminder emails to survey owners/coordinators, providing incentives for respondents, and optimizing the use of online environments by utilizing online teaching aids/methods for examples (Nulty 2008). Of note is to make sure that your participants can actually type on a key board or use dictation software and speech recognition tools when physical limitation and dexterity are issues to be considered, which may be the case depending on the nature of your research or target group of participants.

4.6 Conclusion and Future Directions

This chapter aimed at presenting the pros and cons of using email as an example of Internet-based communication application for collecting qualitative research data and information. We first offered a nine-step process on how to administer an electronic interview, and we then discussed the advantages and disadvantages of such means while drawing from a brief experience in using email to interview older adults for a research study on aging. We now invite the reader to ponder about the use of email in today's exploratory and always evolving research arena as an alternative means to conference calls or face-to-face interviews when time and resources are restricted. There is no question that email as a research tool can be of value. Although misspelling and the lack of punctuation may delay prompt interpretation of the responses, the use of emails as illustrated by our brief study seems to be useful and effective when collecting information from participants who are comfortable with this form of communication. Nonetheless, there is a need for further studies to support, refute, or illuminate these findings.

In all, we believe that email has potential to be a data collection tool since it:

- eliminates the constraints of time and space
- offers cost-effectiveness by eliminating the need for tape recorders, transcription machines, and transcripts
- offers already transcribed textual data
- provides a noncoercive and antihierarchical dialogue enhancing equal opportunity and reciprocity
- may increase response rates

References

Ahern N. Using the internet to conduct research. Nurse Res. 2005;13(2):55–70.
Alzheimer Society of Canada. Reducing caregiver stress.2009; Retrieved 29 Aug 2011 from www.alzheimer.ca/docs/brochure-caring-eng.pdf
Adams J, Neville S. Resisting the 'condom every time for anal sex' health education message. Health Educ J. 2012;71(3):386–94.
Arnett MR, Loewen JM, Romito LM. Use of social media by dental educators. J Dent Educ. 2013;77(11):1402–12.
Australian Bureau of Statistics (ABS). Household use of information technology Australia 2010–11: Catologue no. 8146.0. 2013; Retrieved 30 Aug 2015 from http://www.abs.gov.au/ausstats/abs@.nsf/PrimaryMainFeatures/8147.0
Bell A, Billot J, Crothers C, Gibson A, Goodwin I, Sherman K, Smith P. The internet in New Zealand: 2007–2009. Auckland: Institute of Culture, Discourse and Communication, AUT University; 2010.
Benfield G. Teaching on web: exploring the meanings of silence. UltiBASE e-Journal. 2000; Retrieved 25 Oct 2010 from http://ultibase.rmit.edu.au/Articles/online/benfield1.pdf
Braun V, Clarke V. Successful qualitative research: a practical guide for beginners. London: Sage; 2013.
Brondani MA, MacEntee M, O'Connor D. E-mail as a data collection tool when interviewing older adults. Int J Qual Methods. 2010;10:221–30.

Burkhalter B. Reading race online: discovering racial identity in usenet discussions. In: Smith M, Kollock P, editors. Communities in cyberspace. London: Routledge; 1999. p. 60–75.

Burns P, Jones SC, Iverson D, Caputi P. Riding the wave or paddling in the shallows? Understanding older Australians' use of the internet. Health Promot J Austr. 2012;23:145–8.

Charness N, Parks DC, Sabel BA, editors. Communication, technology and aging: opportunities and challenges for the future. New York: Springer; 2001.

Cheston CC, Flickinger TE, Chisolm MS. Social media use in medical education: a systematic review. Acad Med. 2013;88(6):893–901.

China Highlights. Internet access in China. 2016; Retrieved 20 Jan 2016 from http://www.chinahighlights.com/travelguide/article-internet-access-in-china.htm

Cohen A. Internet insecurity: the identity thieves are out there - and someone could be spying on you. Why your privacy on the net is at risk, and what you can do. 2001; Retrieved 7 Mar 2010 from http://www.time.com/time/archive/preview/0,10987,1101010702-133167,00.html

Constantinides E, Lorenzo-Romero C, Gomez M. Effects of web experience on consumer choice: a multicultural approach. Internet Research. 2010;20:188–209.

Cook C. Email interviewing: generating data with a vulnerable population. J Adv Nurs. 2012; 68(6):1330–9.

Creswell JW. Qualitative inquiry and research design: choosing among five traditions. 4th ed. Thousand Oaks: Sage; 2014.

de Leeuw E, Nicholls WII. Technological innovation in data collection: acceptance, data quality and costs. Sociol Res Online. 1996;1(4.) Retrieved 15 Dec 2009 from http://www.socresonline.org.uk/socresonline/1/4/leeuw.html

EMarketer. Demographic profile – Seniors. 2010; Retrieved 30 Aug 2011 from http://www.emarketer.com/Reports/All/Emarketer_2000738.aspx

Etter JF, Perneger TV. A comparison of a cigarette smokers recruited through the internet or by mail. International Epidemiological Association. 2001;30:521–5.

Flowers LA, Moore JL III. Conducting qualitative research on-line in student affairs. Student Affairs Online. 2003;4(l). Retrieved 2 Sept 2009 from http://www.studentaffairs.com/ejournal/Winter_2003/research.html

Freese J, Rivas S, Hargittai E. Cognitive ability and internet use among older adults. Poetics. 2006;34:236–49.

Funkhouser E, Fellows JL, Gordan VV, Rindal DB, Foy PJ, Gilbert GH. Supplementing online surveys with a mailed option to reduce bias and improve response rate: the National Dental Practice-Based Research Network. J Public Health Dent. 2014;74(4):276–82.

Ginger Software. http://www.gingersoftware.com/punctuation-checker#.VktwJ175LSs

Goodchild JH, Donaldson M. The use of sedation in the dental outpatient setting: a web-based survey of dentists. Dent Implantol Updat. 2011;22:73–80.

Hackworth BA, Kunz MB. Health care and social media: building relationships via social networks. Acad Health Care ManagJournal. 2010;6:55–68.

Hage B. Bridging the digital divide: The impact of computer training, internet, and email use on levels of cognition, depression, and social functioning in older adults. Gerontechnology. 2008;7:117.

Hahn W, Bikson T. Retirees using email and networked computers. Int J Technol Aging. 1989;2:113–23.

Harwood J, editor. Understanding communication and aging. Thousand Oaks: Sage; 2007.

Haythornthwaite C. Online personal networks. New Media Soc. 2000;2:195–226.

Henry RK, Molnar AL. Examination of social networking professionalism among dental and dental hygiene students. J Dent Edu. 2013;77(11):1425–30.

Henry RK, Molnar A, Henry JC. A survey of US dental practices' use of social media. J Contemp Dent Pract. 2012;13:137–41.

Internet Live Stats. Internet users. 2015. Retrieved 12 Nov 2015 from http://www.internetlivestats.com/internet-users/

Johnson E, MacFadden K. SeniorNet's official guide to the web. Emeryville: Lycos Press; 1997.

Jowett A, Peel E, Shaw R. Online interviewing in psychology: reflections on the process. Qual Res Psychol. 2011;8(4):354–69.

King E. The use of the self in qualitative research. In: Richardson JTE, editor. Handbook of qualitative research methods for psychology and the social sciences. Leicester: BPS Books; 1996. p. 175–88.

Kutz D, Shankar K, Connelly K. Making sense of mobile- and web-based wellness information technology: cross-generational study. J Med Internet Res. 2013;15:e83. https://doi.org/10.2196/jmir.2124.

Lewis D, Eysenbach G, Kukafka R, Stavri P, Jimison H. Consumer health informatics: informing consumers and improving health care. New York: Springer; 2005.

Liamputtong P. Qualitative research methods. 4th ed. Melbourne: Oxford University Press; 2013.

Mann C, Stewart F. Internet interviewing. In: Holstein JA, Gubrium JF, editors. Inside interviewing: new lenses, new concerns. Thousand Oaks: Sage; 2003. p. 241–65.

Mariño R, Marwaha P, Barrow S, Baghaie H. Web-based oral health promotion program for older adults. Paper presented at the 10th world congress on preventative dentistry. 2013. Available from: https://iadr.confex.com/iadr/wcpd13/webprogram/Paper182509.html

Mariño R, Habibi E, Au-Yeung W, Morgan M. Use of communication and information technology among Victorian and south Australian oral health profession students. J Dent Educ. 2012;76:1667–74.

Marx M, Libin A, Renaudat K, Cohen-Mansfield J. Barriers encountered in an e-mail tutorial program for computer-illiterate seniors aged 71–96 years. Geron. 2002;2:151–2.

Meho LI. E-mail interviews in qualitative research: a methodological discussion. J Am Soc Inf Sci Technol. 2006;57:1284–95.

Melenhorst AS, Bouwhuis DG. When older adults consider the internet? An exploratory study of benefit perception. Geron. 2004;3:89–101.

McAuliffe D. Challenging methodological traditions: research by e-mail. Qual Rep. 2003;8(1). Retrieved 10 Aug 2009 from http://www.nova.edu/ssss/QR/QR8-1/mcauliffe.html

Merlien Institute. Mobile apps for qualitative research. 2016; Retrieved 3 Jan 2016 from http://mrmw.net/news-blogs/295-a-quick-review-of-mobile-apps-for-qualitative-research

National Center for Educational Statistics. Digest of educational statistics. 2004; Retrieved 29 Aug 2011 from http://nces.ed.gov/programs/digest/d05/tables/dt05_418.asp

Neville S, Adams J, Cook C. Using internet-based approaches to collect qualitative data from vulnerable groups: reflections from the field. Contemp Nurse. 2015;21:1–12.

Nulty D. The adequacy of response rates to online and paper surveys: what can be done? Assessment and Evaluation in Higher Education. 2008;33:301–14.

O'Kane M, Kenne PE, Pearcy H, Morgan T, Ransom G, Devoy K. On the feasibility of automatic punctuation of transcribed speech without prosody or parsing. 1994; Retrieved 30 Aug 2011 from www.assta.org/sst/SST-94-Vol-ll/cache/SST-94-VOL2- Walking inter views -p2.pdf

Patton MQ. Qualitative research and evaluation methods. 4th ed. Thousand Oaks: Sage; 2015.

Pew Internet. The Pew Research Center's Internet & American Life Project. 2010; Retrieved 30 Aug 2011 from http://www.pewinternet.org/Reports/2010/Older-Adults-and-Social-Media/Report.aspx

Plouffe D. The audacity to win: the inside story and lessons of Barack Obama's historic victory. New York, NY: Viking Penguin Group Publications; 2009.

Ramsay M. Social media etiquette: a guide and checklist to the benefits and perils of social marketing. Journal of Database Marketing & Customer Strategy Management. 2010;17:257–61.

Reisenwitz T, Iyer R, Kuhlmeier DB, Eastman JK. The elderly's internet usage: an updated look. J Consum Mark. 2007;24:406–18.

Roy D, Taylor J, Cheston CC, Flickinger TE, Chisolm MS. Social media: portrait of an emerging tool in medical education. Acad Psychiatry. 2016;40(1):136–40.

Scanfield D, Scanfeld V, Larson E. Dissemination of health information through social networks: twitter and antibiotics. Am J Infect Control. 2010;38:182–8.

Selwyn N, Robson K. Using email as a research tool. Social Research Update, 21. 1998; Retrieved 2 Sept 2009 from http://www.soc.surrey.ac.uk/sru/SRU21.html

Serry T, Liamputtong P. The in-depth interviewing method in health. In: Liamputtong P, editor. Research methods in health: foundations for evidence-based practice. 3rd ed. Melbourne: Oxford University Press; 2017. p. 67–83.

Shea V. The core rules of netiquette. 2004; Retrieved 19 Feb 2010 from http://www.albion.com/netiquette/corerules.html

Simeon R. Evaluating domestic and international web-site strategies. Internet Research: Electronic Networking Applications and Policy. 1999;9:297–308.

Smith A. Home broadband 2010. Pew Research Center's Internet & American Life Project. 2010; Retrieved 5 Apr 2011 from http://pewinternet.org/Reports/2010/Home-Broadband-2010.aspx

Statistics and facts on internet usage in Brazil.2016. Retrieved 20 Jan 2016 from http://www.statista.com/topics/2045/internet-usage-in-brazil/

Statistics Canada. The daily. 2002; Retrieved 2 Sept 2010 from http://www.statcan.ca/Daily/English/020611/td020611.htm

Statistics Canada. Internet use among older Canadians. 2001; Retrieved 10 Mar 2009 from http://www.statcan.gc.ca/access_acces/alternative_alternatif.action?l=eng&loc=http://www.statcan.gc.ca/pub/56f0004m/56f0004m2001004-eng.pdf&t=Internet%20Use%20Among%20Older%20Canadians%20%28Connectedness%20Series%29

Statistics Canada. Internet use by individuals, by type of activity (internet users at home). Tables by Subject: Individual and Household Internet Use. 2010; Retrieved 15 Apr 2011 from http://www40.statcan.gc.ca/l01/cst01/comm29a-eng.htm?sdi=internet

Visser J. Dalhousie suspends 13 dentistry students involved in Facebook scandal from clinical activities. National Post, 5 January 2005.

Whyte J, Marlow B. Beliefs and attitudes of older adults toward voluntary use of the Internet: An exploratory investigation. 1999; Retrieved 18 July 2010 from www.csu.edu.au/OZCHI99/short_papers/Whyte.doc

Wikipedia. 2015. https://en.wikipedia.org/wiki/Emoticon

Wong FL, Greenwell M, Gates S, Berkowitz JM. It's what you do! Reflections on the VERBTM campaign. Am J Prev Med. 2008;34:175–82.

Cell Phone Survey

80

Lilian A. Ghandour, Ghinwa Y. El Hayek, and Abla Mehio Sibai

Contents

1 Cellphones in Health Survey Research 1404
2 Methods of Constructing "Dual Frame" Surveys 1405
3 Methodological Considerations When Using Cell Phone Sampling Frames 1407
   3.1 Noncoverage Bias 1408
   3.2 Nonresponse Bias 1410
   3.3 Cost-Efficiency and Feasibility of Cell Phone Surveys 1412
4 Conclusion and Future Directions 1413
References 1413

Abstract

The global rise in cell phones usage has undermined traditional data collection modes, notably landline surveys, and has elicited the development of novel survey methods and designs. Adopting a single landline frame survey is no longer viable, owing to its undesirable implications on response rate, coverage, as well as data representativeness and validity. Subsequently, researchers have developed methods to integrate the cell phone and landline frames, and conduct "dual frame" surveys using either overlapping or nonoverlapping modes of integration. Dual frame surveys have gained popularity as they were shown to enhance the quality of the collected data and improve the validity of national estimates. Of course, cell phone surveys, be it single frame or dual frame, are not void of methodological challenges relevant to sampling frames, participant selection, respondent burden, and collection of reliable and valid data. The evidence concerning the proper implementation of cell phone surveys, as well as the feasibility of a single frame cell phone survey is relatively recent especially

---

L. A. Ghandour (✉) · G. Y. El Hayek · A. Mehio Sibai
Department of Epidemiology and Population Health, Faculty of Health Sciences, American University of Beirut, Beirut, Lebanon
e-mail: lg01@aub.edu.lb; gye01@mail.aub.edu; am00@aub.edu.lb

© Springer Nature Singapore Pte Ltd. 2019
P. Liamputtong (ed.), *Handbook of Research Methods in Health Social Sciences*,
https://doi.org/10.1007/978-981-10-5251-4_27

when youth are the targeted population. Given their proliferation worldwide and the diminishing existing barriers, cell phones are expected to become an inevitable mode for collecting health survey data. Yet, the need remains for contextualizing their feasibility as per each country's settings and circumstances.

> **Keywords**
> Cell phones · Telephone · Bias epidemiology · Validity · Feasibility

1 Cellphones in Health Survey Research

Face-to-face and landline telephone interviews have been the most common data collection methods in health survey research (Nelson et al. 2003; Schofield and Forrester-Knauss 2017; see also ▶ Chap. 32, "Traditional Survey and Questionnaire Platforms"). Face-to-face interviews allow the collection of more detailed information on a wider range of topics (Nelson et al. 2003) and have better response rates (Massey et al. 1997), but they are more costly and time consuming than landline telephone interviews. The latter also allow researchers to reach out to a larger pool of respondents (Anie et al. 1996; Aziz and Kenford 2004) and reduce methodological issues by reducing social desirability, providing greater anonymity (Babor et al. 1990; Schwarz et al. 1991) and minimizing the influence of other household members (Anie et al. 1996).

The traditional landline telephone interviews have been recently challenged by the rise of new communication technologies, namely mobile phones, which have since been used for the collection of health and nonhealth-related survey data (Brick et al. 2007a; Ekman and Litton 2007; Vehovar et al. 2010). Globally, in high as well as low and middle income countries, an increasing number of people have switched to mobile phone use only, reducing the number of households with landline phones (Blumberg and Luke 2014; Liu et al. 2011; McBride et al. 2012). In the United States (USA), nearly 41% of American homes, and particularly two third of adults aged 25–29 years old, do not own a landline telephone (Blumberg and Luke 2014). The World Bank estimates that mobile phone subscriptions grew 40-fold from 1997 (2%) to 2011 (85%) and have either exceeded or are approaching 100% in both developed and developing countries (multiple cell phone lines per subscriber), such as Finland (166%), Italy (158%), Argentina (135%), Bulgaria (141%), Kuwait (175%), Bahrain (128%), Jordan (118%), and Egypt (101%) (World Bank 2013).

The rise in cell phone usage and the number of households that can be solely reached by mobile phones has undermined data collection using landline telephone surveys, specifically through reducing the completeness of the landline phone directories (Blumberg and Luke 2010). From an epidemiological standpoint, this impacts the coverage of landline only phone surveys, and the quality of the data derived using this method (Kempf and Remington 2007; Link et al. 2007; Hu et al. 2010). To elaborate, sampling frames that exclude members or groups of the target population raise concerns that those missing may be different from those included, hence threatening the representativeness of the sample (Lee et al. 2012) and resulting in biased survey estimates, whether health prevalence statistics (Keeter et al. 2007; Blumberg and Luke 2009; Baffour et al. 2016), data on lifestyle behaviors (Ehlen and Ehlen 2007), or data on political attitudes (Keeter et al. 2008).

The rapid increase in wireless phone usage, particularly among the young (Keeter et al. 2007), has rendered landline only surveys not only inaccurate for provision of national data (Lee et al. 2010) but also increasingly inadequate in capturing hard-to-reach young populations (Gundersen et al. 2014). For example, the exclusion of mobile phones from the 2007 CDC Behavioral Risk Factor Surveillance System (BRFSS) in the USA resulted in lower estimates of binge drinking and smoking among young adults (Blumberg and Luke 2009) when compared to those from the household face-to-face National Health Interview Survey (NHIS) (Delnevo et al. 2008). Attempting to address these differences in estimates by weighting methods that adjust for the proportionality of a certain group is not an option, as noncoverage bias is the product of both the degree of noncoverage and differentials between respondents and nonrespondents (Lesser and Kalsbeek 1992). Thus, it has become necessary to integrate cellphone users in landline telephone survey sampling frames.

2 Methods of Constructing "Dual Frame" Surveys

Surveys integrating landline and cellphone users are labeled "dual frame" or "dual-mode" surveys. Dual-mode surveys are feasible but are not void of challenges that are mainly logistical in nature (Brick et al. 2007a, b; Link et al. 2007). There are at least two approaches to integrating mobile phone numbers into landline telephones surveys. One approach uses an overlapping design whereby the two sampling frames are combined (Fig. 1). Such a design requires the availability of two sampling frames, the first including all the residential landline phone numbers and the second all the cell phones numbers; the use of Random Digit Dialing (RDD) for the selection of phone numbers ensures that all phone numbers have equal chances of being selected (The American Association for Public Opinion Research 2010). Obviously, dual users, or households with access to both landline and cell phones, will have a higher chance of being selected, and adjustment for this dual frame multiplicity will follow by composite weighting procedures to generate unbiased results (Blumberg et al. 2009; Best 2010). Such weighting procedures are different for each country, as they are benchmarked against variations in phones distribution in the population (landline only, landline and cell phone, and cell phone only) (The American Association for Public Opinion Research 2010; Barr et al. 2014a). While overlapping frames require a complex weighting strategy, this approach has been particularly favored in risk factor surveillance systems and chronic disease surveys (Hu et al. 2011; Barr et al. 2012; Livingston et al. 2013; Jackson et al. 2014).

Another approach is the nonoverlapping design, better known as the screening dual frame design (Fig. 2). This entails screening the mobile phone sampling frame to select the cell-only users and removing the duplicated units (households with both a landline and cellphone(s)) from the mobile phone sampling frame. This would result in a sampling frame consisting of two separate strata (the mobile only users

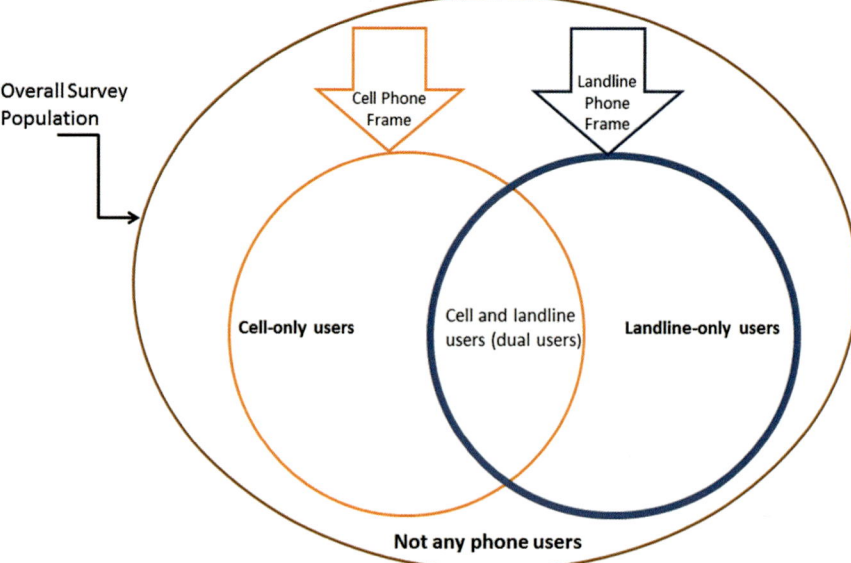

Fig. 1 The overlapping design in a dual frame telephone survey

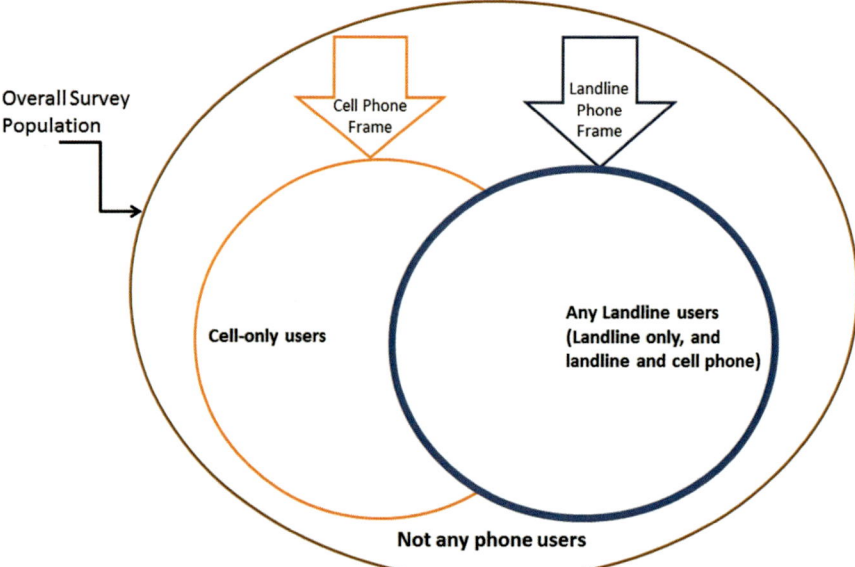

Fig. 2 The dual screening design in a dual frame telephone survey

and landline users), with sampling being comparable to stratified random sample design. The screening process can only be done during data collection, rendering the nonoverlapping design more costly than the overlapping design (The American Association for Public Opinion Research 2010).

Researchers took more interest in the overlapping design as it can result in estimates based on four comparison groups: landline only and landline-cell phone users sampled from landline frame, as well as cell phone only and cell phone-landline users sampled from cell phone frame. As aforementioned, this design has become increasingly popular, notably in national health surveys in Australia (Barr et al. 2014a, b).

Of relevance to integrating mobile phone numbers into telephone surveys is obviously identifying or creating a cell phone sampling frame, which is not as straight forward as a traditional landline sampling frame (Kempf and Remington 2007). In contrast to landline surveys, a cell phone respondent does not represent households, as cell phones are attached to individuals rather than a geographic location leading to further statistical and operational implications (Link et al. 2007). Another major challenge is the issue of establishing the eligibility of the respondent in mobile phone surveys (Brick et al. 2007a; Moura et al. 2011; McBride et al. 2012; Gundersen et al. 2014). This is particularly problematic when call receivers are below 18 years of age (Call et al. 2011) or when they refuse to confirm their age even after the study protocol and the importance of excluding minors from the survey has been clearly explained (McBride et al. 2012). Therefore, the likelihood of rechanneling the call to an eligible subject in landline phone interview is higher (Brick et al. 2007a). Issues of eligibility extend also to cases when the cell phone number is linked to fax machines, business phones, or wireless cards assigned for computers and tablets (Gundersen et al. 2014).

Typically, an assumption is made that cell phones are personal and not shared household devices, meaning there is one-to-one correspondence between the person answering the cellphone and the household (Link et al. 2007). However, cell phone sharing may still be common in various forms (e.g., shared one third of the time, shared half the time), adding to the complexity of establishing whether the cell phone is "shared" or not (Busse and Fuchs 2013). In fact, little and inconsistent evidence on cell phone sharing exists (Brick et al. 2007b; Carley-Baxter et al. 2010; The American Association for Public Opinion Research 2010), and interviewers may need to ask some questions to make sure that the household has not been previously contacted.

3 Methodological Considerations When Using Cell Phone Sampling Frames

Household surveys using face-to-face interviews have been viewed as the gold standard in terms of obtaining national estimates, and other modes of data collection have been consequently compared against them (De Leeuw 2008; Delnevo et al. 2008). In contrast to the relatively good number of studies providing evidence on the agreement in disease and risk factor prevalence rates from landline interviews

Table 1 Agreement (Kappa Statistics) between data obtained from a cell phone survey ($N = 630$) versus a face-to-face interview ($N = 630$): Data from Lebanon collected as part of the Nutrition and Non-communicable Disease Risk Factor Survey nationwide cross-sectional population-based household survey, 2009 (Mahfoud et al. 2015)

Variable	Face-to-face interviews ($N = 630$)		Cell phone survey ($N = 630$)		% Agreement	Kappa statistics
	n	%	n	%		
Current cigarette smoking	245	38.9	239	37.9	95.6	0.91
Reported diabetes	48	7.6	43	6.8	98.3	0.87
Having health insurance	388	61.6	391	62.1	92.2	0.84
Past year alcohol consumption	299	47.5	307	48.7	89.5	0.79
Current waterpipe smoking	159	25.2	170	27.0	91.3	0.77
Reported hypertension	76	12.1	85	13.5	92.9	0.68
Reported heart disease	25	4.0	36	5.7	97	0.67
Reported hyperlipidemia	94	14.9	79	12.5	91.9	0.66

compared to population-based face-to-face interviews (Groves and Kahn 1979; Herzog et al. 1983; Herzog and Rodgers 1988), studies comparing cell phone to face-to-face interviews remain scarce. Mahfoud et al. (2015) examined the reliability of self-reported responses generated by cell phone interviews and face-to-face interviews and observed very high concordance (kappa statistics >0.8) for outcomes such as cigarette smoking and diabetes, as well as substantial agreement ($0.6 < k < 0.8$) for other measures such as water pipe smoking, alcohol consumption, and hypertension (Table 1). The high reliability alleviates concerns on how to handle dual users, in terms of selection and weighting procedures, in dual frame surveys (The American Association for Public Opinion Research 2010).

Overall, the validity and reliability of data obtained from surveys can be compromised by one or more factors including selection of sampling frames, design effect, response rates, and interviewing techniques (Kempf and Remington 2007). Dual frame surveys (e.g., cellphone and landlines) and mixed mode surveys (e.g., face-to-face, phone interviews, and online forms of data collection) may alleviate bias, but they face other challenges such as bias due to mode effects (when the multiple interviews methods used for a single survey influence respondent answers) (Lugtig et al. 2011). When integrating cell phones into phone surveys, an assumption is made that this enhances representativeness and that the data derived from cell phone interviews are both reliable and valid. Below, we cover the main epidemiological biases in cell phone surveys and, when available, methods that have been used to mitigate them or reduce undesirable effects on data quality.

3.1 Noncoverage Bias

The substantial differences between landline and mobile only users that have been found in cross-sectional and longitudinal study designs and in developmental and

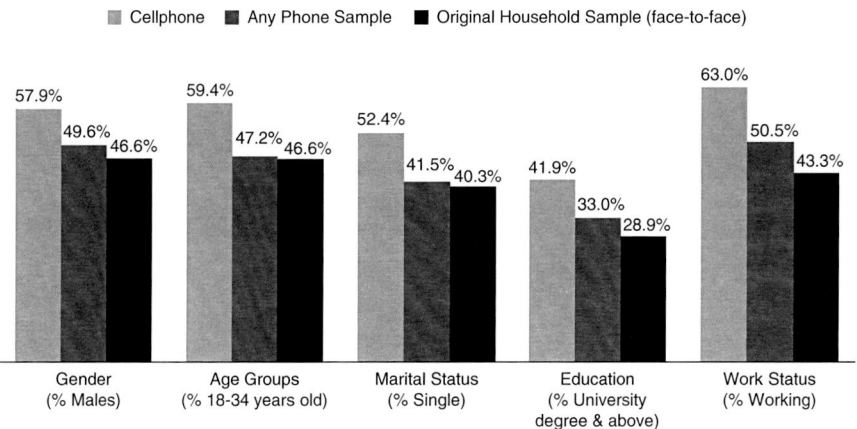

Fig. 3 Comparison of sociodemographic characteristics of cell phone sample ($N = 1381$), any phone sample ($N = 2126$), and face-to-face sample ($N = 2656$): Data from Lebanon collected as part of the Nutrition and Non-communicable Disease Risk Factor Survey nationwide cross-sectional population-based household survey, 2009 (Sibai et al. 2016)

applied research as well (Jackson et al. 2014) necessitate the inclusion of cell phone only users into traditional phone sampling frames. Several studies have noted differences in sociodemographic characteristics, whereby cell phone only users were more likely to be young, males, speak English, to live in rented homes, and to belong to certain ethnicities with low socioeconomic backgrounds than the general population (Blumberg and Luke 2007; Dal Grande and Taylor 2010; Lee et al. 2010, 2012). In other studies, cell phone respondents have also reported higher educational level (Dal Grande and Taylor 2010; Lee et al. 2012). Worth noting is that the evidence has been mostly from the West, with very little data from the less-developed world. One study from Lebanon in the Middle East further stresses these marked differences, showing that cell phones users (irrespective of whether they belong to households with or without landlines) were more likely be male, young, single, of a higher educational level, and part of the labor force (Fig. 3; Sibai et al. 2016).

Differences have also been noted in risk behavior estimates. Compared to landline users, studies have shown that cell phone users (whether cell phone only population, or dual users) have higher rates of heavy drinking; smoking; and risky sexual behavior (Blumberg and Luke 2009; Dal Grande and Taylor 2010); illegal drug use and gambling problems, notably among the young (Barr et al. 2012, 2014a; Livingston et al. 2013); as well as poorer mental well-being and higher incidence of psychological distress (Baffour et al. 2016). Other studies have shown that lying about gambling behavior and attempts to control gambling behavior were significantly more likely to occur within the mobile sample (Jackson et al. 2014), even if compared within the same age category (Keeter et al. 2007). Nonetheless, recent data from the household-based Australian National Health survey showed that mobile-only users were more physically active and had lower prevalence of obesity compared to landline accessible population (Baffour et al. 2016).

Adjustment for age, gender, and geographic location, among other factors responsible for noncoverage and nonresponse, tend to lessen differences in health indicators and prevalence estimates between both landline and cellphone frames (Lee et al. 2010; Hu et al. 2011; Sibai et al. 2016). Nonetheless, adjustment is not sufficient to justify the exclusion of cell phone only users from sampling frames. Hu and colleagues demonstrated the remaining potential noncoverage bias when excluding cell phone only users for 16 health-related outcomes (most notably in the context of alcohol consumption among adults), even after adjustment for demographic differences (Hu et al. 2010). Barr (2008) estimates that, in Australia, when landline phone users drop to below 85%, the difference in health indicators between landline owners only and cell phone users only will become significant and will impact national health statistics.

To sum up, the exclusion of cell phone sampling frame will impact the validity of the survey findings by reducing representativeness of the study. Adjustment for demographic characteristics is unlikely to totally resolve the noncoverage bias issues consequent to its omission.

Taking it a step further, in countries with high penetration rate of cellphone within certain age groups, researchers have benchmarked the quality of data collected solely through a cell phone frame against a national survey that included both landline and cell phone or landline only. As an example, findings on the prevalence of cigarette smoking among young adults aged 18–34 years old were not significantly different between the National Young Adult Health Survey (NYAHS) based on cell phone frame compared to data collected from the BRFSS based on dual frame sampling for the same age group (Gundersen et al. 2014). Similarly, a sexual youth survey targeting women 18–39 years old from Australia reported no differences on almost all variables, including demographic and health-related outcomes between the two sampling frames (Liu et al. 2011). Therefore, the future of having a single sampling frame seems feasible under certain conditions and will tend to reduce the time-length of the survey and the cost per interview, especially that questions related to ownership of telephones would not be needed anymore (Liu et al. 2011; Guterbock et al. 2013; Gundersen et al. 2014).

3.2 Nonresponse Bias

Besides sampling challenges, telephone surveys are increasingly affected by decreasing response rates (Curtin et al. 2005; Carley-Baxter et al. 2010; see also ▶ Chap. 81, "Phone Surveys: Introductions and Response Rates"). Compared to landline telephone surveys, mobile phone surveys are challenged by lower response rates, higher refusal rates, and lower refusal conversion rates (Steeh 2004; Brick et al. 2007a). The 2010 Irish Contraception and Crisis Pregnancy survey (ICCP) found that an overall higher percentage of valid telephone numbers go unanswered after ten contact attempts for the mobile telephone strand (30%) compared to the landline telephone strand (24%) (McBride et al. 2012). Subsequently, lower response rates will occur in the mobile strand as illustrated in the New South Wale Population

Health survey (31.5% vs. 35.1%) (Barr et al. 2012), the gambling behavior survey in Australia (12.7% vs. 22.2%) (Jackson et al. 2014), and in various other surveys across the USA including New Mexico (31.3% vs. 51.9%) and Pennsylvania (23.2% vs. 45.3%) (Link et al. 2007), as well as in the 2011 BRFSS (27.9% vs. 53.0%) (Gundersen et al. 2014).

Factors affecting cooperation and response rates in cell phone surveys are mostly "situational," related to the circumstances in which the respondent had received the call (e.g., at work, shopping, or driving) and less so survey-related (e.g., the topic of the survey) (Brick et al. 2007b; Carley-Baxter et al. 2010). The call is likely to be perceived as a burden by the respondents who may be simultaneously engaged and constrained in other activities when receiving the call, subsequently affecting response rate (Lavrakas et al. 2007). Response rates are also likely to be influenced by the country-specific cell phone charges policy. Mahfoud et al. (2015), for example, reported a relatively high response rate for cellphone interviews (82%), attributing this to the fact that receiving cell phone calls in Lebanon is free of charge. Incentives also play a role. For example, comparing a 10$ to a 5$ incentive leads to higher response rates for the higher incentive provided (25.8% vs. 18.6%) (Brick et al. 2007a). Still, and even with no incentives, McBride et al. (2012) noted "a high-unusual" response rate for both the cell phone (61%) and landline interviews (79%) in the 2010 ICCP survey, accredited to using a telephone system with a blocker caller ID to evade potential back calls that would have entailed additional costs on behalf of the respondents. Of course, mobile telephone owners may be more reluctant to answer a blocked Caller ID or may also become irritated at receiving "missed" calls from an unknown source on their personal mobile phone. Alternatively, respondents may be provided with a toll free hotline or a link to the Institutional Research Ethics Committee and their contact details (Hu et al. 2011; McBride et al. 2012).

One method that has been employed to increase response rates is the refusal conversation procedure through recontacting respondents who refused to participate in the survey upon first contact one final time, giving them an opportunity to reconsider their decision or by scheduling an appointment for the interview (e.g., when incoming calls would not incur a cost on the respondent, or at a more convenient time) (Hu et al. 2011). While interviewers of the BRFSS relied on such a procedure to increase response rates in the cell phone strand (Hu et al. 2011), McBride opted for not using this method halfway through the survey as it resulted in annoyance of respondents (McBride et al. 2012). Often, phone respondents are irritated and inquire about how their telephone numbers were obtained. As such, interviewers will need to stress the importance of surveys for the public good and to explain to the interviewees the principles of RDD and assure that mobile telephone numbers are not being purchased from a database or other external sources (McBride et al. 2012) (see also ▶ Chap. 81, "Phone Surveys: Introductions and Response Rates").

Regardless of methods to enhance response rates, it is important to note that there are different formulas to calculate survey response rates, which may partially explain these observed variations across studies (The American Association for Public Opinion Research 2011).

3.3 Cost-Efficiency and Feasibility of Cell Phone Surveys

One major drawback mentioned in different studies using mobile frames, with or without a landline frame, is the cost incurred to obtain complete interview data, since cell phone numbers typically need to be contacted more often than landlines to establish a final contact or to reconnect after having lost network coverage in the middle of the call (Kuusela et al. 2008; Jackson et al. 2014; Barr et al. 2012).

Some researchers have argued that cell phone surveys are impractical in certain contexts such as in the USA, China, and Canada, where mobile phone surveys induce a cost for the respondent if they are accessing a mobile network other than their service provider (i.e., roaming) or on a prepaid plan that charges for incoming calls (Link et al. 2007; Vehovar et al. 2010; Hu et al. 2011).

Furthermore, additional screening and hence a higher cost is entailed when the study protocol targets subjects based on dual frames whereby longer screening protocols are needed (in contrast to single frames) (Keeter et al. 2007; Gundersen et al. 2014). The evidence has however varied depending on the study context as well as study protocol and design. In the first dual frame survey in Brazil, the cost of cell phone interviewing was 6.6 times that of landline survey, making it inconvenient at the time of the survey (Moura et al. 2011). In the New South Wales Population Health survey in Australia, mobile frames cost was only 2.3 times more than landline frames, rendering the inclusion of a cell phone group manageable (Barr et al. 2012). In the 2007 BRFSS conducted in three states, Pennsylvania, New Mexico, and Georgia, the cost of one complete landline survey interview was approximately 64 USD, increasing to 74 USD for a cell phone survey interview, and approaching 196 USD, when there was the need to screen and select cell phone-only users (Link et al. 2007). Guterbock and colleagues (2013) estimated that cell phone interviews cost overall 50% more than landline phone interviews. Nonetheless, cell phone interviews may still be appealing, useful, and affordable in countries and regions of conflict where access to certain geographical areas is physically challenging and landline interviews are difficult to conduct in the case of mobile population (Sibai et al. 2016).

Contrary to the landline interviews, scheduling calls for cell phone interviews are equally convenient across different time periods, whether on weekdays or during weekends (Brick et al. 2007a). There is no need to worry about the optimal time of the day to contact respondents as the vast majority keep their cell phone turned on all the time (Carley-Baxter et al. 2010). However, as mentioned earlier, cell phone interviews can reach individuals in circumstances that may not be ideal for them to respond to the call or to be fully engaged in the interview (e.g., while shopping). In one study from the USA, it was found that while 56% of mobile phone respondents were at home when undertaking the survey, an additional 14% were driving, 13% were at work, and the others were in public spaces, on holidays, visiting friends and relatives (Brick et al. 2007a). The question of being "fully and cognitively engaged" in the interview process without being distracted by other activities clearly impacts overall data quality (Link et al. 2007). To address these issues, interviewers need to be instructed to be proactive and pick up cues as to whether the respondent might not be in a position to answer the survey and to offer the respondent a callback at a more convenient time (Brick et al. 2007a; Hu et al. 2011; McBride et al. 2012).

Cellphone surveys may need to be shorter than those conducted via landline telephones to ensure high data quality (Lavrakas et al. 2007). Also, there is the issue of privacy that may entail legal consequences and be problematic in some countries where calling private mobile telephones may breach communication legislation and consumer protection acts (Kempf and Remington 2007; Lavrakas et al. 2007; McBride et al. 2012).

4 Conclusion and Future Directions

Cell phone coverage is increasing globally, rendering the use of cell phones a necessity rather than a choice, in order to obtain representative and valid estimates (Link et al. 2007). Yet, it remains imperative to construct or identify, and evaluate both landline and mobile sampling frames before conducting phone surveys to ensure adequate coverage. There is also no "best buy" to sampling mobile phone users; researchers must consider multiple factors, including the survey budget, typical country participation and response rates, as well as legal issues (Kempf and Remington 2007). The use of a mobile phone frame to obtain national data has been deemed proper if particular segments of the population are targeted, such as young adults (Liu et al. 2011; Gundersen et al. 2014). However, published literature still lacks solid and empirical evidence regarding reliability and feasibility of cell phone only as well as dual frame surveys from the middle- and low-income countries, necessitating further contextual research in this field. In the near future, it is expected that the implementation of mobile surveys will become less complex owing to the reduction of existing barriers, prompting researchers to move forward by adopting them, with or without other modes of data collection.

References

Anie KA, Jones PW, Hilton SR, Anderson HR. A computer-assisted telephone interview technique for assessment of asthma morbidity and drug use in adult asthma. J Clin Epidemiol. 1996;49(6): 653–6.

Aziz MA, Kenford S. Comparability of telephone and face-to-face interviews in assessing patients with posttraumatic stress disorder. J Psychiatr Pract. 2004;10(5):307–13.

Babor TF, Brown J, Delboca FK. Validity of self-reports in applied-research on addictive behaviors – fact or fiction. Behav Assess. 1990;12(1):5–31.

Baffour B, Haynes M, Dinsdale S, Western M, Pennay D. Profiling the mobile-only population in Australia: insights from the Australian National Health Survey. Aust N Z J Public Health. 2016;40(5):443–7.

Barr ML. Predicting when declining landline frame coverage will impact on the overall health estimates for the NSW Population Health Survey. 2008. Retrieved from http://www.health.nsw.gov.au/surveys/other/Documents/predicting-when-mobile-only-impacts-2008.pdf.

Barr ML, Van Ritten JJ, Steel DG, Thackway SV. Inclusion of mobile phone numbers into an ongoing population health survey in new South Wales, Australia: design, methods, call outcomes, costs and sample representativeness. BMC Med Res Methodol. 2012;12:177. https://doi.org/10.1186/1471-2288-12-177.

Barr ML, Ferguson RA, Hughes PJ, Steel DG. Developing a weighting strategy to include mobile phone numbers into an ongoing population health survey using an overlapping dual-frame design with limited benchmark information2. BMC Med Res Methodol. 2014a;14:102. https://doi.org/10.1186/1471-2288-14-10.

Barr ML, Ferguson RA, Steel DG. Inclusion of mobile telephone numbers into an ongoing population health survey in New South Wales, Australia, using an overlapping dual-frame design: impact on the time series. BMC Res Notes. 2014b;7:517. https://doi.org/10.1186/1756-0500-7-517.

Best J. First-stage weights for overlapping dual frame telephone surveys. In: Paper presented at the 65th annual conference of the American Association of Public Opinion Research. Chicago; 2010.

Blumberg SJ, Luke JV. Coverage bias in traditional telephone surveys of low-income and young adults. Public Opin Q. 2007;71(5):734–49.

Blumberg SJ, Luke JV. Reevaluating the need for concern regarding noncoverage bias in landline surveys. Am J Public Health. 2009;99(10):1806–10.

Blumberg SJ, Luke JV. Wireless substitution: early release of estimates from the National Health Interview Survey, Jan–June 2010. National Center for Health Statistics, 201(3). 2010. Retrieved from http://www.cdc.gov/nchs/data/nhis/earlyrelease/wireless201005.pdf.

Blumberg SJ, Luke JV. Wireless substitution: early release of estimates from the National Health Interview Survey, July–Dec 2013. 2014. Retrieved from https://www.semanticscholar.org/paper/Wireless-Substitution-Early-Release-of-Estimates-Blumberg-Luke/1cb29f663f0419d8ef7a4e742ee8e1bad1e3f7eb/pdf.

Blumberg SJ, Luke JV, Davidson G, Davern ME, Yu TC, Soderberg K. Wireless substitution: state-level estimates from the National Health Interview Survey, January-December 2007. Natl Health Stat Rep. 2009;14:1–13.

Brick JM, Brick PD, Dipko S, Presser S, Tucker C, Yuan Y. Cell phone survey feasibility in the US: sampling and calling cell numbers versus landline numbers. Public Opin Q. 2007a;71(1):23–39.

Brick JM, Edwards WS, Lee S. Sampling telephone numbers and adults, interview length, and weighting in the California health interview survey cell phone pilot study. Public Opin Q. 2007b;71(5):793–813.

Busse B, Fuchs M. Prevalence of cell phone sharing. Survey Methods: Insights from the Field (SMIF); 2013. http://surveyinsights.org/?p=1019.

Call KT, Davern M, Boudreaux M, Johnson PJ, Nelson J. Bias in telephone surveys that do not sample cell phones: uses and limits of poststratification adjustments. Med Care. 2011;49(4):355–64.

Carley-Baxter LR, Peytchev A, Black MC. Comparison of cell phone and landline surveys: a design perspective. Field Methods. 2010;22(1):3–15.

Curtin R, Presser S, Singer E. Changes in telephone survey nonresponse over the past quarter century. Public Opin Q. 2005;69(1):87–98.

Dal Grande E, Taylor AW. Sampling and coverage issues of telephone surveys used for collecting health information in Australia: results from a face-to-face survey from 1999 to 2008. BMC Med Res Methodol. 2010;10:77. https://doi.org/10.1186/1471-2288-10-77.

De Leeuw ED. Choosing the method of data collection. In: de Leeuw D, Hox JJ, Dillman DA, editors. International handbook of survey methodology. New York: Lawrence Erlbaum Associates; 2008. p. 113–35.

Delnevo CD, Gundersen DA, Hagman BT. Declining estimated prevalence of alcohol drinking and smoking among young adults nationally: artifacts of sample undercoverage? Am J Epidemiol. 2008;167(1):15–9.

Ehlen J, Ehlen P. Cellular-only substitution in the United States as lifestyle adoption – implications for telephone survey coverage. Public Opin Q. 2007;71(5):717–33.

Ekman A, Litton JE. New times, new needs: E-epidemiology. Eur J Epidemiol. 2007;22(5):285–92.

Groves RM, Kahn RL. Surveys by telephone: a national comparison with personal interviews. New York: Academic; 1979.

Gundersen DA, ZuWallack RS, Dayton J, Echeverria SE, Delnevo CD. Assessing the feasibility and sample quality of a national random-digit dialing cellular phone survey of young adults. Am J Epidemiol. 2014;179(1):39–47.

Guterbock T, Peytchev A, Rexrode D. Cell-phone costs revisited: understanding cost and productivity ratios in dual-frame telephone surveys. In: Paper presented at the American Association for Public Opinion Research Conference. Boston; 2013.

Herzog AR, Rodgers WL. Interviewing older adults: mode comparison using data from a face-to-face survey and a telephone resurvey. Public Opin Q. 1988;52(1):84. https://doi.org/10.1086/269083.

Herzog AR, Rodgers WL, Kulka RA. Interviewing older adults: a comparison of telephone and face-to-face modalities. Public Opin Q. 1983;47(3):405. https://doi.org/10.1086/268798.

Hu SS, Balluz L, Battaglia MP, Frankel MR. The impact of cell phones on public health surveillance. Bull World Health Organ. 2010;88(11):799. https://doi.org/10.2471/BLT.10.082669.

Hu SS, Balluz L, Battaglia MP, Frankel MR. Improving public health surveillance using a dual-frame survey of landline and cell phone numbers. Am J Epidemiol. 2011;173(6):703–11.

Jackson AC, Pennay D, Dowling NA, Coles-Janess B, Christensen DR. Improving gambling survey research using dual-frame sampling of landline and mobile phone numbers. J Gambl Stud. 2014;30(2):291–307.

Keeter S, Kennedy C, Clark A, Tompson T, Mokrzycki M. What's missing from national landline RDD surveys? The impact of the growing cell-only population. Public Opin Q. 2007;71(5):772–92.

Keeter S, Dimock M, Christian L. Calling cell phones in'08 pre-election polls. The Pew Research Center for the people & the Press. 2008. Retrieved 18 Dec 2008.

Kempf AM, Remington PL. New challenges for telephone survey research in the twenty-first century. Annu Rev Public Health. 2007;28:113–26.

Kuusela V, Callegaro M, Vehovar V. The influence of mobile telephones on telephone surveys. In: Lepkowski JM, Tucker C, Brick JM, et al., editors. Advances in telephone survey methodology. New York: Wiley; 2008. p. 87–112.

Lavrakas PJ, Shuttles CD, Steeh C, Fienberg H. The state of surveying cell phone numbers in the United States 2007 and beyond. Public Opin Q. 2007;71(5):840–54.

Lee S, Brick JM, Brown ER, Grant D. Growing cell-phone population and noncoverage bias in traditional random digit dial telephone health surveys. Health Serv Res. 2010;45(4):1121–39.

Lee S, Elkasabi M, Streja L. Increasing cell phone usage among Hispanics: implications for telephone surveys. Am J Public Health. 2012;102(6):e19–24.

Lesser VM, Kalsbeek WD. Non-sampling error in surveys. New York: Wiley; 1992.

Link MW, Battaglia MP, Frankel MR, Osborn L, Mokdad AH. Reaching the US cell phone generation comparison of cell phone survey results with an ongoing landline telephone survey. Public Opin Q. 2007;71(5):814–39.

Liu B, Brotherton JM, Shellard D, Donovan B, Saville M, Kaldor JM. Mobile phones are a viable option for surveying young Australian women: a comparison of two telephone survey methods. BMC Med Res Methodol. 2011;11:159. https://doi.org/10.1186/1471-2288-11-159.

Livingston M, Dietze P, Ferris J, Pennay D, Hayes L, Lenton S. Surveying alcohol and other drug use through telephone sampling: a comparison of landline and mobile phone samples. BMC Med Res Methodol. 2013;13:41. https://doi.org/10.1186/1471-2288-13-41.

Lugtig PJ, Lensvelt-Mulders GJL, Frerichs R, Greven A. Estimating nonresponse bias and mode effects in a mixed mode survey. Int J Mark Res. 2011;53(5):669–86.

Mahfoud Z, Ghandour L, Ghandour B, Mokdad AH, Sibai AM. Cell phone and face-to-face interview responses in population-based surveys: how do they compare? Field Methods. 2015;27:39–54.

Massey JT, O'Connor D, Krotki K. Response rates in random digit dialing (RDD) telephone surveys. In: Paper presented at the Proceedings of the American Statistical Association, Section on Survey Research Methods; 1997.

McBride O, Morgan K, McGee H. Recruitment using mobile telephones in an Irish general population sexual health survey: challenges and practical solutions. BMC Med Res Methodol. 2012;12:45. https://doi.org/10.1186/1471-2288-12-45.

Moura EC, Claro RM, Bernal R, Ribeiro J, Malta DC, Morais Neto O. A feasibility study of cell phone and landline phone interviews for monitoring of risk and protection factors for chronic diseases in Brazil. Cad Saude Publica. 2011;27(2):277–86.

Nelson DE, Powell-Griner E, Town M, Kovar MG. A comparison of national estimates from the National Health Interview Survey and the behavioral risk factor surveillance system. Am J Public Health. 2003;93(8):1335–41.

Schofield M, Forrester-Knauss C. Surveys and questionnaires in health research. In: Liamputtong P, editor. Research methods in health: foundations for evidence-based practice. 3rd ed. Melbourne: Oxford University Press; 2017. p. 235–56.

Schwarz N, Strack F, Hippler HJ, Bishop G. The impact of administration mode on response effects in survey measurement. Appl Cogn Psychol. 1991;5(3):193–212.

Sibai AM, Ghandour LA, Chaaban R, Mokdad AH. Potential use of telephone surveys for non-communicable disease surveillance in developing countries: evidence from a national household survey in Lebanon. BMC Med Res Methodol. 2016;16(1):64. https://doi.org/10.1186/s12874-016-0160-0.

Steeh C. A new era for telephone surveys. In: Paper presented at the Annual Conference of the American Association for Public Opinion Research. Phoenix; 2004.

The American Association for Public Opinion Research. New considerations for survey researchers when planning and conducting RDD telephone surveys in the US with respondents reached via cell phone numbers. Deerfield: American Association for Public Opinion Research; 2010.

The American Association for Public Opinion Research. Standard definitions. Final dispositions of case Codes and outcome rates for surveys. 2011. Retrieved: https://www.esomar.org/uploads/public/knowledge-and-standards/codes-and-guidelines/ESOMAR_Standard-Definitions-Final-Dispositions-of-Case-Codes-and-Outcome-Rates-for-Surveys.pdf.

Vehovar V, Berzelak N, Manfreda KL. Mobile phones in an environment of competing survey modes: applying metric for evaluation of costs and errors. Soc Sci Comput Rev. 2010;28(3):303–18.

World Bank. World development indicators. 2013. Retrieved from: http://databank.worldbank.org/data/download/WDI-2013-ebook.pdf.

Phone Surveys: Introductions and Response Rates

81

Jessica Broome

Contents

1 The Relationship Between Interviewer Speech and Vocal Characteristics and Success .. 1418
2 A Model of Interviewer Voice and Speech to Minimize Nonresponse 1420
3 Green Lights and Red Flags 1423
4 Recommendations for Telephone Interviewing Practice 1425
    4.1 Recommendation #1: Train Interviewers to Switch Gears from Conversational Introductions to Standardized Interviews 1426
    4.2 Recommendation #2: Train Interviewer Speech Rates; Consider Implementing Hiring Criteria around Vocal Pitch 1427
    4.3 Recommendation #3: Emphasize Responsiveness to Answerer Concerns 1427
    4.4 Recommendation #4: Train Interviewers to be Aware of and Respond to both "Red Flags" and "Green Lights" from Answerers 1427
5 Conclusion and Future Directions 1428
References 1429

Abstract

As web surveys increase in popularity, the focus in the research industry on telephone surveys continues to decline. However, phone surveys are far from becoming extinct. Limited Internet access among certain populations (including older and lower income groups) makes telephone a preferred methodology when broad cross-sections of a population need to be reached, such as in the health research arena, where large-scale surveys such as the Behavior Risk Factor Surveillance System (BRFS), California Health Interview Survey (CHIS), and the Canadian Community Health Survey (CCHS) rely on telephone surveys. An understudied but critical component of phone surveys is the introduction.

J. Broome (✉)
University of Michigan, Ann Arbor, MI, USA

Sanford, NC, USA
e-mail: Jessica@jessicabroomeresearch.com

Introductions that are effective at convincing sample members to participate can help to improve shrinking response rates in this mode. These declining response rates have the potential to contribute to nonresponse error, and interviewers contribute differentially to nonresponse. Why do some telephone interviewers have better response rates than others? What are key speech and vocal characteristics of interviewers that help their performance, and how can these characteristics be implemented in practice? This chapter will review existing literature on telephone survey introductions, examine components of an ideal introduction, and conclude with suggestions for effective interviewer training in this area.

Keywords

Telephone · Interviewer · Nonresponse · Response rates · Survey · Voice · Speech

1 The Relationship Between Interviewer Speech and Vocal Characteristics and Success

Nonresponse to telephone surveys has been increasing steadily over the past 25 years (Curtin et al. 2005), and declining response rates have the potential to increase nonresponse error (Teitler et al. 2003; Groves et al. 2004; see also ▶ Chaps. 32, "Traditional Survey and Questionnaire Platforms" and ▶ 80, "Cell Phone Survey"). Further, nonresponse rates vary by interviewer (Oksenberg and Cannell 1988; Morton-Williams 1993; O'Muircheartaigh and Campanelli 1999; Snijkers et al. 1999). Uncovering the characteristics and tactics of successful interviewers can help to reduce nonresponse, either by using vocal and personality characteristics as hiring criteria or by training interviewers to adopt characteristics or tactics which have been shown to lead to increased success.

Literature from both survey methodology (Oksenberg et al. 1986) and telemarketing (Ketrow 1990) has found that a pleasing or attractive voice in the initial seconds of a phone call is imperative in extending the interaction. Further, Ketrow (1990) discusses the importance of giving an initial impression of competence, and Oksenberg and colleagues (Oksenberg et al. 1986; Oksenberg and Cannell 1988) found that judges' ratings of phone interviewer competence based on brief recorded excerpts were positively associated with the interviewers' success. This is not to imply that, in survey interview introductions, having a pleasing, competent-sounding voice in the opening statement is enough to guarantee success. However, an interviewer voice that gives listeners a positive first impression may lead to a longer conversation, thus increasing the likelihood of participation.

In contrast to face-to-face interviewers, telephone survey interviewers have just two primary tools which are under their control in their efforts to persuade answerers to participate: what they say (speech) and how they say it (vocal characteristics). A small body of literature (e.g., Sharf and Lehman 1984; Oksenberg et al. 1986; Oksenberg and Cannell 1988; Groves et al. 2007; Conrad et al. 2013) finds relationships between vocal characteristics of interviewers in telephone survey introductions and interviewer success in obtaining interviews. In general, successful interviewers have been ones who spoke louder (Oksenberg et al. 1986; Oksenberg and Cannell 1988; Van der Vaart et al. 2005) and with more falling intonation (Sharf and Lehman 1984; Oksenberg and Cannell 1988). In addition, success has been shown to be correlated both with higher mean fundamental frequency (Sharf and Lehman 1984) and with higher perceived pitch (Oksenberg et al. 1986), as well as variable fundamental frequency (Sharf and Lehman 1984; Groves et al. 2007) and variable pitch (Oksenberg et al. 1986). The terms "pitch" and "fundamental frequency" are often used interchangeably, but a necessary distinction is that fundamental frequency is an acoustic measure of vocal chord vibrations, while pitch is a listener's perception of frequency or how "high" or "low" a voice sounds.

Three recent studies have found nonlinear relationships between success and rate of speech (Groves et al. 2007; Steinkopf et al. 2010; Benkí et al. 2011): contacts with speech that is either overly slow or overly fast tend to be less successful. Benkí et al. (2011) found that contacts with interviewer speech in the range of 3.34–3.68 words per second were the most likely to be successful.

One critical question concerns what underlies these associations; what is it about an interviewer who speaks at a particular rate or with more variable pitch that leads to success, especially given the limited amount of exposure an answerer has to the interviewer's voice before deciding whether or not to participate? Oksenberg et al. (1986) emphasized the importance of an interviewer having a voice that potential respondents find appealing in the first few seconds of a survey interview introduction context, stating that "if vocal characteristics lead the respondent to perceive the interviewer as unappealing, cooperation will be less likely" (p. 99).

Two dimensions of person perception, warmth and competence, have been shown across a range of contexts to be relevant to the development of first impressions of others (Asch 1946; Fiske et al. 2007). Several studies in the literature on interviewer vocal characteristics (Oksenberg et al. 1986; Van der Vaart et al. 2005) suggest that ratings of personal characteristics on these dimensions of person perception are associated with both interviewer response rates and vocal characteristics. These studies involved collecting ratings of several interviewer personality characteristics, which were then successfully reduced to two dimensions interpretable as "warmth" and "competence." Characteristics on the "warmth" dimension included being cheerful, friendly, enthusiastic, polite, interested in her task, and pleasant to listen to. Oksenberg et al. (1986) and van der Vaart et al. (2005) found correlations between high ratings on the warmth dimension and vocal characteristics including variation in pitch, higher pitch, and a faster rate of speech, suggesting that listeners' impressions of interviewer personality are based, at least in part, on physical (acoustic) attributes of interviewers' voices. Characteristics composing the "competence" dimension included being self-assured, educated, intelligent, and professional. Van der Vaart et al. (2005) found that interviewers rated highly on "competence" characteristics tended to have lower pitch.

Importantly, both Oksenberg et al. (1986) and Van der Vaart et al. (2005) found that high ratings on a "warmth" dimension correlated with ratings of judges' willingness to participate. This aligns with Morton-Williams' (1993) finding that warm or "likable" interviewers increased perceived benefits to potential respondents and improved participation rates and also with Cialdini's (1984) "'Liking' Principle of Compliance"; people are more likely to comply with a request from someone they like.

Further, Cialdini (1984) suggests a compliance heuristic based on the principle of authority; requests from an authoritative speaker are more likely to be honored than requests with less authority. Impressions of authoritative characteristics such as competence and confidence, in turn, have been shown to be associated with interviewer success (Oksenberg et al. 1986; Oksenberg and Cannell 1988; Steinkopf et al. 2010).

The initial impression of a phone interviewer's voice as warm and competent may offer the interviewer the proverbial "foot in the door," giving the interviewer an opportunity to tailor their introduction to be more relevant to the individual by keeping a potential respondent (hereafter referred to as a phone "answerer") on the phone longer. Groves and Couper (1998) name "prolonging interaction" as a key strategy of successful interviewers for this reason.

Interviewer responsiveness to sample members has been shown to be effective in persuasion. Campanelli et al. (1997) find that relevant interviewer responses to specific householder concerns, indicating adaptation, are a successful interviewer technique. Groves and McGonagle (2001) are able to quantify this association with their finding that interviewers' cooperation rates in telephone surveys improved after interviewers went through a training program to increase their use of tailoring techniques, specifically by focusing on giving relevant responses to concerns expressed by sample members. Similarly, Pondman (1998) finds a 49% refusal rate among interviewers who were trained in a responsive strategy (compared to a 60% refusal rate among interviewers who did not participate in the training): in response to refusals, rather than asking "why" or repeating the refusal ("You don't want to participate?"), interviewers were instructed to offer to call back if answerers indicated not having time at the moment to take part in the interview, and also to "apply the provision of relevant information about features of the interview in reaction to refusals based on reasons other than time" (p. 75).

2 A Model of Interviewer Voice and Speech to Minimize Nonresponse

Broome (2012) proposes a model (Fig. 1) which suggests that when, excluding "hard-core" nonrespondents (who would not respond no matter what an interviewer said or how she said it), an initial voice which was not scripted could get an interviewer "over the hump" of an interaction, −that is, past an immediate refusal. According to this model, after giving a positive initial impression, an interviewer who was responsive to an answerer would have more success in persuading the answerer to comply with the survey request.

One study by Broome (2015) tested the hypothesis that first impressions, formed in the initial seconds of a telephone interviewer's introduction, are critical in determining the outcome of the contact. This study used a web survey to play

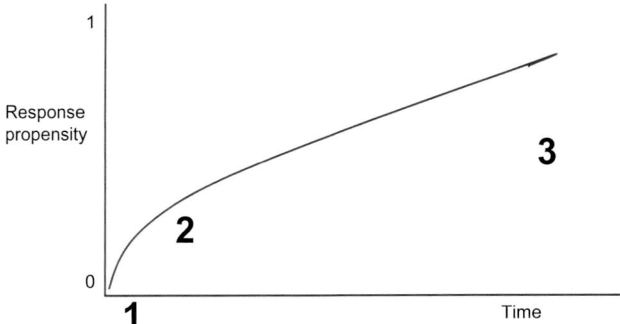

1: Some answerers will not respond– no matter what an interviewer says or how s/he sounds.

2: Initial presentation of a non-scripted voice can get an interviewer "over the hump" of the interaction.

3: Responsiveness for the remainder of the introduction will improve response likelihood.

Fig. 1 A proposed conceptual model of interviewer behavior

recordings of a number of introductions from actual phone studies, for which the outcome was known, to listeners. The study concluded that in fact the only rating by listeners that predicted contact outcome was whether or not the interviewer sounded scripted. Ratings of friendliness, competence, confidence, and other personality traits were *not* predictive of contact outcome. But while ratings of positive personality characteristics were both unrelated to contact success, ratings of an interviewer's scriptedness were significantly *negatively* associated with agreement, indicating that an interviewer who comes across to the listener as less scripted in the initial seconds of a contact has a greater chance of success.

While an initial impression of an interviewer as nonscripted is important, Broome (2012) also found that an interviewer's ability to be responsive to an answerer over the course of the survey introduction is far more important to her success than any initial impression. Through coded transcripts of interviewer/answerer dialogue, Broome explored two components of interviewer responsiveness in this study. The first was whether interviewers addressed answerer concerns. Answerer utterances were coded to indicate if they contained one of thirteen specific concerns, including "bad time," "purpose," or "not interested." Similarly, codes were assigned to interviewer utterances to indicate if the interviewer appropriately addressed a concern (e.g., by responding to "I'm too busy" with "We can call you back later").

The examples below (from actual contacts) show (1) a successful interviewer attempt at responsiveness to a concern and (2) an irrelevant interviewer response to the answerer's concern.

1. Answerer: There's a lot of questions that we probably couldn't even answer.
   Interviewer: Well, it's not a test or anything.
2. Answerer: There's only two of us and my husband's in the shower and I'm in the middle of making dinner.
   Interviewer: Well, this is a very important study.

Answerers may also present what Broome refers to as "conversation starters," which are utterances that are not concerns, but that provide opportunities for interviewers to give a response that demonstrates their attentiveness. Such remarks could be phrased as either comments or questions; they could include observations about the survey task, or they could be peripheral to the task – the point is that the interviewer can respond in a way that shows she has understood and thought about the answerer's comment. As in any conversation, the interviewer can "take the bait" and acknowledge these remarks with a relevant response – what Clark and Schaefer (1989) refer to as "contributing to discourse." Interviewers can also fail to effectively respond to answerers' conversation starters, either by offering content that is not relevant to what the answerer has said or by saying something with no substance. The example below shows a missed opportunity to respond to a conversation starter.

Answerer: The economy? I don't like it and that's the end of it.
Interviewer: Oh. Well.

Interviewers' contributions to discourse can take several forms. Examples of effective interviewer responses to conversation starters are below. In the first, the interviewer's response to the answerer's question show that she is adhering to conversational norms by answering a question addressed to her; her second turn, "Oh, thank you," is an expression of politeness. In the second, the interviewer's response is more substantial and demonstrates attention, adaptation to the answerer's comment, and quick thinking to build rapport with the answerer:

1. Answerer: Um, this is your job right?
   Interviewer: Yes sir it is.
   Answerer: Oh [laughs] oh, ok well we want you to keep your job.
   Interviewer: [laughs] Oh, thank you.
2. Answerer: Just so you know, the next time Nebraska plays Michigan [laughs] we're going to root for Nebraska even though you're giving us fifty bucks.
   Interviewer: That's all right. I'll root for Nebraska if you do the interview. How's that?

Analyses revealed that while overall responsiveness by an interviewer is important, an interviewer's ability to address concerns trumps her responses to conversation starters in persuading an answerer to participate.

If an interviewer can keep a potential respondent on the phone long enough to engage in an exchange, responsiveness is immensely helpful. Returning to the model of interviewer behavior proposed above, Broome (2012) found that interviewers who start out as scripted but are highly responsive as the answerer raises concerns or presents conversation starters are nearly as successful as their counterparts who begin with a low level of scriptedness and then act responsively. Thus, while being less scripted can help interviewers get past the initial "hump," in contacts that survive the initial stage, interviewer responsiveness is crucial to success.

3   Green Lights and Red Flags

The work of an interviewer in being responsive varies greatly. Answerers who ultimately agree tend to express fewer concerns, but produce more conversation starters. In contacts that result in refusal, answerers express a relatively high number of concerns in a shorter time period and often do not give the interviewer a chance to respond. The types of tailoring opportunities (conversation starters and concerns) presented by answerers can offer the interviewer a clue as to where the contact is headed. Concerns about the purpose or content of the survey, or the length of the interview, are most common in contacts where the answerer ultimately agrees to participate, as are conversation starters (Broome 2014). This parallels the finding by Schaeffer and colleagues (2011) that when answerers ask about interview length, acceptance is more likely. This may be because concerns about call timing or survey length are relatively straightforward for interviewers to address:

1. Answerer: But you're calling at a bad time because we have company.
   Interviewer: Oh, oh, I see. Is there a better time that we could call back?
2. Answerer: How many minutes is the survey?
   Interviewer: It takes maybe 10 minutes. It's pretty short.

Sweeping statements of disinterest, such as "I'm not interested" or "I don't want to participate," and personal policies such as "I don't do surveys," are much harder for interviewers to respond to. These statements are the most common concern in refusals (Broome 2014); they should be viewed as red flags and handled with caution. Interviewer responses to these concerns are often ignored and followed or interrupted by hang-ups. In responding to this type of utterance, probing for more information or asking the answerer to elaborate on his or her concerns may not be an interviewer's best strategy, as it prompts answerers to descend into a spiral of negativity. Interviewers may have a better chance of success if they treat and address "I'm not interested" as a concern in and of itself, rather than view it as a symptom of another concern.

Often, interviewers do not have a chance to address "I'm not interested" concerns: Broome found that in 26% of contacts containing an expression of disinterest, the answerer did not say anything after the statement of disinterest before hanging up, showing no willingness to react to an interviewer's response. These statements, then, should be interpreted by interviewers as "red flags" that indicate that unless drastic action is taken, the contact is about to be terminated.

Broome found that statements of disinterest are often presented in combination with another, more easily addressable concern, and interviewers often default to addressing these concerns, rather than the answerer's lack of interest. In the following examples, the interviewer chooses to respond to other issues brought up by the answerers (repeated calls in the first example and a misunderstanding of the purpose of the call in the second):

1. Answerer: I really don't want to take it. So I need you to take me off the list or quit calling here because I don't have time to do a survey. I've already declined. They've called me like three or four times. I told them the last time that I just wasn't interested in doing it.
   Interviewer: Oh, I do apologize ma'am for all the calls. We are actually coming to the end of our study and we really do need representation from your area.
2. Answerer: Yeah well I won't be interested in that. I don't even know what it's about. And then plus I'm tired of telemarketers calling here.
   Interviewer: Oh I completely understand ma'am. You know a lot of times we do get confused with telemarketers. We are not telemarketers.

Interviewers may seek out an addressable concern by asking why the answerer does not want to participate as a means to establish common ground or a mutual understanding with the answerer, but Pondman (1998) found that this type of query prompts answerers to verbalize or repeat their reasons for not wanting to participate and rarely leads to conversions; interviewers at the University of Michigan are advised to "break the habit of asking what the concerns are" (J. Matuzak, March 29, 2011, personal communication). However, this is not an uncommon technique among interviewers.

It is important to note that while it is risky, asking an answerer to specify his or her concerns can be beneficial, as it sometimes prompts the answerer to express a concern that the interviewer can easily address. One technique that may be effective is following the question with a barrage of information intended to assuage myriad possible concerns, as in this example:

Answerer: Uh, you know, I'm not interested in that.
Interviewer: Ok, are there any concerns you have? This study is one of the most important studies in the country. It's looked at by the Federal Reserve Board. Your number was chosen to represent your part of California and you really can't be replaced in the study. It's just general opinion questions.
Answerer: Ok, all right, let's do it.

Instead of probing for a specific concern, other techniques for responding to expressions of disinterest included ignoring, acknowledging, or rejecting them, as in the examples below. It cannot be said with certainty that these responses improve response likelihood; instead, they are presented as options for interviewers to consider in lieu of asking disinterested answerers to elaborate on their concerns.

Here the interviewer bypasses the answerer's statement of disinterest and moves into the household listing:

Answerer: I don't think I'm interested.
Interviewer: We can just do the first part to determine who is eligible and then after that we can just set up an appointment to call back later. It just takes 2 minutes to find out who in your household the computer will pick to participate. So what's your first name?

In the examples below, rather than ignoring the statement of disinterest, the interviewer acknowledges it directly. Faber and Mazlish (1980) suggest that acknowledging and naming negative feelings when they are expressed, rather than rejecting or downplaying them, is an effective tactic for engaging children in distress; it appears that some interviewers employ this technique with potential respondents as well:

1. Answerer: Well, I ain't interested.
   Interviewer: I know you're saying you're not interested, but I'd be more than willing to talk about the study with you right now so you can familiarize yourself with it.
2. Answerer: I just don't want to do it is what it comes down to.
   Interviewer: Yeah, I understand that part. Right.

Sometimes interviewers offer an explanation as to why the answerer's policy of nonresponse does not apply in this situation, as in this example:

Answerer: I just don't like to participate in phone surveys.
Interviewer: Well this is actually not a typical phone survey.

Addressing the lack of interest – that is, treating a statement of disinterest or a nonresponse policy as a legitimate and addressable concern, rather than asking answerers to elaborate on the reasons for their disinterest – may help interviewers to avoid the phenomenon discussed by Pondman (1998), where answerers who are probed to express their reasons for not wanting to participate are less likely to be converted to agreement. Other examples of rebuttals to statements of disinterest included:

1. Interviewer: Most people find this a pretty interesting study and this is really the first time it's ever been done.
2. Interviewer: A lot of people who haven't wanted to do it did participate in it and found it quite interesting.

4 Recommendations for Telephone Interviewing Practice

The findings discussed above, coupled with other findings in the survey methodology literature, have minimal use unless they can be applied to the practice of telephone interviewing to improve response rates. This section will discuss applications of these results for survey practice.

4.1 Recommendation #1: Train Interviewers to Switch Gears from Conversational Introductions to Standardized Interviews

The finding that judgments of an interviewer's scriptedness in the initial seconds of a contact are negatively associated with contact success should be considered by those responsible for hiring, training, and monitoring interviewers. Interviewers should be encouraged to make their speech as natural as possible, through the use of intonation patterns and word selection. Interviewers can be exposed to contacts with both high and low ratings of scriptedness to make clear the difference.

While interviewers may be required to mention particular points in their introduction or even to follow a verbatim introductory script (depending on the institution's policy), they should be trained to sound as conversational as possible, particularly at the start of their introduction. Both Houtkoop-Steenstra and van den Bergh (2000) and Morton-Williams (1993) found that interviewers who were allowed to adapt their introductory script had greater success. Further, work by Conrad et al. (2013) demonstrated that a moderate use of fillers such as "um" and "uh" by interviewers can lead to greater success, possibly because these interviewers sound like they are engaged in a natural conversation, rather than following a script.

It could behoove survey organizations to conduct research around which elements of speech (increased use of fillers such as "um" and "uh," maintenance of natural intonation patterns, and "on the spot" adaptation of scripts) can be taught and then to focus on training interviewers to use these techniques.

Beyond the introduction, the issue of standardized interviewing, and what departures from verbatim interview scripts can mean for data quality, is the subject of much debate. Schober and Conrad (1997) and Conrad and Schober (2000) found clear evidence that "conversational" interviewing, or less rigid adherence to interview scripts, can enhance data accuracy. Along similar lines, Dykema et al. (1997) found that interviewer deviations from standard question wording had minimal impact on respondents' reports. Still, "reading the questions exactly as worded" is a tenet of interview administration which is upheld and enforced in most survey organizations (Broome 2012).

Because emphasizing the need to read questions in a standardized manner may seem in conflict with emphasis on less scripted delivery of introductions, interviewers need to be trained to "wear two hats." It needs to be made explicit to interviewers that there are two distinct (but, arguably, equally important) elements of the phone component of their job, each requiring a different style of speech and interaction. In the introductory or persuasive portion, scriptedness may be a liability, and the ability to "think on one's feet" to respond to answerers is an asset. In contrast, in the interviewing portion, deviating from a script may have ramifications for data quality or, at the very least, will represent a lack of adherence to the organization's procedures. Interviewers should be trained to "switch gears" between these two speech styles and perhaps even be encouraged to acknowledge to respondents that their delivery of the questions will sound different from their introduction.

4.2 Recommendation #2: Train Interviewer Speech Rates; Consider Implementing Hiring Criteria around Vocal Pitch

It may be a worthwhile investment by research organizations to place greater emphasis on interviewers' vocal characteristics. Speech rate in particular is something that can be trained and monitored. Benkí et al. (2011) found that a rate of 3.5 words per second during the introduction is ideal in obtaining cooperation. This rate is fast enough to sound self-assured, but not so fast as to be incomprehensible. Notably, this is higher than the rate of two words per second, which is often suggested anecdotally as an ideal speech rate for delivering interview questions. Just as interviewers may be able to be trained to speak in a less scripted manner during an introduction and more so during an interview, they can be trained to slow down their speech noticeably after the introduction. This type of training could be implemented with the use of software measuring speech rate and displaying it to the interviewer as she talks (similar to speed clocks which show drivers their current speed); interviewers could slow down or speed up their speech in response, and eventually learn how it feels and sounds to speak at a rate of 3.5 (for an introduction) or 2.0 (for question delivery) words per second.

Additionally, Benkí et al. (2011) found lower pitch over the course of a contact to be associated with success. If further research can substantiate these findings, survey organizations may want to limit their hiring to interviewers who are capable of hitting an optimal pitch range (or avoiding a pitch range which is associated with negative impressions and lower success).

4.3 Recommendation #3: Emphasize Responsiveness to Answerer Concerns

While scriptedness and vocal characteristics in the initial seconds of a survey introduction are important, it is important to note that an interviewer's ability to be responsive to answerers is absolutely critical, whether they are addressing concerns or "taking the bait" by responding to conversation starters. Practicing by interviewers of appropriate responses to common answerer concerns should be a top priority in interviewer training; as Groves and McGonagle (2001) demonstrated, interviewers trained in effective responding have greater success.

4.4 Recommendation #4: Train Interviewers to be Aware of and Respond to both "Red Flags" and "Green Lights" from Answerers

Some answerer comments should be viewed as "red flags," or warnings that the contact is about to end. Particular attention should be paid to concerns in the "I'm not interested" category. Interviewers often treat statements in this category as symptoms of a different concern and, instead of addressing the answerer's stated lack of interest, attempt to uncover a more addressable concern, such as a lack of time or worries about privacy. Findings by Pondman (1998) show that asking answerers to elaborate on this type of remark (as in, "May I ask why you don't want to participate?") can have disastrous consequences, sending the answerer on a tirade of negativity and leading to hang-ups at comparable rates as when this question is not asked. Instead, interviewers may have better success by addressing lack of interest in and of itself as a legitimate concern, rather than treating it as a symptom of another concern. Still, statements of disinterest often indicate an impending hang-up and should be treated as red flags by interviewers.

On the other hand, some answerer utterances can be viewed as "green lights," or signals that the answerer is open to participating. Questions about the length or content of the survey are more common in contacts where the answerer ultimately agrees to participate. Similarly, the presentations of conversation starters by answerers, such as questions directed at the interviewer or comments peripheral to the task at hand, are not only more frequent than concerns in agree contacts, but are much more frequent in agree compared to refusal contacts and should be viewed as signs of engagement and likely participation.

Other research has looked at utterances by answerers that may indicate a greater likelihood of agreement. Work by Conrad et al. (2013) found that answerers who use more backchannels such as "mm-hmm" or "I see" seem to be indicating engagement in the conversation. These answerers are more likely to agree with the survey request than those who use fewer backchannels.

Being attuned to green lights in answerer speech, such as backchannels, questions about length or content, or the presentation of more conversation starters relative to the expression of concerns, can help interviewers know when an answerer is likely to agree and adapt their introduction in turn. In such a case, backing off from a "hard sell" may be recommended; however, it may also be advisable for interviewers to gently urge the answerer to begin the interview, rather than schedule a callback.

5 Conclusion and Future Directions

Using the findings from studies on interviewer speech in phone survey introductions can help research organizations inform interviewer training and requirements. Specifically, findings show that less scripted speech in the initial seconds of an introduction can help interviewers to keep potential respondents on the phone. Further, interviewer responsiveness to potential respondents' concerns and conversation starters lead to greater likelihood of participation. Finally, aspects of interviewer speech, such as pitch and rate, may result in higher response rates. Implementing these recommendations can help survey organizations increase efficiencies and potentially reduce nonresponse error.

There is also room for additional research in this area that can add depth to the findings discussed above. Questions for further work may include:

- Does gender, either of interviewer or answerer, interact with the importance of sounding responsive or nonscripted?
- What is the relationship between ratings of scriptedness and actual paralinguistic measures, such as pauses and fillers ("um," "uh")? Is there an ideal rate of fillers and pauses in successful invitations?
- When thinking about "red flags" and "green lights" from answerers in the introduction, is there any specific content (e.g., statements of "I'm not interested") that interviewers can be trained to address?

References

Asch SE. Forming impressions of personality. J Pers Soc Psychol. 1946;9:283–94.

Benkí J, Broome J, Conrad F, Groves R, Kreuter F. Effects of speech rate, pitch, and pausing on survey participation decisions. Paper presented at the 66th annual conference of the American Association for Public Opinion Research, Phoenix. 2011.

Broome J. Vocal characteristics, speech, and behavior of telephone interviewers. Unpublished doctoral dissertation. University of Michigan, Michigan. 2012.

Broome J. How telephone interviewers' responsiveness impacts their success. Field Methods. 2014;27:66–81.

Broome J. First impressions of telephone interviewers. J Off Stat. 2015;31:611–25.

Campanelli P, Sturgis P, Purdon S. Can you hear me knocking: an investigation into the impact of interviewers on survey response rates. London: National Center for Social Research; 1997.

Cialdini RB. Influence: science and practice. New York: Harper Collins; 1984.

Clark HH, Schaefer EF. Contributing to discourse. Cogn Sci. 1989;13:259–94.

Conrad FG, Schober MF. Clarifying question meaning in a household telephone survey. Public Opin Q. 2000;64:1–28.

Conrad F, Broome J, Benkí J, Groves R, Kreuter F, Vannette D, McClain C. Interviewer speech and the success of survey invitations. J R Stat Soc. 2013;176(1):191–210.

Curtin R, Presser S, Singer E. Changes in telephone survey nonresponse over the past quarter century. Public Opin Q. 2005;69(1):87–98.

Dykema J, Lepkowski JM, Blixt S. The effect of interviewer and respondent behavior on data quality: analysis of interaction coding in a validation study. In: Lyberg LE, Biemer P, Collins M, de Leeuw E, Dippo C, Schwarz N, Trewin D, editors. Survey measurement and process quality. New York: Wiley; 1997. p. 311–30.

Faber A, Mazlish E. How to talk so kids will listen & listen so kids will talk. New York: Harper Collins; 1980.

Fiske ST, Cuddy AJC, Glick P. Universal dimensions of social cognition: warmth and competence. Trends Cogn Sci. 2007;11:77–83.

Groves RM, Couper MP. Nonresponse in household interview surveys. New York: Wiley; 1998.

Groves RM, McGonagle K. A theory-guided interviewer training protocol regarding survey participation. J Off Stat. 2001;17:249–65.

Groves RM, Presser S, Dipko S. The role of topic interest in survey participation decisions. Public Opin Q. 2004;68(1):2–31.

Groves RM, O'Hare BC, Gould-Smith D, Benkí J, Maher P. Telephone interviewer voice characteristics and the survey participation decision. In: Lepkowski JM, Tucker C, Brick JM, de Leeuw ED, Japec L, Lavrakas PJ, Link MW, Sangster RL, editors. Advances in telephone survey methodology. New York: Wiley; 2007. p. 385–400.

Houtkoop-Steenstra H, van den Bergh H. Effects of introductions in large-scale telephone survey interviews. Sociol Methods Res. 2000;28(3):281–300.

Ketrow SM. Attributes of a telemarketer's voice and persuasiveness: a review and synthesis of the literature. J Direct Mark. 1990;4:7–21.

Morton-Williams J. Interviewer approaches. Cambridge, UK: Cambridge University Press; 1993.

O'Muircheartaigh C, Campanelli P. A multilevel exploration of the role of interviewers in survey non-response. J R Stat Soc Ser A. 1999;162(3):437–46.

Oksenberg L, Cannell C. Effects of interviewer vocal characteristics on nonresponse. In: Groves RM, Biemer PB, Lyberg LE, Massey JT, Nichols II WL, Waksberg J, editors. Telephone survey methodology. New York: Wiley; 1988. p. 257–72.

Oksenberg L, Coleman L, Cannell C. Interviewers' voices and refusal rates in telephone surveys. Public Opin Q. 1986;50(1):97–111.

Pondman LM. The influence of the interviewer on the refusal rate in telephone interviews. Unpublished doctoral dissertation. Vrije Universiteit, Amsterdam. 1998.

Schaefer NC, Garbarski D, Maynard DW, Freese J. Interactional environments, requests, and participation in the survey interview. Paper presented at the annual meeting of the American Association for Public Opinion Research, Phoenix. 2011.

Schober MF, Conrad FG. Does conversational interviewing reduce survey measurement error? Public Opin Q. 1997;61:576–602.

Sharf DJ, Lehman ME. Relationship between the speech characteristics and effectiveness of telephone interviewers. J Phon. 1984;12(3):219–28.

Snijkers G, Hox J, de Leeuw ED. Interviewers' tactics for fighting survey nonresponse. J Off Stat. 1999;15(2):185–98.

Steinkopf L, Bauer G, Best H. Nonresponse in CATI-surveys. Methods Data Anal. 2010;4(1):3–26.

Teitler JO, Reichman NE, Sprachman S. Costs and benefits of improving response rates for a hard-to-reach population. Public Opin Q. 2003;67:126–38.

Van der Vaart W, Ongena Y, Hoogendoorn A, Dijkstra W. Do interviewers' voice characteristics influence cooperation rates in telephone surveys? Int J Opin Res. 2005;18(4):488–99.

The Freelisting Method

82

Marsha B. Quinlan

Contents

1 Introduction 1432
2 To Freelist or Not 1433
3 Using Written, Electronic, or Oral Interviews 1435
4 Focusing the Domain 1436
5 Freelist Analysis 1438
6 Checking Freelists with Ethnographic Interviews 1440
7 Conclusion and Future Directions 1443
References 1443

Abstract

A freelist is a mental inventory of items an individual thinks of within a given category. Freelists reveal cultural "salience" of particular notions within groups, and variation in individuals' topical knowledge across groups. The ease and accuracy of freelist interviewing, or freelisting, makes it ideal for collecting data on health knowledge and beliefs from relatively large samples. Successful freelisting requires researchers to break the research topic into honed categories. Research participants presented with broad prompts tend to "unpack" mental subcategories and may omit (forget) common items or categories. Researchers should find subdomains to present individually for participants to unpack in separate smaller freelists. Researchers may focus the freelist prompts through successive freelisting, pile sorts, or focus group-interviews. Written freelisting among literate populations allows for rapid data collection, possibly from multiple individuals simultaneously. Among nonliterate peoples, using oral freelists remains a relatively rapid method; however, interviewers must prevent bystanders from "contaminating" individual interviewees' lists. Researchers should cross-

M. B. Quinlan (✉)
Department of Anthropology, Washington State University, Pullman, WA, USA
e-mail: mquinlan@wsu.edu

© Springer Nature Singapore Pte Ltd. 2019
P. Liamputtong (ed.), *Handbook of Research Methods in Health Social Sciences*,
https://doi.org/10.1007/978-981-10-5251-4_12

check freelist responses with informal methods as much as practicable to contextualize and understand the references therein. With proper attention to detail, freelisting can amass high quality data on people's medical understanding, attitudes, and behaviors.

Keywords

Freelist · Free recall · Salience analysis · Systematic data collection · Domain analysis · Rapid Ethnographic Assessment (REA) approaches

1 Introduction

Freelisting is a qualitative, easily quantifiable method. Freelists quickly and easily amass data that (1) identifies items in a cultural domain, or emic category; (2) indicates which of those things are most important, or salient within the culture; and (3) reveals how much variation there is in the knowledge or beliefs in question (Quinlan 2016). In a freelist interview, a respondent simply lists members (things) that they perceive to be part of a domain (e.g., "ways to avoid HIV," "breakfast foods," "reasons to fear hospitals," or "treatments for a cough") in whatever order they come to mind. The resulting lists tap into local knowledge and its variation in a study community. Hence, the method is well suited to find "knowledge, attitudes, and practices" (KAP); ethnomedical beliefs; and some types of prevalence.

Freelisting is a well-established ethnographic method that rests on three assumptions (e.g., Romney and D'Andrade 1964; Henley 1969; Bolton et al. 1980). First, when people freelist, they tend to list terms in order of familiarity. When listing kinship terms, for example, people generally list "mother" before "aunt," and "aunt" before "great-aunt" (Romney and D'Andrade 1964). Second, individuals who know a lot about a subject list more terms than people who know less. For instance, people who can look at an unlabeled map and correctly name many countries also make long freelists of country names (Brewer 1995). And third, terms that most respondents mention indicate locally prominent items: Pennsylvanians list "apple" and "birch" trees more frequently and earlier than they do "orange" or "palm" trees (Gatewood 1983).

Recognition of freelisting as a productive method for health research is increasing. Bayliss et al. (2003), for example, conducted freelisting to find barriers to self-care among Denver patients with chronic comorbid medical problems. In a series of team health studies, Frances Barg successfully employed freelisting on diverse medical topics including women's views of urinary incontinence (Bradway and Barg 2006; Bradway et al. 2010); nutrition concepts and strategies among Philadelphian urban poor (Lucan et al. 2012); parental and pediatrician decision-making for ADHD (Fiks et al. 2011); and perspectives of patients, caregivers, and clinicians on heart failure management (Ahmad et al. 2015). Bolton and Tang employed freelists and associated semi-structured interviewing during postdisaster periods, and recommend freelisting in such situations, in which researchers need an ethnographic method that is "both quick and easy to implement" (2004, p. 97). They argue that such interviewing on local names and common manifestations of health problems, and locally accepted treatments for those problems in affected communities, results in interventions that are more acceptable to local people, thus more effective and sustainable. Finally, freelisting has been used in many studies of medical ethnobotany (e.g., Trotter 1981; Crandon-Malamud 1991; Berlin and Berlin 1996; Nolan and Robbins 1999; Ryan et al. 2000; Nolan 2001, 2004; Finerman and Sackett 2003; Quinlan 2004, 2010; Quinlan and Quinlan 2007; Pieroni et al. 2008; Waldstein 2006; Ceuterick et al. 2008; Giovannini and Heinrich 2009; Mathez-Stiefel et al. 2012; Flores and Quinlan 2014).

In principle, there is a distinction between freelists and open-ended surveying. Freelists inquire about *cultural* domains, while open-ended questions ask for information about the informant (Borgatti 1999). Asking someone to list "medicines *you* use" is an open-ended survey, while asking for "medicines *people here* use" is a true freelist. In practice, it makes sense to consider a freelist a *kind* of survey, and the distinction between freelisting and open-ended surveying may be inconsequential as individuals often answer both open-ended surveys and freelists from a personal perspective. In the over 1,000 oral freelists I conducted on medicines people use in Dominica, it was common, if not typical, for interviewees to respond as though I had asked about a personal attribute, though I asked about a general cultural one. Instead of responding, "People here use...," they replied in the first person singular or plural (e.g., "I'm using..." or "We're using..."). If the egocentric perspective is normal in freelists, it would explain why, in freelists of kin terms, Romney and D'Andrade's (1964) High School informants (who presumably had parents but probably not offspring) listed "mother" and "father" much more than "son" or "daughter," and while several listed grandparental terms, very few listed grandchildren terms. My experience leads me to conclude that individuals' freelists are largely personal or egocentric, although this may vary according to culture or topic. The difference between freelisting and open-ended (freelist styled) surveys may be insignificant for many research questions. However, each researcher must decide whether a tendency to respond personally is an important concern. In either case, one could analyze the lists for salience.

Drawing mostly upon my ethnomedical research in rural Dominica, West Indies, in this chapter, I describe the advantages and obstacles of using freelists. Freelist interviews allowed me to (1) find culturally important illnesses, (2) identify local herbal treatments for those illnesses, and (3) explore sociodemographic variables associated with treatment knowledge. Specifically, I discuss five issues for efficient use of freelists in the field: (1) whether to conduct freelist interviews or not, (2) whether to collect oral or written lists, (3) focusing the domain of each freelist interview, (4) types of freelist analysis, and (5) cross-checking freelists with ethnographic interviewing.

2 To Freelist or Not

Freelists provide inventories and boundaries of cultural domains. In health social science, the freelist method is ideal if one wants to find the most culturally salient knowledge (e.g., cut treatment, mosquito control); attitudes towards, or associations with, an issue or topic (e.g., obesity, vaccinations, violence); or different ways locals do something (e.g., prepare a medicine or a food, decide on healthcare). Freelisted data allows the researcher to discover the relative salience of items across all respondents within a given domain. Salience is a statistic accounting for rank and frequency (e.g., in the domain of English color terms, "red" is more salient – it appears more often and earlier in freelists – than "maroon" [Smith et al. 1995]). Researchers can calculate the mean salience value for all listed items, to reveal the intracultural salience of each term (below). Researchers can also compare individuals' lists to assess who in a community knows more (or less) about a certain domain of knowledge.

A potential shortcoming of freelisting is that inventories may not be as exhaustive as inventories gained through other methods. For example, medical ethnobotanists normally rely on key informant interviews with local plant experts, including "field interviews," i.e., walking through vegetation zones or plots with informants and noting every useful plant found (Alexiades 1996, pp. 65–66). Long interviews with key informants may offer informants visual cues and allow or encourage informants to remember more obscure species. Conducting several long interviews may generate more exhaustive inventories than freelists.

The specificity of domains can limit freelists. For example, my freelists on illnesses that Dominicans "cure" with "bush medicine" did not yield the multiple gynecological conditions for which Dominicans use bush (herbal) medicine (Flores and Quinlan 2014). Dominicans regard childbirth, menstruation, and so on as normal events for healthy women, not illnesses requiring a "cure." Another factor limiting freelists is that they only reflect terms in a respondent's active vocabulary (or lexical command). Informants are able to recognize more items in a domain than they can freelist from memory (Hutchinson 1983). Researchers can, however, maximize freelist output through supplementary prompting (see Brewer 2002; Gravlee et al. 2012).

The advantages of freelisting outweigh the possibility of reduced inventory, in most cases. First, freelists, unlike less-structured interviews, are rapid and simple. They allow for much larger samples in less time. Other rapid interview methods require the researcher to have prior expertise in the domain. Recognition tasks, questionnaires, sorting, and ranking interviews, for example, have predetermined responses built into the instrument (see Bernard 2011). Second, unlike data from less-structured interviewing, freelists are quantifiable. As Handwerker and Borgatti (2014, p. 520) argue, "even simple forms of numerical reasoning add important components to ethnographic research...Reasoning with numbers reveals things you'd otherwise miss." Focused freelists gather every significant or salient item that the population associates with a domain, and freelist data allows one to find areas of consensus or high modality within the community (Boster 1987; D'Andrade 1987; Weller 1987). In addition, an informant's list length is a measure of that person's depth of knowledge or familiarity within a domain (Gatewood 1983, 1984; Borgatti 1990; Brewer 1995; Furlow 2003). Thus, a researcher can use freelists to identify community experts or examine intracultural variation (Quinlan 2000). One can examine freelists' content comparatively. For example, rural Missouri novices in wild plant use listed highly recognizable, ecologically salient species like blackberry and sunflower, while experts listed greater proportions of plain, herbaceous species native to the region, like burdock and plantain (Nolan 2002). In a study of postpartum problems and associated etiologies, signs, and care in Matlab, Bangladesh, Hruschka et al. (2008) found agreement of mentions among local lay women and "traditional birth attendants," while "skilled birth attendants" (i.e., accredited health professionals) had different (biomedical) responses.

3   Using Written, Electronic, or Oral Interviews

In fully literate communities in which the terms sought (e.g., plant names, illnesses, and so on) are in a written language, researchers can provide freelist interview schedules for informants to fill in themselves. Interview sheets simply contain a prompt written above a series of blanks. If the research participants have computer/smartphone and internet access, web-based freelisting surveys are another written option (see Gravlee et al. 2012). These self-administered methods may work well with developed, industrialized populations. Self-administered freelists offer the advantage of privacy, which is good for getting uncontaminated responses that best test for intracultural variation, and privacy is best for potentially touchy topics.

Freelist interviews can also be oral (face-to-face). For example, I collected oral freelists in Dominica. Rural Dominicans vary in literacy. Most herbal medicine names are in French Patois, a largely unwritten Creole language, in which residents struggle (more than they do in English) to spell (sound-out) words. However, collecting oral freelists did complicate the freelist procedure because making oral lists is less formal and less independent than completing a written list. While informally listing plants, participants sometimes called out for help from nearby friends or family who (trying to help) shared various remedies. Every so often, somebody saw another villager doing an interview, approached out of curiosity, and offered suggestions. These occasions were difficult because the freelists should contain the items that one individual knows in the order that they come to mind *for that individual*.

Research assistants and I dealt with shared-answer events in several ways according to the situation. For example, one woman I interviewed had listed two cures for the common cold when the name of a third slipped her mind. Her aunt was about 30 feet away, and the woman called out, "Auntie! What that bush we using for the cold, na?" "*Timayok!*" the aunt screamed back. The woman had already listed *timayok*. "No, the other one!" The aunt yelled, "Hibiscus flower." "Yes, but there's another one again," the woman said to the aunt. The freelister turned to me and said, "Hibiscus flower is good too, you know. We using that plenty for colds." "*Pachuri?*" the aunt called. "THAT is it! *Pachuri*," the woman said to the aunt and me. In this case, I continued with the woman's freelist, leaving a blank space in the third position, writing "hibiscus" in the fourth position, then, filling "*pachuri*" in the third position, because that was the one the woman was thinking of when she involved the aunt. In some cases, we abandoned the freelists for a later time, as it was less than clear in what order the interviewee would have thought of the herb, had he or she not been prompted. Leaving an interview because of compromised data was discouraging and consultants were sometimes miffed when I or another interviewer returned to try again.

We learned to avoid most compromised freelist situations by explaining the project to all the nearby adults, and isolating the informant somewhat by stepping around a corner. Had villagers been writing their own responses, rather than listing them aloud, they still may have called out for memory help, or received suggestions from curious bystanders, but probably less often. Written exercises are inherently more private, but they may not be of higher quality. Gravlee et al. (2012) conducted an experiment to compare university students' freelist responses in face-to-face oral interviews, hand-written, and online freelists. They found that all three modes identified the same salient members of each domain. They also found that the face-to-face interviews had the fewest items per respondent but the most agreement. It is possible that the social nature of the face-to-face mode may elicit slightly more thoughtfulness, concentration, or self-editing.

4 Focusing the Domain

Freelists are an advisable method for accumulating inventories. Yet, they *may* not yield a *total* knowledge. Interviewees commonly forget to list items in a domain or (for expediency) intentionally omit items they know (Brewer 2002). Omissions are, in my experience, most likely if the freelist prompt is broad. Freelist data is ranked so that the order in which people list items reveals psychological or cultural preeminence of items *given a certain prompt*. The more focused the prompt, the more complete the freelist will be for that subject whereas vague, general prompts result in broad, scattered lists of questionable utility (Drawing the line on *how much* to focus one's prompt depends on the research question at hand. Cultural experts could potentially parse out items until each prompt corresponds with only a single item.).

Freelists must deal with a single mental category, called a semantic domain (Weller and Romney 1988; Bernard 2017). If the prompted domain is broad, the inventory in the freelist often consists of clusters of subdomains (mental categories). In an experiment (Quinlan 2016), I asked Ball State University students to complete two freelists on birds. In the first, they named "all the birds you can think of." Responses on these lists were diverse, many listing eagles or (American) bald eagles first, followed by other raptors or by pet bird species, or poultry species, or local wild species, but generally clustering by domains as students unpacked their mental clusters of birds. In the second freelist, done some time later, they named "backyard birds in Indiana." Cardinals (the Ball State mascot) were most salient, followed by robins. One informant stated that "robins" were one of the first birds she learned to name as a child. Indeed, in her freelist of backyard birds, robins were highly ranked, third in a list of 16. In her freelist of *all* birds, however, robins appeared near the list's end, emerging as an "afterthought." Before this informant listed "robin," she listed a series of pet birds (one mental domain for her); followed by a series of colorful, exotic birds such as macaws and toucans; then, some raptors; some poultry birds;

and finally, common local wild birds, including robins. In every case, students listed more local birds with the "backyard bird" prompt than in their "all the birds" lists.

In Dominica, open-ended pilot surveys (e.g., Borgatti 1999), in which I asked consultants to list "all the *bush medicines* [herbal remedies] people use," were similarly too broad. Lists contained clusters of subdomains, usually grouped into treatments for particular ailments, though sometimes grouped by the individuals that grew/used the plant, plant size/shape, and so on. Further, many species – later identified as leading treatments for common illnesses – were missing from these broad lists, presumably because the treatment's subdomain (a particular illness) did not occur to the informant during the interview. Similarly, Saraguro (Ecuadorian Andean) freelisters did not recall several plants in their own home-gardens, omitting scarcer ones (usual in freelists) and also plants that were possibly "too ubiquitous to consider 'interesting'" (Finerman and Sackett 2003, p. 462).

To make freelists most efficient and accurate, it is helpful (for researchers and consultants alike) to narrow the freelist's domain. It is easier, for example, for someone to list "all of the over-the-counter medicines in your house" than "all of the over-the-counter medicines at your pharmacy," and easier still to list "all of the over-the-counter medicines in your medicine cabinet." Asking someone to list *sore throat* medicines that he knows is less daunting and less bother than listing *every* medicine he knows of. Researchers should identify relevant, focused domains, and then conduct freelists on the content of each domain. The researcher thus runs several short, noninvasive interviews, which, when combined, may be more complete than one broad interview.

In cases in which essential categories are not apparent, one might focus domains using either successive freelisting or ethnographic interviews (see below). Successive freelisting is an accurate, efficient method of honing domains. Here, a researcher uses the responses from one freelist as topics for subsequent freelist interviews, yielding related lists of subdomains. Ryan et al. (2000) offer a detailed description of collection and analysis using this technique. I provide one ethnobotanical example.

After conducting the aforementioned unsatisfactory pilot interviews on all *bush medicines*, I conducted a series of freelists focusing on illnesses Dominicans know how to treat with medicinal plants. I used a prompt in the local English Creole (developed with key informants to aid comprehension): "Here in Bwa Mawego [the village], what things they curing with bush medicine?" I collected freelists from a quota sample of 30 adult villagers stratified by age, sex, and village location (see Quinlan 2004), or approximately one fourth of resident adults. These freelists were oral and I wrote informant's responses as they listed the illnesses. Each freelist of illnesses took between 2 and 10 min.

I compiled this data (using the calculations outlined below) and ascertained the most salient treatable illnesses. Twenty-one illnesses terms were highly salient, but focus groups found and consolidated redundant and related salient terms. The final list contained 18 prominent illness domains.

This project's ultimate objective was to find rural Dominicans' customary (or prevalent) medicinal plants. Research assistants and I asked every available adult in the village to freelist bush remedies for each of the 18 illness-domains.

For most people, each freelist of a domain took much less than 1 min. We could usually do all 18 freelists in one sitting with each villager. When we surveyed the whole village with the 18 short freelists, we reinterviewed the individuals who participated in the long *pilot* interviews (in which they named "all" bush medicines). When I summed the separate medicinal species that the former consultants mentioned in the domain-focused exercise, they all had mentioned more species in multiple short freelists than they had in their initial long open-ended survey. Few informants became bored, frustrated, or overwhelmed during the domain-focused freelists because each of the 18 tasks was simple, quick, and different. Numerous people enjoyed their freelisting tasks and returned to the interviewers with their kin and friends who wanted a turn at it. Together, we obtained 1,826 freelists from 126 adults (almost all present in the village), yielding 7,235 total responses (see Quinlan 2004, 2010; Quinlan and Quinlan 2007; Quinlan et al. 2016; Quinlan and Flores in press).

5 Freelist Analysis

Freelist data reveals information about the items people list and the people who list them. The data inherently demonstrates a kind of cultural agreement (Weller and Romney 1988; Furlow 2003). Frequently mentioned items (or species) among individuals indicate common knowledge, or consensus, within the culture. And, the differences in list length and content are measures of intracultural variation.

Salience analysis (or Smith's S) (see Smith 1993) accounts for frequency of mention; however, it is weighted for list position as well. Thus, with my freelists of illnesses people treated with bush medicine, the calculation showed salience estimates for each illness, indicating both the number of people who mentioned the illness, and the order of their responses.

The salience statistic is simple enough to calculate quickly by hand. There are two steps. First, one needs to find salience of listed items (S) for each individual. Here, one ranks items on an individual's list inversely (final item listed equals one and items increase by one moving up the list). Then one divides the rank by the number of items the individual listed (see Table 1). Second, one tabulates a *composite salience value* (CSV) (or mean salience value) for each item listed in all freelists of the domain. Here, one sums all salience scores for that item then divides by the number of informants (see Table 2). ANTHROPAC 4.98 Windows freelist program software (Borgatti 1992) simplifies the entry and salience analysis of freelist data, which is useful with large samples, but which one can also compute in a spreadsheet program, as shown in Tables 1 and 2.

Determining which items are salient is not standardized. Drawing this boundary is a matter of judgment because salience is relative (Quinlan 2016, p. 8). In my experience, there are often visible breaks in the data that make good margins. The first break in salience occurs between items that many people think of and those that only some recall. In Fig. 1 of freelists of Dominican treatments for boils (Quinlan 2000), many people listed *malestomak*, *planté*, and "soft candle" (dripped tallow

Table 1 Weighting salience of items (illnesses) for freelister 1

Illness	Inverted rank	Divide by total listed	Salience (S)
Worms	5	÷ 5	1
Cold	4	÷ 5	0.8
Vomiting	3	÷ 5	0.6
Sore throat	2	÷ 5	0.4
Sprain	1	÷ 5	0.2

Note: For the sake of illustration, this freelist and all lists in Tables 1, 2, and 3 are abbreviated

Table 2 Determining composite salience for three freelisters

Illness	Freelister 1	Freelister 2	Freelister 3	Illness Σ	Composite salience Σ/ n ($n = 3$)
Worms	1	0.875	1	2.875	0.958
Cold	0.8	1	0.5	2.3	0.767
Buttons		0.75	0.75	1.5	0.500
Vomiting	0.6	0.625		1.225	0.408
Pressure		0.5	0.625	1.125	0.375
Inflammation			0.875	0.875	0.292
Cough		0.375	0.35	0.725	0.242
Sore throat	0.4		0.25	0.65	0.217
Asthma		0.25		0.25	0.083
Sprains	0.2			0.2	0.067
Cuts			0.125	0.125	0.042
Shock		0.125		0.125	0.042

Note: Column 2 contains S of freelister 1 (From Table 1). Responses of two other individuals (freelisters 2 and 3) are in columns 3 and 4. For every item (illness), sum individuals' salience scores (Illness Σ) then divide by the sample size (number of freelisters)

candle wax): These were highly salient. Some people listed the subsequent four treatments (*basilik* through *babadin*) – which are somewhat salient and worth inclusion under most circumstances, depending on the researchers objectives. Pepper leaf, *aloz*, and tomato leaf are not very salient, but because they were listed by three or four individuals, they likely are indeed local boil treatments, and though less salient might be retained in inclusive considerations. The straggling items (*zeb kwes* through *miwet*) are either uncommon or a "mistake," as with the informant who freelisted "turnip" as a fruit for Weller and Romney (1988). *Dobla*, though listed by only two individuals like the three least salient items, has higher salience, meaning that it ranked relatively highly among the individuals who listed it. I would not consider it "salient," but worthy of further investigation. Bernard (2011, p. 349) offers as a rule of thumb that items mentioned by 10% of informants be considered salient.

Robbins et al. (2017) improve on Robbins and Nolan (1997) and Ryan et al. (2000) by adding a technique to examine clustering of freelist data, such that one might use freelists to examine cognitive arrangement of emic or etic categories within a domain, across and within freelists of varied lengths. An ethnobotanist might, for example,

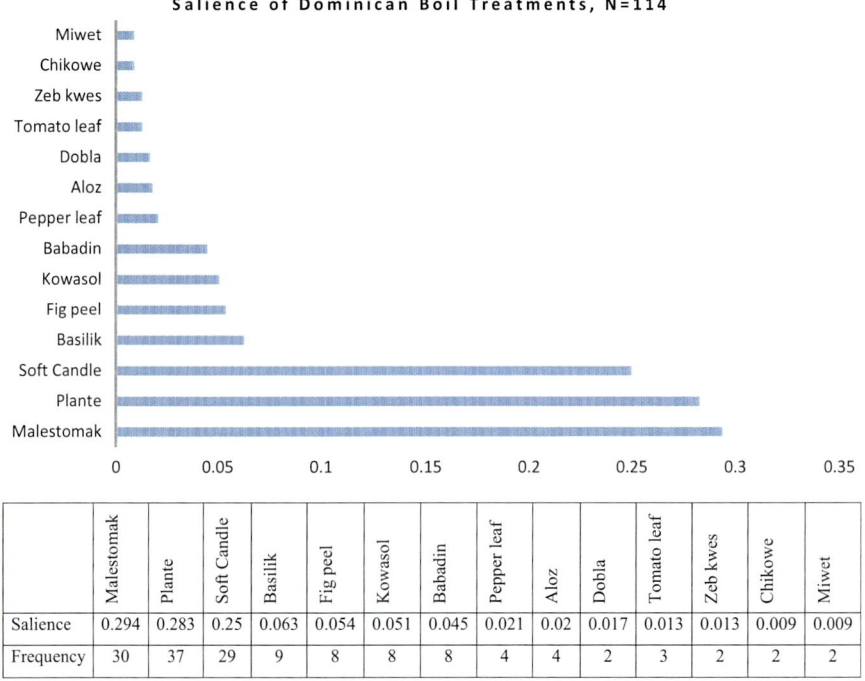

Fig. 1 Dominican treatments for boils. Note: For species names of treatments (mostly given in French Patois), see Quinlan and Flores (in press), or Quinlan (2000)

examine the psychological/cultural salience of plants with perceived humoral qualities, or weeds versus cultivates, or flowering versus nonflowering plants.

In addition to revealing culturally salient items across individuals, freelists measure individuals' expertise within a domain. As mentioned above, knowledgeable people tend to have longer lists. By creating an individual by item matrix one can tabulate items' frequency of mention (Table 3, row totals), and individuals list lengths (Table 3, column totals). With additional sociodemographic data, one can investigate relationships between people's knowledge in a domain and other characteristics such as residence or education-level. I found, for example, that Dominicans' mean list-length of bush medicines positively correlates with age and wealth (Quinlan 2000).

6   Checking Freelists with Ethnographic Interviews

The value of the freelisting technique depends on understanding the cultural domains in question. Informal interviews (such as key informants, focus groups, and so on) in conjunction with freelists permit "ethnographic cross-checking," which

Table 3 Comparing individuals' knowledge

Illness	Freelister 1	Freelister 2	Freelister 3	Illness frequency
Worms	1	1	1	3
Cold	1	1	1	3
Buttons	0	1	1	2
Vomiting	1	1	0	2
Pressure	0	1	1	2
Inflammation	0	0	1	1
Cough	0	1	1	2
Sore throat	1	0	1	2
Asthma	0	1	0	1
Sprains	1	0	0	1
Cuts	0	0	1	1
Shock	0	1	0	1
Total illnesses listed	5	8	8	21

Note: Salience data from Table 2 converted to ones and zeros for the presence or absence on three informants' freelists (1/0 indicate probable presence/absence of knowledge of bush medicine for that illness)

increases accuracy and enhances the depth of ethnographic understanding. Informal ethnography and freelisting can be complementary sources of information (see also Ethnographic method).

The ideal way to find emic domains is to use successive freelisting cross-checked with ethnographic interviews. I recommend getting freelists from a larger sample, then using the salient subdomains of the original as focus group topics. In focus group interviews (Bernard 2017), several local consultants hash out the different *categories* of X (e.g., cold medicines, stomach medicines, and so on) that have been freelisted. Observing focus groups lets an ethnographer witness locals' decision-making rationales and processes (Trotter et al. 2014) (One could omit successive freelists and use only focus groups here. However, freelisted domains are representative of the population, not swayed by charisma or assertiveness of an individual in a focus group.). Initial time spent going through both freelisting and informal interviewing is worthwhile because it expedites the final set of interviews.

Researchers can use ethnographic interviews to accurately "standardize" freelists. Weller and Romney (1988) warn in *Systematic Data Collection* that when freelists consist of phrases or statements, various lists may contain different phrasings of the same concept. A researcher must then use judgment to standardize concepts before tabulating the lists. In unclear cases, it is "desirable" for informants to identify different phrases that represent a single concept (1988, p. 15). Similarly, freelists of terms often contain synonyms and redundant phrasing (e.g., breastfeeding, breastfeeding, nursing, and lactating) for the researcher to cull out. Different terms in a freelist may not be separate entities. Further, one emic term may refer to more than one etic entity.

My first succession of freelists of illnesses yielded 21 highly salient illness domains. Focus groups responded to the terms indicating that several of them were redundant. In their estimation, "cuts" and "sores," though different, belonged "together," as did "prickle-heat" and "buttons" (rashes, pox, and pimples), and "upset stomach" and "vomiting." After much debate, they agreed that a less salient term "arthritis" did not belong with "rheumatism" ("arthritis" is associated with a culture-specific fright-illness in Dominica, while "rheumatism" is not). Here, using focus groups streamlined my interviewing process and lent "emic authority" to the final domains. Without input from the focus groups, I would have performed several superfluous interviews with each subject. Or, if I had deleted redundancies on my own, I would have, despite extensive experience with the local medical system, grouped domains differently.

Final freelists in a succession also contain synonyms and require standardization. Lay people generally use local common names for illnesses, plants, body parts, and so forth. Does "stomach" refer to the single organ or to the general area? Is it synonymous with gut and belly? Plant foods and medicines' common names are often not exact (thus the necessity of Latin species names). Plants often have multiple, distinct-sounding common names that people in a population use interchangeably (e.g., "scallion" and "green onion"), which can be confusing for an outsider.

The problem of several terms for one species multiplies in societies influenced by multiple languages. Anecdotally, with greater Hispanic influence in the USA, *Coriandrum sativum* L. has become a relatively common cooking ingredient, and the Spanish common name "cilantro" appears at least as prevalently as the herb's English common names, "coriander" and "Chinese parsley." One species that may have numerous monikers wherever it is used is *Cannabis sativa* L. In Dominica, it is *kali* in both French Patois and English, *zeb* in Patois, and "marihuana," "weed," "sensi," and "ganja" in English. In addition to single species with multiple names, people identify separate, related species by one generic name (Berlin 1992), such as "begonia" for various species in the genus Begonia. This confusion is not unique to plants, one might wonder whether football means American football or soccer. Thus, conducting interviews in which local consultants identify plants or discuss topics in the lists, in addition to freelisting, is the only way to ensure accuracy of inventories from freelists.

Once terms are standardized and analysis reveals salient terms, ethnographic interviews can fill information about salient items. In medical ethnobotany, for example, informants can describe combinations, preparations, and doses of the salient medicinals, and the circumstances in which each plant might be preferred. For example, among salient Dominican treatments for "worms," *sime kontwa* (*Chenopodium ambrosioides* L.) is a general-purpose vermifuge, while *twef* (*Aristolochia trilobata* L.) is for severe cases, and *kupiyè* (*Portulaca oleracea* L.) is best for children (Quinlan et al., 2002). In another example, "soft candle" wax, the third most salient boil treatment (Fig. 1), is used to plaster a medicinal leaf over a boil, but is not a treatment alone (Quinlan 2000).

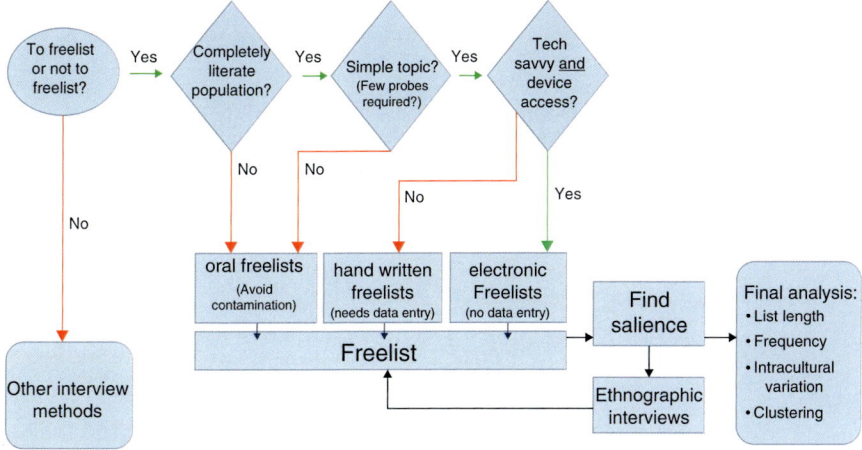

Fig. 2 The ideal freelisting process (modified from Quinlan 2016)

7 Conclusion and Future Directions

Freelisting is a simple, accurate, and quick way to collect data from a large sample of individuals. Freelists reveal the salience of items in the community and variation in knowledge of the domain in question. Among literate communities, written freelists provide privacy and avoid data contamination from spectators, but may be less considered than face-to-face interviews. Interviewers conducting oral freelists should take steps to prevent bystander "contamination." Freelists are especially useful in an iterative or successive process, as outlined in Fig. 2. Generally, researchers should hone domains of freelists tightly: Given broad topical areas, people tend to forget or omit items. They also cluster their responses as they "unpack" their mental subcategories. Omission and clustering of terms may reduce precision of salience estimates.

Using successive freelists factors out mental subdomains from the original topic. Final interviews in the iterative process are fast (and often enjoyable) for informants, and are most complete and accurate for investigators. Researchers should check responses from freelists with informal methods. Determining emic definitions of the terms in a domain is necessary to prevent over- or undercounting responses. Informal methods can also reveal other information about salient items (e.g., how, when, or where the item is appropriate or occurs). With proper attention to detail, freelisting can result in large amounts of high quality sociocultural health data.

References

Ahmad FS, Barg FK, Bowles KH, Alexander M, Goldberg LR, French B, Kangovi S, Gallagher TR, Paciotti B, Kimmel SE. Comparing perspectives of patients, caregivers, and clinicians on heart failure management. J Card Fail. 2015;22(3):210–7.

Alexiades MN. Collecting ethnobotanical data. In: Alexiades MN, editor. Selected guidelines for ethnobotanical research: a field manual. Bronx, NY: New York Botanical Garden; 1996 p. 53–96.

Bayliss EA, Steiner JF, Fernald DH, Crane LA, Main DS. Descriptions of barriers to self-care by persons with comorbid chronic diseases. Ann Fam Med. 2003;1(1):15–21.

Berlin B. Ethnobiological classification. Princeton: Princeton University Press; 1992.

Berlin EA, Berlin B. Medical ethnobiology of the highland Maya of Chiapas, Mexico. Princeton: Princeton University Press; 1996.

Bernard HR. Research methods in anthropology: qualitative and quantitative approaches. 5th ed. Lanham: Altamira/Rowman & Littlefield; 2011.

Bernard HR. Research methods in anthropology: qualitative and quantitative approaches, 6th edn. Lanham, MD: Rowman & Littlefield; 2017.

Bolton P, Tang AM. Using ethnographic methods in the selection of post-disaster, mental health interventions. Prehospital Disaster Med. 2004;19(01):97–101.

Bolton R, Curtis A, Thomas L. Nepali color terms: salience in a listing task. J Steward Anthropol Soc. 1980;12:309–21.

Borgatti SP. Using Anthropac to investigate a cultural domain. Cult Anthropol Methods Newsl. 1990;2(1):8.

Borgatti SP. ANTHROPAC 4.00 methods guide. Columbia: Analytic Technologies; 1992.

Borgatti SP. Elicitation techniques for cultural domain analysis. In: Schensul J, LeCompte M, editors. The ethnographer's toolkit. Walnut Creek: AltaMira Press; 1999. p. 115–51.

Boster JS. Introduction. Am Behav Sci. 1987;31:150–62.

Bradway CW, Barg F. Developing a cultural model for long-term female urinary incontinence. Soc Sci Med. 2006;63(12):3150–61.

Bradway C, Dahlberg B, Barg FK. How women conceptualize urinary incontinence: a cultural model. J Women's Health (Larchmt). 2010;19(8):1533–41.

Brewer D. Cognitive indicators of knowledge in semantic domains. J Quant Anthropol. 1995;5:1047–128.

Brewer D. Supplementary interviewing techniques to maximize output in free listing tasks. Field Methods. 2002;14(1):108–18.

Ceuterick M, Vandebroek I, Torry B, Pieroni A. Cross-cultural adaptation in urban ethnobotany: the Colombian folk pharmacopoeia in London. J Ethnopharmacol. 2008;120(3):342–59.

Crandon-Malamud L. From the fat of our souls: social change, political process, and medical pluralism in Bolivia. Berkeley: University of California Press; 1991.

D'Andrade RG. Modal responses and cultural expertise. Am Behav Sci. 1987;31:194–202.

Fiks AG, Gafen A, Hughes CC, Hunter KF, Barg FK. Using freelisting to understand shared decision making in ADHD: parents' and pediatricians' perspectives. Patient Educ Couns. 2011;84(2):236–44.

Finerman R, Sackett R. Using home gardens to decipher health and healing in the Andes. Med Anthropol Q. 2003;17(4):459–82.

Flores KE, Quinlan MB. Ethnomedicine of menstruation in rural Dominica, West Indies. J Ethnopharmacol. 2014;153(3):624–34.

Furlow CA. Comparing indicators of knowledge within and between cultural domains. Field Methods. 2003;15(1):51–62.

Gatewood JB. Loose talk: linguistic competence and recognition ability. Am Anthropol. 1983;85(2):378–87.

Gatewood JB. Familiarity, vocabulary size, and recognition ability in four semantic domains. Am Ethnol. 1984;11:507–27.

Giovannini P, Heinrich M. Xki yoma' (our medicine) and xki tienda (patent medicine) – interface between traditional and modern medicine among the Mazatecs of Oaxaca, Mexico. J Ethnopharmacol. 2009;121(3):383–99.

Gravlee CC, Bernard HR, Maxwell CR, Jacobsohn A. Mode effects in free-list elicitation: comparing oral, written, and web-based data collection. Soc Sci Comput Rev. 2012;31(1):119–32.

Handwerker WP, Borgatti SP. Reasoning with numbers. In: Bernard HR, Gravelee CC, editors. Handbook of methods in cultural anthropology. Lanham: Rowman & Littlefield; 2014. p. 519–32.

Henley N. A psychological study of the semantics of animal terms. J Verbal Learn Verbal Behav. 1969;8:176–84.

Hruschka DJ, Sibley LM, Kalim N, Edmonds JK. When there is more than one answer key: cultural theories of postpartum hemorrhage in Matlab, Bangladesh. Field Methods. 2008;20(4):315–37.

Hutchinson JW. Expertise and structure of free recall. In: Bagozzi RP, Tybout AM, editors. Advances in consumer research. Ann Arbor: Association for Consumer Research; 1983. p. 585–9.

Lucan SC, Barg FK, Karasz A, Palmer CS, Long JA. Concepts of healthy diet among urban, low-income, African Americans. J Community Health. 2012;37(4):754–62.

Mathez-Stiefel SL, Brandt R, Lachmuth S, Rist S. Are the young less knowledgeable? Local knowledge of natural remedies and its transformations in the Andean highlands. Hum Ecol. 2012;40(6):909–30.

Robbins MC, Nolan JM, Chen D. An Improved Measure of Cognitive Salience in Free Listing Tasks. Field Methods. 2017;1525822X1772672.

Nolan JM. Pursuing the fruits of knowledge: cognitive ethnobotany in Missouri's little Dixie. J Ethnobiol. 2001;21(2):29–51.

Nolan JM. Wild plant classification in little Dixie: variation in a regional culture. J Ecol Anthropol. 2002;6:69–81.

Nolan JM. Ethnobotany in Missouri's little Dixie: cognitive ecology in a regional culture. Lanham: University Press of America; 2004.

Nolan JM, Robbins MC. Cultural conservation of medicinal plant use in the Ozarks. Hum Organ. 1999;58:67–72.

Pieroni A, Sheikh QZ, Ali W, Torry B. Traditional medicines used by Pakistani migrants from Mirpur living in Bradford, Northern England. Complement Ther Med. 2008;16(2):81–6.

Quinlan MB. Bush medicine in Bwa Mawego: Ethnomedicine and medical botany of common illnesses in a Caribbean village. Unpublished doctoral dissertation, University of Missouri-Columbia. An Arbor, MI: UMI Dissertation Publishing. 2000.

Quinlan MB. From the bush: the frontline of health care in a Caribbean village. Belmont: Wadsworth/Cengage Publishing; 2004.

Quinlan MB. Ethnomedicine and ethnobotany of fright, a Caribbean culture-bound psychiatric syndrome. J Ethnobiol Ethnomed. 2010;6(9):1–18.

Quinlan MB, Flores KE. Bush medicine in Dominica: ethnophysiology and medical ethnobotany in a Caribbean horticultural village. Contributions in ethnobiology series. Denton: The Society of Ethnobiology; in press.

Quinlan MB, Quinlan RJ. Modernization and medicinal plant knowledge in a Caribbean horticultural village. Med Anthropol Q. 2007;21(2):169–92.

Quinlan MB, Quinlan RJ, Nolan JM. Ethnophysiology and herbal treatments of intestinal worms in Dominica, West Indies. J Ethnopharmacol. 2002;80(1):75–83.

Quinlan MB, Quinlan RJ, Council SK, Roulette JW. Children's acquisition of ethnobotanical knowledge in a Caribbean horticultural village. J Ethnobiol. 2016;36(2):433–56.

Quinlan M. Considerations for collecting freelists in the field: Examples from Ethobotany. Field Methods 2016;17(3):219–234.

Robbins MC, Nolan JM. An improved measure of cognitive salience in free listing tasks. Cult Anthropol Methods J. 1997;9:8–12.

Romney A, D'Andrade R. Cognitive aspects of English kin terms. Am Anthropol. 1964;66(3.): 146–70. Part 2.

Ryan G, Nolan J, Yoder S. Successive free listing: using multiple free lists to generate explanatory models. Field Methods. 2000;12(2):83–107.

Smith JJ. Using ANTHROPAC 3.5 and a spreadsheet to compute a free list salience index. Cult Anthropol Methods Newsl. 1993;5(3):1–3.

Smith JJ, Furbee L, Maynard K, Quick S, Ross L. Salience counts: a domain analysis of English color terms. J Linguist Anthropol. 1995;5(2):203–16.

Trotter RT II. Remedios cseros: Mexican American home remedies and community health problems. Soc Sci Med. 1981;15B:107–14.

Trotter RT II, Schensul JJ, Kostick KM. Theories and methods in applied anthropology. In: Bernard HR, Gravelee CC, editors. Handbook of methods in cultural anthropology. Lanham: Rowman & Littlefield; 2014. p. 661–95.

Waldstein A. Mexican migrant ethnopharmacology: pharmacopoeia, classification of medicines and explanations of efficacy. J Ethnopharmacol. 2006;108(2):299–310.

Weller SC. Shared knowledge, intracultural variation, and knowledge aggregation. Am Behav Sci. 1987;312:178–93.

Weller S, Romney AK. Systematic data collection. Newbury Park: Sage; 1988.

Solicited Diary Methods

83

Christine Milligan and Ruth Bartlett

Contents

1 Introduction	1448
2 The Solicited Diary: Methodological Underpinnings	1449
3 Approaches to Solicited Diary Method	1451
3.1 What Sorts of Questions Can Solicited Diaries Address?	1452
3.2 Semi and Unstructured Approaches to Solicited Diary Method	1452
4 Solicited Diary Techniques and "Capture" Technologies	1454
5 Analyzing Diary Data	1456
5.1 Analyzing Written Diary Data	1456
5.2 Analyzing Visual Diary Data	1457
6 Strengths and Limitations of Solicited Diary Method	1457
7 Addressing the Limitations of Diary Methods	1459
8 Diary Methods: Practical Examples	1460
9 Conclusion and Future Directions	1461
References	1462

Abstract

To date, solicited diaries have been relatively neglected as a research method within the health and social sciences. Yet, the gathering of chronologically organized diary data can provide unique insights into the life-worlds inhabited by individuals; their experiences, actions, behaviors, and emotions and how these are played out across time and space. Solicited diaries enable informants to actively participate in both recording and reflecting (either in written, oral, or

C. Milligan (✉)
Division of Health Research, Lancaster University, Lancaster, UK
e-mail: c.milligan@lancaster.ac.uk

R. Bartlett
Centre for Innovation and Leadership in Health Sciences, Faculty of Health Sciences, University of Southampton, Southampton, UK
e-mail: r.l.bartlett@soton.ac.uk

© Springer Nature Singapore Pte Ltd. 2019
P. Liamputtong (ed.), *Handbook of Research Methods in Health Social Sciences*,
https://doi.org/10.1007/978-981-10-5251-4_15

visual formats) on their own data. While inevitably these data are reflected upon with a certain research agenda in mind, most qualitative diary methods allow space and time for diarists to depict their own priorities. As such, this research technique affords participants greater control of the data, enabling individuals to not only consider their responses but reveal as little or as much as they feel willing to do so. Solicited diary methods can be used to gather both qualitative and quantitative data, but in this chapter we focus specifically on the use of qualitative diary techniques. We address the methodological underpinnings of these approaches and the contribution they can make to the study of different questions, phenomena, and social problems. We also consider the strengths and limitations of solicited diary methods including debate about the extent to which they are viable techniques for undertaking research with all groups and individuals. Finally, we discuss how the rise of digital technologies is opening up new and exciting approaches for solicited diary techniques, the possibilities for which are only newly being explored.

Keywords

Solicited diaries · Qualitative · Unstructured · Technology · Analysis

1 Introduction

The use of (largely) quantitative solicited diary methods has a history in health research stretching back more than 90 years (Waldron and Eyers 1975). In comparison, the use of qualitative solicited diaries has been far less common – particularly in relation to face-to-face methods such as interviewing and focus groups. But, the design and application of solicited diaries has grown significantly over the last 15 years. Indeed, a recent structured search of key databases identified over 64,000 papers published between 2000 and 2014, using solicited diary methods as part of their research design (Rokkan et al. 2015, p. 203). Of course, this does not differentiate between structured and semi/unstructured diary methods, but it does demonstrate a growing recognition that this technique has the potential to add something different to the repertoire of qualitative methods. Solicited diary method is increasingly being used within research either as the sole method of data collection, or as part of a wider multimethod research design.

So what is a diary, what do we mean by a solicited diary, and how does it differ from the unsolicited diary? Methodologically, the solicited diary is a research technique in which informants actively participate in both recording and reflecting on their own thoughts, feelings, actions, and behaviors. Though written with a specific research agenda in mind, the semi- and unstructured diary also usually enables the diarist to identify his or her own priorities. Rokkan et al. (2015, p. 203) define the diary as "a record of what an individual considers relevant and important in his or her life, for instance; events, activities, interactions, impressions or feelings. It is usually structured by time in some way." A key distinction between the unsolicited and the solicited diary is that the former consists of a diary that the individual chooses to keep voluntarily. He or she has not been asked to keep the diary and there is no specific research or other agenda to its completion – even though some can go on to provide the foundation for important historical research. The record of events and experiences in the internationally recognized diaries of Anne Frank or Samuel Pepys are often cited as key examples of this (Bartlett and Milligan 2015). Solicited diaries, on the other hand, are those that people have been asked to keep for a specific reason – in this case, research. Here, the participant is requested to keep a regular record of his/her thoughts, feelings, experiences, and/or behaviors around a specific topic over a defined period of time.

While we acknowledge the long history of structured diary method in health and social research, in this chapter we focus specifically on the use of in-depth semi and unstructured solicited diary methods. These approaches are designed to encourage the diarist to record a more detailed temporal narrative, often around a loosely structured set of themes constructed by the researcher. The aim is to gain a deeper understanding of a person's actions, experiences, thoughts, and feelings around a specific topic over time. These approaches also allow space for the diarist to record his/her own priorities, so can prove useful for capturing the meaning and weight diarists attach to different events and experiences in their lives (Milligan et al. 2005).

We also consider the methodological underpinnings of solicited diary method in the chapter and the implications for ethics and empowerment. The chapter continues with a discussion of the different approaches to solicited diary design and how it can be used in conjunction with other methods to give unique insights into individuals' personal experiences of specific health and social events. We draw on illustrations from our own and other recent research to demonstrate why researchers have chosen to draw on solicited diary methods as part of their research design and how the method has been applied in the field. We are particularly concerned with the novel insights that can be gained from using this approach.

While our chapter focuses on the potential benefits solicited diary techniques can contribute to a researcher's toolbox, like any research method, it has limitations. Our final section discusses these strengths and limitations in order that researchers are alert to what this might mean for their studies and can plan for it, if incorporating solicited diaries into their research design. We conclude with a brief discussion of the newest developments in applying diary method and future directions.

2 The Solicited Diary: Methodological Underpinnings

There are several methodological underpinnings to diary methods, most of which relate to the immediacy and flexibility of this approach. Firstly, diary methods are favored by researchers for gathering data as, and when, it occurs rather than relying on someone recounting an event or feeling after it has happened (as, for example, with a qualitative interview). Diary methods allow the participant to record an experience in real time (or at least closer to the moment that it occurs). The immediacy of diary method – being able to gain a person's perspective on an event as near to when that event occurs – is important because it can improve the quality of data and open up other ways of knowing about a person's inner world. Indeed, solicited diary methods are often chosen because they have the potential to reduce the recall bias or retrospective censorship that can occur in question-response methods such as interviews (Bartlett and Milligan 2015).

A second methodological reason for using diary methods is because they allow for an extended period of data collection that is relatively unobtrusive. Participants are typically asked to maintain a diary over a certain number of days, weeks, months, or potentially even longer, depending on the phenomenon under investigation and the resources available to the researcher. For example, in the Hull Floods project, as well as undertaking initial in-depth interviews, a panel of 55 people were recruited to keep diaries over a 12–18-month period (Whittle et al. 2010). This allowed for the long-term effects of flooding to be captured in a systematic and impactful way. Hence, diary techniques can "fill in the gaps," offering insights into what happens between longer intervals in longitudinal approaches (Bijker et al. 2015). Being able to capture changes in circumstances and within-person variations over an extended period of time is vital for many topics in the health and social sciences – coping with substance misuse cravings (Cleveland and Harris 2010) and breastfeeding difficulties for first time mothers (Williamson et al. 2012) to name just two. This can be a key reason for adopting diary method techniques.

A third methodological reason for choosing to use diary method is because it facilitates the reporting of sensitive or otherwise "unseen" behaviors. Several researchers have reported on the value of using diary method to research "taboo" or controversial topics such as sexual activity, obesity, and substance misuse (e.g., Allen 2009; Boone et al. 2013). Sleep is another topic that has been researched using diary method, particularly audio diaries, because the method lends itself to eliciting what sleep researchers have called "narratives of the night" (Hislop et al. 2005). Diary methods are accessible tools, then, that can give researchers rich insights into all aspects of human life.

A fourth methodological reason for using diary methods, specifically video-diary method, is because this method generates rich data about how a person looks, sounds, and moves. As one advocate of video-diary method notes, it can make the body "knowable" (Bates 2013). Information about how a participant speaks, behaves, and experiences ill-health can all be gained through video-diary method – and to some extent through audio and photo diary techniques as well. In our view, one of the best examples of using video-diary method for this methodological reason was a study conducted in the United States investigating younger peoples' experience of asthma (Rich et al. 2000). Here, 25 young people were invited to create a video diary of their day-to-day life with asthma. The researchers were able to capture data about the embodied experience of living with asthma that would not have been possible to gather using any other method. For instance, one 18-year-old participant "videotaped herself in rapidly worsening respiratory distress" as she was being driven to hospital by her mother (p. 67). In such circumstances conducting an interview would be neither possible nor ethical. So insights into this kind of bodily experience can only be gained through immediate, accessible, and visual methods such as video-diary method, where the decision to record particular diary data lies firmly with the participant (see also ▶ Chap. 74, "Digital Storytelling Method").

Finally, it is important to note that, solicited diary methods fit well with a participatory research design, where concerns about participants in the research process are paramount (see Higginbotham and Liamputtong 2015; see also ▶ Chaps. 17, "Community-Based Participatory Action Research," and ▶ 17, "Community-Based Participatory Action Research"). Here, researchers might choose to use a diary method because it allows participant's to control the pace and time of data collection in ways that other qualitative methods do not. For example, in the asthma study described above, participants were encouraged to "interview" members of their family about their illness experience on camera, thus giving the young people considerable control of the data production process (p. 55). All of the above highlight reasons why a researcher might choose to use a diary method to investigate a topic.

3 Approaches to Solicited Diary Method

The emergence of participatory research designs has led to a greater emphasizes on the importance of "researching with" rather than "researching on" participants (Higginbotham and Liamputtong 2015). This has resulted in a greater consciousness among many qualitative health and social researchers of the need to find ways of reducing the uneven power relationship that often occurs between the researcher and the researched. These are key issues, but it is also important to consider those other qualities of solicited diary method that are important in deciding whether it should be the method of choice in designing a qualitative study.

As previously indicated, a key feature of solicited diary methods are their focus on how an issue of interest occurs or changes over time. A well-designed diary can yield significant insights into the temporality of an individual's actions or experiences that may not be so accurately gained using other research techniques. The requirement for regular diary contributions made over a defined period of time means that the quality and quantity of data provided is significantly different to that of other research techniques. Indeed, using diary-interview method in a study of scientists' experiences of working in Antarctica, Filep et al. (2015) observed that the diary process appeared to give participants "permission" to be more expressive and creative in giving data in comparison to their interview data, which was less reflective, briefer, and to the point. Bernays et al. (2014), also using diary-interview method in their study of peoples' perceptions of hope after being diagnosed with HIV/AIDS, found an added richness in audio diary data. They noted that because participants made their recordings alone, rather than in the company of a researcher, a very "specific communication space" was created that made it possible to "access resistant narratives that are normatively unsayable" – such as feeling suicidal and hopeless (p. 636). In effect, then, diary methods allow a participant to produce a freer, unencumbered narrative. Hence, they have the potential to offer a more comprehensive picture of an individual's activities, thoughts, and experiences over time.

3.1 What Sorts of Questions Can Solicited Diaries Address?

Before adopting a solicited diary research design, it is important to consider what sort of questions this technique can usefully help to answer – whether as a single method study or as part of a multimethod research design. Following Bolger et al. (2003) we outline a threefold typology of research questions that solicited diary methods can usefully address.

(i) **To gain reliable individual information over time**
Solicited diaries can be useful where accuracy about an individual's experiences, practices, habits, actions, and so on over time are important. They are not only less subject to the vagaries of memory, retrospective censorship, or participants' reframing of data than other methods, they can also be useful where a researcher is interested in uncovering routine, everyday processes and events that may be viewed by the participant as trivial and hence easily forgotten (Milligan et al. 2005; Gill and Liamputtong 2009).

(ii) **To obtain an understanding of within-person change over time and individual differences in such change**
Time may play an important role in understanding embodied responses to internal factors (physical, cognitive, or emotional) or external (social/environmental) factors. Solicited diaries can thus help where research focuses on gathering accurate understandings of how a purposively defined range of an individual's thoughts, feelings, experiences, embodied actions and reactions, etc., to phenomena of interest may (or may not) change over time. These data can then be used to analyze within-person differences in those experiences over time.

(iii) **To conduct causal analysis of between-person changes and individual differences in these changes**
Solicited diaries can help identify underlying causes of change within an individual over time in relation to a phenomena of interest and how these may vary between people. The emphasis of the research here, then, is to uncover those processes that underlie not just within-person variability over time, but *between*-person variability.

3.2 Semi and Unstructured Approaches to Solicited Diary Method

Having confirmed that solicited diaries can usefully contribute to answering the overall aims and objectives of a qualitative study, it is important to consider the form of the diary required. By this we mean, whether a semi or unstructured diary is required and whether written, audio, or visual approaches (or a combination) might be most appropriate. For example, is the study seeking to identify how many times specific events, phenomena, or activities and their variables occur over a prescribed length of time? Is it more concerned with an individual's thoughts, feelings, or the unspoken and often personal actions or experiences that may not be deemed important by the respondent when engaging with more reactive research techniques? Is the solicited diary being used as the sole method within a study or as part of a multimethod research design? Might technology have a useful role to play in the application of the diary and data collection? These sorts of questions are important in any decision about how solicited diaries should be constructed and what instructions should be given to participants.

Whichever format is undertaken, solicited diaries involve participants being given a clear set of guidelines on how to complete the diary and the frequency with which this should be undertaken. In semistructured diaries, participants are given a set of headings or themes linked to the research objectives and asked to record their experiences of a phenomenon, their activities, thoughts, feelings, and so on in relation to those themes at regular intervals over an agreed period of time. In most cases, participants are also given an opportunity to record other issues that may not be part of the predefined themes, but which *they* feel are important in relation to the study. In unstructured approaches, participants are asked only to record their thoughts/experiences/actions around a phenomenon of interest without being given any thematic prompts. Participants then identify their own priorities around the topic in their diary entries without influence from the researcher.

As noted, solicited diaries can either be used as the sole method of data collection, or as part of a multimethod research design. They can be used as a precursor, an adjunct, or a follow-up to survey, interview, observational, or other research techniques. One commonly adopted multimethod approach is that of the diary-interview method or diary-photograph, diary-interview method (DPDIM) pioneered by Zimmerman and Wieder in the 1970s. Here, the diarist acts as a proxy observer, whose regular diary observations of phenomena are followed up by an in-depth interview with the researcher. Gibson (2002), for example, used audio diaries combined with photographic data and interviews to gain a deeper understanding of experiences of transitioning to adulthood among young men with a disability. The researcher was keen to understand the intersectionality of gender, disability, and life-stage identities. Diarists were encouraged to take photographs that reflected their daily lives alongside their regular diary entries. Orban et al. (2012), in a study of how changes in patterns of daily occupations change over time, used the diary data to construct graphs illustrating the sequences of occupations performed by participants. In both these, and similar studies, the diary was used as a source of information and aide-memoire that facilitated subsequent stimulated-recall interviews using the photographs/graphs to help guide discussion. Gibson (2002) also noted that the ease of operating audio-recorders makes the audio-diary particularly suitable for those with physical functional limitations, as well as enabling the capturing of those subtleties of tone that are not possible in written accounts.

Latham (2003) contends that the integration of the nondiscursive (photographs) within DPIM methods facilitates an exploration of tacit knowledge that is often difficult for individuals to write down or verbalize. He suggests that as a research technique, diary methods are both a reportage of events over time and a performance. The subsequent interview then becomes a reperformance or reaccounting of recorded events and phenomena. He thus argues that rather than seeing the different ways in which diarists complete their individual diaries as problematic, we should focus on how diary methods enable us to plug into, and enable, participants' existing narrative resources (Latham 2003, p. 2002).

Finally, Gibson (2002) point out that the subsequent interview in diary interview method is not solely a reperformance or reaccounting of the diary data. Rather, "it provides a situation of co-analysis where the interviewer learns from the participant how the image was created, the motivation for including it and what it represents for them" (p. 387).

4 Solicited Diary Techniques and "Capture" Technologies

In the last decade, the application of technologies in diary-based studies has expanded enormously. While a significant rise in published diary studies involving technologies has occurred since 2000, the majority of these have been published within the last decade. Many of these studies involve the use of "capture" technologies such as cameras, voice recorders, and mobile (smart) phones and are exploratory – qualitative – in design. These approaches to diary studies form the focus of this section.

One of the most well-established applications of capture technology in diary-based research is the video-diary method. This involves participants keeping a diary of an activity or experience using a video camera for an agreed period of time. Researchers have used video diaries to explore such themes as the bodily experiences of running (Bates 2013); children's experiences of various life events (Buchwald 2009), such as running away from home (Edinburgh et al. 2013) and learning in school (Noyes 2004). Video-diary methods privilege action and the visible, and are therefore "ideal device[s], with which to unlock bodily experience and bring the sensuous and affective qualities of embodiment to the screen" (Bates 2013, p. 31). Video diaries are unique and innovative modes of data collection. They are fundamentally different from paper-based diaries, in that they generate moving data and allow for the visualization of phenomena. Video-capture technology is thus contributing to the diversification of diary-based research (see ► Chap. 74, "Digital Storytelling Method").

Video-capture technologies are changing diary methods in a number of ways. First, the data generated by video-diary method gives researchers access to areas and situations that might otherwise be prohibited, or go unseen or unknown. Recall the video-diary study mentioned above of young people living with asthma. One diary entry comprised a short video clip of a cat-allergic patient with a kitten hidden in her bed. Another revealed a dusty construction site outside the participant's home. In both cases, the video data furnished the researcher with an intimate knowledge of a child's private space (bedroom), and in particular, the environmental triggers for their asthma. For the young participant, it meant that they did not have to rely on the spoken word to convey their day-to-day lives as the video footage did that for them. Indeed, it transpired in the subsequent interviews, that the young participants were not even aware that cat hair and dust were triggers. Solicited diaries using capture technologies can thus provide valuable information that the participant may not view as important or may not think to write or verbalize.

Video-capture technologies can also enhance research relations by affording participants greater control over the data production process, than traditional methods allow. Giving the participant the camera means that they make the decisions about the pace and frequency of data collection rather than the researcher. For example, in one video-diary study involving over 70 young girls, the researchers suggested 10 min a day, but most girls chose to film less frequently but for longer (Jackson and Vares 2014). As diary researchers note, participants will keep their diaries in their own way, with the power of selection being handed over to the participant (Bijoux and Myers 2006). This is especially true of video-diary studies, where having a video camera can engender feelings of self-empowerment and control. The evidence suggests this also works well with those people who are not typically seen as being able to take control, such as young children and people with dementia (Capstick 2011).

Another common application of capture technology in diary-based research is the photo diary. This involves participants taking photographs, which capture their experiences and activities, and talking to the researcher about them afterward – a technique known as photo-elicitation (Prosser and Schwartz 1998). The approach originated in anthropology and ethnography and is popular with visual researchers. The following two studies are examples of how photo diaries have been used to investigate areas of academic interest in the health and social sciences:

- Twenty-two secondary school children in New Zealand were issued with disposable cameras to explore sexualities and school culture (Allen 2009).
- Twenty-two parents of preschool aged children were asked to keep a photographic diary to record their child's dietary intake.

Some diary studies use the term photovoice to describe the process of giving participants a camera to record their thoughts and feelings (see, for example, Williams et al. 2016). Photovoice is a similar photographic technique to photo-elicitation, but its roots are firmly in participatory action research (Wang 1999). One practical consideration with any photo-diary method is whether to provide participants with a camera, or ask them to use their own. If you decide to provide participants with a camera, you need to consider whether to provide an analogue (disposable) one, or a digital one. The latter is probably the better of the two options as society becomes increasingly digitalized and allows for ease of transfer of the digital photographs for printing and for analysis. However, both options rely on the researcher having access to sufficient numbers of cameras to disperse to all participants, and therefore can have cost implications. Moreover, each has methodological implications, which may or may not influence your decision. For example, an analogue camera can be easier for some people to operate; plus, the restriction on the amount of images that can be taken, may mean that participants give more thought and consideration to what to photograph (Bartlett and Milligan 2015; see also ▶ Chap. 65, "Understanding Health Through a Different Lens: Photovoice Method" and ▶ 77, "Blogs in Social Research").

5 Analyzing Diary Data

Techniques for analyzing solicited diary data depend on the research objectives, the medium used to gather the data, and the size of the research team. Where a project involves multiple researchers and data gathered using different media (for example, written, audio, and/or visual data), analysis can be complex and require skill and effort to ensure materials can be practically shared and effectively analyzed (Pink 2009). However, for a sole researcher analyzing written diary data, the process is relatively straightforward.

5.1 Analyzing Written Diary Data

Quantitative methods of analysis can be applied to semi- and unstructured diaries, for example, where a researcher may be interested in how many times a diarist refers to a specific phenomenon of interest. Those interested in linguistic issues may also be concerned not just with the temporal narrative, but with the frequency of, and language used. It is also possible to adopt discourse analysis to the diary data (see also ▶ Chap. 50, "Critical Discourse/Discourse Analysis"). This may be particularly relevant for audio diaries where *how* an entry is recorded, and the tone of speech used in the audio diary, may be as important as what is said. In the main, however, the objective within semi- and unstructured diary method is to analyze the free-flowing data, whether written or oral. Following transcription of the data, this can be done using commonly used qualitative data analysis methods such as thematic or constant comparative techniques (see also ▶ Chap. 48, "Thematic Analysis"). Verbatim entries are thus coded, compared, and the data thematically categorized until no new themes emerge (Bartlett and Milligan 2015). The key purpose here is to identify themes emerging either from within an individual's data over time, or across a sample of participants selected on the basis of their similar or differing characteristics. Where appropriate, a well-designed diary with a coherent precoding system (similar to framework analysis) can help to reduce the degree of editing and coding required. While audio diaries produce significant amounts of monologue and often contain background sounds, which may or may not be important to interpret, the basic approach to analyzing these data is the same (Pink 2009). The main difference between written and audio-diary data is the increased amount, and complexity, of the audio data and the researcher's need to consider whether (or which) background sounds and information require interpretation.

While thematic and constant comparative analysis of diary data are useful, it is important to bear in mind that these forms of analysis rely on extracting small sections of diary data and reassigning by code or theme to achieve across person analysis. These approaches can result in the loss of two of the key strengths of diary data: the personal narrative and the temporality of that narrative. For this reason, some researchers reject the thematic approach in favor of narrative analysis that focuses on trying to make sense of the temporal story and conveying the meaning and contextual detail provided by the diary (Thomas 2007; see also ▶ Chap. 49, "Narrative Analysis").

5.2 Analyzing Visual Diary Data

Diaries using visual data require a slightly different analytic approach. The more complex nature of visual diary data, where movement, body-language, setting, as well as the orally delivered diary data means analysis will be more time-consuming, requiring the researcher to do more filtering of what is/not required for the study (Noyes 2004, p. 203). Here, both moving images and spoken words need to be scrutinized and coded. Moreover, visual materials are suggestive, meaning that the researcher's whole body and sensory range is likely to be engaged in the interpretative process (Pink 2009). Hence, as Pink notes, the researcher's reaction to what they see and hear in the diary should be one of the first issues to reflect on in visual diary analysis. Identifying and noting those parts of a diary that the researcher finds most compelling (either positively or negatively), and why, adds another layer of meaning-making to the analysis (Allen 2009).

While photographic diaries produce images that are clearly meaningful to the participant, they too require a process of sifting and coding. This is usually done *with* the participant, selecting a sample for a more in-depth analysis (photo-elicitation/photovoice). Where this is combined with DIM (Diary-interview methods) approaches, the analysis of the photographs in the final interview can be seen as a facilitating coanalysis. In a multimethod study using video diaries, journals, and interviews to examine embodiment and everyday illness, for example, Bates (2013) analyzed her video-diary data in two ways. Firstly, each video diary was transcribed, integrated with the interview and journal data, and analyzed thematically. Secondly, by identifying categories and concepts *across* the different methods, she was able triangulate her interpretation of the video diaries and consider the diary data within a broader context.

The complexity and variety of data produced by video and photographic diaries means that some researchers aim to simplify the process by opting to analyze purposively selected aspects of the visual diary data rather than trying to make sense of the whole. In both Roberts's (2011) and Edinburgh et al.'s (2013) studies, for example, their analysis focused on what their video diarists *said* rather than trying to interpret the performative elements of the diaries; transcribing and analyzing the oral data in much the same way as audio-diary data is analyzed. In such cases, the rationale for using visual diaries may be less about the relevance of visual analysis, and more about the appropriateness of using this particular diary technique for the group or individuals being studied.

6 Strengths and Limitations of Solicited Diary Method

All research methods have their strengths and weaknesses – solicited diary methods are no exception. Critiques suggest they favor more literate and better-educated participants, and so it can be exclusionary to those with poor literacy skills or those with cognitive or physical limitations that hinder their ability to perform the task of diary keeping (see discussions by Meth 2003; Mackrill 2008; Cooley et al. 2014).

Yet, the wealth of highly successful solicited diary-based studies, undertaken with a wide range of individuals with different levels of literacy and so-called vulnerabilities, suggests this critique lacks validity. Solicited diary studies, for example, have been undertaken with people living with dementia, young children, young men with disabilities, older people, survivors of domestic violence, and those from minority ethnic groups, to name but a few (Bartlett and Milligan 2015).

Critiques are often based on a perception that the solicited diary relies solely on the (hand) written word. Yet, oral, visual, and pictorial techniques have often been used as an alternative to the written diary. The key is to take participants' strengths and needs into account and modify the diary method accordingly. Bartlett's (2012) use of audio and photographic diaries to explore the lives of people with dementia; Thomas et al.'s (2015) use of pictorial diaries with slum-dwelling mothers of children who are experiencing diarrhea morbidity in India; and Edinburgh et al.'s (2013) video-diary study of migrant girls who had runaway following sexual exploitation offer just a few examples. These types of studies illustrate how solicited diaries can be creatively adapted to meet the skills of those for whom written diaries may prove exclusionary. Furthermore, as we illustrate here, the growing availability of video and digital technologies means that solicited diary techniques no longer rely solely on gathering hand-written data and paper-based medium. As Cooley et al. (2014) note, semistructured video diaries can be particularly useful for overcoming the difficulties some people face in expressing themselves in written format. It also offers increased depth and freedom of speech with less reliance on researcher/participant rapport (Buchwald 2009).

A further strength of solicited diary methods is that they offer respondents space and time to think and consider questions/themes in private and the freedom to express intense sentiments (Cooley et al. 2014; Filep et al. 2015). Meth's (2003) diary study of domestic violence in South Africa is an excellent example of this. Rather inhibiting women from telling their stories, her diaries enabled space for women to record deeply personal descriptions of often traumatic life experiences. Diaries can thus be revelatory, in that they help researchers understand the embodied and the emotional, and can be used to delve into otherwise unreachable interpretations of social and physical experiences (Filep et al. 2015). Myers (2010) diary study with HIV-positive men, for example, illustrated how solicited diaries provided them with space to express thoughts and feelings that they would have found hard to do face-to-face. The longitudinal approach of diary techniques, and the provision of a situated "safe space" that allows for extensive individual and personal reflection, means that diary methods can provide rich tapestries of narratives with multiple divergent themes (Filep et al. 2015).

Numerous researchers have also noted the therapeutic value of diaries and the potential to enrich the diarists' lives (e.g., Meth 2003; Thomas 2007; Dwyer et al. 2013; Filep et al. 2015). Thomas (2007) maintains that not only can diaries feel therapeutic to the diarist, but they make diarists feel their opinions are valued and valid. Filep et al. (2015) go so far as to note that in their study, one diarist enjoyed the process so much he kept the original and submitted a photocopy.

7 Addressing the Limitations of Diary Methods

Where solicited diaries are used, it is important to bear in mind the action required on the part of the participant to produce the written, audio, or visual data and how this might affect the quality of that data. Diary-keeping relies on the participant's ability and motivation to complete the diary. Where motivation is low, this can result in attrition or gaps in the diary completions. Thus, Bijker et al. (2015) note the importance of personal interaction between the researcher and the diary respondents prior to the start of diary completion, as it creates a commitment to the research that motivates diarists to continue the diary. Milligan et al. (2005) also note that regular contact throughout the project (whether face-to-face, telephone, or electronic media) can help reduce noncompletion, particularly where a diarist may have stopped due to illness, holiday, or other personal circumstances; it can also encourage continuation of diary-keeping once the participant feels able to do so. Finally, Meth (2003) and others (e.g., Ulrich and Grady 2004; Boone et al. 2013) suggest that payment for diary completion can both encourage motivation and demonstrate recognition of the work of diary keeping.

Researchers have also commented that the dynamic or performative elements of producing a video diary or posing for an image as part of diary recording – and the importance of taking this account when analyzing the data. Noyes (2004), for example, in a video-diary study involving primary school children, noted how the participants tended to perform in front of the camera, particularly in the early stages of data collection. Others using video-diary method found that adult participants (particularly men) can be self-conscious about filming themselves (Roberts 2011). However, as (Buchwald 2009) note, just as with audio-taping an interview, the depth of (in this case) diary entries increases as diarists become more comfortable with the camera, and participants to get used to the diary-keeping process.

Others point out that solicited diary methods require the researcher to relinquish control of part of the research process (Mackrill 2008), offering less opportunity for the researcher to check on, and direct, the form and follow of data emerging or to probe interesting issues raised by diarists (Cooley et al. 2014). Critics have also suggested that the irregularity in the way diary data can be produced is a weakness, but this presupposes that other qualitative data produce more consistent and reliable data. Indeed, Mackrill (2008) suggests that *how* the diarist records data, what is and is not recorded, can also be a source of data in itself. Others suggest that there can be a selectivity to what data respondents choose to reveal in their diaries

But as Meth (2003) points out, this is true of any qualitative method. We would also add, that where these issues are important to the study, they can be overcome by combining diary methods with other techniques such as DIM.

A final issue we raise in this section is the potential for diary techniques to raise consciousness within the diarist of events/lack of events that have the potential to cause harm (Milligan 2001; Meth 2003). Because the researcher is not present during diary completion, s/he is unable to provide the same level of emotional support as may be able to do in face-to-face methods. Kenten (2010) has thus suggested that where there may be risk of vulnerabilities being exposed in the diary-keeping process, it may be useful to use a DIM design as the postdiary interview can also be used to debrief the participant, helping to mitigate any potential discomfort caused by the diary-keeping process.

8 Diary Methods: Practical Examples

In this final section, we draw on practical examples of the use of solicited diaries from our own research.

The first study involved the use of solicited diaries as part of a mixed methods study designed to understand the well-being impacts of communal gardening for older people when compared with other social activities (Milligan et al. 2005). The diaries were designed to gather longitudinal data about activities and factors impacting on participants' health and well-being over time. The diary was structured as a brief booklet comprising: instructions for completion and return; a short quality of life survey; and an open text section with narrative prompts. Participants were asked to complete the diaries on a weekly basis and were encouraged to write as much or as little as they wanted – focusing specifically on what was important to *them*. They wrote about thoughts, feelings, and events in their everyday lives that they felt were connected to their health and activities. The project researcher kept a record of all submitted diaries and where a diarist failed to complete, she made contact to investigate why. This turned out to be crucial to the study, facilitating the building of a rapport between the researcher and the participants, providing support and encouragement, and enhancing completion rates. In total, 69 older people aged between 65 and 91 years completed the weekly diaries over a 30 week period. Accounting for attrition and noncompletion, we gathered 1609 diaries in total – a completion rate of 81%. This approach also gave us a better insight into reasons for noncompletion (largely due to illness or holidays) and encouraged continuation of the diary-keeping once the participant either regained their health or returned from holiday.

The diaries revealed not only the ups and downs of older people's health and well-being over a sustained period of time, but also their levels of activity and social connectedness. In particular, they highlighted the importance of family and friends in sustaining and supporting participants' well-being and activity – and in some cases how the stresses and difficulties of family relationships can have an adverse effect on an older person's well-being. "Lucy" (aged 68), for example, lived alone but wrote of the importance of her family in helping her through a period of illness. She wrote in her diary, "I am at last back to my normal fit and happy self. For a time I felt quite low, but thanks to the help and support of my daughter and her family, and to my own determination, I got over this." Conversely, Florence (aged 73) wrote of having a busy and active live despite her health problems, but woven through her diary accounts is a narrative of a difficult relationship with her 42-year-old adult son who lived with her. She wrote of how she found her son's controlling attitude difficult to deal with, and her increasing depression resulting from their regular arguments and inability to live together harmoniously. Over time, her diaries revealed how her activities and social relationships were bound up in strategies to keep her away from what she describes as "the black cloud" hanging over her home (Milligan et al. 2005, p. 1887).

So while how people complete diaries can be very different, from the detailed reflective and emotive entries to the brief factual account. Yet, despite these differences in the length and depth of entries, the diaries gave us detailed longitudinal insights into participants' lives that enabled us to draw out the interconnections between health, environment, and everyday life for older people.

The second study relates to the use of diary interview method to investigate the rise of activism among British people with dementia (Bartlett 2012). Sixteen people with dementia who were engaged in some form of social action or campaigning related to their health condition were recruited to the study. Participants were asked to keep either a written, photo, or audio diary of their activities. Five participants kept a photo diary, three kept a written diary, and one kept an audio diary. The seven remaining participants kept a combination of one or more of the diary methods, but no one chose to keep all three. Everyone kept a diary for about one month.

A range of diary techniques were used in this study to enable people with dementia to participate; it was recognized that not everyone would want or be able to write, or take photographs, or speak into a voice recorder, thus people had a choice. The approach enabled participants to have greater control over the content and pace of data collection than traditional data collection methods allow. For example, the audio diarist took his time to find the right word – he said "worm" instead of "word" – and switched the recorder on and off, presumably to prepare to speak. His finger controlled the record button, not the researcher (Bartlett 2012). Other participants supplemented their diaries with artefacts they had collected during their campaigning activities, such as reports and brochures. Virtually, all the photo diarists took photographs of how and where they spent their leisure time. All of which is to say that diary-interview method enabled people with dementia to not only participate in this study, but to participate on their own terms.

Finally, it is important to note that there were certain drawbacks to using diary interview method in this study involving people with dementia. Notably, the process of keeping a written diary evoked a certain amount of anxiety for some participants; some individuals were not sure whether what they were doing was "right," others did not like reading back what they had written (Bartlett 2012). Researchers are therefore advised to highlight this as a potential risk in the study's participant information sheet.

9 Conclusion and Future Directions

A growing number of researchers are beginning to recognize the potential added value of solicited diary method in social research – whether as a stand-alone method or as one element of a multimethod study. We foresee that this will continue to grow in strength and that, of itself, will trigger further innovative developments.

However, it is perhaps the rise of technologies, especially capture and digital technologies, and their integration into research methods that hold the most promise. These developments mean that researchers will increasingly find themselves having to engage with, and adapt to, the digital culture. The growth of ubiquitous technologies, smart phones, tablets, and so forth means that we are shifting toward a scenario in which the research equipment required for undertaking solicited diary research may be significantly simplified. The capability is now there for research to be undertaken through one device that has multiple capacities and features to facilitate both video, audio, and electronic written diary keeping. Such devices will become an increasingly common part of daily life within many households, particularly in high-income countries. We should not forget, however, that access to, and familiarity with, digital technologies can still raise issues of access for those who are less affluent (including those living in less developed parts of the world) and those whose physical, sensory, or cognitive limitations may require specially adapted equipment.

The acceptability and expectation of self-disclosure practices evidenced through blogging, Facebook, and the rise of the "selfie" is re-energizing the practice of diary keeping but in much more public and interactive ways than we have seen in the past. Weblogs, for example, are simultaneously public and personal diaries. They are intended to be read and to provoke dialogue with others. This presents opportunities for researchers to solicit online diaries or to analyze existing online diary data in ways that have not previously been possible. Arguably, then, the boundaries between solicited and unsolicited diaries are blurring due to web-based applications and practices such as blogging and Facebook, presenting researchers with challenges to the traditional notions of solicited and unsolicited

Finally, given that human relationships are becoming more mobile and interconnected (Urry 2000), it is important that researchers develop and use methods that have some synergy with the day-to-day practices and experiences of their participants. For example, email diary studies, such as Jones's and Woolley's (2014) study of the effects of the London Olympics on daily commuters, are likely to become more commonplace in the future. Whatever direction diary methods take in the future, it is imperative that teaching and texts on research methods move beyond their traditional focus on standard qualitative techniques, to ensure new researchers are aware of the significant potential of adding solicited diary methods to their methodological toolbox.

References

Allen L. "Snapped": researching the sexual cultures of schools using visual methods. Int J Qual Stud Educ. 2009;22(5):549–61.
Bartlett R. Modifying the diary interview method to research the lives of people with dementia. Qual Health Res. 2012;22(12):1717–26.
Bartlett R, Milligan C. What is diary method? London: Bloomsbury Academic; 2015.
Bates C. Video diaries: audio-visual research methods and the elusive body. Vis Stud. 2013;28(1):29–37.

Bernays S, Rhodes T, Jankovic K. Embodied accounts of HIV and hope: using audio diaries with interviews. Qual Health Res. 2014;24(5):629–40.

Bijker RA, Haartson T, Strijker D. How people move to rural areas: insights in the residential search process from a diary approach. J Rural Stud. 2015;38:77–88.

Bijoux D, Myers J. Interviews, solicited diaries and photography: 'New' ways of accessing everyday experiences of place. Grad J Asia-Pacific Stud. 2006;4(1):44–64.

Bolger N, Davis A, Rafaeli E. Diary methods: capturing life as it is lived. Annu Rev Psychol. 2003;54:579–616.

Boone MR, Cook SH, Wilson P. Substance use and sexual risk behavior in HIV positive men who have sex with men: an episode-level analysis. AIDS Behav. 2013;17(5):1883–7.

Buchwald D. Video diary data collection in research with children: an alternative method. Int J Qual Methods. 2009;8(1):12–20.

Capstick A. Travels with a flipcam: bringing the community to people with dementia in a day care setting through visual technology. Vis Stud. 2011;26(2):142–7.

Cleveland HH, Harris KS. The role of coping in moderating within-day associations between negative triggers and substance use cravings: a daily diary investigation. Addict Behav. 2010;35(1):60–63. https://doi.org/10.1016/j.addbeh.2009.08.010.

Cooley SJ, Holland MJG, Cumming J, Novakovic EG, Burns VE. Introducing the use of a semi-structured video diary room to investigate students' learning experiences during an outdoor adventure education groupwork skills course. High Educ. 2014;67:105–21.

Dwyer S, Piquette N, Buckle J, McCaslin E. Women gamblers write a voice. J Group Addict Recover. 2013;8:36–50.

Edinburgh LD, Garcia CM, Saewyc EM. It's called "Going out to play": a video diary study of Hmong girls' perspectives on running away. Health Care Women Int. 2013;34(2):150–68.

Filep CV, Thompson-Fawcett M, Fitzsimmons S, Turner S. Reaching revelatory places: the role of solicited diaries in extending research on emotional geographies into the unfamiliar. Area. 2015;47(4):459–65.

Gibson BE. The integrated use of audio diaries, photography, and interviews in research with disabled young men. Int J Qual Res Methods. 2002;12:382–402.

Gill J, Liamputtong P. Walk a mile in my shoes: researching lived experiences of mothers of children with autism. J Fam Stud. 2009;15(3):309–19. Special issue on "Parenting around the world".

Higginbotham G, Liamputtong P. Participatory qualitative research methodologies in health. Los Angeles: Sage; 2015.

Hislop J, Arber S, Meadows R, Venn S. Narratives of the night: the use of audio diaries researching sleep. Sociol Res Online. 2005;10(4).

Jackson S, Vares T. 'Perfect skin', 'pretty skinny': girls' embodied identities and post-feminist popular culture. J Gend Stud. 2014;1–14. https://doi.org/10.1080/09589236.2013.841573.

Kenten C. Narrating oneself. Forum Qual Soc Res. 2010;11(2). http://www.qualitative-research.net/index.php/fqs/article/viewArticle/1314

Latham A. Research, performance, and doing human geography: some reflections on the diary-photograph, diary-interview method. Environ Plan A. 2003;35(11):1993–2017.

Mackrill T. Solicited diary studies of psychotherapy in qualitative research – pros and cons. Eur J Psychother Couns. 2008;10(1):5–18.

Meth P. Entries and ommisions: using solicited diaries in geographical research. Area. 2003;35(2):195–205.

Milligan C, Bingley A, Gatrell A. Digging deep: using diary techniques to explore the place of health and well-being amongst older people. Soc Sci Med (1982). 2005;61(9):1882–92. https://doi.org/10.1016/j.socscimed.2005.04.002.

Milligan C. Geographies of care: space, place and the voluntary sector. Aldershot: Ashgate; 2001.

Myers J. Moving methods. Area. 2010;42:328–38.

Noyes A. Video diary: a method for exploring learning dispositions. Camb J Educ. 2004;34(2):193–209.

Orban K, Edberg A-K, Erlandsson L-K. Using a time-geographical diary method in order to facilitate reflections on changes in patterns of daily occupations. Scand J Occup Ther. 2012;19(3):249–59.

Pink S. Doing sensory ethnography. London: Sage; 2009.

Prosser J, Schwartz D. Photographs within the sociological research process. In: Prosser J, editor. Image based research: a qualitative sourcebook for researchers. London: Falmer Press; 1998.

Rich M, Lamola S, Amory C, Schneider L. Asthma in life context: Video Intervention/Prevention Assessment (VIA). Pediatrics. 2000;105(3):469–77.

Roberts J. Video diaries: a tool to investigate sustainability-related learning in threshold spaces. Environ Educ Res. 2011;17(5):675–88.

Rokkan T, Phillips J, Lulei M, Poledna S, Kensey S. How was your day? Exploring a day in the life of probation workers across Europe using practice diaries. Eur J Probation. 2015;7(3):201–17.

Thomas F. Eliciting emotions in HIV/AIDS research: a diary-based approach. Area. 2007;39(1):74–82.

Thomas RJ, Karthekayun R, Velusamy V, Kaliappan SR, Kattula D, Muliyal J, Kang G. Comparison of fieldworker interview and pictorial diary method for recording morbidity of infants in semi-urban slums. BMC Public Health. 2015;15:43. https://doi.org/10.1186/s12889-015-1372-7.

Ulrich C, Grady C. Financial incentives and response rates in nursing research. Nurs Res. 2004;53(2):73–4.

Urry J. Sociology beyond societies: mobilities for the twenty-first century. London: Routledge; 2000.

Waldron I, Eyer J. Socioeconomic causes of the recent rise in death rates for 15–24 year olds. Soc Sci Med. 1975;9(7):383–96.

Wang C. Photovoice: a participatory action research strategy applied to women's health. J Women's Health. 1999;8(2):185–92.

Whittle R, Medd W, Deeming H, Kashefi E, Mort M, Twigger Ross C, Walker G, Watson N. After the rain – learning the lessons from flood recovery in Hull. Final project report for Flood, Vulnerability and Urban Resilience: a real-time study of local recovery following the floods of June 2007 in Hull, Lancaster: Lancaster University; 2010.

Williamson I, Leeming D, Lyttle S, Johnson S. 'It should be the most natural thing in the world': exploring first-time mothers' breastfeeding difficulties in the UK using audio-diaries and interviews. Matern Child Nutr. 2012;8(4):434–47. https://doi.org/10.1111/j.1740-8709.2011.00328.x.

Williams S, Sheffield D, Knibb R. A snapshot of the lives of women with polycystic ovary syndrome: a photovoice investigation. J Health Psychol. 2016;21(6):1170–82.

Teddy Diaries: Exploring Social Topics Through Socially Saturated Data

84

Marit Haldar and Randi Wærdahl

Contents

1 Introduction	1466
2 What Are Teddy Diaries?	1466
3 Socially Saturated Naturally Occurring Data	1467
4 The Two Contexts: Home–School Communication and the Comparative Analytical Context	1468
5 Conclusion and Future Directions	1474
References	1476

Abstract

Teddy bears with diaries are common pedagogical tools for home-school collaboration. In this chapter, we use three analytical examples comparing teddy diaries from Norway and China to demonstrate how these diaries give unique access to the display of family life. Because the diaries circulate not only between the school and the family but between families, each family influences the other in how they write their entries. This social process saturates the diaries with the norms, values, and ideas of the social context. Comparing and contrasting diaries from two different contexts adds to the richness of each dataset, as it illuminates the things that we take for granted and the things that are there that we do not talk about. By this methodological demonstration, we wish to challenge two hegemonic positions in qualitative methods and show that you do not have to "be there" to get close to lived life and you do not have to "speak to people" to get trustworthy data about the social. This kind of data is easily assessable for research with the consent of schools and families, and they are easy to initiate in schools or any institutions with groups of some permanence.

M. Haldar (✉) · R. Wærdahl
Department of Social Work, Child Welfare and Social Policy, Oslo and Akershus University College of Applied Sciences, Oslo, Norway
e-mail: marit.haldar@hioa.no; randi.wardahl@hioa.no

> **Keywords**

Display of families · Participatory data · Family life · Comparative analysis · Data saturation · Teddy diaries

1 Introduction

In this chapter, we demonstrate a method for exploring display (Finch 2007) of family life. The significance of this method lies in the combination of using naturally occurring data that have been socially saturated by circulating among families and comparing these data with a second set that has been socially saturated to an equal extent in a different context. When datasets from two of these contexts are compared, each dataset informs the other, enriching and elaborating the data.

Using four examples from teddy diaries produced in two very different contexts, we demonstrate how the combination and comparison of socially saturated naturally occurring data facilitate discovery of taken-for-granted traits of everyday family life. The examples are comparisons between schools and families in the Oslo metropolitan area in Norway and in urban Beijing, China. These contexts are apparently very different, but the comparative context does not have to be as different as this to yield revealing results.

This approach to contemporary family life challenges two hegemonic positions on qualitative methods to obtain trustworthy data: the ethnographical notion of "being there" (Geertz 1973; Clifford and Marcus 1986; Marcus 1995; Moore 1999; Davies 2010; Liamputtong 2013) and the authenticity of the spoken word gathered through personal interviews (Gubrium and Holstein 2002; Kvale 2007; Denzin and Lincoln 2011; Hammersley 2013; Serry and Liamputtong 2017). As a result of these hegemonic positions, texts receive a secondary status as data.

2 What Are Teddy Diaries?

Teddy bears and teddy diaries were introduced as a pedagogical device in Norway after the school reform of 1997 to ease the transition between a student's family and the first year of school. Each new school class gets a teddy bear that will visit every child's home in turn, carrying a diary where the bear's experiences in the children's home are recorded. During the first year in school, the teddy will visit each pupil several times. The diary has an introductory page with a greeting form the teddy bear that usually says something like this:

> Hello, my name is Teddy and I am a special friend of this class. This is my book. I will be very happy if you write and draw in my book about the things you and I experience together. I am sure that your Mommy or your Daddy can help you.
> Warm regards,
> Teddy

The diary entries are written by the children, or by the children together with their parents, and must be read as a negotiated textual description of what the family regards as a valid account of their activities on any given day. Most importantly, the entries represent a choice of topics that they consider worth mentioning to classmates, teachers, and other families. Teachers and classmates share the entries, as well as the families who are next to receive the book and the teddy (Haldar and Wærdahl 2009; Wærdahl and Haldar 2013; Haldar and Engebretsen 2013; Haldar et al. 2015).

3   Socially Saturated Naturally Occurring Data

Naturally occurring data are characterized by being derived from texts, social situations, or processes that are not initiated by a researcher (Geertz 1973; Silverman 2011). Thus, it is typical of such data that they are not communicated to the researcher in a controlled setting, but one that would occur regardless of researcher involvement. An important criterion for natural occurrence is that the researcher's influence on the data is minimal. The researcher's preanalytical questions, glances, or facilitation should not influence them.

Another important feature of naturally occurring data is in their association with the context. The data should exist in their natural context, so context is highly relevant for the data (Graue and Walsh 1998; Haldar and Wærdahl 2009) and may be significant in several ways. What people say depends on the situation and to whom they are talking. What a person writes reflects both where it is written and for whom it is intended. In the case of the teddy diaries, it is crucial that they are filled with stories written by children and their parents in collaboration. Another important trait is that the authors of the diaries are aware that other families, classmates, and teachers will read what they have written. Thus, the natural context for these texts is home–school correspondence. They are simple, probably because they convey everyday events that everyone can recognize. They should engage other children in the class, and they are probably written in some haste, as this is an obligatory task performed from 1 day to the next. These texts would be different if they were written as private diaries, letters to someone's grandmother or replies to a household survey.

Low researcher influence is a prime quality in naturally occurring data, as it indicates low contamination of the "natural." This is not particularly important in the case of the teddy diaries and is in fact almost the reverse. The more that readers and writers have been involved in the contextualization of the data, the more saturated it becomes with the social subject matter we seek (Lareau and Shumar 1996). Every family who writes in the diary and reads the work of other families, every school class and every teacher and even the researcher leave their marks on the text as well as on the interpretations of the text. This social impact on the data is desirable and is a type of influence that researchers are unable to exert on the data on their own.

The diaries contain parent-assisted correspondence between 6-year-olds, written with the knowledge that there is a normative audience of teachers and other parents. All families except the one that receives the diary first know what others have written. All families except the last one know that other families will read what they write. Teddy diaries can, thus, be read as an exchange of normative everyday standards between different homes and between home and the school public. Researcher effect is therefore relatively low, but the impact of the social, cultural, and contextual on the data is very high. What we actually learn from these diaries are topics that a researcher would not necessarily ask about, yet that convey highly saturated information about norms and values and those that are socially accepted (Rose 1989; Dean 2009). We learn what different families choose to publish about small and large events in an ordinary day. Where people are, who does what, who is with whom, how they get along together, or what they do alone can be recorded in the diaries. In many ways, it could be said that the most important qualities of these data are those with which they are imbued by circulation between informants before they reach the researcher. What is exchanged and reinforced by the evaluation of others becomes the most interesting feature of the material. Teddy diaries can be understood as a joint production of what is ordinary (Sacks 1984).

4 The Two Contexts: Home–School Communication and the Comparative Analytical Context

The first context of these diaries is home–school communication. We discovered teddy diaries as mothers when our children were first graders in a Norwegian primary school. In 1997, important school reform in Norway lowered the school admission age to 6 years, and it became crucial to prepare schools for encounters with 6-year-olds. One pedagogical measure was the introduction of a class teddy bear and the teddy diary. We were the first researchers to see them as suitable data material for a sociological analysis of families with children and as providing unique access to the norms and ideals of family life (Haldar and Wærdahl 2009). Because one of us (RW) had planned fieldwork in China, we were intrigued by the possibility of persuading some Chinese families to produce texts about normal life in China, without the researcher even being present. We decided to try the method in Beijing. Chinese teachers welcomed the concept of a class teddy bear equipped with a travel diary was with such great enthusiasm that ten bears went on home visits, and we received reports from nearly 30 families in Beijing for each of the teddies.

As a result, we had two sets of data. The first was from Norway where we contacted two schools in the Oslo area, and were given access to 16 diaries with 319 stories collected in the period 2006–2007. The second was from Beijing, where the process of writing diaries was initiated by one of the authors, driven by the schools and overseen by the teachers. In total, we had ten diaries from China with 284 stories collected from six schools within 3 months in 2006. These collections of diaries gave us analytical access to the normative environments that mold childhood and family ideals in Norway and China.

The other context that is important for the interpretations of teddy diaries is the comparative analytical context (Lamont and Thévenot 2000). By comparing texts written in two contrasting natural contexts, we are able to see what stands out as unique in each of the contexts, as well as obvious facts of which we are oblivious.

This acknowledgment can be transferred to all types of comparisons. The features that stand out as special only do so when we know what constitutes ordinary life and what is taken for granted in a context: The commonplace is often discovered to be common only when something unusual occurs (Lamont and Thévenot 2000).

Below we provide four examples to demonstrate the possibilities of this method. We have chosen examples where comparison between the two datasets has either made the data richer and more interesting or revealed something we might otherwise have overlooked.

Example 1 Developing a collective voice
When teddy diaries are circulating among families, a kind of genre or a script for writing entries is formed. In Norwegian diaries, we see that the posts take the form of a logbook of daily activities. These contributions are stories with a cast of characters who do activities together, often with a rapid sequence of events. A whole range of activities, often outdoors, are listed, with many birthday celebrations, parties, enjoyable moments, or play situations.

> On Saturday we went to the cabin. We had to drive both car and boat. All but Teddy were wearing a life jacket. My cousin, who is twelve and a half years old, came with us. When we arrived we went for a swim, even though the water was cold. We jumped on the trampoline and we played cards. We also went into the woods and we visited some neighbours. On Monday we went home in time for football training. That was good, because Teddy and I were invited to come stay with Anne, and we thought that was great (...).

The expressed ideals are to spend much time outdoors, to be together, to have fun and not least to experience these things as "cosy." These ideals are repeated and reinforced as the teddy diaries circulate between families and the school. Each story almost surpasses the previous one in its cosiness and range of outdoor activities.

When we read the Chinese teddy diaries, we see a different genre evolve. Here, diaries are more contemplative, with thoughts and reflections written down. These stories, like the Norwegian ones, are chronological. The teddy bear is given a warm welcome and is introduced to family members as a valued guest, and it is emphasized that the child is special to be honored by a visit from teddy. This provides an opening to describe the splendid qualities of the child. The stories usually end by reporting how hard it is to say goodbye. Between welcome and farewell, they talk about routines, duties, food, and studies, but also about fun and intimate conversations. Thoughts, feelings, evaluations, and educational advice are the most frequent elements of the Chinese diaries. The circulation of the texts reinforces the ideals of honor, praise, and duty.

> I wish I could bring Teddy home every day (...) when Mum picked me up after school I introduced her to my special friend, Xiao Xiao. Then Mum was happy and proud of me. 'You're awesome! You just have to continue to study hard and work so that Xiao Xiao will come home to you every day!' When we got home I put the teddy bear on the bed, and I took its paw and we agreed that it would help me do my homework. (...) I didn't dare get distracted when Xiao-Xiao was watching me. (...) My Dad and Mum said 'you should be like Xiao Xiao every day, dutiful, that's good'.

Different genres develop in the Norwegian and Chinese diaries, and there is an interesting genre aspect that underlines how circulation works to reinforce values. The stories in the Chinese diaries become increasingly alike as the diaries are filled. This is in great contrast to the Norwegian diaries, where the extraordinary and peculiar are the normative genre. We recognize this from the contrasts between one story and the next. In the Norwegian teddy bear stories, it seems to be of the utmost importance to show that the bear has experienced something very special in the child's home, not just the same as in the one before. This does not mean that norms are absent in the Norwegian stories, while they are present in the Chinese ones. It means that while sameness is a strong norm in the Chinese diaries, uniqueness trumps sameness as a norm in the Norwegian diaries.

Both sets of diaries demonstrate how circulation reinforces the values of the context (Rose 1989; Dean 2009), but it is also interesting to note that these values would be hard to see without the comparative element.

Example 2 The significant routines

The texts provide access to conventional family life. Families most likely deliberately choose not to write about topics that are unpleasant or embarrassing. However, events that a family would like to acknowledge can be described in detail (Gubrium and Holstein 1990; Finch 2007; Morgan 2011).

The obvious cultural content in both the mentioned and the omitted topics is easier to capture through comparison, without which it would be particularly difficult to discover what is missing. It is the unique combination of comparative analysis and socially saturated naturally occurring data that creates this form of data on the obvious. This may be illustrated by the descriptions of obvious activities that are worth mentioning and by those that are omitted.

As a cuddly toy, the teddy bear belongs to the intimate sphere in both Norway and China. In the diaries, we read much about sleeping and bedtime routines. In Norway, sleeping is often described as occurring almost as a result of exhaustion after a day filled with activities, often physical and often outside. One teddy bear put it this way: "Now I'm really tired, so Ola and I are off to bed, a little late because of the birthday party." In another entry, the bear says: "It has been a busy weekend, and I've experienced a lot. Now I'm tired, and Felicia and I are going to bed." Both these quotes say something about the importance of going to bed after an active day. The sleep is well deserved. Going to bed "on time" also seems to be an important norm. In the first case, we see this norm expressed as an exception: *It has become especially late*. In other words, Ola usually goes to bed earlier.

The importance of sleep is also reflected in the Chinese diaries, but the routines around sleeping are different. In the Chinese diaries, they do not go to bed exhausted, but bring certain activities with them into the bed, where there is an element of preparation for the next day. It is common to do learning activities just before or after going to bed. A Chinese child writes:

> I went through my English homework again, so I kept Xiao Xiao in my arms while I went to bed. I told Xiao Xiao that tomorrow I will have a language test, so I should rest well before that.

Another Chinese child writes:

> Last night I kept teddy bear in my arms while I slept. Today when I woke up, I let the bear listen to me reading English. At noon, after lunch, I took the teddy bear with me for an afternoon nap.

Where children sleep and with whom may also be different. As we have seen in the above cases, the teddy often sleeps in the child's bed. However, by comparing sleep routines in the two datasets, we became aware that in the Norwegian diaries it is not uncommon for the whole family to go to bed together in the parents' bed, as in this case:

> Teddy, Mum and I went to bed in Mum and Dad's bed, where we slept well, all together.

In the Chinese diaries, there is no mention of sleeping in the parent's bed, but the teddy sleeping in the arms of the child is common. Moreover, they sleep together several times a day, with the bear taking a nap with the child after school, after dinner, or before homework. Sometimes they sleep in the car on the way home from school, if they have far to travel. In Norwegian diaries, sleeping is never mentioned as anything but a night-time activity.

In both Chinese and Norwegian diaries, sleep is important, but the importance attached to sleep in the two contexts is different. Sleep is primarily described as an ending to the day in both places, but while the Norwegian entries describe sleep as a necessary rest after a long active day, the Chinese stories describe sleep as a preparation for what comes after waking; it occurs not only at night but also at other times of day.

Example 3 Analyzing what is (not) worth writing about
While sleeping is important for teddies in both Norway and China, it is noteworthy that none of the Norwegian children reports resting after school, or before homework. In fact, in the Norwegian diaries, no one talks about doing homework at all. This does not mean that Norwegian children do not do homework; they just do not consider it worth mentioning. Norwegian six-year-olds may have less homework than their Chinese counterparts, and the teddies may go on more weekend visits in Norway. Nevertheless, it is a little surprising that in the context of a home–school communication tool, they do not find it important to mention that they are doing homework, even on weekdays.

The Norwegian diaries never mention homework. However, we did not notice this until we made the comparison with the Chinese diaries. The lack of homework in the former becomes evident when comparing with the latter, in which homework occurs all the time.

> I showed Xiao Xiao to Mum, and she said, 'Why did you bring this home with you?' I said, 'I've been so attentive in class and listened to the teacher, so she rewarded me.' I also showed Xiao Xiao to Daddy, he was also very happy, praised me and said I was good (...). After dinner I wanted to study. I placed Xiao Xiao beside my desk. Xiao Xiao sat quietly and watched me study, so we read English together. Then we should sleep. I washed my face and brushed my teeth with Xiao Xiao beside me. It looked like Xiao Xiao was saying, 'You are a child who is really concerned about hygiene'.

Another virtue that appears, especially in the Chinese diaries, is good hygiene. This is yet another example of how comparing two datasets reveals points that are not mentioned in one of them. In the Norwegian diaries, nobody mentions that they wash or brush their teeth. Obviously, this does not mean that Norwegian children do not; just that it is not seen as important enough to mention. However, the teddy bear is washed in the laundry or has a shower in the stories from both countries.

Comparison sometimes reveals obvious points from what is omitted from the stories. At other times, we understand that something is significant because it is mentioned often and described in detail. The numerous activities and relationships mentioned in the Norwegian diaries are an example. Here, children attend sporting events and competitions, particularly soccer and handball. They go to the cinema and the swimming pool and have grandparents visiting. Overnight visits with a friend are frequent. They regularly travel to their cabins, feed the ducks or other animals, or go hiking in the woods. In general, these are stories about being outdoors frequently, and there are many people in them. Some excerpts from Norwegian diaries illustrate what is typically believed to be important and worth writing about in a Norwegian context:

> On Sunday, Teddy got to taste blueberries. Moreover, we found three orientation posts in the north woods. Teddy enjoyed the walking tour and it was nice that both (maternal) grandmother and grandfather came along for the trip as well as my aunt and cousin.
>
> On Saturday, we picked mushrooms in the woods, during lightning and thunder, but it was okay for Teddy. On Sunday, the rain was really pouring, so we only made a small trip looking for mushrooms in Birkelunden (the city park).

Again, the contrast reveals the importance of things that seem mundane at first glance. In the Chinese diaries, there is no child who is described as sweaty or tired after physical activity, no mischief is described, and there are no overnight guests besides the teddy bear. In the Norwegian diaries, homework is never written about, and there is no brushing of teeth or praising the obedient child, his/her excellence, or virtues. The comparison between the two datasets provides a key to understanding what is meaningful in a given context, from something being described both extensively and in detail, and from being particular to one context when it is absent in the other. Knowledge of the social context is crucial to interpret omissions (Haldar et al. 2013).

Example 4 Bear qualities
Field researcher Cato Wadel (1991) emphasizes informants' awareness of identity when they tell us something, show us something, or give us access to something. He writes that one should be "a sociologist on oneself" because the roles attributed to the researcher in the field indicate important facts about the field (Wadel 1991: 59). Although researchers are not present in the field when we use teddy diaries, and the usual researcher effects are minimal, there is still a figure following the diaries and the texts from home to home, namely the teddy bear (Borovski et al. 2016).

In both China and Norway, the children find it necessary to explain much to an uninformed teddy bear. Moreover, the teddy bear is given many roles, just as a researcher would be. Teddy can be described as a friend, a stranger, a teacher, or someone who needs training. The teddy bear can also be a catalyst of events, feelings, and actions. In this section, we illustrate examples where Teddy is given the role of either instructor or student. Through these descriptions, we gain access to both the ideals and moral imperatives of bringing up children (Bernstein and Triger 2011; Chambers 2012).

In the Chinese texts, we see that when parents write the entries, they may use the teddy bear to provide clear moral advice to the child, or assign the teddy bear an important role as a bearer of good and desirable properties. Here is a Chinese example:

> When I see Teddy's simple and honest look, I think of an old proverb that says: 'To be human means that one must be kind and honest' (...) As the teacher takes care of us, we must take care of the teddy bear; let Teddy grow up together with us and be a healthy, happy, honest, hardworking and good child.

If we compare this with the Norwegian diaries, we never encounter explicit moral advice. However, moral imperatives are still implicit. For example, there are several stories where the teddy bear has eaten too many sweets and suffers from stomachache.

Properties ascribed to the teddy bear can be interpreted as stories about the child's abilities, often in the form of what might be expected from a child or what children are expected to either outgrow or learn. For example, there are teddy bears who are afraid of going to the dentist, teddy bears who are too tired to walk and get to ride on Dad's shoulders, and there are teddy bears that do not finish their food. Not infrequently, the teddy bear is bored – and most of all – tired. To simplify the contrast, we can conclude that being exhausted has a high moral value in both countries. However, while Norwegian children should be exhausted from outdoor activities, Chinese children should be exhausted from doing homework.

Using the teddy to talk about being afraid or refusing to eat is also a way to acknowledge a child's feelings. Because it is not always easy to voice aloud that your child is tired or afraid, or to admit to behavior that is inappropriate for the child's age, the bear adopts the vicarious role of the child for such descriptions. When the class reads aloud from a diary, the other children empathize with the weary frightened teddy, and the child him/herself is not directly affected by the description. In this role, the bear functions as a catalyst for thoughts and feelings the families (children or parents) may not otherwise have talked about. When the bear is told to do things properly, or it has difficulties mastering desired skills, it is no major issue to discuss it, while parents would not describe their children in the same critical way. Here is an example from a Norwegian diary:

> Teddy and Thomas were securely strapped into the back seat, and we headed for the mountains (...) On Sunday, all the children including Teddy set off early. Teddy borrowed some sunglasses and got sunscreen on his nose; the sun is harsh in March! We went skiing with sticky wax under the skis. We did not go particularly fast. Teddy struggled terribly in the snow. We suspect that he has not skied very much before. Maybe his family is Danish?

In the Chinese diaries, we read that teddies are well instructed by the children, with many clear and direct pieces of advice about what the bear should do and how it should behave, often expressed as coming directly from the child. Despite all the well-behaved teddies in the Chinese diaries, there are also bored teddies, impatiently waiting for the child to finish their homework or activities in which bears are not allowed to participate. The latter can be interpreted as examples of venues where it is inappropriate to take a friend or a pet, such as for extra lessons in English. We also find examples of naughty teddies, perhaps reflecting annoying behavior that one would not ascribe to the child in such a text. Here is an example of a Chinese teddy bear who apparently slept a little too long in the morning.

> Today I woke up early and wanted to play with the teddy bear. But when I looked, I saw that the bear was still asleep. I didn't have the heart to disturb him, and thought I would have to wait until he woke up by himself, and then play with him. He is really a lazy little bear. Ah!

Ideas on education and morality can be deduced from such narratives as both direct exhortations and descriptions of desirable properties. However, ideas can also be reflected in how the teddy bear is treated or described according to what are considered good and bad qualities in a teddy bear. The teddy bear can act as a conduit through which to convey clear expectations to a child without appearing authoritarian. Indirectly, the teddy bear takes on the position of the child (James 2007).

Again, we see that comparison adds an extra dimension to interpretation, and that ideals expressed in both datasets become even more apparent in contrast. The ideal of the active and playful Norwegian child is reinforced in contrast to the learning and time-structured Chinese child. Similarly, the obedient Chinese child is reinforced in contrast to the Norwegian child who is always negotiating.

Although the effect of the researcher is minimized in these data, the effect of the teddy bear is significant. The teddy bear is assigned roles, abilities, and expectations, and it is the catalyst for events that do not necessarily occur every day. For example, if the teddy bear is assigned the role of an important guest, there are descriptions of how guests are received. If the role ascribed to the bear is that of a new member of the family, or a close friend, there is access to other kinds of information, and perhaps more intimate everyday events and descriptions.

5   Conclusion and Future Directions

In studying contemporary life through qualitative methods, "being there" is the ethnographer's trademark, and in the sociological toolbox of qualitative methods, personal interviews have achieved a nearly hegemonic position (Silverman 2007; Liamputtong 2013).

In this chapter, we have sought to refute the common idea that truth lies within the individual, as the use of in-depth interviews would suggest. Using simple teddy diaries, we have shown that social research ideals do not have to delve deeply into an individual to reach the "truth." Significant insights into social situations can be found on the surface, in the obvious, and in information easily shared with others. Simple notes from a family's daily lives are rich in socio-cultural awareness of respectable family lives.

We have also demonstrated that the researcher's presence is not necessary. The teddy bear's unspecified nature makes it possible to attribute infinite roles and inclinations, roles that a researcher could never have played. The teddy can be a guest, a little brother or sister, a mischievous child, an ignorant person, someone to be boasted about, or someone to be raised. These assignments to hybrid roles provide important social information.

Most importantly, the data in their entirety show that texts have unused potential. The most common view of texts is that they are not life itself, but they refer to life. However, as we see it, much social life occurs textually, and today probably increasingly so. Social media are full of textual realities and brimming with significant integrative mechanisms and subtle logics of power. Seemingly insignificant texts such as teddy diaries may be very powerful. Our everyday lives are clearly present in such texts.

In this chapter, we have argued that teddy diaries, as naturally occurring data, are a good source of knowledge about the norms, values, and ideals in the social context we wish to examine (Griffin 2007). Circulation and context add richness to the data from teddies and their diaries. The combination and comparison of naturally occurring data help us to discover the familiar. When combined and contrasted, each dataset becomes richer and more nuanced, and together they provide the contrast needed to illuminate the taken for granted.

The diaries contain short simple stories of everyday life, about common knowledge and topics that are easy to discuss. Circulating between readers and writers in a delimited context, reinforcing superficial truths, the diaries are socially saturated with the self-evident. For precisely this reason, they convey significant information about the ideals and taken-for-granted qualities of everyday family life.

Since we discovered Teddy diaries as a rich source of information about families in our own social context, we have both initiated and collected similar diaries from other places in Norway as well as from other communities in other countries around the world. The fact that we can obtain highly socially saturated data with low researcher interference is one of the features that make these data attractive for research. Not having to be there for every interview, observation or participation makes it additionally interesting because we can cover more families without extra researcher efforts. Moreover, by not being there, we give the families an opportunity to choose what to write about, and increase the likelihood of answering questions about their everyday life that we have not even thought about asking in an interview.

This kind of data is easily assessable for research with the consent of schools and families, and they are easy to initiate in any institutions with groups of some permanence. Having had numerous materials to compare with, we have experienced how rich originally "thin descriptions" in a diary may be. As for others aiming to use diaries in their research, we would like to underscore that comparative cases do not need to be transcontinental and very different as in our case here. Two communities close by each other may also be challenged trough comparison that will enrich each dataset. Finally, we find these kinds of data valuable for family research, to reach a negotiated collective voice of a family, and not the perspective of a singular voice as, for example, a parent's understanding of a child.

References

Bernstein G, Triger Z. Over-parenting. UC Davis Law Rev. 2011;44(4):1221–79.
Borovski Lübeck S, Haldar M, Wærdahl R. Bamsens fortellinger om familielivet – hvordan familieidealer reproduseres gjennom skole-hjem-kommunikasjon. [the teddy bear's stories about family life – how family ideals reproduce through home–school communication]. Barnläkaren. 2016;34(1):41–55.
Chambers D. A sociology of family life: change and diversity in intimacy relations. Cambridge: Polity Press; 2012.
Clifford J, Marcus G, editors. Writing culture: the poetics and politics of ethnography. Berkeley: University of California Press; 1986.
Davies J. Introduction: emotions in the field. In: Davies J, Spencer D, editors. Emotions in the field: the psychology and anthropology of fieldwork experience. Stanford: Stanford University Press; 2010. p. 1–34.
Dean M. Governmentality: power and rule in modern society. London: Sage; 2009.
Denzin N, Lincoln Y, editors. The sage handbook of qualitative research. Thousand Oaks: Sage; 2011.
Finch J. Displaying families. Sociology. 2007;41(1):65–81.
Geertz C. The interpretation of cultures: selected essays. New York: Basic Books; 1973.
Graue EM, Walsh DJ. Studying children in context: theories, methods, and ethics. Thousand Oaks: Sage; 1998.
Griffin C. Being dead and being there: research interviews, sharing hand cream and the preference for analysing 'naturally occurring data'. Discourse Stud. 2007;9:246–69.
Gubrium JF, Holstein JA. What is family? Mountain View: Mayfield; 1990.
Gubrium JF, Holstein JA. Handbook of interview research: context and method. Thousand Oaks: Sage; 2002.
Haldar M, Engebretsen E. Governing the liberated child in self-managed family display. Childhood. 2013;21(4):475–87.
Haldar M, Wærdahl R. Teddy diaries: a method for studying the display of family life. Sociology. 2009;43(6):1141–50.
Haldar M, Rueda E, Wærdahl R, Mitchell C, Geldenhuys J. Where are the children? Exploring the boundaries between text and context in the study of place and space in four different countries. Child Soc. 2015;29(1):48–58.
Hammersley M. What is qualitative research? Bloomsbury: Bloomsbury Academic; 2013.
James A. Giving voice to children's voices: practices and problems, pitfalls and potentials. Am Anthropol. 2007;109(2):261–72.
Kvale S. Doing interviews. Los Angeles: Sage; 2007.
Lamont M, Thévenot L. Introduction: toward a renewed comparative cultural sociology. In: Lamont M, Thévenot L, editors. Rethinking comparative cultural sociology. Cambridge: Cambridge University Press; 2000. p. 1–22.
Lareau A, Shumar W. The problem of individualism in family-school policies. Sociol Educ. 1996;69:24–39.
Liamputtong P. Qualitative research methods. 4th ed. Melbourne: Oxford University Press; 2013.
Marcus GE. Ethnography in/of the world system: the emergence of multi-sited ethnography. Annu Rev Anthropol. 1995;24:95–117.
Moore H. Anthropological theory at the turn of the century. In: Moore H, editor. Anthropological theory today. Oxford: Polity Press; 1999. p. 1–23.

Morgan DHJ. Rethinking family practices. Basingstoke: Palgrave Macmillan; 2011.

Rose N. Governing the soul: the shape of the private self. London: Routledge; 1989.

Sacks H. On doing 'being ordinary'. In: Atkinson JM, Heritage J, editors. Structures of social action: studies in conversation analysis. Cambridge: Cambridge University Press; 1984. p. 413–29.

Serry T, Liamputtong P. The in-depth interviewing method in health. In: Liamputtong P, editor. Research methods in health: foundations for evidence-based practice. 3rd ed. Melbourne: Oxford University Press; 2017. p. 67–83.

Silverman D. A very short, fairly interesting and reasonably cheap book about qualitative research. London: Sage; 2007.

Silverman D. Interpreting qualitative data. 4th ed. London: Sage; 2011.

Wadel, C. Feltarbeid i egen kultur. En innføring i kvalitativt orientert samfunnsforskning. [Fieldwork in your own culture. An introduction to qualitative social research] Flekkefjord, SEEK a/s. 1991.

Wærdahl R, Haldar M. Socializing relations in the everyday lives of children. Comparing domestic texts from Norway and China. Childhood. 2013;20(1):115–30.

Qualitative Story Completion: A Method with Exciting Promise

85

Virginia Braun, Victoria Clarke, Nikki Hayfield, Naomi Moller, and Irmgard Tischner

Contents

1	Introduction	1480
2	Theoretical Lens	1481
3	Benefits of SC as a Data Collection Method	1482
4	Suitable Research Topics and Questions in SC Research	1484
5	Stem and Study Design in Story Completion Research	1484
6	Completion Instructions for SC	1487
7	Asking Additional Questions in SC Research	1488
8	Sampling in Story Completion Research	1488
9	Data Collection and Ethical Considerations in Story Completion Research	1489
	9.1 Piloting in SC Research	1490
	9.2 Ethics in SC Research	1490
10	Analyzing Story Completion Data	1491

The chapter is a revised version of Clarke V, Hayfield N, Moller N, Tischner I. Story completion tasks. In: Braun, V, Clarke, V, & Gray, D, editors. (2017) Collecting Qualitative Data: A practical guide to textual, media and virtual techniques. Cambridge: Cambridge University Press. Reprinted with permission of the publishers.

V. Braun (✉)
School of Psychology, The University of Auckland, Auckland, New Zealand
e-mail: V.Braun@auckland.ac.nz

V. Clarke · N. Hayfield
Department of Health and Social Sciences, Faculty of Health and Applied Sciences, University of the West of England (UWE), Bristol, UK
e-mail: Victoria.Clarke@uwe.ac.uk; Nikki2.Hayfield@uwe.ac.uk

N. Moller
School of Psychology, Faculty of Social Sciences, The Open University, Milton Keynes, UK
e-mail: Naomi.Moller@open.ac.uk

I. Tischner
Faculty of Sport and Health Sciences, Technische Universität München, Lehrstuhl Diversitätssoziologie, Munich, Germany
e-mail: Irmgard.Tischner@tum.de

© Springer Nature Singapore Pte Ltd. 2019
P. Liamputtong (ed.), *Handbook of Research Methods in Health Social Sciences*,
https://doi.org/10.1007/978-981-10-5251-4_14

10.1 Story Maps 1491
10.2 Frequency Counts 1493
10.3 Analysis Approaches to Avoid 1493
11 Conclusion and Future Directions 1493
References 1495

Abstract

This chapter introduces the story completion (SC) method of collecting qualitative data, a novel technique that offers exciting potential to the qualitative researcher. SC involves a researcher writing a story "stem" or "cue" – or, more simply put, the start of a story, usually an opening sentence or two – and asking the participants to complete or continue the story. Originally developed as a form of projective test, the use of SC in qualitative research is relatively new. The authors comprise the *Story Completion Research Group*, a group of researchers that have come together to share their experience of using and further developing the method. This chapter explains what SC offers the qualitative researcher – including choices about the "best" epistemological lens and analytic approach for their research question, the potential to collect data about sensitive or taboo topics and to access socially undesirable responses, as well as the possibility of research designs that allow comparisons (for example between male and female respondents). This chapter also provides key guidance, such as what constitutes an appropriate research question, and sampling and design considerations. As a recently developed method, SC has fewer published research studies than some of the other research methods covered in this volume. For this reason, the chapter aims not only to provide a description of the method and recommendations for how best to use it, but also to explore some of the unresolved theoretical and practical questions about SC as well as to suggest future directions for SC.

Keywords

Story-completion · Qualitative methods · Data collection · Innovative

1 Introduction

Story completion (SC) involves a researcher writing a story "stem" or "cue" – or more simply put, the start of a story, usually an opening sentence or two – and asking the participants to complete or continue the story. Most forms of qualitative data collection involve the gathering of direct self-reports, so SC offers the qualitative researcher a radically different approach to data collection, and one that we think holds much potential.

SC originally developed as a form of projective test, for use by psychiatrists and clinical psychologists, to assess the personality and psychopathology of clients (see Rabin 1981). Projective tests involve asking people to respond to ambiguous stimuli, such as inkblots, as in the famous Rorschach inkblot test (Rorschach et al. 1921/1998). The assumption is that because the respondent cannot know unequivocally what the stimulus "is," they have to draw on their own understandings to make sense of it, and "fill in the blanks." In doing so, as the theory of projective tests goes, the participant reveals things about themselves that they may not be conscious of, or would feel uncomfortable revealing if asked directly about. Projective tests are rooted in psychoanalytic theory (Rabin 2001), which assumes that large portions of the self are blocked off to consciousness, and thus unavailable to both clients and clinicians through conventional means, such as self-reported accounts. Projective tests are thought to tap into this "blocked off" information, providing what Murray (1943/1971, p. 1) described as "an x-ray picture of [the] inner self."

Projectives have also been used as a research method, for example, in consumer and business research (e.g., Donoghue 2000; Soley and Smith 2008) and developmental psychology (e.g., Bretherton et al. 2003; George and West 2012). Projectives as a research technique (such as SC) have typically been used in *quantitative* designs, with complex coding systems developed to allow researchers to iron out the variability in individual responses to the projective stimuli, and turn the rich narrative detail into numbers and categories suitable for quantitative analysis.

Because it is assumed that projectives reveal "hidden truths", those who use projective methods in this way rely on a (post)positivist epistemology, taking an essentialist stance on the person and on the data. Such an approach does not sit well with many qualitative researchers, and in the rest of this chapter an alternative approach to using SC is elaborated, one which is grounded more firmly within a qualitative paradigm.

SC was first used in qualitative research in a 1995 study by feminist psychologists Celia Kitzinger and Debra Powell. They used SC to examine how 116 male and female undergraduate students made sense of infidelity in the context of a heterosexual relationship. They suggested that it was not necessary to read the stories as (only) revealing the psychological "truth" of the respondents: "researchers can instead interpret these stories as reflecting contemporary discourses upon which subjects [sic] draw in making sense of experience" (1995, pp. 349–350). This approach to SC is a social constructionist one, rejecting the idea that it is possible to access "real" or "true" feelings or thoughts, and assuming instead that realities are discursively constructed (Burr 2003). Kitzinger and Powell (1995) illustrated the differences between essentialist and constructionist readings of SC data by contrasting two different readings of their data. An essentialist reading would see the data as revealing any gender differences in "attitudes" to infidelity; a social constructionist reading would make sense of the data as replicating various (gendered) discourses about the meanings of infidelity for men and women.

2 Theoretical Lens

We contend that qualitative SC can be used in both essentialist and constructionist qualitative research, and this theoretical and conceptual flexibility makes the SC method eminently adaptable to a range of research questions and approaches to qualitative research! The aim in this chapter is thus to hand researchers the tools from which to choose which theoretical "lens" to apply to their data.

How might different theoretical lenses impact on SC research? Epistemology has implications at both design/data collection phases, but most vitally at the analytic/interpretative phase. In essentialist qualitative SC research, the data are assumed to represent participants' real perceptions of a phenomenon. US psychologists Jennifer Livingston and Maria Testa (2000), for example, used qualitative SC within an *experimental* design in which the female participants were given alcohol, a placebo drink or no drink, to explore women's perceptions of their vulnerability to male aggression in a heterosexual dating scenario. The researchers asked women to imagine *themselves* as the female character in their story and to write in the *first person*; they treated the women's responses as representing their true beliefs about this topic. An example of constructionist SC research can be found in feminist psychologist Hannah Frith's (2013) research on orgasmic "absence", where she treated SC data as capturing the cultural discourses available to participants. Frith's analysis explored how the stories drew on and reinforced various gendered discourses, including women's responsibility to be sexually attractive to maintain men's sexual interest and the notion that men's sexual desire is unbridled and easy to satisfy. Contextualist research, which sits somewhere between essentialism and constructionism, and where multiple truths or situated realities are understood to exist within particular contexts (Braun and Clarke 2013), is also possible using SC. However, at the time of writing, there are no published studies exemplifying this approach.

3 Benefits of SC as a Data Collection Method

Theoretical flexibility is only one benefit that SC offers the qualitative researcher – there are many others:

1. ***SC gives access to a wide range of responses, including socially undesirable ones***: SC offers an alternative to approaches that ask people directly about their views and understandings of a particular topic, instead asking them to write about the *hypothetical* behavior of *others* (Will et al. 1996) or how they would feel in a *hypothetical* situation. When participants are prompted to write hypothetically, they do not have to take ownership of, or justify, their stories in the way they would if they were being asked directly about the topic. Therefore, they are more likely to "relax their guard" and engage with the research topic with less reserve. This gives SC the unusual advantage of breaking down the "social desirability 'barrier' of self-report research" (Moore et al. 1997, p. 372).
2. ***SC ideally suits sensitive topics***: SC also offers a particularly accessible way for participants to take part in research, because it does not necessarily require personal experience of the topic. The use of hypothetical scenario story telling also means participants are slightly "removed" from the topic. This makes SC *especially* useful for exploring sensitive topics – if questioned directly about their *own* experiences, some participants feel uncomfortable, or even unwilling, to

discuss such topics. Sensitive topics that have been explored utilising qualitative SC include orgasmic "absence" (Frith 2013) and sex offending (Gavin 2005).

3. ***SC offers robust and easy-to-implement comparative design options***: This feature of SC can be useful to explore differences between different groups of participants or between different versions of the same story and how they are made sense of. Kitzinger and Powell's (1995) ground-breaking study used a comparative design with unfaithful male and female partners in the story stem, and male and female respondents. Critical psychologists Ginny Braun and Victoria Clarke (2013) similarly used two versions of a story to explore people's perceptions of trans-parenting. The story stem described a parent telling their children that they are uncomfortable living within their assigned gender and want to start the process of "changing sex." Roughly half of the participants completed a male parent (Brian) version and half an otherwise identical female parent (Mary) version. Having two versions enabled the researchers to compare the responses both according to the gender of the parent character and the gender of the participant. This was important because mothers and fathers tend to be perceived very differently in the wider culture, and women tend to be more tolerant of gender diversity and nonconformity than men (Braun and Clarke 2013).

4. ***SC offers scope for methodological innovation***: Qualitative researchers have only recently begun to fully explore the possibilities that SC offers. For example, critical psychologists Nikki Hayfield and Matthew Wood (2014) recently piloted a SC using visual methodologies (Frith et al. 2005) in their research on perceptions of appearance and sexuality. The stem described a dating scenario; once they had completed their stories, participants were directed to the website *Bitstrips* to create a cartoon image of the main character. A preliminary analysis of the images indicated that participants recognized the existence of lesbian and gay appearance norms, in a way which was not necessarily *as* apparent in their written responses. Hence, visual data may provide an anchor for, or "bring to life," textual responses and can also be analyzed in their own right. This allows the potential for different understandings of, insights into, and interpretations of the findings (Frith et al. 2005).

5. ***SC is useful for researching social categories:*** The listed advantages of SC as a method – including the ease of implementing comparative designs – means that it fits well with research focused on understanding the operation of social categories such as gender, race/ethnicity, or sexuality. It enables researchers to explore any divergences in how different social groups make sense of a scenario, *and* whether participants respond differently to variations in, for example, the story character's gender or sexuality.

6. ***SC methods have the advantage of being economical in terms of time and resources:*** In a context of reduced support for social sciences research, it can be an advantage that SC is a thrifty method! Being economical to use also makes SC eminently suited for student research projects. Hard copy stories, for instance, can be handed out to a large group of people and the completed stories returned in 30 minutes or so; online stories can be distributed (and then downloaded) with a few mouse clicks.

Table 1 Examples of existing story completion research

Topic area	Research question/focus	Theoretical framework
Internet infidelity	What are the perceived impacts of cyber-cheating on offline relationships? (Whitty 2005)	Essentialist (perceptions)
Sexual aggression	How do women perceive their vulnerability to sexual aggression in (heterosexual) dating contexts? (Livingston and Testa 2000)	
Infidelity	How do women and men represent unfaithful heterosexual relationships? (Kitzinger and Powell 1995)	Essentialist and constructionist
Sex offending	What cultural narratives do people draw on in stories about child sex offenders? (Gavin 2005)	Constructionist (discursive constructions)
Eating disorders	How are "anorexic" and "bulimic" young women discursively constructed in stories written by young people who do not self-identify as "eating disordered"? (Walsh and Malson 2010)	

4 Suitable Research Topics and Questions in SC Research

The flexibility of SC is one of its key advantages and, accordingly, it can be used to research a broad range of topics. SC is particularly suited to research exploring people's perceptions and understandings and broader social constructions around a topic. However, questions that focus on people's *lived experiences* are not well suited to SC research, because this method does not gather stories about participants' *own* experiences. When developing a research question(s), as in any qualitative project, it is important to ensure it is both focused on a specific topic, but also broad and open-ended – for instance, the research is guided by exploratory "what" or "how" type questions. For example, Kitzinger and Powell (1995, p. 345) aimed to "explore young men's and women's representations of 'unfaithful' heterosexual relationships," and Frith (2013, p. 312) examined "how people account for and explain orgasmic absence during heterosex." These questions are specific enough to guide the research and design, but open enough so that there is plenty of scope for fully exploring participants' responses. It is also important to ensure that the type of question created "fits" with the chosen epistemological approach; "perception" questions tend to be used in essentialist research, whereas "construction" and "representation" questions are most often used in constructionist and critical research. Table 1 provides examples of existing SC studies that demonstrate this.

5 Stem and Study Design in Story Completion Research

The most important design consideration in SC research is the design of the story stem: the "start" of a story that participants are asked to complete. A careful balance needs to be struck between providing the participant with a *meaningful* story stem and leaving enough ambiguity for tapping into their assumptions. Braun and Clarke (2013) suggest six considerations in story stem design:

1. ***Length of the story stem***: How much of the beginning of the story will be written? There are no hard and fast rules here; it depends on the topic and participant group. If the story concerns something likely to be familiar to participants, less detail is necessary for the scenario to be meaningful to them. For a less familiar or more complex topic, such as one focused on the character's psychology, participants may need more detail to understand the scenario that is the focus of the stem. For instance, critical psychologist Irmgard Tischner's (2014) research on constructions of weight-loss used a slightly longer stem: "Thomas has decided that he needs to lose weight. Full of enthusiasm, and in order to prevent him from changing his mind, he is telling his friends in the pub about his plans." Although weight-loss is a familiar topic to most people, the main focus of the research was on social perceptions and interactions around weight-loss *intentions*; this necessitated the story stem including the protagonist's interaction with other people, i. e., him telling his friends about his plans.
2. ***Authentic and engaging scenarios and characters***: Unless the story, its protagonists, and the context resonate with the study participants, it is unlikely they will write a useful story. The stem should engage participants and be easy for them to relate to. Using names and scenarios that sound authentic and believable will help participants imagine or "see" the characters and the scenario, and thus to write a rich and complex story. These details can also cue (potentially not deliberately) certain cultural norms; for example, names may provide cues about ethnicity, class, or religion, which may then shape the stories told.
3. ***Amount of detail***: The most difficult design decisions revolve around the issue of detail in the story stem. Too much detail and direction will potentially limit the variation and richness of the data; not enough could mean the participants will not know "where to take" the story, resulting in data that do not address the research question. Researchers need to design a story stem that stimulates a range of complex and rich stories – otherwise the analysis will not have much to say! To achieve this, give the participants adequate directions by giving them a context or background to the story, and some detail about the characters, and what the topic of the story should be about. At the same time, it is important to avoid overly constraining their responses, by describing the background and characters in too much detail. Participants need to know what their story should be about, but do not give them a suggested plot or ending. Thus, for a study exploring understandings of *motivations* for exercise, for instance, a very open story stem like "Toby decides to become more physically active... What happens next?" may take the stories in too many, and possibly undesired, directions, and not focus on Toby's motivations. On the other hand, giving participants a particular motivation in the story stem (e.g., "Toby wants to develop a six-pack to attract a boyfriend...") is likely to result in a lack of diversity in the data. A better stem for this topic could be "As Toby wipes the sweat off his face and tries to catch his breath, he wonders why he ever thought starting to exercise was a good idea."

4. *Use of deliberate ambiguity:* SC is particularly useful for the exploration of underlying, taken-for-granted assumptions around a topic – for example, the heteronormative assumption that a couple consists of a man and a woman. This can often be achieved by leaving certain elements of the story ambiguous, such as some demographic characteristics of the protagonists (e.g., class, sex, race, sexuality, age). However, if the research question necessitates focusing participants' attention on a particular detail of the story, this should not be left ambiguous.
5. *First or third person:* One design consideration concerns the standpoint the researchers want participants to take. Is it important participants step into the shoes of, and empathize with, one particular protagonist? Or is it better if they assume the position of an omniscient narrator? Although to date qualitative SC has involved mostly third person story stems, first person stems are possible (e.g., Livingston and Testa 2000). These can be useful if it is important for the participants to write from the perspective of a specific character. From a classical projective standpoint, first person SC is assumed to prompt more socially *desirable* responses (Rabin 1981). Therefore, if the researchers want to gain a *broader* range of stories, including socially undesirable responses, it is recommended to use a third person stem. However, as this approach is still new, the impact of participant perspective is also something that can be explored within research design!
6. *A comparative design:* A final design consideration is whether or not to use a comparative design. A comparative design allows exploration and comparison of assumptions made, or perceptions held, about certain social groups or scenarios. If this is a research aim, it is necessary to design versions of the story which reflect the specific differences of interest, and to allocate roughly equal numbers of participants to each of these. It is better not to have too many versions of a story in one study, and to avoid using overly complex designs, because qualitative research is primarily about understanding (potentially complex and dynamic) meaning, rather than compartmentalization – the latter is a specialty of quantitative research designs. Two to six covers the manageable spectrum for comparison for small and medium-sized projects, in terms of both participant recruitment and analysis.

   The other main way comparison can be included in the design involves different participant groups, and exploring the differences between the stories written by people who are, for instance, from different genders, sexualities, generations, or cultural or educational backgrounds. This requires the recruitment of sufficient numbers of participants from each demographic category concerned. For example, counseling psychologist Naomi Moller's (2014) research on perceptions of fat therapists included responses from 18 to 21 year-old undergraduate psychology university students and 16–18 year-old sixth formers. This design made it possible to consider the salience of counselor body weight for the whole group of young people, but also showed how small differences in age and educational experience impacted on the expression of fat stigma. Whereas the stories of both groups clearly reiterated anti-fat cultural narratives, the younger cohort were much more direct in their expression.

6 Completion Instructions for SC

After designing the story stem(s), the researchers need to write completion instructions for participants. In the *participant information sheet*, provide participants with some information about the nature of the task, and what they are expected to do, emphasizing the necessity of writing *a story*. Here is an example from Victoria Clarke's (2014) research on body hair:

> You are invited to complete a story – this means that you read the opening sentences of a story and then write what happens next. There is no right or wrong way to complete the story, and you can be as creative as you like in completing the story! I am interested in the range of different stories that people tell. Don't spend too long thinking about what might happen next – just write about whatever first comes to mind. Because collecting detailed stories is important for my research, you are asked to WRITE A STORY THAT IS AT LEAST 10 LINES/200 WORDS LONG. Some details of the opening sentence of the story are deliberately vague; it's up to you to be creative and 'fill in the blanks'!

Then, ideally just before or after participants are presented with the story stem, provide specific instructions on how they should complete the story. Completion instructions can vary from the broad and open to the more prescriptive and directive. For example, Clarke (2014) instructed participants to simply "read and complete the following story." Another common instruction is to ask participants to write "what happens next." Nikki Hayfield and Matthew Wood's (2014) research on sexuality and appearance provides an example of a more prescriptive approach. Because they wanted participants to focus on the events before, during and after the female character's date, they instructed participants to write their story in three sections. Their story varied by character sexuality (bisexual, lesbian and heterosexual) – these are the instructions for the lesbian version:

Jess is a 21 year old lesbian woman. She has recently met someone, and they have arranged to go on a date.

- Please write about the run-up to the date and how she prepared for it…
- Please write about the date and how it went…
- Please write about what happened next…

(Please feel free to write as much as you like about the characters and as far into the future as you like).

Researchers may also want to provide participants with clear instructions on the length of story they wish them to write, or a time-expectation, to help ensure the stories generate rich, useful data. For example, instruct participants to spend a certain amount of time writing their story (e.g., "please spend at least 10 minutes"), or to write stories of a particular length (e.g., see Clarke's 2014, example above). Such instructions are particularly important for participant groups who are not necessarily highly motivated, such as individuals who take part in order to access particular benefits associated with participation (e.g., research participation for course credit within universities).

7 Asking Additional Questions in SC Research

Although one of the key features of SC is that it provides an indirect approach, some researchers have combined the use of a story stem with a small number of direct questions (in a way that combines some aspects of vignettes, a related data collection approach; see Gray et al. in press). For example, Naomi Moller's (2014) research on perceptions of fat therapists involved asking participants a direct question about the counsellor featured in the story stem: "What weight did you think the counsellor was?" The answers to this question allowed Moller to understand how the participants' defined "fat" – a variable construct – and provided a conceptual anchor for interpreting their stories.

Researchers should also consider whether it is important to ask participants demographic questions beyond the "standard" questions about age, sex/gender, race/ethnicity, sexuality, disability and social class (see Braun and Clarke 2013). Such questions can provide a useful "baseline" for interpreting and contextualizing stories. For example, in her research on body hair, Victoria Clarke (2014) asked a series of questions about whether participants had currently or previously removed or trimmed body hair in particular areas and their reasons for doing so. Given that for women, but increasingly for men too, body hair removal is a dominant social norm (Braun et al. 2013; Terry and Braun 2013), an overview of the participants' own body hair practices provides important information for contextualising the data.

8 Sampling in Story Completion Research

How many participants or stories is the "right" number? In existing SC research, there is a large variation in sample sizes – from 20 (Walsh and Malson 2010) to 234 (Whitty 2005) participants. Sample size depends on a number of factors, including: (a) the complexity of the design – more stories generally require more participants to be able to say something meaningful about each version, especially if researchers intend to make comparisons; (b) the richness of individual stories – richer stories mean fewer participants (note, however, that it may not be possible to predict in advance how rich the stories will be); and (c) the purposes of the research. For a small student project, with a single stem design, and no comparison between different participant groups, a sample size of around 20–40 participants is likely to provide data that are rich and detailed enough for a meaningful analysis. The more comparisons made, the bigger the overall sample will need to be. Braun and Clarke (2013) advise recruiting *at least* ten participants per story stem variation, but to publish the study, journal editors and reviewers may require higher participant numbers than that.

Of course, as with any research, recruiting enough participants can be a challenge, which is why many studies are carried out with a student population. Students, however, are a very specific population, and often not very diverse in terms of demographics. At the same time, students *are* used to discussing and describing ideas in writing, tend to be fairly literate, and thus will not struggle with the task of writing a story (Kitzinger and Powell 1995); the same cannot be assumed for all other participant groups. Think carefully about the needs and expectations of the study participants – busy professionals, for example, may require very clear but short instructions.

Another sampling consideration is determining how many stories each participant will be asked to complete. When using a comparative design with multiple versions of the story stem, one option is to ask participants to complete one version, which is what Clarke (2014) did in her body hair research. Another approach is to get participants to complete more, or all, the versions. In psychologist Helen Gavin's (2005) research on the social construction of sex offenders, for instance, each participant was asked to complete *six* different versions of a story stem. She did so to explore how individual participants' narratives surrounding sex offenders varied when presented with different situations. Asking participants to complete more than one stem may reflect a more pragmatic concern to maximize the number of stories in the data-set. For example, Shah-Beckley and Clarke (2015) were able to halve the number of required participants by asking them to complete two versions of a story stem related to therapists and non-therapists' constructions of heterosex.

One concern when asking participants to respond to multiple story stems is that there may be order effects, which could have different impacts. Participants may write their longest story first, and the richness and quality of data may drop off. However, in Shah-Beckley and Clarke's study (2015), the opposite was true, with participants writing longer stories in response to the second stem. Randomizing order can reduce the risk of systematic impact across different stems. A second concern is whether, through more than one stem, the participant gets "cued in" to what the researcher is interested in, and starts to tell the "right" story. Whether or not this is a concern still needs to be determined, and whether/how it potentially impacts the research likely depends primarily on the specifics of each study.

9 Data Collection and Ethical Considerations in Story Completion Research

When it comes to actually collecting SC data, the key consideration is whether to conduct the study using "paper and pen" completion, or electronically either online using (free or subscription) survey software such as *Qualtrics* (www.qualtrics.com) or *SurveyMonkey* (www.surveymonkey.com), or by emailing the SC to participants as an attachment or in the body of an email (see ▶ Chaps. 76, "Web-Based Survey Methodology," ▶ 78, "Synchronous Text-Based Instant Messaging: Online Interviewing Tool," and ▶ 79, "Asynchronous Email Interviewing Method"). An advantage of hard copy completion is that researchers can hand the SC directly to participants (for example, if researchers recruit on university campuses or at specific events), and, providing they have ethical approval, offer participants a small "reward" (e.g., a chocolate bar) for returning their story. A downside of hardcopy is that researchers then need to manually type up participants' stories ready for analysis.

The key advantage of *electronic* data collection is that responses require little preparation for analysis – emailed stories will need to be cut and pasted and collated in a document; online responses can be downloaded into a document almost instantly (see ▶ Chap. 79, "Asynchronous Email Interviewing Method"). Furthermore, participants can complete the study at a time and place that suits them. However, online SC research that requires participants to have Internet access can limit who can take part; it is the least privileged members of society that tend to have limited or no Internet access (Hargittai 2010), and some groups (such as older participants) *may* be uncomfortable with, or find difficult to use, certain types of technology (Kurniawan 2008). Finally, another important consideration is achieving a good fit between the mode of data collection and the participant group. Researchers do not have to restrict themselves to one mode – it may be most appropriate to ask some participants to complete the study online and others on hard copy.

9.1 Piloting in SC Research

Given the open-ended and exploratory nature of SC research, piloting the stem and instructions to ensure they elicit relevant and useful data is vital (Braun and Clarke 2013). We have often made minor (but transformative) amendments to story stems or instructions following piloting. The nature of SC means that piloting is not generally an onerous task. It is recommended that researchers pilot the stem on the equivalent of 10–20% of the intended final sample; the precise number should be determined in relation to the diversity within the participant group: greater diversity = larger pilot sample. Piloting can be approached in one of two ways: (1) by treating early data collection as a pilot, and using participant responses to judge if the stem and instructions have been interpreted in the way(s) researchers intended; (2) by asking participants to both complete the study *and* comment on the clarity of the instructions and the study design, after completion. If researchers make no (or minimal) changes to the stem following piloting, the pilot data can be incorporated into final sample.

9.2 Ethics in SC Research

As a general rule, SC research raises fewer ethical concerns than research that involves direct interaction with participants and asking them about their personal lives; this is particularly the case for online SC studies that make it even easier for participants to be anonymous and reduce risk for both participants and researchers. However, participant comfort with the topic is still an important ethical consideration, particularly for sensitive topics, and standard accepted ethical practice still needs to be adhered to (e.g. American Psychological Association 2010; British Psychological Society 2009). Researchers should also follow the relevant ethical guidance of their institution and/or professional body (see ▶ Chap. 106, "Ethics and Research with Indigenous Peoples").

10 Analyzing Story Completion Data

To date, two methods have been used to analyze SC data – thematic analysis (TA) (e.g., Livingston and Testa 2000; Frith 2013; Clarke et al. 2015; see ▶ Chap. 48, "Thematic Analysis") and discourse analysis (DA) (e.g. Walsh and Malson 2010; see ▶ Chap. 50, "Critical Discourse/Discourse Analysis"). Following Kitzinger and Powell, TA (Braun and Clarke 2006, 2012) is often slightly adapted from its usual use with self-report data. That is, rather than simply identifying patterns across the stories as a whole, researchers have identified patterns in specific elements of the story (both of these can be thought of as a variant of *horizontal* patterning, in the sense that the patterns intersect the stories). For example, SC research on perceptions of relational infidelity has identified patterns in how both the relationship between primary partners, and between the unfaithful partner and the "other" man/woman, is presented, how infidelity is accounted for, and how the responses to, and consequences of, infidelity are depicted (Kitzinger and Powell 1995; Whitty 2005). This means that SC researchers have identified particular questions they want to ask of the data – either in advance of the analysis, or after data familiarization – and used the techniques of TA to identify patterns in relation to these questions.

As noted above, Kitzinger and Powell (1995) demonstrated that both essentialist and constructionist readings of SC data are possible, and TA has been used to analyse SC data in both essentialist and constructionist ways. Pattern-based DA is also an ideal analytic approach for constructionist approaches to SC (Braun and Clarke 2013). For example, critical psychologists Eleanor Walsh and Helen Malson (2010) used poststructuralist DA (e.g., Wetherell et al. 2001) to interrogate some of the ways in which their participants made sense of anorexia and bulimia, and constituted the causes of, and recovery from, eating disorders. They explored how the participants constructed "dieting" as normal and healthy, for instance, and the ways in which recovery from eating disorder was framed in terms of a return to "normal" dieting rather than (say) a return to unrestricted eating or a lack of concern with body weight.

10.1 Story Maps

In addition to identifying *horizontal* patterning in the data, SC researchers have also examined *vertical* patterning – patterns in how stories unfold. One approach very useful for this type of "narrative" analysis is Braun and Clarke's (2013) story mapping technique that involves distinguishing patterns in the key elements of a story's progression (see also ▶ Chap. 64, "Creative Insight Method Through Arts-Based Research"). Braun and Clarke provide the example of a study exploring perceptions of a young woman "coming out" to her parents as non-heterosexual. The story map for this study identified patterns in: (1) the parent's initial reactions to the coming out; (2) the development of the stories; and (3) the ending or resolution of the stories. After an initial expression of shock, the parents' responses to their daughter coming out were categorized as either (broadly) positive or negative; the

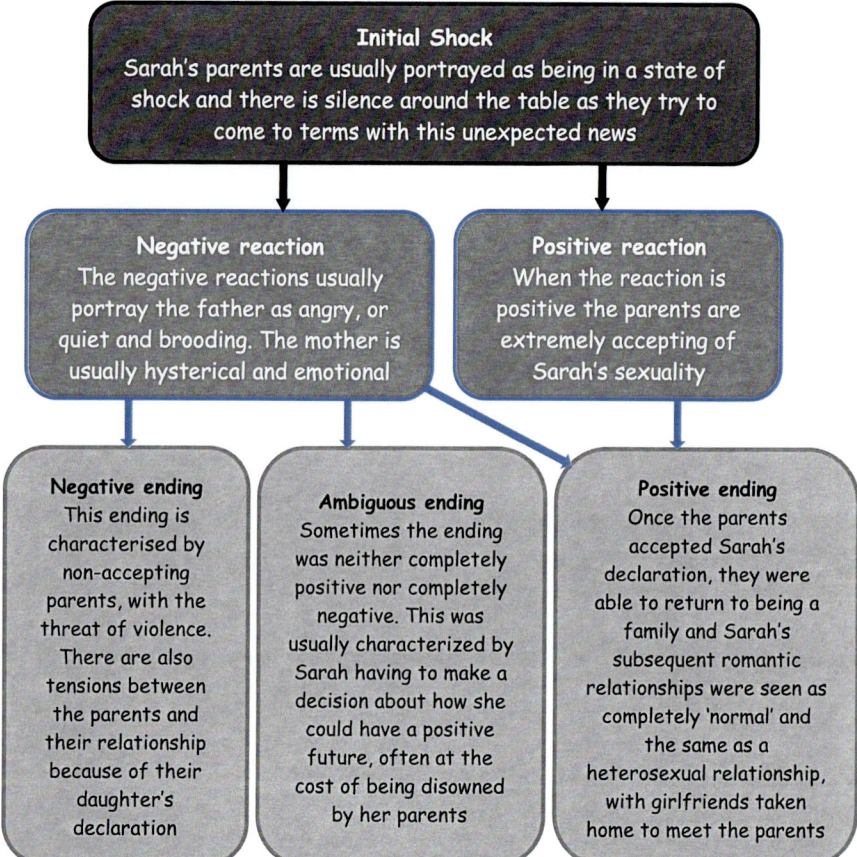

Fig. 1 An example of a story map (Braun and Clarke 2013)

negative reaction stories either ended positively, negatively or ambiguously, and the positive reaction stories always ended positively (see Fig. 1). Depending on the research question and approach, this story mapping technique can be a useful complement to a standard pattern-based analysis (e.g., TA), which helps the analysis to retain a sense of the storied nature of the data. This technique also lightly captures (Western) cultural conventions around story-telling (beginning, middle, end) and the dominance of particular genres (e.g., "happily ever after," "triumph over adversity"), and forces the researcher to think about the ways these are part of data production using SC.

One analytic approach that has yet to be used to analyse SC data, but nonetheless seems particularly apt, is narrative analysis (e.g., Riessman 2007). Narrative techniques could be productively used to identify narratives types and genres, and the structures and styles of particular narrative types, thus extending and developing Braun and Clarke's (2013) story mapping technique.

10.2 Frequency Counts

Researchers who do qualitative research within a qualitative paradigm do not generally recommend the use of frequency counts in the analysis of self-report data, because of the organic and participant-responsive nature of self-report data collection (Braun and Clarke 2013). However, frequency counts *are* often used in the analysis of SC data. For example, in their research on perceptions of infidelity, Kitzinger and Powell (1995) asked how many participants interpreted the female protagonist Claire "seeing someone else" as Claire being unfaithful – a full 10% rejected the implications of infidelity. When asking such concrete questions of the data (and when participants have been set an identical task), reporting numbers or percentages rather than using looser words such as "most" or "some" to capture patterning in the data is both appropriate and potentially analytically informative.

10.3 Analysis Approaches to Avoid

Certain analytic approaches are *not* suited to the analysis of SC data, including approaches such as interpretative phenomenological analysis (Smith et al. 2009) and forms of narrative analysis focused on understanding participants' lived experiences (Riessman 2007). Because participants are not asked for their views directly, and are often asked to write stories about things they may have little or no personal experience of, it is unclear whether SC data tell us anything meaningful about participants' lived experience. Without some big interpretative leaps, SC data would need to be combined with another data source to be suitable for use in research focused on lived experience. Grounded theory has similarly not been used to analyze SC data, and the focus on theory generation and the examination of the social processes and factors that shape particular phenomenon (Charmaz 2006) suggest that it is unlikely to be an appropriate method for analyzing SC data. Finally, approaches centred on the analysis of language practice – such as conversation analysis (e.g., Schegloff 2007) and discursive psychology (e.g., Wiggins and Potter 2010) – are not well suited to SC data. These approaches typically focus on "talk-in-interaction"; the "what" and "how" of "real" talk – both everyday "real" talk and that produced in institutional contexts such as courtrooms or consulting rooms – which is rather different from written, storied data.

11 Conclusion and Future Directions

It is hoped that this chapter conveys the enthusiasm of a group of committed SC researchers. In our view, SC is a method for producing data that provide an exciting, viable, and very accessible, alternative to self-report methods of data collection. SC allows participants control and creativity, and the resulting data can be fun, rich and complex. SC also offers researchers new ways to generate data that provide compelling insights into their chosen topics. This said, the method is still relatively unknown, with as yet a small but growing body of literature evidencing its use and potential. This leaves some questions still to be explored in future research, such as:

1. **What is distinct about SC?** As noted, there are overlaps between SC and vignette research; there are also overlaps between SC and diary research (see ▶ Chap. 83, "Solicited Diary Methods") – another often written solicited account. Points of overlap and of difference between these forms of data collection could further be delineated.
2. **What sample size is best?** To date, studies have used quite varied sample sizes. Ideal sample sizes, and whether or not a higher N offers more convincing and useful data/results, remain important questions to keep exploring.
3. **Does SC work well with a wide range of sample populations?** So far, SC has predominantly been used with UG students, who are educated, literate, and used to putting thought into textual form. How well SC works with different populations thus remains an important question. For instance, does it work well for less educated populations? How might it work in populations with different story-telling traditions? Could it also work effectively as an oral method, such as in clinical contexts, or with children?
4. **What are the factors that explain why SC does not always work?** Sometimes, SC data can, for instance, be flat, restricted in scope, or very brief. Some of the potential causes have been discussed, such as over-specifying the stem; another possibility to consider is whether even a hypothetical scenario may prove threatening to participants in some way, thereby restricting the data. For instance, in Shah-Beckley and Clarke's (2015) study, potential concerns about their own professional competence may have been evoked in the respondents, leading to "flat" data. Are there other consistent aspects to consider in designing effective stems?
5. **What can and should be done with story "refusal"?** Story refusal is when the participant either effectively ignores the stem topic, or offers up a humorous and/or fantastical story – such as accounts of a (hairy) yeti in our hair removal research, or of a monster/therapist eating the storyteller/client in our research on perceptions of fat therapists. These stories exemplify the creative potential that participants have with SC research – and can provide great pleasure to read as a researcher – and in some cases, the discursive or thematic aspects may resonate with the more conventional stories provided (e.g. that it is monstrous to be hairy/fat). But the theory of what such data mean still requires some thought.
6. *Finally,* **what is the potential of other analytic approaches to SC data?** What is the cost and the benefit of not focusing on the *storied* aspect of this approach? What potential is there in using narrative analysis to provide quite different insights to those offered by TA and DA?

These questions are important ones for the development of SC, but already, the evidence indicates that SC offers a valuable alternative to existing methods of qualitative data collection. The existence of questions offers researchers the opportunity to explore and innovate. Have fun with it!

References

American Psychological Association. Ethical principles of psychologists and code of conduct. 2010. Available: http://www.apa.org/ethics/code/.

Braun V, Clarke V. Using thematic analysis in psychology. Qual Res Psychol. 2006;3:77–101.

Braun V, Clarke V. Thematic analysis. In: Cooper H, Camic PM, Long DL, Panter AT, Rindskopf D, Sher KJ, editors. APA handbook of research methods in psychology, Research designs: Quantitative, qualitative, neuropsychological, and biological, vol. 2. Washington, DC: American Psychological Association; 2012. p. 57–71.

Braun V, Clarke V. Successful qualitative research: a practical guide for beginners. London: Sage; 2013.

Braun V, Tricklebank G, Clarke V. 'It shouldn't stick out from your bikini at the beach': meaning, gender, and the hairy/hairless body. Psychol Women Q. 2013;37(4):478–93.

Bretherton I, Oppenheim D, Emde RN, the MacArthur Narrative Working Group. The MacArthur story stem battery. In: Emde RN, Wolfe DP, Oppenheim D, editors. Revealing the inner worlds of young children: the MacArthur story stem battery and parent-child narratives. New York: Oxford University Press; 2003. p. 381–96.

British Psychological Society. Code of ethics and conduct. Leicester: British Psychological Society; 2009.

Burr V. Social constructionism. 2nd ed. London: Psychology Press; 2003.

Charmaz K. Constructing grounded theory: a practical guide through qualitative analysis. Thousand Oaks: Sage; 2006.

Clarke V. Telling tales of the unexpected: using story completion to explore constructions of non-normative body hair practices. Paper presented at Appearance Matters 6, 1–2 July 2014. Bristol; 2014.

Clarke V, Braun V, Wooles K. Thou shalt not covet another man? Exploring constructions of same-sex and different-sex infidelity using story completion. J Community Appl Soc Psychol. 2015;25(2):153–66.

Donoghue S. Projective techniques in consumer research. J Fam Ecol Consum Sci. 2000;28:47–53.

Frith H. Accounting for orgasmic absence: exploring heterosex using the story completion method. Psychol Sex. 2013;4(3):310–22.

Frith H, Riley S, Archer L, Gleeson K. Editorial. Qual Res Psychol. 2005;2(3):187–98.

Gavin H. The social construction of the child sex offender explored by narrative. Qual Rep. 2005;10 (3):395–415.

George C, West ML. The adult attachment projective system, attachment theory and assessment in adults. New York: The Guildford Press; 2012.

Gray D, Royall B, Malson H. Hypothetically speaking: using vignettes as a stand-alone qualitative method. In: Braun V, Clarke V, Gray D, editors. Collecting qualitative data: a practical guide to textual, media and virtual techniques. Cambridge: Cambridge University Press; in press.

Hargittai E. Digital na(t)ives? Variation in internet skills and uses among members of the 'net generation'. Sociol Inq. 2010;80(1):92–113.

Hayfield N, Wood M. Exploring sexuality and appearance using story completion and visual methods. Paper presented at Appearance Matters 6, 1–2 July 2014. Bristol; 2014.

Kitzinger C, Powell D. Engendering infidelity: essentialist and social constructionist readings of a story completion task. Fem Psychol. 1995;5(3):345–72.

Kurniawan S. Older people and mobile phones: a multi-method investigation. Int J Hum Comput Stud. 2008;66(12):889–901.

Livingston JA, Testa M. Qualitative analysis of women's perceived vulnerability to sexual aggression in a hypothetical dating context. J Soc Pers Relat. 2000;17(ni):729–41.

Moller N. Assumptions about fat counsellors: findings from a story-completion task. Paper presented at Appearance Matters 6, 1–2 July 2014. Bristol; 2014.

Moore SM, Gullone E, Kostanski M. An examination of adolescent risk-taking using a story completion task. J Adolesc. 1997;20:369–79.

Murray HA. Thematic apperception test manual. Cambridge, MA: Harvard University Press; 1971. (Original work published 1943)

Rabin AI, editor. Assessment with projective techniques. New York: Springer Publishing Company; 1981.

Rabin AI. Projective techniques at midcentury: a retrospective review of an introduction to projective techniques by Harold H. Anderson and Gladys L. Anderson. J Pers Assess. 2001;76(2):353–67.

Riessman CK. Narrative methods for the human sciences. Thousand Oaks: Sage; 2007.

Rorschach H, Lemkau P, Kronenberg B. Psychodiagnostics: a diagnostic test based on perception. 10th rev. edn. Switzerland: Verlag; 1921/1998. Huber.

Schegloff EA. Sequence organisation in interaction: a primer in conversation analysis. New York: Cambridge University Press; 2007.

Shah-Beckley I, Clarke V. Exploring constructions of sexual refusal in heterosexual relationships: a qualitative story completion study. Manuscript under submission. 2015.

Smith JA, Flowers P, Larkin M. Interpretative phenomenological analysis: theory, method and research. London: Sage; 2009.

Soley L, Smith AL. Projective techniques for social science and business research. Shirley, NY: The Southshore Press; 2008.

Terry G, Braun V. To let hair be, or to not let hair be? Gender and body hair removal practices in Aotearoa/New Zealand. Body Image. 2013;10(4):599–606.

Tischner I. Gendered constructions of weight-loss perceptions and motivations. Paper presented at Appearance Matters 6, 1–2 July 2014. Bristol; 2014.

Walsh E, Malson H. Discursive constructions of eating disorders: a story completion task. Fem Psychol. 2010;20(4):529–37.

Wetherell M, Taylor S, Yates S. Discourse theory and practice: a reader. London: Sage; 2001.

Whitty MT. The realness of cybercheating: men's and women's representations of unfaithful internet relationships. Soc Sci Comput Rev. 2005;23(1):57–67.

Wiggins S, Potter J. Discursive psychology. In: Willig C, Stainton-Rogers W, editors. The sage handbook of qualitative research in psychology. London: Sage; 2010. p. 73–90.

Wilkinson S. Focus groups: a feminist method. Psychol Women Q. 1999;23:221–44.

Will V, Eadie D, MacAskill S. Projective and enabling techniques explored. Mark Intell Plan. 1996;14:38–43.